Geophysical Monograph Series

180 **Arctic Sea Ice Decline: Observations, Projections, Mechanisms, and Implications** *Eric T. DeWeaver, Cecilia M. Bitz, and L.-Bruno Tremblay (Eds.)*
181 **Midlatitude Ionospheric Dynamics and Disturbances** *Paul M. Kintner, Jr., Anthea J. Coster, Tim Fuller-Rowell, Anthony J. Mannucci, Michael Mendillo, and Roderick Heelis (Eds.)*
182 **The Stromboli Volcano: An Integrated Study of the 2002–2003 Eruption** *Sonia Calvari, Salvatore Inguaggiato, Giuseppe Puglisi, Maurizio Ripepe, and Mauro Rosi (Eds.)*
183 **Carbon Sequestration and Its Role in the Global Carbon Cycle** *Brian J. McPherson and Eric T. Sundquist (Eds.)*
184 **Carbon Cycling in Northern Peatlands Andrew J. Baird,** *Lisa R. Belyea, Xavier Comas, A. S. Reeve, and Lee D. Slater (Eds.)*
185 **Indian Ocean Biogeochemical Processes and Ecological Variability** *Jerry D. Wiggert, Raleigh R. Hood, S. Wajih A. Naqvi, Kenneth H. Brink, and Sharon L. Smith (Eds.)*
186 **Amazonia and Global Change Michael Keller, Mercedes** *Bustamante, John Gash, and Pedro Silva Dias (Eds.)*
187 **Surface Ocean–Lower Atmosphere Processes** *Corinne Le Quèrè and Eric S. Saltzman (Eds.)*
188 **Diversity of Hydrothermal Systems on Slow Spreading Ocean Ridges** *Peter A. Rona, Colin W. Devey, Jérôme Dyment, and Bramley J. Murton (Eds.)*
189 **Climate Dynamics: Why Does Climate Vary?** *De-Zheng Sun and Frank Bryan (Eds.)*
190 **The Stratosphere: Dynamics, Transport, and Chemistry** *L. M. Polvani, A. H. Sobel, and D. W. Waugh (Eds.)*
191 **Rainfall: State of the Science** *Firat Y. Testik and Mekonnen Gebremichael (Eds.)*
192 **Antarctic Subglacial Aquatic Environments** *Martin J. Siegert, Mahlon C. Kennicut II, and Robert A. Bindschadler*
193 **Abrupt Climate Change: Mechanisms, Patterns, and Impacts** *Harunur Rashid, Leonid Polyak, and Ellen Mosley-Thompson (Eds.)*
194 **Stream Restoration in Dynamic Fluvial Systems: Scientific Approaches, Analyses, and Tools** *Andrew Simon, Sean J. Bennett, and Janine M. Castro (Eds.)*
195 **Monitoring and Modeling the Deepwater Horizon Oil Spill: A Record-Breaking Enterprise** *Yonggang Liu, Amy MacFadyen, Zhen-Gang Ji, and Robert H. Weisberg (Eds.)*
196 **Extreme Events and Natural Hazards: The Complexity Perspective** *A. Surjalal Sharma, Armin Bunde, Vijay P. Dimri, and Daniel N. Baker (Eds.)*
197 **Auroral Phenomenology and Magnetospheric Processes: Earth and Other Planets** *Andreas Keiling, Eric Donovan, Fran Bagenal, and Tomas Karlsson (Eds.)*
198 **Climates, Landscapes, and Civilizations** *Liviu Giosan, Dorian Q. Fuller, Kathleen Nicoll, Rowan K. Flad, and Peter D. Clift (Eds.)*
199 **Dynamics of the Earth's Radiation Belts and Inner Magnetosphere** *Danny Summers, Ian R. Mann, Daniel N. Baker, and Michael Schulz (Eds.)*
200 **Lagrangian Modeling of the Atmosphere** *John Lin (Ed.)*
201 **Modeling the Ionosphere–Thermosphere** *Jospeh D. Huba, Robert W. Schunk, and George V Khazanov (Eds.)*
202 **The Mediterranean Sea: Temporal Variability and Spatial Patterns** *Gian Luca Eusebi Borzelli, Miroslav Gacic, Piero Lionello, and Paola Malanotte-Rizzoli (Eds.)*
203 **Future Earth—Advancing Civic Understanding of the Anthropocene** *Diana Dalbotten, Gillian Roehrig, and Patrick Hamilton (Eds.)*
204 **The Galápagos: A Natural Laboratory for the Earth Sciences** *Karen S. Harpp, Eric Mittelstaedt, Noémi d'Ozouville, and David W. Graham (Eds.)*
205 **Modeling Atmospheric and Oceanic Flows: Insights from Laboratory Experiments and Numerical Simulations** *Thomas von Larcher and Paul D. Williams (Eds.)*
206 **Remote Sensing of the Terrestrial Water Cycle** *Venkat Lakshmi (Eds.)*
207 **Magnetotails in the Solar System** *Andreas Keiling, Caitríona Jackman, and Peter Delamere (Eds.)*
208 **Hawaiian Volcanoes: From Source to Surface Rebecca** *Carey, Valerie Cayol, Michael Poland, and Dominique Weis (Eds.)*
209 **Sea Ice: Physics, Mechanics, and Remote Sensing** *Mohammed Shokr and Nirmal Sinha (Eds.)*
210 **Fluid Dynamics in Complex Fractured-Porous Systems** *Boris Faybishenko, Sally M. Benson, and John E. Gale (Eds.)*
211 **Subduction Dynamics: From Mantle Flow to Mega Disasters** *Gabriele Morra, David A. Yuen, Scott King, Sang Mook Lee, and Seth Stein (Eds.)*
212 **The Early Earth: Accretion and Differentiation** *James Badro and Michael Walter (Eds.)*
213 **Global Vegetation Dynamics: Concepts and Applications in the MC1 Model** *Dominique Bachelet and David Turner (Eds.)*
214 **Extreme Events: Observations, Modeling and Economics** *Mario Chavez, Michael Ghil, and Jaime Urrutia-Fucugauchi (Eds.)*
215 **Auroral Dynamics and Space Weather** *Yongliang Zhang and Larry Paxton (Eds.)*
216 **Low-Frequency Waves in Space Plasmas** *Andreas Keiling, Dong-Hun Lee, and Valery Nakariakov (Eds.)*
217 **Deep Earth: Physics and Chemistry of the Lower Mantle and Core** *Hidenori Terasaki and Rebecca A. Fischer (Eds.)*
218 **Integrated Imaging of the Earth: Theory and Applications** *Max Moorkamp, Peter G. Lelievre, Niklas Linde, and Amir Khan (Eds.)*
219 **Plate Boundaries and Natural Hazards** *Joao Duarte and Wouter Schellart (Eds.)*
220 **Ionospheric Space Weather: Longitude and Hemispheric Dependences and Lower Atmosphere Forcing** *Timothy Fuller-Rowell, Endawoke Yizengaw, Patricia H. Doherty, and Sunanda Basu (Eds.)*
221 **Terrestrial Water Cycle and Climate Change Natural and Human-Induced Impacts** *Qiuhong Tang and Taikan Oki (Eds.)*
222 **Magnetosphere–Ionosphere Coupling in the Solar System** *Charles R. Chappell, Robert W. Schunk, Peter M. Banks, James L. Burch, and Richard M. Thorne (Eds.)*
223 **Natural Hazard Uncertainty Assessment: Modeling and Decision Support** *Karin Riley, Peter Webley, and Matthew Thompson (Eds.)*
224 **Hydrodynamics of Time-Periodic Groundwater Flow: Diffusion Waves in Porous Media** *Joe S. Depner and Todd C. Rasmussen*
225 **Active Global Seismology** *Ibrahim Cemen and Yucel Yilmaz (Eds.)*
226 **Climate Extremes** *Simon Wang*
227 **Fault Zone Dynamic Processes** *Marion Thomas*
228 **Flood Damage Survey and Assessment: New Insights from Research and Practice** *Daniela Molinari, Scira Menoni, and Francesco Ballio (Eds.)*
229 **Water–Energy–Food Nexus—Principles and Practices** *P. Abdul Salam, Sangam Shrestha, Vishnu Prasad Pandey, and Anil K Anal (Eds.)*
230 **Dawn–Dusk Asymmetries in Planetary Plasma Environments** *Stein Haaland, Andrei Rounov, and Colin Forsyth (Eds.)*
231 **Bioenergy and Land Use Change** *Zhangcai Qin, Umakant Mishra, and Astley Hastings (Eds.)*
232 **Microstructural Geochronology: Planetary Records Down to Atom Scale** *Desmond Moser, Fernando Corfu, James Darling, Steven Reddy, and Kimberly Tait (Eds.)*
233 **Global Flood Hazard: Applications in Modeling, Mapping, and Forecasting** *Guy Schuman, Paul D Bates, Giuseppe T Aronica, and Heiko Apel (Eds.)*

Geophysical Monograph 234

Pre-Earthquake Processes

A Multidisciplinary Approach to Earthquake Prediction Studies

Dimitar Ouzounov
Sergey Pulinets
Katsumi Hattori
Patrick Taylor
Editors

This Work is a co-publication between the American Geophysical Union and John Wiley and Sons, Inc.

AGU 100
ADVANCING EARTH AND SPACE SCIENCE

WILEY

This Work is a co-publication between the American Geophysical Union and John Wiley & Sons, Inc.

This edition first published 2018 by John Wiley & Sons, Inc., 111 River Street, Hoboken, NJ 07030, USA and the American Geophysical Union, 2000 Florida Avenue, N.W., Washington, D.C. 20009

Published under the aegis of the AGU Publications Committee
Brooks Hanson, Senior Vice President, Publications
Lisa Tauxe, Chair, Publications Committee
For details about the American Geophysical Union visit us at www.agu.org.

Wiley Global Headquarters
111 River Street, Hoboken, NJ 07030, USA

For details of our global editorial offices, customer services, and more information about Wiley products visit us at www.wiley.com.

Library of Congress Cataloging-in-Publication data is available.

ISBN: 9781119156932

Cover image: Dr. Katsumi Hattori, Chiba University, Japan
Cover design: Wiley

Set in 10/12pt Times by SPi Global, Pondicherry, India

Printed in the United States of America

10 9 8 7 6 5 4 3 2 1

CONTENTS

CONTRIBUTORS

Tomokazu Asano
Hayakawa Institute of Seismo
Electromagnetics, Co. Ltd
University of Electro-Communications (UEC)
Incubation Center
Chofu, Tokyo, Japan

Chieh-Hung Chen
Department of Earth and Environment Sciences
National Chung Cheng University
Taiwan

Yuh-Ing Chen
Graduate Institute of Statistics
National Central University
Taoyuan, Taiwan

Andrea Dadomo
GEOINVEST Srl
Piacenza, Italy

Dmitry Davidenko
Space Research Institute
Russian Academy of Sciences
Moscow, Russia

Carolina Filizzola
Institute of Methodologies of Environmental Analysis
National Research Council
Tito Scalo, Italy

Ching-Chou Fu
Institute of Earth Sciences
Academia Sinica
Taipei, Taiwan

Nicola Genzano
School of Engineering
Basilicata University
Potenza, Italy

Peng Han
Southern University of Science and Technology
Shenzhen, China

Katsumi Hattori
Graduate School of Science
Chiba University
Chiba, Japan

Masashi Hayakawa
Hayakawa Institute of Seismo
Electromagnetics, Co. Ltd
University of Electro-Communications (UEC)
Incubation Center
Chofu, Tokyo, Japan;
University of Electro-Communications
Chofu, Tokyo, Japan

Yi-Ying Ho
Institute of Earth Sciences
Academia Sinica
Taipei, Taiwan

Zaisen Jiang
Key Laboratory of Earthquake Prediction
Institute of Earthquake Science
China Earthquake Administration
Beijing, China

Menas C. Kafatos
Center of Excellence in Earth Systems, Modeling and Observations (CEESMO)
Chapman University
Orange, CA, USA

Alexander Karelin
IZMIRAN
Russian Academy of Sciences
Moscow, Russia

Cheng-Ling Kuo
Institute of Space Science
National Central University
Taoyuan, Taiwan

Lou-Chuang Lee
Institute of Earth Sciences
Academia Sinica
Taipei, Taiwan

Mei Li
China Earthquake Networks Center
China Earthquake Administration
Beijing, China

Mariano Lisi
School of Engineering,
Basilicata University
Potenza, Italy

Jann-Yenq (Tiger) Liu
Institute of Space Science
National Central University
Taoyuan, Taiwan

Jie Liu
China Earthquake Networks Center
China Earthquake Administration
Beijing, China

Giovanni Martinelli
ARPAE Emilia-Romagna
Reggio Emilia, Italy

George Minadakis
Institute of Geodynamics
National Observatory of Athens
Athens, Greece

Toshiyasu Nagao
Institute of Oceanic Research and Development
Tokai University, Japan

Katerina Orfanogiannaki
Institute of Geodynamics
National Observatory of Athens
Athens, Greece

Dimitar Ouzounov
Center of Excellence in Earth Systems, Modeling and Observations (CEESMO)
Chapman University
Orange, CA, USA

Gerassimos Papadopoulos
Institute of Geodynamics
National Observatory of Athens
Athens, Greece

Michel Parrot
LPC2E/CNRS
Orléans, France

Antonella Peresan
National Institute of Oceanography and Experimental Geophysics
OGS-CRS
Udine, Italy

Nicola Pergola
Institute of Methodologies for Environmental Analysis of the National Research Council
Tito Scalo, Italy

Sergey Pulinets
Space Research Institute
Russian Academy of Sciences
Moscow, Russia

Alexander Rozhnoi
Institute of Physics of the Earth
Russian Academy of Sciences
Moscow, Russia

Peter Shebalin
Institute of Earthquake Prediction Theory and Mathematical Geophysics
Russian Academy of Sciences
Moscow, Russia

Xuhui Shen
Institute of Crustal Dynamics
China Earthquake Administration
Beijing, China

Tzay-Chyn Shin
Central Weather Bureau
Taiwan

Maria Solovieva
Institute of Physics of the Earth
Russian Academy of Sciences
Moscow, Russia

Patrick Taylor
NASA GSFC
Greenbelt, MD, USA

Valerio Tramutoli
School of Engineering
Basilicata University
Potenza, Italy

Yi-Ben Tsai†
Department of Earth Sciences
National Central University
Zhongli District, Taoyuan City, Taiwan

Seiya Uyeda
Japan Academy
Japan

Hui Wang
Key Laboratory of Earthquake Prediction
Institute of Earthquake Science
China Earthquake Administration
Beijing, China

Hong-Yuan Yen
Department of Earth Sciences
National Central University
Zhongli District, Taoyuan City, Taiwan

Huaizhong Yu
China Earthquake Networks Center
China Earthquake Administration
Beijing, China

Guomin Zhang
Key Laboratory of Earthquake Prediction
Institute of Earthquake Science
China Earthquake Administration
Beijing, China

Yongxian Zhang
China Earthquake Networks Center
China Earthquake Administration
Beijing, China

PREFACE

Our current knowledge of the pre-earthquake phenomena has developed over past 25 centuries, originating from the earliest scientific observation: the first scientific treatise to report comprehensively about earthquake-associated phenomena is generally recognized to be *Meteorologica*, written by Aristotle in 340 BC in ancient Greece. He presented the first description of earthquakes and other weather phenomena, including his theory of "pneuma" (breath), a phenomenon emanating before earthquakes. This was followed by continuous observations of atmospheric signs preceding large earthquakes in Europe. At the end of the nineteenth century *Alexander von Humboldt* [1897] studied the air circulation, similar to the "respiration of Earth" phenomenon, associated with earthquake processes originally proposed by Russian academician *V. Vernadsky* [1912]. During the same period of great science discoveries, John Milne—an English seismologist and geologist who invented the first modern seismograph—published his first edition of *Earthquakes and Other Movements* [*Milne*, 1913] with some novel ideas about warning of major earthquakes. Milne published the first quantitative analysis of atmospheric signals associated with seismicity based on 387 earthquakes observed in northern Japan, where he found that sinusoidal curves of the means of monthly temperature were generally a little in advance of the crest of the waves indicating the earthquake arrival [*Milne*, 1913]. Later in the seventh edition [*Milne and Lee*, 1939], they wrote: "It has been suggested that the small foreshocks indicated the removal of obstacles on a fault plane and that the occurrence of such a series of little shocks might be taken as a warning of other impending disasters" and: "Again, it has been suggested that the quake might be indicated by changes in the velocity which seismic waves are propagated."

Over the past 15–20 years there has been a major revival of interest on pre-earthquake studies in Japan, Russia, China, Europe, and Taiwan. Recent large earthquakes in Asia and Europe have highlighted the work being conducted in these regions on the various studies searching for earthquake precursors, either as forecasts or predictions. One result has been the organizing of a special session on pre-earthquake processes during the Fall Meetings of American Geophysical Union (AGU). Topics included in this book were initially presented at the sessions on pre-earthquake studies, organized by the co-editors of this book, during AGU Fall Meetings since 2004. In this volume we highlight 20 of these studies, which are presented as current research and progress reports on our understanding of pre-earthquake investigations. They are globally distributed, with the majority concerning China, Japan, Russia, Taiwan, Italy, France, and Greece. They describe the various parameters being observed and their relationship to pre-seismic activity. Some of these monitored parameters include: observing crustal geochemical fluids and gases; ultra-low frequency (ULF) magnetic signals; atmospheric effects including ionospheric total electron content (TEC) measurements; and several recording seismicity in regions experiencing earthquakes. In addition there are several national case histories.

Despite the fact that this volume represents only a portion of some of the global research in earthquake prediction being conducted they illustrate the scope and diversity of the investigations being carried out. This volume is intended to show the variety of parameters, including seismic, atmospheric, geochemical, and historical, involved in this important field of research and to bring knowledge and awareness of these investigations to the broader geosciences community.

This book was designed to introduce an interdisciplinary approach towards pre-earthquake studies. Our goal is to document the latest progress made in studying pre-earthquake processes and provide a snapshot of the latest international development on a wide range of research. The peer-reviewed studies can be cited in future journal publications and hopefully this will help satisfy unconvinced scientists about the existence of physical phenomena preceding earthquakes.

It is hoped that this volume will serve as an initial point for those not familiar with earthquake precursor studies and may point to the direction for future research. We expect that this book will encourage others to study pre-earthquake phenomena and continue developing an interdisciplinary physical approach in exploring our changing planet and improving earthquake prediction, as well as facilitating the education of new generations of scientists.

REFERENCES

Humboldt, A., Von (1897), *COSMOS: A Sketch of the Physical Description of the Universe*, New York, 462 pp.

Milne, J. (1913), *Earthquakes and Other Movements*, 1st edn, London, 210 pp.

Milne, J. and Lee, A. W. (1939), *Earthquakes and Other Movements*, 7th edn, Kegan Paul, Trench, & Co., Ltd, London, 244 pp.

Vernadsky, V. (1912), About the gas exchange of Earth crust, *Russ. Acad.Sci. St Petersburg*, *6*, 2141–2162. (In Russian.)

Dimitar Ouzounov, *Chapman University, Orange CA, USA*
Sergey Pulinets, *Space Research Institute, RAS, Moscow, Russia*
Katsumi Hattori, *Chiba University, Chiba, Japan*
Patrick Taylor, *NASA GSFC, Greenbelt, MD, USA*
January 2018

ACKNOWLEDGMENTS

We would like to thank Dr. Ritu Bose, AGU book editor from Wiley, for the great enthusiasm and support, for without this the project would not have been possible.

The co-editors would like to thank the 50 reviewers who donated their time to review the 20 chapters.

Part I: Historical Development of Pre-Earthquake Phenomena Studies

1

International Cooperation in Pre-Earthquake Studies: History and New Directions

Seiya Uyeda[1] **and Toshiyasu Nagao**[2]

ABSTRACT

Earthquake prediction must specify the time, epicenter, and size of an impending earthquake with useful accuracy. Among the long-, intermediate-,and short-term predictions, only the latter is meaningful for directly protecting human lives and social infrastructures. Globally, seismologists running national earthquake prediction projects have long been pessimistic about short-term prediction. In recent years, however, positive signs for short-term predictability are emerging from diverse scientific communities. This article is a summary of the history and new directions of international cooperation in earthquake prediction research.

Modern earthquake seismology may be said to have started in California by the advent of elastic rebound theory after the 1906 *M* 8.3 San Francisco earthquake [*Reid*, 1910]. The theory stipulates that rocks on opposite sides of a fault subjected to shear stress slowly deform until their strength is exceeded, thereby producing an earthquake. Stress is accumulated over many years but is released instantly. This idea has since been the theoretical guideline for earthquake generation. In the following several decades, however, no serious scientific attempt for earthquake prediction had been undertaken. In the 1960s, national projects for earthquake prediction started in several countries, including Japan, former USSR, China, Taiwan, and the USA. The reason why this happened more or less simultaneously is not clear. Perhaps it was because scientific activity recovered following World War II. This development was aided by the Worldwide Standardized Seismograph Network (WWSSN) for nuclear blast detection [*Bolt*, 1976].

[1] *Japan Academy, Japan*

[2] *Institute of Oceanic Research and Development, Tokai University, Japan*

In Japan, the report "Prediction of earthquakes—progress to date and plans for further development," known as the "Blue Print" [*Tsuboi et al.*, 1962] appeared in 1962. Its content was essentially empirical but covered all possible items for monitoring, such as crustal movements, tides, seismicity, seismic wave velocity, active faults, geochemistry, and water-level changes in boreholes, including even geomagnetism and geoelectric currents. Actual funding for earthquake prediction started in 1965 and has continued until now through consecutive 5-year plans. Russia has a similar long history of earthquake prediction, dating back to the days of the former USSR. In China, the China Earthquake Bureau (CEA) has been responsible for earthquake prediction since the late 1960s. In the USA, the National Earthquake Hazards Reduction Program (NEHRP) was proposed in 1965, but it was not until 1977 that it was authorized by Congress. Taiwan, before the 1999, Chi-Chi (or 921) *M* 7.3 earthquake, relied largely on fire fighters, police, and military at emergencies or natural disasters. After the 1999 earthquake, they established a permanent search-and-rescue team and an early-warning system.

The International Commission on Earthquake Prediction was set up 1967 under the auspices of the International Union of Geodesy and Geophysics

Pre-Earthquake Processes: A Multidisciplinary Approach to Earthquake Prediction Studies, Geophysical Monograph 234, First Edition. Edited by Dimitar Ouzounov, Sergey Pulinets, Katsumi Hattori, and Patrick Taylor.

(IUGG). The IUGG is the world's largest organization of geophysics. Optimism prevailed globally [*Press*, 1975] in the early 1970s due, for instance, to the apparent success of the dilatancy diffusion model [*Sholtz et al.*, 1973], which appeared to explain all the precursory phenomena such as anomalous uplift, seismic-wave velocity change, etc. Dilatancy is the volume increase due to generation of microfractures under high stress and diffusive influx of water into them was considered as the cause of lowering of rock strength, facilitating earthquakes. The success of long-, intermediate-, and short-term predictions of the 1975 *M* 7.3 Haisheng earthquake also boosted optimism, but that optimism proved ephemeral. The seismic-wave velocity changes predicted by the dilatancy diffusion model were not supported by accurate measurements in California [*McEvilly and Johnson*, 1974] and the CEA failed to predict the 1976 *M* 7.8 Tangshan earthquake [*Chen et al.*, 1988]. In the USA, the Parkfield earthquake, predicted to occur before 1993, did not occur until 2004 [*Langbein et al.*, 2005]. In fact, no internationally accepted prediction by any national earthquake prediction projects proved successful. This led to a totally pessimistic outlook, not only by scientific and social communities but also by governments, and this pessimism [*Evernden*, 1982] has essentially lasted until today. The 2011 *M* 9 Tohoku-Oki earthquake occurred during the seventh five-year prediction research program, after which the seismological communities in Japan completely gave up official short-term earthquake prediction research.

During the same period, mainly in the 1970s, the plate tectonics concept became established. As is well established, according to plate tectonics theory, Earth's surface layer is composed of about ten slowly moving ($<10\,cm\,year^{-1}$) plates ($<150\,km$ thick) and their interactions cause the whole range of tectonic processes observed, from mountain building to earthquakes [e. g. *Uyeda*, 1977; *Turcotte and Schubert*, 1982]. For geoscientists, including seismologists naturally, plate tectonics was an eye opener towards a more global view, as seismologists became to see that seismicity in their own country was part of a global network of activity. This helped promote "international cooperation."

In spite of this general trend, because they could not identify reliable precursors, seismologists maintained a negative attitude towards earthquake prediction. In Japan, most resources of national projects were spent on improvements to seismic networks rather than on acquisition of precursory phenomena data, which usually requires monitoring of nonseismic physical quantities. Throughout this long period of general pessimism, however, some fervent scientists in many countries continued their positive efforts. For instance, the VAN method (named after its originators P. Varotsos, K. Alexopoulos, and K. Nomicos) of short-term prediction has been practiced in Greece over the past several decades. This method is based on the observation of low-frequency transient changes of Earth's electric field, called seismic electric signals (SES), that precede earthquakes, enabling forecast of the parameters (epicenter, magnitude, and time window) of impending earthquakes [*Varotsos and Alexopoulos*, 1984; *Varotsos et al.*, 1986].

As described by M. Hayakawa in Chapter 2, the main focus of these efforts was on the electromagnetic precursors. Electromagnetic study of earthquakes had been initiated in Japan in the early 1980s [*Gokhberg et al.*, 1982] using very low-freqency (VLF) and LF electromagnetic emissions. It should be noted, however, that the seismo-electromagnetic groups in Japan were funded only once, just after the 1995 Kobe arthquake, but this could be considered as the start of extensive precursor studies [*Nagao et al.*, 2002], and may have inspired seismo-electromagnetic studies in other countries such as Taiwan, Russia, and Italy. Two Japanese institutions were asked to conduct feasibility studies of electromagnetic effects on short-term earthquake prediction for a period of 5 years (1996–2001): (a) RIKEN (Institute of Physical and Chemical Research) and (b) NASDA (National Space Development Agency of Japan, now JAXA) in the Earthquake Integrated Frontier framework of the Science and Technology Agency (now MEXT). The RIKEN project was led by S. Uyeda and the NASDA project by M. Hayakawa.

The RIKEN group installed about 40 VAN type geoelectric stations all over Japan. They found geoelectric signals before $M>5$ earthquakes [*Uyeda et al.*, 2000], but they also found that Japan is not suitable for the VAN method because of noise from DC electric trains. The stations did, however, detect very significant precursors of DC and ultra-low frequency (ULF) electric and magnetic radiation before the volcano-seismic activity in 2000 in the Izu island region where there is no DC-driven electric trains [*Uyeda et al.*, 2002].

Nowadays active research is carried out using the ULF to VHF bands in many countries. Among others, a possible new type of preseismic change of the ionosphere [*Heki*, 2011; *Kamogawa and Kakinami*, 2013; *Heki and Enomoto*, 2015] may be noteworthy. Lithosphere–atmosphere–ionosphere coupling (LAIC) [*Kuo et al.*, 2014; *Pulinets et al.*, 2015) and thermal radiation anomalies [*Ouzounov et al.*, 2011] are also important approaches in this field, and these will be thoroughly treated in later chapters of this book.

In 1996, CNES (Centre National d'Etudes Spatiales) began to develop microsatellites: DEMETER (detection of electro-magnetic emissions transmitted from earthquake regions), launched in 2004, is the pioneer of this international project. Using DEMETER, it was found that nighttime VLF emission markedly decreases in the 4 h before $M>4.8$ earthquakes [*Nemec et al.*, 2008]. China, in cooperation with Italy, plans to build a fleet of electromagnetic satellites by launching one every few years from 2017.

In July 1999, the general assembly of the IUGG was held in Birmingham, UK. In one of its sessions, extremely active discussion was held not only on electromagnetic signals but also on their physical mechanisms and it was decided to establish a Working Group on Electromagnetic Study of Earthquakes and Volcanoes (EMSEV). The EMSEV was founded by the IUGG in 2001 as an Inter-Association Working Group belonging to IAGA (International Association of Geomagnetism and Aeronomy), IASPEI (International Association of Seismology and Physics of Earth's Interior) and IAVCEI (International Association of Volcanology and Chemistry of Earth's Interior). This "Inter-Association" Working Group clearly establishes EMSEV as an international multidisciplinary team.

Establishing EMSEV was the dawn of both new earthquake prediction research and "International Cooperation". The first meeting of EMSEV was in 2002 in Moscow. Since then, international EMSEV meetings have been held every 2 years: France (2004), India (2006), Romania (2008), USA (2010), Japan (2012), Poland (2014), and China (2016). The next meeting will be in Italy (2018): For details see http://emsev-iugg.org/emsev/. The EMSEV Working Group have also organized special sessions at IUGG international meetings: Japan (2003, Sapporo), Italy (2007, Pergia), Australia (2011, Melbourne), and Czech Republic (2015, Prague), as well as at a diverse array of international meetings organized by IAGA, IAVCEI, and IASPEI (Iceland, Japan, Mexico, South Africa, Hungary) and regional workshops (Philippines, Kyrgyzstan, Greece, France).

In conclusion, we would like to draw your attention to the recent remarkable revival of seismology in earthquake prediction research, which has now emerged from the shadows of electromagnetic research. Examples include the study on earthquake generation using spatial organization of foreshocks [*Lippiello et al.*, 2012] and the possible role of slow earthquakes on the generation of huge earthquakes [*Obara and Kato*, 2016]. Perhaps, now is the time to discard the long-held pessimism and combine all our forces to venture towards transforming precursor information into practical earthquake prediction.

REFERENCES

Bolt, B. (1976), *Nuclear Explosions and Earthquakes: The Parted Veil*, W.H. Freeman, San Francisco, 309 pp.

Chen, Y., Tsoi, K., Chen, F., Gao, G., Zou, Q., and Chen, Z. (1988), *The Great Tangshan Earthquake of 1976*, Pergamon, Tarrytown, 153 pp.

Evernden, J. (1982), Earthquake prediction: What we have learned and what we should do now, *Bull. Seismol. Soc. Am.*, *72*, 343–349.

Gokhberg, M. B., Morgounov, V., Yoshino, T., and Tomizawa, I. (1982), Experimental measurement of electromagnetic emissions possibly related to earthquake in Japan, *J. Geophys. Res.*, *87* (B9), 7824–7828.

Heki, K. (2011), Ionospheric electron enhancement preceding the 2011 Tohoku-Oki earthquake, *Geophys. Res. Lett.*, *38*, L17312, doi:10.1029/2011GL047908.

Heki, K. and Enomoto, Y. (2015), *Mw* dependence of the preseismic ionospheric electron enhancements, *J. Geophys. Res. Space Phys.*, *120*, 7006–7020, doi:10.1002/2015JA021353.

Kamogawa, M. and Kakinami, Y. (2013), Is an ionospheric electronenhancement preceding the 2011 Tohoku-Oki earthquake a precursor? *J. Geophys. Res. Space Phys.*, *118*, 1751–1754, doi:10.1002/jgra.50118.

Kuo, C. L., Lee, L. C., and Huba, J. D. (2014), An improved coupling model for the lithosphere–atmosphere–ionosphere system, *J. Geophys. Res. Space Phys.*, *119*, 3189–3205, doi:10.1002/2013JA019392.

Langbein, J., Dreger, D., Fletcher, J., Harde-beck, J. L., Hellweg, M., Ji, C., Johnston, M., Murray, J. R., Nadeau, R. M., and Rymer, M. (2005), Preliminary report on the 28 September 2004, 6.0 Parkfield, California earthquake, *Seismol. Res. Lett.*, *76*, 10–26.

Lippiello, E., Marzocchi, W., de Arcangelis, L., and Godano, C. (2012), Spatial organization of foreshocks as a tool to forecast large earthquakes, *Sci. Rep.*, *2*, 846.

McEvilly, T. V. and Johnson, L. R. (1974), Stability of *P* and *S* velocities from central California quarry blasts, *Bull. Seismol. Soc. Am.*, *64*, 342–353.

Nagao, T., Enomoto, Y., Fujinawa, Y., Hata, M., Hayakawa, M., Huang, Q., Izutsu, J., Kushida, Y., Maeda, K., Oike, K., Uyeda, S., and Yoshino, T. (2002), Electromagnetic anomalies associated with 1995 KOBE Kobe earthquake, *J. Geodyn.*, *33*, 401–411.

Nemec, F., Santolık, O., Parrot, M., and Berthelier, J. J. (2008), Spacecraft observations of electromagnetic perturbations connected with seismic activity, *Geophys. Res. Lett.*, *35*, L05109, doi:10.1029/2007GL032517.

Obara, K. and Kato, A. (2016), Connecting slow earthquakes to huge earthquakes, *Science*, *353*(6296), 353–357.

Ouzounov, D., Pulinets, S. Romanov, A., Romanov, A., Tsybulya, K. Davidenko, D., Kafatos, M., and Taylor, P. (2011), Atmosphere-ionosphere response to the *M* 9 Tohoku earthquake revealed by multi-instrument space-borne and ground observations: Preliminary results, *Earthquake Sci.*, *24*, 557–564.

Press, F. (1975), Earthquake prediction, *Sci. Am.*, *232*(5), 14–23.

Pulinets, S., Ouzounov, D., Karelin, A., and Davidenko, D. (2015), Physical bases of the generation of short-term earthquake precursors: a complex model of ionization-induced geophysical processes in the lithosphere–atmosphere–ionosphere–magnetosphere system, *Geomagn. Aeron.*, *55*, 540–558.

Reid, H. F. (1910), *The Mechanics of the Earthquake*, Report of the State Earthquake Investigation Commission, Vol. *II*, Carnegie Institute, Washington, DC, 192 pp.

Sholtz, C., Sykes, L., and Aggarwal, Y. (1973), Earthquake prediction: a physical basis. *Science*, *181*, 803–810.

Tsuboi, C., Wadati, K., and Hagiwara, T. (1962), *Prediction of Earthquakes – Progress to Date and Plans for Further Development*, Report of Earthquake Prediction Research

Group, Japan, Earthquake Research Institute, University of Tokyo, 21 pp.

Turcotte, D. and Schubert, G. (1982), *Geodynamics—Applications of Continuum Physics to Geological Problems*, John Wiley & Sons., Chichester, 450 pp.

Uyeda, S. (1977), *The New View of the Earth—Moving Continents and Moving Oceans*, W. H. Freeman and Co., San Francisco, CA, 217 pp.

Uyeda, S., Nagao, T., Orihara, Y., Yamaguchi, T. and Takahashi, I. (2000), Geoelectric potential changes: possible precursors to earthquakes, *Proc. Nat. Acad. Sci. USA*, *97*(9), 4561–4566.

Uyeda, S., Hayakawa, M., Nagao, T., Molchanov, O., Hattori, K., Orihara, Y., Gotoh, K., Akinaga, Y., and Tanaka, H. (2002), Electric and magnetic phenomena observed before the volcano-seismic activity 2000 in the Izu Island Region, Japan, *Proc. Nat. Acad. Sci. USA*, *99*(11), 7352–7355.

Varotsos, P. and Alexopoulos, K. (1984), Physical properties of the variations of the electric field of the earth preceding earthquakes, I & II. *Tectonophysics*, *110*, 73–125.

Varotsos, P., Alexopoulos, K., Nomicos, K. and Lazaridou, M. (1986), Earthquake prediction and electric signals. *Nature (London)*, *322*, 120.

2

Earthquake Precursor Studies in Japan

Masashi Hayakawa

ABSTRACT

The purpose of this chapter is to review earthquake precursor studies in Japan. Among the long-, medium-, and short-term earthquake predictions that result from the study of precursors, the most meaningful are the short-term predictions because of their immediate effect on human lives. Very few investigations of earthquake precursors in Japan were undertaken before the 1995 Kobe *M* 6.9 earthquake, but extensive studies started after that event. The most important finding during the past two decades is that many of earthquake precursors are electromagnetic rather than seismological. Two Japanese frontier projects during the years of 1996–2001, in particular, have contributed very much to the progress of seismo-electromagnetics. Electromagnetic phenomena in possible association with earthquakes have been reviewed here with special reference to those achievements made during and after the frontier projects. Stimulated by the success of these Japanese frontier projects, national precursor studies projects devoted to earthquake prediction have been developed in different countries. In addition, several precursors are described (not only electromagnetic, but also of ground movements) that were associated with the *M* 9 Tōhoku mega-earthquake of 2011. Future directions of earthquake precursor studies and short-term earthquake prediction are considered, and conclusions are presented.

2.1. INTRODUCTION

In the context of earthquake prediction, people use the same terminology in totally different ways, and this results in confusion. Earthquake prediction can be classified into the following three types depending on the timescale considered [e.g., *Uyeda*, 2013; *Hayakawa*, 2015]: long-, medium-, and short-term prediction. Long-term prediction involves a timescale of hundreds to thousands years, which can be studied on the basis of plate tectonics, activity of earthquakes, anecdotal records, fault records, archeological survey, etc. Medium-term prediction involves a few decades to a few years, and uses databases of seismicity and crustal movements. Using an earthquake catalog it is possible to evaluate the long- and medium-term occurrence probability of a large earthquake in a certain area during a prescribed period. Long- and medium-term predictions, however, are mere statistical forecasts, and should not be considered as real predictions.

Only short-term earthquake prediction based on reliable earthquake precursors is of real value, but it must specify three important parameters of an impending earthquake: the time (when), the position or epicenter (where), and the prospective magnitude (how big—a few days to 1 week before an earthquake). Although success is very difficult to achieve, short-term earthquake prediction is the only useful and meaningful form for protecting human lives and social infrastructures. The purpose of this chapter is to review earthquake precursor studies initiated after the 1995 Kobe *M* 6.9 earthquake, then to emphasize the importance of electromagnetic phenomena in short-term earthquake prediction, and finally to present future directions of earthquake prediction study in Japan.

Before presenting earthquake precursor studies in Japan, it is necessary to briefly consider the history of general earthquake prediction in the field of seismology in Japan. The essential points that emerge from previous

Hayakawa Institute of Seismo Electromagnetics, Co. Ltd, University of Electro-Communications (UEC), Incubation Center, Chofu, Tokyo, Japan; University of Electro-Communications, Chofu, Tokyo, Japan

Pre-Earthquake Processes: A Multidisciplinary Approach to Earthquake Prediction Studies, Geophysical Monograph 234, First Edition. Edited by Dimitar Ouzounov, Sergey Pulinets, Katsumi Hattori, and Patrick Taylor.

reviews on earthquake prediction in Japan [*Rikitake*, 2001a,b; *Uyeda*, 2012, 2013, 2015] are mentioned briefly. A national earthquake prediction project in the field of seismology started in 1965 and has continued to the present through consecutive 5-year plans. Unfortunately the approach involved in this project does not include short-term earthquake prediction. The general situation of this project for earthquake prediction has remained nearly the same, even after the 1995 Kobe earthquake and the 2011 Tōhoku *M* 9 mega-earthquake. Consequently precursor studies for earthquake prediction in Japan have been carried out extensively only after the 1995 Kobe earthquake, and we are very grateful to have been funded by the government during the years of 1996–2001 in the form of frontier projects (the only such governmental funding to date).

2.2. EARTHQUAKE PRECURSOR STUDIES IN JAPAN

2.2.1. General Description of Precursors

The purpose of this chapter focuses on earthquake precursor research by Japanese colleagues, and where necessary on important contributions by workers in other countries. Specifically, the following account is based mainly on my recent book [*Hayakawa*, 2015].

It goes without saying that short-term prediction absolutely requires study of earthquake precursors. There have been reports of many kinds of earthquake precursors, from ancient Greek times to the present [e.g., *Rikitake*, 2001a; *Molchanov and Hayakawa*, 2008; *Uyeda et al.*, 2008; *Hayakawa and Hobara*, 2010]. Earthquake precursor data can include geodetic signals such as tilt, Global Positioning System (GPS) data, hydrological data such as level, temperature, and chemistry of groundwater, electromagnetic fluctuations in various frequencies, emission of radon and other ionized gases, and anomalous animal behavior [*Rikitake*, 2001a]. Seismological events such as foreshocks and preseismic quiescence also can form precursor data. However, a majority of the reported earthquake precursor data found during the past few decades have been proven to be nonseismological, and such nonseismological (mainly electromagnetic) measurements were mainly undertaken after the 1995 Kobe earthquake [*Hayakawa*,1999, 2009, 2012, 2013; *Hayakawa and Molchanov*, 2002; *Pulinets and Boyarchuk*, 2004; *Molchanov and Hayakawa*, 2008]. As previously mentioned, however, these data were never treated seriously by the long-running Japanese seismological earthquake prediction project.

Figure 2.1 is a conceptual picture of different seismo-electromagnetic phenomena measured by different radiofrequency techniques, and Figure 2.2 summarizes a history of worldwide seismo-electromagnetic studies considered to be significant. The first two observational items in Figure 2.2 refer to the lithospheric effect, or direct effect of the lithospheric pre-earthquake phenomena. The third item is a seismo-atmospheric effect, and the last three items refer to the ionospheric effect. The first geoelectric current measurement has the longest history, including recent achievements of the Greek VAN method (named after its originators P. Varotsos, K. Alexopoulos, and K. Nomicos)[*Varotsos*, 2005]. The second, the study of ULF (ultra-low frequency: < 1 Hz) electromagnetic emissions started with the Spitak *M* 6.9 [*Kopytenko et al.*, 1990; *Molchanov et al.*, 1992] and Loma Prieta *M* 7.1 [*Fraser-Smith et al.*, 1990] earthquakes, and is of extreme importance in short-term earthquake prediction studies. Seismo-atmospheric perturbation, the third item in Figure 2.2, was discovered associated with the 1995 Kobe earthquake [*Kushida and Kushida*, 2002]. The last item of ionospheric perturbation has a relatively short history. Since the convincing evidence of ionospheric perturbation associated with the 1995 Kobe earthquake was found using subionospheric very-low-frequency (VLF 3–30 kHz) propagation [*Hayakawa et al.*, 1996a], there has been widespread use of VLF–LF (30–300 kHz) networks all over the world, including Europe, India, Russia, and South America. Stimulated by the discovery of ionospheric perturbations associated with the Kobe earthquake, many scientists devoted attention to the upper ionosphere (such as the F region) [*Pulinets and Boyarchuk*, 2004], because VLF–LF waves can monitor specifically the lowest part of the ionosphere. Statistical correlations have recently been established between the ionospheric perturbations (both in the lower and upper ionosphere) and earthquakes [*Liu*, 2009; *Hayakawa et al.*, 2010], which provide the potential to attempt to forecast any earthquake as a practical goal. The French satellite DEMETER (detection of electromagnetic emissions transmitted from earthquake regions), dedicated to the study of seismo-electromagnetics, was launched in 2004, and much scientific success has been achieved in the study of how the ionosphere is disturbed due to the lithospheric pre-earthquake effect [*Parrot*, 2012, 2013].

2.2.2. Japanese Research Activity

One more important point related to Figure 2.2 is that, as noted earlier, the seismo-electromagnetic group in Japan were funded only once just after the 1995 Kobe earthquake, and this is considered to be the commencement of extensive precursor studies in Japan [*Nagao et al.*, 2002]. Two institutions were asked to carry out a feasibility study of electromagnetic effects for short-term earthquake prediction over 5 years (1996–2001) in the framework of Earthquake Integrated Frontier by the former Science and Technology Agency (now MEXT

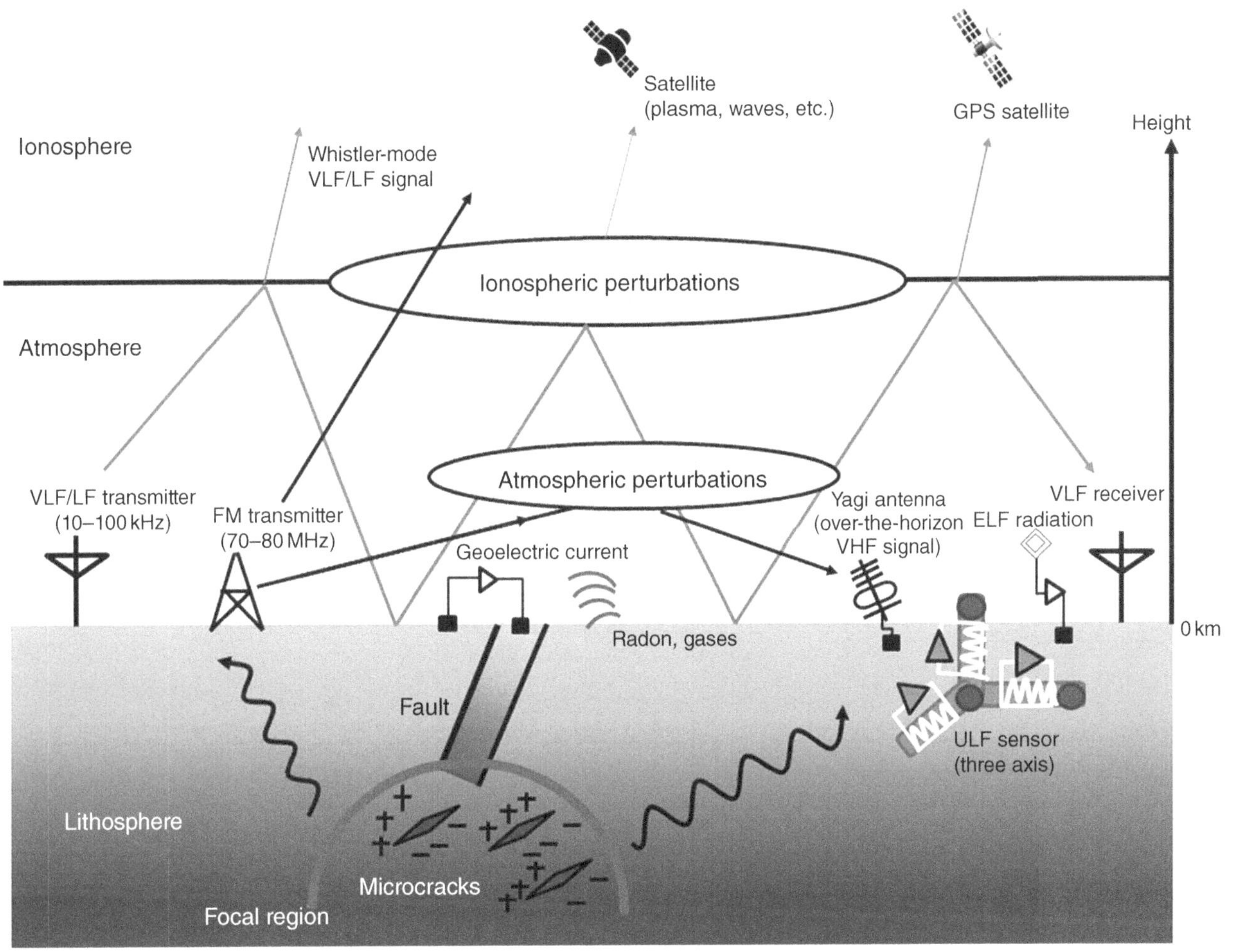

Figure 2.1 Conceptional general view of electomagnetic phenomena in possible association with earthquakes and different radiofrequency techniques to measure those electromagnetic effects.

(Ministry of Education, Culture, Sports, Science and Technology)): (a) RIKEN (Institute of Physical and Chemical Research) and (b) former NASDA (National Space Development Agency of Japan, presently JAXA). S. Uyeda led the RIKEN frontier project while the author led the NASDA frontier project.

The RIKEN group devoted their interests to installing VAN geoelectric measurement stations as in Greece, and finally about 40 stations were established all over Japan. They found geoelectric changes before four out of five $M > 5$ earthquakes [*Uyeda et al.*, 2004]. However, another important conclusion was that the presence of DC electric trains in Japan made it nearly impossible to observe the VAN geoelectric precursors in Japan, which has led them to abandon many geoelectric potential measurement stations and to concentrate their efforts where there are no DC trains. An outstanding finding was that they detected very significant precursors of DC and ULF electric and magnetic radiation before the volcano-seismic activity in 2000 in the Izu island region [*Uyeda et al.*, 2002a]. In addition, 19 anomalous changes in the telluric current were identified during the monitoring conducted on Kozu-shima Island about 170 km south of Tokyo from 14 May 1997 to 25 June 2000. *Orihara et al.* [2012] showed by rigorous statistics that correlation with nearby earthquakes was clearly beyond chance. Also in the Izu island region, anomalous changes in the ULF range (0.01 Hz), starting from a few months before the 2000 major volcano-seismic swarm activity, were observed in both geoelectric and geomagnetic fields. The changes culminated immediately before nearby M 6 earthquakes [*Uyeda et al.*, 2004]. Since then, Japanese colleagues have paid a lot of attention to ULF electromagnetic radiation before an earthquake [*Molchanov and Hayakawa*, 2008]. *Hayakawa et al.* [1996b] reported the third event of seismogenic ULF radiation prior to the 1993 Guam M 7.8 earthquake, a feature also reported by *Fraser-Smith et al.* [1990] and *Kopytenko et al.* [1990] for the earlier Spitak M 6.8 and Loma Prieta M 6.9 earthquakes. In the collaboration of the RIKEN and NASDA Frontier

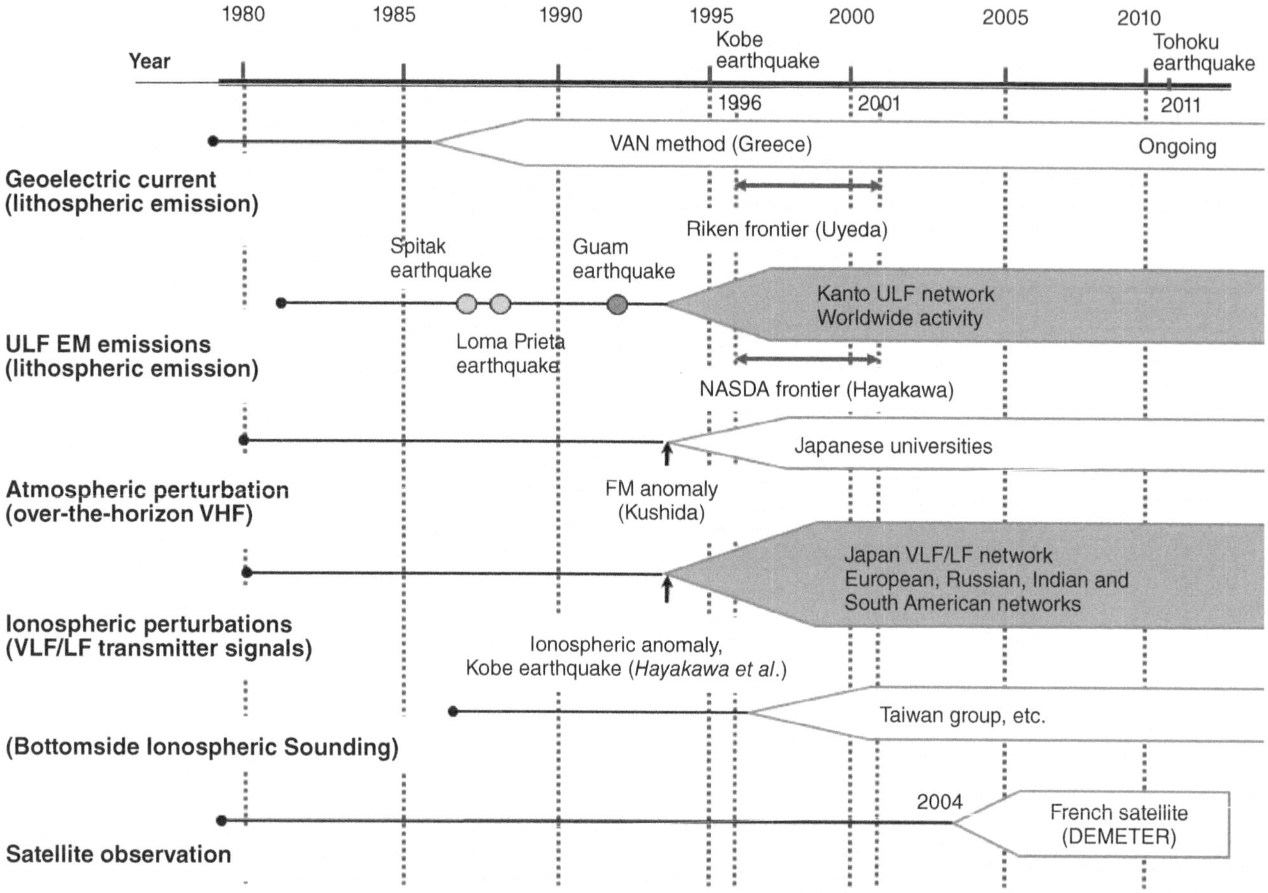

Figure 2.2 History of seismo-electromagnetic studies (including lithospheric and atmospheric effects, and the ionospheric signature.

Projects, a network to observe seismogenic ULF electromagnetic emissions was established in the Tokyo area, comprising a few stations in both Izu and Boso (Chiba) peninsulas as well as others in other areas, and this ULF network was successfully operated for about 10 years [*Molchanov and Hayakawa*, 2008; *Hattori*, 2013]. New signal processing techniques to detect ULF radiation have been developed; for example, for the first time fractal analysis was applied to the 1993 Guam earthquake by *Hayakawa et al.* [1999], subsequently followed by many such studies around the world. Recently, *Hattori* [2013] found statistically on the basis of 10-year data in Tokyo that the probability of occurrence of ULF emissions is higher before rather than after the earthquake.

In the meantime the NASDA Frontier Project team expanded the observation region as much as possible in order to complement its own interests [*Hayakawa et al.*, 2004]. Not only were ULF electromagnetic emissions from the lithosphere targeted in collaboration with the RIKEN team, but the NASDA team also targeted atmospheric and ionospheric signatures of earthquakes by making full use of ground- and satellite-based observations. As for seismo-atmospheric effects, *Fukumoto et al.* [2001] studied the characteristics of over-the-horizon VHF (very high frequency 30–300 MHz) transmitter signals as found by *Kushida and Kushida* [2002]. Although they found that the over-the-horizon VHF signals really were observed prior to an earthquake, their direction-finding results suggested that the reception of those signals is not due to ionospheric reflection, as suggested by *Kushida and Kushida* [2002], but due to atmospheric perturbation. *Fujiwara et al.* [2004] made a statistical study of VHF atmospheric effects as well. *Yasuda et al.* [2009] further extended this VHF study by developing a new interferometric direction finder. This new direction finding enabled them to distinguish between natural VHF seismogenic noises and over-the-horizon VHF signals. Based on those observed characteristics, *Hayakawa et al.* [2007] have proposed a mechanism to explain the seismo-atmospheric perturbation due to an increase in surface temperature, groundwater lifting, and gas emanation, leading to a change in atmospheric refraction index. Moriya and his colleagues established an extensive network for the over-the-horizon VHF signals at many stations in Hokkaido, and *Moriya et al.* [2010] have made a statistical study based on the long-term data on the over-the-horizon VHF signals and found a close correlation of VHF signals with earthquakes. A further seismo-atmospheric effect, ELF (extremely low frequency: 1–10 Hz) pulsive radiation prior to earthquakes, has been detected based on long-term observations in Kamchatka [*Schekotov et al.*, 2007]. Thus the polarization of ELF radiation is suggested as a new parameter among the seismogenic precursor activity. This parameter and the corresponding direction finding have been used to detect such seismogenic ELF radiation before an earthquake [*Hayakawa et al.*, 2012; *Schekotov et al.*, 2013a].

Use of satellites has been one of the most important aspects of the NASDA, aiming to make it possible to monitor features of the Earth's surface, such as the near-surface temperature variation, soil moisture variations, etc., prior to earthquakes. The advanced very high resolution radiometer (AVHRR) onboard the National Oceanic and Atmospheric Administration (NOAA) satellite has provided thermal images of the abnormal enhancement of the outgoing infrared radiation above seismo-active regions in central Asia, China, and Japan [*Tronin et al.*, 2002].

Detecting ionospheric perturbation was the primary concern of the NASDA Frontier Project, during which a network of observing VLF–LF transmitter signals was established (the so-called UEC (University of Electro-Communications) network), comprising seven stations in Japan. Signals from two Japanese transmitters (with call signs of JJY (40 kHz, Fukushima) and JJI (22.2 kHz, Miyazaki) and three transmitters in other countries (NWC (19.8 kHz, Australia), NPM (21.4 kHz, Hawaii) and NLK (24.8 kHz, Jim Creek, Washington State)) are simultaneously received at each station. This VLF–LF network has contributed much to case studies of different large earthquakes occurring in Japan and around the world: 2003 Tokachi-oki *M* 8.3earthquake [*Shvets et al.*, 2004], 2004 Mid-Niigata prefecture *M* 6.8 earthquake [*Hayakawa et al.*, 2006], 2004 Sumatra *M* 9.1 earthquake [*Horie et al.*, 2007], 2007 Niigata Chuetsu-oki *M* 6.6 earthquake [*Hayakawa et al.*, 2008], 2009 L'Aquila *M* 6.9 earthquake [*Rozhnoi et al.*, 2009], and 2010 Haiti *M* 7 earthquake [*Hayakawa et al.*, 2011a]. Statistical studies also utilized this network to examine the close correlation of the subionospheric signal anomaly (ionospheric perturbation) with shallow earthquakes [*Rozhnoi et al.*, 2004; *Maekawa et al.*, 2006; *Hayakawa et al.*, 2010].

Another discovery was made by *Schekotov et al.* [2006] following analysis of a few years of data in Kamchatka, which revealed depression of the ULF magnetic field (horizontal component) before an earthquake as a result of the lower ionospheric perturbation. That is, this new type of anomaly is attributed to the enhanced absorption in the lower ionosphere of conventional ULF emissions of magnetospheric origin, and thus is likely to be the same effect as the VLF–LF ionospheric perturbation.

Lastly, satellite observations of seismogenic phenomena are considered. Initially full use was made of the data observed by the intercosmos (IK)-24 satellite. *Molchanov et al.* [1993] detected ELF–VLF noise anomalies before earthquakes. Furthermore, data on the ionospheric plasma observed by the same satellite was used to study the possible influence of seismicity by gravity waves on the equatorial ionospheric anomaly [*Molchanov et al.*, 2002a,b]. The present author had been involved in the initial planning of the French satellite DEMETER, which was launched in 2004. The satellite data were kindly provided to other interested scientists. *Molchanov et al.* [2006] used the VLF wide-band data to detect the whistler-mode signals from any VLF–LF transmitters, and found that the signal intensity decreased significantly above the epicenter of several earthquakes, including the 2004 Sumatra earthquake, which was consistent with NASDA's ground-based VLF observation. *Muto et al.* [2008] confirmed the conclusion of *Molchanov et al.* [2006]. Studies using satellite data have continued. *Rozhnoi et al.* [2012] studied the spectral broadening of VLF transmitter signals in order to examine seismo-ionospheric irregularities, and *Rozhnoi et al.* [2015] have since compared the satellite VLF data with the corresponding ground-based VLF data to study the seismo-ionospheric response.

2.2.3. Precursors for the 2011 Tōhoku Earthquake

Although most were recognized only retrospectively, there were in fact precursors to the 2011 Tōhoku earthquake. Several reports have now been published on these precursors, mainly by Japanese colleagues, and are listed below.

1. Seismicity: *Nagao et al.* [2014] demonstrated that seismic activation started in about mid-2009 and culminated at the main shock.
2. Electromagnetic phenomena:

   (a) Lithospheric effects. *Kopytenko et al.* [2012] investigated the magnetic field variations at three observatories (Esashi, Mizusawa, and Kakioka) over the 11-year period of 1 January 1 2000 through 31 January 2011. They found a medium-term anomaly beginning around 3 years before the earthquake, and also a short-term precursor in the frequency range of 0.033–0.01 Hz was observed starting from 22 February 2011. Another type of anomalous daily variation of ULF magnetic data was observed approximately 2 months prior to the main shock [*Xu et al.*, 2013]. Unfortunately, these conventional analysis results seemed to require further study, because the anomaly in geomagnetic and ULF radiations is not so intense. After taking these into account,

Hayakawa et al. [2015] performed a critical, so-called natural time, analysis of the ULF data for the 2011 Tōhoku earthquake, and found that the horizontal magnetic field component (f = 0.03–0.05 Hz) fulfilled all criticality conditions for 3–5 March 2011, a few days before the shock, which is likely to be a short-term precursor.

(b) Atmospheric effects. *Ohta et al.* [2013] observed ELF atmospheric radiation on the basis of measurements at three stations in the Nagoya area: Nakatsugawa (Gifu prefecture), Shinojima (in Mikawa Bay), and Izu. It was found that ELF pulsive radiation in the frequency range from 1 to 10 Hz was reliably recorded on 6 March as a precursor to the earthquake. Further confirmation on its seismic origin was provided by the observational fact that the azimuths of the radiation source from all observatories coincided approximately with the region of the forthcoming earthquake.

Schekotov and Hayakawa [2015] extensively investigated the 5-year ULF data at Kakioka around the 2011 Tōhoku earthquake. The sequence of spectra obtained was compared with the evolution of seismicity, which resulted in detection of radiation in the vertical component. The ULF radiation data exhibit seasonal variations with a winter maxima, but can be observed to increase when approaching the moment of the earthquake and to decrease after that. As this radiation seems to be correlated with atmospheric parameters, the authors considered that it could not be caused by a subsurface source, but rather a possible source could be atmospheric discharges.

(c) Ionospheric signatures. *Hayakawa et al.* [2012, 2013a,b] studied the data obtained from their VLF–LF observation network and found a very convincing short-term precursor to the 2011 Tōhoku earthquake. A remarkable anomaly characterized by a decrease in the nighttime amplitude and an enhancement in amplitude fluctuation was detected on 5 and 6 March along the propagation from the NLK (Jim Creek, USA) transmitter to Chofu (together with similar signatures at other stations of Kochi and Kasugai (Nagoya)). The corresponding anomaly was also detected in observations in Russia.

Schekotov et al. [2013a,b] and *Hayakawa et al.* [2013b] studied the depression of ULF (f = 0.03–0.05 Hz) horizontal magnetic field variations of magnetospheric origin at various distances (300–1300 km) from the earthquake epicenter. They found a remarkable depression on 6 March at Kakioka (closest to the earthquake epicenter). *Schekotov et al.* [2013a] statistically correlated this kind of depression with seismicity.

The above two phenomena, subionospheric VLF anomaly and ULF field depression, can be consistently explained by a single cause, perturbation in the lower ionosphere [*Hayakawa et al.*, 2013b]. Another issue of current debate is the imminent variation of ionospheric contents (about 40 h) before this earthquake [*Heki*, 2011; *Kamogawa and Kakinami*, 2013], but further study is needed.

3. Land movements. Another highly promising development is the detection of preseismic land movements using GPS data [e.g., *Chen et al.*, 2013]. *Kamiyama et al.* [2014] investigated GPS data on land movements extensively for a period of about 10 years before the 2011 Tōhoku earthquake, trying to detect medium- and short-term as well as imminent preseismic precursors. They found a latitudinal trend toward the south whereby variation in land movement stopped around 6 months before the earthquake, showing a flat trend thereafter. The westward longitudinal trend in land movement changes also ceased about 2 months before the earthquake. So, some preseismic movements of the crust might have occurred several months before the earthquake. However, the clearest anomaly in crustal movements detected by *Kamiyama et al.* [2014] is the short-term precursor before the earthquake. The longitudinal movement change exhibited a sharp increase over 6–10 March, and stayed flat, in readiness to rebound following the activity of 3–6 March. As mentioned previously, 3–6 March has been reported to be full of electromagnetic signatures. In this sense, it is of crucial importance to combine the seismic (land movement) information with electromagnetic signatures for further understanding of the mechanism of seismo-electromagnetic phenomena.

2.3. FUTURE DIRECTION OF EARTHQUAKE PREDICTION

As of 2015, the main organizations of the earthquake prediction program, including the Seismological Society of Japan and the MEXT Headquarters for Earthquake Research Promotion, maintain an official position that they neither can or want to make any short-term prediction. Over the past two decades, however, significant new directions in precursor research have prevailed concerning short-term earthquake prediction, with the most typical example being electromagnetic phenomena in possible association with earthquakes. Extensive progress has been achieved in this new scientific field since the 1995 Kobe earthquake, as mentioned previusly. The task of short-term earthquake prediction is therefore no longer the preserve of seismologists but can transferred to other researchers with different skill sets who are eager and capable of undertaking this task. It is crucial in this context to understand that short-term earthquake prediction will only be possible by conducting thorough scientific research into precursor phenomena. Taking into account both the long-held resistance of the seismological research community and the enormous progress of seismo-electromagnetics, in 2014 the "Earthquake Prediction Society of Japan" was established, in which the focus is on scientific studies of any type of precursor phenomena, including mechanical (seismic), electromagnetic, and macroscopic approaches.

There are many challenges ahead, but I do feel very optimistic about the future of earthquake prediction. First it will be necessary to address the technical problems listed below.

1. Long-term continuous observations. There are numerous electromagnetic precursors detectable using radiofrequency techniques, but the most important task is to carry out long-term (at least 5 years, but preferably 10 years or so) observations in order to establish a clear statistical correlation between any precursor and earthquakes. The most serious problem in continuous observations is the presence of data gaps owing to malfunction of equipment, so special care has to be paid to the importance of continuous observations of any observational item. It is only from long-term continuous observations, however, that a high probability exists for discovering a new phenomenon in seismogenic studies. Anomalies in Schumann resonances (SR) were a new discovery made by *Hayakawa et al.* [2005], but subsequently an additional discovery was made, also in the ELF band, i.e., unusual SR-like line emissions observed in possible association with earthquakes [*Ohta et al.*, 2009; *Hayakawa et al.*, 2011b], which can be interpreted in terms of excitation of gyrotropic waves in a thin layer of the lower ionosphere [*Sorokin and Hayakawa*, 2014]. Lastly, it is important to note the observation of acoustic emissions [*Gorbatikov et al.*, 2002] as a prominent earthquake precursor.

2. Development of highly sensitive equipment. In any frequency range, as discussed in this chapter, detectors for seismogenic electromagnetic signatures are the first step towards distinguishing seismogenic effects with smaller intensity from other interference noise. Initially, therefore, it is necessary in any frequency band to develop highly sensitive detectors. For example, a new DC geopotential detector free from the problem of contact potential with the ground, a new type of ULF sensor (with a different protocol), various kinds of direction-finding systems in ULF [*Kopytenko et al.*, 2002], ELF–VLF, VHF, and VLF–LF Doppler-shift observations [*Asai et al.*, 2011], and a VHF interferometer direction finder [*Yasuda et al.*, 2009]. Development of such sensitive receivers will surely provide new findings and discoveries of seismogenic effects. Another important direction of development is to reduce the cost of high performance equipment. For example, to establish a dense VLF network all over Japan would require 20–30 (at least) VLF receivers, which would be achievable if the unit price could be reduced.

3. Signal processing techniques. As for new measurement techniques, developing new signal processing techniques for long-term big data is an essential requirement. In the fields of DC/ULF and VLF–LF it is obvious how important new signal processing techniques were in the study of earthquake prediction. Much progress has since been achieved in the vital field of signal processing techniques, including wavelet analysis [*Alperovich et al.*, 2001], principal component analysis (independent component analysis) [*Gotoh et al.*, 2002], and fractal analysis [e.g., *Hayakawa and Ida*, 2008] to study nonlinear processes in the lithosphere and various kinds of direction-finding determination. *Varotsos* [2005] proposed a new conceptual analysis method (similar, in principle, to fractal analysis), a natural time method, in order to specify the date of an impending earthquake with a sufficient accuracy. In the future it will be necessary to consider how important this kind of critical analysis is, in further confirming the result by conventional statistical analysis or identifying any precursor imbedded in the noisy data [*Hayakawa et al.*, 2015]. Further development of new signal processing methods is a necessity.

4. Ability of private companies to release earthquake prediction information. In addition to the above progress in new directions for tackling technical problems, there are promising signs in these endeavors arising from the private sector. Short-term earthquake prediction should not be regarded as the sole domain of national projects or national governments. A few private ventures working on earthquake prediction with the help of scientists have been established, and there are local authorities in high earthquake/tsunami risk areas also involved in this research.

In addition to the above technical problems of data collection and analysis, a final goal of seismo-electromagnetic study is to develop better understanding of the physical mechanisms of different kinds of seismogenic effects (seismogenic emissions, seismo-atmospheric and seismo-ionospheric perturbations, and plasma and electromagnetic signatures in the ionosphere). It is obvious that elucidation of those physical mechanisms will also help much in the improvement of earthquake prediction. There are difficult but challenging problems to be solved in the determination of physical mechanisms, which can be summarized as listed below.

1. Extensive coordination of numerous observations. Seismo-electromagnetic phenomena are known to take place in an extremely wide range of regions, not only in the lithosphere, but also in the atmosphere and ionosphere. The most difficult, but challenging, topic for seismo-electromagnetics (earthquake predictability) is the mechanism of why and how the ionosphere is perturbed prior to an earthquake (i.e., LAIC—lithosphere–atmosphere–ionosphere coupling) is extremely difficult

to comprehend. One possible way for coordinating the various observational items is to establish a field test site (such as Kamchatka as a part of the collaboration between Russians and Japanese [*Gladychev et al.*, 2001; *Uyeda et al.*, 2002b]). The primary requirement for such a test field site is that it must be located at a place free from anthropogenic interference noise. The second requirement is to observe simultaneously as many electromagnetic phenomena in a wide frequency band as possible, as in Figure 2.2, in order to achieve extensive coordination of observational data, which will enable us to observationally verify any particular hypothesis. A few hypotheses for the LAIC [*Surkov and Hayakawa*, 2014; *Sorokin et al.*, 2015] have been proposed previously, including: (a) electrostatic channel [*Freund*, 2009], in which positive hole charge carriers in crustal rocks play the crucial role; (b) chemical channel [*Pulinets and Ouzounov*, 2011], in which radon emanating from the ground is the main player; and (c) atmospheric oscillation channel [*Hayakawa et al.*, 2004; *Molchanov and Hayakawa*, 2008; *Korepanov et al.*, 2009], in which the perturbation in the pre-earthquake Earth surface (such as land motion, temperature and pressure changes, or emanation of ionized gas, etc.) excites the atmospheric oscillations traveling up to the ionosphere. Those who are interested in more details on these LAIC mechanisms should consult the recent monograph by *Sorokin et al.* [2015]. A new direction is the fusion of electromagnetic precursor data with mechanical data such as seismicity and land movements measured by GPS (as done in *Kamiyama et al.* [2014]), because the main agent of LAIC is definitely located in the lithosphere.

2. Detailed study on the mechanisms of generation of seismogenic emissions and also on the physical mechanism of LAIC. Generation mechanisms of electromagnetic emissions in a wide frequency range (from DC/ULF to VHF or even higher) and also mechanisms of how and why both the atmosphere and ionosphere are perturbed by a pre-earthquake fracture effect, are poorly understood at the moment, even though various hypotheses have been proposed for each item of the seismogenic effects [e.g., *Hayakawa*, 2009, 2012, 2013; *Hayakawa and Molchanov*, 2002; *Pulinets and Boyarchuk*, 2004; *Molchanov and Hayakawa*, 2008; *Surkov and Hayakawa*, 2014]. Elucidation on which hypothesis is more plausible or acceptable will be possible with the help of improved radiofrequency techniques and multidisciplinary coordination of different kinds of seismogenic phenomena, together with elaboration of theoretical considerations. The final point to be considered here is the necessity of satellite observations, since it is obvious how effective satellite results are for the scientific study of LAIC [*Parrot*, 2012, 2013]. In near future, Japan has a strong desire to launch its own satellite dedicated to earthquake prediction study (like the French DEMETER satellite [*Parrot*, 2012, 2013]).

2.4. CONCLUSION

The important conclusions derived from this review can be summarized as follows.

1. Enormous progress in the studies of earthquake precursors has been achieved in Japan (and also in other countries) during the past two decades since the 1995 Kobe earthquake, and it has been found that many of earthquake precursors are not seismological, but mainly electromagnetic. Several electromagnetic precursors have been identified, and some have been demonstrated to statistically correlate with earthquakes, including ionospheric perturbations not only in the lower part but also in the upper F region. Those electromagnetic phenomena are likely to be of essential importance in short-term earthquake prediction.

2. Short-term earthquake prediction has essentially two aspects. One is purely the academic science (or obtaining the statistical correlation of any precursor with earthquakes on the basis of long-term observation and elucidation of the associated mechanisms of those precursors), and the other is the actual attempt of earthquake forecast with the use of such precursors. It is strongly recommended here that both directions should be pursued in future, especially in Japan, where private sectors are very active in this research field.

3. Once a statistical correlation of any earthquake precursor with earthquakes has been established, the most important next step is to elucidate the generation mechanism of each earthquake precursor. Then, the main goal of scientific research should be to understand much better the mechanism of LAIC. This woul be possible only with the coordinated measurements of multiple parameters, including the lithospheric, atmospheric and ionospheric. The simultaneous use of data on crustal movements is also required.

4. As earthquake precursor studies will not be supported by the government in Japan, at least in the coming few years, the important role of private sectors in the actual earthquake prediction/forecast should be emphasized. In their earthquake forecast, the most required outcome will be an increase in its accuracy, which should rely on the above-mentioned adoption of multiparameter and multidisciplinary observations.

ACKNOWLEDGMENTS

The author is grateful to many collaborators not only in Japan, but also in other countries. Also he thanks the editors of this book for giving him an opportunity to write this review.

REFERENCES

Asai, S., Yamamoto, S., Kasahara, Y., Hobara, Y., Inaba, T., and Hayakawa, M. (2011), Measurement of Doppler shifts of short-distance subionospheric LF transmitter signals and seismic effects, *J. Geophys. Res.*, doi:10.1029/2010JA016055.

Alperovich, L., Zheludev, V., and Hayakawa, M. (2001), Application of a wavelet technique for the detection of earthquake signatures in the geomagnetic field, *Nat. Hazards Earth System Sci.*, *1*, 75–81.

Chen, C.H., Wen, S., Liu, J.Y., Hattori, K., Han, P., Hobara, Y., Wang, C. H., Yeh, T. K., and Yen. H. Y. (2013), Surface displacements in Japan before the 11 March 2011 *M*.9.0 Tōhoku-Oki earthquake, *J. Asian Earth Sci.*, *80*, 165–171.

Fraser-Smith, A. C., Bernardi, A., McGill, P. R., Ladd, M. E., Helliwell, R. A., and Villard Jr., O. G. (1990), Low frequency magnetic field measurements near the epicenter of the Loma-Prieta earthquake, *Geopys. Res. Lett.*, *17*, 1465–1468.

Freund, F. (2009), Stress-activated positive hole charge carriers in rocks and the generation of pre-earthquake signal, in M. Hayakawa (ed.), *Electromagnetic Phenomena Associated with Earthquakes*, pp. 41–96, Transworld Research Network, Trivandrum, India.

Fujiwara, H., Kamogawa, M., Ikeda, M., Liu, J. Y., Sakata, H., Chen, Y. I., Ofuruton, H., Muramatsu, S., Chuo, Y. J., and Ohtsuki. Y. H. (2004), Atmospheric anomalies observed during earthquake occurrences, *Geophys. Res. Lett.*, *31*, L17110.

Fukumoto, Y., Hayakawa, M., and Yasuda, H. (2001), Investigation of over-horizon VHF radio signals associated with earthquakes, *Nat. Hazards Earth System Sci.*, *1*, 107–112.

Gladychev, V., Baransky, L., Schekotov, A., Fedorov, E., Pokhotelov, O., Andreevsky, S., Rozhnoi, A., Khabazin, Y., Belyaev, G., Gorbatikov, A., Gordeev, E., Chebrov, V., Sinitsin, V., Lutikov, A., Yunga, S., Kosarev, G., Surkov, V., Molchanov, O., Hayakawa, M., Uyeda, S., Nagao, T., Hattori, K., and Noda. Y. (2001), Study of electromagnetic emissions associated with seismic activity in Kamchatka region, *Nat. Hazards Earth System Sci.*, *1*, 127–136.

Gorbatikov, A. V., Molchanov, O. A., Hayakawa, M., Uyeda, S., Hattori, K., Nagao, T., Tanaka, H., Nikolaev, A. V., and Maltsev. P. (2002), Acoustic emission possibly related to earthquakes, observed at Matsushiro, Japan and its implications, in M. Hayakawa and O. A. Molchanov (eds), *Seismo Electromagnetics (Lithosphere–Atmosphere–Ionosphere Coupling)*, pp. 1–10, Terra Scientific Publishing, Tokyo.

Gotoh, K., Akinaga, Y., Hayakawa, M., and Hattori, K. (2002), Principal component analysis of ULF geomagnetic data for Izu islands earthquakes in July 2000, *J. Atmos. Electr.*, *22*(1), 1–12.

Hattori, K. (2013), ULF geomagnetic changes associated with major earthquakes, in M. Hayakawa (ed.), *Earthquake Prediction Studies: Seismo Electromagnetics*, pp. 129–152, Terra Scientific Publishing, Tokyo.

Hayakawa, M. (ed.) (1999), *Atmospheric and Ionospheric Electromagnetic Phenomena Associated with Earthquakes*, Terra Scientific Publishing, Tokyo, 996 pp.

Hayakawa, M. (ed.) (2009), *Electromagnetic Phenomena Associated with Earthquakes*, Transworld Research Network, Trivandrum, India, 279 pp.

Hayakawa, M. (ed.) (2012), *The Frontier of Earthquake Prediction Studies*, Nihon-Senmontosho-Shuppan, Tokyo, 794 pp.

Hayakawa, M. (ed.) (2013), *Earthquake Prediction Studies: Seismo Electromagnetics*, Terra Scientific Publishing, Tokyo, 168 pp.

Hayakawa, M. (2015), *Earthquake Prediction with Radio Techniques*, John Wiley & Sons, Singapore, 294 pp.

Hayakawa, M. and Molchanov, O. A. (eds) (2002), *Seismo Electromagnetics: Lithosphere–Atmosphere–Ionosphere Coupling*, Terra Scientific Publishing, Tokyo, 477 pp.

Hayakawa, M., and Ida, Y. (2008), Fractal (mono- and multi-) analysis for the ULF data during the 1993 Guam earthquake for the study of prefracture criticality, *Curr. Dev. Theory Appl. Wavelets*, *2*(2), 159–174.

Hayakawa, M. and Hobara, Y. (2010), Current status of seismo-electromagnetics for short-term earthquake prediction, *Geomatics, Nat. Hazards Risk*, *1*(2), 115–155.

Hayakawa, M., Molchanov, O. A., Ondoh, T., and Kawai, E. (1996a), The precursory signature effect of the Kobe earthquake on VLF subionospheric signals, *J. Comm. Res. Lab.*, *43*, 169–180.

Hayakawa, M., Kawate, R., Molchanov, O. A., and Yumoto, K. (1996b), Results of ultra-low-frequency magnetic field measurements during the Guam earthquake of 8 August 1993, *Geophys. Res. Lett.*, *23*, 241–244.

Hayakawa, M., Itoh, T., and Smirnova, N. (1999), Fractal analysis of ULF geomagnetic data associated with the Guam earthquake on August 8, 1993, *Geophys. Res. Lett.*, *26*(18), 2797–2800.

Hayakawa, M., Molchanov, O. A., and NASDA/UEC Team (2004), Summary report of NASDA's earthquake remote sensing frontier project, *Phys. Chem. Earth*, *29*, 617–625.

Hayakawa, M., Ohta, K., Nickolaenko, A. P., and Ando, Y. (2005), Anomalous effect in Schumann resonance phenomena observed in Japan, possibly associated with the Chi-chi earthquake in Taiwan, *Ann. Geophysicae*, *23*, 1335–1346.

Hayakawa, M., Ohta, K., Maekawa, S., Yamauchi, T. Ida, Y., Gotoh, T., Yonaiguchi, N., Sasaki, H., and Nakamura, T. (2006), Electromagnetic precursors to the 2004 Mid Niigata Prefecture earthquake, *Phys. Chem. Earth*, *31*, 356–364.

Hayakawa, M., Surkov, V. V., Fukumoto, Y., and Yonaiguchi, N. (2007), Characteristics of VHF over-horizon signals possibly related to impending earthquakes and a mechanism of seismo-atmospheric perturbations, *J. Atmos. Solar-Terr. Phys.*, *69*, 1057–1062.

Hayakawa, M., Horie, T., Yoshida, M., Kasahara, Y., Muto, F., Ohta, K., and Nakamura, T. (2008), On the ionospheric perturbation associated with the 2007 Niigata Chuetsu-oki earthquake, as seen from subionospheric VLF/LF network observations, *Nat. Hazards Earth System Sci.*, *8*, 573–576.

Hayakawa, M., Kasahara, Y., Nakamura, T., Muto, F., Horie, T., Maekawa, S., Hobara, Y., Rozhnoi, A. A., Solivieva, M., and Molchanov, O. A. (2010), A statistical study on the correlation between lower ionospheric perturbations as seen by subionospheric VLF/LF propagation and earthquakes, *J. Geophys. Res.*, *115*, A09305.

Hayakawa, M., Raulin, J. P., Kasahara, Y., Bertoni, F. C. P., Hobara, Y., and Guevara-Day, W. (2011a), Ionospheric perturbations in possible association with the 2010 Haiti earthquake, as based on medium-distance subionospheric VLF propagation data, *Nat. Hazards Earth System Sci.*, *11*, 513–518.

Hayakawa, M., Y. Hobara, K. Ohta, J. Izutsu, A. P. Nickolaenko, and V. M. Sorokin (2011b), Seismogenic effects in the ELF Schumann resonance band, *IEEJ Trans. Fund. Mater.*, *131*(9), 684–690.

Hayakawa, M., Hobara, Y., Yasuda, Y., Yamaguchi, H., Ohta, K., Izutsu, J., and Nakamura, T. (2012), Possible precursor to the March 11, 2011, Japan earthquake: Ionospheric perturbations as seen by subionospheric very low frequency/low frequency propagation, *Ann. Geophys.*, *55*(1), 95–99.

Hayakawa, M., Hobara, Y., Rozhnoi, A., Solovieva, M., Ohta, K., Izutsu, J., Nakamura, T., and Kasahara Y. (2013a), The ionospheric precursor to the 2011 March 11 earthquake based upon observations obtained from the Japan-Pacific subionospheric VLF/LF network, *Terr. Atmos. Ocean. Sci.*, *24*(3), 393–408, doi: 10.3319/TAO.2012.12.14.01 (AA).

Hayakawa, M., Rozhnoi, A., Solovieva, M., Hobara, Y., Ohta, K., Schekotov, A., and Fedorov, E. (2013b), The lower ionospheric perturbation as a precursor to the 11 March 2011 Japan earthquake, *Geomatics, Nat. Hazards Risk*, *4*(3), 275–287, doi:org/10.1080/19475705.2012.751938.

Hayakawa, M., Schekotov, A., Potirakis, S., and Eftaxias, K. (2015), Criticality features in ULF magnetic fields prior to the 2011 Tōhoku earthquake, *Proc. Japan Acad. Ser. B*, *91*, 25–30, doi: 10.2183/pjab.91.25.

Heki, K. (2011), Ionospheric electron enhancement preceding the 2011 Tōhoku-Oki earthquake, *Geophys. Res. Lett.*, *38*, L17312.

Horie, T., Yamauchi, T. Yoshida, M., and Hayakawa, M. (2007), The wave-like structures of ionospheric perturbation associated with Sumatra earthquake of 26 December 2004, as revealed from VLF observation in Japan of NWC signals, *J. Atmos. Solar-Terr. Phys.*, *69*, 1021–1028.

Kamiyama, M., Sugito, M., Kuse, M., Chekotov, A., and Hayakawa, M. (2014), On the precursors to the 2011 Tōhoku earthquake: crustal movements and electromagnetic signatures, *Geomatics, Nat. Hazards Risk*, doi:10.1080/19475705.2014.937773.

Kamogawa, M. and Kakinami, Y. (2013), Is an ionospheric electron enhancement preceding the 2011 Tōhoku-Oki earthquake a precursor?, *J. Geophys. Res. Space Phys.*, *118*, 1751–1754.

Kopytenko, Yu. A., Matiashvili, T. G., Voronov, P. M., Kopytenko, E. A., and Molchanov, O. A. (1990), *Discovering of ultra-low-frequency emissions connected with Spitak earthquake and its aftershock activity with the data of geomagnetic pulsations observations at Dusheti and Vardzija*, IZMIRAN, Moscow, Preprint N3 (888), 27 pp.

Kopytenko, Yu. A., Ismaguilov, V. S., Hattori, K., Voronov, P. M., Hayakawa, M., Molchanov, O. A., Kopytenko, E. A., and Zaitev, D. B. (2002), Monitoring of the ULF electromagnetic disturbances at the station network before EQ in seismic zones of Izu and Chiba peninsulas (Japan), in M. Hayakawa and O. A. Molchanov (eds), *Seismo Electromagnetics (Lithosphere–Atmosphere–Ionosphere Coupling)*, pp. 11–18, Terra Scientific Publishing, Tokyo.

Kopytenko, Yu. A., Ismaguilov, V. S., Hattori, K., and Hayakawa, M. (2012), Anomaly disturbances of the magnetic fields before the strong earthquake in Japan on March 11, 2011, *Ann. Geophys.*, *55*(1), 101–107, doi: 10.4401/ag-5260.

Korepanov, V., Hayakawa, M., Yampolski, Y., and Lizunov, G. (2009), AGW as a seismo-ionospheric coupling responsible agent, *Phys. Chem. Earth*, *34*, 485–495.

Kushida, Y. and Kushida, R. (2002), Possibility of earthquake forecast by radio observations in the VHF band, *J. Atmos. Electr.*, *22*, 239–255.

Liu, J. Y. (2009), Earthquake precursors observed in the ionospheric F-region, in M. Hayakawa (ed.), *Electromagnetic Phenomena Associated with Earthquakes*, pp. 187–204. Transworld Research Network, Trivandrum, India.

Maekawa, S., Horie, T., Yamauchi, T., Sawaya, M., Ishikawa, M., Hayakawa, M., and Sasaki, H. (2006), A statistical study on the effect of earthquakes on the ionosphere, based on the subionospheric LF propagation data in Japan, *Ann. Geophysicae*, *24*, 2219–2225.

Molchanov, O. A. and Hayakawa, M. (2008), *Seismo Electromagnetics and Related Phenomena: History and Latest Results*, Terra Scientific Publishing, Tokyo, 189 pp.

Molchanov, O. A., Kopytenko, Yu. A., Voronov, P. M., Kopytenko, E. A., Matiashvili, T. G., Fraser-Smith, A. C., and Bernardi, A. (1992), Results of ULF magnetic field measurements near the epicenters of the Spitak ($Ms = 6.9$) and Loma Prieta ($Ms = 7.1$) earthquakes: comparative analysis, *Geophys. Res. Lett.*, *19*, 1495–1498.

Molchanov, O. A., Mazhaeva, O. A., Goliavin, A. N., and Hayakawa, M. (1993), Observations by the Intercosmos-24 satellite of ELF-VLF electromagnetic emissions associated with earthquakes, *Ann. Geophysicae*, *11*, 431–440.

Molchanov, O. A., Hayakawa, M., Afonin, V. V., Akentieva, O. A., and Mareev, E. A. (2002a), Possible influence of seismicity by gravity waves on ionospheric equatrial anomaly from data of IK-24 satellite 1. Search for idea of seismo-ionosphere coupling, in M. Hayakawa and O. A. Molchanov (eds), *Seismo Electromagnetics (Lithosphere–Atmosphere–Ionosphere Coupling)*, pp. 275–285, Terra Scientific Publishing, Tokyo.

Molchanov, O. A., Hayakawa, M., Afonin, V. V., Akentieva, O. A., Mareev, E. A., and Trakhtengerts, V. Yu. (2002b), Possible influence of seismicity by gravity waves on ionospheric equatrial anomaly from data of IK-24 satellite 2. Equatorial anomaly and small-scale ionospheric turbulence, in M. Hayakawa and O. A. Molchanov (eds), *Seismo Electromagnetics (Lithosphere–Atmosphere–Ionosphere Coupling)*, pp. 287–296, Terra Scientific Publishing, Tokyo.

Molchanov, O. A., Rozhnoi, A., Solovieva, M., Akentieva, O., Berthelier, J. J., Parrot, M., Lefeuvre, F., Biagi, P. F., Castellana, L., and Hayakawa, M. (2006), Global diagnostics of the ionospheric perturbations related to the seismic activity using the VLF radio signals collected on the DEMETER satellite, *Nat. Hazards Earth System Sci.*, *6*, 745–753.

Moriya, T., Mogi, T., and Takada, M. (2010), Anomalous pre-seismic transmission of VHF-band radio waves resulting from large earthquakes, and its statistical relationship to magnitude of impending earthquakes, *Geophys. J. Int.*, *180*, 858–870.

Muto, F., Yoshida, M., Horie, T., Hayakawa, M., Parrot, M., and Molchanov, O. A. (2008), Detection of ionospheric perturbations associated with Japanese earthquakes on the basis of reception of LF transmitter signals on the satellite DEMETER, *Nat. Hazards Earth System Sci.*, *8*, 135–141.

Nagao, T., Enomoto, Y., Fujinawa, Y., Hata, M., Hayakawa, M., Huang, Q., Izutsu, J., Kushida, Y., Maeda, K., Oike, K., Uyeda, S., and Yoshino, T. (2002), Electromagnetic anomalies associated with 1995 Kobe earthquake, *J. Geodyn.*, *33*, 477–487.

Nagao, T., Orihara, Y., and Kamogawa, M. (2014), Precursory phenomena possibly related to the 2011 *M*9.0 off the Pacific coast of Tōhoku earthquake, *J. Disaster Res.*, *9*, 303–310.

Ohta, K., Izutsu, J., and Hayakawa, M. (2009), Anomalous excitation of Schumann resonances and additional anomalous resonances before the 2004 Mid-Niigata prefecture earthquake and the 2007 Noto Hantou earthquake, *Phys. Chem. Earth*, *34*(6 & 7), 441–448.

Ohta, K., Izutsu, J., Chekotov, A., and Hayakawa, M. (2013), The ULF/ELF electromagnetic radiation before the 11 March 2011 Japanese earthquake, *Radio Sci.*, *48*, 589–596, doi:10.1002/ rds.20064.

Orihara, T., Kamogawa, M., Nagao, T., and Uyeda, S. (2012), Pre-seismic anomalous telluric current signals observed in Kozu-shima Island, Japan, *Proc. Nat. Acad. Sci. USA*, *109*, 19125–19128.

Parrot, M. (2012), Satellite observation of ionospheric disturbance before large earthquakes, in M. Hayakawa (ed.), *The Frontier of Earthquake Prediction Studies*, pp. 738–764, Nihon-Senmontosho-Shuppan, Tokyo.

Parrot, M. (2013), Satellite observation of ionospheric perturbations related to seismic activity, in M. Hayakawa (ed.), *Earthquake Prediction Studies: Seismo Electromagnetics*, pp. 1–16, Terra Scientific Publishing, Tokyo.

Pulinets, S. A. and Boyarchuk, K. (2004), *Ionospheric Precursors of Earthquakes*, Springer, Berlin, 315 pp.

Pulinets, S. and Ouzounov, D. (2011), Lithosphere–atmosphere–ionosphere coupling (LAIC) model—a unified concept for earthquake precursors validation, *J. Asian Earth Sci.*, *41*(4–5), 371–382.

Rikitake, T. (2001a), *Predictions and Precursors of Major Earthquakes: The Science of Macroscopic Anomalous Phenomena*, Terra Scientific Publishing Company, Tokyo, 198 pp.

Rikitake, T. (2001b), *Earthquake Prediction: Achievements and Future*, Nihon-Senmontosho-Shuppan, Tokyo, 617 pp. (In Japanese.)

Rozhnoi, A., Solovieva, M. S., Molchanov, O. A., and Hayakawa, M. (2004), Middle latitude LF (40 kHz) phase variations associated with earthquakes for quiet and disturbed geomagnetic conditions, *Phys. Chem. Earth*, *29*, 589–598.

Rozhnoi, A., Solovieva, M., Molchanov, O., et al. (2009), Anomalies in VLF radio signals prior to the Abrusso earthquake($M=6.3$) on 6 April 2009, *Nat. Hazards Earth System Sci.*, *9*, 1727–1732.

Rozhnoi, A., Solovieva, M., Biagi, P. F., Schwingenschuh, K., and Hayakawa, M. (2012), Low frequency signal spectrum analysis for strong earthquakes, *Ann. Geophys.*, *55*(1), 181–186, doi: 10.4401/ag-5076.

Rozhnoi, A., Solovieva, M., Parrot, M., et al. (2015), VLF/LF signal studies of the ionospheric response to strong seismic activity in the far Eastern region combining the DEMETER and ground-based observations, *Phys. Chem. Earth*, Parts A/B/C, https://doi.org/10.1016/j.pce.2015.02.005.

Schekotov, A. and Hayakawa, M. (2015), Seismo-meteo-electromagnetic phenomena observed during a 5-year interval around the 2011 Tōhoku earthquake, *Phys. Chem. Earth*, Parts A/B/C, doi:10.106/j.pce.2015.01.010.

Schekotov, A., Molchanov, O., Hattori, K., Fedorov, E., Gladyshev, V. A., Belyaev, G. G., Chebrov, V., Sinitsin, V., Gordeev, E., and Hayakawa, M. (2006), Seismo-ionospheric depression of the ULF geomagnetic fluctuations at Kamchatka and Japan, *Phys. Chem. Earth*, *31*, 313–318.

Schekotov, A. Y., Molchanov, O. A., Hayakawa, M., Fedorov, E. N., Chebrov, V. N., Sinitsin, V. I., Gordeev, E. E., Belyaev, G. G., and Yagova, N. V. (2007), ULF/ELF magnetic field variations from atmosphere induced by seismicity, *Radio Sci.*, *42*, RS6S90, doi:10.1029/2005RS003441.

Schekotov, A., Fedorov, E., Molchanov, O. A., and Hayakawa, M. (2013a), Low frequency electromagnetic precursors as a prospect for earthquake prediction, in M. Hayakawa (ed.), *Earthquake Prediction Studies: Seismo Electromagnetics*, pp. 81–99, Terra Scientific Publishing, Tokyo.

Schekotov, A., Fedorov, E., Hobara, Y., and Hayakawa, M. (2013b), ULF magnetic field depression as a possible precursor to the 2011/3.11 Japan earthquake, *J. Atmos. Electr.*, *33*(1), 41–51.

Shvets, A. V., Hayakawa, M., and Maekawa, S. (2004), Results of subionospheric radio LF monitoring prior to the Tokachi ($M=8$, Hokkaido, 25 September 2003) earthquake, *Nat. Hazards Earth System Sci.*, *4*, 647–653.

Sorokin, V. and Hayakawa, M. (2014), Plasma and electromagnetic effects caused by the seismic-related disturbances of electric current in the global circuit, *Mod. Appl. Sci.*, *8*(4), 61–83.

Sorokin, V., Chumyrev, V., and Hayakawa, M. (2015), *Electrodynamic Coupling of Lithosphere-Atmosphere-Ionosphere of the Earth*, Nova Publishers, New York, 355 pp.

Surkov, V. and Hayakawa, M. (2014), *Ultra and Extremely Low Frequency Electromagnetic Field*, Springer, Tokyo, 486 pp.

Tronin, A. A., Hayakawa, M., and Molchanov, O. A. (2002), Thermal IR satellite data application for earthquake research in Japan and China, *J. Geodyn.*, *33*, 477–487.

Uyeda, S. (2012), Earthquake prediction in Japan, in M. Hayakawa (ed.), *The Frontier of Earthquake Prediction Studies*, pp. 3–13, Nihon-Senmontosho-Shuppan, Tokyo. (In Japanese.)

Uyeda, S. (2013), On earthquake prediction in Japan, *Proc. Japan Acad. Ser. B*, *89*, 391–400.

Uyeda, S. (2015), Current affairs in earthquake prediction in Japan, *J. Asian Earth Sci.*, https://doi.org/10.1016/j/jseas.2015.07.006

Uyeda, S., Hayakawa, M., Nagao, T., Molchanov, O., Hattori, K., Orihara, Y., Gotoh, K., Akinaga, Y.,and Tanaka, H. (2002a), Electric and magnetic phenomena observed before the volcano-seismic activity 2000 in the Izu island region, Japan, *Proc. Nat. Acad. Sci. USA*, *99*(11), 7352–7355.

Uyeda, S., Nagao, T. Hattori, K., Noda, Y., Hayakawa, M., Miyaki, K., Molchanov, O., Gladychev, V., Baransky, L., Schekotov, A., Belyaev, G., Fedorov, E., Pokhotelov, O., Andreevsky, S., Rozhnoi, A., Khabazin, Y., Gorbatikov, A., Gordeev, E., Chebrov, V., Lutikov, A., Yunga, S., Kosarev, G., and Surkov, V. (2002b), Russian–Japanese complex geophysical observatory in Kamchatka for monitoring of phenomena connected with seismic activity, in M. Hayakawa and O. A. Molchanov (eds), *Seismo Electromagnetics (Lithosphere–Atmosphere–Ionosphere Coupling)*, pp. 413–419, Terra Scientific Publishing, Tokyo.

Uyeda, S., Nagao, T., and Tanaka, H. (2004), A report from the RIKEN International Frontier Research Project on Earthquakes (IFREQ). *Terr. Atmos. Ocean. Sci.*, *15*, 269–310.

Uyeda, S., Nagao, T., and Kamogawa, M. (2008), Short-term earthquake prediction: Current status of seismo-electromagnetics, *Tectonophysics*, *470*, 205–213.

Varotsos, P. (2005), *The Physics of Seismic Electric Signals*, Terra Scientific Publishing, Tokyo, 338 pp.

Xu, G., Han, P., Huang, Q., Hattori, K., Febriani, F., and Yamaguchi, H. (2013), Anomalous behavior of geomagnetic diurnal variations prior to the 2011 off the Pacific coast of Tōhoku earthquake (*Mw*9.0), *J. Asian Earth Sci.*, *77*, 59–65.

Yasuda, Y., Ida, Y. Goto, T., and Hayakawa, M. (2009), Interferometric direction finding of over-horizon VHF transmitter signals and natural VHF radio emissions possibly associated with earthquakes, *Radio Sci.*, *44*, RS2009, doi:10.1029/2008RS003884.

3

Pre-Earthquake Observations and Their Application in Earthquake Prediction in China: A Review of Historical and Recent Progress

Hui Wang[1], Yongxian Zhang[2], Jie Liu[2], Xuhui Shen[3], Huaizhong Yu[2], Zaisen Jiang[1], and Guomin Zhang[1]

ABSTRACT

Developments in observing and interpreting pre-earthquake phenomena in mainland China are summarized in this paper. The investigated pre-earthquake phenomena include changes in crustal deformation and seismicity, as well as geoelectric, geomagnetic, and groundwater properties. These phenomena are related to crustal deformation changes in and near seismogenic zones prior to an earthquake mainshock. Some general characteristics of pre-earthquake phenomena are listed below, based on case studies of earthquakes that have occurred in mainland China in recent decades.

1. Pre-earthquake phenomena display two different patterns of temporal changes: gradual changes over several months or years and rapid changes over a few days or months.

2. Significant variations in pre-earthquake phenomena usually occur prior to large earthquakes. The pre-earthquake phenomena can be characterized by the amplitude and duration of the anomalous changes, as well as by the spatial and temporal distribution of these changes.

3. Pre-earthquake phenomena are usually observed close to the epicenters of mainshocks. The pre-earthquake phenomena associated with three earthquakes that occurred along the southeastern margin of the Tibetan Plateau in recent years are re-examined. Retrospective studies show that, although preliminary mechanical models provide possible physical explanations for these pre-earthquake phenomena, it is very difficult to use these models to predict earthquakes.

3.1. INTRODUCTION

"银川地震, 每岁小动,民习以为常, 大约春冬居多, 如井水忽混浊, 炮声闪长, 群犬围吠, 即防止患。如若秋多雨水, 冬时未有不震者。"

-*17* 世纪的《银川小志》

"Small earthquakes occur in Yinchuan every year. The residents are accustomed to it. The small earthquakes mostly occur in winter and spring. If wells foul, earth sounds persist and dogs bark, you should be careful, and there may be an earthquake. In particular, extra rain in the autumn would indicate that earthquakes are coming in the winter."

[From the local chronicles of Yinchuan city recorded in the seventeenth century.]

China is one of the few countries in the world with rich and abundant historical records of earthquakes. The above statement is typical of historical documents that describe pre-earthquake phenomena in China. In this quotation, multiple types of pre-earthquake phenomena are described, including small earthquakes, ground

[1]*Key Laboratory of Earthquake Prediction, Institute of Earthquake Science, China Earthquake Administration, Beijing, China*

[2]*China Earthquake Networks Center, China Earthquake Administration, Beijing, China*

[3]*Institute of Crustal Dynamics, China Earthquake Administration, Beijing, China*

Pre-Earthquake Processes: A Multidisciplinary Approach to Earthquake Prediction Studies, Geophysical Monograph 234, First Edition. Edited by Dimitar Ouzounov, Sergey Pulinets, Katsumi Hattori, and Patrick Taylor.

motion, water-well level changes, earth sounds, and abnormal animal activity. Moreover, a perceived correlation is mentioned, namely, that the timing of earthquake occurrence might be related to weather and seasonal changes. Similar historical accounts have been found in numerous other local chronicles. These documents record the earliest stages of earthquake prediction concepts based on observations of pre-earthquake phenomena.

Earthquake hazards in China are very serious. The Chinese catalog of historical earthquakes includes more than 1000 disastrous events with $M \geq 6$ since AD 23. At least 13 of these events were particularly deadly ($M \geq 8$) [*Ming et al.*, 1995]. The 1556 Huaxian *M* 8 earthquake reportedly killed 830,000 people and remains the deadliest earthquake in human history. Modern earthquakes in China have also caused huge losses. The 1976 Tangshan earthquake (*Ms* 7.8) destroyed the entire city of Tangshan and killed ~240,000 people. The 2008 Wenchuan earthquake (*Ms* 8.0) killed more than 70,000 people and injured hundreds of thousands, and the property losses caused by the earthquake have been valued at more than 800 billion RMB. Reducing earthquake hazards by predicting earthquakes has always been one of the national goals of China.

Past huge earthquake disasters led to studies of pre-earthquake phenomena, which were expected to be one of the most important approaches to earthquake prediction in China after 1966. The pre-earthquake phenomena investigated included a wide variety of physical or chemical phenomena that have been observed prior to some earthquakes at least and that might be related to the nucleation of seismogenic processes. These anomalous phenomena represent abnormal changes relative to background conditions, and they occur at different times during the complex process of earthquake generation.

During the past several decades, the practice of using observations of pre-earthquake phenomena to predict earthquakes has primarily met with controversy and failures, except for a small number of successes [*Jackson*, 2004; *Yue*, 2005; *Wang et al.*, 2006; *Chen*, 2010; *Chen and Wang*, 2010]. In the mid-1970s, seismologists were confident that earthquake prediction would be achieved within a short time. This confidence came in part from the prediction in China of the 1975 Haicheng *Ms* 7.3 earthquake. Two official middle-term predictions were made based on observations of pre-earthquake phenomena. On the day of the mainshock, action taken by scientists and government officials saved thousands of lives. This sequence of events effectively constituted an imminent prediction [*Wang et al.*, 2006]. However, another devastating earthquake that occurred 18 months later, the Tangshan *Ms* 7.8 earthquake, forced seismological societies to acknowledge the difficulties involved in earthquake prediction.

The goals of this paper are to survey the published scientific literature that addresses mainland China and to summarize the basic characteristics of recorded pre-earthquake phenomena. Our report summarizes the results that may be used in future earthquake prediction research.

3.2. STUDIES OF PRE-EARTHQUAKE PHENOMENA IN MAINLAND CHINA

3.2.1. Selection of Pre-Earthquake Phenomena

To select reliable pre-earthquake phenomena, two stages of evaluation have been proposed [*Wyss*, 1997b]. First, the pre-earthquake phenomena must have been reported in scientific publications as having occurred prior to some earthquakes at least. All records of pre-earthquake phenomena should be sourced from credible, controlled, and calibrated experiments. Additionally, the times of the detailed observations of the pre-earthquake phenomena should be documented. Second, the pre-earthquake phenomena should be precisely defined. The quantitative values of the pre-earthquake phenomena, the methodology and data processing used in interpreting the observations, and the signal to noise ratio should be clear. Finally, accepted physical models that explain the occurrence of the phenomena should exist [*Scholz*, 1968, 1977].

Based on these criteria, acceptable pre-earthquake phenomena can be recorded at different stages prior to earthquakes that have not yet occurred (Figure 3.1). The phenomena should relate to crustal deformation occurring in and near the seismogenic area prior to the mainshock earthquakes. These phenomena are as follows [*Zhang et al.*, 2013]:

1. Crustal deformation. According to Reid's elastic rebound theory, earthquakes are the consequence of accumulated crustal deformation. The horizontal displacements and vertical displacements produced by crustal movement change during the stages prior to mainshock earthquakes. One of the most significant pre-earthquake phenomena is the acceleration of the rate of crustal deformation and changes in the deformation direction.

2. Seismicity. This phenomenon is already measured by seismic networks and usually consists of characteristic spatial patterns, including seismic doughnuts and gaps, quiescence, frequent background seismicity, foreshocks, and variations in the *b*-value. Some reported seismic parameters, such as focal mechanisms and P- and S-wave velocities, change prior to earthquakes.

3. Geoelectricity and geomagnetism. Pre-earthquake geoelectric and geomagnetic effects include localized changes in electric and magnetic fields. There are two types of pre-earthquake changes related to these phenomena: temporal parameters, such as electrical resistivity, and spatial parameters, including geoelectric and geomagnetic fields. Changes in infrared emissions, another type of

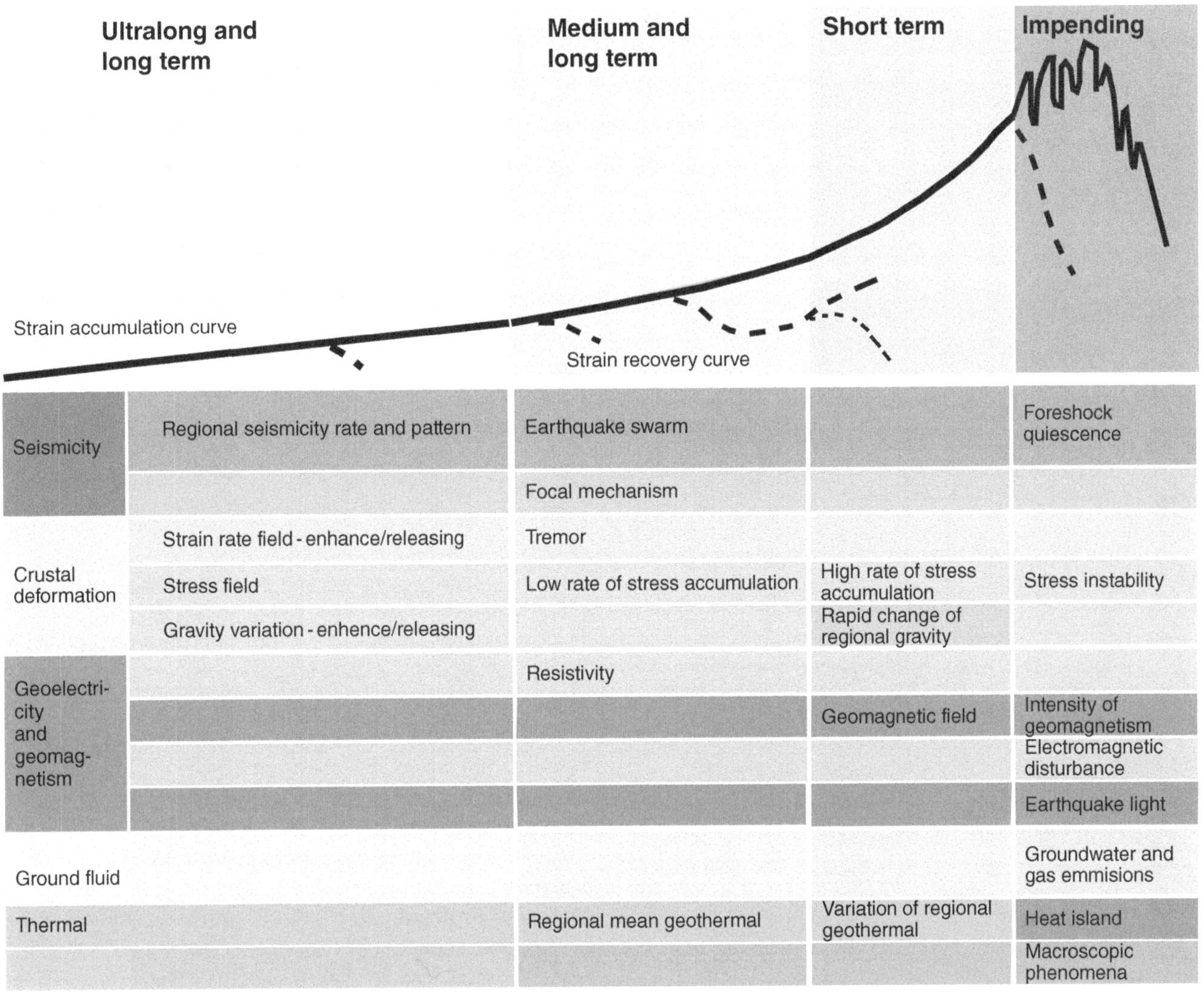

Figure 3.1 Conceptual model showing the relationship between the strain accumulation curve and pre-earthquake phenomena during different stages of earthquake nucleation.

geoelectric effect, from the ground prior to earthquakes have been observed by satellites.

4. Ground fluids. Groundwater is sensitive to changes in crustal stresses and can exhibit various pre-earthquake phenomena, including changes in water level, quality, content, and temperature. Additionally, various gases can be emitted from the earth.

3.2.2. The Earthquake Monitoring System in Mainland China

A multidisciplinary earthquake monitoring system was constructed in mainland China during the 1970s and the 1980s, and it provides the fundamental infrastructure for the study and use of pre-earthquake phenomena. Digital modernization of the system was carried out during the Ninth Five-Year Plan (1996–2000). After the 2008 Wenchuan *Ms* 8.0 earthquake, the monitoring system was updated further. Several new national projects, such as the Global Navigation Satellite System (GNSS) project, the Digital Earthquake Observation Networks project, the Geophysical Background Field project, and the Geophysical Field Monitoring project, as well as experimental sites for studying earthquake prediction, have been proposed to enhance monitoring capabilities in China [*Huang et al.*, 2017].

By 2012, the total number of permanent seismic stations operating in mainland China was more than 1000 (Figure 3.2a). Other permanent stations that make observations of pre-earthquake phenomena have also been deployed and contain strainmeters, gravimeters, magnetometers, earth electricity instruments, and water-level gauges, as well as other types of instruments (Figure 3.2b). The variations in seismicity, crustal deformation (including fixed crustal deformation and borehole strain), geoelectricity, geomagnetism, gravity and underground fluids

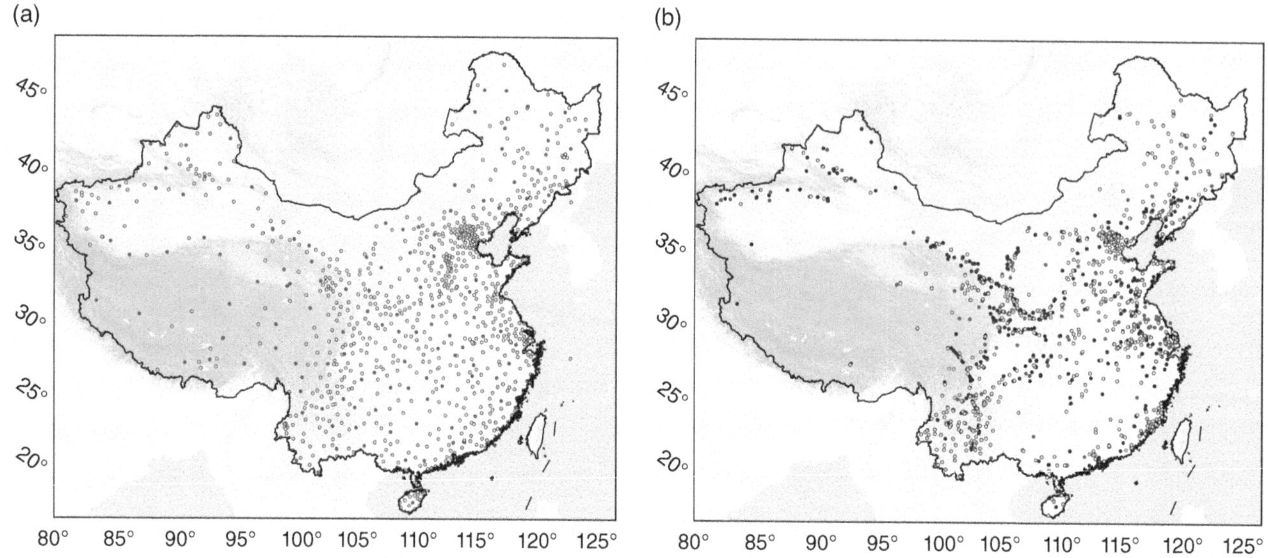

Figure 3.2 Maps showing the distribution of observation stations for monitoring pre-earthquake phenomena in mainland China. (a) China National Digital Seismic Network. Red circles represent national seismic stations; gray circles represent regional seismic stations. (b) Other observation stations that recorded pre-earthquake phenomena. Red circles represent geomagnetic observation stations; blue circles represent crustal deformation observation stations; gray circles represent groundwater observation stations. (*See electronic version for color representation.*)

Table 3.1 Statistical description of the multidisciplinary pre-earthquake phenomena monitoring system operating in mainland China

Methods	Number of stations	Number of instruments
Gravity	39	42
Deformation	246	561
Geomagnetism	162	366
Geoelectricity	139	222
Underground fluid	453	1101
Accessible observation	354	416
Total	1393	2708

(including parameters related to groundwater dynamics and chemistry), and other factors have been observed. More than 1000 permanent stations for observing pre-earthquake phenomena are in operation, and the number of deployed instruments is more than 2000 (Table 3.1). In addition to the permanent observation stations, portable survey lines are also periodically installed. For example, in 2013, mobile crustal-level gauges were installed over a total length of 10,439 km, 4430 mobile gravity stations were established, and 1308 geomagnetic recordings were conducted.

The Crustal Motion Observation Network of China (CMONC) was constructed in 1998–2001 and consists of 27 continuous observation stations and 1056 mobile observation stations [*Li et al.*, 2012]. The network was expanded between 2008 and 2011. A total 233 new continuous observation stations and more than 1000 new mobile observation stations were deployed. The network provides a precise crustal velocity field over all of mainland China (Figure 3.3).

This multidisciplinary earthquake monitoring system provides abundant information. The amount of original data and professional products produced by the pre-earthquake phenomena monitoring network occupies a data volume of more than 1.2 TB [*C. Liu et al.*, 2015]. The volume of original data and data products increases by 400 GB year^{-1}. These data provide ever more detailed information on processes operating inside the Earth. For example, during 2008–2011, the completeness of the catalog of earthquake magnitudes recorded by the China Earthquake Networks Center (CENC) improved to 1.6 from the previous value of 2.6 over most parts of mainland China, except for parts of Tibet and Xinjiang where fewer stations were deployed due to difficult natural conditions [*Mignan et al.*, 2013].

3.2.3. Summarized Characteristics of Pre-Earthquake Phenomena

Based on a comprehensive quantitative analysis of earthquake case studies, some preliminary characteristics of pre-earthquake phenomena in mainland China have been reviewed [*Ma et al.*, 1982; *Jiang et al.*, 2009; *Chen*, 2010]. The scientific methods and working routines for earthquake prediction over the long, intermediate, short and imminent terms were assessed based on investigations

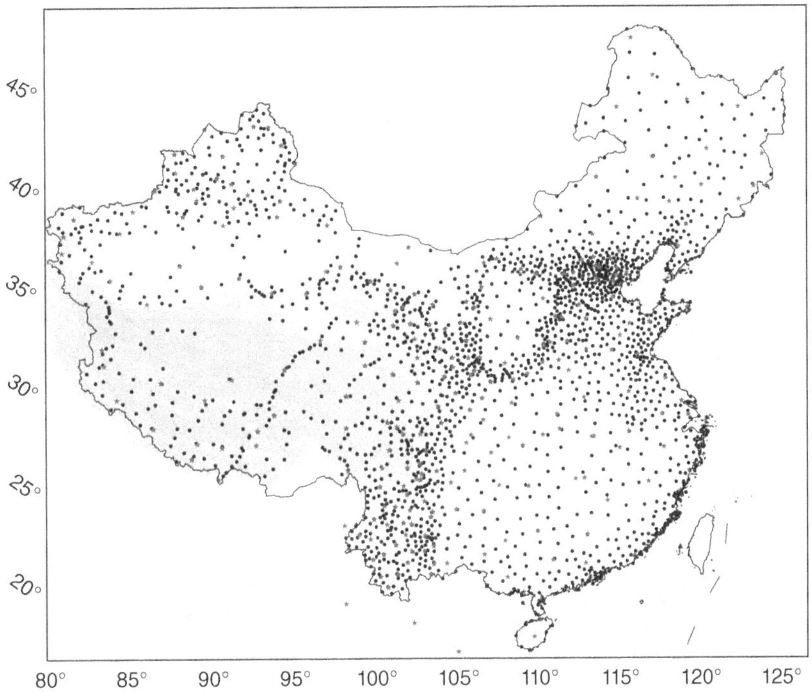

Figure 3.3 Crustal Movement Observation Network in mainland China. Red stars indicate continuous GPS stations; blue circles show GPS campaign stations. (*See electronic version for color representation.*)

of the relationships between the reviewed characteristics and potential future earthquakes. Over the next 10–15 years, the pre-earthquake phenomena that provide useful information over the long to intermediate term could be applied to identify key regions in which to improve the monitoring. Such reviews also can be applied to estimate seismic tendencies over the entire subcontinent of China in annual national-scale assessments. Experiments involving the assessment of the seismic hazard in China have been proposed based on these pre-earthquake phenomena. Case-study-based evaluations of earthquake predictions and assessments of time-dependent seismic hazards within specified areas have provided interesting test examples for practical earthquake predictions.

3.2.3.1. Earthquake Cases in China

Since 1966, more than 1200 earthquakes > M 5.0 have occurred in mainland China in areas covered by the network monitoring pre-earthquake phenomena. Pre-earthquake phenomena prior to some events have been investigated and recorded in a series of books named *Earthquake Cases in China*. This series of books includes ten volumes covering 1966–2006 and is accessible online (http://www.csi.ac.cn/publish/main/828/850/index.html). More than 200 earthquakes have been reported in these books, and the information provided for each earthquake includes the geological background, the level of seismic hazard, and the seismic parameters, such as the seismic activity and focal mechanism, as well as other pre-earthquake phenomena that were noted prior to each event [*Zhang et al.*, 1988, 1990a,b; *Z. Zhang et al.*, 1999, 2000; *Chen et al.*, 2002a,b,c, 2008; *Jiang et al.*, 2014]. The uniform format and specifications of these reports make them valuable data for the comprehensive study of pre-earthquake phenomena in mainland China [*Zhang et al.*, 2013].

We analyzed the basic spatiotemporal characteristics of pre-earthquake phenomena based on *Earthquake Cases in China* from 1986 to 2006, relying on the more reliable records from these years. More than 3,000 observations of pre-earthquake phenomena were reported as having preceded approximately 270 earthquakes (Figure 3.4), of which at least 22 had magnitudes > 7.0, 59 had magnitudes of ~6, and nearly 200 had magnitudes of 4–5. The recorded pre-earthquake phenomena can be classified into 75 different physical parameters (Table 3.2). A detailed introduction to these potential pre-earthquake phenomena and their application to earthquake predictions in China was provided by *Huang et al.* [2017].

3.2.3.2. Frequency Variations of Pre-Earthquake Phenomena

Although pre-earthquake phenomena are common prior to moderate or large earthquakes, they differ in terms of their frequencies prior to different earthquakes. A total of 75 physical parameters of pre-earthquake phenomena were selected for observation by the national

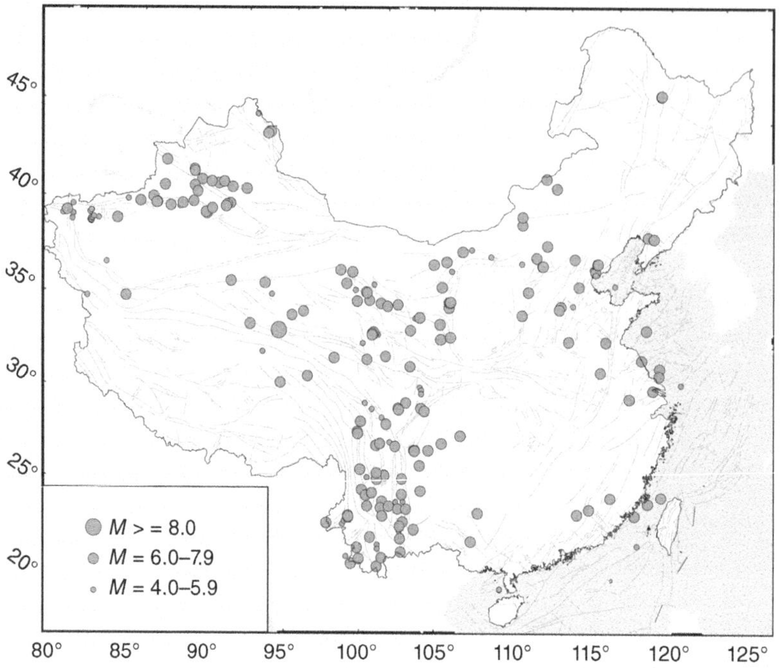

Figure 3.4 Distribution map of earthquakes in the book series *Earthquake Cases in China* during 1966–2006.

monitoring network, but changes in only 44 physical parameters were frequently recorded in *Earthquake Cases in China* from 1986 to 2006. The other 31 pre-earthquake phenomena reportedly occurred fewer than five times.

Figure 3.5 shows the frequency of occurrence of these 44 pre-earthquake phenomena prior to the recorded earthquakes. The parameters or processes associated with pre-earthquake phenomena that occurred with the greatest frequency included (a) earthquake frequency, (b) tilting, (c) seismic gaps, (d) the radon content of groundwater, (e) *b*-values and (f) the water level in wells. These six factors exhibited pre-earthquake changes in 53%, 53%, 45%, 45%, 44% and 38% of the total number of earthquakes, respectively. Hence, these six types of pre-earthquake phenomena appear prior to at least half of all recorded earthquakes.

3.2.3.3. Various Changes in Pre-Earthquake Phenomena

Most pre-earthquake phenomena can be divided into two categories according to their initial time of onset. One type of pre-earthquake phenomenon changes slowly, relative to the pre-earthquake level, and the changes usually last at least 1 year. This type of pre-earthquake phenomenon can be noted at an intermediate time prior to the main earthquake. The other type of pre-earthquake phenomenon changes rapidly over a few days to 1–2 months prior to the subsequent earthquake. If the duration is short and the change is rapid, then this type of pre-earthquake phenomenon mainly appears as a short-term event prior to the mainshock. The different patterns of change may be regarded as an indication of the different stages that occur during the earthquake nucleation process.

Some pre-earthquake phenomena, such as crustal motion and changes in baseline length, the geoelectric field, the geomagnetic field, radon content, and water temperature, have been recorded more frequently as short-term phenomena than as intermediate- or imminent-term phenomena in regions with an epicentral distance of less than 100 km. Additionally, water-temperature changes occurred frequently during the short-term period in regions < 300 km and > 100 km from the epicenter of an approaching mainshock. Fluid chemistry changes also appeared frequently during the short-term period in regions < 300 km and more than 200 km from the epicenter of an approaching mainshock [*Zheng et al.*, 2006]. Consequently, water temperature and fluid chemistry observations may provide additional short-term information on the earthquake nucleation process. At the same time, electromagnetic and crustal temperature changes mainly appear within 3 months of an earthquake. These changes may provide additional information on impending earthquakes.

3.2.3.4. Characteristics of Pre-Earthquake Phenomena and Their Spatiotemporal Distributions

Pre-earthquake phenomena are usually concentrated around seismogenic structures. The scale of the distribution of the recorded pre-earthquake phenomena is proportional to the magnitude of the future earthquake. Pre-earthquake phenomena are sometimes localized

Table 3.2 List of physical parameters of pre-earthquake phenomena observed in mainland China

Subject	Pre-earthquake phenomena items
Seismology	Seismic band, earthquake distribution (temporal, spatial, quiescence or activation), seismic gap (segment), precursory earthquake (swarm), foreshock activity, seismic window, earthquake deficiency, earthquake-swarm activity, induced foreshock, earthquake clustering, earthquake migration.
	Earthquake frequency, *b*-value, strain release (energy release), *h*-value, time interval between earthquakes, modulation ratio of small earthquakes, wave velocity (wave velocity ratio), coseismic stress drop τ, sign-contradiction ratio of P-wave first motions, ambient stress τ_0, quality factor (*Q* value), composite fault plane solution of small earthquakes, amplitude ratio, seismic waveform, total area of fault planes ($\Sigma(t)$), seismic coda wave sustained time ratio τ_H/τ_V attenuation coefficient a of seismic coda wave, S-wave polarization, seismic intensity factor M_f-value, degree of seismic inhomogeneity (GL value), η value, *D* value of seismicity, comprehensive index *A* of seismic activity, degree of seismic concentration *C*, degree of seismic activity *S*, microtremor.
	Earthquake rhythm, seismicity entropy (Q^t, Q^N, $Q\Sigma$), exponential of earthquake situation (*A* (*b*)) value, attenuation rate *p*, special concentration *C*1, degree of seismic activity γ, fractal dimension of magnitude capacity D^0, algorithmic complexity (*C*(*n*), *Ac*), parameter of seismic gap (σH), band concentration *CB*, "*E*, *N* and *S*" elements, fuzzy degree of seismic activity *Sy*, seismic information efficiency (I_q), number of areas of earthquake occurrence (*A* value)
Deformation	Tilt (pendulum, tube)
	Fixed leveling (leveling of short route), leveling (leveling of long route), mobile leveling, sea leveling
	Fixed baseline (short baseline), mobile baseline
	Fault creep, ranging, roving gravity
	Borehole strain (piezo-capacity strain, vibrating-string strain), volumetric strain, strain of fault (extensometer)
	Borehole stress (electric induction stress, steel-string stress)
Geoelectricity and geomagnetism	Vertical magnetic intensity (*Z* variation of geomagnetism, low-point drift of daily variation of geomagnetism, distortion of daily variation of geomagnetism, amplitude difference of geomagnetism), induced magnetic effects (geomagnetic transfer functions)
	Total intensity of geomagnetism
	Magnetic declination
	Roving geomagnetism
	Apparent resistivity (ρs), spontaneous potential (V_{SP})
	Electromagnetic disturbance
	Thermal infrared (outgoing longwave radiation (OLR), brightness temperature), atmospheric electric potential
Ground water	Well-water level, lake-water level, water temperature, (spring) water-flow quantity, variation of oil well, radon content in groundwater, air, and soil, mercury content in groundwater, air, and soil; CO_2, He, H_2, O_2, N_2, Ar, H_2S, CH_4, total amount of gas in groundwater; SiO_2; K^+, Li^+, Ca_2^+, Mg_2^+. CL^-. F^-. HCO_3^-, SO_4^{2-}; total water hardness, pH value, water conductivity, microscopic electricity in groundwater
Comprehensiveness	Comprehensive statistic (anomalous item number, ratio of anomalous station items), precursory information entropy (*H*)
Others	Macroscopic phenomena, ground temperature, atmospheric pressure, drought, waterlogging

Source: *Institute of earthquake Science and China Earthquake Networks Center* [2007].

within a fault zone related to particular geological structures. Figure 3.6 shows the distances over which pre-earthquake phenomena have been recorded for *M* 5, *M* 6, and *M* 7 earthquakes. From this figure, we can see that the pre-earthquake phenomena associated with *M* 5 earthquakes were mainly recorded within epicentral distances of < 200 km. For *M* 6 earthquakes, the pre-earthquake phenomena occur within 300 km of the epicenter, and for *M* 7 earthquakes, the pre-earthquake phenomena occur within 100–500 km of the epicenter, reflecting a wider distribution.

The temporal pattern shows that the pre-earthquake phenomena were mainly recorded within a 5-year period prior to each earthquake. Most of the pre-earthquake phenomena were recorded within 1 month to 2 years prior to the eventual mainshock, indicating that intermediate-term pre-earthquake phenomena are more pronounced than long-term phenomena. The number of reported instances of intermediate-term pre-earthquake phenomena is the largest, while the numbers of reported instances of short-term and impending phenomena are the least. Despite these details, the time at which pre-earthquake phenomena become observable prior to earthquakes is nearly proportional to the magnitude of the future earthquake. For *M* 5 earthquakes, most of the pre-earthquake phenomena were recorded within the 10

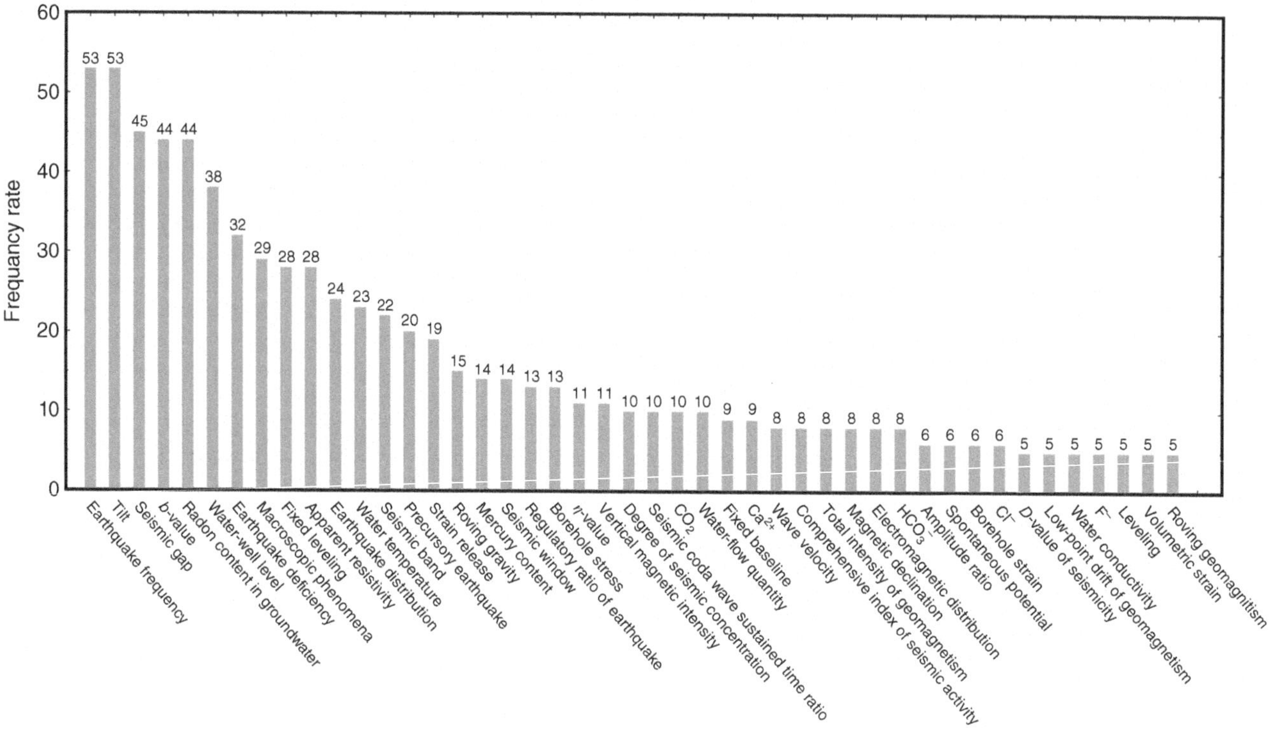

Figure 3.5 Frequency of pre-earthquake phenomena that appeared prior to eventual mainshock earthquakes during 1986–2006, according to the book series *Earthquake Cases in China*.

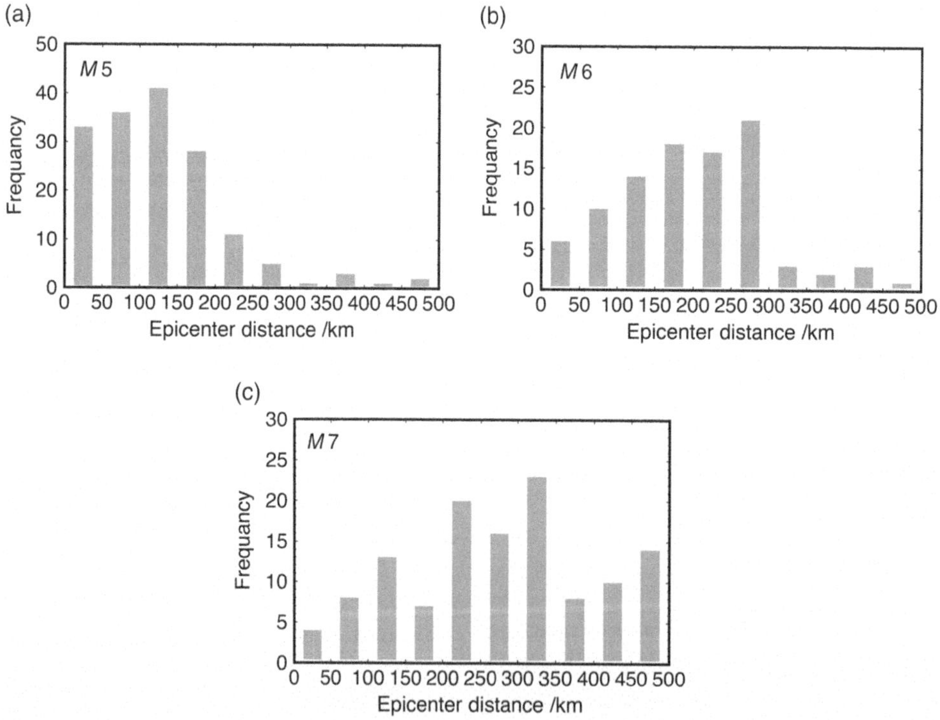

Figure 3.6 Rates of pre-earthquake phenomena for *M* 5, *M* 6, and *M* 7 earthquakes, based on the book series *Earthquake Cases in China*.

days prior to each earthquake. For *M* 6 earthquakes, most of the pre-earthquake phenomena were recorded within 1 month of the earthquake. For *M* 7 earthquakes, the temporal relation is less certain due to the small number of occurrences of events of this magnitude.

In addition, the spatial pattern of intense pre-earthquake phenomena may also change during the different stages before the mainshock. In the long-term and intermediate-term stages, pre-earthquake phenomena tend to appear first in the seismogenic zone then in the surrounding region. In the short-term and impending stages, pre-earthquake phenomena may be concentrated in the seismogenic region [*Ma et al.*, 1982]. The changing patterns of pre-earthquake phenomena may provide another indicator to estimate the place and time of an approaching mainshock.

3.2.3.5. Preliminary Empirical Relationship Between the Spatiotemporal Characteristics of Pre-Earthquake Phenomena and the Magnitude of the Eventual Earthquake

Some preliminary characteristics of pre-earthquake phenomena, including their frequency, variety, changes in amplitude, duration, and spatial extent, show positive correlations with the magnitudes of the eventual mainshock earthquakes. In general, the pre-earthquake phenomena tend to be more pronounced for larger earthquakes. In addition, the number of physical parameters affected by the pre-earthquake phenomena, as well as their duration and spatial extent, will be larger as the magnitude of the eventual earthquake increases. For example, the average numbers of pre-earthquake phenomena recorded prior to *M* 5, *M* 6 and *M* 7 earthquakes are 7.0, 10.3 and 20.3, respectively, based on the *Earthquake Cases in China* book series. This observation provides one of the fundamental facts for estimating the magnitudes of approaching mainshock earthquakes.

Based on case studies of earthquakes in China, some preliminary statistical relationships between the characteristics of pre-earthquake phenomena and the magnitudes of the subsequent mainshock earthquakes have been established. Usually, the duration of the pre-earthquake phenomena and the spatial distribution of the area affected by pre-earthquake phenomena exhibit a logarithmic relationship with the magnitude of the eventual mainshock earthquake. Table 3.3 shows selected statistical relationships between the durations over which the pre-earthquake phenomena can be observed and the magnitudes of the subsequent earthquakes.

The statistical relationships between the amplitudes of pre-earthquake phenomena and the magnitudes of the eventual mainshock earthquakes are complicated. The amplitudes of pre-earthquake phenomena are usually larger closer to the seismogenic zones, and they may attenuate with increasing epicentral distance. This complexity may be due to differences in the tectonic environments around seismogenic zones. Although the amplitudes of pre-earthquake phenomena have been revealed to be larger when the eventual mainshock is larger, the quantitative statistical relationship between them remains unclear. The complexity reflects the nonlinear nature of the process of earthquake nucleation.

Table 3.3 Statistical relationship between the duration of pre-earthquake phenomena and the magnitude of the eventual mainshock based on *Earthquake Cases in China*

Method	Statistical relationship[a]
Vp/*Vs*	$\log T = 0.34M + 0.63$
Seismicity	$\log T = 0.60M - 1.72$
b-value	$\log T = 0.20M + 1.32$
Hydrogeochemistry	$\log T = 0.506M - 0.984$
Resistivity	$\log T = 0.34M - 0.13$
Surface deformation	$\log T = 0.46M - 1.26$
Drought	$\log T = 0.67M - 1.27$
Surface stress	$\log T = 0.19M + 1.11$

[a] *T* is duration of trend pre-earthquake phenomena item, *M* is the magnitude of forthcoming main shock

3.2.4. Pre-Earthquake Phenomena Observed from Space

Space-based technology provides new methods to monitor the Earth, and these methods feature a global scope, high dynamic range, continuous observations, and more opportunities to record information on pre-earthquake phenomena. These advantages of space-based technology might lead to improvements in our understanding of the laws governing earthquake preparation and occurrence. At present, satellite technologies, as represented by the GNSS and satellite-based infrared instruments, have played a significant role in earthquake monitoring.

The CMONC provides detailed direct information on crustal movement and dynamic changes using global positioning system (GPS) measurements [*Li et al.*, 2012]. The GPS observations show that crustal deformation around seismogenic zones is enhanced over a relatively large region and that this enhancement is consistent with strain accumulation on seismogenic structures over the course of several years prior to major earthquakes. For example, the 2001 *Ms* 8.1 Kokoxili earthquake occurred in a region with elevated shear strain [*Jiang et al.*, 2003], and the 2008 Wenchuan earthquake occurred at the edge of a zone of large E–W compressive stresses [*Wu et al.*, 2015]. The high degree of locking and slip deficit in a seismogenic fault may change the regional crustal deformation pattern in the vicinity of the seismogenic zone over the short term [*Wu et al.*, 2015].

Remote sensing monitoring has also become a new aspect of earthquake monitoring. Studies of satellite-based measurements of infrared-related pre-earthquake phenomena have been performed in China since the 1990s [*Xu et al.*, 1991]. Different characteristics of infrared radiation have been used to investigate pre-earthquake phenomena. These characteristics include the brightness temperature (BT), the flux of outgoing longwave radiation (OLR), and the surface latent heat flux (SLHF). More than 100 earthquakes have been studied from the perspective of multiparameter infrared variations, and the results show some general characteristics of these parameters. For example, abnormal variations in these parameters appeared before approximately 50% of earthquakes with magnitudes above 5.0. The anomalies mainly appeared within 2 weeks prior to the earthquakes. The anomalous area was different for each parameter. The BT anomalies are usually isolated and have a wide distribution, whereas the OLR and SLHF anomalies occur near the epicenter and are always distributed along the seismogenic fault. These infrared anomalies are more likely to be detected before earthquakes with magnitudes above 6.0 [*Shen et al.*, 2013].

Satellite-based electromagnetic measurements have also been used to observe other types of pre-earthquake phenomena. The main methods used to detect ionospheric anomalies include ground-based vertical and slant measurements of the ionosphere, the inversion of GPS observations to determine total electron content (TEC), observation of occultation events, and in situ satellite observations. Many studies have noted pre-earthquake ionospheric anomalies for selected large earthquakes [*Le et al.*, 2015]. Data on global earthquake cases also show that these pre-earthquake phenomena typically appear several days prior to the mainshock [*Le et al.*, 2015]. These pre-earthquake phenomena are distributed over a relatively large area in the ionosphere and are more pronounced prior to earthquakes above M 7 [*Shen et al.*, 2013].

However, the areas of pre-earthquake phenomena observed from space are usually larger than those observed from ground stations. For example, several days before the 2008 Wenchuan earthquake, abnormal TEC values in the ionosphere could be observed even in south China and spanned a distance of approximately 11° [*Ma and Wu*, 2012]. Even so, statistics show a positive relationship between most pre-earthquake phenomena observed from space and seismic activity. Further work is needed to perform full temporal and spatial scans that include nonseismogenic areas and aseismic periods.

3.2.5. Some Mechanical Theories of Intraplate Earthquakes in China

Some mechanical theories that describe the nucleation process of intraplate earthquakes have been developed based on pre-earthquake phenomena observations made in China. The results show that most pre-earthquake phenomena are related to crustal deformation that occurs near seismogenic structures prior to the eventual mainshock earthquakes.

Models of rock mechanics experiments illustrate four stages of the behavior of seismogenic faults during the process of earthquake nucleation. These stages reflect stability, metastability, metainstability, and instability. Gradual changes in pre-earthquake phenomena are usually observed during the stability and metastability stages, and rapid changes in pre-earthquake phenomena usually occur during the metainstability stage [*Ma*, 2016]. Earthquakes occur in the instability stage. Such models provide an explanation for the existence of the two kinds of pre-earthquake phenomena.

Conceptual and numerical models have also been proposed. One type of model is based on the genesis of a single seismic event, whereas the other type of model proposes that strong earthquakes in China are related to the interaction of several events. Models of a single seismogenic zone, such as the composite seismogenic model [*Kuo et al.*, 1973], the expansion-creeping model [*Niu*, 1978], and the strong body earthquake-generating model [*Mei*, 1995, 1996a,b], have been proposed since 1970. The composite seismogenic model assumes that the seismogenic zone consists of two types of mechanical units: stress-accumulating units, which are locked high-strength regions, and stress-accommodating units, which are creeping low-strength regions. In this model, the tectonic stress load is transferred from the stress-accommodating unit to the stress-accumulating unit. The stress concentration increases in the stress-accumulating unit until the earthquake occurs. Quantitative simulations based on these models show the rock dilatancy and indicate when the accumulated stress reaches a threshold value. Continuous stress accumulation produces fault creep and accumulated rupture, leading to unstable rupture in the locking units. The model also provides a mechanism that explains both the rapid and longer-term changes seen in pre-earthquake phenomena. Updated versions of this model have been applied to explain the nucleation process of the 2008 Wenchuan Ms 8.0 earthquake [*Zhang et al.*, 2010].

Models of the interactions among strong earthquakes are primarily numerical. The tectonic -block seismogenic model uses a networked system that consists of mechanical springs, dampers, and sliders to simulate the nonlinear process of earthquake nucleation in a fault system [*Zhang et al.*, 1995]. As a first-order approximation of complex fault networks, the model represents some features of the interactions among strong earthquakes in mainland China. The modeling results also provide useful insights into the mechanisms underlying major earthquakes and the evolution of pre-earthquake phenomena.

3.3. SOME CASE STUDIES OF PRE-EARTHQUAKE PHENOMENA OBSERVED OVER THE PAST DECADE

Although abundant data on pre-earthquake phenomena have been collected in the past 50 years in mainland China, it is still difficult to apply them to predict earthquakes [*Wang et al.*, 2006; *Chen*, 2009; *Chen and Wang*, 2010; *Rhoades*, 2010]. Controversies and failures still occur, although there have been a small number of successes in the application of pre-earthquake phenomena [*Wu*, 1999; *G. Zhang et al.*, 1999; *Yue*, 2005; *Wu and Jiang*, 2006; *Chen*, 2010; *Zhuang and Jiang*, 2012].

In the past 10 years, many moderate and strong earthquakes have occurred along the southeastern margin of the Tibetan Plateau. They provide excellent examples that permit examination of the relationship between pre-earthquake phenomena and mainshock earthquakes. Here, we re-examine the pre-earthquake phenomena that occurred prior to three earthquakes in the region, i.e., the Wenchuan *Ms* 8.0 earthquake, the Lushan *Ms* 7.0 earthquake and the Ludian *Ms* 6.5 earthquake (Figure 3.7).

3.3.1. The 2008 Wenchuan *Ms* 8.0 Earthquake

On 12 May 2008, a devastating earthquake occurred on the central-northern segment of the Longmen Shan fault in the southeastern Tibetan Plateau. This earthquake produced a rupture length greater than 400 km [*Xu et al.*, 2009] and caused the greatest amount of damage of any earthquake after the Tangshan earthquake in 1976. The central-northern segment of the Longmen Shan fault was determined to be one of the key national regions for seismic monitoring and earthquake protection in 2005 [*Zhang et al.*, 2006].

Many reports have discussed pre-earthquake phenomena prior to the 2008 Wenchuan event. *Ma and Wu* [2012] provided a collection of the relevant literature. The pre-earthquake phenomena that occurred prior to the Wenchuan earthquake were characterized by continuous trends lasting at least several decades, whereas rapidly evolving changes were rarely detected.

The most significant feature of the pre-earthquake phenomena that occurred prior to the Wenchuan earthquake was a seismic gap on the Longmen Shan fault. Although this fault can generate major earthquakes, historical and instrumental records showed a lack of seismic activity along the fault [*Ming et al.*, 1995; *Division of Earthquake Monitoring and Prediction*, 1999]. The largest historically recorded earthquake to have occurred on this fault is the 1970 Dayi *M* 6.2 earthquake. Studies of paleoseismicity have shown that the recurrence interval of major earthquakes on the Yingxiu-Beichuan fault is on the scale of millennia and that activity on the fault might represent a significant seismic hazard [*Densmore et al.*, 2007]. The spatial distribution of small earthquakes along the Longmen Shan fault showed that the Mianzhu–Maoxian–Jiangyou–Pingwu segment had low *b*-values and relatively high stress concentrations from 1970 to 1999 [*Yi et al.*, 2006]. The earthquakes with magnitudes above M_L 4 that occurred from September 2001 to March 2007 formed a seismic doughnut around the Longmen Shan fault, covering an area of more than 522,400 km^2.

Level measurements on the Xianshuihe fault, which is connected to the Longmen Shan fault, also showed significant changes prior to the Wenchuan earthquake. Observations collected at the Xuxu, Goupu, Zhuwo, and Laoqianning stations showed that the curves connecting the locations of the epicenters during 2002–2008 deviated from the long-term left-lateral shear trend from the mid-1980s to 2001 along the Xianshuihe fault (Figure 3.8). Numerical models showed that interseismic locking on the Xianshuihe fault could increase the loading rate on the Longmen Shan fault [*Luo and Liu*, 2010]. The deviations in the long-term trends present in the level observations on the Xianshuihe fault were regarded as significant pre-earthquake phenomena prior to the Wenchuan earthquake.

Another significant pre-earthquake phenomenon involved gravity variations [*Zhu et al.*, 2010]. A combination of absolute and relative gravity measurements from 1998 to 2000 showed that the Longmen Shan fault represented a boundary between the western and eastern parts of Sichuan province. The gravity anomaly in the western part of Sichuan province was negative, whereas the anomaly in the northern part of Sichuan province was positive. The gravity difference between these two parts of Sichuan province was more than 90×10^{-8} m s^{-2}. During 2000–2002, the gravity difference between the western part of Sichuan province and the northern part of Sichuan province increased. A large gravity gradient formed along the Wenchuan–Maerkang region prior to the Wenchuan earthquake (Figure 3.9).

Unfortunately, although a unique anomalous enhancement in ionospheric electron content in southern China 3 days prior to the 2008 Wenchuan earthquake has been reported by many researchers [*Le et al.*, 2015], other rapid changes reflecting pre-earthquake phenomena prior to the earthquake were insignificant [*Chen and Wang*, 2010]. This lack of observed changes may have resulted from the lack of a dense network of monitoring stations and studies of the region before 2008.

3.3.2. The 2013 Lushan *Ms* 7.0 Earthquake

The *Ms* 7.0 (*Mw* 6.6) Lushan earthquake occurred on 20April 2013, 5 years after the 2008 Wenchuan *Ms* 8.0 earthquake. This large-magnitude event ruptured the southern segment of the Longmen Shan fault. The stress transfer produced by the 2008 Wenchuan earthquake

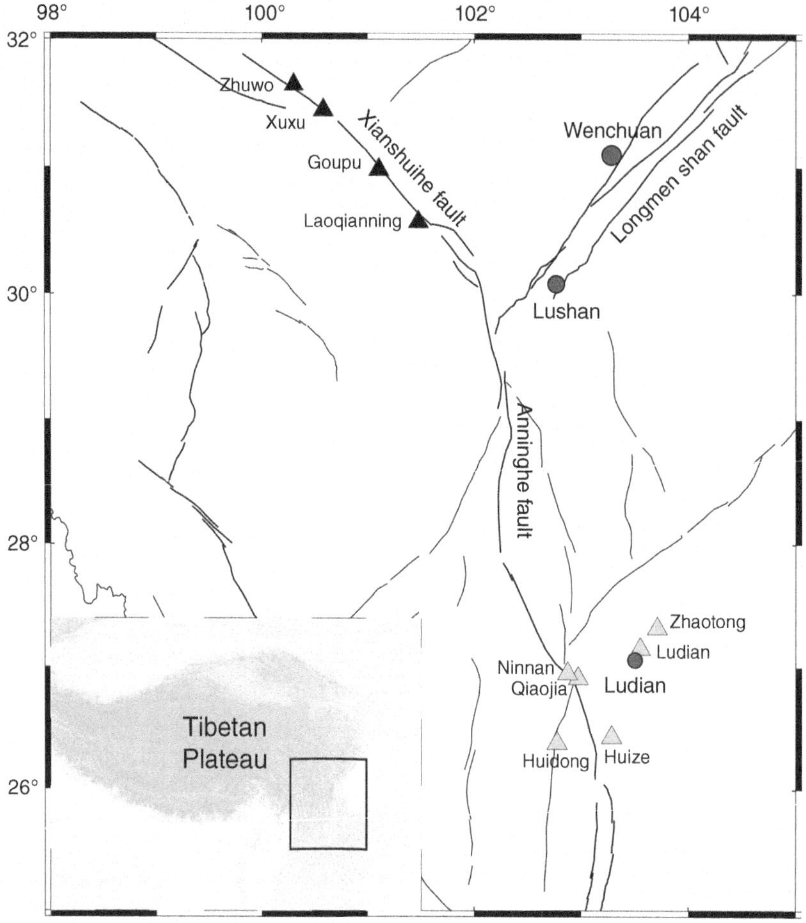

Figure 3.7 Simplified tectonic map of the southeastern margin of the Tibetan Plateau. Red dots show the epicenters of the 2008 Wenchuan *Ms* 8.0 earthquake, the 2013 Lushan *Ms* 7.0 earthquake and the 2014 Ludian *Ms* 6.5 earthquake; blue triangles show the locations of level-measurement stations; pink triangles show the locations of regional groundwater-observation stations. (*See electronic version for color representation.*)

resulted in an increase in the Coulomb failure stress change (ΔCFS) in the southern segment of the Longmen Shan fault after 2008 [*Parsons et al.*, 2008; *Toda et al.*, 2008], indicating an increase in the regional seismic hazard.

Gradual changes reflecting pre-earthquake phenomena include seismicity and gravity variations. Variations in *b*-values along the southern Longmen Shan fault zone show that the fault segment between Baoxing and Chengdu reached high stress levels during 1980–2013 [*Yi et al.*, 2008]. Moreover, a region with increasing ΔCFS was produced by the 2008 Wenchuan earthquake [*Parsons et al.*, 2008; *Toda et al.*, 2008]. Gravity measurements in 2010–2012 showed that the local gravity varied in and near the potential source area. The anomalous variations that were noted near Baoxing, Tianquan, Kangding, Luding, and Shimian were as large as 100×10^{-8} m s^{-2}. The dynamic pattern of regional gravity changes and the pattern of cumulative gravity measurements appeared 2– 3 years before the *Ms* 7.0 Lushan earthquake [*Zhu et al.*, 2013]. Based on comprehensive analysis of the changes in multiple pre-earthquake phenomena, a scientific earthquake prediction was made by *Su et al.* [2014] on 22 February 2013, approximately 2 months before the mainshock on 20 April 2013.

Rapid changes in pre-earthquake phenomena prior to the mainshock were noted after the data from the regional monitoring network were re-examined. Thermal BT anomalies appeared on 15 January 2013, and continued through 19 April 2013. The pre-earthquake phenomena migrated from the Longmen Shan fault and the Xiansuihe fault to the intersection zone between these two faults, near the epicenter of the later earthquake [*Xie et al.*, 2015]. Ground temperature changes were observed at the Kangding station starting on 30 January 2013, and these changes may also have been related to the stress changes in the seismogenic structure associated with the Lushan earthquake [*Chen et al.*, 2013]. Electromagnetic changes

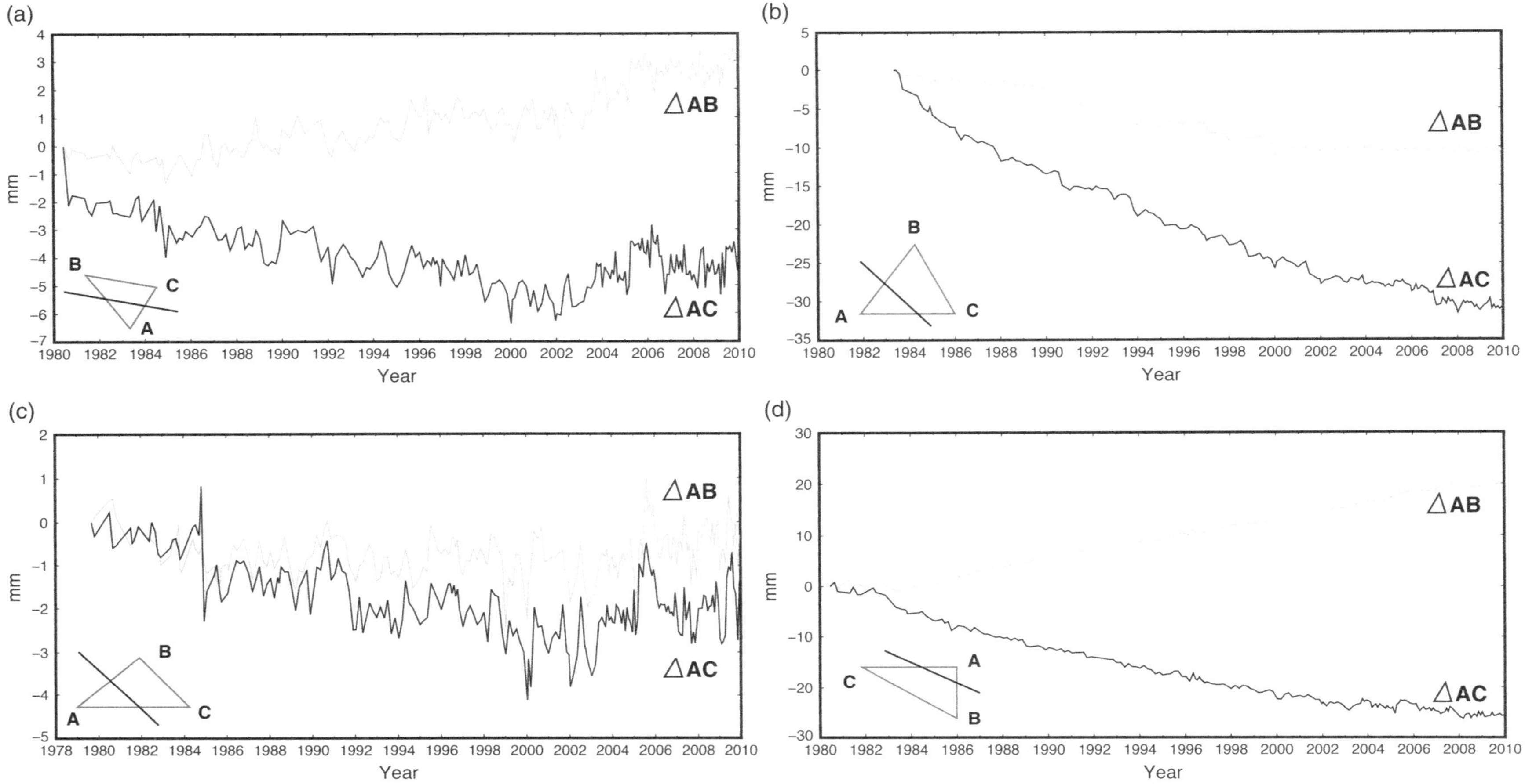

Figure 3.8 Curves of observations collected at fixed baseline stations along the Xianshuihe fault. The black line and gray line show the motion of the two baselines in different directions. The data shown were collected at the following stations: (a) Xuxu, (b) Goupu, (c) Laoqianning, (d) Zhuwo. Station locations are shown in Figure 3.7.

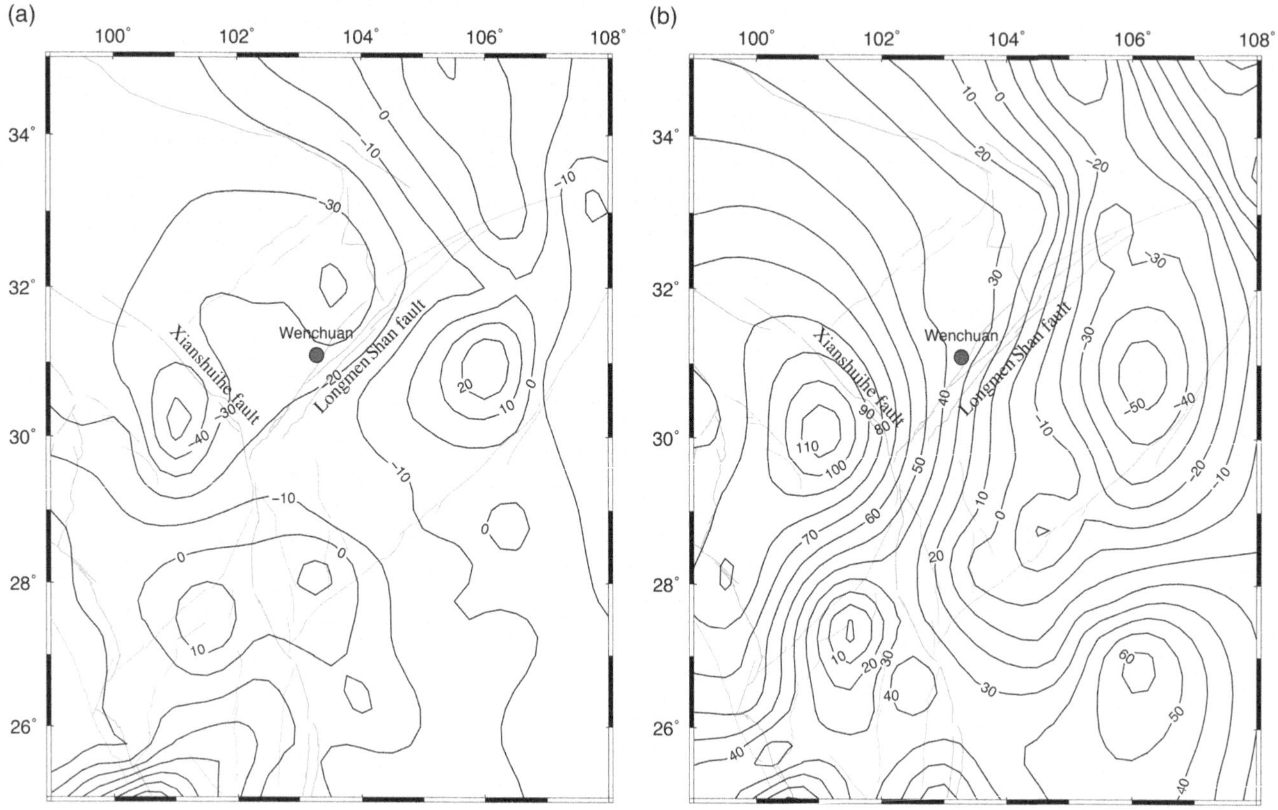

Figure 3.9 Contours of changes (in μGal) before the 2008 Wenchuan earthquake: (a) in 1998–2000; (b) in 2000–2002. Gray lines show major active faults. Red circle shows the epicenter of the 2008 Wenchuan earthquake. (*See electronic version for color representation.*)

before the Lushan earthquake were also recorded at stations around the seismogenic zone. The signal with the greatest lead time was obtained 21 days prior to the mainshock [*Ma et al.*, 2013]. Geoelectric field variations, mainly diurnal waveform distortions and changes in the frequency spectrum characteristics, generally appeared 1–2 months before the earthquake [*An et al.*, 2013]. However, the observed rapid changes in pre-earthquake phenomena were far from the epicenter of the Lushan earthquake.

3.3.3. The 2014 Ludian *Ms* 6.5 Earthquake

The *Ms* 6.5 Ludian earthquake occurred on 3 August 2014, striking Ludian county in Yunnan province, southwestern China [*Cheng et al.*, 2015]. The epicenter of the earthquake was located in the boundary zone between the Tibetan Plateau and the South China block, where the Zhaotong and the Lianfeng faults occur near the southern end of the Mabian fault and other small-scale sinistral faults [*Wen et al.*, 2013]. The earthquake occurred within one of the key regions for seismic monitoring and earthquake protection in the national annual-scale consultation during 2008–2014.

The largest measured earthquake that has taken place in the middle segments of the eastern boundary of the Sichuan–Yunnan tectonic block was a *Ms* 7.1 earthquake that occurred in 1974. The occurrence cycle of *M* 6 earthquakes appears to be quasi-periodic, and the recurrence interval is 17–19 years in this region. Because the last previous event was the 1995 Wuding *M* 6.5 earthquake, the seismic risk in the region in 2014 was determined to be high, based on the characteristic earthquake model [*Nishenko and Buland*, 1987].

The GPS data also show significant changes prior to the event. Figure 3.10 depicts the regional GPS velocity field and crustal shear strain rates. During 2007–2013, sinistral strike-slip movement dominated the shear strain pattern along the eastern boundary of the Sichuan–Yunnan tectonic block. The localized shear stress along the boundary is consistent with the intense seismicity [*Wen et al.*, 2008]. However, the regional crustal shear strain rate increased significantly during 2011–2013, which might have been due to the high degree of locking and slip deficit in the seismogenic zone of the Ludian earthquake [*Wu et al.*, 2015].

Variations in groundwater near seismogenic zones represent a significant type of rapid change associated

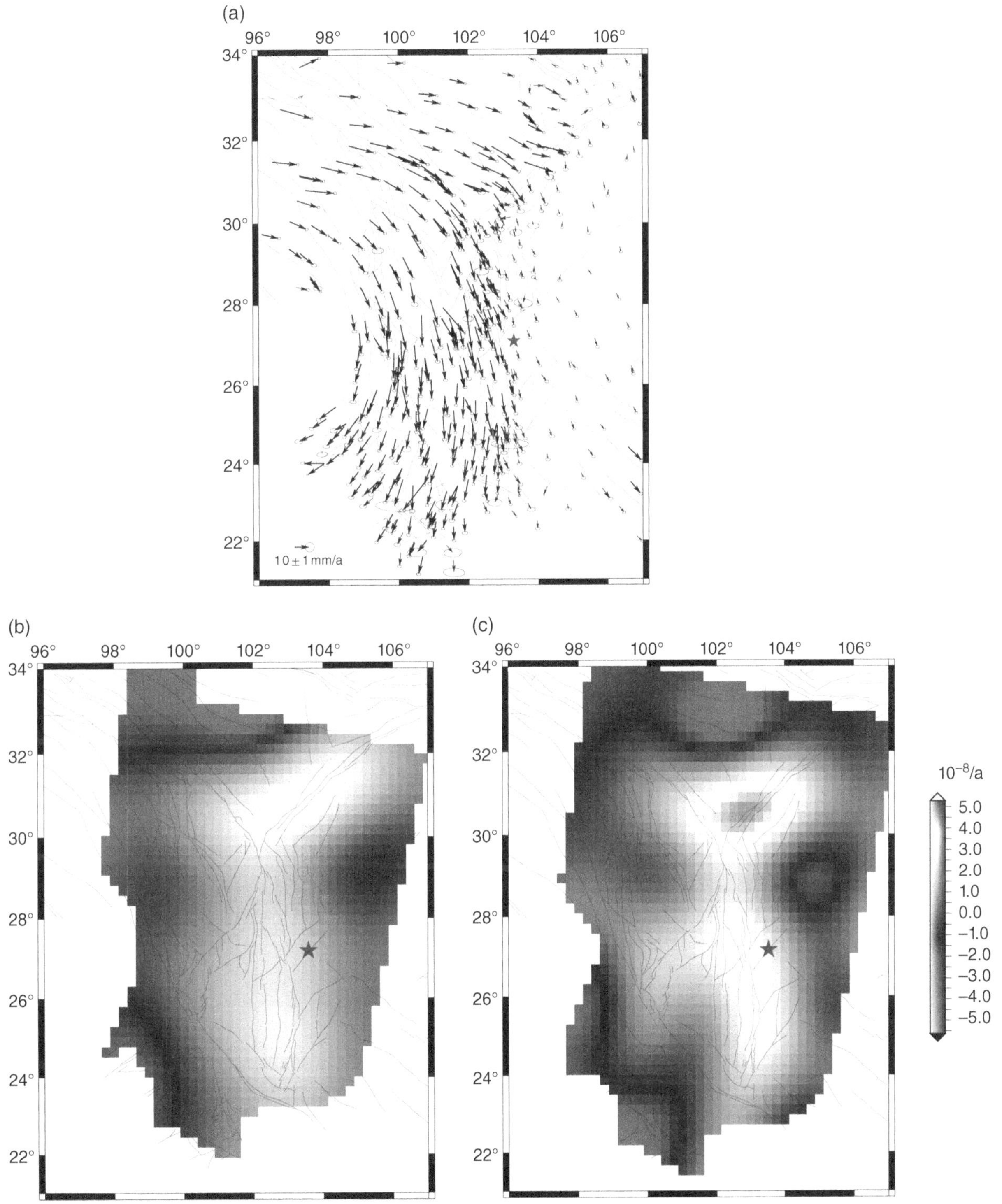

Figure 3.10 The GPS velocity field and regional shear-strain-rate fields prior to the Ludian earthquake. The red star represents the epicenter of the Ludian earthquake. (a) GPS site velocities relative to stable southern China during 2007–2013; (b) shear strain rate field during 2007–2011; (c) shear strain rate field during 2011–2013. (*See electronic version for color representation.*)

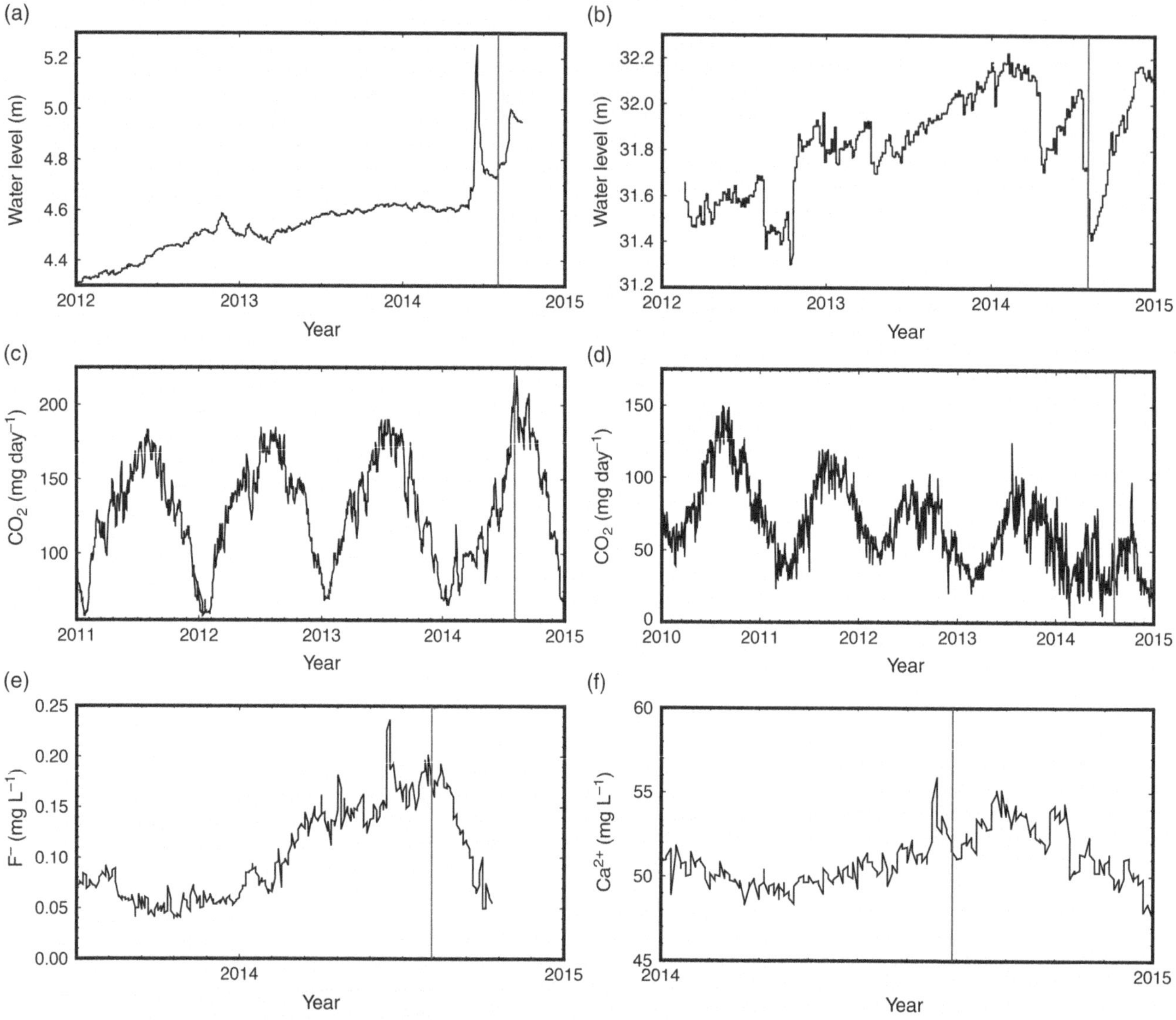

Figure 3.11 Curves of observations from fixed groundwater observation stations near the epicenter of Ludian earthquake. (After Liu *et al.* [2015b].) Red lines show the time of occurrence of the Ludian earthquake. The data shown were collected at the following stations: (a) Zhaotong, (b) Huize, (c) Ninnan, (d) Huidong, (e) Ludian, (f) Qiaojia. Station locations are shown in Figure 3.7. (*See electronic version for color representation.*)

with pre-earthquake phenomena. Rapid groundwater changes occurred 6 months before the earthquake within the seismogenic zone [*Y. Liu et al.*, 2015]. For example, the water levels in wells at the Zhaotong station and the Huize station changed significantly. Rapidly increasing well water levels were observed at the Zhaotong station in June 2014, two months prior to the mainshock (Figure 3.11a). Similar drastic changes, with an amplitude reaching 0.59 m, occurred at the Huize station in July 2014 (Figure 3.11b). Emissions of CO_2 at the Ningnan station reached a peak value in July 2014, a few days before the mainshock (Figure 3.11c), but the emissions at the Huidong station were at the lowest value observed during the preceding 4 years in May 2014 (Figure 3.11d). The CO_2 emissions at the Ludian station increased continuously after February 2014 but started to decrease after the mainshock (Figure 3.11e). The concentrations of Ca^+ measured in the groundwater at the Qiaojia station changed rapidly in July 2014 (Figure 3.11f). All these stations are located near the seismogenic zone of the Ludian earthquake. These reported rapid changes might provide some useful insights into the nucleation process of the Ludian earthquake.

3.3.4. Summary

The 2008 Wenchuan earthquake, the 2013 Lushan earthquake, and the 2014 Ludian earthquake all occurred along the eastern margin of the Tibetan Plateau. However, the characteristics of the pre-earthquake phenomena preceding each event differed. The observed changes associated with the pre-earthquake phenomena prior to the Ludian earthquake were rapid and significant near the epicenter, whereas the rapid changes that occurred prior to the Wenchuan earthquake and the Lushan earthquake occurred far from the epicenters of these events. Hence, the relationship between pre-earthquake phenomena and these two events remains unclear. Such differences may be due to differences in the nucleation mechanisms of these earthquakes [*Ma*, 2016]. The Wenchuan and Lushan earthquakes both occurred along the Longmen Shan fault and were thrust events, while the Ludian earthquake was mainly a strike-slip event.

3.4. DISCUSSION

3.4.1. Some Key Problems in the Observations of Pre-Earthquake Phenomena

The multidisciplinary earthquake monitoring system is still sparse in mainland China. The average distance between stations is approximately 100 km. This distance is far larger than the nucleation zone of a major earthquake [*Ohnaka*, 2000]. Many pre-earthquake phenomena can be observed in certain isolated stations near the epicenter. At the same time, the spatial distribution of the monitoring network is uneven. The network is denser in eastern China and sparser in western China due to the limitations imposed by the environment and the requirements for additional data in regions with greater population. However, from a scientific perspective, the network is insufficient because the seismicity in western China is more severe than that in eastern China.

The mechanisms responsible for pre-earthquake phenomena remain unclear. Although pre-earthquake phenomena are related to crustal deformation that occurs in and near seismogenic areas prior to mainshock earthquakes, the nature of this relationship is still being explored. In particular, the relationship between reported pre-earthquake phenomena and the eventual earthquake is sometimes controversial. For example, anomalous tremor signals were widely detected by broadband seismographs 1–3 days before the mainshock of the 2008 Wenchuan earthquake [*Hu and Hao*, 2008]. The controversy over whether these signals represented pre-earthquake phenomena or were simply noise caused by unrelated factors lasted for several years [*Fu et al.*, 2009; *Hu et al.*, 2010].

In addition, human activities associated with rapid urbanization have also disturbed earthquake monitoring observations in mainland China in recent years. Some established earthquake monitoring stations have had to be abandoned due to engineering projects and construction. Human activities also produce noise that may hide weak signals related to the earthquake nucleation process. Optimization of the spatial distribution of multidisciplinary earthquake monitoring systems, as well as the development of observation technologies and earthquake prediction methods, are necessary.

3.4.2 Difficulties and Potential Future Improvements in the Application of Observations of Pre-Earthquake Phenomena to Predict Earthquakes

Although we have made considerable efforts to observe and comprehensively analyze pre-earthquake phenomena in the past [*Huang et al.*, 2017], the observability of pre-earthquake phenomena remains limited, and the application of these observations in the prediction of earthquakes remains extremely difficult.

The stations that observe pre-earthquake phenomena are located on the Earth's surface, whereas seismic ruptures are typically initiated at depths of 10–20 km. We cannot directly observe the nucleation process of earthquakes in the deeper reaches of the crust. Due to the limitations of the available observation technology, we have to observe pre-earthquake phenomena by using sparsely distributed instruments located on the surface of the Earth or in the shallow parts of the crust. Incomplete, inadequate and sometimes imprecise observations provide insufficient data to reliably identify pre-earthquake phenomena that are directly related to earthquakes.

The scale of the distribution of pre-earthquake phenomena is usually larger than the scale of the eventual earthquake rupture. As shown in Figure 3.6, the pre-earthquake phenomena associated with M 5, M 6, and M 7 earthquakes have mainly been observed within 200 km, 300 km, and 500 km, respectively, of the earthquake epicenters. Although this apparent relationship may be partly caused by the regions analyzed in the relevant case studies, which were limited to 200 km, 300 km, and 500 km from the epicenters of M 5, M 6, and M 7 earthquakes, respectively [*Zhang et al.*, 2013], the observations clearly reflect an exponential relationship between magnitude and rupture length. On the other hand, the average rupture lengths for magnitude 5, 6, 7, and 8 earthquakes are approximately 5 km, 15 km, 40 km, and 100 km, respectively [*Wells and Coppersmith*, 1994]. It is difficult to determine the location of the epicenter of a major event based only on recorded observations of pre-earthquake phenomena. Furthermore, distinguishing the pattern of pre-earthquake phenomena preceding large earthquakes from that preceding some smaller earthquakes is quite challenging. The pre-earthquake phenomena associated with the 2008 Wenchuan Ms 8.0 earthquake support this statement. The earthquake occurred on the central and northern segments of the Longmen Shan fault, and the length of the rupture zone was greater than 400 km. Prior to the event, many pre-earthquake phenomena appeared throughout the Sichuan-Yunnan tectonic block, while fewer appeared along the Longmen Shan fault [*Chen*, 2010]. Such spatial patterns have led to controversial interpretations of the relationship between pre-earthquake phenomena and earthquakes on the Longmen Shan fault. Prior to this disastrous event, some authors identified two regions with high seismic hazard levels around the Longmen Shan fault. One of these regions was located along the northern part of the fault, and the other was located along the southern part of the fault. The seismic potential of the Longmen Shan region was underestimated.

The recurrence intervals of major earthquakes are typically much longer than the times over which earthquake-monitoring systems have been in operation. Slow tectonic loading in midcontinental areas results in long recurrence intervals and few repeat major intraplate earthquakes. The recurrence interval of major earthquakes in mainland China is usually more than 1000 years [*Xu and Deng*, 1996; *Liu-Zeng et al.*, 2007; *Lin and Guo*, 2008; *Harkins et al.*, 2010; *Liu et al.*, 2011]. It is very difficult to obtain information, especially on pre-earthquake phenomena, that covers the whole earthquake process from initial stress accumulation to rupture. At the same time, different geological structures and crustal environments are likely to produce different spatiotemporal patterns of pre-earthquake phenomena. The collection of a sufficient number of cases over a short time is extremely difficult. In addition, our knowledge about the pre-earthquake phenomena is mainly based on moderate earthquakes, which occur more frequently. Thus, we may be able to obtain more scientific data on pre-earthquake phenomena related to moderate earthquakes than those related to large earthquakes.

Acceptable physical models that explain the occurrence of pre-earthquake phenomena are limited. The initiation and occurrence of earthquakes are the results of a self-organizing geological evolution process that is constrained by the complex geodynamics of the deep crust [*Geller et al.*, 1997; *Wyss*, 1997a]. Many processes, such as stress transfer from surrounding plates, local mantle convection, and interaction between faults and blocks, affect this process. It is typically believed that the seismicity within seismic zones reflects an active-quiet-active pattern. This pattern indicates that the observed pre-earthquake phenomena may be related to a seismic sequence rather than a single earthquake. For example, the recurrence interval of magnitude 6 earthquakes at the Parkfield site on the San Andreas Fault was thought to be 22 ± 3 years [*Bakun and Lindh*, 1985]. However, the predicted earthquake did not occur in approximately 1988, as expected. Instead, it occurred in 2004, 16 years later. One possible explanation for this delay involves the effects of a stress shadow produced by several previous earthquakes nearby [*Jackson and Kagan*, 2006]. Such complex geodynamic behavior would result in different patterns of pre-earthquake phenomena. We know little about such complex geophysical processes, and previously proposed physical models for earthquake nucleation and earthquake development are rough approximations. Although these theories are plausible and provide some understanding of the earthquake nucleation process, many parameters in the models remain poorly known.

Objective, credible, sufficient, and continuous data on earthquake processes represent the scientific foundation of earthquake prediction. New technology applied to the Earth sciences, such as space-based observations, broadband seismographs, deep-hole seismic detection [*Niu et al.*, 2008], ocean observation technology, and active observations involving artificially controlled sources, all provide background information in support of high-density, broadband dynamic, wide-coverage, multiparameter earthquake data. These technologies yield new scientific data that permit us to advance our understanding of earthquakes. At the same time, new technologies often yield data that challenge the ways in which we interpret earthquake processes and force us to examine these processes in new and different ways. The construction of four-dimensional fields that display multiple variables, such as crustal deformation and stress–strain relationships, as well as geophysical and geochemical parameters, is a goal worth pursuing [*Plesch et al.*, 2007; *Shi et al.*, 2014; *Shaw et al.*, 2015]. The observation and application of pre-earthquake phenomena will face new opportunities and challenges in the future.

3.5. CONCLUSIONS

Based on earthquakes that have occurred in China, some preliminary characteristics were derived from observations of pre-earthquake phenomena. The following major conclusions could be drawn from this work.

1. The temporal patterns of pre-earthquake phenomena vary prior to the eventual mainshock earthquakes. Some types of pre-earthquake phenomena are observed several months to several years prior to major earthquakes. They usually evolve slowly and steadily, exhibiting deviations from a long-term trend. Other types of pre-earthquake phenomena tend to change rapidly a few days to a few months prior to major earthquakes.

2. Larger variations associated with pre-earthquake phenomena are usually recorded prior to large earthquakes. In general, more types of pre-earthquake phenomena are recorded before large earthquakes. Moreover, the duration of these phenomena is longer, and the spatial extent over which they can be observed is larger.

3. The spatial distributions of pre-earthquake phenomena are uneven. Pre-earthquake phenomena are typically noted close to the epicenter of the eventual earthquake, and they are concentrated along the seismogenic fault. Before earthquakes of magnitude 7 or larger, pre-earthquake phenomena tend to occur across regions with varying diameters around the epicenters, whereas they tend to occur close to the epicenter in the short term and impending term.

4. Pre-earthquake phenomena are very complex. Although we have observed many pre-earthquake phenomena and some preliminary features of these data have been noted in the past 50 years, the relationship between these phenomena and the approaching mainshock remains controversial. Although some earthquakes exhibited significant pre-earthquake phenomena, many other earthquakes were not accompanied by significant pre-earthquake phenomena. Furthermore, many significant changes resembling pre-earthquake phenomena have occurred but were not succeeded by any earthquake. The controversy associated with pre-earthquake phenomena also highlights the difficulty of predicting earthquakes in practice.

ACKNOWLEDGMENTS

We thank the anonymous reviewers and the editor, Dimitar Ouzounov, for their constructive comments, which improved the manuscript. We also thank Dr. Lei Zhang for providing the groundwater data associated with the Ludian earthquake and Prof. Yiqing Zhu for providing the gravity data associated with the Wenchuan earthquake. This work was supported by the Basic Research Project of the Institute of Earthquake Science, CEA (Grant 2014IES010206) and by the National Natural Science Foundation of China (Grant 41774111).

REFERENCES

An, Z., Du, X., Tan, D., et al. (2013), Study on the geoelectric field variation of Sichuan Lushan *Ms* 7.0 and Wenchuan *Ms* 8.0 earthquake, *Chin. J. Geophys.*, *56*(11), 3868–3876, doi:10.6038/cjg20131128. (In Chinese.)

Bakun, W. H. and Lindh, A. G. (1985), The Parkfield, California, earthquake prediction experiment, *Science*, *229*(4714), 619–624.

Chen, Q. (2010), Earthquake prediction in China: Discussions after the 2008 Wenchuan earthquake, *Quaternary Sciences*, *30*(4), 721–735, doi:10.3969/j.issn.1001-7410.2010.04.08. (In Chinese.)

Chen, Q. and Wang, K. (2010), The 2008 Wenchuan earthquake and earthquake prediction in China, *Bull. Seismol. Soc. Am.*, *100*(5B), 2840–2857, doi:10.1785/0120090314.

Chen, Q., Zhen, D., and Che, S. (2002a), *Earthquake Cases in China (1992–1994)*, Seismological Press, Beijing. (In Chinese.)

Chen, Q., Zheng, D. and Huang, W. (2008), *Earthquake Cases in China (2000–2002)*, Seismological Press, Beijing. (In Chinese.)

Chen, Q., Zhen, D., and Gao, R. (2002b), *Earthquake Cases in China (1997–1999)*, Seismological Press, Beijing. (In Chinese.)

Chen, Q., Zhen, D., Liu, G., et al. (2002c), *Earthquake Cases in China (1995–1996)*, Seismological Press, Beijing. (In Chinese.)

Chen, S., Liu, P., Liu, L., et al. (2013), A phenomenon of ground temperature change prior to Lushan earthquake observed in Kangding. *Seismol. Geol.*, *35*(3), 634–640, doi:10.3969/j.issn.0253-4967.2013.03.017. (In Chinese.)

Chen, Y. (2009), Earthquake prediction: retrospect and prospect, *Sci. China Ser. D—Earth Sci.*, *39*(12), 1633–1658. (In Chinese.)

Cheng, J., Wu, Z., Liu, J., et al. (2015), Preliminary report on the 3 August 2014, *Mw* 6.2/*Ms* 6.5 Ludian, Yunnan–Sichuan border, Southwest China, earthquake, *Seismol. Res. Lett.*, *86*(3), 750–763, doi:10.1785/0220140208.

Densmore, A. L., Ellis, M. A., Li, Y., et al. (2007), Active tectonics of the Beichuan and Pengguan faults at the eastern margin of the Tibetan Plateau, *Tectonics*, *26*(TC4005), doi:10.1029/2006TC001987.

Division of Earthquake Monitoring and Prediction, C.S.B. (1999), *Catalog of Chinese Historical Strong Earthquakes (1912–1990 Ms>4.7)*, China Seismological Press, Beijing. (In Chinese.)

Fu, R., Wan, K., Chong, J., et al. (2009), Earthquake auspice or other factor?—Discussion with authors of the paper "The short-term anomalies detected by broadband seismographs before the May 12 Wenchuan earthquake, Sichuan, China,". *Chin. J. Geophys.*, *52*(2), 584–589. (In Chinese.)

Geller, R. J., Jackson, D. D., Kagan, Y. Y., et al. (1997), Earthquake cannot be predicted, *Science*, *275*, 1616.

Harkins, N., Kirby, E., Shi, X., et al. (2010), Millennial slip rates along the eastern Kunlun fault: Implications for the dynamics of intracontinental deformation in Asia, *Lithosphere*, *2*(4), 247–266, doi:10.1130/L85.1.

Hu, X. and Hao, X. (2008), The short-term anomalies detected by broadband seismographs before the May 12 Wenchuan earthquakes, Sichuan, China. *Chin. J. Geophys.*, *61*(6), 1726–1734. (In Chinese.)

Hu, X., Hao, X., and Xue, X. (2010), The analysis of the non-typhoon-induced microseisms before the 2008 Wenchuan earthquake. *Chin. J. Geophys.*, *53*(12), 2875–2886, doi:10.3969/j.issn.00001-5733.2010.12.011. (In Chinese.)

Huang, F., Li, M., Ma, Y., et al. (2017), Studies on earthquake precursors in China: A review for recent 50 years, *Geod. Geodyn.*, doi:10.1016/j.geog.2016.12.002.

Institute of Earthquake Science and China Earthquake Networks Center, C.E.N. (2007), *Specification for Earthquake Case Summarization*, Vol. DB/T 24-2007, China Earthquake Administration, Beijing. (In Chinese.)

Jackson, D. D. (2004, Earthquake prediction and forecasting, in R.S.J. Sparks and C.J. Hawkesworth (eds), *The State of the Planet: Frontiers and Challenges in Geophysics*, pp. 335–348, American Geophysical Union, Washington, DC.

Jackson, D. D. and Kagan, Y. Y. (2006), The 2004 Parkfield earthquake, the 1985 prediction, and characteristic earthquakes: lessons for the future, *Bull. Seismol. Soc. Am.*, *96*(4B), S397–S409, doi:10.1785/0120050821.

Jiang, H., Fu, H., Yang, M., et al. (2014), *Earthquake Cases in China (2003–2006)*, Seismological Press, Beijing. (In Chinese.)

Jiang, H., Miao, Q., WU, Q., et al. (2009), Analysis on statistical features of precursor based on earthquake cases in China mainland, *Acta Seismol. Sin.*, *31*(3), 245–259. (In Chinese.)

Jiang, Z., Zhang, X., Zhu, Y., et al. (2003), Regional tectonic deformation background before the Kunlun *Ms* 8.1 earthquake (In Chinese.), *Sci. China Ser. D—Earth Sci.*, *33*(suppl.), 163–172.

Kuo, T.-K., Chin, P.-Y., Hsu, W.-Y., et al. (1973), Preliminary study on a model for the development of the focus of an earthquake. *Chin. J. Geophys.*, *16*(3), 43–48. (In Chinese.)

Le, H., Liu, J., Zhaoa, B., et al. (2015), Recent progress in ionospheric earthquake precursor study in China: A brief review, *J. Asian Earth Sci.*, *114*(Part 2), 420–430, doi:10.1016/j.jseaes.2015.06.024.

Li, Q., You, X., Yang, S., et al. (2012), A precise velocity field of tectonic deformation in China as inferred from intensive GPS observations, *Sci. China Earth Sci.*, *55*(5), 695–698.

Lin, A. and Guo, J. (2008), Nonuniform slip rate and millennial recurrence interval of largeearthquakes along the eastern segment of the Kunlun Fault, northern Tibet, *Bull. Seismol. Soc. Am.*, *98*(6), 2866–2878, doi:10.1785/0120070193.

Liu, C., Li, Z., Wang, J., et al. (2015. Progress in production and service of the earthquake preursory observation networks in China, *J. Seismol. Res.*, *38*(2), 313–319. (In Chinese.)

Liu, M., Stein, S., and Wang, H. (2011), 2000 years of migrating earthquakes in North China: how earthquakes in mid-continents differ from thoese at plate boundaries, *Lithosphere*, *3*(2), 128–132, doi:10.1130/L129.1.

Liu, Y., Ren, H., Zhang, L., et al. (2015. Underground fluid anomalies and the precursor mechanisms of the Ludian *Ms* 6.5 earthquake. *Seismol. Geol.*, *37*(1), 307–318, doi:10.3969/j.issn.0253-4967.2015.01.024. (In Chinese.)

Liu-Zeng, J., Klinger, Y., Xu, X., et al. (2007), Millennial recurrence of large earthquakes on the Haiyuan fault near Songshan, Gansu Province, China, *Bull. Seismol. Soc. Am.*, *97*(1), 14–34, doi: 10.1785/0120050118.

Luo, G. and Liu, M. (2010), Stress evolution and fault interactions before and after the 2008 Great Wenchuan earthquake, *Tectonophysics*, *491*, 127–140, doi:10.1016/j.tecto.2009.12.019.

Ma, J. (2016), On "whether earthquake precursors help for prediction do exist," *Chin. Sci. Bull.*, *61*, 409–414, doi:10.1360/N972015-01239. (In Chinese.)

Ma, Q., Fang, G., Li, W., et al. (2013), Electromagnetic anomalies before the 2013 Lushan *Ms* 7.0 earthquake, *Acta Seismol. Sin.*, *35*(5), 717–730. (In Chinese.)

Ma, T. and Wu, Z. (2012), Precursor-like anomalies prior to the 2008 Wenchuan Earthquake: a critical-but-constructive review, *Int. J. Geophys.*, *2012*(583097), doi:10.1155/2012/583097.

Ma, Z., Fu, Z., Zhang, Y., et al., (1982), *Earthquake prediction: Nine major earthquakes in China (1966–1976)*, Seismological Press, Beijing. (In Chinese.)

Mei, S. (1995), On the physical model of earthquake precursor fields and the mechanism of precursors' time–space distribution—origin and evidences of the strong body earthquake-generating model, *Acta Seismol. Sin.*, *8*(3), 337–349.

Mei, S. (1996a), On the physical model of earthquake precursor fields and the mechanism of precursors' time-space distribution (III) - anomalies of seismicity and crustal deformation and their mechanisms when a strong earthquake is in preparation, *Acta Seismol. Sin.*, *9*(2), 223–234.

Mei, S. (1996b), On the physical model of earthquake precursor fields and the mechanism of precursors′ time-space distribution (II)—How do the stress and strain fields evolve while a strong earthquake is in preparation and how does the evolution affect the seismicity and earthquake precursors? *Acta Seismol. Sin.*, *9*(1), 1–12.

Mignan, A., Jiang, C., Zechar, J. D., et al. (2013), Completeness of the Mainland China earthquake catalog and implications for the setup of the China Earthquake Forecast Testing Center, *Bull. Seismol. Soc. Am.*, *103*(2A), 845–859, doi:10.1785/0120120052.

Ming, Z. Q., Hu, G., Jiang, X., et al., (1995), *Catalog of Chinese Historic strong earthquakes from 23 AD to 1911*, Seismological Publishing House, Beijing. (In Chinese.)

Nishenko, S. P. and Buland, R. (1987), A generic recurrence interval distribution for earthquake forecasting, *Bull. Seismol. Soc. Am.*, *77*(4), 1382–1399.

Niu, F., Sliver, P. G., Daley, T. M., et al. (2008), Preseismic velocity changes observed from active source monitoring at the Parkfield SAFOD drill site, *Nature*, *454*, 204–208, doi:10.1038/nature07111.

Niu, Z. (1978), On the theory of precursors of tectonic earthquake—dilatancy–creep model of earthquake source development, *Chin. J. Geophys.*, *21*(3), 199–212. (In Chinese.)

Ohnaka, M. (2000), A physical scaling relation between the size of an earthquake and its nucleation zone size, *Pure Appl. Geophys.*, *157*, 2259–2282.

Parsons, T., Ji, C., and Kirby, E. (2008), Stress changes from the 2008 Wenchuan earthquake and increased hazard in the Sichuan basin, *Nature*, *454*(7203), 509–510, doi:10.1038/nature07177.

Plesch, A., Shaw, J. H., Benson, C., et al. (2007), Community fault model (CFM) for southern California, *Bull. Seismol. Soc. Am.*, *97*(6), 1793–1802, doi:10.1785/0120050211.

Rhoades, D. A. (2010), Lessons and questions from thirty years of testing the precursory swarm hypothesis, *Pure Appl. Geophys.*, *167*, 629–644, doi:10.1007/s00024-010-0071-7.

Scholz, C. H. (1968), The frequency-magnitude relation of microfracturing in rock and its relation to earthquakes, *Bull. Seismol. Soc. Am.*, *58*(1), 399–415.

Scholz, C. H. (1977), A physical interpretation of the Haicheng earthquake prediction, *Nature*, *267*, 121–124.

Shaw, J. H., Plesch, A., Tape, C., et al. (2015), Unified structure representation of the southern California crust and upper mantle, *Earth Planet. Sci. Lett.*, *415*, 1–15, doi:10.1016/j.epsl.2015.01.016.

Shen, X., Zhang, X., Hong, S., et al. (2013), Progress and development on multi-parameters remote sensing application in earthquake monitoring, *Earthquake Sci.*, *26*(6), 427–437, doi:10.1007/s11589-013-0053-9.

Shi, Y., Zhang, B., Zhang, S., et al. (2014), On numerical earthquake prediciton, *Earthquake Sci.*, *27*(3), 319–335, doi:10.1007/s11589-014-0082-z.

Su, Q., Yang, Y., Zheng, B., et al. (2014), A review of the thinking and process about prediction of Lushan M7.0 earthquake on Apr.20, 2013. *Seismol. Geol.*, *36*(4), 1077–1093, doi:10.3969/j.issn.0253-4967.2014.04.012. (In Chinese.)

Toda, S., Lin, J., Meghraoui, M., et al. (2008), 12 May 2008 $M = 7.9$ Wenchuan, China, earthquake calculated to increase failure stress and seismicity rate on three major fault systems, *Geophys. Res. Lett.*, *35*(L17305), doi:10.1029/2008GL034903.

Wang, K., Chen, Q., Sun, S., et al. (2006), Predicting the 1975 Haicheng earthquake, *Bull. Seismol. Soc. Am.*, *96*(3), 757–795, doi:10.1785/0120050191.

Wells, D. L. and Coppersmith, K. J. (1994), New empirical relationships among magnitude, rupture length, rupture width, rupture area, and surface displacement, *Bull. Seismol. Soc. Am.*, *84*(4), 974–1002.

Wen, X., Ma, S., Xu, X., et al. (2008), Historical pattern and behavior of earthquake ruptures along the eastern boundary of the Sichuan–Yunnan faulted-block, southwestern China, *Phys. Earth Planet. Inter.*, *168*, 16–36, doi:10.1016/j.pepi.2008.04.013.

Wen, X., Du, F., Yi, G., et al. (2013), Earthquake potential of the Zhaotong and Lianfeng fault zones of the eastern Sichuan-Yunnan border region, *Chin. J. Geophys.*, *56*(10), 3361–3372, doi:10.6038/cjg20131012. (In Chinese.)

Wu, Y., Jiang, Z., Zhao, J., et al. (2015), Crustal deformation before the 2008 Wenchuan *Ms* 8.0 earthquake studied using GPS data, *J. Geodyn.*, *85*, 11–23, doi:10.1016/j.jog.2014.12.002.

Wu, Z. (1999), Seismological problems of the statistical test of earthquake precursors-A review on the current discussions about earthquake prediction (II), *Earthquake Res. China*, *15*(1), 14–22. (In Chinese.)

Wu, Z. and Jiang, C. (2006), Performance evaluation and statistical test of candidate earthquake precursors: Revisit in the perspective of geodynamics—critical review on the recent earthquake prediction debate, *Earthquake Res. China*, *22*(3), 236–241. (In Chinese.)

Wyss, M. (1997a), Cannot earthquakes be predicted? *Science*, *278*, 487–490, doi:10.1126/science.278.5337.487.

Wyss, M. (1997b), Second round of evaluations of proposed earthquake precursors, *Pure Appl. Geophys.*, *149*(1), 3–16, doi:10.1007/BF00945158.

Xie, T., Zheng, X., Kang, C., et al. (2015), Posiible thermal brightness temperature anomalies associated with the Lushan (China) *M* 7.0 earthquake on 20 April 2013. *Seismol. Geol.*, *37*(1), 149–161, doi:10.3969/j.issn.0253-4967.2015.01.012. (In Chinese.)

Xu, X. and Deng, Q. (1996), Nonlinear characteristics of paleoseismicity in China, *J. Geophys. Res. Solid Earth*, *101*(B3), 6209–6231.

Xu, X., Qiang, Z., and Lin, C. (1991), Earthquake satellite thermal infrared anomalies and surface temperature anomaly, *Chin. Sci. Bull.*, *36*(4), 291–294. (In Chinese.)

Xu, X., Wen, X., Yu, G., et al. (2009), Coseismic reverse- and oblique-slip surface faulting generated by the 2008 *Mw* 7.9 Wenchuan earthquake, China, *Geology*, *37*(6), 515–518, doi:10.1130/G25462A.1.

Yi, G., Wen, X., Wang, S., et al. (2006), Study on fault sliding behaviors and strong-earthquake risk of the Longmenshan–Minshan fault zones from current seismicity parameters, *Earthquake Res. China*, *22*(2), 117–125. (In Chinese.)

Yi, G., Wen, X., and Su, Y. (2008), Study on the potential strong-earthquake risk for the eastern boundary of the Sichuan–Yunnan active faulted-block, China. *Chin. J. Geophys.*, *51*(6), 1719–1725. (In Chinese.)

Yue, M. (2005), Thoughts about the strategy for the development of earthquake prediction. *Recent Dev. World Seismol.*, *5*, 7–21. (In Chinese.)

Zhang, G., Geng, L., Zhang, Y., et al. (1995), Seismogenic model of groups of earthquakes in tectonic block and analysis for some features of earthquake precursory, *Acta Seismol. Sin.*, *17*(1), 1–10. (In Chinese.)

Zhang, G., Zhu, L., Song, X., et al. (1999. Predictions of the 1997 strong earthquakes in Jiashi, Xinjiang, China, *Bull. Seismol. Soc. Am.*, *89*(5), 1171–1183.

Zhang, G., Fu, Z., Wang, X., et al. (2006), Stuy on determination of the national significant seismic monitoring and protection regions, *Earthquake Res. China*, *22*(3), 209–221. (In Chinese.)

Zhang, P., Wen, X., Shen, Z., et al. (2010), Oblique, high-angle, listric-reverse faulting and associated development of strain: the Wenchuan earthquake of May 12, 2008, Sichuan, China, *Ann. Rev. Earth Planet. Sci.*, *38*, 353–382, doi:10.1146/annurev-earth-040809-152602.

Zhang, Z., Luo, G., Li, H., et al., (1988), *Earthquake Cases in China (1966–1975)*, Seismological Press, Beijing. (In Chinese.)

Zhang, Z., Luo, G., Li, H., et al., (1990a), *Earthquake Cases in China (1976–1980)*, Seismological Press, Beijing. (In Chinese.)

Zhang, Z., Luo, G., Li, H., et al., (1990b), *Earthquake Cases in China (1981–1985)*, Seismological Press, Beijing. (In Chinese.)

Zhang, Z., Zhen, D., and Xu, J., (1999b), *Earthquake Cases in China (1986–1988)*, Seismological Press, Beijing. (In Chinese.)

Zhang, Z., Luo, G., Li, H., et al. (2000), *Earthquake Cases in China (1989–1991)*, Seismological Press, Beijing. (In Chinese.)

Zhang, Z., Chen, Q., and Zhen, D. (2013), Summary of earthquake case studies in China, Seismological Press, Beijing. (In Chinese.)

Zheng, Z., Zhang, G., He, K., et al. (2006), Statistics and anlayses of the anomalalies of earthquake cases in Chinses mainland, *Earthquake*, *26*(2), 29–37. (In Chinese.)

Zhu, Y., Zhan, F.B., Zhou, J., et al. (2010), Gravity Measurements and Their Variations before the 2008 Wenchuan Earthquake, *Bull. Seismol. Soc. Am.*, *100*(5B), 2815–2824, doi:10.1785/0120100081.

Zhu, Y., Wen, X., Sun, H., et al. (2013), Gravity changes before the Lushan, Sichuan, $Ms = 7.0$ earthquake of 2013, *Chin. J. Geophys.*, *56*(6), 1887–1894, doi:10.6038/cjg20130611. (In Chinese.)

Zhuang, J., and Jiang, C. (2012), Scoring annual earthquake predictions in China, *Tectonophysics*, *524–525*, 155–164, doi:10.1016/j.tecto.2011.12.033.

4

Multidisciplinary Earthquake Precursor Studies in Taiwan: A Review and Future Prospects

Yi-Ben Tsai[†,1], Jann-Yenq(Tiger) Liu[2], Tzay-Chyn Shin[3], Hong-Yuan Yen[1], and Chieh-Hung Chen[4]

ABSTRACT

Taiwan is situated in a unique plate-tectonic environment. It is intensely compressed by the Philippine Sea plate on its east side and by the Eurasian plate on its west. These active tectonic processes have resulted in a collision zone encompassing the main part of Taiwan, which lies between two subduction zones: one in its northeastern offshore region, and another in its southern offshore region. Frequent shallow earthquakes take place inland, with many disastrous ones striking the densely populated areas in western Taiwan, as well as the Longitudinal Valley in eastern Taiwan. The Taiwan earthquake science community recognized early on the significance of these seismotectonic and societal impacts, launching the first multidisciplinary earthquake prediction research program in late 1970s, as a cooperative effort between the Institute of Earth Sciences (IES) and the University of Southern California (USC), jointly funded by the National Science Council and the US Geological Survey. The program comprised seismological, geodetic, gravitational, geomagnetic, and geochemical components. After the 1999 *M* 7.6 Chi-Chi earthquake, a new series of multidisciplinary research programs focused on the integrated study of Taiwan earthquake precursors, code named iSTEP-1, iSTEP-2 and iSTEP-3, were launched. These programs included the earlier components, as well as adding a seismo-electromagnetic component and statistical validation methodology. These programs are described in the first section. At least seven different types of precursors have been positively identified to be associated with the Chi-Chi earthquake under the above-mentioned programs, as well as by other researchers. These precursors include seismicity, crustal deformation, groundwater level, geomagnetic, atmospheric, and ionospheric changes, with premonitory times ranging from years, months, down to days. Among them, the ionospheric and electromagnetic precursors appear to be most promising. These findings are summarized in the second section. Finally, we present our prospects for future multidisciplinary earthquake precursor studies in Taiwan. In consideration of the seismotectonic and societal impacts, several island-wide advanced monitoring networks have been installed and operated by dedicated technical staff and researchers. In light of the positive findings from the Chi-Chi earthquake, Taiwan arguably could be considered as a highly promising multidisciplinary test ground for earthquake precursors. It is hoped that the earthquake science community, through persistent and diligent work, would be successful in identifying positive precursors in years, months and/or days to benefit people in Taiwan prior to future disastrous earthquakes. In the long term, sufficient empirical precursors observed in Taiwan may help to unravel a fundamental question in earthquake science: "*Are there earthquake precursors that can lead to useful predictions?*"

[1] *Department of Earth Sciences, National Central University, Zhongli District, Taoyuan City, Taiwan*
[2] *Institute of Space Science, National Central University, Taoyuan, Taiwan*
[3] *Central Weather Bureau, Taiwan*
[4] *Department of Earth and Environment Sciences, National Chung Cheng University, Taiwan*

Pre-Earthquake Processes: A Multidisciplinary Approach to Earthquake Prediction Studies, Geophysical Monograph 234, First Edition. Edited by Dimitar Ouzounov, Sergey Pulinets, Katsumi Hattori, and Patrick Taylor.

4.1. INTRODUCTION

Taiwan is situated in the boundary zone between the Eurasian and Philippine Sea plates, as shown in Figure 4.1a. Figure 4.1b shows that the Philippine Sea plate subducts northward under the Eurasian plate along the Ryukyu Trench in northeastern Taiwan and its offshore region, whereas the South China Sea basin subducts eastward under the Phillippine Sea plate along the Manila Trench in southern Taiwan and its offshore region. Intense compression takes place across the main part of Taiwan as a result of the Eurasian plate moving east-southeastward at 33.5 mm year^{-1} on its west side and by the Philippine Sea plate moving northwestward at 49.8 mm year^{-1} on its east, as shown in Figure 4.2a. The bulk of the island is logjamed and being steadily uplifted to form a series of N–S-trending high mountain ranges in its central region, as shown in Figure 4.2b.

The active plate-tectonic processes have resulted in frequent seismicity, as revealed by instrumental observations since 1897 [*Hsu*, 1971; *Tsai et al.*, 1977; *Wu*, 1978, *Tsai*, 1986]. Figure 4.3a shows the epicenters of 1989 M 5.0+ earthquakes that occurred during 1900–2014 [*Chang et al.*, 2016]. This figure shows that most inland earthquakes are shallow. Many damaging earthquakes have occurred since historic times, some of which caused great loss of life and structural damage. Figure 4.3b shows the strong-shaking areas and the fatalities caused by the five most disastrous earthquakes that struck densely populated western Taiwan in the 20th century.

Faced with this situation, the people in Taiwan realize that the potential for devastating earthquake disasters is ever increasing with time, as rapid urbanization and infrastructure development continues. Obviously, the benefits would be enormous if predictions of major earthquakes near population centers or critical facilities could be made and handled properly. Both the general public and government in Taiwan are genuinely interested in earthquake prediction out of concerns for the high potential of seismic hazards.

Equally significant, the earthquake science community recognized early on, given the high risk of seismotectonic and societal impacts, the necessity for pursuing the difficult goal of earthquake prediction. The first multidisplinary earthquake prediction research program was

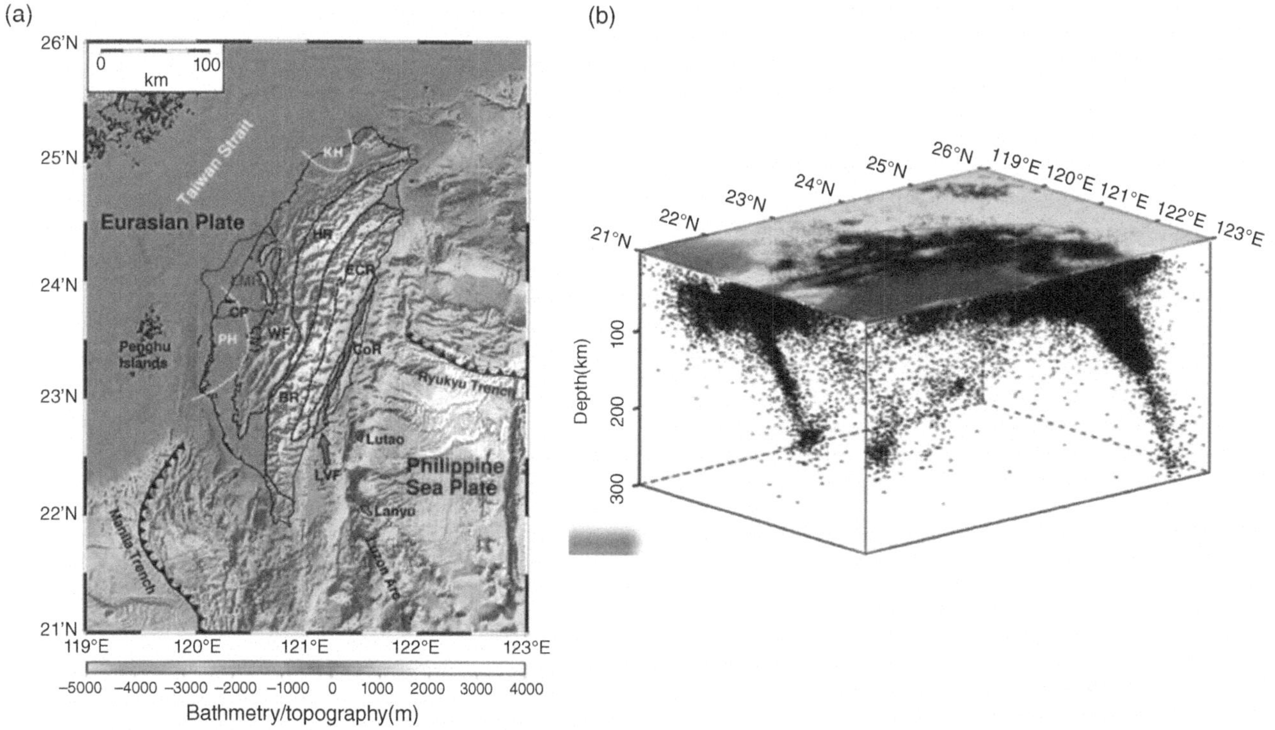

Figure 4.1 (a) Tectonic plates near Taiwan [*Hsieh et al.*, 2014]. (b) Three-dimensional projection of seismicity 1994–2015. CP, Coastal Plain; WF, Western Foothills; HR, Hsuehshan Range; BR, Backbone Range; ECR, Eastern Central Range; CoR, Coastal Range; LVF, Longitudinal Valley Fault; PH, Peikang High (a geometric basement high); KH, Kuanyin High (a geometric basement high); LMH, Lukang magnetization high; Manila and Ryukyu Trenches are drawn on the basis of bathymetry only. (Source: Reproduced with permission of C.T. Lee.)

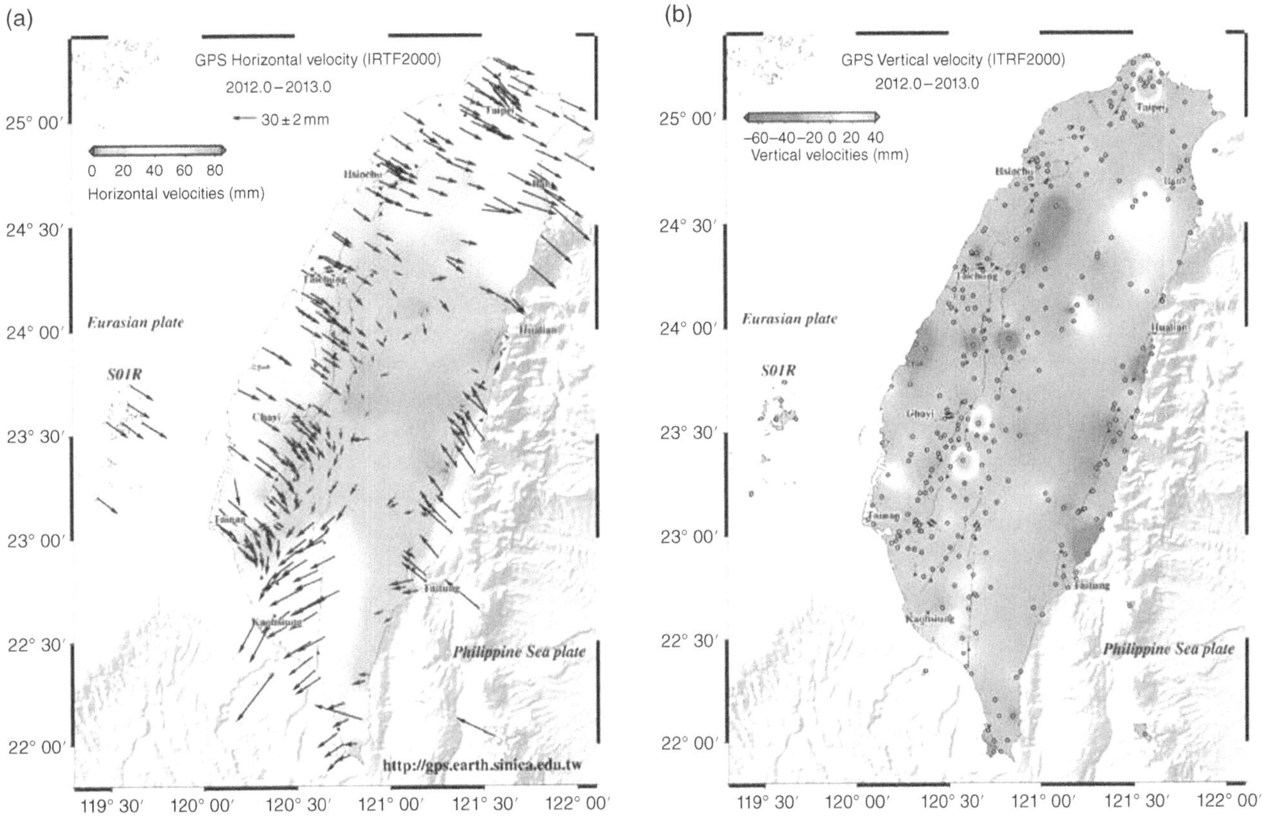

Figure 4.2 (a) GPS horizontal and (b) vertical velocities in Taiwan: S01R (23.66°N, 119.59°E) is the reference station of the GPS network. (Source: http://gps.earth.sinica.edu.tw)

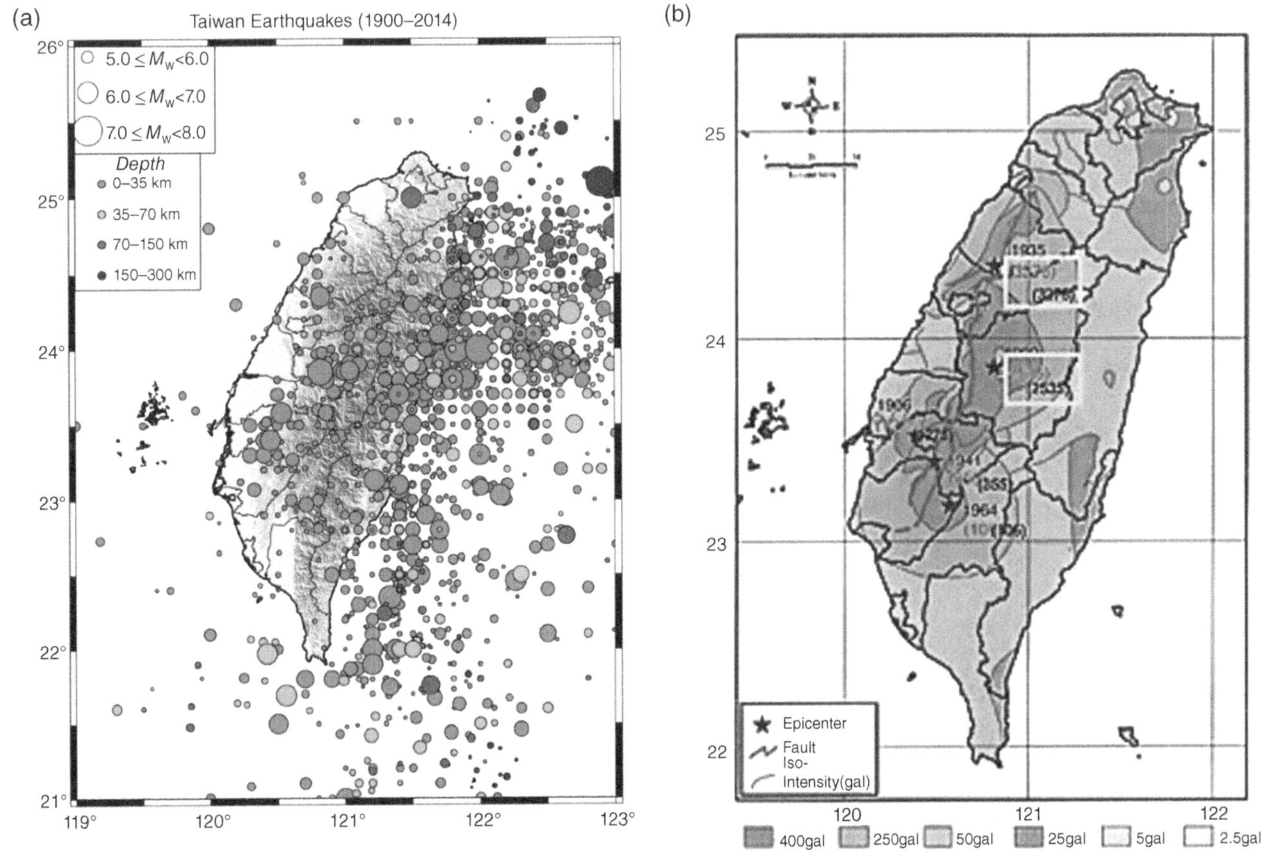

Figure 4.3 (a) Epicenters of *M* 5.0+ earthquakes in Taiwan, 1900–2014 [*Chang et al.*, 2016]. (b) Intensive shaking areas of the five most disastrous earthquakes in western Taiwan during the 20th century. (Source: Reproduced with permission of Elsevier.)

launched in 1979 as a cooperative effort between the Institute of Earth Sciences (IES) and University of Southern California (USC), with joint funding by the National Science Council and US Geological Survey (USGS). Over the years a small yet dedicated group of scientists have persistently made efforts to identify positive earthquake precursors in Taiwan.

After the 1999 *M* 7.6 Chi-Chi earthquake, a new series of multidisciplinary research programs focused on the integrated study of Taiwan earthquake precursors, code named iSTEP-1, iSTEP-2 and iSTEP-3, were established. Thus far, multiple types of precursory phenomena have been identified positively against long-term temporal and/or spatial pre-earthquake backgrounds, especially those pertaining to the Chi-Chi earthquake.

In the following we will first give a concise summary of these multidisciplinary research programs. Then we will review the multiple types of precursors that have been positively identified as being associated with the 1999 *M* 7.6 Chi-Chi earthquake. This will be followed by future prospects for multidisciplinary earthquake precursor studies in Taiwan.

4.2. FIRST MULTIDISCIPLINARY EARTHQUAKE PREDICTION RESEARCH PROGRAM IN TAIWAN (1979–1984)

As mentioned above, the first multidisciplinary earthquake prediction research program in Taiwan began in 1979 [*Tsai et al.*, 1983]. It was prompted partly by a sudden nearly threefold increase in the monthly microearthquake occurrences during 1977–1978, relative to the previous activity levels since 1973. T. L. Teng and Y. B. Tsai served as the co-principal investigators (PI). The program was composed of five multidisciplinary components:

1. study of both spatial and temporal changes in microearthquake activities (PI: J. H. Wang);
2. repeated trilateration surveys to detect crustal deformation across the Longitudinal Valley in eastern Taiwan and the Lanyang Plain in northeastern Taiwan (PI: S. B. Yu);
3. repeated microgravity surveys to detect vertical crustal deformation along linear profiles traversing the Longitudinal Valley, Lanyang Plain, as well as the 1906 and 1946 earthquake faults in western Taiwan (PI: Y. H. Yeh);
4. repeated measurements by proton magnetometers to detect possible changes in geomagnetic total intensity due to tectonic stress at 22 sites located throughout Taiwan—two continuous-recording proton magnetometers were installed as reference sites, one in northern and one in southern Taiwan (PI: Y. H. Yeh).
5. weekly sampling of geothermal water to monitor changes in radon content at five hot springs in northern Taiwan (PI: K. K. Liu).

Tsai et al. [1983] reported the observational results were and some are highlighted here. Figure 4.4a shows some of the observational sites mentioned above. Figure 4.4b shows the trilateration survey results across the Lanyang Plain and Longitudinal Valley in eastern Taiwan. These findings in horizontal crustal strain rates have been fully confirmed by contemporary global positioning system (GPS) measurements in recent years. Figure 4.5a correlates the weekly radon sampling data at four hot springs with microearthquake activity in northern Taiwan from 1980 to 1983.

Even though some observational results were encouraging, they were far from conclusive. On the other hand, three of the above five program components have benefited from upgraded instrumentation and new methodology over the years to make major impacts on our understanding of the Taiwan tectonic environment. For example, the IES staff working on trilateration surveys were quick to gain expertise by collaborating with their more experienced USGS counterparts. This enabled them to pick up the GPS methodology early on to start monitoring island-wide crustal deformation in Taiwan [*Yu and Chen.*, 1994].

As another example of the extended impact of the IES–USC program, a detailed gravity anomaly map of Taiwan was obtained by using a LaCoste-Romberg model-D gravimeter [*Yen et al.*, 1995]. It was later refined with recalculated terrain corrections (Figure 4.5b). This map was used recently to constrain the three-dimensional crustal structure models for Taiwan [*Yen and Hsieh*, 2010].

As a third example of the extended impact of the IES–USC program, Figure 4.6a shows a detailed magnetic anomaly map of Taiwan [*Yen et al.*, 2009]. It reveals several prominent features of tectonic significance. The map was used recently to obtain the Curie-point surface at depth, as well as to estimate the surface heat flow distributions in Taiwan [*Hsieh et al.*, 2014].

4.3. INTEGRATED SEARCH FOR TAIWAN EARTHQUAKE PRECURSORS (iSTEP) PROGRAM (2002–2006)

The biggest breakthrough of the IES–USC program took place following the disastrous 1999 *M* 7.6 Chi-Chi earthquake. The number of continuous-recording geomagnetic stations were increased from two to eight to form a network covering the whole of Taiwan, as shown in Figure 4.6b, since the first eartquake prediction research program in early 1980s. The recordings by this netwok over the years before the Chi-Chi earthquake showed that the geomagnetic field in Taiwan generally followed the trends observed at the reference station Lunping (LP) in northern Taiwan.

Surprisingly, upon inspecting the continuous records of this geomagnetic network shortly after the earthquake,

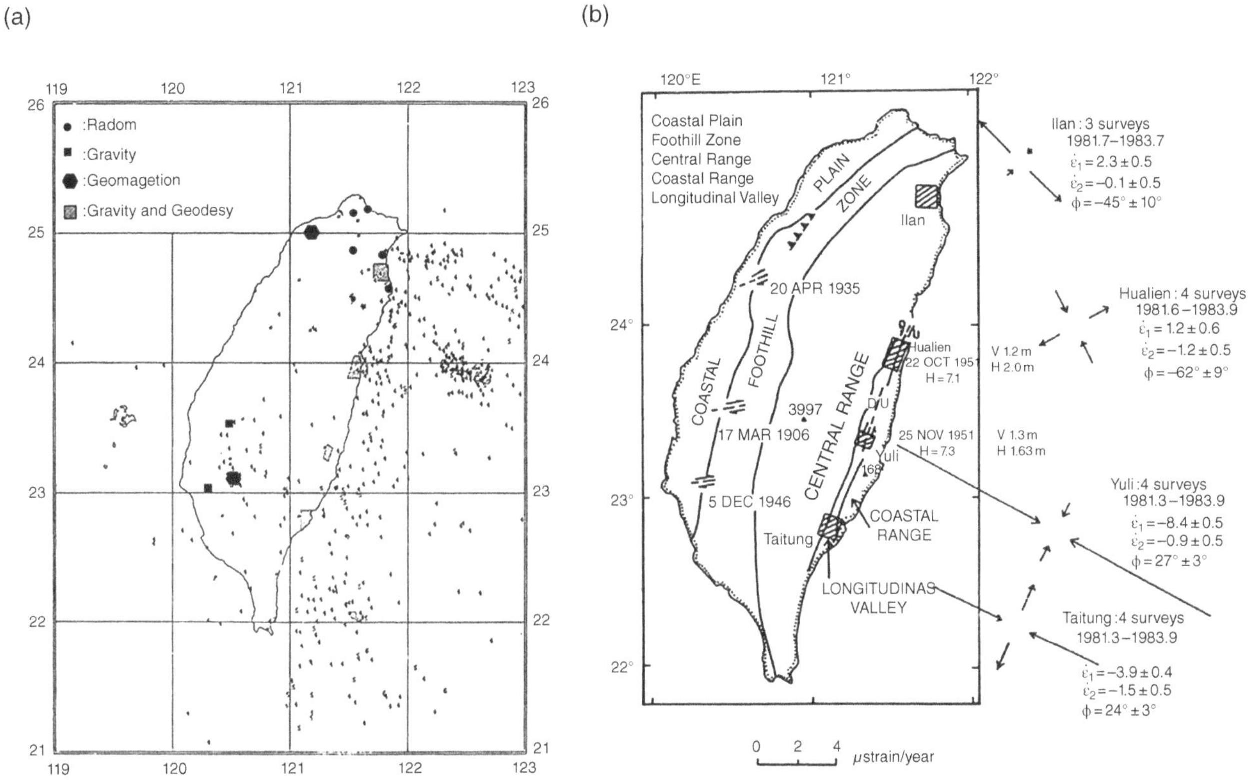

Figure 4.4 (a) Observational sites. (b) Triangulation results in eastern Taiwan from the multidisciplinary earthquake prediction research program in Taiwan, 1979–1984 [*Tsai et al.*, 1983].

it was found that major disturbances of over 200 nT, relative to the undisturbed station LP, took place at station Liyutan (LY), which is located only about 6 km to the north of the northern end of the Chelungpu fault, at least 1 month prior to the earthquake. These disturbances subsided about 1 month after the earthquake [*Yen et al.*, 2004].

Another type of clear electromagnetic precursors in the ionosphere several days before large earthquakes in Taiwan was identified at about the same time. They were based on corroborating total electron content (TEC) data obtained at an ionosonde station and six GPS receiving stations [*Liu et al.*, 2000, 2001a]. These two positive observations prompted us to think that major earthquakes in Taiwan might have other types of identifiable precursors.

At about the same time, a physical process was proposed to explain the electromagnetic signals observed both on the ground and in the ionosphere, which were thought to be associated with earthquakes [*Freund et al.*, 1999, 2004; *Freund*, 2000]. According to this process, igneous rocks contain positive hole pairs (PHP), i.e., dormant electronic charge carriers. The PHP could be activated by microfracturing and/or dislocation movement. Upon activation, the PHP would release highly mobile charge carriers in the form of positive holes that can conduct through rocks. Once generated, these charges would spread out from the source volume to form huge electric currents.

These new developments in observations, as well as on a feasible physical mechanism, eventually led us to organize a multidisciplinary program, code named iSTEP, to undertake integrated search for Taiwan earthquake precursors from April 2002 to March 2006. Initiation of this program was also inspired by the International Frontier Research on Earthquakes (IFREQ) [*Uyeda et al.*, 2004] and the Earthquake Remote Sensing Frontier Project (ERSFP) [*Hayakawa et al.*, 2004] projects that were established after the disastrous 1995 Kobe (Japan) earthquake. Since then, close collaborations in this area among scientists in Taiwan and Japan have been very productive.

The iSTEP program was funded by the Ministry of Education. It was set to achieve the following scientific goals:

1. to identify credible precursory phenomena of earthquakes, including those of seismological, geodetic, geomagnetic, and ionospheric nature;

2. to characterize feasible physical mechanisms underlying the identified precursory phenomena, in order to provide quantitative interpretations;

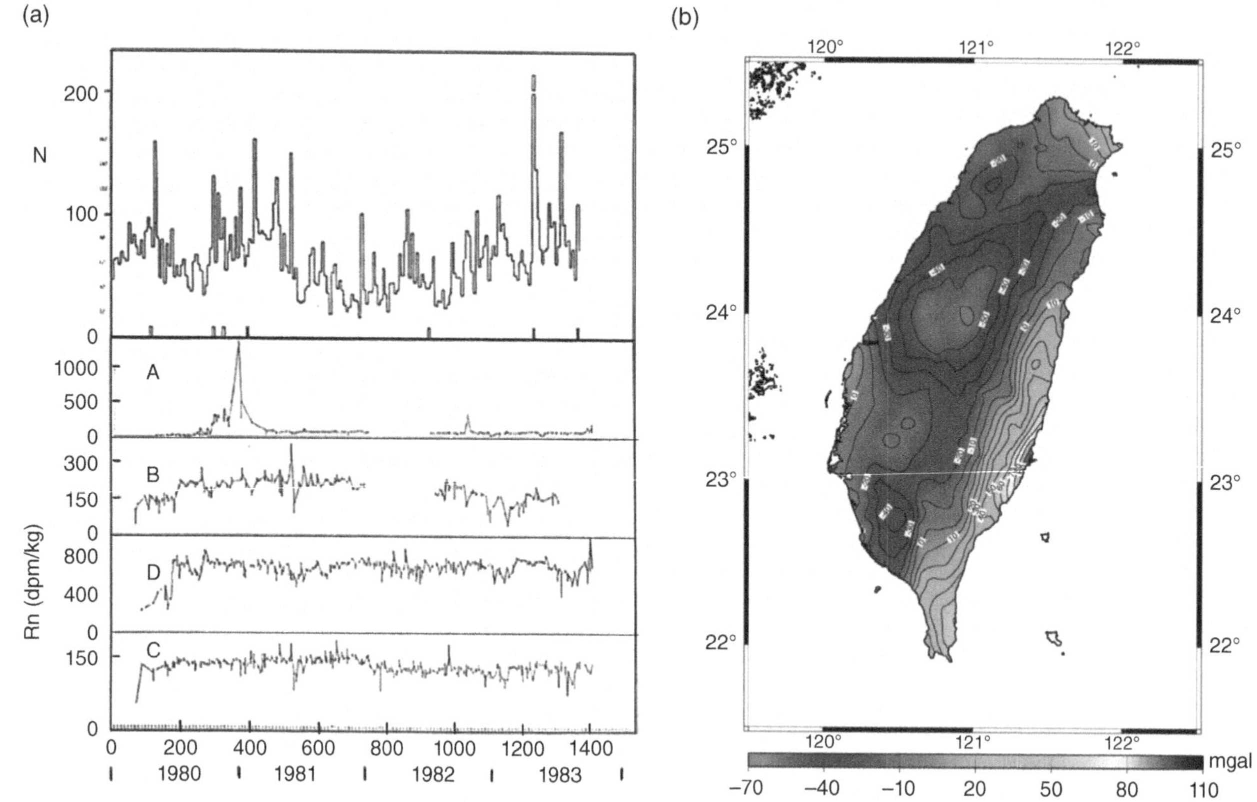

Figure 4.5 (a) Comparison of temporal variations of radon content at four hotsprings with seismicity in northern Taiwan [*Tsai et al.*, 1983]. (b) Gravity anomaly map of Taiwan [*Yen and Hsieh*, 2010].

3. to correlate the identified precursory phenomena with existing or emerging earthquake preparation models, in order to lay a foundation for earthquake prediction.

Striving to achieve above goals and taking into account the strengths of our research team, the program was composed of an overall project with five subprojects, focusing studies on seismological, geomagnetic and gravitational variations, ground surface deformation, and ionospheric variations, as well as statistical validation of these variations. Figure 4.6b shows some of the observational stations for the program. The five major components of iSTEP, as shown in Figure 4.7, are briefly described below.

1. The overall aim of the project was to coordinate the activities and integrate the findings of the multidisciplinary subprojects through biweekly project meetings. It also organized international workshops to promote exchange and cooperation with scientists from abroad (PIs: Y. B. Tsai and J. Y. Liu).

2. Identification of potential precursors by seismological data (PIs: K. F. Ma, Y. B. Tsai). Taiwan is endowed with a long and consistent earthquake catalog since 1900. It is also densely covered by advanced seismic networks which can provide high-quality digital waveform data. Several tasks were undertaken to take advantage of these data.

A. Spatial and temporal variations of seismicity rate: Significant changes relative to normal background seismicity were observed in several large earthquakes. Another item was to study changes in the *b*-value.

B. Seismic velocity changes: rock properties might change prior to major earthquakes, resulting in discernable seismic velocity changes. For this task, we compiled and made use of the vast amount of arrival-time residuals at individual seismic stations from routine earthquake location operations by the Central Weather Bureau Seismic Network (CWBSN) of Taiwan.

C. Changes in shear-wave splitting: shear-wave splitting due to anisotropy might take place in a critical part of the crust before major earthquakes [*Crampin*, 1998]. We were also contemplating possible correlation between this phenomenon with electromagnetic disturbances on the ground and above in the ionosphere.

D. Stress transfer and triggering of aftershocks: once the location, magnitude, and focal mechanism of a major earthquake became available, stress changes could be quickly calculated to forecast aftershock distribution for timely hazard mitigation.

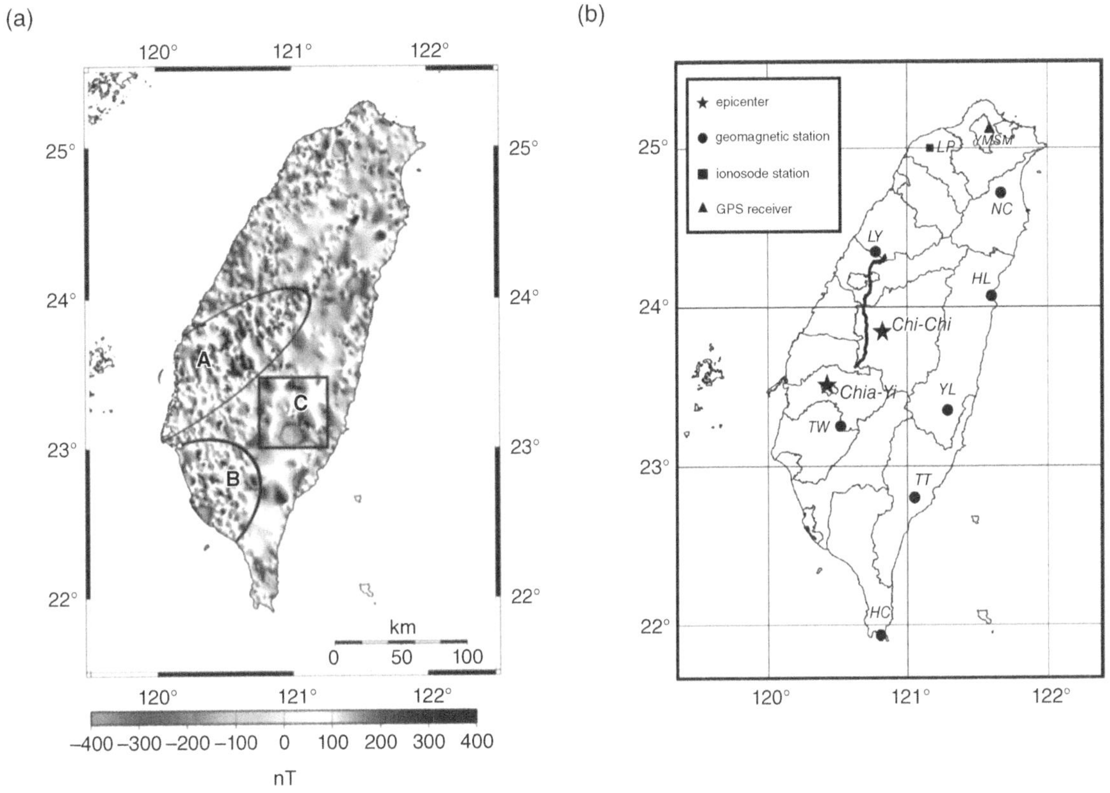

Figure 4.6 (a) Geomagnetic anomaly map of Taiwan [*Yen et al.*, 2009]. (b) Observational stations under iSTEP. LP, Lunping (25.00°N, 121.17°E) is the ionosode station; NC, Neicheng (24.72°N, 121.68°E), LY, Liyutan (24.35°N, 120.78°E), HL, Hualien (24.07°N, 121.61°E), YL, Yuli (23.35°N, 121.29°E), TW, Tsengwen (23.25°N, 120.53°E), TT, Taitung (22.80°N, 121.06°E) and HC, Hengchun (21.94°N, 120.81°E) are the geomagnetic stations; YMSM is the Yangmingshan GPS station. [*Tsai et al.*, 2006].

3. Identification of potential precursors by geomagnetic and gravity data (PI: H. Y. Yen). Inspired by the finding of large geomagnetic disturbances before and after the Chi-Chi earthquake, as mentioned above, we planned to analyze the whole data set obtained by the eight-station continuous-recording geomagnetic network, in order to search for possible precursors related to other large earthquakes in Taiwan. The existing geomagnetic network and ionosonde station would also be upgraded and expanded. In addition, repeated microgravity surveys would be made along perpendicular profiles across known active faults to detect possible crustal deformation, and subsurface mass redistribution before and after major earthquakes.

4. Identification of potential precursors in crustal deformation by interferometric synthetic aperture radar (InSAR) techniques (PIs: K.S. Chen, C.P. Chang). Use of satellite differential InSAR for monitoring crustal deformation related to earthquakes has advanced significantly in recent years [*Massonnet et al.*, 1993]. This subproject focused on several tasks.

A. To integrate very high frequency (VHF) and GPS atmospheric and ionospheric observations into radar interferometry for path-lenth corrections in order to improve accuracy of determining the terrain height and its changes.

B. To detect pre-, co-, and postseismic ground-surface movements.

C. To correlate radar interferograms with temporal and spatial distributions of seismo- electromagnetic phenomena in the atmosphere and ionosphere.

D. To process and analyze the thermal infrared (IR) satellite imageries to search for possible earthquake precursors.

5. Identification of potential precursors by ionospheric data (PI: J. Y. Liu). Although seismogenic zones are situated deep in the crust, earthquakes may be preceded by electric, electromagnetic, or luminous phenomena observable above ground that might help indicate impending seismic activity [*Chen et al.*, 2004; *Freund et al.*, 2004; *Hayakawa et al.*, 2004; *Liu et al.*, 2004a,b; *Uyeda et al.*, 2004]. In this subproject we aimed to systematically search through ionospheric variations of all large earthquakes in Taiwan to see whether there are useful precursors. Both ionosonde and GPS data would be used.

6. Statistical studies of potential earthquake precursors (PI: Y. I. Chen). Extensive and reliable data on earthquake

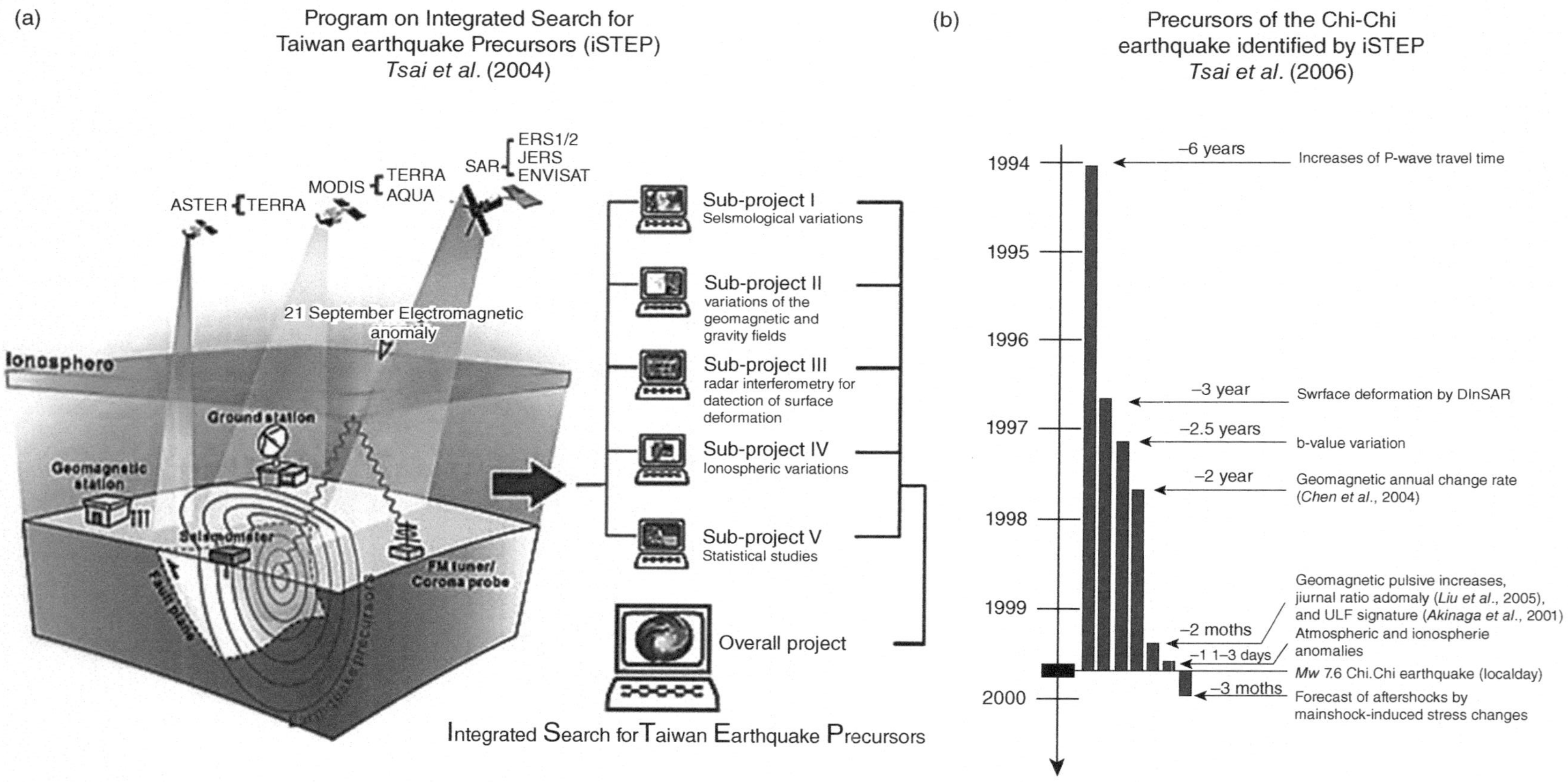

Figure 4.7 (a) The main component and five subcomponent projects [*Tsai et al.*, 2004]. (b) Precursors of the *M* 7.6 Chi-Chi earthquakee identified as of 2006 under the multidisplinary "integrated Search for Taiwan Earthquake Precursors" (iSTEP) program [*Tsai et al.*, 2006].

precursors are expensive to collect and might take a long time to accumulate [*Vere-Jones*, 1995]. Fortunately, we have obtained a substantial amount of data on anomalous fluctuations of the critical plasma frequency (*fo*F2) and TEC of the ionosphere, which were positively correlated with large earthquakes in Taiwan area [*Liu et al.*, 2000, 2001, 2004a,b]. These anomalies revealed not only temporal but also spatial information about the earthquakes to follow. In this subproject we aimed to develop appropriate statistical process control (SPC) techniques [*Stoumbos et al.*, 2000] to identify variations of the observed precursory parameters. In order to validate systematically the reliablity of precursory parameters, the Monte Carlo simulations [*Musson*, 1997; *Stark*, 1997] of some resonable prediction processes based on either the occurrence of precursors or previous earthquake catalogs would be compared with the earthquakes of interest. The generalized linear model [*McCullogh*, 2000] would be further introduced to define the relationship between the observed precursors and associated earthquake parameters.

Results of the iSTEP program have been published in a 2004 speial issue of TAO (*Terrestrial, Atmospheric and Oceanic Sciences Journal*; Vol. 15, No. 3), as well as in other mainstream scientific journals.

4.4. FOLLOW-UP iSTEP PROGRAMS

The intial iSTEP program was funded during 2002–2006. Under this program, an integrated ground-based seismo-electromagnetic observation system, including eight networks composed of magnetometers, electrode arrays, corona probes, FM tuners, Doppler sounding systems, ionosondes, GPS receivers, and all-sky cameras, has been deployed to cover the whole of Taiwan. This system has been routinely monitoring potential earthquake precursors in the lithosphere, atmosphere, and ionosphere. The data would be crucial for finding possible lithosphere–atmosphere–ionosphere coupling (LAIC) mechanisms. Several statistical analysis methods have been developed to validate the observed anomalies as credible precursors. The statistical results showed that the ionospheric TEC anomalies derived from ground-based GPS receivers, due to worldwide availability, appear to be the most promising short-term credible precursors.

Since the initial iSTEP program (now called iSTEP-1), two follow-up programs focusing on seismo-electromagnetic precursory phenomena have been undertaken, as described by *Liu et al.* [2016a,b]. The iSTEP-2 project (PI: J. Y. Liu) was carried out during 2006–2012. This project required a longer time period for data collection and analysis through the addition of satellite observations. Another aim of the project was to develop physical and statistical models. The integrated ground-based observation system established remained operational without interruption, even though this project had not been officially funded, rather it was supported as part of other basic ionospheric research projects.

Under iSTEP-2, many new obsrvations were obtained. They included possible seismo-lithospheric precursors of Earth's surface magnetic field and the GPS surface deformation, seismo-atmospheric precursors based on infrasound signals, and seismo-ionospheric precursors (SIPs) in the electron density profile, the electron temperature, ion density, and neutral temperature probed by satellites. The TEC distributions in the global ionosphere map (GIM) were routinely published, with a 2- or 4-day time delay. This allowed us to monitor temporal SIPs at a specific location, and conduct spatial analysis to discriminate the SIPs from global effects, such as solar flares, magnetic storms, etc.

Statistical analyses for detecting both temporal and spatial precursors in the ionospheric TEC were developed. Meanwhile, ionospheric model simulations were also introduced in order to find causal mechanisms to explain the observed SIPs.

Currently, the iSTEP-3 (2012-2016, PI: J. Y. Liu) program, focusing on the SIP study, is formally funded. The program consists of a main project and three subprojects. The main project continues to operate the integrated ground-based observation system, develops physical models, and compares model simulations with observed precursors. The main themes of the three subprojects are:

1. to develop near real-time GIMs with a 4-h time delay for worldwide SIP monitoring;
2. to monitor lithospheric, atmospheric, and ionospheric precursors and to find possible coupling mechanisms among them;
3. to make earthquake hazard assessments using the observed precursors.

Multiple types of precursory phenomena associated with several major earthquakes in Taiwan, as well as in other regions of the world, have been identified under the iSTEP-1, iSTEP-2, and iSTEP-3 programs. A concise summary of several types of positive precursors identified from observations of the 1999 *M* 7.6 Chi-Chi earthquake under these multidisciplinary iSTEP programs, as well as by other researchers, is presented below.

4.5. TYPES OF POSITIVE PRECURSORS IDENTIFIED FROM OBSERVATIONS OF THE 1999 *M* 7.6 CHI-CHI EARTHQUAKE

The Chi-Chi earthquake, named after the town where its epicenter was located, struck central Taiwan on 21 September 1999. The earthquake was caused by rupture of the N–S-striking Chelungpu fault for a total length of about 100 km, as shown in Figure 4.6b. With a magnitude

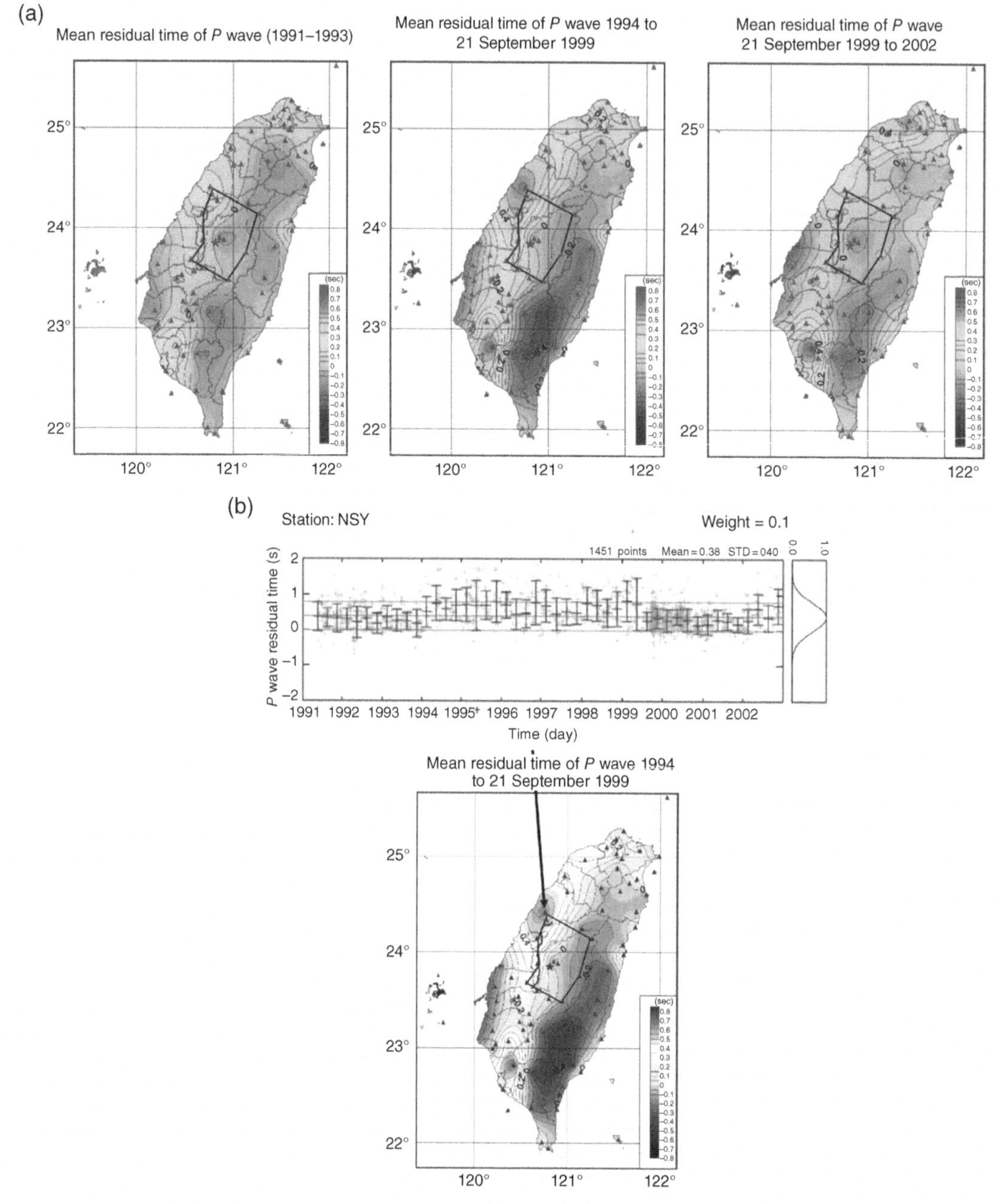

Figure 4.8 (a) Spatial variations of *P*-wave arrival-time residuals. (b) Temporal variations of *P*-wave arrival-time residuals during three periods before and after the 1999 *M* 7.6 Chi-Chi earthquake in Taiwan [*Lee and Tsai*, 2004].

of M_L 7.3 (CWBSN) and M_W 7.6 (USGS), it was the largest inland earthquake to occur in Taiwan in the 20th century. More than 2500 people were killed or missing, and over 100,000 residential housing units collapsed.

Fortuitously, owing to its central location and availability of data from several pre-existing observational networks, multiple types of precursory phenomena associated with the earthquake have been retrospectively identified thus far under the iSTEP programs, as well as by other researchers. These precursory phenomena included seismological and seismicity variations, ground-surface deformation, groundwater-level changes, electromagnetic field changes, and ionospheric variations. Figure 4.7b shows the time lines of precursors, as identified under the original iSTEP program, ranging from 6 years for increase of *P*-wave arrival-time residuals down to several days in atmospheric and ionospheric anomalies [*Tsai et al.*, 2006]. Some of the major findings are briefly summarized below.

4.5.1. Increase of *P*-Wave Arrival-Time Residuals

The *P*-wave arrival-time residuals are byproducts of routine earthquake location by the CWBSN. *Tsai et al.* [2004] analyzed these data from 1991 to 2002 to identify distinct

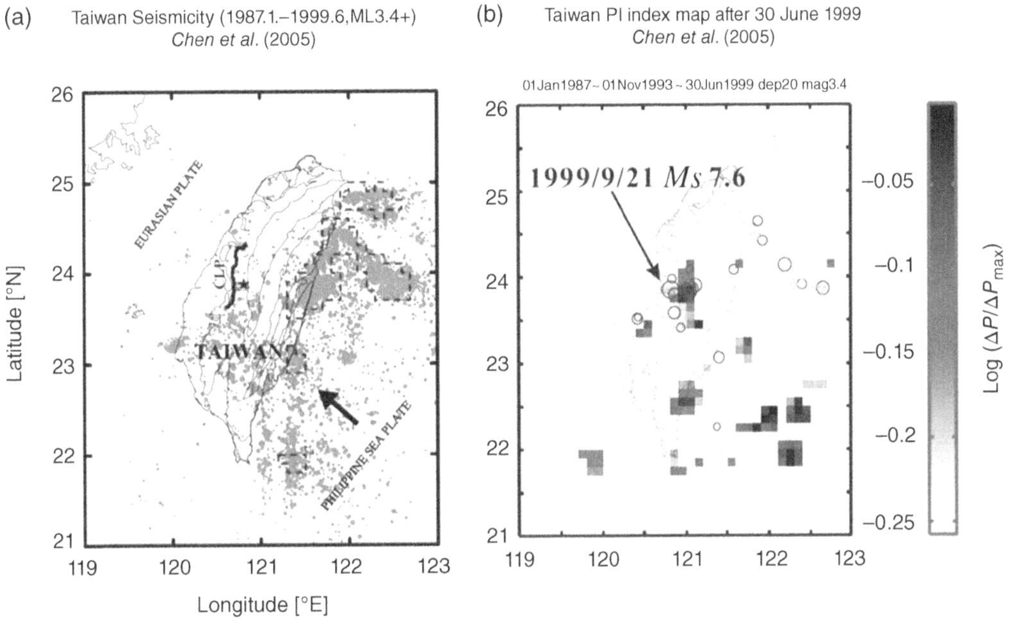

Figure 4.9 (a) Distribution of Taiwan seismicity (M 3.4+, January 1987 to June 1999); (b) Distribution of pattern informatics (PI) index of Taiwan seismicity during 3 months before the 1999 M 7.6 Chi-Chi earthquake [*Chen et al.*, 2005]. CLP, the Chelungpu fault. The red dashed lines denote the top 5% active area of historic coarse-graining intensity of earthquakes. Circles represent earthquakes with $M \geq 6$ that occurred after June 1999 and inverted triangles represent earthquakes with $M \geq 6$ that occurred between November 1993 and June 1999. Colored pixels (hotspots) represent areas with large seismicity change caused by both seismic activation and quiescence, indicating high probability for future large events. (Source: Reproduced with permission of Elsevier.) (*See electronic version for color representation.*)

changes in the spatial distribution of the mean *P*-wave arrival-time residuals over two time periods before and one period after the Chi-Chi earthquake, i.e. 1991–1993, 1994 to 20 September 1999, 21 September 1999, as shown in Figure 4.8a. From the figure we can see the mean *P*-wave residuals at stations within about 40 km east of the Chelungpu fault changed very little both before and after the earthquake. In contrast, the mean *P*-wave residuals at stations immediately west of the Chelungpu fault increased significantly beginning 1994, about 6 years before the earthquake.

Figure 4.8b shows the temporal variations of *P*-wave arrival-time residuals at San-Yi (NSY; 24.41°N, 120.77°E) seismic station, located at central Taiwan, from 1991 to 2002. From the top panel we can see both the mean and standard deviation of the *P*-wave arrival-time residuals during the 1994 to 20 September 1999 period were significantly greater than those of the preceding or following period. The anomalous zone, bounded by the black borders marked with seismic stations, suggests a decrease of *P*-wave velocity beneath the zone, probably due to dilatancy or development of cracks in the crust [*Nur*, 1972].

4.5.2. Seismicity Changes

Chen et al. [2005] used a pattern informatics method to detect locations of anomalous seismic activity preceding the Chi-Chi earthquake. Figure 4.9a shows the epicenters of M_L 3.4+ earthquakes used in their study. It was found that the area near epicenter of the Chi-Chi earthquake exhibited signatures of anomalous activity related to the seismic activiation and quiescence over a time span of about 6 years before the earthquake, as shown in Figure 4.9b.

Wu and Chen [2007] used the CWBSN seismicity data with M_L 2.0+ and focal depth less than 35 km from 1994 to 2005 to study seismicity reversal patterns for the Chi-Chi earthquake. Figure 4.10a shows the epicenter distribution of the seismicity data set used. Figure 4.10b shows the monthly occurrence of events for the whole data set from 1994 to 2005. It can be seen from the figure that the monthly occurrence of events was anomalously low over the whole Taiwan region for nearly 9 months from January 1999 to the time just before the Chi-Chi earthquake on 21 September 1999.

The *b*-value of the Gutenberg–Richter relation [*Gutenberg and Richter*, 1944] was also calculated for the three regions spanning the north, central, and south segments of Chelungpu fault, as shown in the upper left panel of Figure 4.11a. A complete catalog of M_L 2.0+ earthquakes with focal depth shallower than 40 km was used to obtain the temporal variations of the *b*-value from 1 January 1994 to 31 August 1999 in each of the three regions, as shown in the upper right panel of Figure 4.11a. The *b*-values were computed using a 100-event window sliding by 10 events. The results showed

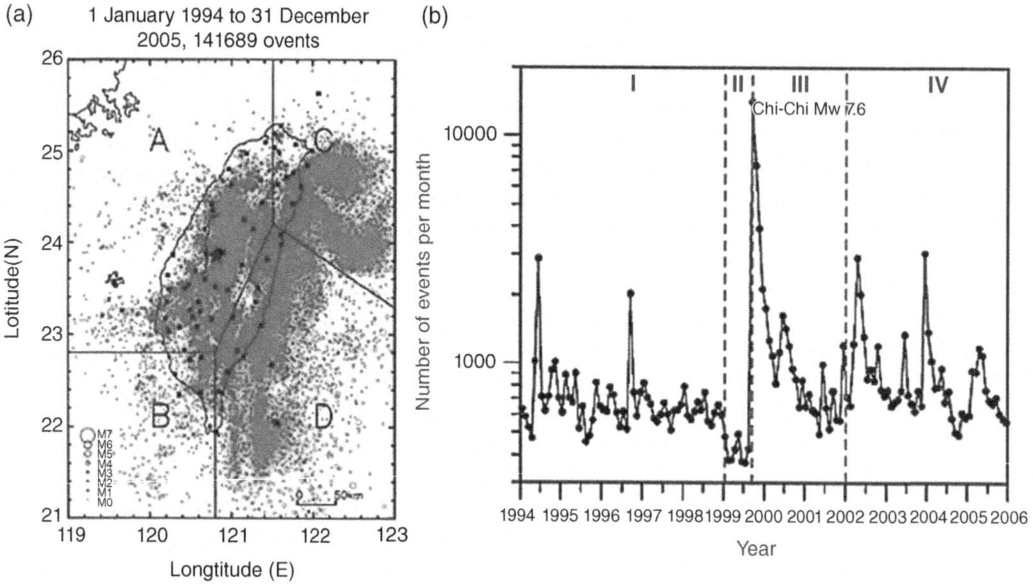

Figure 4.10 (a) Distribution of Taiwan seismicity, 1994–2005. (b) Time series of the monthly earthquake occurrence, 1994–2005 [*Wu and Chen*, 2007]. (Source: Reproduced with permission of American Geophysical Union.)

that the *b*-value started to increase steadily in 1995 and reached its peak in 1997 for all three regions. Thereafter, the *b*-value dropped back sharply in regions A and C, whereas it remained high in region B until 31 August 1999, just before the Chi-Chi earthquake.

We further analyzed the spatial variations of the *b*-value by applying a dense-gridding technique [*Wiemer and Katsumata*, 1999], which takes overlapping grids centering at nodes 5 km apart with 20 km in radius. The results are shown in the two lower left panels of Figure 4.11a for 1994–1995 and 1997, respectively. We then calculated the standardized *b*-value change, *Z*, for each grid by the method of *Shi and Bolt* [1982]. A grid was considered to have anomalous change if the *Z*-value was greater than 1.282 or < -1.282. The spatial variations of the *Z*-value, as shown in the lower right panel of Figure 4.11a, indicate a significant *b*-value increase to the west of northern Chelungpu fault together with a significant *b*-value decrease to the east of central Chelungpu fault just north of the Chi-Chi earthquake epicenter.

In summary, the above studies suggested that precursory temporal and spatial variations in seismicity and *P*-wave velocity ranging from 6 years to 9 months before the Chi-Chi earthquake, could be positively identified.

4.5.3. Ground Surface Deformation Observed by DInSAR Images

Carefully chosen ERS-2 radar images were used to identify possible precursory ground surface deformation in the areas near the Chelungpu fault before the Chi-Chi earthquake [*Chang et al.*, 2004]. Figure 4.11b shows the DInSAR interferograms of the areas near the Chelungpu fault for three consecutive years before and one year after the Chi-Chi earthquake. We can see from the figure that ground-surface deformation began at least 3 years before the earthquake in the areas immediately to the west of northern segment of the Chelungpu fault where clear coseismic surface-deformation patterns were observed.

4.5.4. Groundwater-Level Changes

Coseismic and preseismic groundwater-level changes associated with the Chi-Chi earthquake were studied by *Chia et al.* [2001], *Wang et al.* [2001], and *Chen et al.* [2012], by using the groundwater-level data recorded at many monitoring stations densely distributed in the Choshuichi alluvial fan, as shown in the left panel of Figure 4.12a. Each groundwater monitoring station consists of five wells ranging from 14 to 300 m in depth and equipped with a screen covering from 6 to 36 m in length. Each well only monitors the groundwater level of one particular aquifer, which is primarily composed of gravel or sand. The groundwater levels in these wells are measured by piezometric sensors and recorded automatically at an hourly sampling rate [WRA, 2001]. The stations with significant changes associated with the Chi-Chi earthquake are marked by red open circles in the right panel of Figure 4.12a (see electronic version for color representation.)

Recently, *Chen et al.* [2015a] showed definitive coupling of premonitory anomalous phenomena between the long-term groundwater levels recorded at 42 of these monitoring stations and seismicity, using the time-varying surface

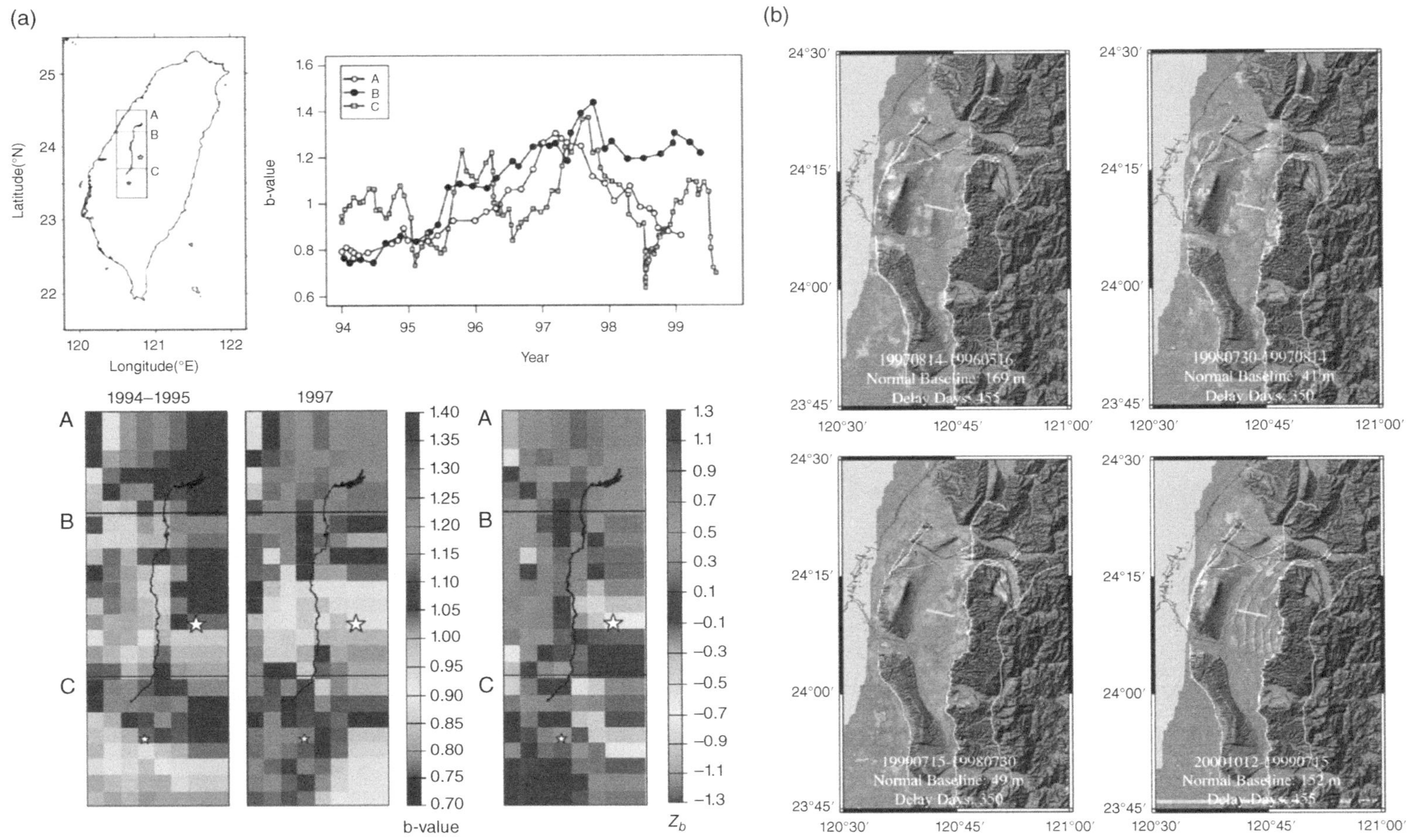

Figure 4.11 (a)Temporal and spatial variations of the *b*-value near the Chelungpu fault zone before the *M* 7.6 Chi-Chi earthquake. (b) Temporal variations of InSAR images in areas near the Chelungpu fault [*Tsai et al.*, 2005].

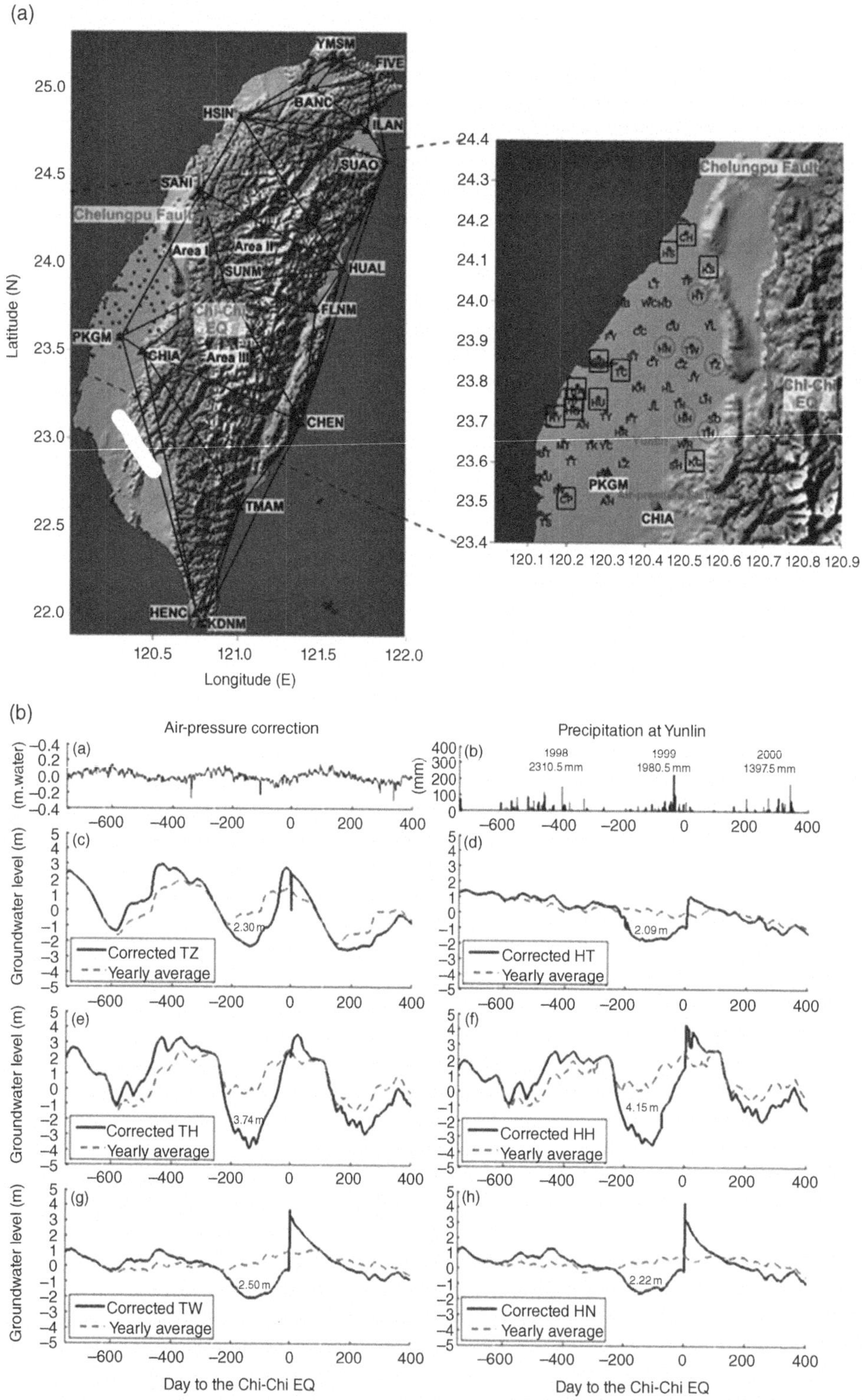

Figure 4.12 (a) GPS stations in Taiwan (left) and groundwater-level monitoring wells in central Taiwan (right). (b) Groundwater level records at six wells before and after the *M* 7.6 Chi-Chi earthquake with (top) barometric and precipitation records in central Taiwan [*Chen et al.*, 2015a]. Abbreviations YMSM, FIVE, BANC, HISN, ILAN, SUAO, SANI, HUAL, SUNM, FLNM, PKGM, CHIA, CHEN, TMAM, HENC, and KDNM stand for the GPS receiving stations. Groundwater stations with significant anomalies related to the Chi-Chi earthquake are denoted by red open circles, those without by blue open rectangles. (*See electronic version for color representation.*)

strain patterns derived from 16 GPS stations throughout Taiwan before the Chi-Chi earthquake. As shown in Figure 4.12b, the prominent noise-free groundwater-level anomalies at six stations were consistently comprised of a sequence of decreasing, flat, and rising phases for about 6 months prior to the earthquake. It is interesting to point out that these wells were located in the footwall block of the Chelungpu fault directly west of the Chi-Chi earthquake epicenter.

Figure 4.13a shows that these sequential changes in groundwater level agreed very well with changes in the strain-rate patterns computed from the GPS data over the areas surrounding the Chelungpu fault, as shown in Figure 4.13a. These coupling agreements show that, in addition to compression, tension also could be generated before occurrence of a major thrust earthquake. Figure 4.13b shows the GPS records at station San-Yi (SANI), located just 6 km north from the northern end of Chelungpu fault. By comparing the records between two consecutive 6-month periods, we can clearly see that the N–S and vertical components had changed direction during the 6 months immediately prior to the earthquake. This suggests that continuous GPS recordings could be used to monitor premonitory temporal changes in crustal movement before major earthquakes.

4.5.5. Geomagnetic Fluctuations

Observations of the geomagnetic fluctuations associated with earthquakes were part of the first earthquake prediction research program in Taiwan in the late 1970s. An island-wide geomagnetic network of eight stations, each equipped with a proton precession magnetometer with a 0.1 nT sensitivity and continuous recording sysyem, was completed in 1988. Figure 4.6b shows that all stations are located in areas of frequent seismicity, except the LP station in northwestern Taiwan. Accordingly, this station was chosen as a reference for the network.

Soon after the Chi-Chi earthquake, the geomagnetic data recorded by the network were analyzed [*Yen et al.*, 2004]. Figure 4.14a shows the total intensity record at LP station from mid-August to November 1999. Two magnetic storms were observed: one at about 2 days after the Chi-Chi earthquake and another about 1 month later at about the same time that the *M* 6.2 Chia-Yi earthquake occurred. The LP station was located about 130 km and 185 km, respectively, north of the epicenters of these two earthquakes. Apparently, each station was far away enough so that its surrounding geomagnetic field was not disturbed by both earthquakes.

Figure 4.14b shows the differences in geomagnetic total intensity between the LY station and the reference LP station from mid-August to November 1999. The figure shows that the difference fluctuated significantly, with the largest amplitude reaching up to about 200 nT. The disturbances at the LY station located near the northern end of the Chelungpu fault persisted from mid-August, through the Chi-Chi earthquake and up to the time of Chia-Yi earthquake; thereafter it returned to the normal quiet level.

Such large anomalous geomagnetic disturbances are probably the result of a large ground current generated by accumulation and release of crustal stress that subsequently leads to severe fault ruptures at the time of earthquakes [*Freund*, 2000]. This phenomenon is quite consistent with previous observations of the ULF anomalies by *Akinaga et al.* [2001], as well as that of the diurnal geomagnetic anomalies noted by *Liu et al.* [2006].

4.5.6. Atmospheric and Ionospheric Electromagnetic Anomalies

Atmospheric and ionospheric electromagnetic phenomena possibly associated with earthquakes have been reported extensively in the literature [e.g. *Hayakawa and Fujinawa*, 1994; *Hayakawa*, 1999]. Major advances in this area have been made in recent years by *Liu et al.* [2004a,b, 2015]. Two outstanding findings related to the Chi-Chi earthquake are highlighted below.

It has been observed that anomalies in ionospheric electron density could take place a few days before strong earthquakes [*Pulinets et al.*, 1994; *Liu et al.*, 2000, 2001]. *Liu et al.* [2004a,b] examined the data recorded by a local ionosonde and a ground-based GPS receiver in Taiwan during 1994–2003 (see Figure 4.14a). They found that the electron density at the F2 peak (NmF2) and the TEC measured by the GPS receiver decreased significantly during 10:00–20:00 hours local time a few days before most of the $M \geq 6.0$ earthquakes in Taiwan.

Figure 4.15a shows the observed ionospheric TEC time history during September 1999, as compared with the 15-day running median as well as the corresponding upper and lower bounds at ±1I QR (interquartile range) levels. The figure shows that the TEC level recorded by GPS anomalously decreased below the corresponding lower bound during 10:00–20:00 hours local time on 17 and 18 September 1999, i.e., 4 and 3 days, respectively, before the Chi-Chi earthquake.

It has been known that atmospheric conductivity could be changed by the earthquake preparation process. A network of lightning detectors in Taiwan is continuously monitoring occurrences of lightning due to cloud-to-ground (C-G) discharges. The lightning occurrences 15 days both before and after the Chi-Chi earthquake are shown in Figure 4.16a. It is apparent that the lightning occurrences increased drastically on 17 September 1999, 4 days before the earthquake. From Figure 4.16b we can see that the C-G discharges on this day occurred mainly near the southern end of the Chelungpu fault. The coincidental occurrence in time of enhanced atmospheric C-G lightnings and reduced ionospheric TEC level recorded by GPS is highly suggestive that a possible coupling mechanism exists between these two phenomena.

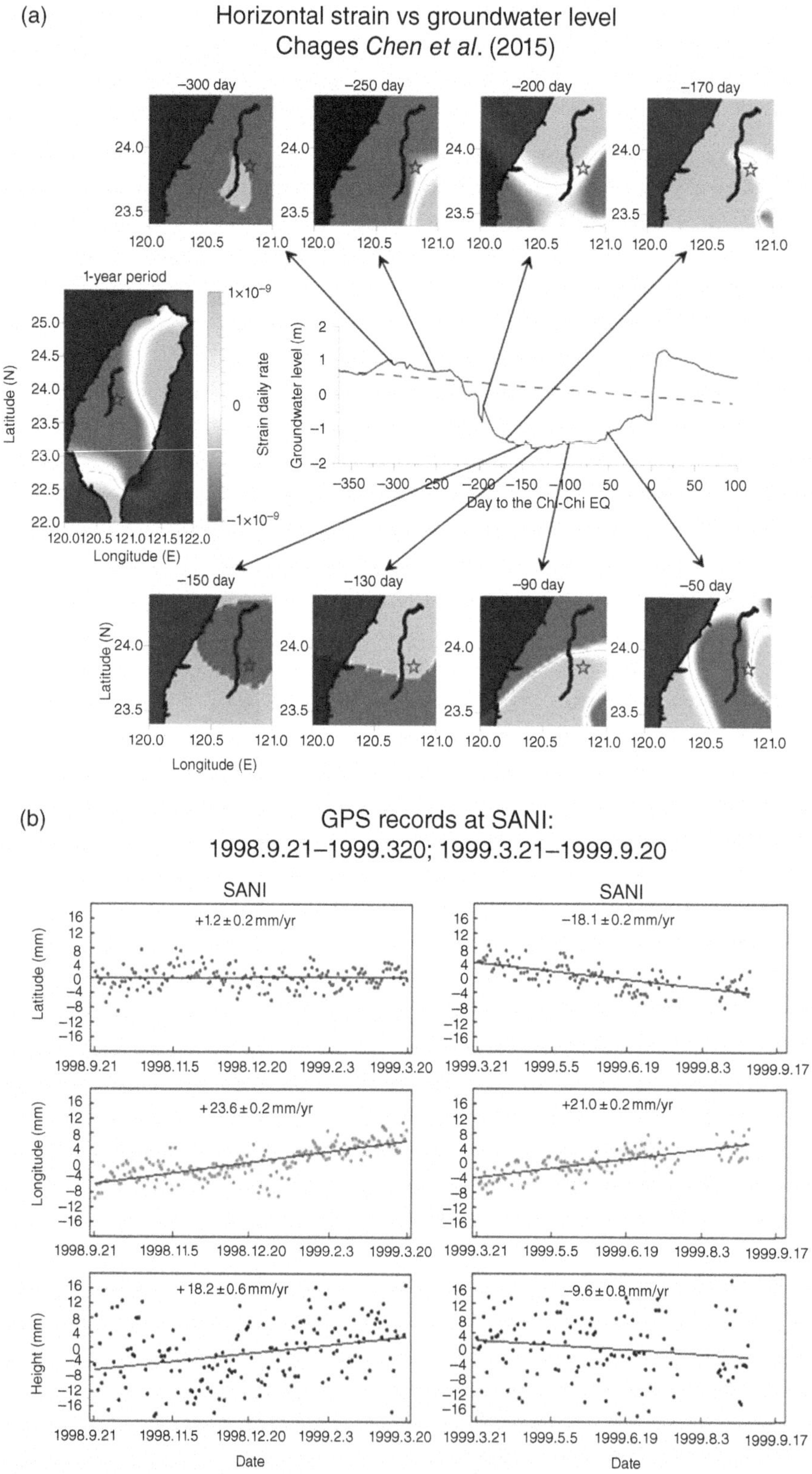

Figure 4.13 (a) Correlation between temporal variations of horizontal-strain patterns in areas surrounding the Chelungpu fault in Central Taiwan and a groundwater-level record before the 1999 *M* 7.6 Chi-Chi earthquake [*Chen et al.*, 2015a]. (b) Segments of a GPS record at San-Yi (SANI; 24.41°N, 120.77°E) site during (left) 21 September 1998 to 20 March 1999 and (right) 21 March 1999 to 20 September 1999. Note the changes in the directions of motion in the (top) N–S and (bottom) vertical components.

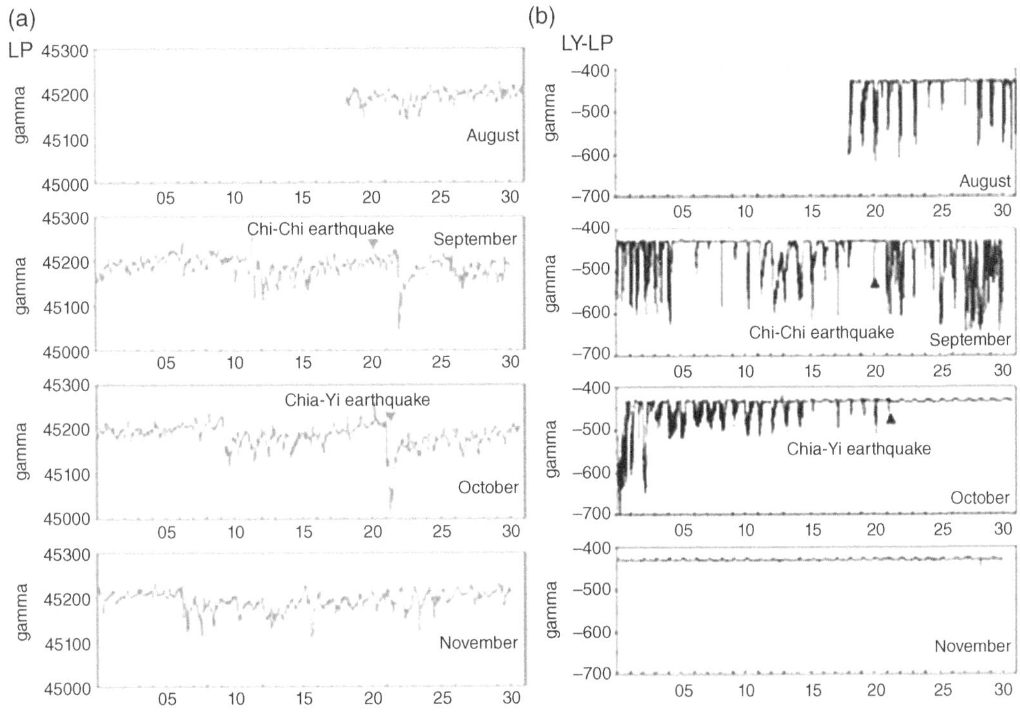

Figure 4.14 (a) Geomagnetic total intensity (GTI) records at the reference station LP. (b) The difference in GTI between stations Lunping (LP) and Liyutan (LY) located near the northern end of the Chelungpu fault before and after the 1999 *M* 7.6 Chi-Chi earthquake in Taiwan [*Yen et al.*, 2004].

It is encouraging that multiple types of precursory phenomena were positively identified associated with the *M* 7.6 Chi-Chi earthquake using reliable instrumental data. They include seismological, geodetic, groundwater level, geomagnetic, atmospheric, and ionospheric anomalies. Some of the observations corroborated with each other, such as those between the GPS crustal strain and the groundwater-level changes, as well as those between the bursts of atmospheric C-G lightning and the decreases of the ionospheric TEC level observed by a local ionosonde and a ground-based GPS receiver in Taiwan.

The observed premonitory times, ranging from about 6 years, through several months, and down to a few days, could provide long-term, to intermediate-term, to short-term precursors. Most of these observations were also spatially indicative of the likely source zone of the Chi-Chi earthquake. Further studies are needed to synthysize all this information in order to estimate the magnitude of impending earthquakes.

4.6. POSITIVE PRECURSORS IDENTIFIED WITH SEVERAL RECENT MODERATE EARTHQUAKES IN TAIWAN

At the end of iSTEP-1, a *M* 7.0 doublet earthquake occurred on 26 December 2006 in southwestern offshore region of Taiwan. They provided a chance to test our methods for identifying earthquake precursors. Again, the TEC anomalies were observed 4 days before the earthquakes [*Liu et al.*, 2006, 2008; *Hsiao et al.*, 2009, *C. Y. Liu et al.*, 2011]. Meanwhile, the orientations of residual ground displacements, which were retrieved from the GPS data after removing the long-term plate motion, semi-annual and annual components, were shown to align with the direction of the maximum compression of the earthquakes [*Chen et al.*, 2011]. Alignment of the pre-earthquake residual displacements widely covered the whole island, leading to further understanding of the earthquake preparation zone.

About 10 years after the Chi-Chi event, the *M* 6.2 Ming-Jian earthquake occurred on 5 November 2009 in almost the same region. The residual displacements were again found to align with the maximum compression axis of the earthquake about 20 days earlier. The residual displacements became randomly oriented at the terminal stage of fault failure, as expected by the damage–mechanic models [*Chen et al.*, 2014].

The *M* 6.4 Jiashian earthquake occurred in southern Taiwan on 4 March 2010. Data on GPS residual displacements showed that the crust with northeastern movement was resisted in the areas to the southwest of the epicenter about 10 days before the earthquake [*Chen et al.*, 2013a]. Similarly, two *M* 6.0 earthquakes occurred in central Taiwan on 27 March and 2 June 2013. Changes in the

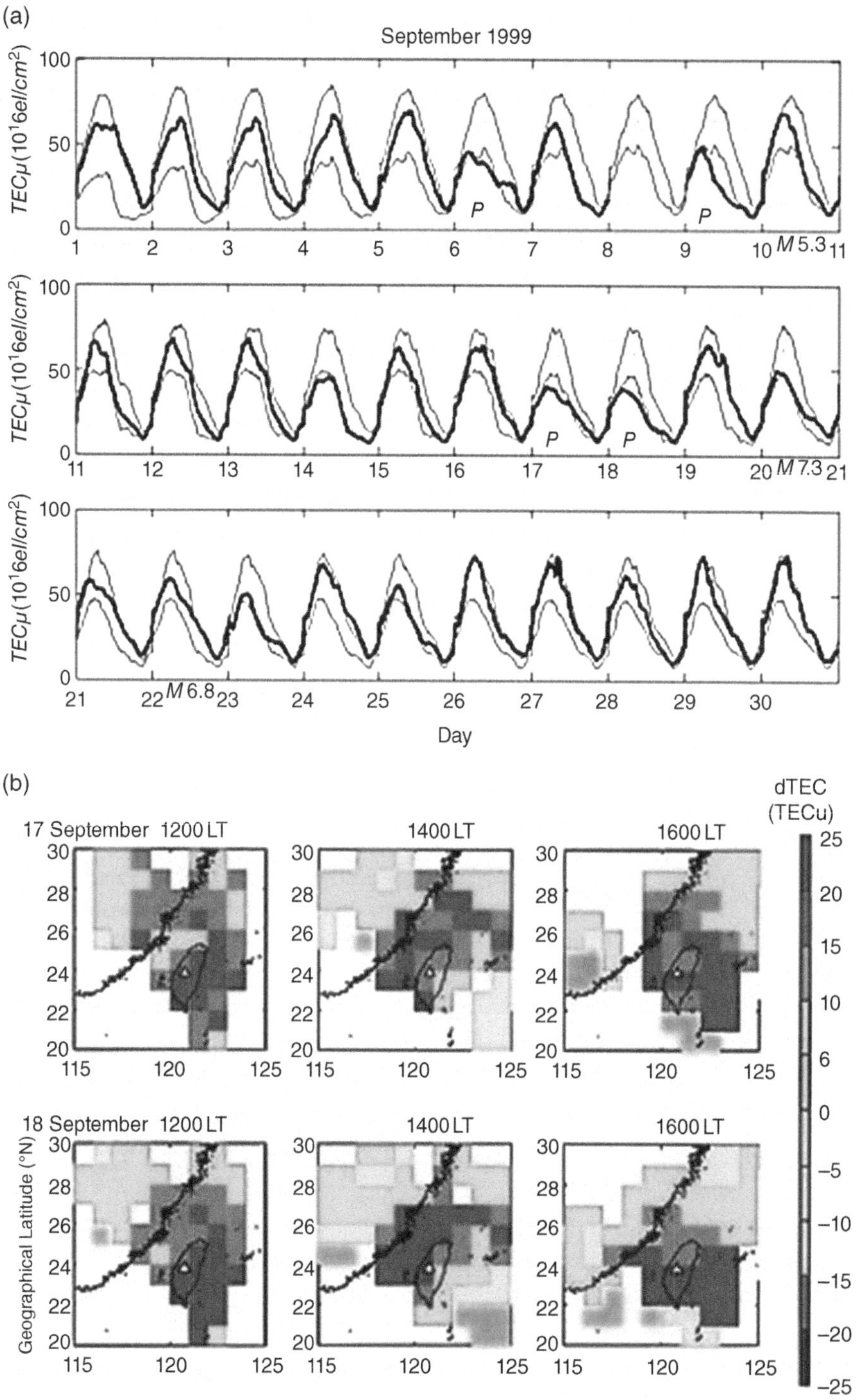

Figure 4.15 (a) Daily variations of ionospheric total electron content (TEC) in September 1999; (b) Spatial variations of TEC on 17 September 1999 [*Liu et al.*, 2004b]; LT, local time; P, seismo-ionospheric precursor of the GPS TEC.

GPS residual displacements from alignment to disorderly were again observed [*Chen et al.*, 2012].

In addition, seismo-conductivity anomalies associated with these earthquakes were observed by using the three-component geomagnetic data recorded at three stations [*Chen et al.*, 2015c]. They were obtained through the depth–frequency relationship derived from the skin effect in the magnetic transfer function, as proposed by *Chen et al.* [2013b]. Appearance of the seismo-conductivity anomalies in the terminal stage of fault failure, as determined by the disorientation of residual GPS displacement, suggests that variations of conductivity could change along with the

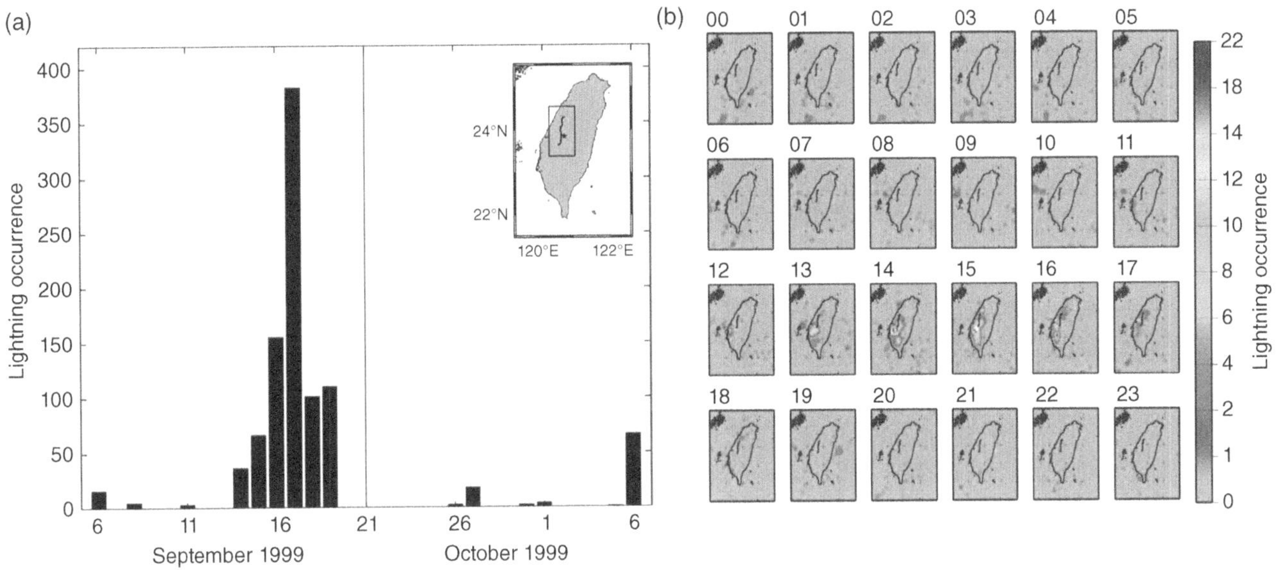

Figure 4.16 (a) Daily lightning occurrences during 6 September 1999 to 6 October 1999 around the Chelungpu fault area. (b) Spatial patterns of hourly lightning occurrences on 17 September 1999 in the Taiwan region.

stress accumulation and reach a maximum just before the earthquakes. In order to take advantage of these repeatable observations with the highly reliable TEC anomalies, studies on feasible LAIC mechanisms are underway.

In recent years, several multiparameter continuous soil-gas and gamma-ray monitoring stations have been deployed in Taiwan to study possible seismo-geochemical variations in both space and time [*Chyi et al.*, 2005; *Walia et al.*, 2013]. These stations are stratigetically located near active faults in southwestern Taiwan [*Yang et al.*, 2005, 2006], the Longitudinal Valley of eastern Taiwan [*Fu et al.*, 2009, 2015, 2017a; *Kuo et al.*, 2010], and northern Taiwan [*Fu et al.*, 2017b]. The geochemical earthquake precursors observed to date are presented in Chapter 11 of this monograph [*Fu and Lee*, 2018].

4.7. FUTURE PROSPECTS ON MULTIDISCIPLINARY EARTHQUAKE PRECURSOR STUDIES IN TAIWAN

As reviewed above, the earthquake science community in Taiwan recognized early on, given the prevailing high risk of severe seismotectonic and societal impacts, the necessity of pursuing the difficult goal of earthquake prediction. The first multidisplinary earthquake prediction research program was launched in 1979, jointly funded by the National Science Council and the US Geological Survey. Since then, a small yet dedicated group of scientists have persistently made efforts to identify possible earthquake precursors. As a result, multiple types of earthquake precursory phenomena have been positively observed, especially those pertaining to the 1999 *M* 7.6 Chi-Chi earthquake.

Looking to the future, it is encouraging that several advanced multidisciplinary real-time monitoring networks, in addition to the iSTEP-3 observational systems, are currently available for making long-term observations to identify positive earthquake precursors in Taiwan. These netwoks are listed below:

1. the short-period, broadband, and strong-motion seismic networks operated by the CWBSN (Figure 4.17a);
2. the broadband seismic network operated by the IES (Figure 4.17b);
3. the continuous-recording GPS network operated by the CWBSN (Figure 4.18a);
4. the total-intensity geomagnetic network operated by the CWBSN (Figure 4.18b);
5. the groundwater-level monitoring network operated by the CWBSN (Figure 4.19a);
6. the groundwater-level monitoring network operated by the Water Resource Agency (WRA) (Figure 4.19b);
7. the geochemical and borehole strainmeter networks operated by the Central Geological Survey (CGS) (Figure 4.20);
8. the gamma-ray network operated by IES (Figure 4.21a).
9. the geoelectric-potential netwok operated by the National central University (NCU) (Figure 4.21b);
10. the three-component geomagnetic network operated by the IES, National Central University (NCU), and Chung Cheng University (CCU) (Figure 4.22).

Networks referred to above are either directly operated by the technical staff of relevant government agencies, such as the CWB, CGS, and WRA, or by research staff in academic institutions, such as the IES, NCU, and CCU,

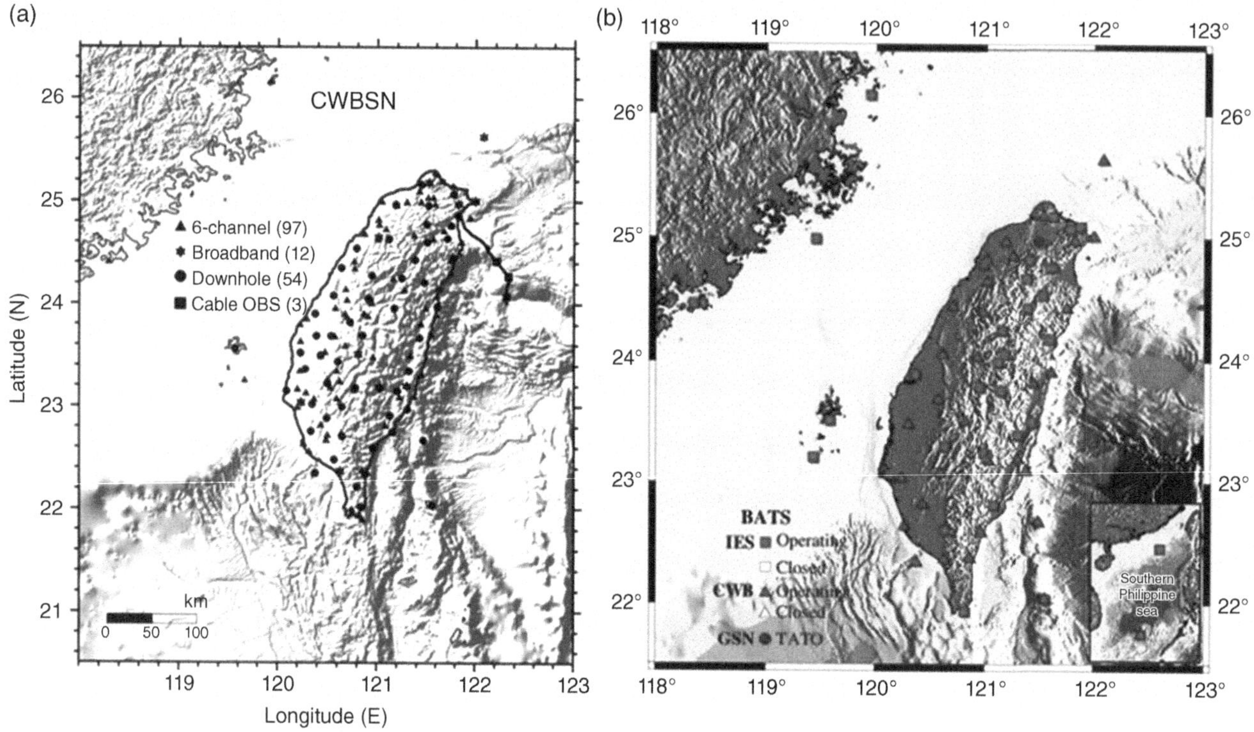

Figure 4.17 (a) Map of the Central Weather Bureau Seismic Network of Taiwan. (Source: http://cwb.gov.tw) (b) Map of the Broadband Array in Taiwan for Seismology (BATS) network. (Source: http://tec.earth.sinica.edu.tw)

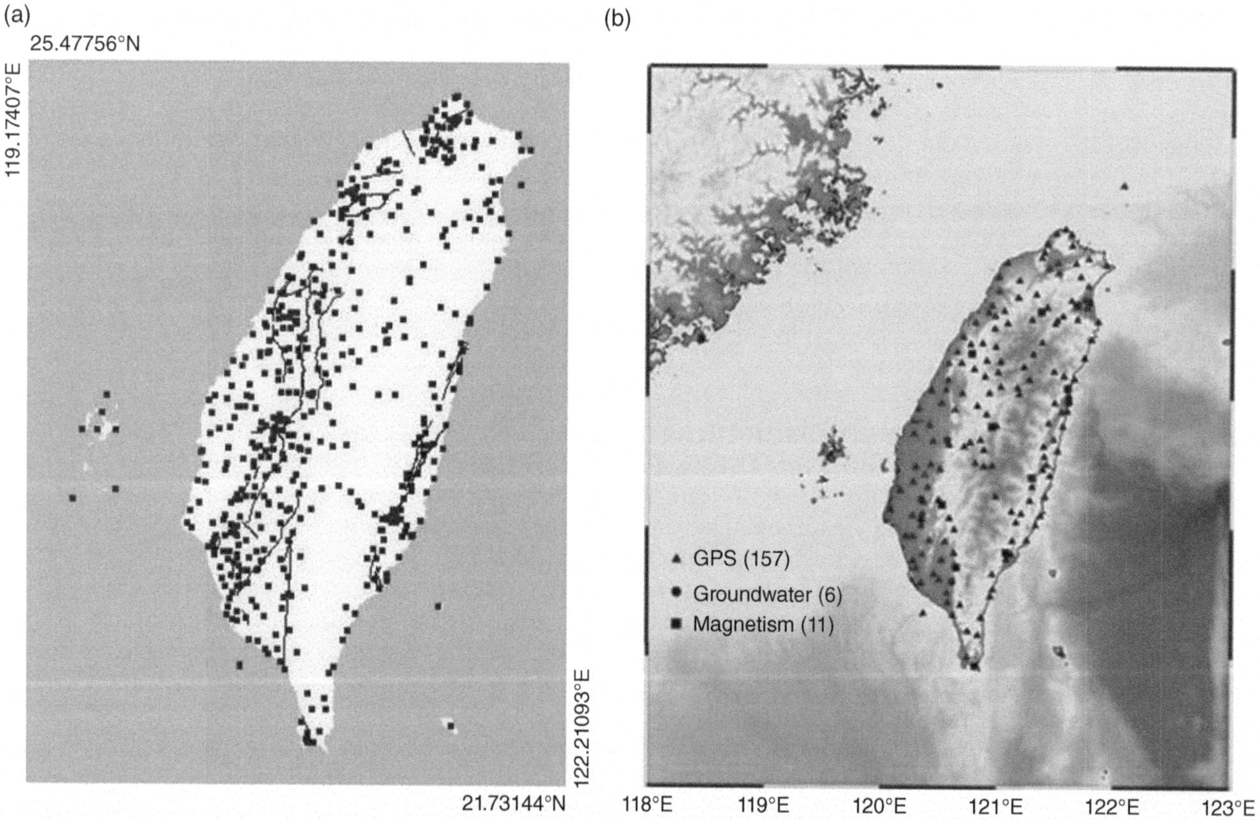

Figure 4.18 (a) Map of continuous GPS stations in Taiwan. (Source: http://gps.earth.sinica.edu.tw) (b) Map of GPS geomagnetic and groundwater monitoring stations in the CWBSN (Source: http://cwb.gov.tw).

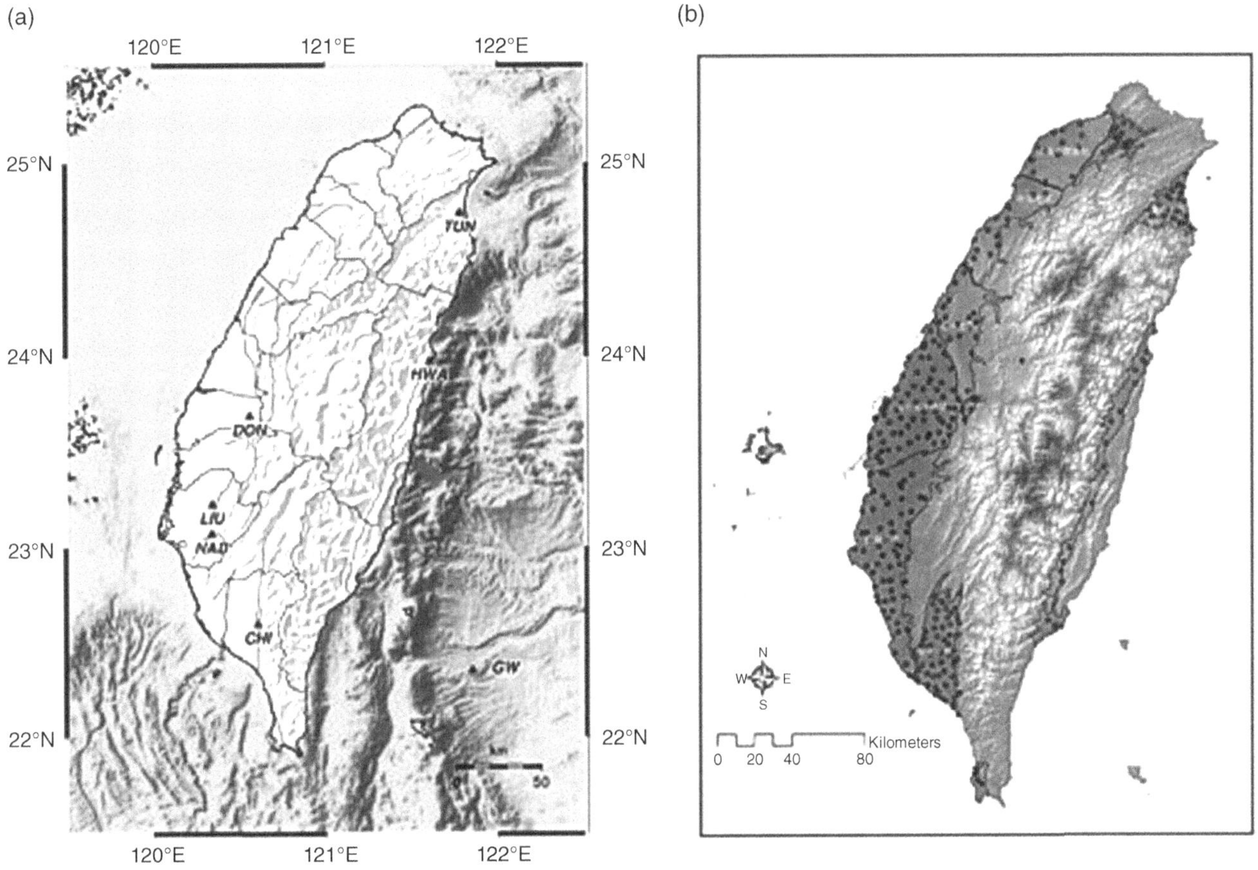

Figure 4.19 (a) Map of the CWBSN groundwater-level monitoring stations [*Shin et al.*, 2013]. (b) Map of WRA groundwater-level monitoring stations [*WRA*, 2001]. TUN, MWA, DON, LIU, NAU, CHI, and GW are groundwater-level monitoring station locations.

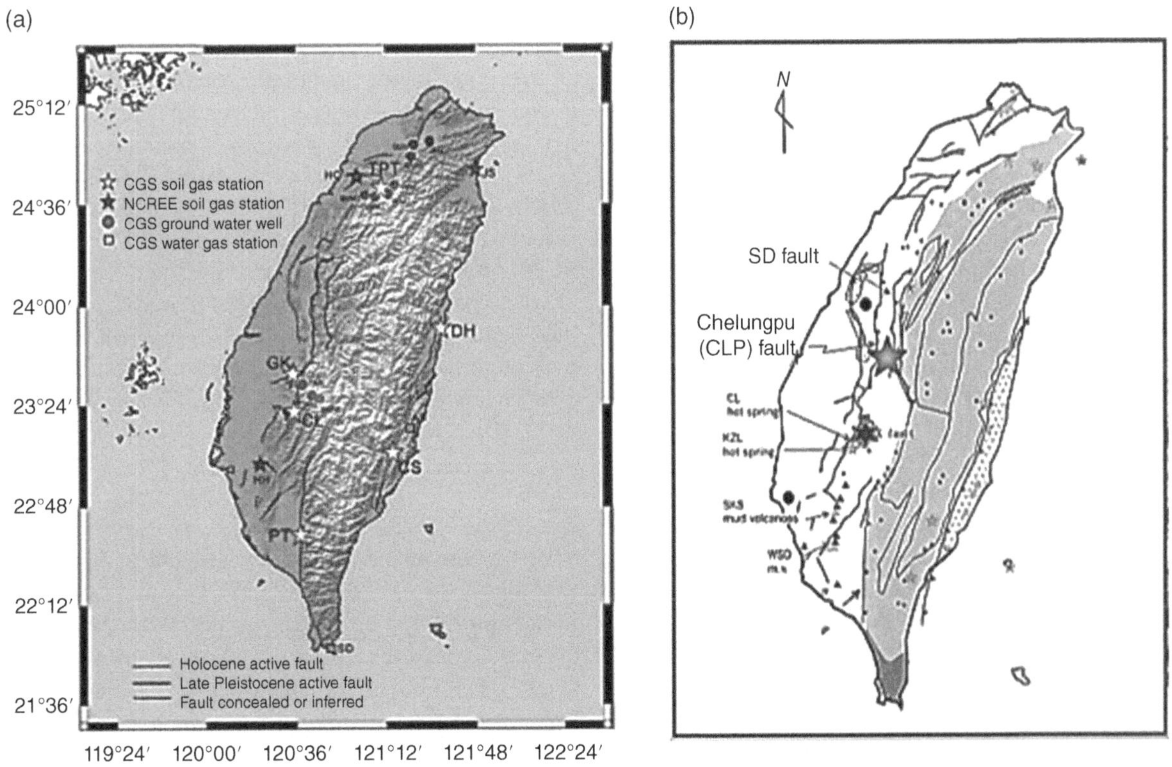

Figure 4.20 (a) Map of the Central Geological Survey gas and fluid geochemical monitoring stations. (b) Automatic fluid geochemical monitoring stations on the Chukou fault [*Yang et al.*, 2015].

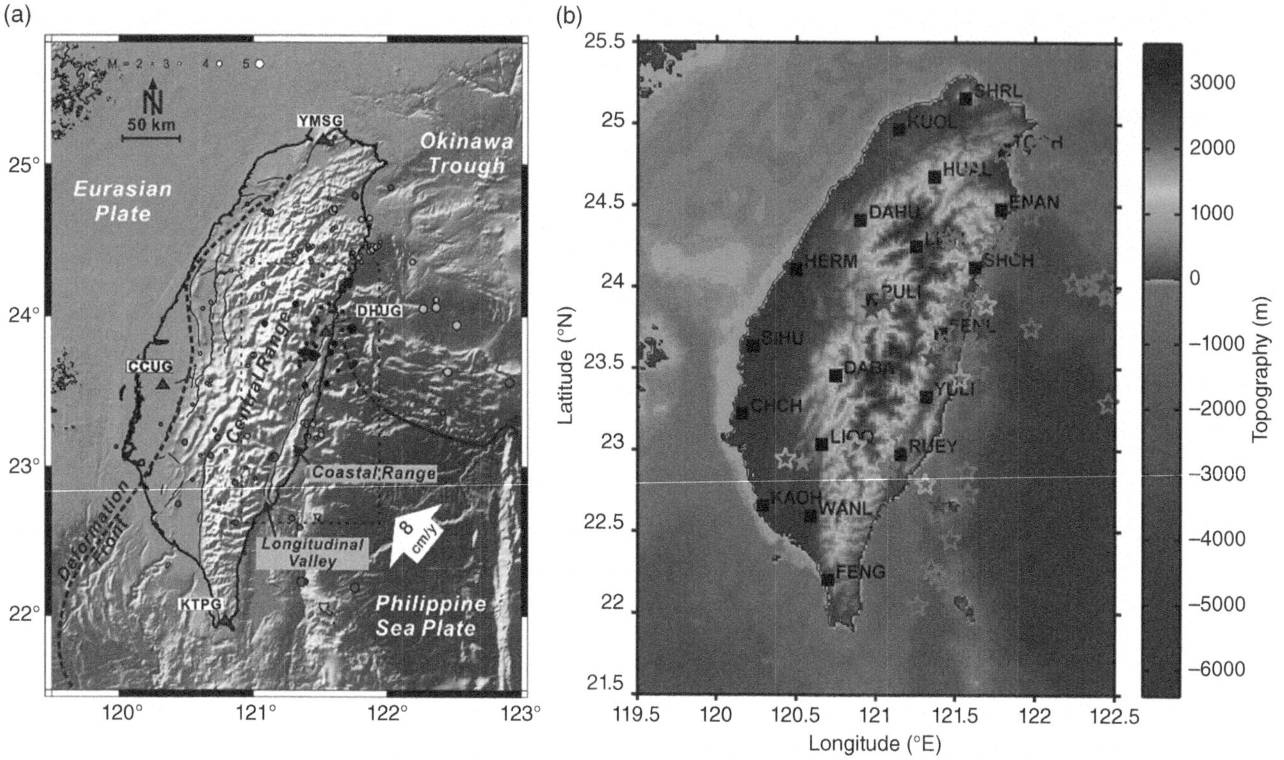

Figure 4.21 (a) Map of gamma-ray monitoring stations operated by the Institute of Earth Sciences [*Fu et al.*, 2015]. (b) Map of geoelectric monitoring stations operated by the National Central University [*C. C. Chen et al.*, 2015]. YMSG, DHUG, CCUG, and KTPG denote gamma-ray monitoring stations.

through funding by the Ministry of Science and Technology. Moreover, the CWB has taken the lead to organize annual workshops to exchange findings on earthquake precursors using the observational data obtained by these networks.

In summary, Taiwan is situated in a highly active seismotectonic environment and is constantly subject to frequent small and large shallow inland earthquakes, making it a favorable location for gathering data on these phenomena. This is further compounded by dedicated scientific endeavors backed by long-term government support and genuine public interest, out of concerns for seismic hazards. Endowed with these favorable natural, societal, and scientific conditions, Taiwan could arguably qualify as a highly promising test ground for multidisciplinary earthquake precursor studies. This has been amply highlighted by positive identification of multiple types of premonitory phenomena associated with the 1999 M 7.6 Chi-Chi earthquake.

Since earthquake prediction became a subject of scientific study in the mid-1970s [*Press*, 1975; *Rikitake*, 1976], the question, "*Are earthquakes predictable?*" has been hotly debated in the earthquake science community since then. Responses to this question range widely, from very negative to very positive, with moderate responses in between. Perhaps a more pragmatic question would be: "*Are there earthquake precursors that can lead to useful predictions?*" [*Seife*, 2005].

With the benefit of several unique seismotectonic situations, scientific potentential, and societal support, it is hoped that the earthquake science community, through persistent and diligent work, will be successful in identifying positive precursors at timescales of years, months and/or days, that would inform the people of Taiwan prior to future disastrous earthquakes. In the future, sufficient empirical precursors positively observed in Taiwan may help to establish an underlying physics-based mechanisms to form the foundation for the ultimate goal of earthquake prediction [*Aki*, 1995; *Freund*, 2000; *Kanamori and Brodsky*, 2004; *Pulinets*, 2004; *Kim et al.*, 2012; *Kuo et al.*, 2014].

ACKNOWLEDGMENTS

We are deeply indebted to Drs Ta-Liang Teng and Jack Evernden who initiated, established, and supported the first multidisciplinary earthquake prediction research program in Taiwan. We owe most sincere gratitude to our colleagues for their dedicated efforts. The studies reported here were supported over the years by the Presidential Office (through IES), Ministries of Economic Affairs

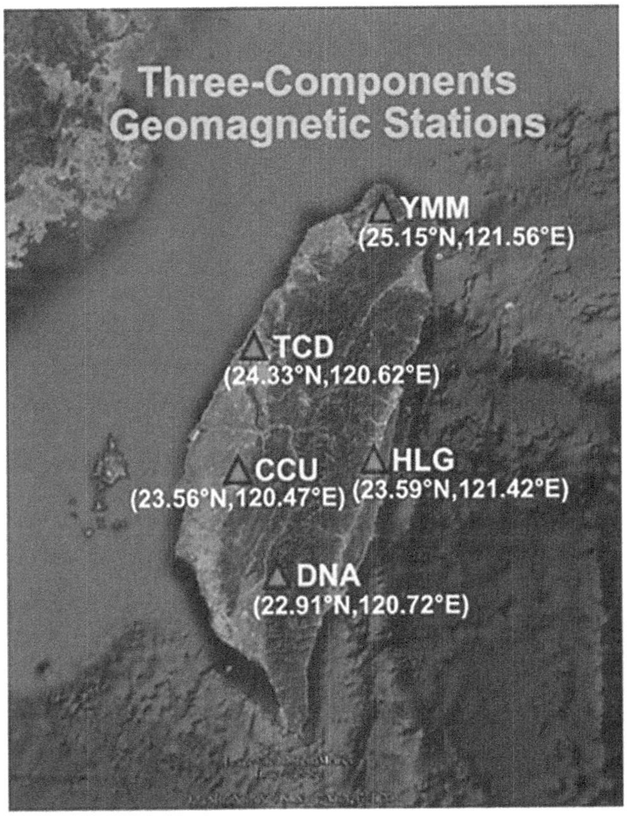

Figure 4.22 The three-component geomagnetic stations in Taiwan operated by the Institute of Earth Science and National Chung Cheng University. YMM, TCD, CCU, HLG, and DNA are three-component geomagnetic stations.

(through WRA), Education (through universities), Interior Affairs (through CGS), Science and Technology (formerly National Science Council, through IES and universities), and Transportation (through CWB), as well as by the US Geological Survey. Insightful comments and suggestions by the two anonymous reviewers are greatly appreciated. Editorial comments by Ms S. Halloran and the members of her writing class have been helpful. The first author, Professor Yi-Ben Tsai passed away peacefully in San Francisco at 09:30 LT on 31 January 2018. In the last moment, Professor Tsai excepted this chapter as his final work. JYL and HYY continued finalizing the revision. The authors wish to thank Dr. Chun-Rong Chen at NCU for his editing and cartography skills.

REFERENCES

Aki, K. (1995), Earthquake prediction, societal implications, in *U.S. National Report to IUGG, 1991–1994, Rev. Geophys.*, *33*(Suppl.).

Akinaga, Y., Hayakawa, M., Liu, J. Y. Yumoto, K., and Hattori, K. (2001), A precursory ULF signature for the Chi-Chi earthquake in Taiwan, *Nat. Hazards Earth Syst. Sci.*, *1*, 33–36.

Chang, C. P., Chen, K. S., Wang, C. T., Yen, J. Y., Chang, T. Y., and Lin, C. W. (2004), Application of space-borne radar interferometry on crustal deformations in Taiwan: A prospective from the nature of events, *Terr. Atmos. Ocean Sci. J.*, *15*(3), 523–543.

Chang, W.Y., Chen, K.P., and Tsai, Y.B. (2016), An updated and refined catalog of earthquakes in Taiwan (1900–2014) with homogenized M_w magnitudes, *Earth, Planets, Space*, *68*(45), 1–9.

Chen, C. C., Rundle, J. B., Holliday, J. R., Nanjo, K. Z., Turcotte, D. L., Li, S. C., and Tiampo, K. F. (2005), The 1999 Chi-Chi, Taiwan earthquake as a typical example of seismic activation and quiescence, *Geophys. Res. Lett.*, *32*, L22315.

Chen, C. C., Chen, H. C., Hsu, H. L., Hsiao, W. C., Lin, C. C. and Chen, C. Y. (2015), *A Study of Geoelectric Phenomena Related to Earthquakes*, MOTC-CWB-104-E-2 Annual Technical Report, 139–177, Central Weather Bereau, Taipei. (In Chinese.).

Chen, C. H., Yeh, T. K., Liu, J. Y., Wang, C. H., Wen, S., Yen, H. Y. and Chang, S. H. (2011), Surface deformation and seismic rebound: Implications and applications, *Surv. Geophys.*, *32*, 291–313.

Chen, C. H., Wang, C. H., Wen, S., Yeh, T. K., Lin, C. H., Liu, J. Y., Yen, H. Y., Lin, C., Rau, R. J., and Lin, T. W. (2012), Anomalous frequency characteristics of groundwater level before major earthquakes in Taiwan, *Hydrol. Earth Syst. Sci.*, *17*, 1693–1703.

Chen, C. H., Wen, S., Yeh, T. K., Wang, C. H., Yen, H. Y., Liu, J. Y., Hobara, Y., and Han, P. (2013a), Observation of surface displacements from GPS analyses before and after the Jiashian earthquake (M=6.4) in Taiwan, *J. Asian Earth Sci.* *62*, 662–671.

Chen, C. H., Hsu, H. L., Wen, S., Yeh, T. K., Chang, F. Y., Wang, C. H., Liu, J. Y., Sun, Y. Y., Hattori, K., Yen, H. Y. and Han, P. (2013b), Evaluation of seismo-electric anomalies using magnetic data in Taiwan, *Nat. Hazards Earth Syst. Sci.*, *13*, 1–8.

Chen, C. H., Tang, C. C., Cheng, K. C. Wang, C. H., Wen, S., Lin, C. H., Wen, Y. Y., Meng, G., Yeh, T. K., Jan, J. C., Yen, H. Y., and Liu, J. Y. (2015a), Groundwater–strain coupling before the 1999 *Mw* 7.6 Taiwan Chi-Chi earthquake, *J. Hydrol.*, *524*, 378–384.

Chen, C. H., Lin, C. H., Hsu, H. L., Wang, C. H., Lee, L. C., Han, P., Wen, S., and Chen, C. S. (2015b), Evaluating the March 27, 2013 M6.2 earthquake hypocenter using momentary high-conductivity materials, *Terr. Atmos. Ocean Sci. J.*, *26*, 1–9.

Chen, C. H., Lin, C. H., Wang, C. H., Liu, J. Y., Yeh, T. K., Yen, H. Y., and Lin, T. W. (2015c), Potential relationships between seismo-deformation and seismo-conductivity anomalies, *J. Asian Earth Sci.*, *114*, 327–337.

Chen, Y. I., Liu, J. Y., Tsai, Y. B., and Chen, C. S. (2004), Statistical tests for pre-earthquake ionospheric anomaly, *Terr. Atmos. Ocean Sci. J.*, *15*(3), 385–396.

Chia, Y., Wang, Y. S., Chiu, J. J., and Lin, C. W. (2001), Changes of groundwater level due to the 1999 Chi-Chi earthquake in the Choshui River alluvial fan in Taiwan, *Bull. Seismol. Soc. Am.*, *91*, 1062–1068.

Chyi, L. L., Quick, T. J., Yang, T. F., and Chen, C. H. (2005), Soil gas radon spectra and earthquakes, *Terr. Atmos. Ocean Sci. J.*, *16*, 763–774.

Crampin, S.(1998), Shear-wave splitting in a critical crust, *Rev. Inst. Fr. Petrol.*, *53*, 749–763.

Freund, F. (2000), Time-resolved study of charge generation and propagation in igneous rocks, *J. Geophys. Res.*, *105*, 11001–11019.

Freund, F., Gupta, A., Butow, S. J., and Tenn, S. (1999) Molecular hydrogen and dormant charge carriers in minerals and rocks, in M. Hayakawa (ed.), *Atmospheric and Ionospheric Electromagnetic Phenomena Associated With Earthquake*, pp. 859–871, Tokyo, Terra Scientific Publishers.

Freund, F. T., Takeuchi, A., Lau, B. W. S., Post, R., Keefner, J., Mellon, J., and Al-Manaseer, A. (2004), Stress-induced changes in the electrical conductivity of igneous rocks and the generation of ground currents, *Terr. Atmos. Ocean Sci. J.*, *15*(3), 437–467.

Fu, C. C. and Lee, L. C. (2018), Continuous monitoring of fluid and gas geochemistry for seismic study in Taiwan, D. Ouzounov, S. Pulinets, K. Hattori, and P. Taylor (eds), *Pre-Earthquake Processes: A Multi-disciplinary Approach to Earthquake Prediction Studies*, pp. 414, Geophysical Monograph 234, American Geophysical Union, Washington, DC, and John Wiley & Sons, Inc., Hoboken, NJ.

Fu, C. C., Yang, T. F., Walia, V., Liu, T. K., Lin, S. J., Chen, C. H., and Hou, C. S. (2009), Variations of soil- gas compositionaround the active Chihshang Fault in a plate suture zone, eastern Taiwan, *Radiat. Meas.*, *44*, 940–944.

Fu, C. C., Wang, P. K., Lee, L. C., Lin, C. H., Chang, W. Y., Giuliani, G., and Ouzounov, D. (2015), Temporal variation of gamma rays as a possible precursor of earthquake in the Longitudinal Valley of eastern Taiwan, *J. Asian Earth Sci.*, *114*(2), 362–372.

Fu, C. C., Yang, T. F., Tsai, M. C., Lee, L. C., Liu, T. K., Walia, V., Chen, C. H., Chang, W. Y., Kumar, A., and Lai, T. H. (2017a), Exploring the relationship between soil degassing and seismic activity by continuous radon monitoring in the Longitudinal Valley of eastern Taiwan, Chemical Geology, dx.doi.org/10.1016/j.chemgeo.2016.12.042.

Fu, C.C., Yang, T. F., C.C. Chen, Lee, L. C., Y.M. Wu, Liu, T. K., Walia, V., Kumar, A., and Lai, T. H. (2017b), Spatial and temporal anomalies of soil gas in northern Taiwan and its tectonic and seismic implications, *J. Asian Earth Sci.*, *144*, 64–77.

Gutenberg, B. and Richter, C. F. (1944), Frequency of earthquakes in California, *Bull. Seismol. Soc. Am.*, *34*, 185–188.

Hayakawa, M. (ed.) (1999), *Atmospheric and Ionospheric Electromagnetic Phenomena Associated with Earthquakes*, TERRAPUB, Tokyo, 997 pp.

Hayakawa, M. and Fujinawa, Y. (eds) (1994), *Electromagnetic Phenomena Related to Earthquake Prediction*, TERRAPUB, Tokyo.

Hayakawa, M., Molchanov, O. A., and NASDA/UEC Team (2004), Achievements of NASDA's Earthquake remote sensing frontier project, *Terr. Atmos. Ocean Sci. J.*, *15*(3), 311–327.

Hsiao, C.C., Liu, J. Y., Oyama, K. I., Yen, N. L., Wang, Y. H., and Miau, J. J. (2009), Ionospheric electron density anomaly prior to the December 26, 2006 *M* 7.0 Pingtung earthquake doublet observed by FORMOSAT-3/COSMIC, *Phys. Chem. Earth*, *34*, 474–478.

Hsieh, H. H., Chen, C. H., Lin, P. Y., and Yen, H. Y., (2014), Currie point depth from spectral analysis of magnetic data in Taiwan, *J. Asian Earth Sci.*, *90*, 26–33.

Hsu, M. T., (1971), Seismicity of Taiwan and some related problems, *Bull. Intern. Inst. Seism. Earthq. Eng.*, *8*, 41–60.

Kanamori, H. and Brodsky, E. E. (2004), The physics of earthquakes, *Rep. Prog. Phys.*, *67*, 1429–1496.

Kim, V. P., Liu, J. Y., and Hegai, V. V. (2012), Modeling the pre-earthquake electrostatistic effect on the F region ionosphere, *Adv. Space Res.*, *50*, 1524–1533.

Kuo, C. L., Lee, L. C., and Huba, J. D. (2014), An improved coupling model for the lithosphere-atmosphere-ionosphere system, *J. Geophys. Res. Space Phys.*, *119*, 3189–3205.

Kuo, T., Lin, C., Chang, G., Fan, K., Cheng, W., and Lewis, C. (2010), Estimation of aseismic crustal strain using repetitive radon precursors of 2003 *M* 6.8, 2006 *M* 6.1, and 2008 *M* 5.0 earthquakes in eastern Taiwan, *Nat. Hazards*, *53*, 219–228.

Lee, C.P. and Tsai, Y. B. (2004), Variations of P-wave travel-time residuals before and after the 1999 Chi-Chi, Taiwan, earthquake, *Bull. Seismol. Soc. Am.*, *94*, 2348–2365.

Liu, C. Y., Liu, J. Y., Wen, W. S., Li, J. Z., Xia, Y. Q., and Cui, X. Y. (2011), An integrated study of anomalies observed before four major earthquakes: 2004 Sumatra *M* 9.3, 2006 Pingtung *M* 7.0, 2007 Chuetsu Oki *M* 6.8, and 2008 Wenchuan *M* 8.0, *J. Asian Earth Sci.*, *41*, 401–409.

Liu, J. Y., Chen, Y. I., Tsai, Y. B., and Chuo, Y. J. (2000), Seismo-ionospheric signatures prior to $M \geq 6.0$ Taiwan earthquakes, *Geophys. Res. Lett.*, *27*, 3113–3116.

Liu, J. Y., Chuo, Y. J., and Chen, Y. I. (2001), Ionospheric GPS TEC perturbations prior to the 20 September 1999, Chi-Chi earthquake, *Geophys. Res. Lett.*, *28*, 1383–1386.

Liu, J. Y., Chen, Y. I., Chuo, Y. J., and Tsai, H. F. (2001b), Variations of ionospheric total electron content during the Chi-Chi earthquake, *Geophys. Res. Lett.*, *28*(7), 1383–1386.

Liu, J. Y., Chuo, Y. J., Shan, S. J., Tsai, Y. B., Chen, Y. I. Pulinets, S. A., and Yu, S. B. (2004a), Seismo-ionospheric anomalies monitored by GPS TEC, *Ann. Geophys.*, *22*, 1585–1593.

Liu, J. Y., Chen, Y. I., Jhuang, H. K., and Lin, Y. H. (2004b), Ionospheric *fo*F2 and TEC anomalous days associated with *M* 5.0+ earthquakes in Taiwan during 1997–1999, *Terr. Atmos. Ocean Sci. J.*, *15*(3), 371–383.

Liu, J. Y., Chen, S. W., Chen, Y. C., Yen, H. Y., Chang, C. P., Chang, W. Y., Tsai, L. C., Chen, C. H., and Yang, W. H. (2008), Seismo-ionospheric precursors of the 26 December 2006 *M* 7.0 Pingtung earthquake doublet, *Terr. Atmos. Ocean Sci. J.*, *19*, 751–759.

Liu, J.Y., Chen, Y. I., Huang, C. H., Ho, Y. Y., and Chen, C. H. (2015), A statistical study of lightning activities and $M > 5.0$ earthquakes in Taiwan during 1993–2004, *Surv. Geophys.*, *36*, 851–859.

Liu, J. Y., Tsai, Y. B., Chen, C. H., Chen, Y. I., and Yen, H. Y. (2016a), Integrated search for Taiwan earthquake precursors, *IEEJ Transaction*, *136*, 214–220, doi: 10.1541/ieejfms.136.214.

Liu, J. Y., Yen, H. Y., Tsai, Y. B., Chen, C. H., and Chen, Y. I. (2016b), A summary of iSTEP (integrated Study and Test of Earthquake Precursor) projects (Abstract), *Japan Geoscience Union Meeting 2016*, 22–26 May, Chiba, Japan.

Massonnet, D., Rossi, M., Carmona, C., Adragna, F., Peltzer, G., Feigl, K., and Rabaute, T. (1993), The displacement field of the Lander raethquake mapped by radar interferometry, *Nature*, *364*, 138–142.

McCullogh, C. E. (2000), Generalized linear models, *J. Am. Stat. Assoc.*, *95*, 975–978.

Musson, R. M. W. (1997), Testing earthquake prediction results statistically, *Seismol. Res. Lett.*, *68*, 944–946.

Nur, A. (1972), Dilatancy, pore fluids, and premonitory variations in *ts*/*tp* travel times, *Bull. Seism. Soc. Amer.*, *62*, 1217–1222.

Press, F. (1975), Earthquake prediction, *Sci. Am.*, *232*(5), 14–23.

Pulinets, S. (2004), Ionospheric precursors of earthquakes: Recent advances in theory and practical applications, *Terr. Atmos. Ocean Sci. J.*, *15*(3), 413–435.

Pulinets, S.A., Legen'ka, A. D., and Alekeseev, V. A. (1994) Pre-earthquake ionospheric effects and their possible mechanisms, in *Dusty and Dirty Plasmas, Noise and Chaos in Space and in Laboratory*, pp. 545–557, Plenum Publishing, New York.

Rikitake, T., (1976), *Earthquake Prediction*, Elsevier, 357 pp.

Seife, C. (2005), What is the universe made of? *Science*, *309*(5731), 78–102.

Shi, Y. and Bolt, B. A. 1982. The standard error of the magnitude–frequency *b* value, *Bull. Seismol. Soc. Am.*, *72*, 1677–1687.

Shin, T.C., Chang, C. H., Pu, H. C., Lin, H. W., and Leu, P. L. (2013), The geophysical database management system in Taiwan, *Terr. Atmos. Ocean Sci. J.*, *24*, 11–18.

Stark, P. B. (1997), Earthquake predictions: the null hypothesis, *Geophys. J. Int.*, *131*, 495–499.

Stoumbos, Z. G., Reynolds, M. R. Jr., Ruan, T. P., and Woodall, W. H. (2000), The state of statistical process control as we proceed into the 21st century. *J. Am. Stat. Assoc.*, *95*, 992–998.

Tsai, Y. B. (1986), Seismotectonics of Taiwan, *Tectonophysics*, *125*, 17–37.

Tsai, Y. B., Teng, T. L., Chiu, J. M. and Liu, H. L. (1977), Tectonic implications of the seismicity in the Taiwan region, *Mem. Geol. Soc. China*, *2*, 13–41.

Tsai, Y. B., Teng, T. L., Yeh, Y. H., Yu, S. B., Liu, K. K., and Wang, J. H. (1983), Status of earthquake prediction research in Taiwan, ROC, *Bull. Inst. Earth Sci.*, *3*, 1–26.

Tsai, Y. B., Liu, J. Y., Ma, K. F., Yen, H. Y., Chen, K. S., Chen, Y. I., and Lee, C. P. (2004), Preliminary results of the iSTEP program on integrated search for Taiwan earthquake precursors, *Terr. Atmos. Ocean Sci. J.*, *15*(3), 545–562.

Tsai, Y. B., Liu, J. Y., Ma, K. F., Yen, H. Y., Chen, K. S., Chen, Y. I., and Lee, C. P. (2006), Precursory phenomena associated with the Chi-Chi earthquake in Taiwan as identified under the iSTEP program, *Phys. Chem. Earth*, *31*, 365–377.

Uyeda, S., Nagao, T., and Tanaka, H. (2004), A report from RIKEN international frontier research project on earthquakes (IFREQ), *Terr. Atmos. Ocean Sci. J.*, *15*(3), 269–310.

Vere-Jones, D. (1995), Forecasting earthquakes and earthquake risk, *Int. J. Forecast.*, *11*, 503–538.

Walia, V., Yang, T. F., Lin, S. J., Kumar, A., Fu, C. C., Chiu, J. M., Chang, H. H., Wen, K. L., and Chen, C. H., (2013), Temporal variation of soil gas compositions for earthquake surveilance in Taiwan, *Radiat. Meas.*, *50*, 154–159.

Wang, C. Y., Cheng, I. H., Chin, C. V., and Yu, S. B. (2001), Coseismic hydrologic response of an alluvial fan to the 1999 Chi-Chi earthquake, Taiwan, *Geology*, *29*, 831–834.

Wiemer, S. and Katsumata, K. (1999), Spatial variability of seismicity parameters in aftershock zones, *J. Geophys. Res.*, *104*, 13135–13151.

WRA, (2001), *Report of Monitoring, Investigating and Analyzing of Land Subsidence in Taiwan (1/4)*, Water Resources Agency, Ministry of Economic Affairs, Taipei, Taiwan. (In Chinese.)

Wu, F. T. (1978), Recent tectonics of Taiwan, in S, Uyeda, R. W. Murphy and K. Kobayashi (eds), *Geodynamics of the Western Pacific, Advances in Earth and Planetary Sciences, J. Phys. Earth*, *26*(Suppl.), 265–299.

Wu, Y. M. and Chen, C. C. (2007), Seismic reversal pattern for the 1999 Chi-Chi, Taiwan, *Mw* 7.6 earthquake, *Tectonophysics*, *429*, 125–132.

Yang, T. F., Walia, V., Chyi, L. L., Fu, C. C., Chen, C. H., Liu, T. K., Song, S. R., Lee, C. Y., and Lee, M. (2005), Variations of soil radon and thoron concentrations in a fault zone and prospective earthquakes in SW Taiwan, *Radiat. Meas.*, *40*, 496–502.

Yang, T. F., Fu, C. C., Walia, V., Chen, C. H., Chyi, L. L., Liu, T. K., Song, S.R., Lee, M., Lin, C. W., and Lin, C. C. (2006), Seismo-geochemical variations in SW Taiwan: multi-parameter automatic gas monitoring results, *Pure Appl. Geophys.*, *163*(4), 693–709.

Yang, T. F., Chen, C. H., Fu, C. C., Liu, C. K., Hilton, D., Walia, V., Lin, S. J., and Lai, T. H. (2015), Fluid geochemical observations of earthquake precursors, presented at the *Workshop on Earthquake Activities and Precursors*, 11 August, Central Weather Bereau, Taipei. (In Chinese.)

Yen, H. Y. and Hsieh, H. H. (2010), A study on the compatibility of 3-D seismic velocity structures with gravity data of Taiwan, *Terr. Atmos. Ocean Sci. J.*, *21*, 897–904.

Yen, H. Y., Yeh, Y. H., Lin, C. H., Chen, K. J., and Tsai, Y. B., (1995), Gravity survey of Taiwan, *J. Phys. Earth*, *43*, 685–696.

Yen, H. Y.,Chen, C. H., Yeh, Y. H., Liu, J. Y., Lin, C. R. and Tsai, Y. B., (2004), Geomagnetic fluctuations during the 1999 Chi-Chi earthquake in Taiwan, *Earth Planets Space*, *56*, 39–45.

Yen, H. Y., Chen, C. H., Hsieh, H. H., Lin, C. R., Yeh, Y. H., Tsai, Y. B., Liu, J. Y., Yu, G. K., and Chen, Y. R. (2009), Magnetic survey of Taiwan and its preliminary interpretations, *Terr. Atmos. Ocean Sci. J.*, *20*, 309–314.

Yu, S. B. and Chen, H. Y. (1994), Global positioning system measurements of crustal deformation in the Taiwan arc-continent collision zone, *Terr. Atmos. Ocean Sci. J.*, *5*, 477–498.

5

Contributions to a History of Earthquake Prediction Research

Giovanni Martinelli

ABSTRACT

Over the past 25 centuries a large amount of data regarding precursory phenomena of seismic events have been gathered and recorded by observers, whether professional, casual, or indirect. Such data have been systematized over the past 120 years and have become the subject of systematic applied research in the past 30 years. Yet, the scientific community has displayed a reluctance to value these historical paths as a means for a better understanding of the process of gathering and applying knowledge to the field of earthquake prediction research. In this chapter the possible reasons why there have been difficulties in the advancement of research into precursory earthquake phenomena are reexamined. In addition, some consequences of the indirect observation of natural phenomena with time are indicated. Some possible development paths are also reexamined for a better understanding of the history of earthquake prediction.

5.1. INTRODUCTION

Earthquake prediction can still be regarded as being among the most controversial scientific theories. Experimental evidence has often been encouraging but later turned out to be disappointing. Thus, research connected with earthquake prediction has been characterized by the absence of great projects aimed at gathering experimental data that are typical of other fields. This limited number of large-scale research projects is certainly due to the complexity of this subject. Time-series research can be useful for understanding cognitive processes characterized by diffused and fragmentary information, such as in earthquake prediction. The awareness that chronologically identified events are reported in precise social and economic contexts can eventually lead to an understanding of the importance of the history of research into earthquake precursors. This history is mostly made up of indirect information.

Research into earthquake prediction has been characterized by a poor organization of information, even though data on earthquake precursors have been present in other fields of study where systematic measurements of experimental data are vital. A great deal of indirect data (water level, pendulum, meteorological, astronomic, electromagnetic emission, and magnetic anomalies [*Breiner and Kovach*, 1967] have been recorded by researchers who have been working in other subject areas. Scientific heritage can be found globally, in particular, in Europe, where the tradition of scientific observatories and laboratories has provided significant data records.

Only since the end of World War II, however, has there been a global distribution of uniform seismological instrumentation. Hitherto, encouraging results had been obtained in just a few test areas where research projects into earthquake prediction were carried out. Earthquake prediction means the establishment, in advance, of a number of parameters that define when a seismic event will occur in time and space, as well as its epicenter and magnitude. The essential characteristics of the main lines of past research into earthquake prediction can be found in two different approaches to the problem: statistical and deterministic prediction.

Statistical prediction means the calculation of the probability that an earthquake of a certain magnitude will occur in a given location within a certain period of time.

ARPAE Emilia-Romagna, Reggio Emilia, Italy

Pre-Earthquake Processes: A Multidisciplinary Approach to Earthquake Prediction Studies, Geophysical Monograph 234, First Edition. Edited by Dimitar Ouzounov, Sergey Pulinets, Katsumi Hattori, and Patrick Taylor.

If assessments of this kind are to produce tangible results, it is necessary not only to study the historical seismicity of the area under investigation, but also to evaluate geological factors on both regional and local scales [*Gelfand et al.*, 1972; *Caputo et al.*, 1980; *Keilis-Borok*, 1990]. Such an approach is particularly useful for studying regions of high seismic activity and for assessing seismic hazards. Deterministic prediction, on the other hand, is based on the understanding of established physical laws that relate to precursors and the actual occurrence of seismic events. The empirical approach to the problem of earthquake prediction—that is, the attempt to establish phenomenological laws—has led many scholars and researchers to suggest specific relationships between precursors and earthquakes, such as *Bendandi* [1931] who utilized astronomical parameters, as reported by *Fidani* [2012], and *Lorenzini* [1898] who utilized water-level data, as described by *Albarello et al.* [1991].

5.2. 600 BC TO AD 1500: THE FIRST EMPIRICAL OBSERVATIONS OF POSSIBLE EARTHQUAKE PRECURSORS

This period was dominated by Aristotle's ideas on the origin of earthquakes. In his view, earthquakes were generated by underground winds [*Liner*, 1997; *Guidoboni*, 1998]. A number of Greek philosophers believed that earthquakes were generated by meteorological factors, while Aristotle (384–322 BC) in *Meteorologica* [*Aristotle*, 1982] described the possible origin of earthquakes and reached the conclusion that seismic events were generated by underground winds [*Howell*, 1986; *Liner*, 1997]. The Aristotelian theory of earthquakes lasted for about 1700 years, a record in the history of scientific theories [*Guidoboni*, 1998].

Earthquake prediction studies began in the Mediterranean area and spread into central Europe, the Euro-Asian Continent, the Far East in ways and at times reflected in the history of earthquake theories and seismic instruments [*Martinelli*, 1998]. There are a number of examples of earthquake predictions in the ancient literature, but the prediction issued was often based on guesswork or conjecture in order to fit prevailing general disaster beliefs. Only in a few cases was it possible to identify a causal relationship between a phenomenon defined as "precursory" and an earthquake.

An early example worthy of mention is that of Pherecydes of Syros (600–550 BC), who is thought to have been a teacher of Pythagoras. Cicero mentions in his *De divinatione* the case of Pherecydes, who predicted an earthquake when he saw that the water from a well that was usually fully filled had disappeared before an earthquake [*Schinche*, 1875]. In the same period Anaximander of Miletus (610–546 BC) warned the Spartans to be careful because an earthquake was looming. The earthquake occurred a few days later but no information is available about the forecasting methods [*Liner*, 1997]. There are nine volcanoes in the Aegean Sea and five of them have erupted during historical times [*Siebert and Simkin*, 2002]. Hence, the observation of natural phenomena could have affected the ideas of the ancient Greek philosophers. Current research into water-level changes in aquifers prior to earthquakes shows that Pherecydes had identified a phenomenon that still seems to play a useful part in earthquake predictions [*Roeloffs*, 1988].

Throughout the Middle Ages, philosophers and scholars turned their attention to investigating the cause of earthquakes rather than trying to predict them. There were times when phenomena now considered to be precursory were observed and recorded, but the main reason for doing so was to provide evidence in support of this or that general theory of the origin of earthquakes [e.g. *Buoni*, 1571]. In line with the philosophical and religious attitudes of the time, scholars turned their attention to other phenomena, such as different kinds of disasters, miracles, etc., which accompanied earthquakes and seemed to be linked to them. The predominant theory also throughout the Middle Ages was that of Aristotle and his subterranean wind theory. The first critique of Aristotle's ideas came from Georg Bauer (known by the name of Agricola, 1494–1555), who refused to believe the Aristotelean seismology of signs forecasting earthquakes and thought that earthquakes could not be predicted [*Guidoboni*, 1998].

5.3. THE PERIOD 1500–1800: THE DIFFUSION OF GREEK KNOWLEDGE

In the Renaissance, the direct observation of natural phenomena became very important, and a new attitude could be found in natural philosophy. A. J. Buoni, a doctor from Ferrara, refers to bubbling gas and the clouding of water in wells before and during the 1570 Ferrara earthquake. This piece of evidence is recorded in his treatise *Del terremoto* [Buoni, 1571], which contains a good deal of observational and historical information, including a reference to Nicolò Cardano's observation of similar phenomena. In connection to this, it is worth noting that anomalous gas emissions as well as the occurrence of geophysical and geochemical anomalies in subterranean water are currently the subject of intense research activity [*Wakita et al.*, 1988].

In the 16th and 17th centuries scientific theories became more important, and partly as a result of the printing press and the availability of data and the abandonment of the Aristotelian theory of earthquakes, historical records of earthquakes reflected this more favorable cultural climate. In this connection it is worth noting that the 17th century saw a substantial exchange of scientific information between Jesuit fathers and Chinese authorities.

Of particular importance in this respect was Father Matteo Ricci's journey there for cultural and scientific purposes. He established a good diplomatic relationship with the Chinese authorities, and "initiated" cultural exchanges.

Father Matteo Ricci designated as his successor Father Nicola Longobardo, whose name was adapted by the Chinese to Long Huamin. Longobardo was a native of Caltagirone, Sicily. In 1597 he arrived in Peking, and died there in 1655. His work on *The Interpretation of Earthquakes* was published there in 1626 [*Longobardo*, 1626; *Martinelli*, 1998]. In this work he sets out clearly those precursors of which he was aware, such as anomalous gas emissions from the earth, the clouding of water in wells, and a change in the taste of water prior to earthquakes. These phenomena are now considered to be indicators of geochemical changes in ground fluids [e.g. *Wakita et al.*, 1988; *Martinelli*, 1998], possibly linked to earthquakes. He also mentioned the occurrence of exceptionally high tides, which are now considered to be indicators of possible crustal deformation. Father Longobardo attributed the appearance of precursory phenomena to "subterranean gas pressure," and he also considered certain meteorological conditions and the appearance of certain cloud formations to be seismic precursors, stressing the importance of the work of Aristotle, especially the *Meteorologica* [*Aristotle*, 1982].

Another historical source printed in China in 1663 and known as the *Longde County Annals*, quoted by *Ma Zonjin et al.* [1990], reports the description of precursory phenomena almost exactly as described by Longobardo in 1626 [*Martinelli*, 1998]. This feature could be interpreted as confirmation of the presence of Greek thought in the Chinese cultural environment. All the listed phenomena are the subject of modern earthquake prediction research [*Rikitake*, 1982; *Sultankhodzhaev*, 1984]. It is also interesting that modern Chinese earthquake prediction research considers the observation of meteorological phenomena to be important [*Lu Dajiong*, 1988]. Lu Dajiong's work clearly demonstrates the survival of traces of Aristotelian thought, as brought to China with great effect by the Jesuit missions.

Until the end of the 18th century, however, theories about the causes of earthquakes excluded the inclusion of precursor studies within proper research programs [*Santoli*, 1783], with the result that they remained little more than curiosities of nature.

5.4. THE PERIOD 1800–1920: THE BEGINNING OF THE SCIENTIFIC METHOD AND SYSTEMATIC OBSERVATION

In the 1870s the Italian chemist Demetrio Lorenzini published water-level data recorded in a well in northern Italy [*Albarello et al.*, 1991]. He corresponded with Michele Stefano de Rossi [1879] and related water-level variations to local earthquakes. As well as Michele Stefano de Rossi (1834–1898), in the nineteenth century we can find the scientific achievements of many outstanding scholars, such as Robert Mallet (1810–1881), Timoteo Bertelli (1826–1905), Giuseppe Mercalli (1850–1914), and John Milne (1850–1913). Such scientists observed and cataloged many phenomena that were considered precursors of earthquakes [e.g. *de Rossi*, 1879]. Their scientific contribution included many geophysical parameters that have been examined in modern times within the advanced fields of geomagnetism, gravimetry, and observations of slow deformation in Earth's crust, applied geochemistry. and hydrology.

In particular, in the period 1870–1888 Michele Stefano de Rossi and Timoteo Bertelli set up the first instrumental network throughout Italy, comprising 20 "tromometers" (instruments based on the principle of a pendulum) capable of detecting slow movements of crustal deformation [*Ferrari et al.*, 2000], and a possible precursor of a M 5.2 was detected (Figure 5.1). On the occasion of the March 1872 earthquake in central Germany, much information was garnered by German scholars [*Oeser*, 1992]. Among these, the anomalous behavior of a compass that was observed near Wroclaw just before the earthquake is worth mentioning. A similar observation was made in central Italy in 1873 [*Serpieri*, 1873; *Caputo*, 1995]. Many similar observations were reported as precursory phenomena of geomagnetic type [*Taramelli and Mercalli*, 1888; *Ferrari*, 1991].

Further investigations that marked the late nineteenth and the early twentieth centuries concerned lightning phenomena observed before, during, and after earthquakes [*Galli*, 1906, 1910]. Such research stressed a possible relation with electromagnetic phenomena or with the uprising of deep originated fluids. In the early 20th century, partly as a result of the 1909 Messina earthquake, research activity into the subject became more intense. *Mondello* [1909] reported successful experiments in detecting self-potential anomalies before the occurrence of some earthquakes. *Alfani* [1909] reported that Father Maccioni had twice succeeded in detecting radio signals as precursors of earthquakes using a "coherer"—an instrument capable of detecting electromagnetic emissions, devised by Calzecchi-Onesti in 1887. In 1909 Father A. Maccioni published the first impressive results of experiments to predict earthquakes using slightly modified versions of the "coherer" [*Maccioni*, 1909]. Again in 1909, the Italian engineer Antonio Prati improved the device and patented a version meant for commercial use. Here we have what is probably the first attempt to produce a device capable of practical application in studying precursory phenomena for earthquake prediction purposes. *Ungania* [1924] suggested that information concerning

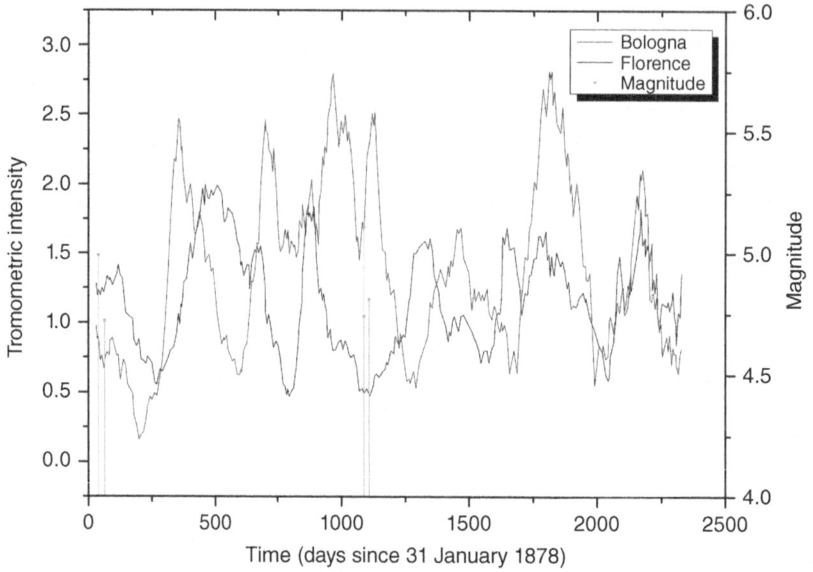

Figure 5.1 Tromometric measurements (angular displacements by a pendulum-based instrument) recorded in Bologna are in black. Measurements recorded in Florence, which is 100 km away, are indicated in blue. Before the seismic event characterized by *M* 5.2, the time series are uncorrelated [*Ferrari et al.*, 2000]. (*See electronic version for color representation.*)

impending earthquakes obtained by means of the "coherer" should be broadcast in Italy.

At roughly the same period, research of a similar nature was carried out in Japan, involving a network of observatories for monitoring electromagnetic emissions. Of particular importance was the founding of the Japanese Seismological Society after the Yokohama earthquake of 1880. British interest in this kind of research—an example is Robert Mallet's fundamental study of the 1857 earthquake in southern Italy [*Mallet*, 1862]—was "exported" to Japan, since that country had more to offer for experiments. The man responsible for this cultural transfer was John Milne. He worked in Japan from 1876 to 1895, actively supporting earthquake precursor research [*Milne*, 1890], and after the *M* 8 earthquake of Nobi in 1891 Aikitu Tanakadate (1856–1952) started geomagnetic field measurements [*Uyeda*, 2013].

Almost at the same time Giuseppe Mercalli carried out research into variations in electric and magnetic parameters that precede earthquakes [*Taramelli and Mercalli*, 1888; *Lapina*, 1953]. The interest all over Europe in the study and assessment of electromagnetic phenomena can be testified to by the activity of geophysical observatories fitted out with equipment for continuous monitoring of such phenomena [*Chapman*, 1930]. Similar investigations are in progress today as part of modern research projects [*Fraser-Smith et al.*, 1990; *Bella et al.*, 1994].

After the Nobi earthquake of 1891, the Imperial Earthquake Investigation Committee was set up to investigate seismic and volcanic phenomena and to limit their effects. The need for a multidisciplinary approach to earthquake prediction studies was stressed first of all by the Russian prince B. B. Galitzin, who drafted a complete research program in 1911 for earthquake prediction [*Galitzin*, 1960], which involved:

1. the study of the frequency and magnitude of seismic events and of the features of the recording of seismic oscillations;
2. the study of the propagation rate of seismic waves in order to evaluate the state of tension in seismically active areas;
3. geodetic measurements aimed at discovering slow deformation in Earth's crust;
4. gravimetric measurements;
5. the study of the condition of springs and of wells, and study of the composition of the gases in Earth's crust.

B. B. Galitzin's attention to the behavior of fluids in Earth's crust may be due to his direct observations; nevertheless, his research was certainly influenced by the remarkable correspondence he had with many European geophysicists. He also visited Italy and exchanged ideas with local scientists. In Europe, the interest in the behavior of Earth's fluids connected with the prediction of seismic events can be found in a great number of historical sources [*de Rossi*, 1879; *Mercalli*, 1883; *Baratta*, 1901; *Moreira et al.*, 1993]. In particular, Michele Stefano De Rossi reported he had contemporarily observed crustal deformation phenomena and significant variations in the flow rate and temperature of geofluids offshore from the Ischia Island (Italy) *M* 5.8 earthquake of 1883 [*Luongo et al.*, 2011] and his findings had almost definitely been read by B. B.Galitzin.

5.5. RESEARCH STARTING IN THE MODERN AND CONTEMPORARY AGES: THE FORMER USSR AND CHINA

The Aschabad earthquake in 1948 fostered systematic research into earthquake precursors in the former USSR [*Zarkov*, 1983]. In the region of Garm (Tadjikistan) a test area was equipped in 1949 and fundamental research was carried out there in order to predict seismic events. In 1953, G. A. Gamburtsev suggested a new research program on precursors, and this was later carried out mostly in Uzbekistan after the Tashkent earthquake in 1966. In 1969 Semenov observed in the Garm region that the compression wave velocity decreased for a period of time preceding some local earthquakes. In 1956, Y. Muminov had already started to record chemical parameters in the waters of local springs in order to check variations in the composition of thermo-mineral waters being used therapeutic purposes.

The research unit led by A. N. Sultankhodzhaev carried out further research into geochemical precursors in thermo-mineral waters exploited by wells of Uzbezkistan, which produced interesting results [*Sultankhodzhaev*, 1984]. Given the quality and number of publications, it is possible to state that most of the research into geochemical and geophysical earthquake precursors that have been carried out in many nations has certainly been inspired by the pioneering research carried out in central Asia in the 1960s, 1970s, and 1980s [*Abdullabekov*, 1991; *Kissin et al.*, 1993; *Gokhberg et al.*, 1995]. It is in the past few decades that the most encouraging results on these topics have been accomplished and confirmed after a validation procedure [Wyss, 1991]. *Semenov* [1969] and coworkers in a test area of Tadzhikistan observed that the compression wave velocity decreased before some local earthquakes. This phenomenon was later recognized as a dilatancy effect [*Scholz et al.*, 1973]. All present earthquake prediction research projects are based on the dilatancy theory proposed by *Scholz et al.* [1973], or on modified versions of it [*Mjachkin et al.*, 1975]. These theories try to provide an organic explanation of the contemporaneous occurrence of a number of different precursory phenomena [e.g. *Ponomarev et al.*, 1997].

Characteristic of the former Soviet Union's research into earthquake prediction is the strong and quantified multidisciplinary approach. Indeed, the entire Academy of Sciences of the USSR has been actively involved in this topic and active cooperation still characterizes the findings of geophysicists, geochemists, mathematicians, seismologists, etc. Thus, in the former Soviet Union the geochemical approach has produced, broadly speaking, very interesting results, but such features would not always be recognized in some countries, where a strictly seismological approach seems to prevail. Those interesting results from a geochemical approach concur with previous reports by *Carapezza et al.* [1980] and *Dall'Aglio* [1995] (see also *Mavlianova* [1983] and *Barsukov et al.* [1984].

Influence of European scientific thought can also be found in China, where a massive earthquake prediction research program was launched in the 1960s and has produced encouraging results. In particular, in 1956 Fu Chengyi formulated a research project regarding earthquake prediction that started in 1966 after the Xingtai earthquake [*Ma Zonjin et al.*, 1990]. The main guidelines of the Chinese research program were similar to those of the USSR, but a particular characteristic was attention paid to the empirical observation of a number of natural phenomena, such as animal behavior and meteorological phenomena that are believed to be earthquake precursors. Alongside this there is an automatic instrumental measurement network of a variety of parameters, using advanced technology [*Wan Dikun*, 1993; *Huang et al.*, 2017]. In particular, a network for groundwater monitoring was set up in China. The network consists of monitoring over 600 thermal springs and wells (depth range 0.1–2.0 km). Possible precursory variations in water levels were observed before some strong earthquakes [*Huang et al.*, 2004; *Shi et al.*, 2013], but validation procedures are being assessed so as to avoid the issue of false alarms [*Saegusa*, 1999].

5.6. MODERN EARTHQUAKE PREDICTION STUDIES IN THE UNITED STATES AND JAPAN

Whitcomb et al. [1973] duplicated in California the experiments of *Semenov* [1969] on velocity variations of compression waves before shocks (Figure 5.2), while *Castle et al.* [1974] reported crustal deformation preceding earthquakes in California. Wiss [1977] interpreted the observed phenomena as being caused by dilatancy, but some failures in predicting earthquakes in California in the same years induced a skeptical reaction. Skepticism about the potential dangers stemming from possibly false or missed alarms before earthquakes greatly limited research into earthquake precursors in the USA in the following decade [*NRC*, 1975; *Hough*, 2010] but new research started in California in 1985 in the Parkfield area. The test area of Parkfield, California, has been instrumental for monitoring precursory phenomena [*Bakun*, 1990]. The scientific method followed is very similar to that proposed by *Galitzin* in 1911. As a multidisciplinary approach it can be considered as a recovery of the working tradition of the Geophysical Observatories Network in Europe during the 19th century.

In 1927, Shiratoi first detected radon changes in hot springs in Japan before some earthquakes [*Shiratoi*, 1927]. Reports on fluid monitoring of seismic precursor parameters also appeared in the 1940s and 1950s [*Imamura*,

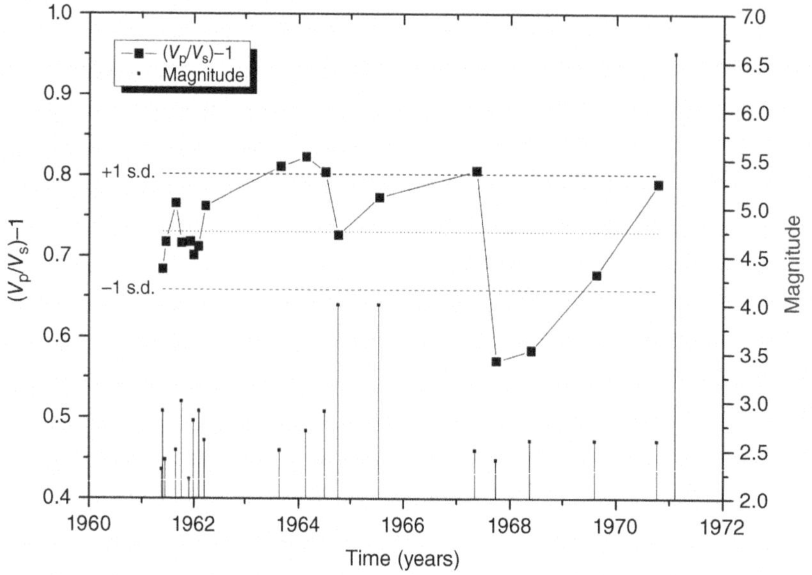

Figure 5.2 V_p/V_s measurements studied by *Whitcomb et al.* [1973] in California. Before a *M* 6.7 event, a significant variation was observed in the period 1967–1969.

1947; *Hatuda*, 1953; *Okabe*, 1956]. Japanese technological interest in this topic increased after the publication of a comprehensive report on earthquake prediction by *Tsuboi et al.* [1962], commonly called the "blueprint". The "blueprint" proposed guidelines for a research activity program with the aim of earthquake prediction in the territory of Japan. The program started some years later giving priority to the detection of crustal movements, detection of seismic activity, geomagnetic monitoring, and laboratory rock mechanics [*Rikitake* 1966]. Scarce attention was paid to fluid monitoring in the initial stages of the Japanese program. Intense research activity in this field began in the late 1970s with the establishement of the Laboratory for Earthquake Chemistry, where remarkable scientific results were achieved [e.g. *Wakita et al.*, 1980]. Some recent failures concerning earthquake prediction in Japan have prompted local authorities to avoid mentioning the word 'prediction' in recent documents. *Uyeda* [2013] criticized the apparent lowering of interest in short-term prediction in Japan, although the multidisciplinary approach to research has been reaffirmed.

5.7. RECENT EARTHQUAKE PREDICTION STUDIES IN EUROPE AND IN OTHER WORLD SITES

In recent decades important results have been obtained in Turkey, where an earthquake prediction project has been carried out jointly by Turkish and German research institutes. The basic characteristic of this joint project was its multiparameter approach, involving the simultaneous observation and analysis of multiple precursors. This feature, too, goes back to the working tradition of geophysical observatories of the 19th century. The results achieved so far have confirmed the suitability of this approach, and the scientific communities involved has been largely supportive [see, e.g., *Schindler et al.*, 1993].

The example provided by this joint Turkish and German research has also influenced other research projects in Greece, Italy, Germany, Iceland, India, New Zealand, and South Africa. Since the second half of the 1980s, the scientific debate concerning earthquake prediction has been strongly affected by the progress made in Greece, Japan, and France [*Varotsos and Kulhànek*, 1993] concerning deterministic earthquake prediction through the study of precursory phenomena of electrical field variations. Seismologists, as a scientific discipline, have accepted precursory phenomena approach to the problem with skepticism [*Burton*, 1985; *Geller*, 1996], drawing attention to the apparent fortuitousness of predictions that were actually guesses by researchers of electric-field precursors. Total denial of the validity of this type of research may, however, unfairly cast doubt on analogous valid research results, which proved to be uniquivocal and to share common features thus representative of a consolidated state of the art [*Sobolev*, 1975]. Contributions from the latter side of this debate include *Di Bello et al.* [1994] and *Jouniaux et al.* [1994]. Both these works clearly proved the nonfortuitousness of the occurrence of electric-type precursor signals possibly related to release of electric charges during crustal strain.

5.8. NEW PARAMETERS AND NEW MONITORING TECHNIQUES

Several research studies published over the past 30 years on earthquake precursors were carried out by means of satellite techniques [e.g. *Gorny et al.*, 1988; *Quiang et al.*, 1997; *Pulinets and Ouzunov*, 2011; *Lisi et al.*, 2016]. In particular, studies have detected anomalous space–time transients in the thermal infrared radiation emitted by the Earth, possibly due to the expulsion of underground gases induced by crustal deformation processes during the earthquake preparatory stage. Crustal deformation processes may be accompanied by fluid expulsion when available, thus a new research field of fluid monitoring has been opened. Ground-based and satellite-based joint measurements will probably better explain the observed phenomena. These satellite techniques are characterized by relatively low costs. These features are promoting an increase in the use of methods based on remote sensing techniques, thus allowing universities and research centers to launch independent research projects without the need for large-scale infrastructures. Together with the contemporary diffusion of personal electronic devices capable of instantaneous monitoring, the prospects for future research developments in these various environmental parameters looks promising.

5.9. BIOGRAPHICAL INFORMATION ON SOME SCIENTISTS INVOLVED IN EARTHQUAKE PREDICTION RESEARCH

History of sciences teaches that human memory is relatively short, thus some basic information about outstanding scientists who actively promoted research into earthquake prediction are reported below.

• De Rossi Michele Stefano (Rome, 1834; Rocca di Papa, 1898) started, along with Father Timoteo Bertelli, the monitoring of crustal deformation in the Italian peninsula. He became the director of the Geodynamic Observatory of Rocca di Papa. In 1874 he founded the scientific journal *Bullettino del Vulcanesimo Italiano* in which copious amounts of earthquake precursory information was reported. His ideas were adopted by Prince Boris Borisovich Galitzin.

• Prince Galitzin Boris Borisovich (Saint Petersburg, 1862–1916) invented the first electromagnetic seismograph. In 1911 he became the Chairman of the International Association of Seismology. In the same year he published a list of parameters for earthquake prediction. The list also included various nonseismic parameters and is still used in many research projects worldwide that also include advanced monitoring techniques.

• Gamburtsev Grigori Aleksandrovic (Saint Petersburg, 1903; Moscow, 1955) was director of the Institute Physics of the Earth in Moscow in the period 1949–1955. In 1947 he led the first earthquake prediction research project in the former USSR. It was also the first post-World War II project in the world and the observed parameters were in the list compiled by B.B. Galitzin.

• Rikitake Tsuneji (Tokyo, 1921–2004) was affiliated to the Earthquake Research Institute from 1942. He was Professor at the University of Tokyo and is considered one of the founders of modern research into earthquake prediction in Japan. His name has been linked to studies on geomagnetism and to the dissemination of ideas on earthquake prediction thanks to his well-known books.

• Keilis-Borok Vladimir (Moscow, 1921; Culver City, 2013) was the founder of the International Institute of Earthquake Prediction and Mathematical Geophysics in Moscow. He was President of the International Union of Geodesy and Geophysics in the period 1987–1991. He set up new algorithmic methods for earthquake prediction within the framework of so-called "pattern recognition." The method, together with similar methods developed by other researchers, is presently still deemed to be encouraging and is used in many sites around the world.

5.10. CONCLUSIONS

Many examples of scientific research into earthquake prediction allow us to trace paths of historical investigation into this topic. There are possible ways to further investigate the links between earthquake prediction research and decision-making. If the scientific publications regarding seismic event precursors are carefully examined, the process of scientific growth could be considered as a chronologically coherent trajectory; this actually was not the case as intense debates have always surrounded research into earthquake prediction. Hence, the scientific geophysical environment has been split into two distinct parts: seismologists on the one hand, and geophysicists and geochemists concerned with nonseismic observations, on the other. Nothwithstanding the normal scientific dynamics resulting from belonging to one "school" rather than another, economics also plays a role in such occasions, e.g., the involvement of insurance companies. The assessments of probability of seismic occurrence are carried out on the basis of catalogs of recorded events [*Freeman*, 1932]. However, nowhere in the world has seismological research into earthquake prediction ever reached outcomes so good as to assert itself above other methodologies concerned with the question of defense against earthquakes. Up to now, however, only a small amount of funding has been assigned to research using other methodologies, as compared with the seismological approach. The present cutbacks to funding allocated to research that characterizes a certain number of developed nations, may lead us to assume little likelihood of research into earthquake prediction in the near future. A new, expanding economic cycle like the one that characterized the 1960s will probably be needed to reverse the present trend. Hence, for now the richest historical stage, at least for ground-based monitoring techniques, belongs firmly to the past: from 1975 to 1990, when the scientific production reached its best outcomes, both in terms of quality and quantity. New monitoring satellite-based techniques could contribute to lowering the costs of such research and to creating the conditions for a greater scientific and social impact.

REFERENCES

Abdullabekov, K. N. (1991), *Electromagnetic Phenomena in the Earth's Crust*, Balkema, Rotterdam, 150 pp.

Albarello, D., Ferrari, G., Martinelli, G., and Mucciarelli, M. (1991), Well-level variation as a possible seismic precursor: a statistical assessment from Italian historical data, *Tectonophysics*, *193*, 385–395, doi: 10.1016/0040-1951(91) 90347-U

Alfani, G. (1909), *Il terremoto sarà preveduto? Lo strumento del Padre Maccioni*, Corriere della Sera, 9 maggio.

Aristotle (1982), *Meteorologica*, P. Louis (ed.), Paris.

Bakun, W. H. (1990), The Parkfield, California earthquake prediction experiments, in *Prediction of Earthquakes, Occurrence and Ground Motion, Proceedings of the ECE/UN Seminar*, C. S. Oliveira (ed.), pp. 681–693, Lisbon.

Baratta, M. (1901), *I terremoti d'Italia. Saggio di storia geografia e bibliografia sismica italiana*, Torino, 951 pp. (antistatic reprint, Sala Bolognese 1979).

Barsukov, V. L., Varshal G. M., Garanin A. B., and Serebrennikov, V. S. (1984), Hydrochemical precursors of earthquakes, in *Earthquake Prediction, Proceedings of the International Symposium on Earthquakes Prediction*, pp. 169–180, UNESCO, Paris.

Bella, F., Biagi, P. F., Caputo, M., Della Monica, G., Ermini, A., Plastino, W., and Sgrigna, D. (1994), Electromagnetic background and preseismic anomalies recorded in the Amare Cave (Central Italy), in *Electromagnetic Phenomena Related to Earthquake Prediction*, M. Hayakawa and Y. Fujinawa (eds), pp. 181–192, Tokyo.

Bendandi, R.(1931), *Un principio fondamentale dell'Universo*. S.T.E, Faenza, 300 pp.

Breiner, S. and Kovach, R. (1967), Local geomagnetic events associated with displacements on the San Andreas Fault, *Science*, *158*, 116–118.

Buoni, A. J. (1571), *Del terremoto: Dialogo di Jacopo Antonio Buoni medico ferrarese distinto in quattro giornate*, Modena.

Burton, P. W. (1985), Electrical earthquake prediction, *Nature*, *315*, 370–371.

Caputo, M. (1995), Introduzione al Convegno, in *Terremoti in Italia*, pp. 3–5, Atti dei Convegni Lincei, Accademia Nazionale dei Lincei, Roma.

Caputo, M., Keilis-Borok, V.I., Oficerova, E., Ranzman, E., Rotwain, I., and Solovieff, A. (1980), Pattern recognition of earthquake-zone areas in Italy, *Phys. Earth Planet. Inter.*, *21*, 305–320.

Carapezza, M., Nuccio, P. M., and Valenza, M. (1980), Geochemical precursors of earthquakes, in *High Pressure Science and Technology*, B. Vodar and P. H. Marteau (eds), pp. 90–103, Pergamon Press, New York.

Castle, R. O., Alt, J. N., Savage, J. C., and Balazs, E. I. (1974), Elevation changes preceding the San Fernando Earthquake of February 9, 1971, *Geology*, *5*, 61–66.

Chapman, S. (1930), A note on two apparent large temporary local magnetic disturbances possibly connected with earthquakes, *Terr. Mag. Atmosph. Elec.*, *35*, 81–83.

Dall'Aglio, M. (1995), Ricerche sui premonitori geochimici dei terremoti. Passato, presente e futuro, in *Terremoti in Italia*, pp. 295–300, Atti dei Convegni Lincei, Accademia Nazionale dei Lincei, Roma.

De Rossi, M.S. (1879), *La meteorologia endogena*, Vol. *1*, Milano.

Di Bello, G., Lapenna, V., Satriano, C., and Tramutoli, V. (1994), Self-potential time series analysis in a seismic area of Southern Italy: first results, *Ann. Geofisica*, *37*, 1137–1148.

Ferrari, G. (1991), The 1887 Ligurian earthquake: a detailed study from contemporary scientific observations, *Tectonophysics*, *193*, 131–139.

Ferrari, G., Albarello, D., and Martinelli, G. (2000), Tromometric measurements as a tool for crustal deformation monitoring, *Seismol. Res. Lett.*, *71*, 562–569, doi: 10.1785/gssrl.71.5.562

Fidani, C. (2012), The Raffaele Bendandi earthquake warnings based on planetary positions, *New Concepts Global Tecton. Newslett.*, *65*, 47–54.

Fraser-Smith, A. C., Bernardi, A., Mc Gill, P. R., Bowen, M. M., Ladd, M. E., Helliwell, R. A., and Villard, O. G. Jr. (1990), Low-frequency magnetic field measurements near the epicenter of the *Ms* 7.1 Loma Prieta earthquake, *Geophys. Res. Lett.*, *17*, 1465–1468.

Freeman, J. R. (1932), *Earthquake Damage and Earth Insurance*, McGraw Hill, New York, 904 pp.

Galitzin, B. B. (1960), *Izbrannie Trudiy (Selected Works)*, Vol. *2*, Izd-vo, AN SSSR, Moscow.

Galli, I. 1906, Fenomeni luminosi nei terremoti, in *Atti del Congresso dei Naturalisti Italiani*, pp. 264–277, Milano.

Galli, I. (1910), Raccolta e classificazione di fenomeni luminosi osservati nei terremoti, *Boll. Soc. Sismol. Ital.*, *14*, 221–448.

Gelfand, I. M., Guberman, Sh. A., Izvekova, M. L., Keilis-Borok, V. I., and Ranzman E. Ia. (1972), Criteria of high seismicity determined by pattern recognition, in A. R. Ritsema (ed.), *The Upper Mantle*, *Tectonophysics 13*, 415–422.

Geller, R. (ed.) (1996), Debate on "VAN," *Geophys. Res. Lett.*, *23*, 1291–1452.

Gokhberg, M. B., Morgounov, V. A., and Pokhotelov, O. A. (1995), *Earthquake Prediction: Seismo-Electromagnetic Phenomena*, Gordon and Breach Publishers, OPA, Amsterdam, 193 pp.

Gorny, V. I., Salman, A. G., Tronin, A. A., and Shilin, B. B. (1988), The earth outgoing IR radiation as an indicator of seismic activity, *Proc. Acad. Sci. USSR*, *301*, 67–69.

Guidoboni, E. (1998), Earthquakes, theories from antiquity to 1600, in G. A. Good (ed.), *Sciences of the Earth, An Encyclopaedia of Events, People, and Phenomena.*, Vol.*1*, pp. 197–205), Garland Publishing, New York.

Hatuda, Z. (1953), Radon content and its change in soil air near the ground surface, *Memoires of the College of Science*, pp. 285–306, series B XX, University of Kyoto.

Hough, S. E. (2010), *Predicting the Unpredictable. The Tumultuous Science of Earthquake Prediction*, Princeton University Press. Princeton, 260 pp.

Howell, B. F. (1986), History of ideas on the cause of earthquakes. *EOS (Trans. Am. Geophys. Union)*, *67*, 1323–1326.

Huang, F., Jian, C., Tang, Y., Xu, G., Deng, Z., Chi, G-C., and Ferrar C. D. (2004), Response changes of some wells in the mainland subsurface fluid monitoring network of China, due to the September 21, 1999, *Ms* 7.6 Chi-Chi earthquake, *Tectonophysics*,*390*,217–234,doi:10.1016/j.tecto.2004.03.022.

Huang, F., Li, M., Ma, Y., Han, Y., Tian, L., Yan, W., and Li, X. (2017), Studies on earthquake precursors in China: A review for recent 50 years, *Geod. Geodyn.*, http://dx.doi.org/10.1016/j.geog.2016.12.002

Imamura, G. (1947), Report on the observed variation of the Toghiomata hot spring immediately before the Nagano earthquake of July 15, 1947, *Kagaku*, *11*, 16–17.

Jouniaux, L., Lallemant, S., and Pozzi, J. P. (1994), Changes in the permeability, streaming potential and resistivity of a claystone from the Nankai prism under stress, *Geophys. Res. Lett.*, *21*, 149–152.

Keilis-Borok, V. I. (ed.) (1990), *Intermediate-Term Earthquakes Prediction: Models, Algorithms, Worldwide Tests*, Special Issue, *Phys. Earth Planet. Inter.*, *61*, 1–139.

Kissin, I. G., Belikov, W. M., Ishankuliyev, G. A., Wang Chengmin, Zhang Wei, Dong Shouyu, Jia Huazhou, and Dikun Wan (1993), High amplitude hydrogeologic precursors of earthquakes in seismic regions of the former Soviet Union and China: a comparative analysis, *J. Earthquake Pred. Res.*, *2*, 89–103.

Lapina, M. I. (1953), Geomagnetism and seismic phenomena, *Izv. AN SSSR, Ser. Geof.*, *5*, 393 (in Russian).

Liner, C.L. (1997), *Greek Seismology-Being an Annotated Sourcebook of Earthquake Theories and Concepts in Classical Antiquity*, Samizdat Press, 135 pp.

Lisi, M., Filizzola, C., Genzano, N., Paciello, R., Pergola, N., and Tramutoli, V. (2016), Reducing atmospheric noise in RST analysis of TIR satellite radiances for earthquakes prone areas satellite monitoring, *Phys. Chem. Earth*, *85–86*, 87–97, doi:10.1016/j.pce.2015.07.013

Longobardo, N. (Long Huamin) (1626), *Interpretazione del terremoto*, Italian translation by G. Matteucig, English translation by Yue Xiaozhu and Lu Dajiong, Centro Studi Internazionale "Before Day", Napoli 1988, 57 pp.

Lorenzini, D.(1898), *Guida dei Bagni della Porretta e dintorni*, Bologna.

Lu Dajiong (1988), *Impending Earthquake Prediction*, Nanjing.

Luongo, G., Carlino, S., Cubellis, E., Delizia, I., and Obrizzo, F. (2011), *Casamicciola Milleottocentottantatre*, Bibliopolis, Napoli, 282 pp.

Maccioni, A. (1909), Nuova scoperta nel campo della sismologia, *Atti della R. Accad. Fisiocrit. Siena*, *1*, 435–444.

Mallet, R. (1862), *Great Neapolitan Earthquake of 1857. The First Principles of Observational Seismology*, *2* Vols, Chapman and Hall, London. (Anastatic reprint (1987) in E. Guidoboni and G. Ferrari (eds), *Mallet's Macroseismic Survey on the Neapolitan Earthquake of 16th December, 1857*, 4 Vols, ING-SGA, Bologna.)

Martinelli, G. (1998), Earthquakes, prediction, in G. A. Good (ed.), *Sciences of the Earth, An Encyclopedia of Events, People, and Phenomena*, pp. 192–196.,Garland Publishing, Inc. New York.

Mavlianova, G. A. (1983), *Hydrogeoseismological Precursors of Earthquakes*, Uzbekh Soviet Socialist Republic, Tashkent.

Ma Zonjin, Fu Zhengxiang, Zhang Yingzhen, Wang Chengmin, Zhang Guomin, and Lin Defu (1990), *Earthquake Prediction—Nine Major Earthquakes in China (1966–1976)*, Beijing.

Mercalli, G. (1883), *Vulcani e fenomeni vulcanici in Italia*, Milano, 374 pp. (Anastatic reprint, Sala, Bolognese 1980.)

Milne, J. (1890), Earthquakes in connection with electric and magnetic phenomena, *Trans. Seismol. Soc. Jpn*, *15*, 135–163.

Mjachkin, V. I., Brace, W. F., Sobolev, G. A., and Dietrich J. H. (1975), Two models for earthquake forerunners, *Pure Appl. Geophys.*, *113*, 169–181.

Mondello, U. (1909), *Sulla presenza di onde elettromagnetiche precorritrici del sismo*, Pubblicazioni dell'Osservatorio di Ardenza al Mare 4, Livorno, 15 pp.

Moreira, V. S., Marques, J. S., Cruz, J. F., and Nunes, J. C. (1993), Review of the historical seismicity in the Gulf of Cadiz area before the 1 November 1755 earthquake. An intermediate report, in M. Stucchi (ed.) *Historical Investigation of European Earthquakes*, Vol.*1*, pp. 225–235, Materials of the CEC Project "Review of Historical Seismicity in Europe," with the collaboration of J. Vogt, Milano,.

NRC Panel on the Public Policy Implications of Earthquake Prediction of the Advisory Committe on Emergency Planning (1975), *Earthquake Prediction and Public Policy*, National Academy of Sciences,Washington, DC, 142 pp.

Oeser, E. (1992), Historical earthquake theories from Aristotle to Kant, in R. Gutdeutsch, G. Grünthal, and R. Musson Abhand (eds), *Historical Earthquakes in Central Europe*, pp. 11–31, Geologische Bundesanstalt, Wien.

Okabe, S. (1956), Time variation of the atmospheric radon-content near the ground surface with relation to some geophysical phenomena, *Memories of the College of Science*, pp. 99–115, Series A 28, University of Kyoto.

Ponomarev, A. V., Zavyalov, A. D., Smirnov, V. B., and Lockner, D. A. (1997), Physical modeling of the formation and evolution of seismically active fault zones, *Tectonophysics*, *227*, 57–81.

Pulinets, S.A. and Ouzunov, D. (2011), Lithosphere–atmosphere–ionosphere coupling (LAIC) model-an unified concept for earthquake precursors validation, *J. Asian Earth Sci.*, *41*, 371–382, doi:10.1016/j.jseaes.2010.03.005

Quiang, Z. I., Xu, X. D., and Dian, C. D. (1997), Thermal infrared anomaly precursor of impending earthquakes, *Pure Appl. Geophys.*, *149*, 159–171.

Rikitake, T. (1966), A five year plan for earthquake prediction research in Japan, *Tectonophysics*, *3*, 1–15.

Rikitake, T. (1982), *Earthquake Forecasting and Warning*, Tokyo.

Roeloffs, E. A. 1988, Hydrologic precursors to earthquakes: a review, *Pure Appl. Geophys.*, *126*, 177–209.

Saegusa, A. (1999), China clamps down on inaccurate warnings, *Nature*, *97*, 284.

Santoli, V. M. (1783), *De mephiti et vallibus anxanti*, Napoli, 107 pp.

Schinche, T. (1875), *De fontibus librorum Ciceronis qui sunt de divinatione*, Jena.

Schindler, C., Balderer, W., Gerber, W., and Imbach, E. (1993), The Marmara Poly-Project: tectonics and recent crustal movements revealed by space-geodesy and their interaction with the circulation of groundwater, heat flow and seismicity in Northwestern Turkey, *Terra Nova*, *5*, 164–173.

Scholz, C. H., Sykes L. R., and Aggarwal, Y. P. (1973), Earthquake prediction: a physical basis, *Science*, *181*, 803.

Semenov, A. M. (1969), Variations in the travel-time of transverse and longitudinal waves before violent earthquakes. *Izv. Acad. Sci. USSR, Phys. Solid Earth*, *4*, 245–248.

Serpieri, A. (1873), Rapporto delle osservazioni fatte sul terremoto avvenuto in Italia la sera del 12 marzo 1873, *Suppl. Meteorol. Ital.*, *1872*, 45–83.

Shi, Z., Wang, G., and Liu, C. (2013), Advances in research on earthquake fluids hydrogeology in China: a review, *Earthquake Sci.*, *26*, 415–425, 10.1007/s11589-014-0060-5

Shiratoi, K. (1927), The variation of radon activity of hot springs, *Sci. Rep. Tohoku Imp. Univ. Ser. III*, *16*, 614–621.

Siebert, L., and Simkin, T. (2002), *Volcanoes of the World: an Illustrated Catalog of Holocene Volcanoes and their Eruptions*. Smithsonian Institution, Global Volcanism Program Digital Information Series, GVP-3, (http://www.volcano.si.edu).

Sobolev, G. A. (1975), Application of electric method to the tentative short-term forecast of Kamchatka earthquake, *Pure Appl. Geophys.*, *113*, 229–235.

Sultankhodzhaev, A. N. (1984), Hydrogeosismic precursors to earthquakes, in *Proceedings of the International Symposium on Earthquake Prediction*, pp. 181–191, UNESCO, Paris.

Taramelli, T. and Mercalli, G. (1888), Il terremoto ligure del 23 febbraio 1887, *Ann. Uff. Centr. Meteorol. Geodin. Ital., Ser. II*, *8*(1886), 331–626.

Tsuboi, C., Wadati, K., and Hagiwara, T. (1962), *Prediction of earthquakes – progress to date and plans for further development*, Report of the Earthquake Prediction Research Group of Japan, Earthquake Research Institute of Tokyo.

Ungania, E. (1924), *Presismofono Ungania. Unico apparecchio preavvisatore dei terremoti, segnalatore delle perturbazioni elettromagnetiche*, Bologna.

Uyeda, S. (2013), On Earthquake prediction in Japan, *Proc. Jpn Acad. Ser. B, Phys. Biol. Sci.*, *89*, 391–400, doi: 10.2183/pjab.89.391

Varotsos, P. and Kulhànek, O. (eds) (1993), Measurements and theoretical models of the Earth's electrical field variations related to earthquakes, *Tectonophysics*, *224*(2–3), 1–228.

Wakita, H., Nakamura, Y., Notsu, K. Noguchi, M., and Asada, T. (1980), Radon anomaly: possible precursor of the 1978 Izu-Oshima-Kinkai earthquake, *Science*, *207*, 882–883.

Wakita, H., Nakamura, Y., and Sano, Y. (1988), Short-term and intermediate-term geochemical precursors, *Pure Appl. Geophys.* *126*, 267–278.

Wan Dikun (1993), China's national seismic well-network for observation of groundwater behaviour (water level and hydrogeochemistry) and typical earthquake cases, *J. Earthquake Pred. Res.*, *2*, 1–16.

Whitcomb, J. H., Garmany, J. D., and Anderson, D.L. (1973), A bare precursory change in seismic body-wave velocities occurring before the earthquake in San Fernando, California, *Science*, *180*, 632–635.

Wiss, M. (1977), Interpretation of the southern California uplift in terms of the dilatancy hypothesis. *Nature*, *266*, 805–808.

Wyss, M. (ed.) (1991), *Evaluation of Proposed Earthquake Precursors*, American Geophysical Union, Washington, DC, 94 pp.

Zarkov, V. N. (1983), *Vnutrennee stroenie zemli i planet*, Moscow, 404 pp. (In Russian.)

Part II: Latest Physical Models and Concepts of Pre-Earthquake Processes

6

Lithosphere-Atmosphere-Ionosphere-Magnetosphere Coupling —A Concept for Pre-Earthquake Signals Generation

Sergey Pulinets[1], Dimitar Ouzounov[2], Alexander Karelin[3], and Dmitry Davidenko[1]

ABSTRACT

The physics of earthquake genesis at the latest stage of a seismic cycle has always been the subject of scientific interest and discussion. In such discussion and in model development, however, only solid Earth processes have been contemplated. Development of satellite technologies brought new insight to the problem: new types of anomalies in the atmosphere and ionosphere were discovered that appear a few days/weeks before the main shock within the zone of earthquake preparation. These anomalies were thoroughly investigated, and results of this research have demonstrated a high statistical confidence warrants their identification as short-term earthquake precursors. In this chapter we seek to clarify how information and energy are transported from underground to the upper and lower layers of the atmosphere, including near-Earth space. Owing to the interdisciplinary character of the model, it is very difficult to create a common code for all its physical and spatial domains. That is why the material presented here should be considered conceptual rather than modeling output. As an example, if in the boundary layer plasmachemistry instrumentation is applied, for ionospheric anomalies the electromagnetism approach is adopted. This conceptual approach, however, proved to be valid not only for earthquakes but also for other natural and technological disasters where air ionization occurs.

6.1. INTRODUCTION

The physical and chemical processes happening during the latest stage of the earthquake cycle have been studied for decades in several countries using the ground-based measurements [e.g., *Scholz et al.*, 1973; *Kanamori and Anderson*, 1975; *Dobrovolsky*, 1991; *Sobolev*, 1993; *Zavyalov*, 2006; *Wiemer and Schorlemmer*, 2007]. Such careful studies, however, led to the establishment of scaling laws connecting the spatial distribution of short-term precursors or foreshock activity with the magnitude of the forthcoming earthquake [*Dobrovolsky et al.*, 1979; *Bowman et al.*, 1998; *Toutain and Baubron*, 1998]. This area has been identified as the "earthquake preparation zone" [*Dobrovolsky et al.*, 1979] or "critical zone" [*Bowman et al.*, 1998], and its radius formally is determined as the maximum distance from the earthquake epicenter where the short-term precursors could still be detected. This determination remained as an abstract conception until the moment it could be verified by remote sensing techniques. For the first time, it was possible to recognize this area from a satellite and to identify it as a physical entity that could be studied rather than as an abstract concept. In Figure 6.1 the spatial distribution of a ground-surface thermal anomaly is depicted. It was recorded on 1 April 2009, 5 days before the L'Aquila, Italy, *M* 6.3 earthquake on 6 April 2009 [*Pulinets et al.*, 2013]. The thermal infrared (TIR) anomaly recorded by MODIS [*Pergola et al.*, 2010] lies within the circles determined by *Dobrovolsky et al.* [1979] (blue circle), and *Bowman et al.* [1998] (red circle).

[1]*Space Research Institute, Russian Academy of Sciences, Moscow, Russia*

[2]*Center of Excellence in Earth Systems, Modeling and Observations (CEESMO), Chapman University, Orange, CA, USA*

[3]*IZMIRAN, Russian Academy of Sciences, Moscow, Russia*

Pre-Earthquake Processes: A Multidisciplinary Approach to Earthquake Prediction Studies, Geophysical Monograph 234, First Edition. Edited by Dimitar Ouzounov, Sergey Pulinets, Katsumi Hattori, and Patrick Taylor.

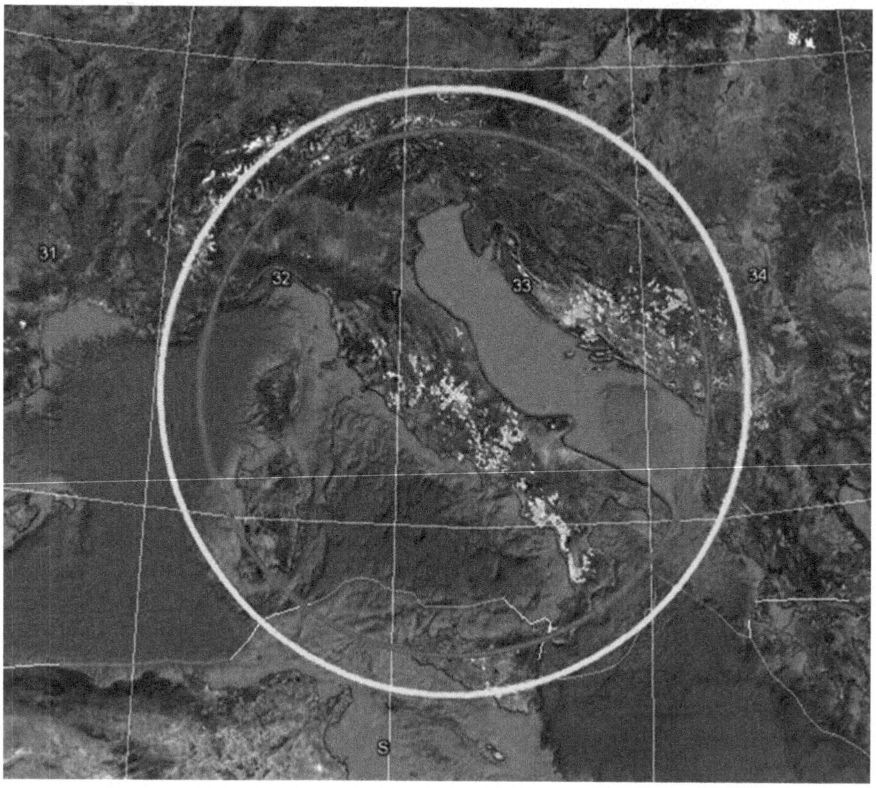

Figure 6.1 Thermal infrared (TIR) anomaly registered 5 days before the L'Aquila *M* 6.8 earthquake of 6 April 2009 (yellow and red). Blue circle, the earthquake preparation zone zccording to *Dobrovolsky et al.* [1979]; red circle, the critical area according to *Bowman et al.* [1998]. The earthquake epicenter indicated by the red cross. (Source: *Pulinets et al.* [2013]. Reproduced with the permission of Sergey Pulinets.) (*See electronic version for color representation.*)

Such TIR anomalies (identified by yellow and red areas) ideally fit within the earthquake preparation zone, and this zone has been verified using the spatial distribution of a variety of precursors for different earthquakes of different magnitudes [*Liu et al.*, 2009]. From this, we can conclude that the earthquake preparation zone is the area within which we may observe the processes of lithosphere–atmosphere–ionosphere–magnetosphere coupling (LAIMC). To determine the whole chain of processes involved we need to unite such disciplines as seismology, rock mechanics, nuclear physics, microphysics, plasmachemistry, dusty plasma, thermodynamics, meteorology, atmospheric electricity, radiowave propagation, ionospheric and magnetospheric physics, solar–terrestrial physics. As this would be too much of an undertaking for one chapter we will concentrate on the main interfaces between the geophysical components.

There have been many studies of the different aspects of the LAIMC framework during past two decades, but up to now a complete description of a LAIMC model has not been presented. Based on publications to date, research into LAIMC can be divided into two groups: (a) attempts to explain the preseismic thermal anomalies [*Qiang et al.*, 1991; *Tramutoli et al.*, 2009], and (b) modeling seismo-ionospheric coupling [*Sorokin et al.*, 2000; *Klimenko et al.*, 2011; *Kuo et al.*, 2011, 2014, 2015; *Enomoto*, 2012; *Namgaladze*, 2013; *Sorokin and Hayakawa*, 2013; *Hegai et al.*, 2015]. The only author who tries to unite both the thermal and ionospheric anomalies is *Freund* [2013], but his modeling is based on the solid body effects observed only over the land, while preseismic anomalies take place both over the land and over the ocean, making universal use of this model problematic.

6.2. LITHOSPHERE–ATMOSPHERE INTERFACE

If imagining how the lithosphere can interact with the atmosphere, the natural starting point would be to consider something that comprises the atmosphere, i.e., gases. However, how do the gases released from Earth's crust know that an earthquake is approaching if those gases are always released? It is well known that at the latest stage of the seismic cycle crustal deformation is not elastic but brittle, leading to asperity formation and the opening of new ways for migration of gases in the crust. The Gutenberg–Richter relation, also known as the frequency–magnitude relation (FMR),

$$\log N(M) = a - b \cdot M \tag{6.1}$$

describes the character of seismic activity, where N is the number of earthquakes with magnitude $\geq M$ and a and b are local constants. At the latest stage of earthquake preparation this relationship is characterized by a decreasing b-value. *Schorlemmer et al.* [2004] claimed "lower than average b-values characterize locked patches of faults (asperities), from which future mainshocks are more likely to be generated." This provided a direct indication of the physical interpretation of the FMR. *Turcotte* [1997] interpreted FMR as a power law (fractal) scaling between the number of earthquakes with rupture areas greater than a given value and the rupture area itself, which has a spatial fractal dimension $D{\sim}2b$. *Turcotte and Malamud* [2002] interpret the low b-values as a probable strong and homogeneous stress field near an asperity. Thus the decrease in b-value is equivalent to a decrease in fractal dimension, which can be interpreted as consolidation and clustering of seismic activity (observed experimentally), and augmentation of crack formation, leading to a higher level of radon emanation before earthquakes as a result of changes in migration pathways in Earth's crust [*Khilyuk et al.*, 2000].

In the Figure 6.2 the b-value and fractal dimension D_2 can be seen to decrease before the Kobe, Japan, M 6.9 earthquake on 17 January 1995 [*Enescu and Itu*, 2003], and variations in radon gas levels were detected in Kobe area [*Igarashi et al.*, 1995]. Also apparent in Figure 6.2 is that the period of increased seismic activity before the Kobe earthquake starts following a period of seismic quiescence (SQ) coincident with growth of the fractal dimension. Therefore, we can contemplate that during periods of decreased b-value magnitude an increase in radon levels will be observed, as happened before the Kobe earthquake. Unfortunately, there have been very few reliable measurements of radon levels in earthquake prone areas in recent years. It is hoped that new, more sophisticated measurements of radon using gamma-spectrometry will shed more light on this phenomena [*Fu et al.*, 2015]

According to *Pulinets et al.* [2015] radon is the main source of boundary-layer modification through air ionization. Analogous to the effects of galactic cosmic rays on the formation of condensation nuclei at the tropopause [*Lee et al.*, 2003; *Svensmark et al.*, 2007], radon produces large hydrated ion clusters that lead to the formation of preseismic thermal anomalies before an earthquake because of latent heat emmission, and to preseismic ionospheric anomalies through local modification of boundary-layer electric conductivity, which has an effect on the ionosphere through electromagnetic coupling in the global atmospheric electric circuit. These two consequences of air ionization form two main branches of the model: generation of thermal anomalies through the geochemical–thermal interface and generation of ionospheric and magnetospheric anomalies through the geochemical–electromagnetic interface.

6.3. GEOCHEMICAL-THERMAL INTERFACE

This interface transforms the geochemical emanation of radon into the heat generation machine through the so-called ion induced nucleation (IIN) process [*Laakso et al.*, 2002; *Kathmann et al.*, 2005]. This is caused by an avalanche-type cumulative characteristic of the processes involved in combination with the environmental conditions, from which we can select three main factors.

1. As shown by *Yasuoka et al.* [2006], the radon emanation before an earthquake follows the power-law of the Benioff strain [*Ben-Zion and Lyakhovsky*, 2002], which can be approximated by a log-periodic oscillation model [*Sornette and Sammis*, 1995]. This sharp augmentation of radon emanation leads to fast growth in the ion production rate.
2. The semidecay period of radon is 3.8 days, so in contrast to ionization by coronal discharge where only the running value of the ionization source is taken into account, the earlier emanated radon does not disappear but is stored in the area, so has a cumulative effect on ionization activity.
3. During nighttime, owing to temperature inversion, the mixing height of the boundary layer is very low (100–300 m) [*Eresmaa et al.*, 2012], therefore, both the radon as the source of ionization and newly formed ions are concentrated at the ground surface.

As shown by *Boyarchuk et al.* [2006], the higher the ionization rate and the greater the ion's concentration, the more stable are the newly formed ion clusters, which gives them opportunity to grow to the larger sizes of up to 1–3 μm [*Pulinets and Ouzounov*, 2011]. Obviously, the effectiveness of thermal energy release will depend on the number of cluster ions and their size, because the latent heat released will be the product of these two values. Bearing in mind items 1–3 mentioned above, we can contemplate an ion production rate of order of $q_i \leq 10^{10}\,m^{-3}\,s^{-1}$ [*Chernogor*, 2012]. According to recent measurements of radon in southern Lebanon [*Kobeissi et al.*, 2015] this value can reach $10^{11}\,m^{-3}\,s^{-1}$. If the ion cluster grows up to 1 μm size (such a particle will contain 4×10^{10} water molecules) and using the latent heat constant $U_0 \sim 40.68 \times 10^3\,J\,mol^{-1}$ ($1\,mol = 6.022 \times 10^{23}$) we obtain a thermal energy release near $27\,W\,m^{-3}\,s^{-1}$ for $q_i = 10^{10}\,m^{-3}\,s^{-1}$.

Figure 6.3 demonstrates schematically the geochemical interface, which transforms the radioactive gaseous flux into an exothermic reactor. To estimate the effectiveness of the thermal energy release because of air ionization the software "PLASER" was used, which primarily was developed for modeling of the local kinetics of laser plasma media [*Karelin*, 2007]. This software complex permits creation of a model with selected media relaxation using a set of characteristics,

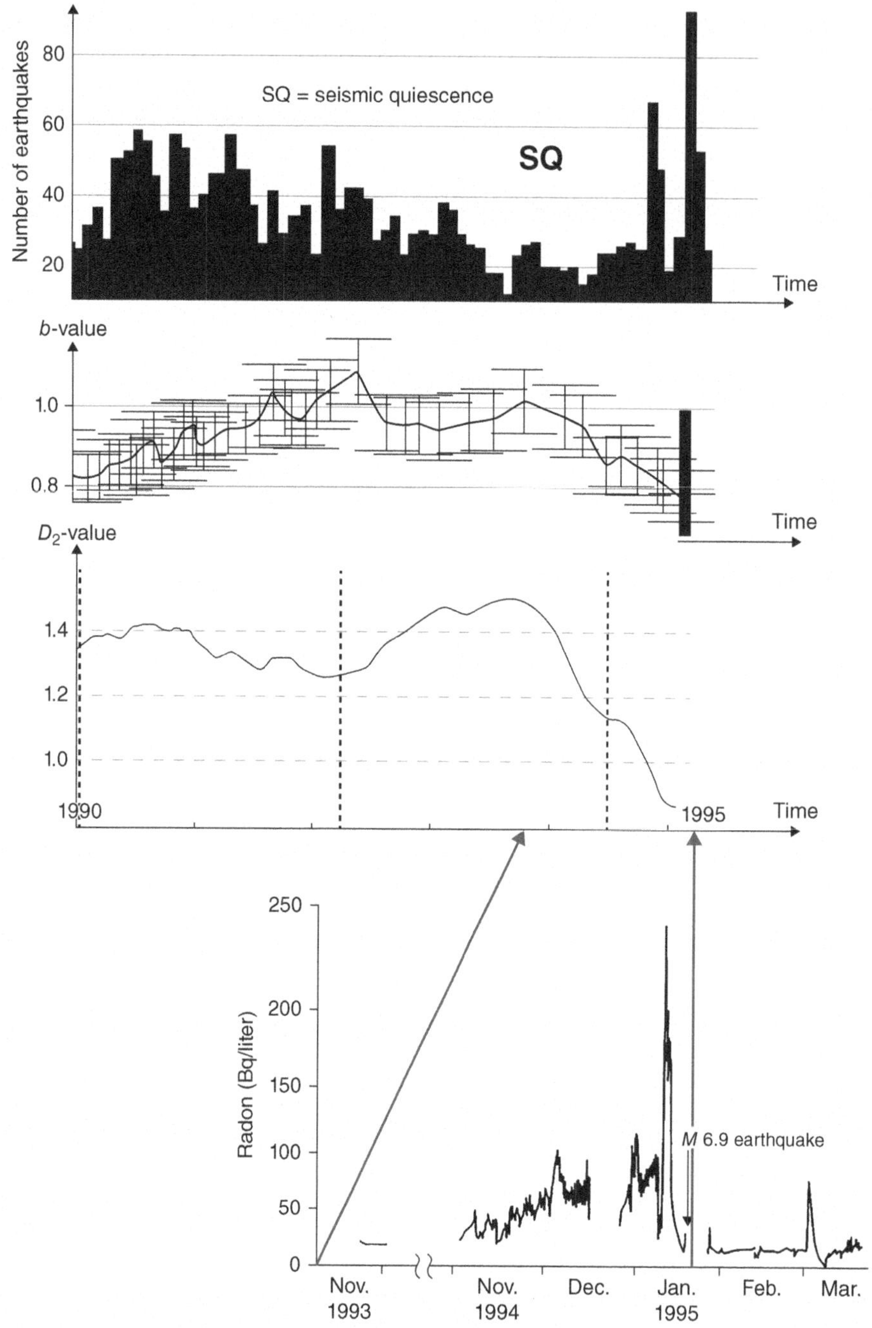

Figure 6.2 From top to bottom: seismic activity in the Kobe area for the period 1990–1995; *b*-value variations for the same period of time; fractal dimension D_2 for seismic activity in the Kobe area; radon activity (in water) for the period from November 1993, the time correspondence is indicated by arrows.

plasmachemical reactions, and other necessary parameters under the action of a rigid ionizer. The model includes provision of quantitative calculation of the ion composition and population of electron state of the atoms, molecules, and ions N_i, as well as the electron T_e and gas T_g temperatures in the nonstationary regime in accordance with kinetic equations. For the process considered, the main source of energy is water vapor in the atmosphere. Figure 6.4 shows the calculated increase of the air temperature as a function of relative humidity calculated using the PLASER software.

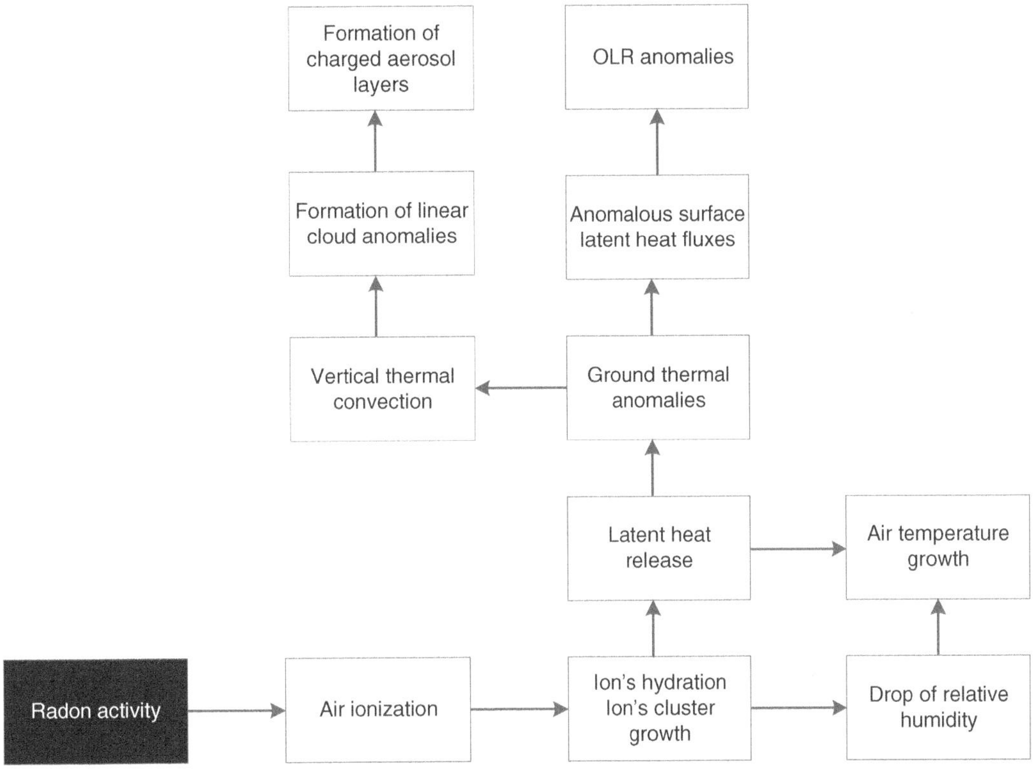

Figure 6.3 Schematic presentation of the geochemical–thermal interface.

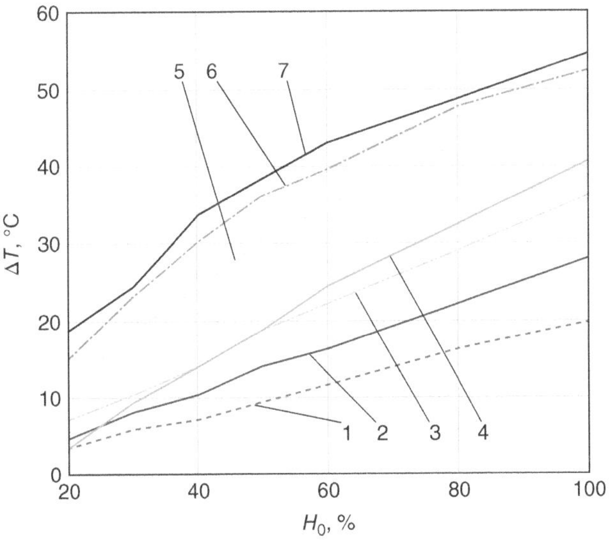

Figure 6.4 Dependence of the possible air heating ΔT on the initial relative humidity H_0 for different initial air temperatures $T = 5$ (1), 10 (2), 15 (3), 20 (4), 25 (5), 30 (6), and 35 (7) °C under $f = 40$ (cm^3·s)$^{-1}$ and $\tau = 10$ h.

The complete set of plasmachemical reactions as a result of ionization was considered by *Pulinets and Boyarchuk* [2004], here we present only the main final negative and positive ion clusters: $NO_3^-\cdot(H_2O)_n$, $NO_3^-\cdot(HNO_3)_n(H2O)_m$, and $H_3O^+\cdot(H_2O)_m$, which can be regarded with a high level of probability as the main ions of the troposphere.

Owing to thermal energy release into the atmosphere, several kinds of thermal anomalies are observed in the atmosphere.

1. Thermal anomaly registered in the very near ground layer of the atmosphere, usually measured in terms of brightness temperature and known as the TIR (transient infrared) anomaly.
2. Thermal anomaly registered in the lower part of the planetary boundary layer (PBL) of the atmosphere in the form of an air temperature increase (usually accompanied by a decrease in relative humidity), known as a meteorological anomaly.
3. Thermal anomaly derived from the column water vapor content known as the surface latent heat flux (SLHF), which is usually registered at altitudes 2–5 km above the ground surface.
4. Thermal anomaly registered at the so-called top of atmosphere or top of clouds, i.e., at an altitude close to the tropopause (10–15 km), and known as outgoing long-wave infrared radiation (OLR).

These anomalies can be separated into two categories by their physical manifestation: 1 and 4 are measured as electromagnetic emissions within infrared bands 2–7 μm and 8–14 μm; 2 and 3 can be referred to as thermodynamic phenomena and usually are measured using microwave detectors. Readers are referred to Chapter 15 where the nature and morphology of thermal anomalies are described in greater detail [*Ouzounov et al.*, 2018a].

6.4. GEOCHEMICAL-ELECTROMAGNETIC INTERFACE

It is natural to assume that intensive ionization will essentially change the near-ground air conductivity and electric properties of the global atmospheric electric circuit within the earthquake preparation zone. Usually in the literature, however, only one possibility of change due to air ionization is considered, i.e., increase in air conductivity. This phenomenon is not so simple though because we should also consider the ion's mobility, which strongly depends on the ion mass. This parameter could vary within an interval more than four orders of magnitude, and the resulting conductivity will depend on the partial concentration of the light and heavy ions. In the Table 6.1 it can be seen how an ion's mobility can vary [*Hirsikko*, 2011]; for example, an ion's mobility can decrease by four orders of magnitude with increasing size up to nearly 80 nm. In the case of ion cluster formation before earthquakes, particles of order 1000 nm can be expected, which would add at least one order of magnitude decrease to ion mobility.

Therefore, we can propose at least three different options for the behavior of boundary layer conductivity in response to ionization:

1. a sharp increase in concentration of light ions, leading to an increase in boundary layer conductivity;
2. a balance between the partial concentration of the light and heavy cluster ions in a situation that appears as if there is no ionization, when conductivity is the same as without ionization.
3. formation of a cluster of large ions through ion-induced nucleation processes with a concentration greater than that of small ions (sometimes completely removing the light ions), what dramatically decreases air conductivity.

Total vertical current, I, can be expressed as:

$$I = e\left(\sum n^+\mu^+ + \sum n^-\mu^-\right)E = \sigma E \qquad (6.2)$$

where n^+ and n^- are the concentration of positive and negative ions respectively, μ^+ and μ^- are mobility of positive and negative ions respectively, E is the vertical electric field in the atmosphere, e the electric charge of an electron, and σ the total air conductivity.

In fair-weather conditions there are no additional sources or sinks of the vertical current between the ionosphere and ground, so the low air conductivity will determine the total electric current in the global atmospheric electric circuit, and consequently the value of ionosphere potential over the earthquake preparation zone. Thus we can consider variations of the ionosphere potential as an indicator of changes in the boundary layer and troposphere conductivity. The presence of this effect was demonstrated by *Markson* [2007] by analysing the variations in the global atmospheric electric circuit recorded during 50 years of observations. The most pronounced effect

Table 6.1 Mobility and diameter of cluster ions.

Ion cluster size/density	Analyzer	Fraction	Mobility cm^2V^{-1} s^{-1}	Diameter nm
Small	IS_1	N_1/P_1	2.51–3.14	0.36–0.45
	IS_1	N_2/P_2	2.01–2.51	0.45–0.56
	IS_1	N_3/P_3	1.60–2.01	0.56–0.70
	IS_1	N_4/P_4	1.28–1.60	0.70–0.85
Large	IS_1	N_5/P_5	1.02–1.28	0.85–1.03
	IS_1	N_6/P_6	0.79–1.02	1.03–1.24
	IS_1	N_7/P_7	0.63–0.79	1.24–1.42
	IS_1	N_8/P_8	0.50–0.63	1.42–1.60
Intermediate	IS_1	N_9/P_9	0.40–0.50	1.6–1.8
	IS_1	N_{10}/P_{10}	0.32–0.40	1.8–2.0
	IS_1	N_{11}/P_{11}	0.25–0.32	2.0–2.3
	IS_2	N_{12}/P_{12}	0.150–0.293	2.1–3.2
	IS_2	N_{13}/P_{13}	0.074–0.150	3.2–4.8
	IS_2	N_{14}/P_{14}	0.034–0.074	4.8–7.4
Light	IS_2	N_{15}/P_{15}	0.016–0.034	7.4–11.0
	IS_3	N_{16}/P_{16}	0.0091–0.0205	9.7–14.8
	IS_3	N_{17}/P_{17}	0.0042–0.0091	15–22
Heavy	IS_3	N_{18}/P_{18}	0.00192–0.00420	22–34
	IS_3	N_{19}/P_{19}	0.00087–0.00192	34–52
	IS_3	N_{20}/P_{20}	0.00041–0.00087	52–79

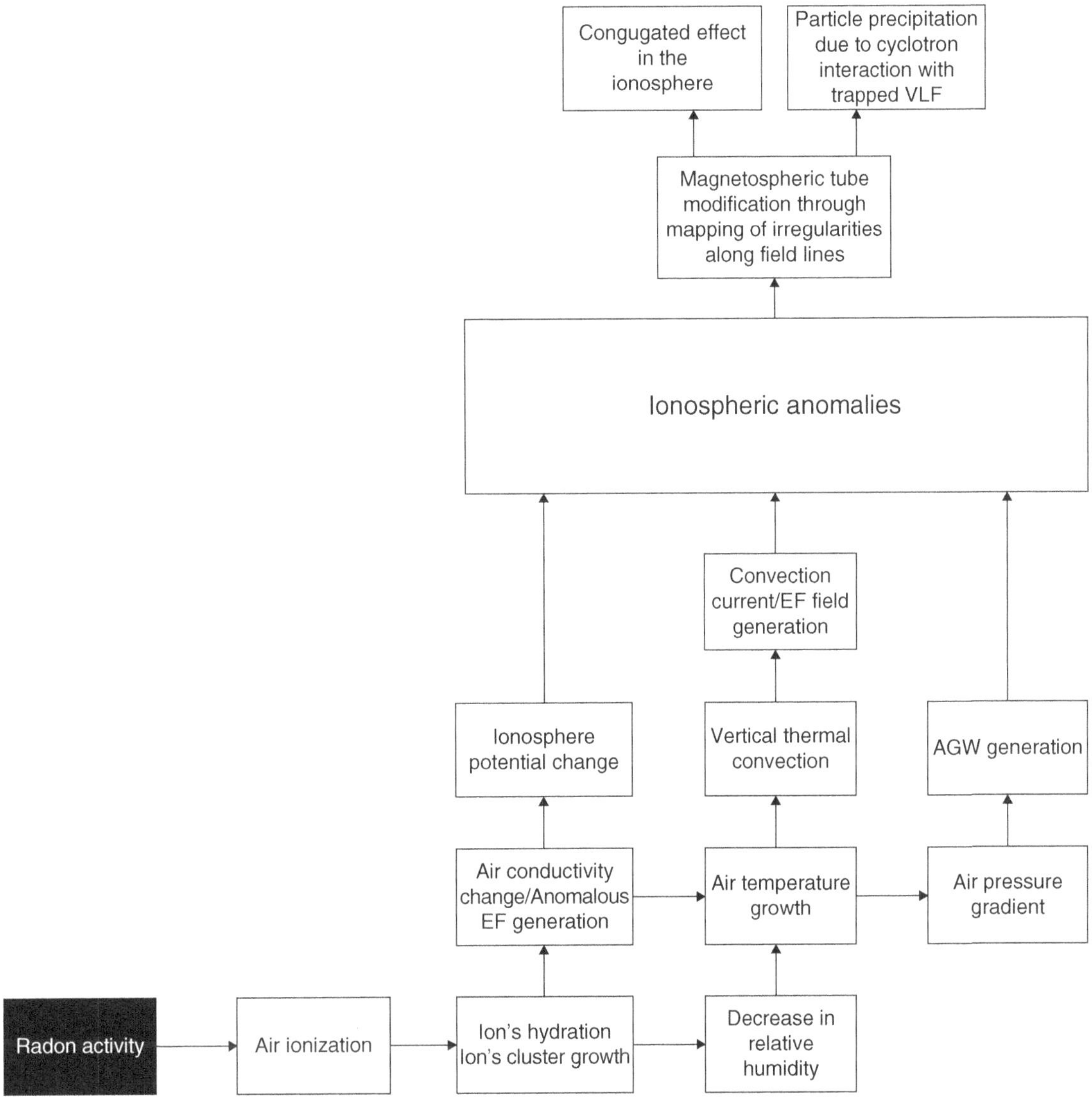

Figure 6.5 Schematic presentation of the geochemical–electromagnetic interface: VLF, very low frequency; EF, electric field; AGW, acoustic gravity waves.

was registered during the period of atmospheric nuclear weapon testing. It is natural to expect that local variations of ionosphere potential will cause variations in electron concentration and of temperature in the ionosphere over the region of ionospheric potential modification.

The phenomena of seismo-ionospheric anomalies can be considered from another point of view, by taking into account electric charge separation related to the natural vertical electric field and the convection electric generator; *Pulinets and Boyarchuk* [2004] and *Morozov* [2006] have considered both phenomena. In these conditions, aside from the downward-directed natural vertical electric field, which has a near-ground surface intensity of 100–150 V m^{-1}, an additional field can be generated that sometimes reaches a value more than 1000 V m^{-1} [*Vershinin et al.*, 1999]. Field reversals can happen (as confirmed experimentally), thus the electric field could be directed upwards. In this situation we could resolve the problem of anomalous electric field penetration from the ground surface into the ionosphere. In the flowchart of the geochemical–electromagnetic interface presented in the Figure 6.5 both possibilities are presented, and will be discussed later in more detail.

The phenomenon of electric field penetration into the ionosphere has been a matter of intensive scientific debate since the pioneering work of *Park and Dejnakarintra* [1973, 1977]. They considered the phenomenon of electric field penetration into the ionosphere emanating from thunderstorm clouds. Using their approach, *Kim and Hegai* [1997] considered the problem just for a seismogenic source of an electric field on the ground surface, which was extended in *Pulinets et al.* [1998]. Under nighttime conditions with a value for an electric field on the ground surface of 1000 V m^{-1}, their most optimal estimation provided a value of 0.7 mV m^{-1} in the ionosphere. In their latest work [*Hegai et al.*, 2015], for a nighttime ionosphere during the solar minimum and inside large-scale plasma bubbles, they estimated the value of an electric field perpendicular to the geomagnetic field as 0.2 and 1.0 mV m^{-1}, respectively.

There are many studies modeling the observed seismo-ionospheric effects in the ionosphere [*Namgaladze*, 2013; *Sorokin and Hayakawa*, 2013; *Kuo et al.*, 2011, 2014; and see the reviews by *Hayakawa et al.* [2018] and *Kuo et al.* [2018]. In order to reproduce the experimentally observed ionospheric anomalies that occur before earthquakes, the models manually introduce a so-called external electric field or external current of several orders (from 4 to 5) of magnitude larger than natural values, but the origin external fields are speculative without valid experimental support. *Denisenko* [2015] reviewed all these modeling studies and clearly demonstrated their physical inconsistency. Even accepting the approach of *Hegai et al.* [2015], he still maintains that penetration of an electric field into the ionosphere is negligibly small.

So in concluding this discussion we should acknowledge the lack of consensus in support of electric field penetration from ground level into the ionosphere. In such a situation the correct approach is to focus on experimental results so that modeling objectives correspond to physical reality. As direct measurements of an electric field in the ionosphere may also not be reliable because of the problem of sensor calibration, the most appropriate source of information on anomalies in the ionosphere before strong earthquakes is provided by measurements of an ion's drift velocity. In the most recent paper, *Liu and Chao* [2016] used data from the Ionospheric Plasma and Electrodynamics Instrument (IPEI) onboard the ROCSAT-1 satellite to estimate the electric field in the ionosphere through determination of an ion's drift velocity in the equatorial ionosphere. Their estimation for a period of 1–5 days before the *M* 6.8 earthquake in Taiwan on 31 March 2001 was 0.91 mV m^{-1} of the west-directed zonal electric field (instead of the expected 0.51 mV m^{-1} east-directed zonal electric field in the afternoon hours for the reference days without the seismo-ionospheric disturbance). This effect will be explained below in discussion of Figure 6.6. In the meantime we compare in Table 6.2 estimations of the seismogenic electric field in ionosphere calculated by the different models. From these data it would appear that the model of *Hegai et al.* [2015] most adequately represents reality.

To understand the results of *Liu and Chao* [2016] let us turn to Figure 6.6a, which consists from two halves (left and right) and portrays the longitudinal (zonal) cross-section of the equatorial ionosphere. The geomagnetic field is directed perpendicular to the plane of the figure. The panels on the left of the figure correspond to increased air conductivity, whereas those on the right correspond to the reverse situation when concentration of heavy ion clusters prevails, and air conductivity decreases. In the initial stages under weak ionization levels the light ions will prevail in the boundary layer of the atmosphere, which will lead to a general increase in the bulk conductivity of the atmosphere and consequently a decrease in ionospheric potential relative to the ground (left panel of Figure 6.6a). If this situation happens during afternoon hours when the east-directed electric field forming the equatorial anomaly is present (white arrows in Figure 6.6a), the ionospheric potential decreases (concave curve at the top of the schematic presentation of the atmosphere–ionosphere interface). The ionosphere is a highly conductive media and will not tolerate local decrease in potential, but will try to maintain equipotentiality by inducing an electric field directed to the center of the anomaly (grey arrows). In these conditions, the "grey field" (seismogenic) will be added to the "white field" (natural) on the west side and subtracted on the east side of the conductivity anomaly. Thus in the situation in the left panel of Figure 6.6a, the vertical **ExB** drift velocity will increase to the west due to the overall addition of the natural and seismogenic electric fields in the west-side area of the conductivity anomaly, and will decrease in the east-side of conductivity anomaly because of the overall subtraction of natural and seismogenic fields created by area of the conductivity anomaly. In these circumstances the level of equatorial anomaly development (ratio of electron concentration in the crests of the anomaly to the concentration in the trough of the anomaly) will increase to the west of the conductivity anomaly center, and to the east of the conductivity anomaly center the equatorial anomaly will be inhibited. The three-dimensional image of this configuration is shown in Figure 6.6b.

Another scenario arises when the ionization rate is very high, relative humidity is sufficient to create large ion clusters, and calm-weather conditions prevail, as large clouds of aerosol-size heavy ion clusters can form in this situation. This condition will lead to the decreased air conductivity anomaly creation. Amplification of the equatorial anomaly on the east side of conductivity anomaly will be observed in this situation, with inhibition of the equatorial anomaly to the west-side of decreased air conductivity anomaly. Usually, during the precursory period the left panel corresponds to the early stage at the start of increased radon emanation and formation of large ion clusters, and the situation depicted in the right panel corresponds to the start of the so-called "ion's aging" stage. This interpretation is supported by the experimental results in the upper panel of the figure, where the differential global ionosphere

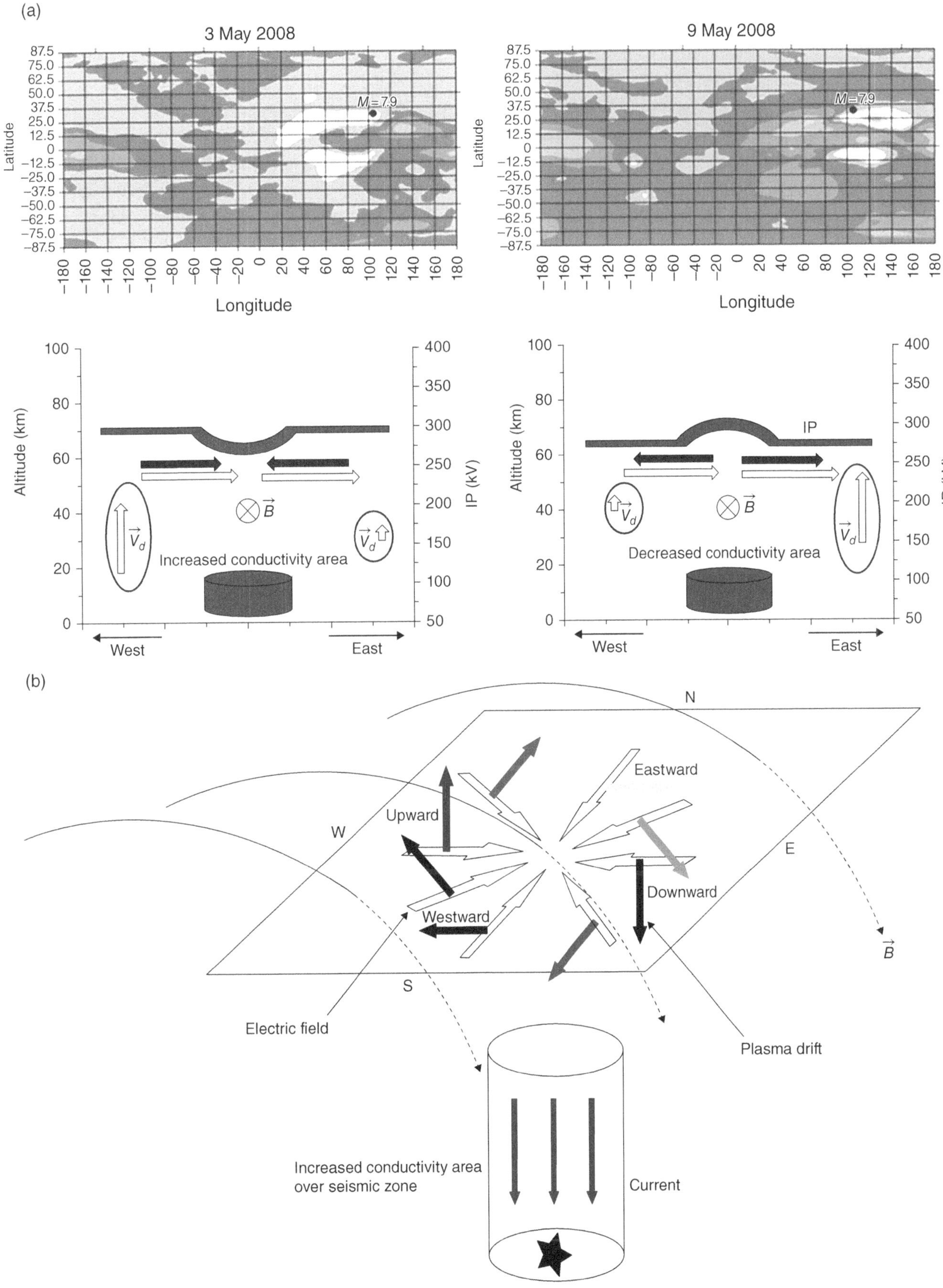

Figure 6.6 (a) Bottom panel: schematic conception of atmosphere–ionosphere coupling through the global atmospheric electric circuit; left panel: for a condition of increased air conductivity; right panel: for a condition of decreased air conductivity. IP, ionosphere potential. Upper panel: the differential maps obtained from the global ionosphere maps GPS total electron concentration data for the period before the Wenchuan earthquake on 12 May 2008; left panel: two-dimensional distribution obtained on 3 May 2008; right panel: two-dimensional distribution obtained on 9 May 2008. (b) three-dimensional presentation of the left panel of Figure 6.6a for the case of increased air conductivity.

maps (GIM) for the Wenchuan *M* 8 earthquake on 12 May 2008 in China are shown: the left side demonstrates the differential GIM map obtained on 3 May 2008, while the right side corresponds to the 9 May 2008, a few days after the process of nucleation started.

What is the relation between this figure and the results of *Liu and Chao* [2016]? Their case corresponds to the east part of the left panel and to the west part of the right panel of Figure 6.6, i.e. to the case of equatorial anomaly inhibition. Considering the importance of this validation of the proposed LAIMC concept, we quote the text from [*Liu and Chao*, 2016]:

> From the IGRF (international geomagnetic reference field) model, we find the **B** field at 600 km altitude over the epicenter is 0.32×10^{-4} nT with the magnetic dip $I = 33.4$ deg. Since *Vpar* is parallel to the magnetic field, no electric field would be produced. By contrast, for the Seismo-Ionospheric Precursor (SIP) day, $V_{PerM} = -11.12$ m/s, $E_{SIP} = 0.36$ mV/m in the westward direction while for the reference day, $V_{PerM} = 17.05$ m/s, $E_0 = 0.55$ mV/m in the eastward direction. Therefore, the seismo-generated electric field can be calculated by $E_{sg} = E_{SIP} - E_0 = 0.91$ mV/m in the westward direction.

As can be seen from the experimental results the seismogenic electric field by absolute value is larger than the natural east-directed electric field, so it not only reduces the upward drift of the plasma, but converts it into downward drift ($V_{PerM} = -11.12$ m/s).

A quite natural question arises: if the seismogenic electric field is so effective in modification of the equatorial anomaly, what effect would we observe without the natural east-directed electric field, for example, in local morning hours? We answer this using the data on electron concentration measurements onboard the French DEMETER satellite for the same Wenchuan earthquake [*Ryu et al.*, 2014]. Figure 6.7 presents the latitudinal distributions of the morning (1010 LT (local time)) passes of the DEMETER satellites for 2 days of observations (4 and 5 May 2008). As can be seen, most of distributions quite naturally do not show the two-hump structure characteristic of the developed equatorial anomaly, and it is only the pass closest to epicenter (20515) that records the two-hump equatorial anomaly structure at 10 am local time. This observation confirms the validity of the proposed physical model, and indicates suitable instruments for detection of precursory signals by monitoring for possible development of an equatorial anomaly at the appropriate hours local time when in natural conditions it should not develop.

At present, therefore, we have two conceptions of electromagnetic coupling: (a) strong anomalous electric field penetration into the ionosphere and (b) boundary-layer-conductivity variations. The question then arises as to whether they are mutually exclusive, and if so which is correct, or whether both occur; the latter seems closer to reality. This conclusion can be drawn when consideration is given to local time temporal dynamics of the planetary boundary layer (PBL; the lowest 1–2 km of the atmosphere, i.e., the region most directly influenced by the exchange of momentum, heat, and water vapor at Earth's

Table 6.2 Seismogenic electric field in the ionosphere estimated by different models.

Model	Electric field
Hegai et al., 2015	0.2–1 mV m^{-1}
Sorokin and Hayakawa, 2013	10 mV m^{-1}
Namgaladze, 2013	2–10 mV m^{-1}
Kuo et al., 2011; 2014	1–7 mV m^{-1}
Denisenko, 2015	1 μV m^{-1}

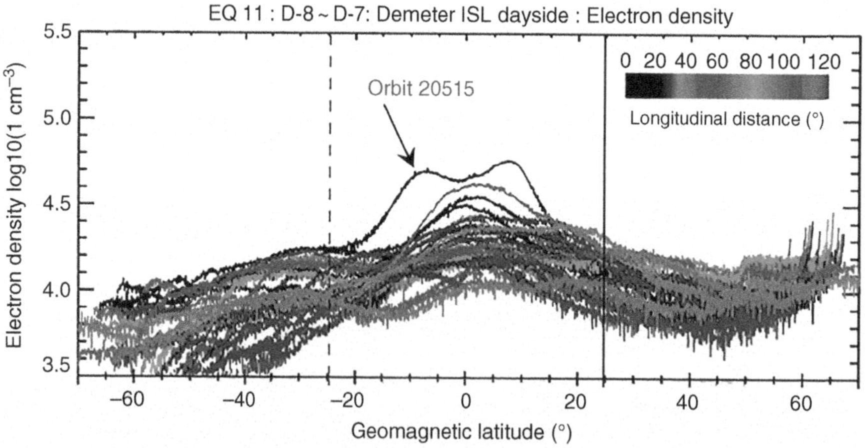

Figure 6.7 Latitudinal (geomagnetic latitude) profiles of morning 10 am local time half-orbits of the DEMETER satellite for 4–5 May 2008. The two-hump curve (red) represents the satellite orbit closest to the epicenter $\Delta_{long} = 5.5°$, orbit 20515. (Source: Modified from *Ryu et al.* [2014].) (*See electronic version for color representation.*)

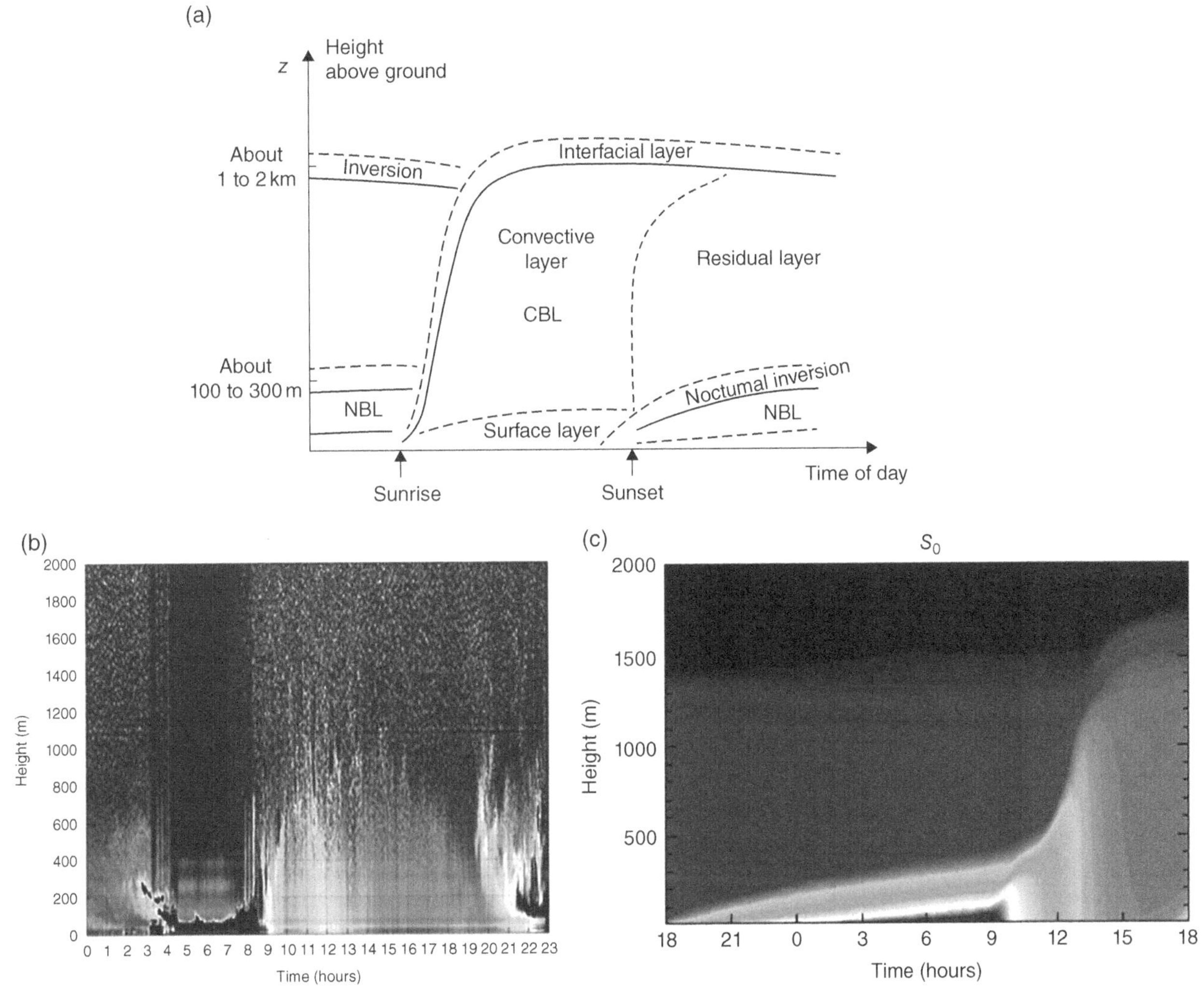

Figure 6.8 (a) Schematic presentation of the atmospheric boundary layer and its daily variations: CBL and NBL are the convective and nocturnal boundary layers respectively. (b) Daily variations of aerosol distribution after *Eresmaa et al.* [2012]. (c) Upper border of the atmospheric boundary layer recorded by radon in atmosphere. (Source: Adapted from *Eresmaa et al.* [2012], American Meteorological Society, and *Vinuesa et al.* [2007], Copernicus Publications.)

surface [*Kaimal and Finningan*, 1994]), as presented in Figure 6.8a. When looking at the structure (at least at quiet fair-weather conditions) we can observe the variable shape and behavior of the PBL with local time: the very narrow (100–300 m) layer at night the sharp expansion at sunrise, and reaching an altitude near 2 km in daytime. Thus at nighttime conditions the large-ion clusters/aerosols would not be able to rise into the atmosphere, and radon would stay near the ground surface, augmenting the level of ionization and the ion production rate.

This proposition is strongly supported by experimental data. The upper edge of the PBL can be monitored by LiDAR [*Eresmaa et al.*, 2012], with the help of radon as a tracer [*Vinuesa et al.*, 2007]. At nighttime (Figure 6.8c) radon is confined to the ground, especially under temperature inversion conditions, while during daytime it spreads over the whole boundary layer up to 2 km altitude. According to modeling results, the maximum of radon concentration is reached just before sunrise. Bearing in mind the presence of a natural electric field and varying mobilities of the positive and negative ions, we can expect separation of electric charges and generation of charges in space over the entire area of the earthquake preparation zone (for *M* 7 the radius of this zone is 1000 km). Theoretical estimations show [*Redin et al.*, 2014] that in conditions of low turbulence a negative space charge can form over the area of increased ionization in the presence of aerosols. Concentration of large negative ions N_- essentially exceeds concentration of large positive ions N_+ as well as concentration of light ions n_- and n_+ in the electrode layer. It is necessary to note that in relation to the concentration of light ions, starting at altitude 0.5 m the concentration of negative light ions n_- also exceeds the concentration of positive light ions n_+. Calculations also show that the value of electric field starting from 0.5 m height exceeds the value of the natural electric field equal to 100 V m^{-1} by up to an order of magnitude. Therefore, favorable conditions are created at nighttime for the electric field penetration mechanism to operate.

In daytime conditions the large-ion clusters rise up to an altitude near 2 km (Figure 6.8b) together with radon, which will continue to produce the nucleation centers and further formation of the large charged clusters. According to *Gringel* [1986] "The first 2 km of the atmosphere contribute about 50 percent and the first 13 km about 95 percent to the total columnar resistance." Thus during daytime we may expect the "air conductivity mechanism" to be working effectively (Figure 6.6).

Returning to other possible mechanisms of atmosphere–ionosphere coupling, we note also the possibility of the effect on the ionosphere of acoustic gravity waves (AGW) [*Hayakawa and Molchanov*, 2007]. One plausible source able to generate the air pressure oscillation could be large-scale ground-surface thermal anomalies [*Genzano et al.*, 2007]. As a possible indicator of this coupling process, the stratification of the lower ionosphere before earthquakes could serve such a purpose, and has been observed by oblique ionospheric sounding [*Blaunstein and Plohotniuc*, 2008]. Nevertheless, in the upper ionosphere we observe stationary long lasting (at least 4 h) anomalies over the earthquake preparation zone, which do not indicate any wave activity in the ionosphere before an earthquake. If more solid experimental evidence of AGW effects in the ionosphere could be obtained, it would be possible to consider this mechanism more carefully in future publications.

In Figure 6.5, presenting the geochemical–electromagnetic interface, furthest from the ground surface are the magnetospheric anomalies resulting from ionospheric irregularities penetrating to the magnetosphere through geomagnetic field lines due to their high conductivity. Several studies [*Sorokin et al.*, 2000; *Pulinets et al.*, 2002] have considered the formation of field-aligned irregularities. This modification of magnetospheric tubes is manifested experimentally by a so-called conjugated effect in ionospheric precursors first observed by Intercosmos-19 satellite in records of three phenoma: very low frequency (VLF) emissions and particle precipitation [*Ruzhin and Larkina*, 1996], and formation of large-scale irregularities of electron concentration registered by the topside sounder [*Pulinets and Legen'ka*, 1997]. Later, the DEMETER satellite recorded conjugated effects in the formation of plasma bubbles in the crest of an equatorial anomaly, extra-low frequency (ELF) noises ($f < 500$ Hz), local electron concentration, and electron and ion temperature [Pulinets, 2012]. It is natural to expect that conjugated effects also would be recorded by GPS total electron content (TEC) observations [*Liu et al., 2010*; *Pulinets et al.*, 2010] (see the reviews by *Parrot and Li* [2018] and *Liu et al.* [2018], chapters 17 and 18 of this book respectively).

Such strong evidence of conjugated effects in the magnetosphere is further proof of the electromagnetic nature of seismo-ionospheric–magnetospheric coupling rather than wave-like disturbances.

6.5. THE SYNERGY OF EARTHQUAKE PRECURSORS AND INTEGRATED PARAMETERS

The most important factor to be borne in mind is that all physical precursors (at least considered within the framework of the LAIMC model) are not independent; they are elements of an open complex nonlinear dissipative system. Earthquake precursors should be considered from the point of view of synergetics, and their comprehensive analysis should reveal a consecutive direction in the process of earthquake preparation, the so-called "arrow of time" [*Eddington*, 1928] indicating the approach of a system to its critical point. Sometimes such a consecutive direction could be detected when multiparameter analysis shows the time delays of one variable increasing in reaction to another, thereby demonstrating the temporal chain of the processes. Such a chain was demonstrated by *Pulinets et al.* [2015] with the temporal/altitude development of several precursors before the L'Aquila *M* 6.3 earthquake on 6 April 2009 in Italy (Figure 6.9). The oblique dashed line in the figure may be interpreted as the arrow of time. We can conclude that a multiparameter approach provides not only an extensive collection of data on different pre-earthquake anomalies, but also is an instrument for detection of their synergy and for demonstration of their affiliation with seismo-tectonic development, expressed by a decrease in *b*-value [*Papadopoulos et al.*, 2010] and entropy dynamics expressed as the negative logarithm of *b* [*De Santis et al.*, 2011].

Some of the major earthquakes, such as the Sumatra *M* 9.1 earthquake of 2004, the Wenchuan *M* 8 earthquake of 2008, the L'Aquila *M* 6.3 earthquake of 2009, and the Tohoku *M* 9 earthquake of 2011, have attracted the attention of a majority of scientists studying earthquakes and an incredible amount of experimental data has been collected, both satellite and ground-based measurements. These events provide the experimental background for the development of precursor generation mechanisms based on collected experimental data. A problem, however, is that these data were collected and interpreted after earthquake had happened. If the desire

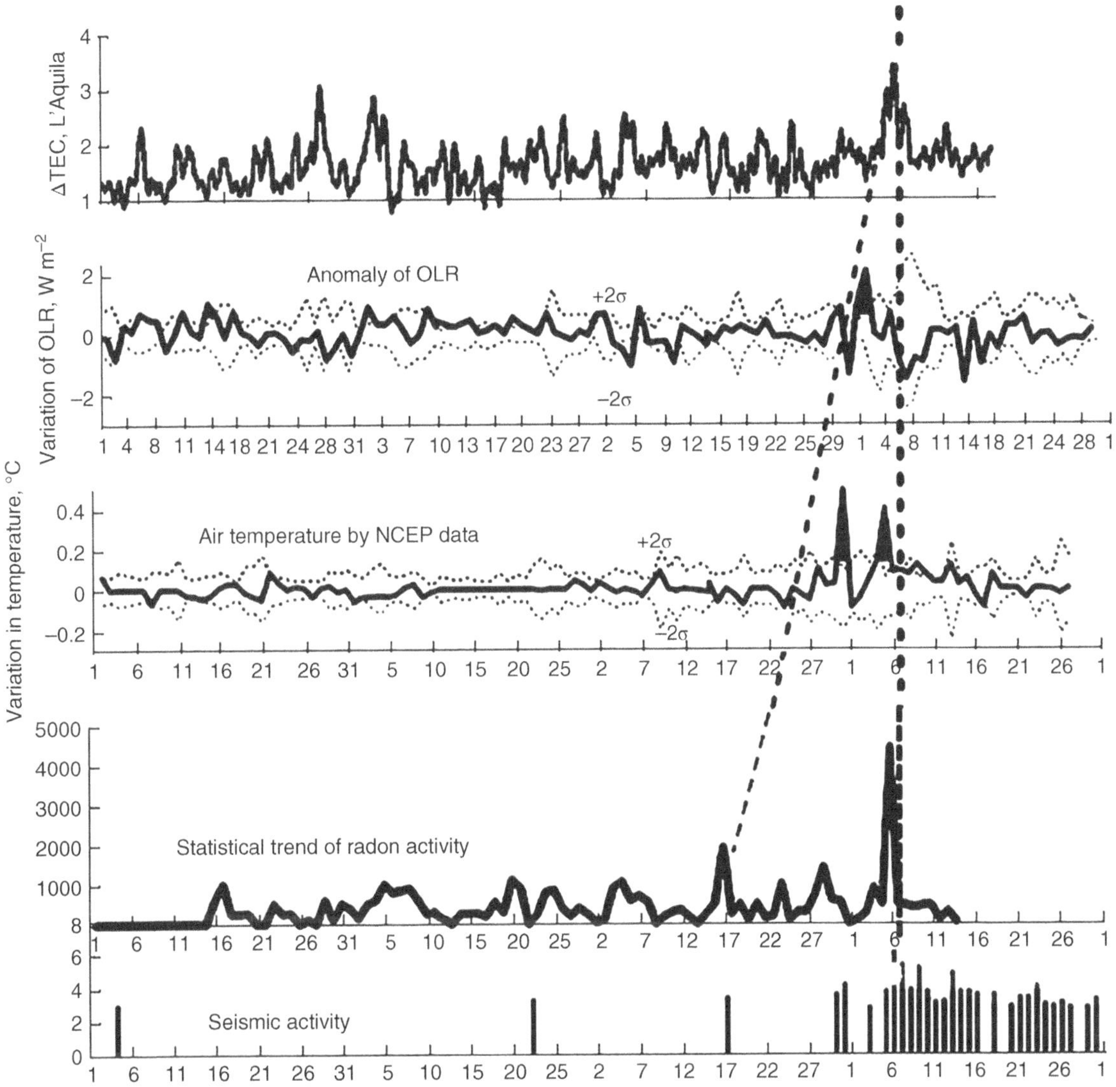

Figure 6.9 Temporal dynamics of seismic activity, radon release, and variations of atmospheric and ionospheric parameters before the L'Aquila earthquake: OLR, outgoing longwave radiation. (Source: National Centers for Environmental Protection.),

is for short-term forecasting, we need real-time multiparameter monitoring and immediate interpretation.

It is practically impossible, however, to collect the large set of data for all parameters in real-time globally, especially if we consider that ground-based measurements exist only in specific locations or geophysical observatories. We therefore should use the advantage of the operability and global coverage of satellite remote sensing and teleseismic measurements. It is difficult to control so many parameters simultaneously in real time, thus the so-called integral parameters should be selected that characterize the state of the complex system. Usually the fractal dimension variation is used to monitor the state of a complex system. *Enescu and Itu* [2003] demonstrated for the Kobe earthquake a clear decrease in fractal dimension D_2. *Turcotte and Malamud* [2002] showed that $D \sim 2b$, so, instead of calculating the fractal dimension the *b*-value variations could be monitored. It is difficult to strictly monitor underground activities, such as formation of asperities and cracks, in real time, but the *b*-value is just such an integrated parameter that shows changes in the Earth's crust and can be monitored by recording seismic flow in real time.

As one can be seen from Figure 6.2, the decrease in *b*-value correlated with radon emanation increase, which stimulates the process of generation of short-term precursors in the atmosphere and ionosphere [*Pulinets et al.*, 2015]. It is practically impossible to monitor radon variations remotely, especially from satellite, but *Pulinets et al.* [2015] proposed a new integral parameter (ACP, atmospheric chemical potential) that could be a good proxy for radon variations, i.e., "correction of chemical potential of the water vapor in atmosphere (ACP)." Its temporal behavior is very similar to radon behavior before an earthquake: the local absolute maximum occurs within the time interval of 1 month before the earthquake (usually 20–3 days before the main shock), and the local minimum is reached −3 to +2 days around the time of the main shock. The typical temporal behavior of ACP in the Kamchatka peninsula area is demonstrated in the top panel of Figure 6.10; the statistical distribution obtained for more than 10 years of Kamchatka earthquakes analysis with $M \geq 6$ for the main maximum of ACP leading time is shown in the middle panel of Figure 6.10, and the main minimum leading time statistical distribution for the same set of experimental data is shown in the bottom panel [*Pulinets et al.*, 2016]. The Integral character of the ACP can be explained if several important factors are taken into consideration.

1. Its value is derived from the air temperature and relative humidity time series—increase in air temperature and decrease in relative humidity is characteristic of ionization processes in the atmosphere.
2. Taking into account that air temperature and relative humidity are an integrated characteristic over a sufficiently large area, they characterize the level of ionization over the area, and not at a single point as is the case with radon measurements.
3. This parameter shows such a striking self-similarity for different earthquakes in different parts of the globe that it worthy of discussion in more detail.

The presence of ions in the atmosphere creates the possibility for molecules of water vapor to join to these ions through a hydration process that is different from condensation. In the process of condensation the chemical potential is equal to the latent heat, which is $Q = 40.683\,\mathrm{kJ\,mol^{-1}}$ or $U_0 = 0.422\,\mathrm{eV\,mol^{-1}}$. The process of evaporation/condensation, which is a phase transition of the first order, always takes place during chemical potential equality. Newly formed ions created by radon ionization, however, have a different chemical potential that should be taken into account. In a one-component approximation we introduce the correction to the chemical potential ΔU that takes into account this circumstance. In this case the real chemical potential can be expressed as:

$$U(t) = U_0 + \Delta U \cdot \cos 2t \tag{6.3}$$

where U_0 is the chemical potential for pure water, $U(t)$ is the chemical potential where the ionization and hydration is taken into account, and $\cos 2t$ takes into account the daily changes of the solar radiation.

Then we can express the air relative humidity in the following form:

$$H(t) = \frac{\exp\left(-U(t)/k\cdot T\right)}{\exp\left(-U_0/k\cdot T\right)} = \exp\left(\frac{U_0 - U(t)}{k\cdot T}\right) = \exp\left(-\frac{0.032\cdot\Delta U\cdot\cos^2 t}{(k\cdot T)^2}\right) \tag{6.4}$$

where k is the Boltzman constant and T is the air temperature. The increase in water molecule chemical potential ΔU that can be derived from (6.4) indicates the strength of the nucleation process, which can be used as an indicator of an impending earthquake, as shown by *Boyarchuk et al.* [2010]. In turn, air temperature and relative humidity can be determined by satellite remote sensing with the help of microwave sounder data, and this information can be incorporated in real-time assimilative modeling of the atmosphere.

6.6. SUMMARY AND CONCLUSIONS

The proposed physical framework provides a realistic opportunity to realize multiparameter monitoring of short-term precursors. The combination of seismic data and remote sensing data (mainly in the form of integrated parameters) makes it possible to detect globally the earthquake preparation period within an interval of time 1 month to a few days before a strong ($M > 6$) earthquake in quasireal time. Using modeling, practical applications were developed to automatically detect the different kinds of precursors that could be used in prospective earthquake monitoring [*Ouzounov et al.*, 2018b].

To conclude it would not be helpful to compare the approach proposed here with existing models, such as those mentioned in Introduction. In order to simplify our investigation we considered using only two groups of precursors, thermal and ionospheric, with the help of various criteria as summarized in Table 6.3; LAIMC, the concept proposed here, is also listed in Table 6.3. From the Table 6.3 it is possible to conclude that the proposed concept appears to apply universally because it is able to explain the physical mechanism and morphologic properties of the main short-term precursors in the atmosphere and the ionosphere, both over the land and the ocean. Nevertheless, a generalized numerical code describing the entire model in its completeness is still lacking, but its derivation is just matter of time and modeling innovation.

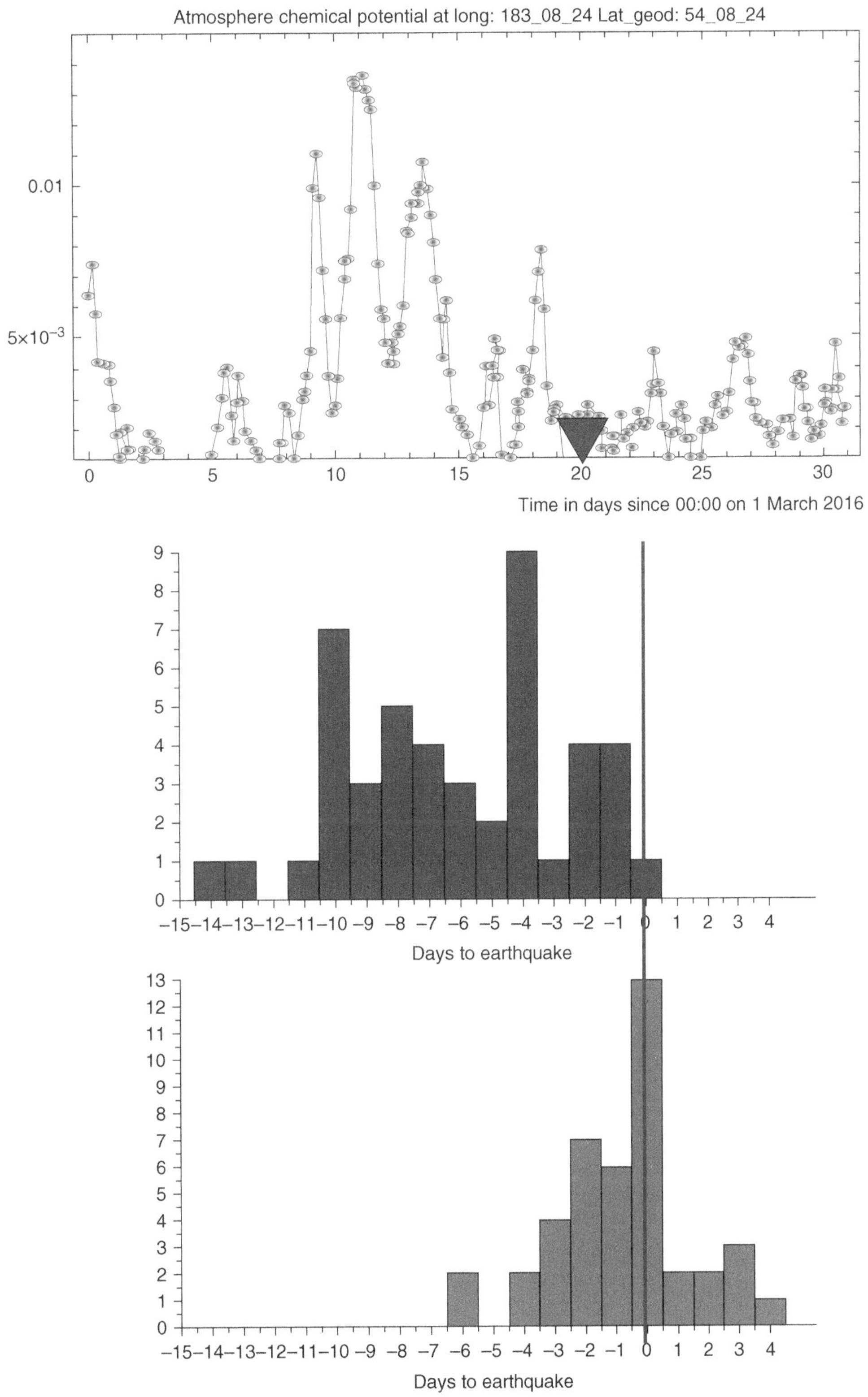

Figure 6.10 Upper panel: correction of chemical potential time series around the time of the *M* 6.7 earthquake near the east coast of Kamchatka on 20 March 2016. Lower panel: histogram of main maximum temporal position of the chemical potential in relation to the day of the main shock for 11 years of observation in Kamchatka area for earthquakes with magnitude *M*>6 (red), and the same for the main minimum chemical potential (blue). (*See electronic version for color representation.*)

Table 6.3 Model comparison

Model	Main source	Thermal precursors	Ionospheric precursors	Experimental data verification	Physical correctness	Existing numeric simulations	Land	Ocean
Freund, 2013	Stress-activated p-holes, surface positive charge	+	+	Need very high surface electric field, not recorded, separated clouds of positive charge, not recorded	Absolutely wrong mechanism of coupling with ionosphere	No numeric estimation of thermal flux, no numeric model of coupling with ionosphere	+	–
Kuo et al., 2011; 2014	External currents, possibly explained by the Freund model	–	+	Correct presentation of ionospheric anomalies and formation of plasma bubbles	Use of external currents four to six orders of magnitude greater than natural ones, not recorded. In second version undetermined negative ions flux not recorded	Correct simulation within the selected flux-tube, starting from introduction of external currents	+	–
Sorokin et al., 2000; *Sorokin and Hayakawa*, 2013	External currents in atmosphere	–	+	One-dimensional calculation of conjugated effect, use of very old experimental data for comparison	Use of external currents four to six orders of magnitude higher than natural ones	One-dimensional numeric simulation of ionospheric effects	+	–
Namgaladze, 2013	External currents in atmosphere	–	+	Correct presentation of ionospheric anomalies including conjugated effect	Use of external currents four to six orders of magnitude higher than natural ones	Correct three-dimensional simulation, starting from introduction of external currents	+	–
Klimenko et al., 2011	External electric fields in ionosphere	–	+	Correct presentation of ionospheric anomalies including conjugated effect	Use of external electric fields matching to obtain correct ionospheric effect	Correct three-dimensional simulation, starting from introduction of external electric field fields	+	+
Hegai et al., 2015	Vertical electric field on the ground surface	–	+	Correct presentation of isolated ionospheric anomaly	Correct one-dimensional calculation using the preset value of vertical electric field on the ground surface	Correct one-dimensional calculation using the preset value of vertical electric field	+	–

Qiang et al., 1991; *Tramutoli et al.*, 2009	Fluxes of greenhouse gases from underground	Only thermal infrared at ground surface, no latent heat, no outgoing longwave radiation	–	Experimental measurements of increased fluxes of greenhouse gases exist	Some doubts exist on the possibility of air temperature increase of large areas, no effect on decrease in humidity	No simulations	+	+
Enomoto, 2012	Fluxes of deep gases negatively electrified	–	+	Electromagnetic effects possibly connected with the proposed fluxes of electrified gases exist	Reversed magnetotelluric induction mechanism is used	Some estimates exist by *Vanhamäki et al.* (2005)	–	+
LAIMC	Air ionization by increased fluxes of radon from underground	+ All four types of thermal anomalies	+	All effects verified by the existing experimental observations of all types of precursors	Physical mechanisms based on fundamental principles of atmospheric and ionospheric physics	Numeric simulations separately for atmospheric and ionospheric effects exist	+	+

ACKNOWLEDGMENTS

The authors thank to the International Space Science Institute (Bern, Beijing) for their support of the international project Validation of Lithosphere–Atmosphere–Ionosphere–Magnetosphere Coupling (LAIMC), which promotes the concept of studying interaction between geospheres by utilizing spaceborne multi-instrument observations.

REFERENCES

Ben-Zion, Y. and Lyakhovsky, V. (2002), Accelerated seismic release and related aspects of seismicity patterns on earthquake faults, *Pure Appl. Geophys.*, *159*, 2385–2412, 2002.

Blaunstein, N. and Plohotniuc, E. (2008), *Ionosphere and Applied Aspects of Radio Communication and Radar*, CRC Press, 600 pp.

Bowman, D. D., Ouillon, G., Sammis, C. G., Sornette, A., and Sornette, D. (1998), An observation test of the critical earthquake concept, *J. Geophys. Res.*, *103*, 24,359–24,372.

Boyarchuk, K. A., Karelin, A. V., and Shirokov, R. V. (2006), *The Basic Model of the Ionized Atmosphere Kinetics*, VNIIEM Publications, Moscow, 320 pp.

Boyarchuk, K. A., Karelin, A. V., and Nadolski, A. V. (2010), Statistical analysis of the chemical potential correction value of the water vapor in atmosphere on the distance from earthquake epicenter, *Issues Electromech.*, *116*, 39–46.

Chernogor, L. F. (2012), *Physics and Ecology of Catastrophes*, Kharkov National University Publications, Kharkov, 556 pp.

De Santis, A., Cianchini, G., Favali, P., Beranzoli, L., and Boschi, E. (2011), The Gutenberg–Richter Law and entropy of earthquakes: Two case studies in Central Italy, *Bull. Seismol. Soc. Am.*, *101*, 1386–1395.

Denisenko, V. V. (2015), Estimate for the strength of the electric field penetrating from the Earth's surface to the ionosphere. *Russ. J. Phys. Chem. B*, *9*, 789–795.

Dobrovolsky, I. P. (1991), *The Theory of Tectonic Earthquake Preparation*, IPE RAS Publications, Moscow, 224 pp.

Dobrovolsky, I. P., Zubkov, S. I., and Myachkin, V. I. (1979), Estimation of the size of earthquake preparation zones, *Pure Appl. Geophys.*, *117*, 1025–1044.

Eddington, A. (1928), *The Nature of the Physical World*, Cambridge University Press, Cambridge, 361 pp.

Enescu, B. and Itu, K. (2003), Values of *b* and *p*: their variations and relation to physical processes for earthquakes in Japan, *Ann. Disast. Prev. Res. Inst., Kyoto Univ.*, *46B*.

Enomoto, Y. (2012), Coupled interaction of earthquake nucleation with deep Earth gases: a possible mechanism for seismo-electromagnetic phenomena, *Geophys. J. Int.*, *191*, 1210–1214.

Eresmaa, N., Härkönen, J., Joffre, S. M., Schultz, D. M., Karppinen, A., and Kukkonen, J. (2012), A three-step method for estimating the mixing height using ceilometer data from the Helsinki testbed. *J. Appl. Meteorol. Climatol.*. *51*, 2172–2187.

Freund, F. (2013), Earthquake forewarning—a multidisciplinary challenge from the ground up to space, *Acta Geophys.*, *61*(4), 775–807.

Fu, C.-C., Wang, P.-K., Lee, L.-C., Lin, C.-H., Chang, W.-Y., Giuliani, G., and Ouzounov, D. (2015), Temporal variation of gamma rays as a possible precursor of earthquake in the Longitudinal Valley of eastern Taiwan, *J. Asian Earth Sci.*, *114*(2), 362–372.

Genzano, N., Aliano, C., Filizzola, C., Pergola, N., and Tramutoli, V. (2007), A robust satellite technique for monitoring seismically active areas: The case of Bhuj–Gujarat earthquake, *Tectonophysics*, *431*, 197–210.

Gringel, W. (1986), Electrical structure from 0 to 30 kilometers, in *The Earth's Electrical Environment*, pp. 166–182, National Academy Press, Washington, DC.

Hayakawa, M. and Molchanov, O. (2007), Seismo-electromagnetics as a new field of radiophysics: electromagnetic phenomena associated with earthquakes, *Radio Sci. Bull.*, *320*, 8–17.

Hayakawa, M., Asano, T., Rozhnoi, A., and Solovieva, M. (2018), Very-low- to low-frequency sounding of ionospheric perturbations and possible association with earthquakes, in D. Ouzounov, S. Pulinets, K. Hattori, and P. Taylor (eds), *Pre-Earthquake Processes: A Multi-disciplinary Approach to Earthquake Prediction Studies*, pp. Geophysical Monograph 234, American Geophysical Union, Washington, DC, and John Wiley & Sons, Inc., Hoboken, NJ, 277–304.

Hegai, V. V., Kim, V. P., and Liu, J. Y. (2015), On a possible seismomagnetic effect in the topside ionosphere, *Adv. Space Res.*, *56*, 1707–1713.

Hirsikko, A. (2011), *On formation, growth and concentrations of air ions*, Academic dissertation, Report series of aerosol science, No. 125(2011), Helsinki Universtiy, 64 pp.

Igarashi, G., Saeki, S., Takahata, N., Sumikawa, K., Tasaka, S., Sasaki, Y., Takahashi, M., and Sano, Y. (1995), Groundwater radon anomaly before the Kobe earthquake in Japan, *Science*, *269*, 60–61.

Kaimal, J. C. and Finnigan, J. J. (1994), *Atmospheric Boundary Layer Flows: Their Structure and Measurement*, Oxford University Press, New York, NY.

Kanamori, H. and Anderson, D. L. (1975), Theoretical basis of some empirical relations in seismology, *Bull. Seismol. Soc. Am.*, *65*, 1073–1095.

Karelin, A. V. (2007), *The Physical Background of the Laser-Reactor*, VNIIEM Publications, Moscow, 157 pp.

Kathmann, S.M., Schenter, G. K., and Garrett, B. C. (2005), Ion-induced nucleation: The importance of chemistry, *Phys. Rev. Lett.*, *94*, doi: 10.1103/PhysRevLett.94.116104

Khilyuk, L. F., Chillingar, G. V., Robertson, J. O. Jr., and Endres, B. (2000), *Gas Migration. Events Preceding Earthquakes*, Gulf Publishing Company, Houston, TX, 390 pp.

Kim, V. P. and Hegai, V.V. (1997), On possible changes in the midlatitude upper ionosphere before strong earthquakes, *J. Earthquake Pred. Res.*, *6*, 275–280.

Klimenko, M. V., Klimenko, V. V. Zakharenkova, I. E. Pulinets, S. A. Zhao, B., and Tzidilina, M. N. (2011), Formation mechanism of great positive disturbances prior to Wenchuan earthquake on May 12, 2008, *Adv. Space Res.*, *48*, 488–499.

Kobeissi, M. A., Gomez, F., and Tabet, C. (2015), Measurement of anomalous radon gas emanation across the Yammouneh Fault in southern Lebanon: A possible approach to earthquake prediction, *Int. J. Disaster Risk Sci.*, *6*, 250–266.

Kuo, C. L., Huba, J. D., Joyce, G., and Lee, L. C. (2011), Ionosphere plasma bubbles and density variations induced by preearthquake rock currents and associated surface charges, *J. Geophys. Res*, *116*, article ID A10317.

Kuo, C. L., Lee, L. C., and Huba, J. D. (2014), An improved coupling model for the lithosphere–atmosphere–ionosphere system, *J. Geophys. Res. Space Phys.*, *119*, 3189–3205, doi:10.1002/2013JA019392

Kuo, C. L., Ho, Y. Y., and Lee, L. C. (2018), Electrical Coupling Between the Ionosphere and Surface Charges in the Earthquake Fault Zone, in D. Ouzounov, S. Pulinets, K. Hattori, and P. Taylor (eds), *Pre-Earthquake Processes: A Multi-disciplinary Approach to Earthquake Prediction Studies*, pp. Geophysical Monograph 234, American Geophysical Union, Washington, DC, and John Wiley and Sons, Inc., Hoboken, NJ.

Laaks, L., Mäkelä, J. M., Pirjola, L., and Kulmala, M. (2002), Model studies on ion-induced nucleation in the atmosphere, *J. Geophys. Res.*, *107*(D20), 4427, doi:10.1029/2002JD002140

Lee, S.-H., Reeves, J. M., Wilson, J. C., Hunton, D. E., Viggiano, A. A., Miller, T. M., Ballenthin, J. O., and Lait, L. R. (2003), Particle formation by ion nucleation in the upper troposphere and lower stratosphere, *Science*, *301*(5641), pp. 1886–1889.

Liu, J.-Y. and Chao, C.-K. (2016), An observing system simulation experiment for 1 FORMOSAT-5/AIP detecting seismo-ionospheric precursors, *Terr. Atmos. Ocean. Sci.*, *28*(2), doi: 10.3319/TAO.2016.07.18.01(EOF5)

Liu, J. -Y., Chen, Y. I., Chen, C. H., Liu, C. Y., Chen, C. Y., Nishihashi, M., Li, J. Z., Xia, Y. Q., Oyama, K. I., Hattori, K., and Lin, C. H. (2009), Seismoionospheric GPS total electron content anomalies observed before the 12 May 2008 *Mw* 7.9 Wenchuan earthquake, *J. Geophys. Res. Space Phys.*, *114*, A04320, doi:10.1029/2008JA013698

Liu, J.-Y, Hattori, K., and Chen, Y. (2018), Application of total electron content derived from the Global Navigation Satellite System for detecting earthquake precursors, in D. Ouzounov, S. Pulinets, K. Hattori, and P. Taylor (eds), *Pre-Earthquake Processes: A Multidisciplinary Approach to Earthquake Prediction Studies*, pp., Geophysical Monograph 234, American Geophysical Union, Washington, DC, and John Wiley and Sons, Inc., Hoboken, NJ.

Markson, R. (2007), The global circuit intensity: its measurement and variation over the last 50 years, *Bull. Am. Meteorol. Soc.*, doi:10.1175/BAMS-88-2-223, 223–241

Morozov, V. N. (2006), The influence of convective current generator on the global current, *Nonlin. Processes Geophys.*, *13*, 243–246.

Namgaladze, A. A. (2013), Earthquakes and global electrical circuit, *Russ. J. Phys. Chem. B*, *7*(5), 589–593.

Ouzounov, D., Pulinets, S., Kafatos, M., and Taylor, P. (2018a), Thermal radiation anomalies associated with major earthquakes, in D. Ouzounov, S. Pulinets, K. Hattori, and P. Taylor (eds), *Pre-Earthquake Processes: A Multidisciplinary Approach to Earthquake Prediction Studies*, pp., Geophysical Monograph 234, American Geophysical Union, Washington, DC, and John Wiley and Sons, Inc., Hoboken, NJ.

Ouzounov, D., Pulinets, S., Hattori, K., Liu, J.-Y., and Han, P. (2018b), Multiparameter assessment of pre-earthquake atmospheric signals, in D. Ouzounov, S. Pulinets, K. Hattori, and P. Taylor (eds), *Pre-Earthquake Processes: A Multidisciplinary Approach to Earthquake Prediction Studies*, pp., Geophysical Monograph 234, American Geophysical Union, Washington, DC, and John Wiley and Sons, Inc., Hoboken, NJ.

Papadopoulos, G. A., Charalampakis, M., Fokaefs, A., and Minadakis, G. (2010), Strong foreshock signal preceding the L'Aquila (Italy) earthquake (*Mw* 6.3) of 6 April 2009, *Nat. Hazards Earth Syst. Sci.*, *10*, 19–24.

Park, C. G. and Dejnakarintra, M. (1973), Penetration of thundercloud electric fields into the ionosphere and magnetosphere, 1. Middle and auroral latitudes. *J. Geophys. Res.*, *84*, 960–964.

Park, C. G. and Dejnakarintra, M. (1977), Thundercloud electric fields in the ionosphere, in Electrical Processes in Atmospheres, Dolezhalek, H. and Reiter, R., eds., Steinkopff, Darmstadt, pp. 544–551.

Parrot, M. and Li, M. (2018) Statistical analysis of the ionospheric density recorded by DEMETER during seismic activity, in D. Ouzounov, S. Pulinets, K. Hattori, and P. Taylor (eds), *Pre-Earthquake Processes: A Multidisciplinary Approach to Earthquake Prediction Studies*, pp., Geophysical Monograph 234, American Geophysical Union, Washington, DC, and John Wiley and Sons, Inc., Hoboken, NJ.

Pergola, N., Aliano, C., Coviello, I., Filizzola, C., Genzano, N., Lacava, T., Lisi, M., Mazzeo, G., and Tramutoli, V. (2010), Using RST approach and EOS-MODIS radiances for monitoring seismically active regions: a study on the 6 April 2009 Abruzzo earthquake, *Nat. Hazards Earth Syst. Sci.*, *10*, 239–249, doi:10.5194/nhess-10-239-2010

Pulinets, S. (2012) Low-latitude atmosphere-ionosphere effects initiated by strong earthquakes preparation process, *Int. J. Geophysics*, *2012*, article ID 131842, doi:10.1155/2012/131842

Pulinets, S. A. and Boyarchuk, K. A. (2004), *Ionospheric Precursors of Earthquakes*, Springer, Berlin, 315 pp.

Pulinets, S. A. and Legen'ka, A. D. (1997), First simultaneous observations of the topside density variations and VLF emissions before the Irpinia earthquake, November, 23, 1980 in magnetically conjugated regions, in *Proceedings of International Workshop on Seismo Electromagnetics*, pp. 56–59, University of Electro-Communications Publications, Chofu, Japan.

Pulinets, S. and Ouzounov, D. (2011), Lithosphere–atmosphere–Ionosphere Coupling (LAIC) model - an unified concept for earthquake precursors validation, *J. Asian Earth Sci.*, *41*, 371–382.

Pulinets, S. A., Khegai, V. V., Boyarchuk, K. A., and Lomonosov, A. M. (1998), Atmospheric electric field as a source of ionospheric variability, *Phys. Uspekhi*, *41*, 515–522.

Pulinets, S. A., Boyarchuk, K. A., Hegai, V. V., and Karelin, A. V. (2002), Conception and model of seismo-ionosphere–magnetosphere coupling, in M. Hayakawa and O.A. Molchanov (Eds), *Seismo-Electromagnetics: Lithosphere–Atmosphere–Ionosphere Coupling*, pp. 353–361, Terra Scientific Publishing, Tokyo.

Pulinets S.A., Bondur, V. G., Tsidilina, M. N., and Gaponova, M. V. (2010), Verification of the concept of seismoionospheric relations under quiet heliogeomagnetic conditions, using the Wenchuan (China) earthquake of May 12, 2008, as an example, *Geomag. Aeron.*, *50*, 231–242.

Pulinets, S. A., Tramutoli, V., Genzano, N., and Yudin, I. A. (2013), TIR anomalies scaling using the earthquake preparation zone concept, *2013 AGU Meeting of the Americas*, Paper NH42A-06, Cancun, Mexico, 14–17 May.

Pulinets, S. A., Ouzounov, D. P., Karelin, A. V., and Davidenko, D. V. (2015), Physical bases of the generation of short-term earthquake precursors: A complex model of ionization-induced geophysical processes in the lithosphere–atmosphere–ionosphere–magnetosphere system, *Geomagn. Aeron.*, *55*(4), 540–558.

Pulinets, S., Ouzounov, D., Davidenko, D., and Petrukhin, A. (2016), Multiparameter monitoring of short-term earthquake precursors and its physical basis. Implementation in the Kamchatka region, *E3S Web of Conferences 11*, 00019, doi: 10.1051/e3sconf/20161100019

Qiang, Z. J., Xu, X. D., and Dian, C. G. (1991), Thermal infrared anomaly precursor of impending earthquakes, *Chin. Sci. Bull.*, *36*, 319–323.

Redin, A., Kupovykh, G., Kudrinskaya, T., and Boldyreff, A. (2014), Surface layer electrodynamic structure under severe aerosol pollution, *XV International Conference on Atmospheric Electricity*, P-07-11, 15–20 June, Norman, OK, 6 pp.

Ruzhin, Yu. Ya. and Larkina, V. I. (1996), Magnetic conjugation and time coherency of seismo-ionosphere vlf bursts and energetic particles, *Proceedings of 13th International Wroclaw Symposium on Electromagnetic Compatibility*, pp. 645–648, Wroclaw.

Ryu, K., Parrot, M., Kim, S. G., Jeong, K. S., Chae, J. S., Pulinets, S., and Oyama, K.-I. (2014), Suspected seismo-ionospheric coupling observed by satellite measurements and GPS TEC related to the *M* 7.9 Wenchuan earthquake of 12 May 2008, *J. Geophys. Res. Space Phys.*, *119*, 10,305–10,323.

Scholz, C. H., Sykes, L. R., and Aggarwal, Y. P. (1973), Earthquake prediction: A physical basis, *Science*, *181*, 803–809.

Schorlemmer, D., Wiemer, S., and Wyss, M. (2004), Earthquake statistics at Parkfield: 1. Stationarity of *b*-values, *J. Geophys. Res.*, *109*, B12307, doi:10.1029/2004JB003234

Sobolev, G. A. (1993), *The Basics of Earthquake Prediction*, Nauka Publications, Moscow, 314 pp.

Sornette, D. and Sammis, C. G. (1995), Complex critical exponents from renormalization group theory of earthquakes: implications for earthquake predictions, *J. Phys. I France*, *5*, 607–619.

Sorokin, V. and Hayakawa, M. (2013), Generation of seismic-related DC electric fields and lithosphere–atmosphere–ionosphere coupling, *Modern Appl. Sci.*, *7*(6), 1–25.

Sorokin, V. M, Chmyrev, V. M., and Hayakawa, M. (2000), The formation of ionosphere-magnetosphere ducts over the seismic zone, *Planet. Space Sci.*, *48*, 175–180.

Svensmark, H., Pedersen, J. O. P., Marsch, N. D., Enghoff, M. B., and Uggerhøj, U. I. (2007), Experimental evidence for the role of ions in particle nucleation under atmospheric conditions. *Proc. Royal Soc. A*, *463*, 385–396.

Toutain, J.-P. and Baubron, J.-C. (1998), Gas geochemistry and seismotectonics: a review. *Tectonophysics*, *304*, 1–27.

Tramutoli, V., Aliano, C., Corrado, R., Filizzola, C., Genzano, N., Lisi, M., Lanorte, V., Tsamalashvili, T., and Pergola, N. (2009), Abrupt change in greenhouse gases emission rate as a possible genetic model of TIR anomalies observed from satellite in earthquake active regions, in *Proceedings of the 33rd International Symposium on Remote Sensing of the Environment*, Vol. 2, pp. 567–570, Stresa, Italy.

Turcotte, D. L. (1997), *Fractals and Chaos in Geology and Geophysics*, Cambridge University Press, Cambridge, 414 pp.

Turcotte, D. L. and Malamud, B. D. (2002), Earthquakes as a complex system, in W. H. K. Lee, H. Kanamori, P. Jennings, and C. Kisslinger (eds), *International Handbook of Earthquake and Engineering Seismology Part A*, pp. 209–227 Academic Press, New York

Vershinin, E. F., Buzevich, A. V., Yumoto, K., Saita, K., and Tanaka, Y. (1999), Correlations of seismic activity with electromagnetic emissions and variations in Kamchatka region, in M. Hayakawa (ed.), *Atmospheric and Ionospheric Electromagnetic Phenomena Associated with Earthquakes*, pp. 513–517, Terra Scientific Publishing Company, Tokyo.

Vinuesa, J.-F., Basu, S., and Galmarini, S. (2007), The diurnal evolution of ^{222}Rn and its progeny in the atmospheric boundary layer during the Wangara experiment. *Atmos. Chem. Phys.*, *7*, 5003–5019.

Wiemer, S. and Schorlemmer, D. (2007), ALM: An Asperity-based Likelihood Model for California, *Seismol. Res. Lett.*, *8*(1),134–140.

Yasuoka, Y., Igarashi, G., Ishikawa, T., Tokonami, S., and Shinogi, M. (2006), Evidence of precursor phenomena in the Kobe earthquake obtained from atmospheric radon concentration, *Appl. Geochem.*, *21*, 1064–1072.

Zavyalov, A. D. (2006), *The Theory of Middle-Term Earthquake Forecast. Basics, Methodology, Realization*, Nauka Publications, Moscow, 256 pp.

7

Electrical Coupling Between the Ionosphere and Surface Charges in the Earthquake Fault Zone

Cheng-Ling Kuo[1], Yi-Ying Ho[2], and Lou-Chuang Lee[2]

ABSTRACT

Several lithosphere–atmosphere–ionosphere coupling (LAIC) models have been proposed to explain observed precursors before strong earthquakes, such as atmospheric conductivity anomalies and ionospheric total electron content (TEC) variations. These coupling models include (a) radon ionization leading to charged aerosols and change of load resistance in the global atmospheric electric circuit, (b) an electric coupling model of stressed rock–atmosphere–ionosphere based on experimental evidence of stressed rock currents, and (c) ionosphere dynamics with an imposed zonal electric field. Here, we summarize our modeling of stressed-rock–atmosphere–ionosphere coupling for the LAIC process. A physical mechanism of electrical coupling between the atmosphere and surface charge in the fault region is proposed. We also discuss other possible seismic anomalies before and after the earthquake and suggest that in future electric field measurements are used to validate the possible existence of electric activities associated with an earthquake preparation region.

7.1. PAST RESEARCH ON LITHOSPHERE-ATMOSPHERE-IONOSPHERE COUPLING MODELS

Prior to the 22 May 1960 M >9 Chile earthquake, the critical frequency of the F2 layer (*fo*F2) on 20 May was slightly below its median value for a few hours in a quiescent period of geomagnetic activity [*Foppiano et al.*, 2008]. Before the 1964 Alaska M 9.2 earthquake, magnetic disturbance and ionospheric irregularities were observed in the magnetometer and ionosonde data [*Moore*, 1964; *Davies and Baker*, 1965]. In addition, related ground-based observations of earthquake precursors have been reported, including ionospheric sounding [*Liu et al.*, 2001; *Pulinets and Boyarchuk*, 2004; *Liu et al.*, 2009], DC electric field emissions at ultralow to very frequencies (ULF–VHF) [*Varotsos*, 2006; *Schekotov et al.*, 2007; *Smirnova and Hayakawa*, 2007; *Surkov and Hayakawa*, 2007; *Uyeda et al.*, 2009], amplitude and phase anomalies of subionospheric VLF/LF signals from powerful transmitters [*Hayakawa*, 2007; *Hayakawa et al.*, 2010, 2012], thermal anomaly [*Ouzounov and Freund*, 2004; *Ouzounov et al.*, 2006; *Pulinets et al.*, 2006] and total electron content (TEC) variations before strong earthquakes [*Liu et al.*, 2000, 2001, 2004, 2009, 2010; *Chuo*, 2002; *Pulinets and Legen'ka*, 2003; *Klimenko et al.*, 2011; *Zolotov et al.*, 2011; *Namgaladze et al.*, 2012]. Observations from both ground-based instruments and satellite data have shown the presence of related electromagnetic perturbations in the atmosphere and ionosphere over the earthquake fault region several days before the main shock.

Scientific payloads onboard satellites have also recorded wave and plasma disturbances before earthquake events [*Chmyrev et al.*, 1989; *Parrot et al.*, 1993; *Parrot*, 1995; *Akhoondzadeh et al.*, 2010; *Li and Parrot*, 2013]. The observed electromagnetic perturbations can be traced back along the magnetic field line to the region

[1]Institute of Space Science, National Central University, Taoyuan, Taiwan
[2]Institute of Earth Sciences, Academia Sinica, Taipei, Taiwan

Pre-Earthquake Processes: A Multidisciplinary Approach to Earthquake Prediction Studies, Geophysical Monograph 234, First Edition. Edited by Dimitar Ouzounov, Sergey Pulinets, Katsumi Hattori, and Patrick Taylor.

of earthquake activities [*Pulinets and Legen'ka*, 2003; *Pulinets et al.*, 2003; *Pulinets and Boyarchuk*, 2004; *Liu et al.*, 2009]. These precursor phenomena include the presence of electron density fluctuations in the ionosphere above the seismic regions [*Liu et al.*, 2009; *Kakinami et al.*, 2010], the change of ionic composition and temperature of plasma in the ionosphere [*He et al.*, 2010; *Oyama et al.*, 2011], and the oscillation of height profile of the ionospheric F region [*Afraimovich et al.*, 1999; *Astafyeva and Heki*, 2009; *Cahyadi and Heki*, 2013].

Several LAIC models have been proposed to explain these precursors, such as atmospheric conductivity anomalies and ionospheric TEC variations before strong earthquakes. Radon ionization may lead to charged aerosols and a change of load resistance, thus causing TEC changes in the global atmospheric electric circuit [*Pulinets et al.*, 2000, 2003, 2006; *Pulinets and Legen'ka*, 2003; *Pulinets and Boyarchuk*, 2004; *Pulinets*, 2009; *Pulinets and Ouzounov*, 2011]. *Kuo et al.* [2011, 2014, 2015] proposed electric coupling of the stressed-rock–atmosphere–ionosphere, as shown in Figure 7.1. The coupling model is adapted from the global atmospheric electric circuit, with the current system in the concentric capacitor between lithosphere and ionosphere [*Wilson*, 1921].

The current source in the lithosphere can be caused by the stressed rock [*Freund*, 2000, 2002, 2010, 2013; Freund and Sornette, 2007; *Freund et al.*, 2009], as shown in Figure 7.2. In our electric LAIC model, we assume the presence of a "dynamo (battery)" that provides the current source flowing out of the battery through lithosphere to the atmosphere, and the ionosphere. However, some [e.g., *Denisenko*, 2015] have suggested that the dynamo currents cannot flow out of the lithosphere to the atmosphere based on a two-charge-layer model. We take the opportunity in this chapter to point out the differences between these two models, which are illustrated in Figure 7.3.

In Figure 7.3 there are two charge layers embedded in the lithosphere in model (a), while in model (b) there exists a dynamo (battery) current source. For convenience we consider only one-dimensional models in Figure 7.3. In model (a), the positive charge layer is located at $\mathbf{z}=-\mathbf{z}_1$ and the negative at $\mathbf{z}=-\mathbf{z}_2$. Electric fields are directed away from the positive charge layer (red color) and toward the negative charge layer (blue color). As a result, the atmosphere electric field is $E=E_+ + E_-=0$ for $z>0$ where E_+ and E_- are electric fields from the positive and negative charge layers, respectively. The two charge layers cannot produce electric fields and hence currents in the atmosphere. It is noted that a charge in the lithosphere will flow away in a conductive lithosphere with a decay time $t_{decaly}=\varepsilon_0/\sigma_{litho}=10^{-10}$ s, where the vacuum permittivity $\varepsilon_0=8.85\times10^{-12}$ F m^{-1} and $\sigma_{litho}=0.01$ S m^{-1}.

In model (b) there is a dynamo (battery) that provides a current source. The dynamo current (J_D) inside the

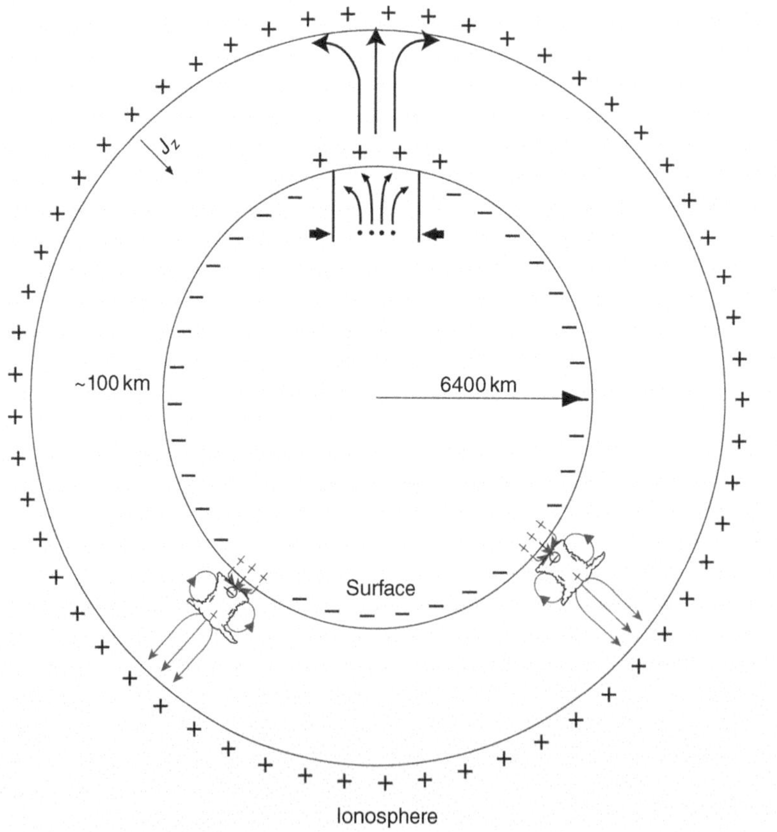

Figure 7.1 Diagram to demonstrate electric coupling between lithosphere–atmosphere–ionosphere.

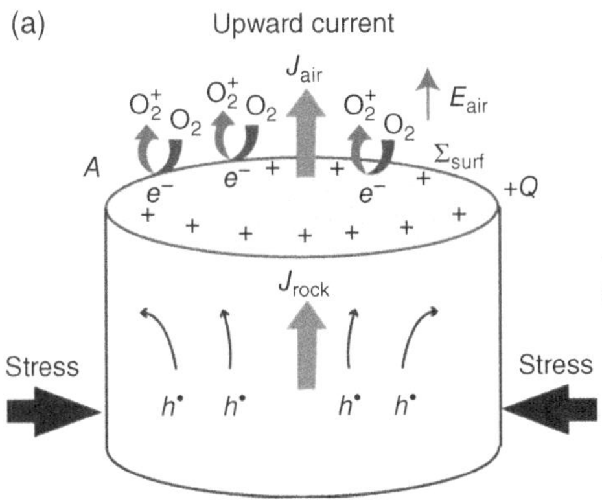

Figure 7.2 As rocks are subjected to forces of stress in an earthquake preparation region, deformation of the lattice structure in rocks can produce electronic charge carriers (positive holes, h•) and electric currents (J_{rock}). The positive holes diffuse from the highly stressed region into the unstressed region, repel each other electrostatically, and push toward the rock surface. The field-ionization ($O_2 \rightarrow O_2^+ + e^-$) occurring at the surface, in particular at sharp points, can produce the O_2^+ ions. The surface charge density Σ_{surf} and total charge amount Q are accumulated by positive charges from the rock surface and O_2^+ ions in the air. The upward electric field $\mathbf{E}_{air}$ associated with the positive surface charges will drive the upward current J_{air} [*Kuo et al.*, 2014]. (Source: Reproduced with permission of John Wiley and Sons.)

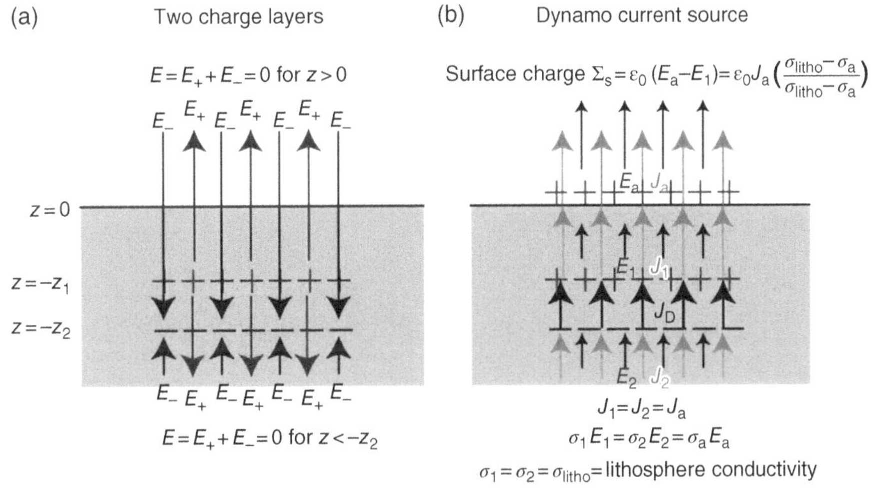

Figure 7.3 (a) Model with two charge layers imbedded in the lithosphere. (b) Model with dynamo (battery) current source.

battery leads to charge separation, with the appearance of two charge layers. However, to maintain the charge layers, continuous charging from the battery is required. In a conductive lithosphere, the currents will flow out of the dynamo through the lithosphere to the atmosphere. However, the air conductivity in the lithosphere near the ground at $z = 0$ km is low (2×10^{-14} S m^{-1}) and the lithosphere conductivity is high (1×10^{-2} S m^{-1}).

Additional conductivity due to extra surface charge density should be considered [e.g., *Kuo et al.*, 2011]. Surface charge density can be derived from $\Sigma_s = \varepsilon_0 \left(E_a - E_1\right) = \varepsilon_0 J_a \left(\frac{\sigma_{litho} - \sigma_a}{\sigma_{litho}\sigma_a}\right)$, where σ_a is much smaller than σ_{litho}. The surface charge density in equation (7.12) is approximated as $\Sigma_s \approx \varepsilon_0 J_a / \sigma_a = 4.4 \times 10^{-7}$ Cm^{-2} for $J_a = 1$ nA m^{-2} and $\sigma_a = 2 \times 10^{-14}$ S m^{-1}. For considering an additional conductivity in previous work [*Kuo et al.*, 2011, equation 9], the surface charge density provides an additional conductivity $\sigma_1 = en_+\mu_+ = \frac{\Sigma_s}{h}\mu_+$ where the surface charge density is corrected as $\Sigma_s = \sqrt{\left(\varepsilon_0 h J_a\right)/\mu_+}$ [*Kuo et al.*, 2011, equation 10], n_+ and μ_+ are density and mobility of positive ions (2.5×10^{-4} V m^2), and h is the estimated characteristic height of surface charge layer. For this case, the characteristic height of the surface charge layer for the additional conductivity $\sigma_1 = 2 \times 10^{-13}$ S m^{-1}

is estimated to be 55 m. It is well known that in the atmosphere with a space-varying conductivity profile, induced space charges are present to maintain the current continuity [*Pasko*, 2013].

The imposed zonal electric field associated with seismic-related events may lead to TEC variations [*Zolotov et al.*, 2011, 2012; *Namgaladze et al.*, 2012; Kuo et al., 2014]. The processes between LAIC have drawn more and more attention from the scientific community [*Molchanov et al.*, 2004; *Pulinets and Boyarchuk*, 2004; *Klimenko et al.*, 2011; *Sorokin and Hayakawa*, 2013; *Ouzounov et al.*, 2014; *Sorokin et al.*, 2015], but more multi-instrument experiments and spaceborne observations are required to validate the possible LAIC process [*Ouzounov et al.*, 2014].

In the following sections we calculate the electrical coupling between the atmosphere current and surface charge in the fault region in the ionosphere and the ionospheric dynamics in the presence of atmospheric current is calculated, as well as simulate TEC variations at magnetic latitudes of 7.5°, 15°, 22.5°, and 30°. In section 7.4 we compare preseismic TEC changes for 2011 Tohoku-Oki earthquake with observations of the Japanese GEONET. In section 7.5 we compare our model results with DEMETER observations of the relationship between the preseismic plasma density and ion velocity variations for a Chile earthquake. Finally, we discuss other possible seismic anomalies before and after the earthquake.

7.2. IONOSPHERE DYNAMICS IN THE PRESENCE OF AN ATMOSPHERIC ELECTRIC CURRENT

The atmospheric electric current source shown in Figure 7.4 may cause ionospheric dynamics [*Kuo et al.*, 2011, 2014; *Kuo and Lee*, 2015, 2017]. Here, we focus on the effect of an upward-driven current on the ionosphere. The source of the current in the atmosphere may be stressed rock or charged aerosols injected into the atmosphere by soil gases associated with seismic events, which can be examined using an electric coupling model of the ionosphere and atmosphere.

7.2.1. Upward Current into the Ionosphere

The upward current **J** has a parallel component of $\mathbf{J}_{\parallel}$ along the magnetic field line in the z' axis direction and a perpendicular component of $\mathbf{J}_{\perp}$ in the y' axis direction. To simplify equation (7.1), we can ignore the parallel electric field because the parallel conductivity $\sigma_{\parallel}$ is much greater than the Pedersen conductivity σ_P and the Hall conductivity σ_H, thus the local perpendicular current $\mathbf{J}_{\perp}$ is expressed as $\mathbf{J}_{\perp} = (J_{x'}, J_{y'})$ and the perpendicular electric field $\mathbf{E}_{\perp} = (E_{x'}, E_{y'})$. For the perpendicular current,

$$\mathbf{J}_{\perp} = \begin{pmatrix} J_{x'} \\ J_{y'} \end{pmatrix} = \begin{pmatrix} \sigma_P & \sigma_H \\ -\sigma_H & \sigma_P \end{pmatrix} \begin{pmatrix} E_{x'} \\ E_{y'} \end{pmatrix} \tag{7.1}$$

In our case we have $J_{x'} = 0$, $J_{y'} = J_{\perp}$, and $\mathbf{J}'_{\perp} = (0, J_{\perp})$. The electric field can be obtained from equation (7.1) as

$$E_{x'} = \frac{-\sigma_H}{\sigma_P^2 + \sigma_H^2} J_{\perp} \tag{7.2a}$$

$$E_{y'} = \frac{-\sigma_P}{\sigma_P^2 + \sigma_H^2} J_{\perp} \tag{7.2b}$$

For an upward (downward) current from stressed rock in the lithosphere, as shown in Figure 7.4, the component of electric field $E_{x'}$ in equation (7.2a) is in an eastward

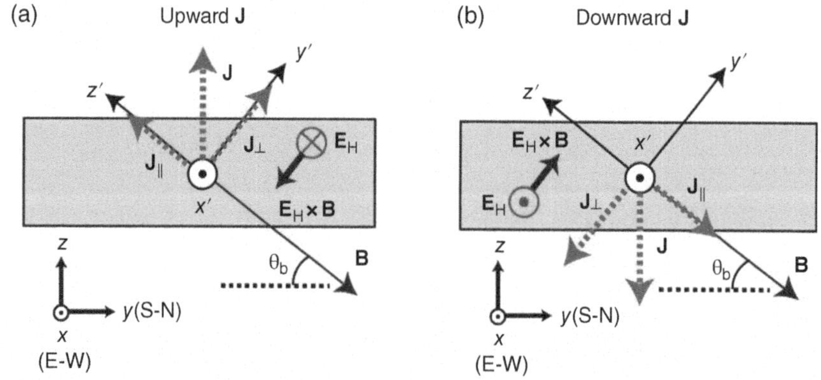

Figure 7.4 The effect of Pedersen and Hall conductivity. The rotated coordinates (x', y', z') are shown by the black line with arrows. The positive direction of the x' axis and y' axis are, respectively, in the eastward and northward direction in the local coordinates of Earth. For an upward current density **J**, the parallel and perpendicular components of current density **J** are along the z' axis and y' axis, respectively. The perpendicular (a) upward and (b) downward current $\mathbf{J}_{\perp}$ leads to the presence of a Pedersen electric field ($\mathbf{E}_P$) along the $+y'$ ($-y'$) axis and the Hall electric field ($\mathbf{E}_H$) along the $+x'$ ($-x'$) axis [*Kuo et al.*, 2014].

(westward) direction, and the component of electric field $E_{y'}$ in equation (7.2b) has an upward (downward) direction. We term $E_{x'}$ as the Hall electric field, $\mathbf{E}_H = E_{x'}\hat{\mathbf{x}}'$, and $E_{y'}$ as the Pedersen electric field, $\mathbf{E}_P = E_{y'}\hat{\mathbf{y}}'$. Note that the average Hall conductivity σ_H is observed to be about ten times the Pedersen conductivity σ_P at nighttime and daytime (see the website http://wdc.kugi.kyoto-u.ac.jp/ionocond/sigcal/index.html) at the altitude of layer E (85–120 km) of the ionosphere. Therefore, the Hall electric field $\mathbf{E}_H$ is about ten times the Pedersen electric field $\mathbf{E}_P$. The Hall electric field $\mathbf{E}_H$ will lead to a downward $\mathbf{E}\times\mathbf{B}$ ionospheric plasma motion as the ionospheric dynamo source.

7.2.2. Atmospheric Current Effect on the Ionosphere Model

We use the current flowing from the atmosphere into the ionosphere as the driving force, like a neutral wind effect in the ionosphere. The imposed electric field in the ionosphere SAMI3 model is $\mathbf{E}_{ext}$ (e.g., $\mathbf{E}_P$ and $\mathbf{E}_H$), described below. First, we use the generalized coordinates (x_1, x_2, x_3) in the ionosphere to express the terms in the conservation equation of current density, $\nabla\cdot\mathbf{J}=0$ [*Richmond*, 1995],

$$\mathbf{E} = -\nabla\Phi = -\frac{1}{W}\sum_{i=2}^{3}\mathbf{a}_i\frac{\partial\Phi}{\partial x_i} \tag{7.3}$$

$$\nabla\cdot\mathbf{J} = \frac{1}{W}\sum_{i=2}^{3}\frac{\partial(\mathbf{a}_i\cdot\mathbf{J})}{\partial x_i} = 0 \tag{7.4}$$

where the volume element is

$$W = \left(\nabla x_2\times\nabla x_3\cdot\nabla x_1\right)^{-1}, \tag{7.5}$$

and the area vectors $\mathbf{a}_i$ are defined as

$$\mathbf{a}_i = W\nabla x_1. \tag{7.6}$$

The generalized coordinates (x_1, x_2, x_3) one-to-one correspond to the dipolar coordinates (s, p, φ), where s is parallel to the magnetic field, and p and φ are perpendicular to the magnetic field. The pointing vector is written as $\mathbf{r}=\mathbf{r}(s, p, \varphi)$, and the curvilinear scale factors are defined in the following expressions. The dipolar coordinate (s, p, φ) is orthogonal,

$$h_s = \left|\frac{\partial\mathbf{r}}{\partial s}\right| = \left(|\nabla s|\right)^{-1} = \frac{r^3}{R_E^3}\frac{1}{\Delta} = \frac{1}{b_s} \tag{7.7}$$

$$h_p = \left|\frac{\partial\mathbf{r}}{\partial p}\right| = \left(|\nabla p|\right)^{-1} = \frac{R_E\sin^3\theta}{\Delta} = \frac{1}{b_s}\frac{R_E^2}{p^2}\frac{1}{r\sin\theta} \tag{7.8}$$

$$h_\varphi = \left|\frac{\partial\mathbf{r}}{\partial\varphi}\right| = \left(|\nabla_\varphi|\right)^{-1} = r\sin\theta \tag{7.9}$$

where, $\Delta=(1+3\cos^2)^{1/2}$ and $b_s=\mathbf{B}/B_0$. The volume element in equation (7.5) is $W=h_s h_p h_\varphi$. One of the area vectors in equation (7.6) exemplifies that $\mathbf{a}_1=h_p h_\varphi \mathbf{e}_s$ where $\mathbf{e}_s = \frac{\partial\mathbf{r}}{\partial s}/\left|\frac{\partial\mathbf{r}}{\partial s}\right|$. Equation (7.3) without the x_1 component of the electric field is due to the assumption of equal potential along magnetic-field lines, i. e., $\frac{\partial\Phi}{\partial x_1}=0$. The current term $\mathbf{J}$ of equation (7.4) is replaced with current density in equation (7.10),

$$\mathbf{J}_\perp = \sigma_p\left(\mathbf{E}-\mathbf{E}_{ex}+\mathbf{V}_n\times\mathbf{B}\right)+\sigma_h\mathbf{b}\times\left(\mathbf{E}-\mathbf{E}_{ex}+\mathbf{V}_n\times\mathbf{B}\right) + \sigma_{pi}\frac{m_i}{e}\mathbf{g}+\sigma_{hi}\mathbf{b}\times\frac{m_i}{e}\mathbf{g} \tag{7.10}$$

and equation (7.4) is integrated with respect to s along the magnetic field line from the one hemisphere in the ionosphere to the opposite hemisphere. Under the assumption of equal potential along the magnetic field lines in the ionosphere model, the three-dimensional equation (7.4) can be reduced to a two-dimensional partial differential equation for electric potential, Φ,

$$\frac{\partial}{\partial p}C_{pp}\frac{\partial\Phi}{\partial p}-\frac{\partial}{\partial p}C_{p\varphi}\frac{\partial\Phi}{\partial p}-\frac{\partial}{\partial p}C_{\varphi p}\frac{\partial\Phi}{\partial p}+\frac{\partial}{\partial p}C_{\varphi\varphi}\frac{\partial\Phi}{\partial\phi} = \frac{\partial S_p}{\partial p}+\frac{\partial S_\varphi}{\partial\phi} \tag{7.11}$$

where

$$C_{pp} = \int\left(\frac{\mathbf{a}_p\cdot\mathbf{a}_p}{W}\right)\sigma_p ds = \int\frac{h_s h_\varphi}{h_p}\sigma_p ds = p\int\frac{\Delta}{b_s}\sigma_p ds \tag{7.12a}$$

$$C_{p\varphi} = C_{\varphi p} = \int\left(\frac{\mathbf{b}\cdot\mathbf{a}_p\times\mathbf{a}_\varphi}{W}\right)\sigma_p ds = \int h_s\sigma_h ds = \int\frac{\sigma_s}{b_s}ds \tag{7.12b}$$

$$C_{\varphi\varphi} = \int\left(\frac{\mathbf{a}_\varphi\cdot\mathbf{a}_\varphi}{W}\right)\sigma_p ds = \int\frac{h_s h_p}{h_\varphi}\sigma_p ds = \frac{1}{p}\int\frac{1}{b_s\Delta}\sigma_p ds \tag{7.12c}$$

The Pedersen and Hall conductivities for ions and electrons are

$$\sigma_p = \frac{n_i e}{B}\frac{\nu_{in}/\Omega_i}{1+\nu_{in}^2/\Omega_i^2}+\frac{n_e e}{B}\frac{\nu_{en}/\Omega_e}{1+\nu_{en}^2/\Omega_e^2} \tag{7.13a}$$

$$\sigma_h = \frac{n_i e}{B}\frac{1}{1+\nu_{in}^2/\Omega_i^2}+\frac{n_e e}{B}\frac{1}{1+\nu_{en}^2/\Omega_e^2} \tag{7.13b}$$

where n_i and n_e are ion and electron densities; Ω_i and Ω_e are ion gyrofrequency and electron gyrofrequency; v_{in} and v_{en} are ion and electron collision frequencies with neutral particles.

The source terms of equation (7.11) are

$$
\begin{aligned}
S_p &= \int \mathbf{a}_p \cdot \left[\sigma_p(-\mathbf{E}_{\text{ext}} + \mathbf{V}_n \times \mathbf{B}) + \sigma_h \mathbf{b} \times (-\mathbf{E}_{\text{ext}} + \mathbf{V}_n \times \mathbf{B}) \right. \\
&\quad \left. + \sigma_{pi} \frac{m_i}{e}\mathbf{g} + \sigma_{hi}\mathbf{b} \times \frac{m_i}{e}\mathbf{g} \right] ds \\
&= \int h_s h_\varphi \left(-\sigma_p E_{\text{ext},p} - \sigma_h E_{\text{ext},\varphi} + \sigma_p B V_{n,\varphi} \right. \\
&\quad \left. + \sigma_p B V_{n,p} + \sigma_{pi} \frac{m_i}{e} g_p \right) \mathrm{d}s \\
&= \int \frac{r \sin\theta}{b_s} \left[-(\sigma_p E_{\text{ext},p} - \sigma_h E_{\text{ext},\varphi}) \right. \\
&\quad \left. + (\sigma_p B V_{n,\varphi} + \sigma_h B V_{n,p}) + \sigma_{pi} \frac{m_i}{\mathrm{e}} g_p \right] \mathrm{d}s \\
&= -F_{pE_{\text{ext}}} + F_{pV_n} + F_{pg}
\end{aligned} \tag{7.14}
$$

where

$$F_{pE_{\text{ext}}} = \int \frac{r\sin\theta}{b_s} (\sigma_p E_{\text{ext},p} + \sigma_h E_{\text{ext},\varphi}) \mathrm{d}s \tag{7.15a}$$

$$F_{pV_n} = \int \frac{r\sin\theta}{b_s} (\sigma_p B V_{n,p} + \sigma_h B V_{n,\varphi}) \mathrm{d}s \tag{7.15b}$$

$$F_{pg} = \int \frac{r\sin\theta}{b_s} \left(\sigma_{pi} \frac{m_i}{\mathrm{e}} g_p \right) \mathrm{d}s \tag{7.15c}$$

and

$$
\begin{aligned}
S_\varphi &= \int \mathbf{a}_\varphi \cdot \left[\sigma_p(-\mathbf{E}_{\text{ext}} + \mathbf{V}_n \times \mathbf{B}) + \sigma_h \mathbf{b} \times (-\mathbf{E}_{\text{ext}} + \mathbf{V}_n \times \mathbf{B}) \right. \\
&\quad \left. + \sigma_{pi} \frac{m_i}{e}\mathbf{g} + \sigma_{hi}\mathbf{b} \times \frac{m_i}{e}\mathbf{g} \right] ds \\
&= \int h_s h_p \left(-\sigma_p E_{\text{ext},\varphi} - \sigma_h E_{\text{ext},p} - \sigma_p B V_{n,p} \right. \\
&\quad \left. + \sigma_h B V_{n,\varphi} + \sigma_{hi} \frac{m_i}{\mathrm{e}} g_p \right) \mathrm{d}s \\
&= \int \frac{R_E \sin^3\theta}{b_s \Delta} \left[-(\sigma_p E_{\text{ext},\varphi} + \sigma_h E_{\text{ext},p}) \right. \\
&\quad \left. + (-\sigma_p B V_{n,p} + \sigma_h B V_{n,\varphi}) + \sigma_{hi} \frac{m_i}{\mathrm{e}} g_p \right] \mathrm{d}s \\
&= -F_{\varphi E_{\text{ext}}} + F_{\varphi V_n} + F_{\varphi g}
\end{aligned} \tag{7.16}
$$

where

$$F_{\varphi E_{\text{ext}}} = \int \frac{R_E \sin^3\theta}{\Delta b_s} (\sigma_p E_{\text{ext},\varphi} + \sigma_h E_{\text{ext},p}) \mathrm{d}s \tag{7.17a}$$

$$F_{\varphi V_n} = \int \frac{R_E \sin^3\theta}{\Delta b_s} (-\sigma_p B V_{n,p} + \sigma_h B V_{n,\varphi}) \mathrm{d}s \tag{7.17b}$$

$$F_{\varphi g} = \int \frac{R_E \sin^3\theta}{\Delta b_s} \left(\sigma_{hi} \frac{m_i}{\mathrm{e}} g_p \right) \mathrm{d}s. \tag{7.17c}$$

The Pedersen and Hall conductivities for ions are

$$\sigma_{pi} = \frac{n_i e}{\Omega_i} \left(\frac{v_{in}/\Omega_i}{1 + v_{in}^2/\Omega_i^2} \right) \tag{7.18a}$$

$$\sigma_{hi} = \frac{n_i e}{\Omega_i} \left(\frac{1}{1 + v_{in}^2/\Omega_i^2} \right) \tag{7.18b}$$

The imposed electric fields $\mathbf{E}_{ext}$ in equations (7.14) and (7.16) are driving forces terms (S_P and S_φ) for solving electric potential equation (7.11) in the ionosphere of the SAMI3 model; the imposed electric field $\mathbf{E}_{ext}$ is solved by the current flowing into the ionosphere, as described in this chapter.

7.3. RESULTS

7.3.1. Ionospheric Electric Field Caused by an Atmospheric Upward/Downward Current

Figure 7.5 shows the vector field (white arrows) and magnitude of electric field E_{85km} at an altitude of 85 km produced by the upward current with $J_{max} = 100$ nA m^{-2} and $\sigma_H = 10\sigma_P$. Note that the average Hall conductivity σ_H is observed to be about ten times the Pedersen conductivity σ_P (see website, http://wdc.kugi.kyoto-u.ac.jp/ionocond/sigcal/index.html). In the following calculations, we set $\sigma_P = 1 \times 10^{-6}$ S m^{-1} in the daytime and $\sigma_P = 2 \times 10^{-7}$ S m^{-1} in the nighttime for an altitude of 85 km. Note that the model computation is conducted for an earthquake preparation region located at magnetic latitudes of 7.5°, 15°, 22.5°, and 30°.

7.3.2. TEC Variations for Upward/Downward Currents at Magnetic Latitudes of 7.5°, 15°, 22.5°, and 30°

Figure 7.6 shows contour plots of ionospheric electron density and velocity vectors in the meridional plane above the source region of the current: (a) at magnetic latitude 15° for upward currents, (b) at 15° for downward currents, (c) at 30° for upward currents, and (d) at 30° for downward currents. The white dashed lines in Figure 7.6 indicate the magnetic field lines from the earthquake region to its conjugated point in the opposite hemisphere. For upward currents at magnetic latitude 15° and 30°, the westward electric field, shown in Figures 7.6a and 7.6c, causes downward $\mathbf{E} \times \mathbf{B}$ ionospheric plasma motion along the magnetic field from the source region of the current. Downward ionospheric plasma velocities are also shown by black lines attached to a dot in Figures 7.6a and 7.6c,

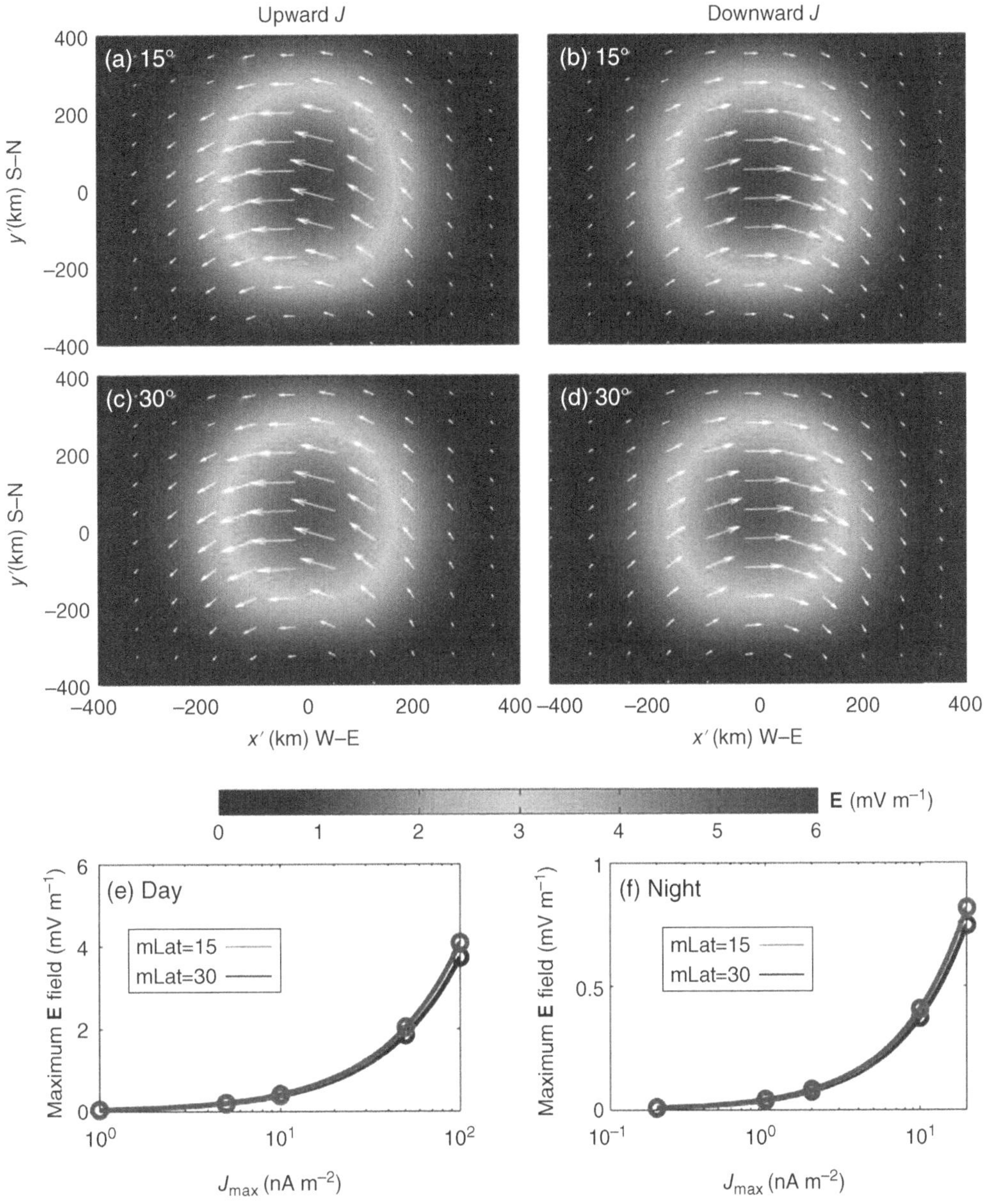

Figure 7.5 The vector electric field (white arrows) and contours of the electric field magnitude produced by (a) upward and (b) downward currents at magnetic latitude 15° and by (c) upward and (d) downward currents at magnetic latitude 30°. The Hall conductivity is assumed to be ten times the Pedersen conductivity $10\sigma_P \approx \sigma_H$ in the E layer of the ionosphere. (e) The magnitude of the electric field is plotted as a function of the current for magnetic latitude 15° and 30° in the daytime. (f) The magnitude of the electric field is plotted as a function of the current with magnetic latitude 15° and 30° in the nighttime [*Kuo et al.*, 2014]. mLat, magnetic latitude. (Source: Reproduced with permission of John Wiley and Sons.)

where the direction of motion is away from the dot. For downward currents with magnetic latitude 15° and 30°, the eastward electric field causes upward **E** × **B** ionospheric plasma motion, as shown in Figures 7.6b and 7.6d. For upward (downward) current at latitude 15°, the maximum of plasma velocity is 133 m s^{-1} (120 m s^{-1}). For upward (downward) current at latitude 30 °, the maximum of plasma velocity is 193 m s^{-1} (160 m s^{-1}).

We show the TEC variations (%) for upward (downward) currents at magnetic latitude 7.5°, 15°, 22.5° and 30° in Figures 7.7a, 7.7c, 7.7e, and 7.7g (Figures 7.7b, 7.7d, 7.7f, and 7.7h), respectively. For upward currents, the westward electric field causes downward **E** × **B** plasma motion from slightly higher latitude (north) to lower latitude (south), as indicated in Figure 7.4a. The plasma moves across the magnetic line from its starting point latitude at

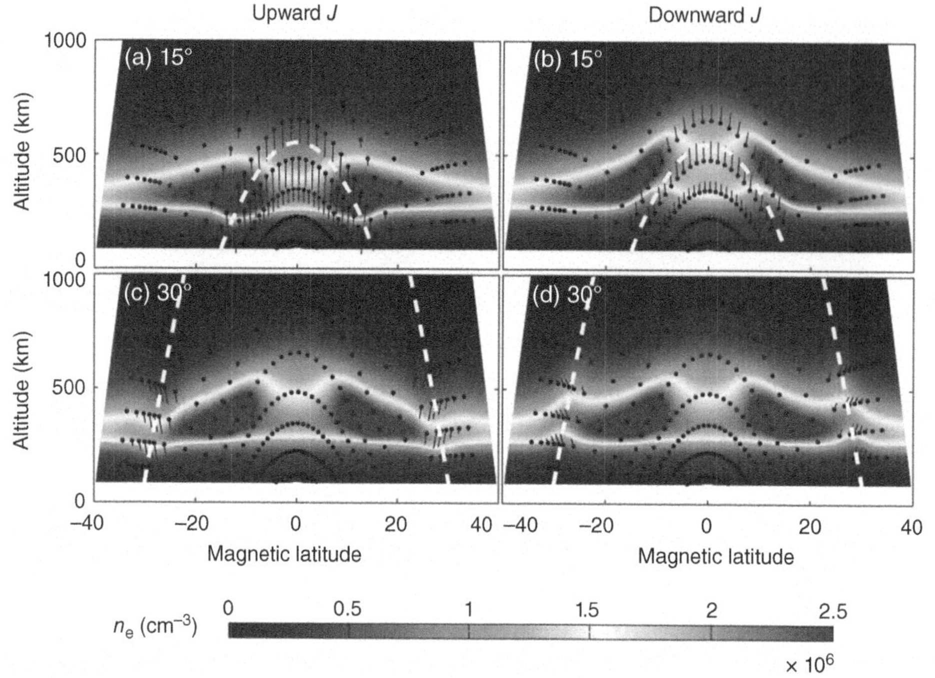

Figure 7.6 The velocities and contour plots of electron density in the meridional plane above the current source region at magnetic latitude 15° for (a) upward currents (westward electric field) and (b) for downward currents (eastward electric field), and at 30° for (c) upward and (d) downward currents. Black lines attached to a black dot indicate velocities of ionospheric plasma, where the direction of motion is away from the dot [*Kuo et al.*, 2014]. (Source: Reproduced with permission of John Wiley and Sons.)

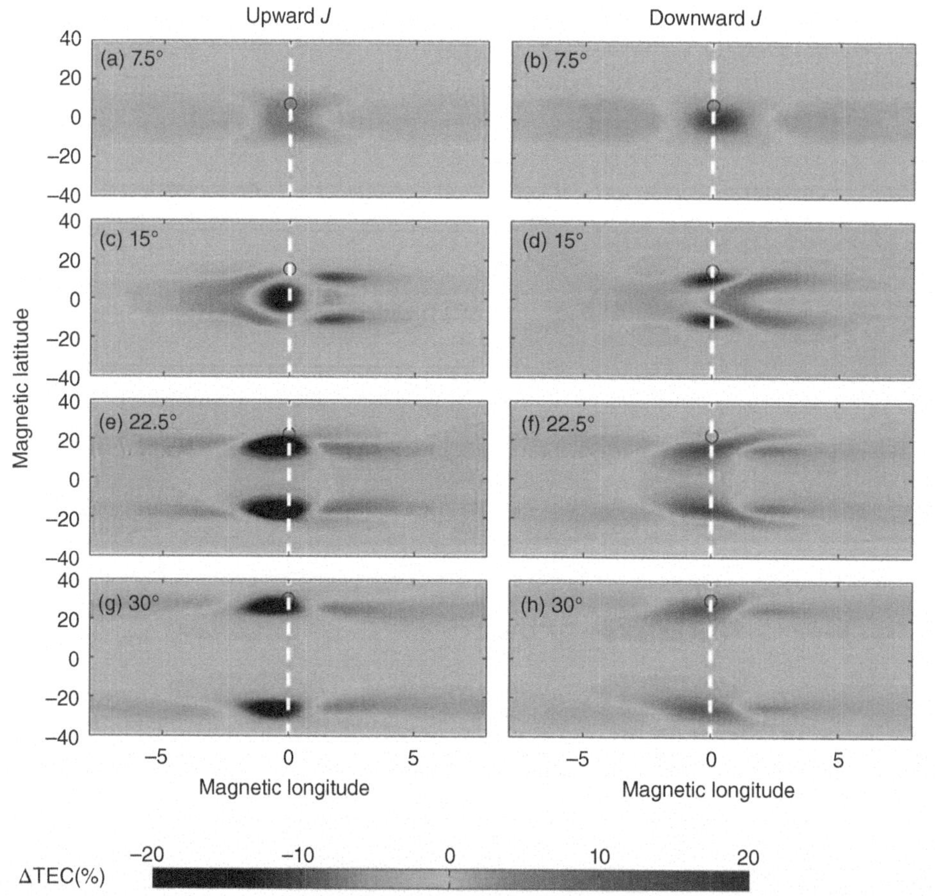

Figure 7.7 TEC variations (%) for the case with upward (downward) current are shown in the left (right) column. The panels from up to down are results for magnetic latitude 7.5°, 15°, 22.5°, and 30°, respectively. The open circle indicates the location of the current source (earthquake epicenter) [*Kuo et al.*, 2014]. (Source: Reproduced with permission of John Wiley and Sons.)

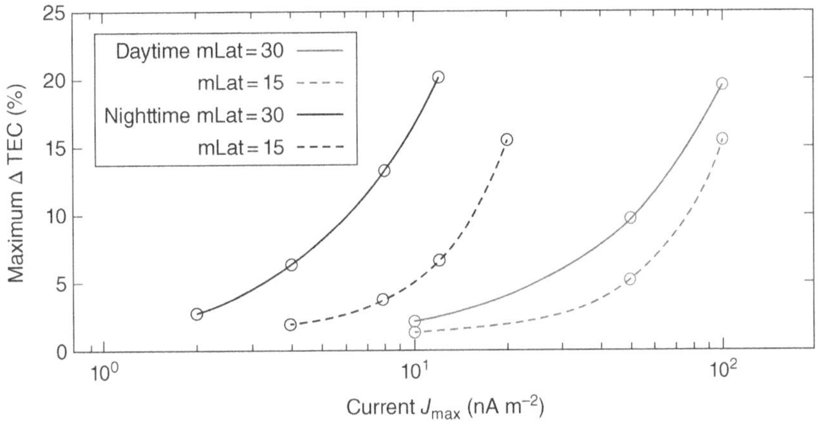

Figure 7.8 Maximum ΔTEC (%) varies with source current density J_{max} (nA m^{-2}) where the solid (dashed) lines are for ΔTEC at latitude 30°(15°). The blue (black) lines are for daytime (nighttime) ionospheric conditions [*Kuo et al.*, 2014]. mLat, magnetic latitude. (Source: Reproduced with permission of John Wiley and Sons.) (*See electronic version for color representation.*)

the foot of the diagram, e.g., the dashed line in Figure 7.6a for 15° and in Figure 7.6c for 30°. Along the magnetic line from its starting point latitude from northern to southern hemisphere, plasma transport from a higher (lower) density region can lead to an increase (decrease) inTEC.

Figure 7.8 shows that the maximum percentage of TEC change varies with dynamo current J_{max} at magnetic latitude 15° and 30° for daytime and nighttime. In order to generate $\Delta_{TEC} \cong 2$–20%, a current density J_{max} of 1–10 nA m^{-2} (10–100 nA m^{-2}) is required for the nighttime (daytime) ionosphere. *Freund et al.* [2009] reported that a pressure of 10 MPa when applied to rock could generate a current density of 100 nA m^{-2}, which is comparable to the J_{max} required generating $\Delta_{TEC} \cong 20\%$ in the daytime ionosphere. On the other hand, it remains to be observed that the injection of charged aerosols into the atmosphere [*Sorokin and Hayakawa*, 2013] can generate a current density of 1–100 nA m^{-2}.

7.3.3. Nighttime Plasma Bubbles Triggered by Atmospheric Current

At nighttime, at the higher magnetic latitude 30°, the decrease (increase) of TEC is similar to the results shown in Figure 7.7g (7.7h), but the decrease of TEC near the source region will not cause the formation of plasma bubbles. At the lower magnetic latitude 15° shown in Figure 7.7c (7.7d), however, the depletion region of the electron density caused by the westward (eastward) electric field will trigger the occurrence of plasma bubbles, as shown in Figure 7.9 (Figure 7.10). The westward electric field leads to two depletion regions of electron density on both sides of the fault region, while electron density depletion is produced in the middle region in the case of an eastward electric field. Figure 7.9 (Figure 7.10) shows the double (single) plasma bubbles caused by the westward (eastward) electric field. The plasma bubble is generated by the depletion region of electron density within 20° of latitude from the equator of the ionosphere.

In Figure 7.10 the downward dynamo current drives an eastward electric field in the ionosphere. The eastward electric field locally causes an upward plasma motion in the lower E region. The upward motion of low-density plasma triggers the Rayleigh–Taylor instability, which occurs with a heavy fluid (F layer) on top of a light fluid (E layer). The buoyant force lifts the low-density plasma (bubble) up.

7.4. CASE STUDY A: PRESEISMIC TEC CHANGES ASSOCIATED WITH THE TOHOKU-OKI EARTHQUAKE

7.4.1. The Ionospheric TEC Anomaly Before the 2011 Tohoku-Oki Earthquake

Heki [2011] reported that the Japanese GPS dense network detected a precursory positive TEC, with ΔTEC~3 TECU (TEC units), at~40 min before the Tohoku-Oki earthquake (*Mw* 9.0). Similar preseismic TEC anomalies were also observed associated with the 2010 Chile earthquake (*Mw* 8.8), the 2004 Sumatra-Andaman (*Mw* 9.2), and the 1994 Hokkaido-Toho-Oki (*Mw* 8.3). In this section we apply our improved LAIC model to compute the TEC variations and compare the simulation results with the reported TEC observations. For the Tohoku-Oki earthquake simulations, we assumed that the stressed associated current started~40 min before the earthquake, linearly increased and reached its maximum magnitude at the time of the earthquake main shock. It is suggested that a dynamo current density of ~25 nA m^{-2} is required to produce the observed ΔTEC~3 TECU.

With the aid of the Japanese dense GPS GEONET observation network (http://www.gsi.go.jp) a possible earthquake precursor anomaly could be detected for the 11March 2011 Tohoku-Oki earthquake [*Heki*, 2011; *Heki*

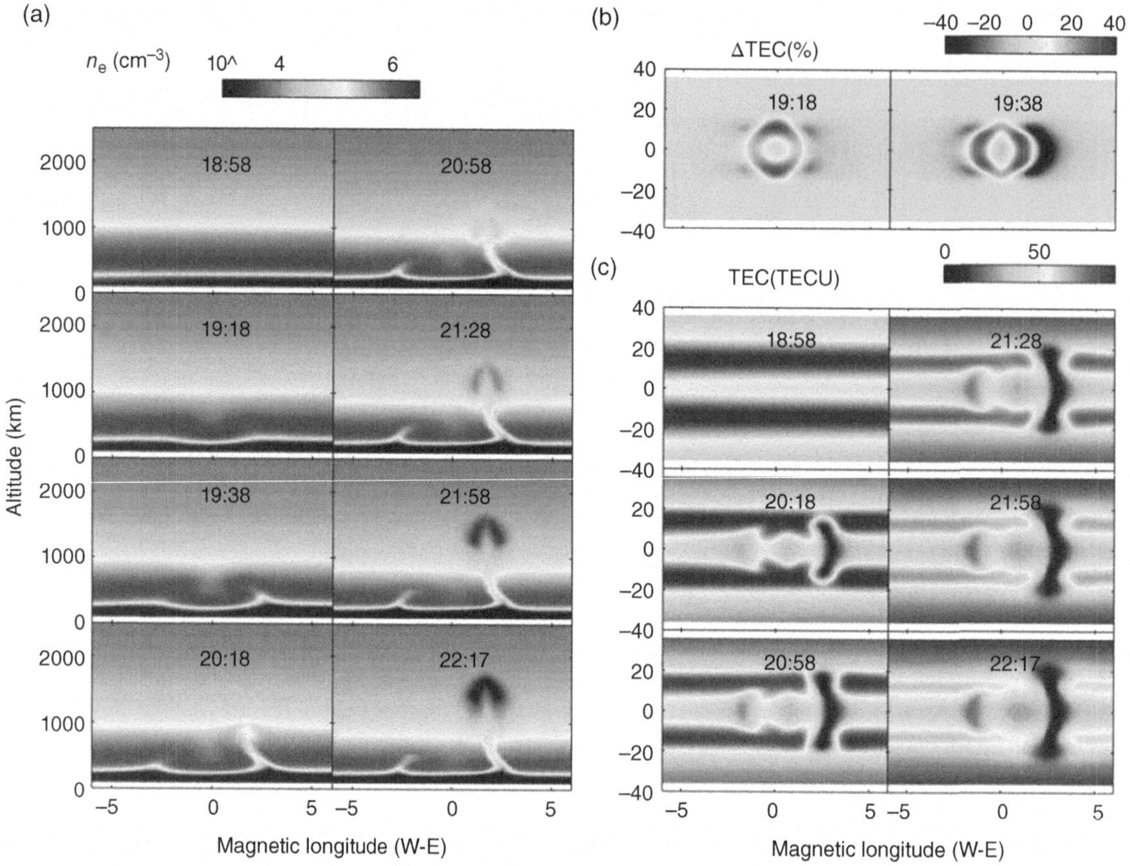

Figure 7.9 Double plasma bubbles are triggered by zonal westward electric field (upward current) at low magnetic latitude 15°: (a) contour plots of electron density n_e in the equatorial plane at different times for the nighttime case; (b) contour plots of ΔTEC during applied westward electric field; (c) contour plots of TEC during formation of plasma bubble [*Kuo et al.*, 2014]. (Source: Reproduced with permission of John Wiley and Sons.)

and Enomoto, 2013]. *Heki* [2011] used GPS-TEC data to find a clear precursory positive ionospheric TEC anomaly over the epicentral region. The TEC variations started at ~40 min before the earthquake and reached nearly 10% of the background TEC. At the time of the main shock (5:46 UT), eight GPS satellites had visibility there [*Heki*, 2011]. The coseismic ionospheric disturbances (CIDs) can be seen by the GPS satellites as irregular TEC changes caused by acoustic waves ~ 10 min after the earthquake, and ionospheric oscillations caused by atmospheric waves or internal gravity waves 40–80 min after the earthquake.

7.4.2. Comparisons Between Modeling Results and the Japanese GEONET TEC Observations

The 2011 Tohoku-Oki earthquake had a fault region of ~ 450 km in length and ~ 200 km in width along the Japan Trench where the Pacific Plate subducts beneath northeast Japan, as modeled by *Heki* [2011]. The fault region orientation has an azimuth angle ~ 30° from north, with its epicenter at 38.3°N, 142.4°E. It is assumed that the maximum current density J_{max} = 25 nA m^{-2} increases linearly from zero to its maximum value in the 40 min period (UT 05:06–05:46) before the main shock, as shown in Figure 7.11, since the increase in TEC started 40 min before the earthquake (UT 05:46), and the region of TEC increase grew in area to reach its maximum ΔTEC.

In comparison with the simulation results of *Kuo et al.* [2014], the modeling results show the presence of an eastward (westward) electric field for downward (upward) dynamo current flowing from the atmosphere into the ionosphere. At magnetic latitude 30°, close to the epicenter, the imposed eastward (westward) electric field causes a nearly upward-northward or downward-southward direction of E × B motion for ionospheric plasma (shown later in Figure 7.12a). For the nearly upward-northward plasma-motion direction with eastward electric field caused by the downward current, the **E** × **B** motion drives the ionospheric plasma from the higher density region to the lower density region, enhancing

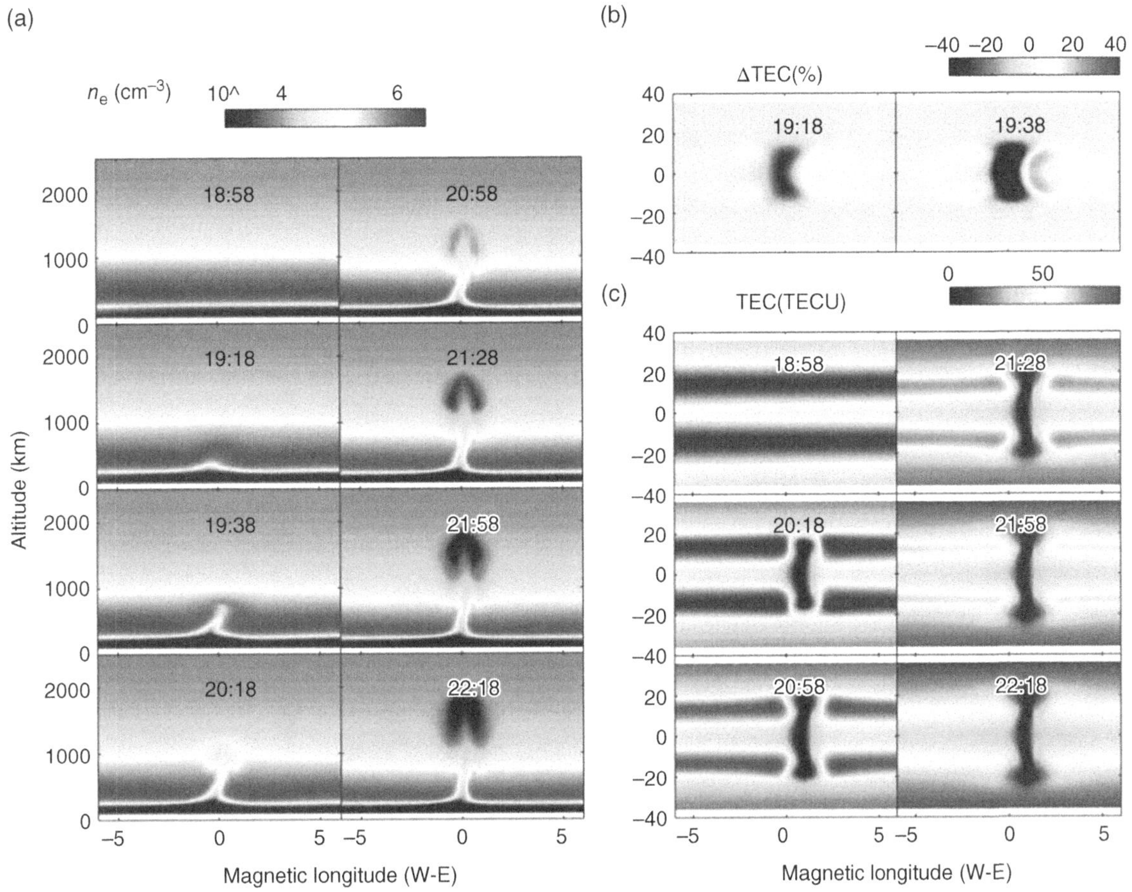

Figure 7.10 A single plasma bubble is triggered by zonal eastward electric field (downward current) at low magnetic latitude 15°: (a) contour plots of electron density n_e in the equatorial plane at different times for the nighttime case; (b) contour plots of ΔTEC during applied westward electric field; (c) contour plots of TEC during formation of plasma bubble [*Kuo et al.*, 2014]. (Source: Reproduced with permission of John Wiley and Sons.)

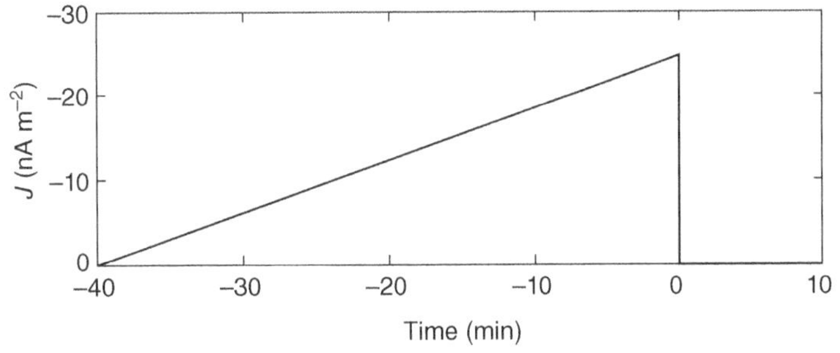

Figure 7.11 Maximum current density linearly increases from zero to its maximum value in the 40-min period (UT 05:06–05:46) before the main shock [*Kuo et al.*, 2015]. (Source: Reproduced with permission of Terrestrial, Atmospheric and Oceanic Sciences.)

plasma density (Figure 7.12c) and increasing TEC (Figure 7.12b). Hence, we choose the downward current with the eastward electric field as our dynamo current.

The typical dynamo current J_{max} value used in the calculations is 10–100 nA m^{-2}, corresponding to ΔTEC up to 1–7 TECU in the daytime case, as indicated by solid lines in Figure 7.8. In the nighttime case, the smaller dynamo current (1–10 nA m^{-2}) value can lead to similar ΔTEC values. In our calculation, the dynamo current equaled the ionospheric conductivity multiplication and

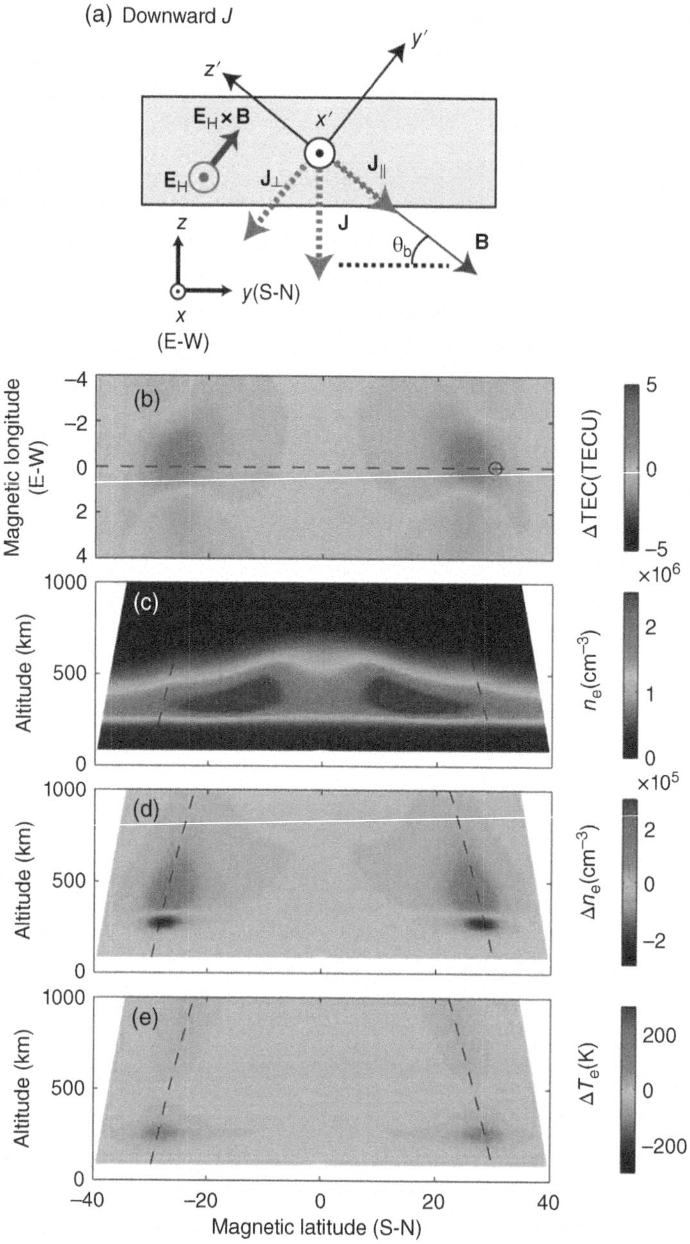

Figure 7.12 An ionospheric anomaly caused by downward current at magnetic latitude 30°; (a) the downward current leads to the presence of an eastward electric field and the resulting **E** × **B** motion enhances the ionospheric plasma density; (b) contour plots of ΔTEC in TECU where the open circle indicates the source region; (c) contour plots of electron density n_e in the meridional planes; (d) contour plots of electron density variations Δn_e in the meridional planes; (e) temperature variations in the meridional planes [*Kuo et al.*, 2015]. (Source: Reproduced with permission of Terrestrial, Atmospheric and Oceanic Sciences.)

created an electric field. The typical daytime ionospheric conductivity is ten times that of the nighttime conductivity. Therefore, greater current density is required to reach the equivalent ΔTEC for the daytime ionosphere.

Figures 7.13a–7.13c show the observed TEC variations at subionospheric point of more than 1000 ground GPS sites in the Japanese GEONET and their corresponding ΔTEC measurements are indicated by the colored dots in TECU. Figures 7.13d–7.13f show the TEC contour lines from the simulation. Figures 7.13g–7.13i show the TEC contours, where the color code indicates the TECU value. The applied eastward electric field

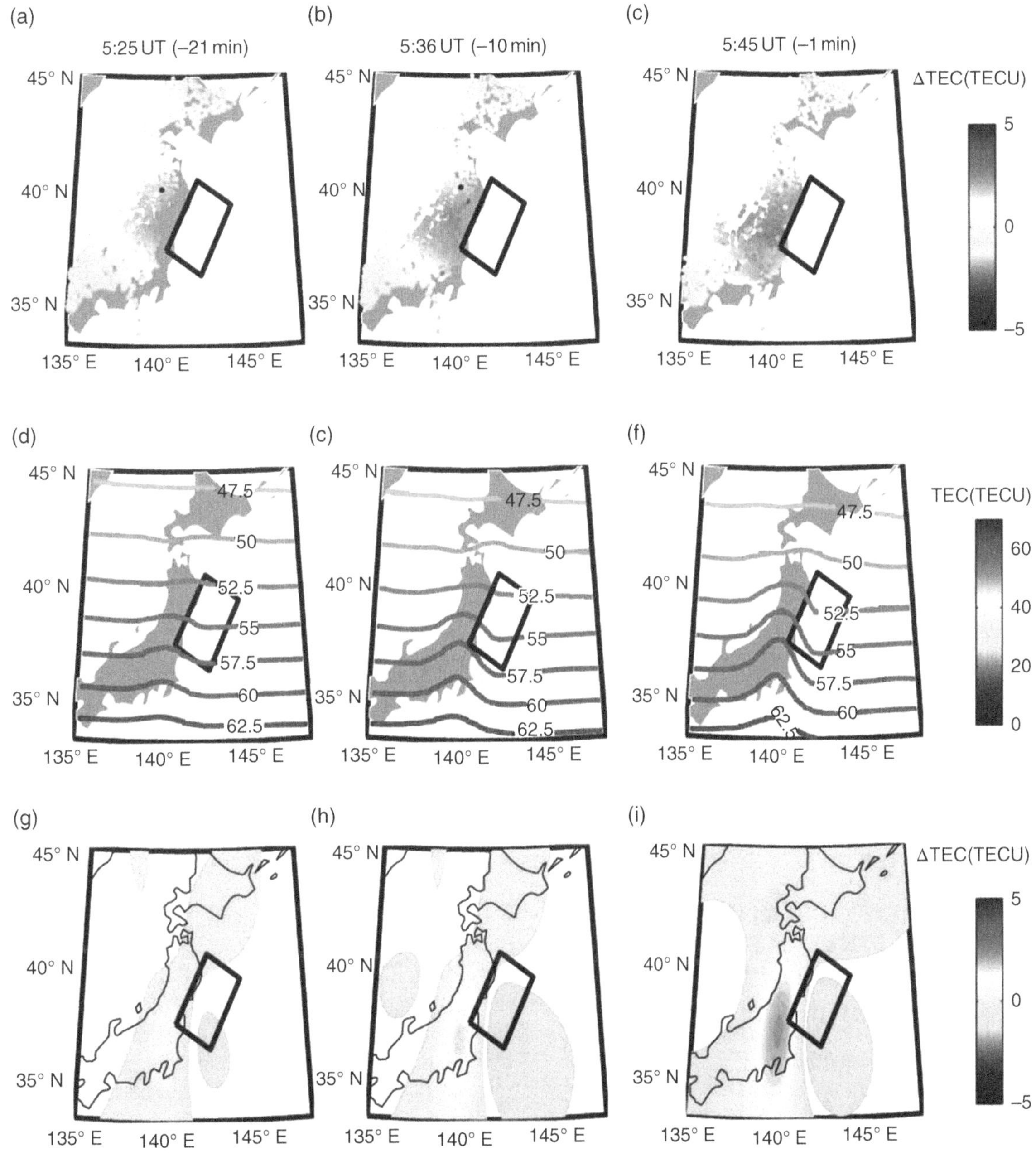

Figure 7.13 The observed results of ΔTEC from the Japanese GEONET where color code indicates the magnitude of TEC in a time sequence of (a) 21 min, (b) 10 min, and (c) 1 min before the main shock of the earthquake. The corresponding TEC contour lines from our simulation results are plotted in (d), (e), and (f). The corresponding ΔTEC from our simulation results are plotted in (g), (h), and (i) [*Kuo et al.*, 2015]. (Source: Reproduced with permission of Terrestrial, Atmospheric and Oceanic Sciences.) (*See electronic version for color representation.*)

leads to an upward $\mathbf{E}\times\mathbf{B}$ motion and an increase in TEC. The ΔTEC shown in Figures 7.13g– 7.13i can be used to compare with the measured ΔTEC results in Figures 7.13a–7.13c.

Figure 7.14 shows a comparison of ΔTEC profiles from modeling results (red dots) with the observations (blue dots) in TECU 1 min before the main shock. Figures 7.14a–714c are the geolongitude profiles at 139°, 140°, and 141° respectively, and Figures 7.14d–7.14f are the geolatitude profiles at 36°, 38°, and 40° respectively. The modeling results with $J_{max} \approx 25$ nA m^{-2} approximately match the observations.

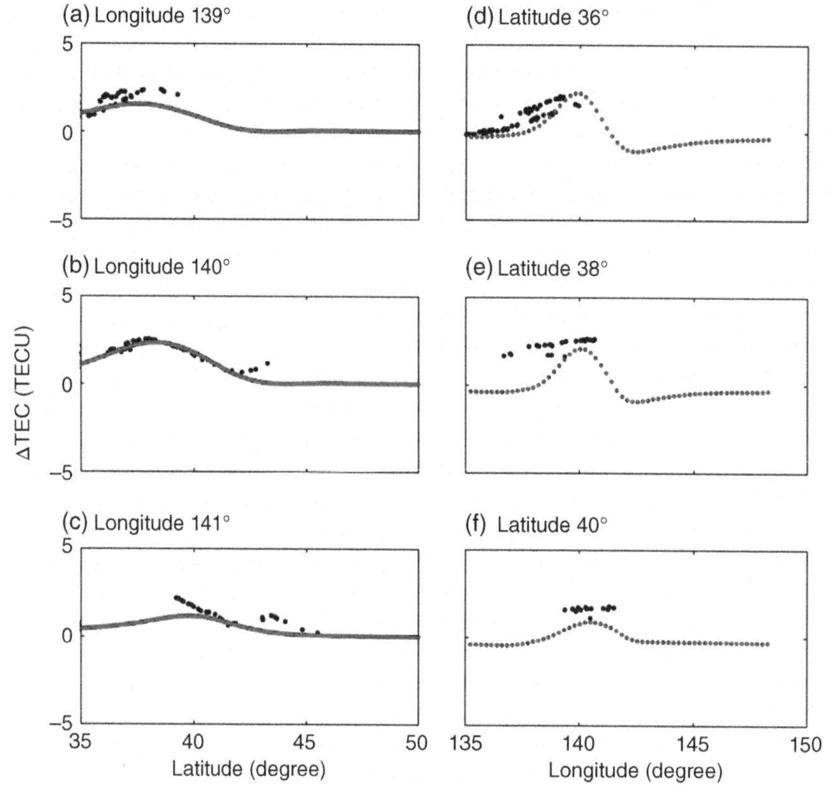

Figure 7.14 Comparison of modeling results (red dots) with observed ΔTEC (blue dots) in TECU at UT 05:45, 1 min before the main shock: (a) profile at longitude 139°, (b) profile at longitude 140°, (c) profile at longitude 141°, (d) profile at latitude 36°, (e) profile at latitude 38°, and (f) profile at latitude 40° [*Kuo et al.*, 2015]. (Source: Reproduced with permission of Terrestrial, Atmospheric and Oceanic Sciences.) (*See electronic version for color representation.*)

7.5. CASE STUDY B: PRESEISMIC PLASMA DENSITY AND ION VELOCITY VARIATION FOR CHILE EARTHQUAKE

7.5.1. The Plasma Density and Velocity Anomaly Before the 2010 Chile Earthquake

The French satellite DEMETER is designed to detect the ionospheric parameter variations before and after earthquakes. Due to its 98° inclination and sun-synchronous orbit, DEMETER provides data within 65° invariant latitude at 10:30 and 22:30 local time (LT) [*Parrot et al.*, 2006]. DEMETER has detected and recorded designated variables continuously for the years 2004–2010. This section investigates the variations of the plasma density and velocity before the 2010 Chile earthquake. The Chile earthquake (27 February 2010) was one of the largest earthquakes that occurred during the DEMETER mission. The epicenter was located at 35.91°S, 72.73°W, with a magnitude of 8.8. The geomagnetic latitude of the epicenter is 25.74°S. *Ho et al.* [2013] reported that DEMETER detected density enhancement during the nighttime before Chile earthquake, which we focus on here.

Figure 7.15 displays the temporal density and perpendicular velocity variations from 30 days before to 10 days after the Chile earthquake. The F10.7 index (solar radio flux at 10.7 cm; 2800 MHz) and the planetary K-index (Kp; an indicator of disturbances in the Earth's magnetic field) indicate that solar and geomagnetic activities are relatively quiet during this period. Therefore, we can pay attention to the local ionospheric variations that might relate to the earthquake. We select the density and velocity data points at the nearest location in each analysis day. The previous 30-day data of a certain observation day are utilized to set up the reference of anomalies [*Liu et al.*, 2011].

$$\mathrm{UB = M + 1.5UQ - M} \quad \text{(upper bound)} \tag{7.19a}$$

$$\mathrm{LB = M - 1.5M - LQ} \quad \text{(lower bound)} \tag{7.19b}$$

where UB, LB, M, UQ, and LQ are the upper bound, lower bound, 30-day running median, upper quartile, and lower quartile, respectively. For the Chile earthquake, the results show that the plasma density increases 9–19 and 28 days before the earthquake. Meanwhile, the plasma perpendicular velocity increases 9, 10, 14, 15, 17, and 25 days before the earthquake. Density and velocity therefore

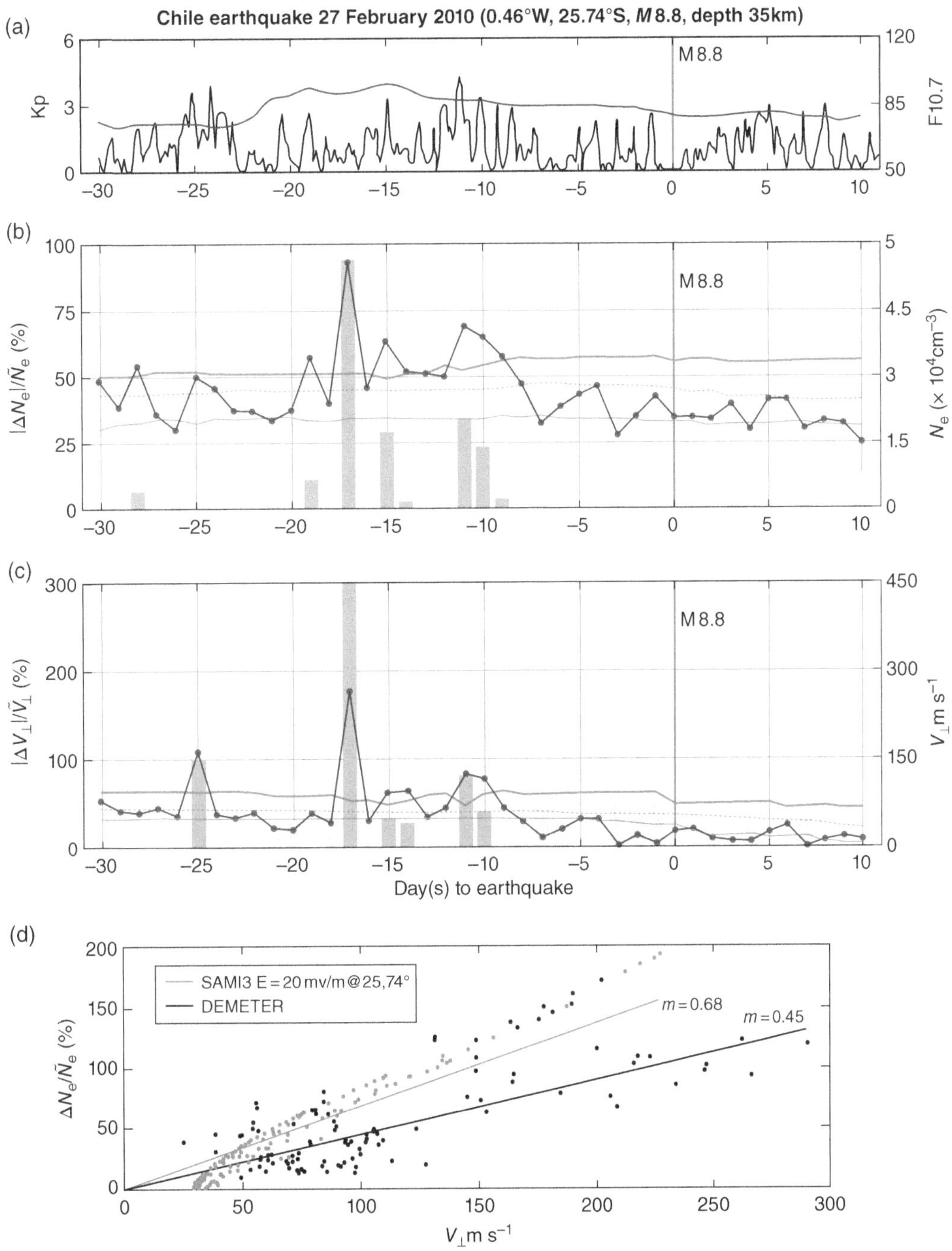

Figure 7.15 Variations of space environments and the ionosphere before and after the Chile earthquake. (a) The geomagnetic index Kp and the solar activity index F10.7. (b and c) Temporal profile of electron density and ion velocity respectively. The red line represents the observations, and the blue and dotted lines the upper and lower bounds and the median. The pink and the light blue bars denote the percentage variations of electron density ($|\Delta N_e|N_e$) and perpendicular velocity ($|\Delta V_\perp|V_\perp$). The geomagnetic latitude of the epicenter is 25.74°S. (d) Comparison between the simulation (blue) and the observation (black) results before the 2010 Chile earthquake. The dots denote the perpendicular velocity and the density variations. The black (blue) best-fit line passes through the origin from DEMETER observations (SAMI3 simulations), and m is the slope. (*See electronic version for color representation.*)

increase over almost the same period. The bars (Figure 7.15) show the percentage density variations given by

$$\frac{|\Delta N_e|}{\tilde{N}_e} = \frac{N_e - \mathrm{UB}}{\tilde{N}_e} \times 100\% \ (\text{upper anomaly})$$
$$\frac{|\Delta N_e|}{\tilde{N}_e} = \frac{\mathrm{LB} - N_e}{\tilde{N}_e} \times 100\% \ (\text{lower anomaly}) \quad (7.20)$$

where N_e, UB, LB, and $\tilde{N}_e$ are the observation, related upper and lower bounds, and median of the electron density, respectively. The percentage velocity variations are in the same form

$$\frac{|\Delta V_\perp|}{\tilde{V}_\perp} = \frac{V_\perp - \mathrm{UB}}{\tilde{V}_\perp} \times 100\% \ (\text{upper anomaly})$$
$$\frac{|\Delta V_\perp|}{\tilde{V}_\perp} = \frac{\mathrm{LB} - V\perp}{\tilde{V}_\perp} \times 100\% \ (\text{lower anomaly}) \quad (7.21)$$

The most pronounced enhancement of the parameters appears on 10 February (17 days before the earthquake) when percentage variations (Figure 7.15) reach 93% in density and 306% in velocity.

To check whether the anomalous distributions are the result of ionospheric seasonal variation, we compare the increased anomalous strength of density and velocity for the same seasons from 2008 to 2010 (Figure 7.16). The density and perpendicular/parallel velocity anomalous strengths are defined as

$$\Delta N_e = N_e - \mathrm{UB}(N_e) \quad (7.22)$$

$$\Delta V_\perp = V_\perp - \mathrm{UB}(V_\perp), \Delta V_\parallel = V_\parallel - \mathrm{UB}(V_\parallel) \quad (7.23)$$

The anomalous strengths 30 days before and after the earthquake are displayed for comparison. The density strengths (ΔN_e) in 2010 are larger than those in 2008 and 2009. In 2010, all density anomalies and 98.77% of

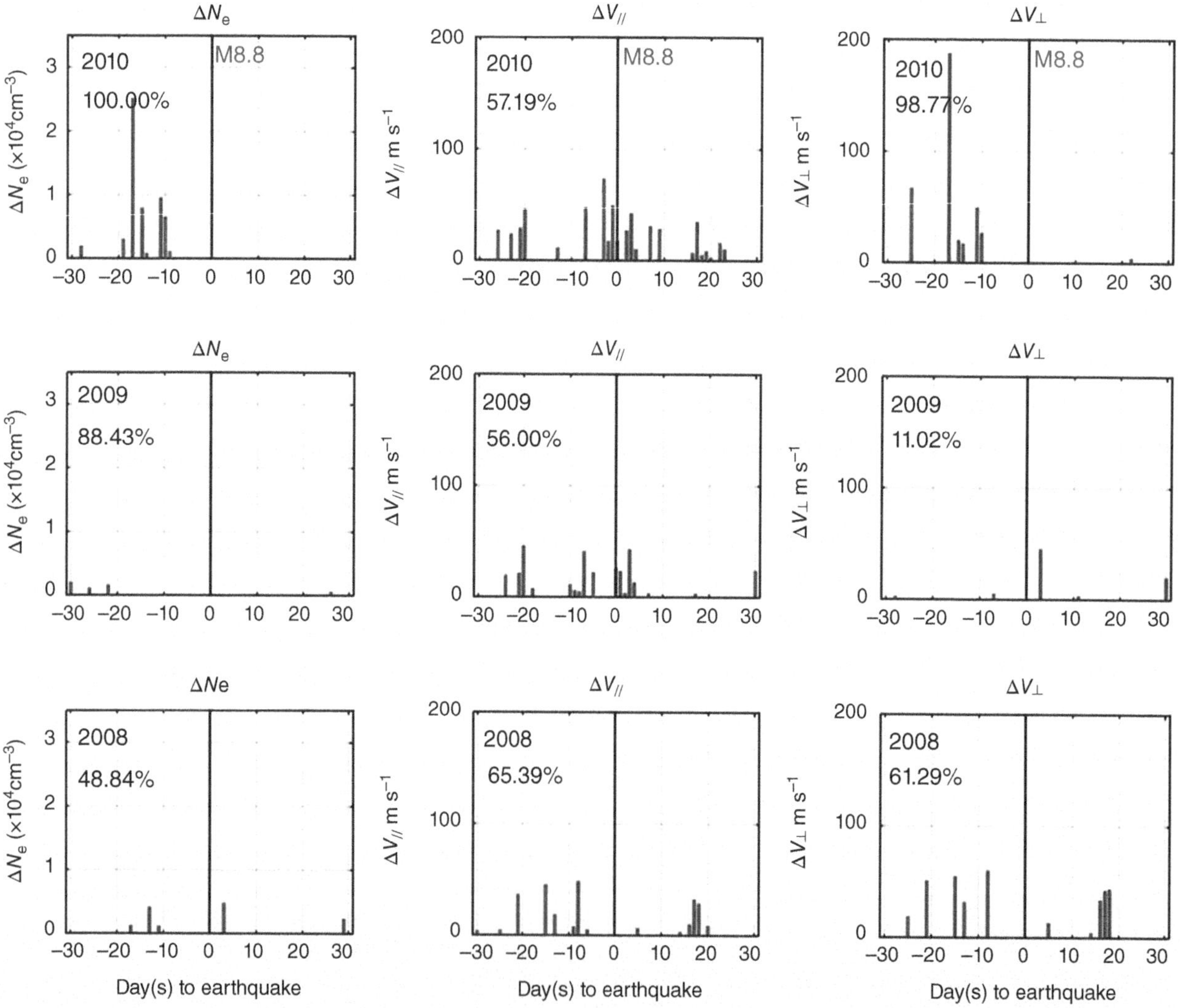

Figure 7.16 Comparison of increased anomalies in 2008, 2009, and 2010 (lower to upper panels). From left to right the panels display anomalies of electron density, parallel velocity, and perpendicular velocity. The red bars show the values of ΔN_e, $\Delta V_\perp$ and $\Delta V_{\parallel}$. (*See electronic version for color representation.*)

perpendicular velocity anomalies occur before the day of the earthquake. By contrast, only 61% of perpendicular velocity anomalies occur before the day of the earthquake in 2008, and 11% perpendicular velocity variations occurred during the same period in 2009. On the other hand, the parallel velocity does not show an obvious variance in its occurrence before and after the day of the earthquake in these 3 years. These data suggest that increased density and perpendicular velocity might relate to the 2010 Chile earthquake but that parallel velocity is not affected by the event.

Insight into plasma density changes can be gained from consideration of the perpendicular velocity data, that is $\mathbf{E}\times\mathbf{B}$ motion. Figure 7.15d shows the relations between the increased density variation $\left(\frac{|\Delta N_e|}{\tilde{N}_e}\right)$ and the related perpendicular velocity ($V_\perp$) 1–30 days before the Chile earthquake observed at an altitude of 670 km by DEMETER within ±2.5° latitude and ±7.8° longitude of the epicenter. Each black dot denotes a pair of $\frac{|\Delta N_e|}{\tilde{N}_e}$ and $V_\perp$. The best-fit (black) line (slope $m=0.45$) passes through the origin where we assume that there is no density variation when the perpendicular velocity is zero.

We have also checked other earthquake events during the DEMETER mission. Figures 7.17 and 7.18 show the density and velocity variations before and after the Samoa Islands earthquake (187.9°E, 15.4°S, M 8.1) on 9 September 2009 and the Kermadec Islands earthquake (183.7°E, 27.4°S, M 7.0) on 18 February 2009, respectively. The geomagnetic latitude for the Samoa Islands earthquake is 16.86°S and for the Kermadec Islands earthquake is 29.26°S. Similar to the 2010 Chile earthquake, these two earthquakes also occurred in quiescent geomagnetic conditions in the southern hemisphere, and we use the same criterion to identify their anomalies. For the 2009 Samoa Islands earthquake, although there is an 8-day data gap before the earthquake, we still find that density obviously increases 4–12 days before the earthquake. During this period, decreased anomalies of perpendicular velocity occur. Percentage density variation $\left(\frac{|\Delta N_e|}{N_e}\right)$ reached 80% and percentage velocity variation $\left(\frac{|\Delta V_\perp|}{\tilde{V}_\perp}\right)$ reached 344%. For the 2009 Kermadec Islands earthquake, there are both increased and decreased density anomalies on 4, 6, and 20–24 days before the earthquake. Velocity anomalies appear intermittently during this period. Percentage density variation $\left(|\frac{|\Delta N_e|}{\tilde{N}_e}\right)$ and percentage velocity variation $\left(\frac{|\Delta V_\perp|}{\tilde{V}_\perp}\right)$ reached 40% and 141% respectively. Figures 7.17 and 7.18 shows relations between the increased density variation $\left(\frac{|\Delta N_e|}{\tilde{N}_e}\right)$ and the related perpendicular velocity ($V_\perp$) 1–30 days before these two earthquakes. The slope of the best-fit line is $m=0.41$ for the Samoa Islands earthquake and $m=0.66$ for the Kermadec Islands earthquake. Comparing with the Chile earthquake, it is interesting that the slopes increase with the geomagnetic latitude.

7.5.2. Comparison Between DEMETER Observations and Coupling Model Results

The numerically modeled movement of ionospheric plasma described in section 7.3 is utilized here. We take the geophysical conditions for the most pronounced anomalous day (10 February) before the Chile earthquake: F10.7 = 91, F10.7A = 82.69, $A_p = 3$. Simulations include different imposed electric fields at the bottom side of the ionosphere and the latitude of the epicenter (25.74°S). The results indicate that a larger electric field leads to both larger density variation and velocity. The slope m also increases with increasing electric field E: $m=0.43$ for $E=0$ mV m^{-1}, $m=0.59$ for $E=15$ mV m^{-1}, $m=0.68$ for $E=20$ mV m^{-1}, and $m=0.73$ for $E=25$ mV m^{-1} (Figure 7.19). The simulation results with the closest maximum of perpendicular velocity to that observed are selected for comparison. Thus we chose the $E=20$ mV m^{-1} ($m=0.68$) results to compare with the observations for the Chile earthquake (Figure 7.18d), and the $E=10$ mV m^{-1} simulation results to compare with the observations ($m=0.82$, Figure 7.19) for the Kermadec Islands earthquake. The slope value of the simulation results is larger than that of the observations, indicating that a larger velocity in observations than in simulations is required to trigger the same density variation.

The satellite data provide only in situ observations at an altitude of 670 km, whereas simulations provide information along an altitudinal range. Figure 7.20a shows the density and velocity distributions $N_e(0)$, $N_e(E)$, ΔN_e, and $V_\perp$ in the meridional plane of the imposed electric field at the epicenter of 25.74°S. Figure 7.20b shows the relations of the density variation and the perpendicular velocity at different altitudes. The maximum of density variations reaches 190%, 80%, and 124% at 670 km, 400 km, and 200 km, respectively. It shows a positive correlation at altitudes of 400 km and 670 km. The density variation percentage at 670 km is twice that at 400 km because of the larger background density of the EIA (equatorial ionization anomaly). However, the velocity at 670 km is three times that at 400 km due to the lower collision frequency in the higher ionosphere. The slope at 400 km is larger than that at 670 km. The correlation is not strong with the data for 200 km altitude.

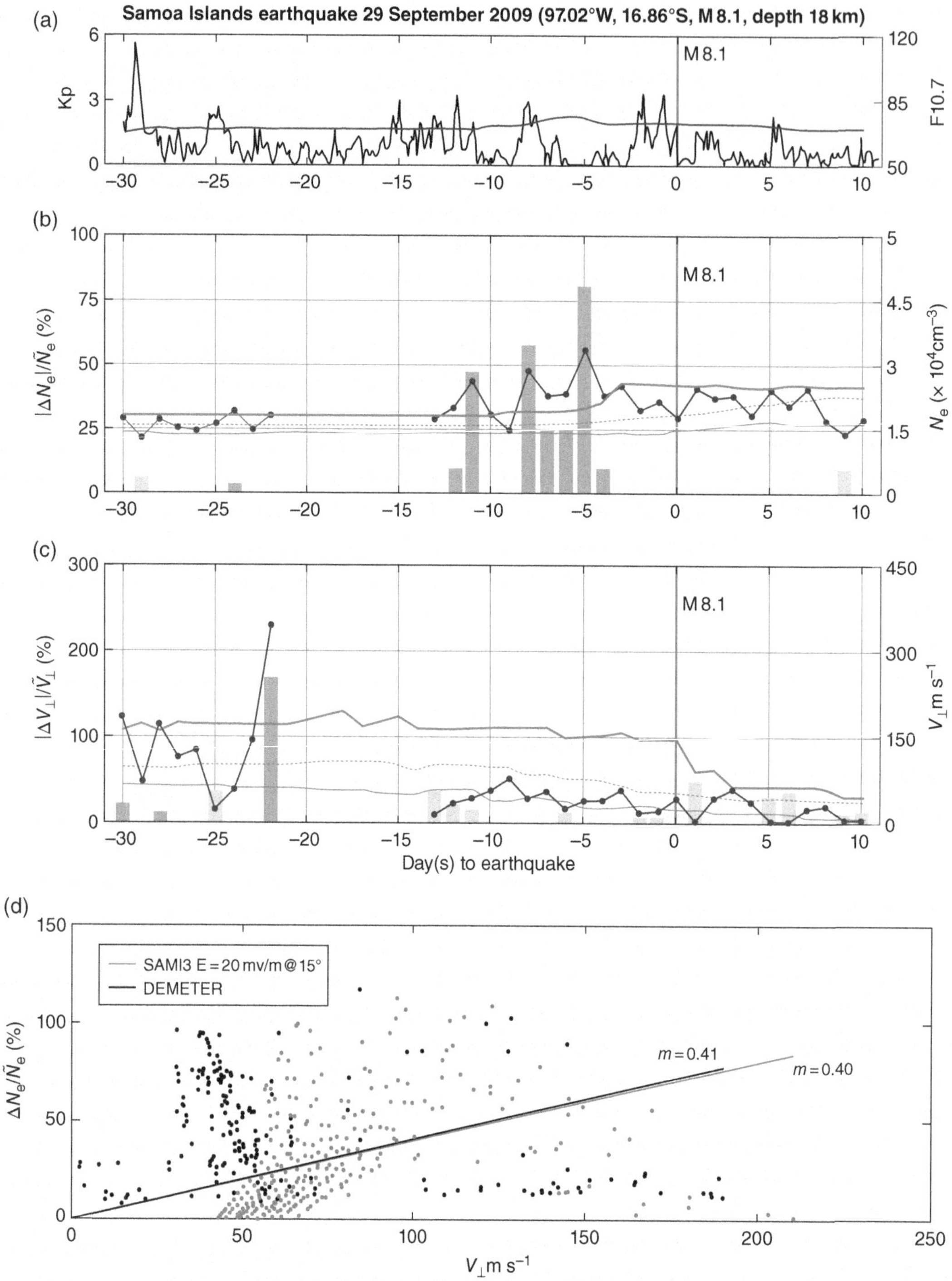

Figure 7.17 Variations of space environments and the ionosphere before and after the 2009 Samoa Islands earthquake. The geomagnetic latitude of the epicenter is 16.86°S.

Observations for the three individual earthquakes show that the slope value increases with the geomagnetic latitude. To gain insight on this observation we ran simulations by imposing an eastward electric field ($E = 20$ mV m^{-1}) at various geomagnetic latitudes from 10° to 35° (Figure 7.21). The data are chosen for a simulation time when the velocity reaches its maximum value. The electron density, density variation, parallel and perpendicular velocity in the meridional plane are shown for six cases at latitudes of 10°, 15°, 20°, 25.54°, 30°, and 35°. The distributions of density variation and velocity are field aligned. The relation of the density variation and perpendicular

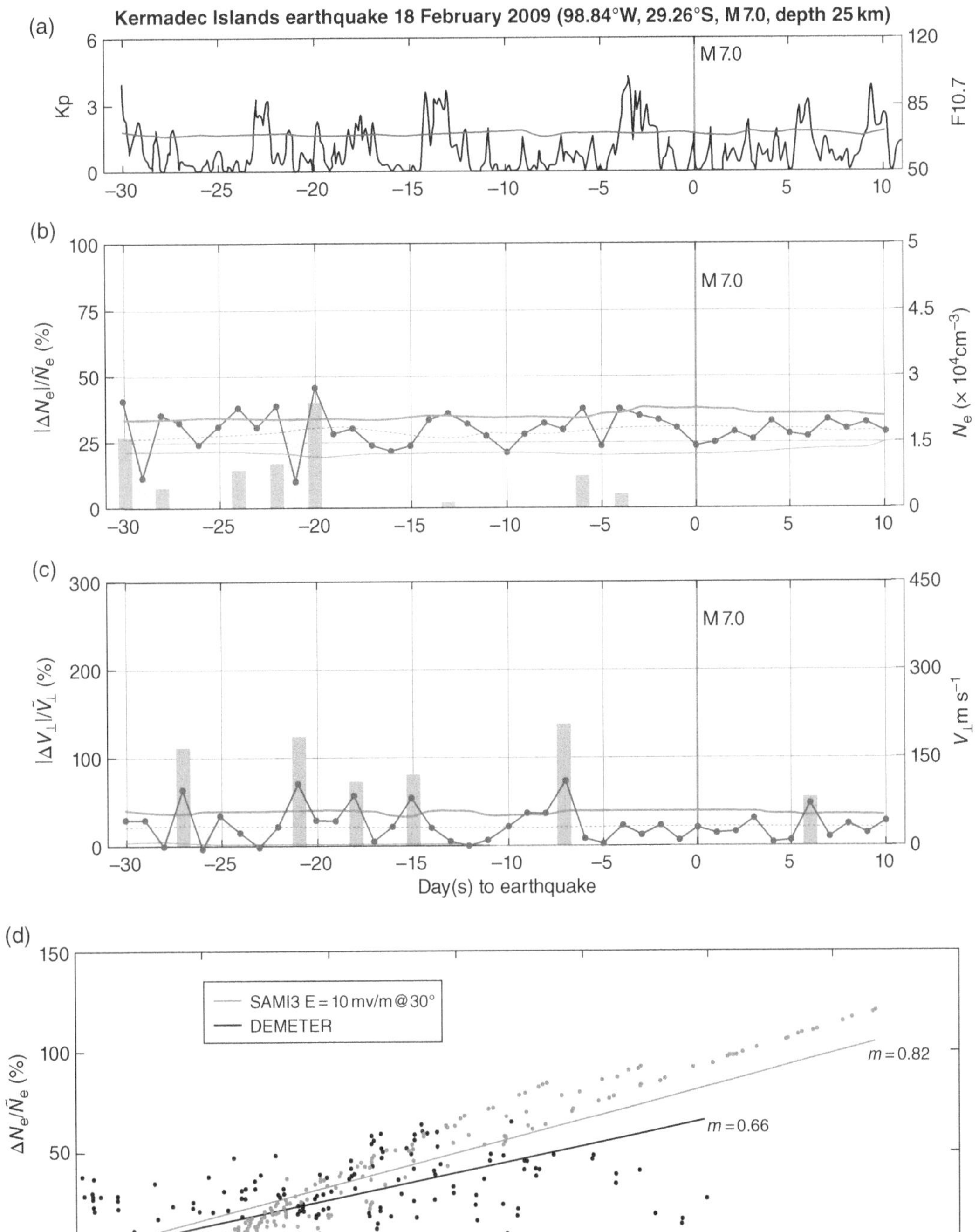

Figure 7.18 Variations of space environments and the ionosphere before and after the 2009 Kermadec Islands earthquake. The geomagnetic latitude of the epicenter is 29.26°S.

velocity at the satellite altitude (670 km) is shown in Figure 7.22. In general, slope increases with the latitude except at 20° latitude. The geomagnetic field line of 20° latitude passes through the EIA region. The **E** × **B** velocity uplifts the plasma of the EIA, leading to a larger density variation for the 20° case. For the 10° and 15° cases, the upper altitude of the magnetic field line passing through the source region is about 200 km for a source at 10° latitude and 460 km for a source at 15° latitude. As shown in Figure 7.21, the field-aligned density cannot reach the satellite altitude. The density variation and perpendicular velocity are not large at altitude 670 km for these two cases.

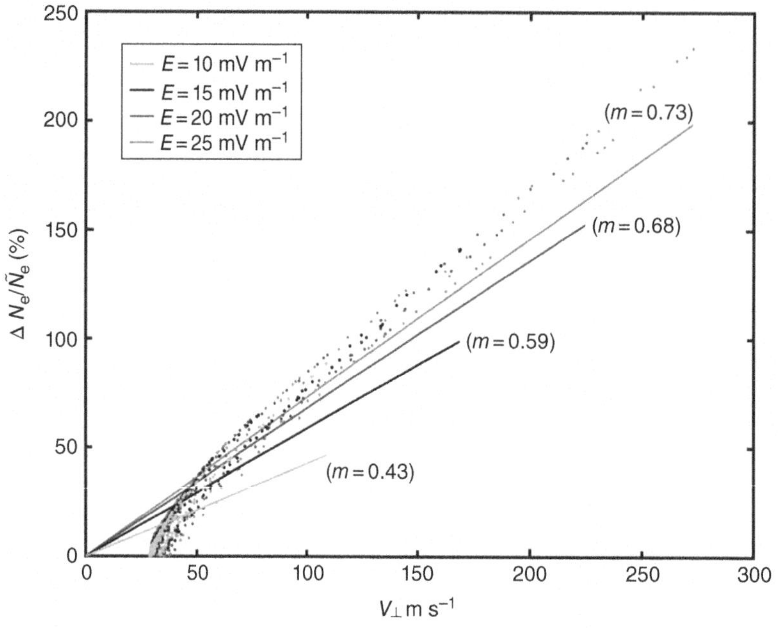

Figure 7.19 Relations between perpendicular velocity and density variation with different imposed electric fields at the epicenter latitude. The dots denote perpendicular velocity and density variation. The lines are best-fit lines and *m* is slope.

The DEMETER observations show that not only density but also perpendicular velocity change before large earthquakes: 2010 Chile, 2009 Kermadec Islands, and 2009 Samoa Islands earthquakes; For each earthquake we ran the simulations under similar geomagnetic conditions. Although the slopes are not exactly the same, the density variations and perpendicular velocity show positive relations in both observations and simulations. Furthermore, the slopes become larger when the latitudes increase in both observations and simulations. We can speculate the plasma motion from the simulation results. In our analysis, the perpendicular velocity can be treated as the result of triggering by the Hall electric field, which was imposed on the bottom side of the ionosphere.

7.6. DISCUSSION

7.6.1. LAIC Through Acoustic-Gravity Waves

Another possible LAIC mechanism is that of acoustic-gravity waves, which propagate from the sea/land boundary into the ionosphere. In doing so they cause disturbance of ionospheric plasma, which is a widely accepted phenomenon occurring the aftershock period of an earthquake [*Calais and Minster*, 1995; *Davies and Archambeau*, 1998; *Astafyeva and Afraimovich*, 2006; *DasGupta et al.*, 2006; *Astafyeva et al.*, 2011]. Recent numerical studies demonstrate the significant ionospheric variations caused by acoustic-gravity waves generated by the sea/land movement in comparison with GPS-measured TEC variations [*Shinagawa et al.*, 2007, 2013].

Acoustic-gravity waves, however, may not explain all ionospheric anomalies in the absence of significant sea/land movement before an earthquake. Simulation results show the percentage of TEC decrease can be as large as 20% for a tsunami with a velocity of 10 m s^{-1}, but this does not explain a precursor in the ionosphere (5% TEC increase 40 minutes before the main shock [*Heki*, 2011]) without a significant sea/land movement. Therefore, an alternative possible mechanism may be needed to account for the observed TEC increase of the ionosphere, e.g., electric coupling between the lithosphere, atmosphere and ionosphere. In comparison with gravity-acoustic waves, electric coupling processes can quickly affect the atmosphere and ionosphere.

7.6.2. Earthquake Lights

Other interesting events before an earthquake are associated light and electric-related activities [*Mizutani et al.*, 1976; *Lockner et al.*, 1983; *Derr*, 1986; *Derr and Persinger*, 1990; *St-Laurent et al.*, 2006; *Derr et al.*, 2011; *Heraud and Lira*, 2011]. Lights have been reported before large earthquakes ($M \geq 6$), e.g.: in the mid-1960s, lights were observed associated with a series of earthquakes before the Matsushiro earthquake on 4 December 1965 in Nagano, Japan; a few hours before the Cross Sound earthquake of 1 July 1973 (*M* 6.7); and the Pisco, Peru,

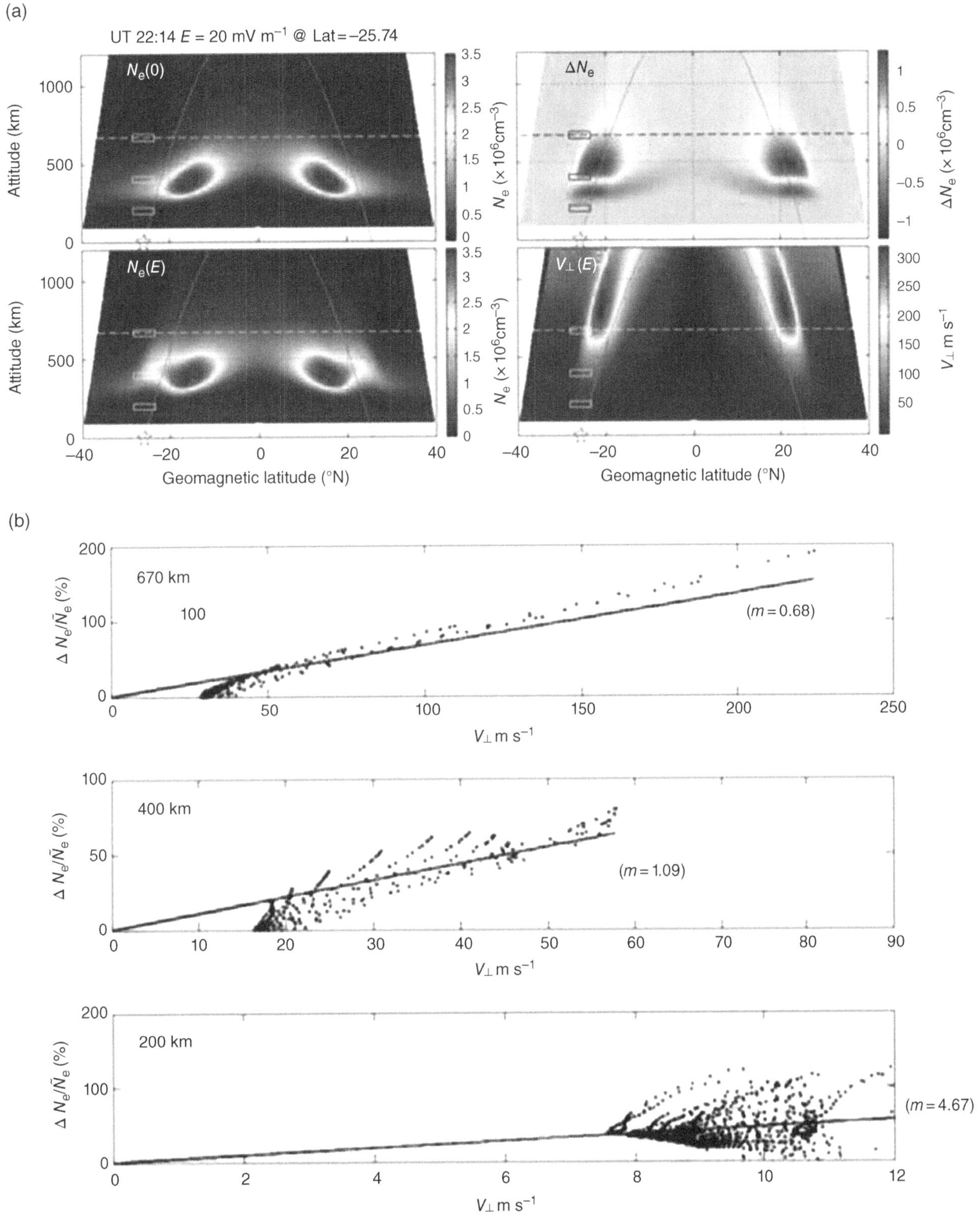

Figure 7.20 The ionospheric density and velocity distributions caused by an imposed eastward electric field (20 mV m^{-1}). (a) Density and velocity distributions in the meridional plane. The two left-hand panels are the density without and with an electric field. The two right-hand panels are the density variation and the velocity perpendicular to the geomagnetic field. (b) Relations between perpendicular velocity and density variations at a different altitude. From top to bottom panels the data represent 670 km, 400 km and 200 km.

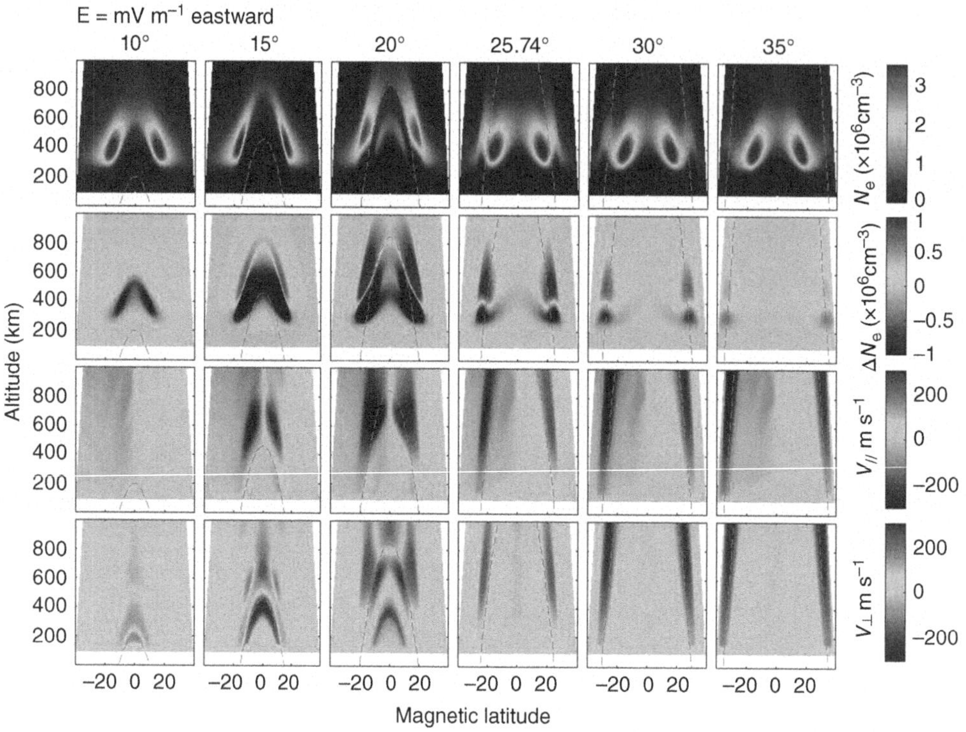

Figure 7.21 Simulation results with an eastward electric field (E = 20 mV m) at geomagnetic latitudes 5°, 10°, 15°, 20°, 25°, and 30°. From top to bottom the panels are electron density, density variation, parallel velocity, and perpendicular velocity.

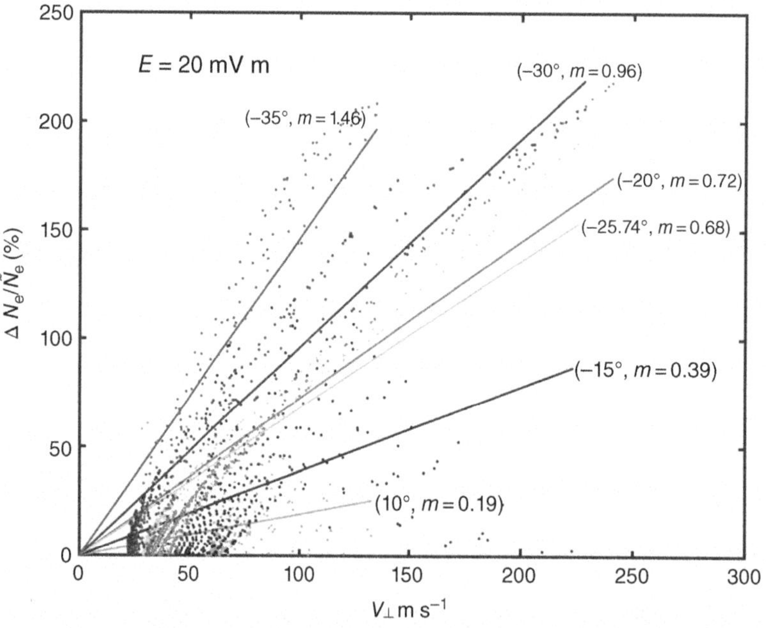

Figure 7.22 Relation between perpendicular velocity and density variation with an imposed electric field (20 mV m^{-1}) at different latitudes. The dots correspond to the data points of the perpendicular velocity and the density variation. The lines are their best-fit lines and m is the slope.

earthquake of August 2007 (*M* 8) [see website, http://news.nationalgeographic.com/news/2014/01/140106-earthquake-lights-earthquake-prediction-geology-science/]. For related electric activities, *Liu et al.* [2015], using statistical studies, also reported incidents of lightning activity near the earthquake region.

7.6.3. Decrease of Potential in Air (φ_{air}) Versus Dynamo Potential (φ_{dynamo})

Due to high resistivity (low conductivity) in the ambient atmosphere, decrease in integrated potential in air, φ_{air}, can be very large, where

$$\varphi_{air} = \int_0^{85\text{km}} E_z \text{d}z. \qquad (7.24)$$

Let φ_{dynamo} be the voltage across the dynamo (battery). If φ_{air} is much larger than φ_{dynamo}, the current flow from the dynamo source through the atmosphere to the lower ionosphere may be significantly reduced. There are several ways to resolve this issue as discussed below.

1. Current carried by charged radon ions from lithosphere through the lower part of atmosphere can reduce the decrease in potential across the atmosphere [*Sorokin and Hayakawa*, 2013; *Kuo et al.*, 2014].
2. Induced charges and large electric fields near Earth's surface can lead to ionizations of air molecules and reduced air resistivity, and hence φ_{air} [*Kuo et al.*, 2011].
3. An electric static field can be built up in the atmosphere and lower ionosphere from charges near Earth's surface, which are injected by the lithosphere dynamo current. The electrostatic electric field built up in the lower ionosphere can drive the motion of the ionosphere, leading to variations of electron density and TEC variations.
4. A large decrease in potential in the lower atmosphere or accumulation of charges near Earth's surface may lead to electric breakdown, forming filamentary current channels (lightning) to facilitate the passing of dynamo currents to the ionosphere.
5. A combination of electric field and filamentary current paths may also occur and lead to TEC variations.

7.7. SUMMARY

It is widely accepted that prediction of earthquakes seems impossible, especially among earthquake research scientists. Nevertheless, as Table 7.1 shows there are a significant number of reported precursor phenomena. The latter include: air ionization at the ground-to-air interface, changes in electrical conductivity of the soil, geoelectric and geomagnetic anomalies in Earth's crust, ionospheric perturbations, ULF and ELF and radiofrequency emissions, anomalous infrared emissions from an earthquake preparation zone, anomalous fog/haze/cloud formation, and unusual animal behavior [*Thériault et al.*, 2014].

Apart from the stressed-rock-activated electric current over Earth's surface, radon released from stressed crust can cause ionizing radiation in the atmosphere, where secondary charged particles (electrons and ionized particles) form. In addition, gamma rays produced in the radiation

Table 7.1 Possible anomalies associated with Earthquake activities.

Time relative to earthquake	Time period of precursor event	Anomalies observed	Proposed causal mechanism
Before	A few months	Radon and groundwater level	
	Several days	Earthquake lights [*Mizutani et al.*, 1976; *Lockner et al.*, 1983; *Derr*, 1986; *Derr and Persinger*, 1990; *St-Laurent et al.*, 2006; *Derr et al.*, 2011; *Heraud and Lira*, 2011] and lightning [*Liu et al.*, 2015]	Stressed-rock current [*Freund*, 2010]; atmospheric model [*Pasko*, 2013]
		VLF/ULF anomalies [*Hayakawa*, 2007; *Hayakawa et al.*, 2010, 2012]	Air ionization at the ground-to-air interface
		Thermal infrared anomalies [*Ouzounov and Freund*, 2004; *Ouzounov et al.*, 2006; *Pulinets et al.*, 2006]	Stressed-rock current [*Freund*, 2010]
		Ionospheric anomalies (see references for this chapter)	Radon ionization and conductivity change [*Pulinets and Boyarchuk*, 2004]
	Tens of minutes before	TEC increment [*Heki*, 2011]	Atmosphere–ionosphere coupling model [*Kuo et al.*, 2011, 2014]
After	Immediately	Coseismic acoustic-gravity waves [*Calais and Minster*, 1995; *Davies and Archambeau*, 1998; *Astafyeva and Afraimovich*, 2006; *DasGupta et al.*, 2006; *Astafyeva et al.*, 2011]	Vertical ground motion the cause of TEC variation [*Shinagawa et al.*, 2013]

process also cause atmospheric ionization. Charged particles in the atmosphere are quickly neutralized, however, in the absence of a charge-separation process, or the attachment of ion clusters, aerosols, or suspected dust grains ionization region is involved.

Experimental evidence of current generation in stressed rock is required as it could provide a plausible explanation of the uncertain relationship between nonseismic pre-earthquake electromagnetic events and actual seismic events [*Freund*, 2010]. In this endeavor we suggest setting up ground observation of a DC electric field or use VLF noise anomalies to monitor electric activities in the atmosphere and lithosphere. Finally, we consider the magnitude of a dynamo current flowing into the atmosphere and ionosphere. With $J_{max} \cong 200$ nA m^{-2}, we have $E = 8$ mV m^{-1} in the daytime ionosphere at a magnetic latitude of $-30°$, corresponding to $V_{\perp} = 200$ m s^{-1}, which has often been observed by DEMETER.

ACKNOWLEDGMENTS

This research is supported by the Ministry of Science and Technology grants MOST 104-2111-M-008-023, MOST 104-2111-M-001-003-MY2, MOST 104-2119-M-001-004, and MOST 104-2811-M-001-081. This research is also supported by grant MOTC-CWB-104-E-02 from the Center Weather Bureau.

REFERENCES

Afraimovich, E. L., Perevalova, N. P., Plotnikov, A. V., and Uralov, A. M. (1999), The shock-acoustic waves generated by earthquakes, *Ann. Geophys.*, *19*(4), 395–409, doi:10.5194/angeo-19-395-2001

Akhoondzadeh, M., Parrot, M., and Saradjian, M. R. (2010), Electron and ion density variations before strong earthquakes ($M>6.0$) using DEMETER and GPS data, *Nat. Hazards Earth Syst. Sci.*, *10*(1), 7–18, doi:10.5194/nhess-10-7-2010.

Astafyeva, E. I. and Afraimovich, E. L. (2006), Long-distance traveling ionospheric disturbances caused by the great Sumatra-Andaman earthquake on 26 December 2004, *Earth Planets Space*, *58*, 1025–1031.

Astafyeva, E. and Heki, K. (2009), Dependence of waveform of near-field coseismic ionospheric disturbances on focal mechanisms, *Earth Planets Space*, *61*, 939–943.

Astafyeva, E., Lognonné, P., and Rolland, L. (2011), First ionospheric images of the seismic fault slip on the example of the Tohoku-oki earthquake, *Geophys. Res. Lett.*, *38*(22), doi:10.1029/2011GL049623

Cahyadi, M. N. and K. Heki (2013), Ionospheric disturbances of the 2007 Bengkulu and the 2005 Nias earthquakes, Sumatra, observed with a regional GPS network, *J. Geophys. Res. Space Phys.*, *118*(4), 1777–1787, doi:10.1002/jgra.50208.

Calais, E. and Minster, J. B. (1995), GPS detection of ionospheric perturbations following the January 17, 1994, Northridge Earthquake, *Geophys. Res. Lett.*, *22*(9), 1045–1048, doi: 10.1029/95gl00168.

Chmyrev, V. M., Isaev, N. V., Bilichenko, S. V., and Stanev, G. (1989), Observation by space-borne detectors of electric fields and hydromagnetic waves in the ionosphere over an earthquake centre, *Phys. Earth Planet. Inter.*, *57*(1–2), 110–114, doi:10.1016/0031-9201(89)90220-3

Chuo, Y. (2002), The ionospheric perturbations prior to the Chi-Chi and Chia-Yi earthquakes, *J. Geodyn.*, *33*, 509–517, doi:10.1016/S0264-3707(02)00011-X

DasGupta, A., Das, A., Hui, D., Bandyopadhyay, K. K., and Sivaraman, M. R. (2006), Ionospheric perturbations observed by the GPS following the December 26th, 2004 Sumatra-Andaman earthquake, *Earth Planets Space*, *58*, 167–172.

Davies, J. B. and Archambeau, C. B. (1998), Modeling of atmospheric and ionospheric disturbances from shallow seismic sources, *Phys. Earth Planet. Inter.*, *105*, 183–199.

Davies, K. and Baker, D. M. (1965), Ionospheric effects observed around the time of the Alaskan Earthquake of March 28, 1964, *J. Geophys. Res.*, *70*, 2251–2253, doi:10.1029/JZ070i009p02251

Denisenko, V. V. (2015), Estimate for the strength of the electric field penetrating from the Earth's surface to the ionosphere, *Russ. J. Phys. Chem. B*, *9*: 789, doi:10.1134/S199079311505019X.

Derr, J. S. (1986), Rock mechanics: Luminous phenomena and their relationship to rock fracture, *Nature*, *321*(6069), 470–471.

Derr, J. S. and Persinger, M. A. (1990), Luminous phenomena and seismic energy in the Central United States, *J. Scient. Explor.*, *4*(1), 55–69.

Derr, J. S., St-Laurent, F., Freund, F. T., and Thériault, R. (2011), Earthquake lights, in *Encyclopedia of Earth Sciences Series, Encyclopedia of Solid Earth Geophysics*, H. K. Gupta (ed.), pp. 165–167, Springer.

Foppiano, A. J., Ovalle, E. M. Bataille, K., and Stepanova, M. (2008), Ionospheric evidence of the May 1960 earthquake over Concepción?, *Geofís. Int.*, *47*(3), 179–183.

Freund, F. (2000), Time-resolved study of charge generation and propagation in igneous rocks, *J. Geophys. Res.*, *105*, 11001–11020, doi:10.1029/1999JB900423

Freund, F. (2002), Charge generation and propagation in igneous rocks, *J. Geodyn.*, *33*, 543–570, doi:10.1016/S0264-3707(02)00015-7

Freund, F. (2010), Toward a unified solid state theory for pre-earthquake signals, *Acta Geophys.*, *58*(5), 719–766, doi:10.2478/s11600-009-0066-x

Freund, F. (2013), Earthquake forewarning—a multidisciplinary challenge from the ground up to space, *Acta Geophys.*, *61*(4), 775–807, doi:10.2478/s11600-013-0130-4

Freund, F. and Sornette, D. (2007), Electro-magnetic earthquake bursts and critical rupture of peroxy bond networks in rocks, *Tectonophysics*, *431*, 33–47.

Freund, F. T., Kulahci, I. G., Cyr, G., Ling, J., Winnick, M., Tregloan-Reed, J., and Freund, M. M. (2009), Air ionization at rock surfaces and pre-earthquake signals, *J. Atmos. Terr. Phys.*, *71*, 1824–1834.

Hayakawa, M. (2007), VLF/LF radio sounding of ionospheric perturbations associated with earthquakes, *Sensors*, *7*(7), 1141–1158, doi:10.3390/s7071141

Hayakawa, M., Kasahara, Y., Nakamura, T., Muto, F., Horie, T., Maekawa, S., Hobara, Y., Rozhnoi, A. A.,

Solovieva, M., and Molchanov, O. A. (2010), A statistical study on the correlation between lower ionospheric perturbations as seen by subionospheric VLF/LF propagation and earthquakes, *J. Geophys. Res. Space Phys.*, *115*(A9), A09305, doi:10.1029/2009ja015143
Hayakawa, M., Kasahara, Y., Endoh, T., Hobara, Y., and Asai, S. (2012), The observation of Doppler shifts of subionospheric LF signal in possible association with earthquakes, *J. Geophys. Res. Space Phys.*, *117*(A9), A09304, doi:10.1029/2012ja017752
He, Y., Yang, D., Zhu, R., Qian, J., and Parrot, M. (2010), Variations of electron density and temperature in ionosphere based on the DEMETER ISL data, *Earthquake Sci.*, *23*(4), 349–355, doi:10.1007/s11589-010-0732-8
Heki, K. (2011), Ionospheric electron enhancement preceding the 2011 Tohoku-Oki earthquake, *Geophys. Res. Lett.*, *38*(17), L17312, doi: 10.1029/2011gl047908
Heki, K. and Enomoto, Y. (2013), Preseismic ionospheric electron enhancements revisited, *J. Geophys. Res. Space Phys.*, *118*(10), 6618–6626, doi:10.1002/jgra.50578
Heraud, J. A. and Lira, J. A. (2011), Co-seismic luminescence in Lima, 150 km from the epicenter of the Pisco, Peru earthquake of 15 August 2007, *Nat. Hazards Earth Syst. Sci.*, *11*, 1025–1036.
Ho, Y.-Y., Jhuang, H.-K., Su, Y.-C., and Liu, J.-Y. (2013), Seismo-ionospheric anomalies in total electron content of the GIM and electron density of DEMETER before the 27 February 2010 M8.8 Chile earthquake, *Adv. Space Res.*, *51*(12), 2309–2315, doi:https://doi.org/10.1016/j.asr.2013.02.006
Kakinami, Y., Liu, J.-Y., Tsai, L.-C., and Oyama, K.-I. (2010), Ionospheric electron content anomalies detected by a FORMOSAT-3/COSMIC empirical model before and after the Wenchuan Earthquake, *Int. J. Remote Sens.*, *31*(13), 3571–3578, doi:10.1080/01431161003727788
Klimenko, M. V., Klimenko, V. V., Zakharenkova, I. E., Pulinets, S. A., Zhao, B., and Tsidilina, M. N. (2011), Formation mechanism of great positive TEC disturbances prior to Wenchuan earthquake on May 12, 2008, *Adv. Space Res.*, *48*(3), 488–499, doi:10.1016/j.asr.2011.03.040
Kuo, C. L. and Lee, L. C. (2015), Ionospheric plasma dynamics and instability caused by upward currents above thunderstorms, *J. Geophys. Res. Space Phys.*, *120*(4), 3240–3253, doi:10.1002/2014JA020767
Kuo, C. L. and Lee, L. C. (2017), Reply to comment by B. E. Prokhorov and O. V. Zolotov on "An improved coupling model for the lithosphere–atmosphere–ionosphere system", *J. Geophys. Res. Space Phys.*, *122*, doi:10.1002/2016JA023579
Kuo, C. L., Huba, J. D., Joyce, G., and Lee, L. C. (2011), Ionosphere plasma bubbles and density variations induced by pre-earthquake rock currents and associated surface charges, *J. Geophys. Res. Space Phys.*, *116*(A10), A10317, doi:10.1029/2011ja016628
Kuo, C. L., Lee, L. C., and Huba, J. D. (2014), An improved coupling model for the lithosphere–atmosphere–ionosphere system, *J. Geophys. Res. Space Phys.*, *119*(4), 3189–3205, doi:10.1002/2013JA019392
Kuo, C.-L., Lee, L.-C., and Heki, K. (2015), Preseismic TEC changes for Tohoku-Oki earthquake: Comparisons between simulations and observations, *Terr. Atmos. Ocean. Sci.*, *26*(1), 63–72, doi:10.3319/TAO.2014.08.19.06
Li, M. and Parrot, M. (2013), Statistical analysis of an ionospheric parameter as a base for earthquake prediction, *J. Geophys. Res. Space Phys.*, *118*(6), 3731–3739, doi:10.1002/jgra.50313
Liu, J. Y., Chen, Y. I., Pulinets, S. A., Tsai, Y. B., and Chuo, Y. J. (2000), Seismo-ionospheric signatures prior to $M >= 6.0$ Taiwan earthquakes, *Geophys. Res. Lett.*, *27*, 3113–3116, doi:10.1029/2000GL011395
Liu, J. Y., Chen, Y. I., Chuo, Y. J., and Tsai, H. F. (2001), Variations of ionospheric total electron content during the Chi-Chi earthquake, *Geophys. Res. Lett.*, *28*, 1383–1386, doi:10.1029/2000GL012511
Liu, J. Y., Chuo, Y., Shan, S., Tsai, Y., Chen, Y., Pulinets, S., and Yu, S. (2004), Pre-earthquake ionospheric anomalies registered by continuous GPS TEC measurements, *Ann. Geophys.*, *22*(5), 1585–1593, doi:10.5194/angeo-22-1585-2004
Liu, J. Y., Chen, Y. I., Chen, C. H., Liu, C. Y., Chen, C. Y., Nishihashi, M., Li, J. Z., Xia, Y. Q., Oyama, K. I., Hattori, K., and Lin, C. H. (2009), Seismoionospheric GPS total electron content anomalies observed before the 12 May 2008 Mw7.9 Wenchuan earthquake, *J. Geophys. Res.*, *114*, 04320, doi:10.1029/2008JA013698
Liu, J. Y., Tsai, H. F., Lin, C. H., Kamogawa, M., Chen, Y. I., Huang, B. S., Yu, S. B., and Yeh, Y. H. (2010), Coseismic ionospheric disturbances triggered by the Chi-Chi earthquake, *J. Geophys. Res.*, *115*(A8), A08303, doi:10.1029/2009ja014943
Liu, J. Y., Le, H., Chen, Y. I., Chen, C. H., Liu, L., Wan, W., Su, Y. Z., Sun, Y. Y., Lin, C. H., and Chen, M. Q. (2011), Observations and simulations of seismoionospheric GPS total electron content anomalies before the 12 January 2010 M7 Haiti earthquake, *J. Geophys. Res. Space Phys.*, *116*(A4), doi:10.1029/2010JA015704.
Liu, J. Y., Chen, Y. I., Huang, C. H., Ho, Y. Y., and Chen, C. H. (2015), A statistical study of lightning activities and $M \geq 5.0$ earthquakes in Taiwan during 1993–2004, *Surv. Geophys.*, *36*(6), 851–859, doi:10.1007/s10712-015-9342-2
Lockner, D. A., Johnston, M. J. S., and Byerlee, J. D. (1983), A mechanism to explain the generation of earthquake lights, *Nature*, *302*, 28–33.
Mizutani, H., Ishido, T., Yokokura, T., and Ohnishi, S. (1976), Electrokinetic phenomena associated with earthquakes, *Geophys. Res. Lett.*, *3*, 365–368.
Molchanov, O., Fedorov, E., Schekotov, A., Gordeev, E., Chebrov, V., Surkov, V., Rozhnoi, A., Andreevsky, S., Iudin, D., Yunga, S., Lutikov, A., Hayakawa, M., and Biagi, P. F. (2004), Lithosphere–atmosphere–ionosphere coupling as governing mechanism for preseismic short-term events in atmosphere and ionosphere, *Nat. Hazards Earth Syst. Sci.*, *4*(5/6), 757–767, doi:10.5194/nhess-4-757-2004
Moore, G. W. (1964), Magnetic disturbances preceding the 1964 Alaska Earthquake, *Nature*, *203*(4944), 508–509, doi:10.1038/203508b0
Namgaladze, A. A., Zolotov, O. V., Karpov, M. I., and Romanovskaya, Y. V. (2012), Manifestations of the earthquake preparations in the ionosphere total electron content variations, *Natural Sci.*, *4*(11), 848–855, doi:10.4236/ns.2012.411113.
Ouzounov, D. and Freund, F. (2004), Mid-infrared emission prior to strong earthquakes analyzed by remote sensing data,

Adv. Space Res., *33*(3), 268–273, doi:10.1016/S0273-1177(03)00486-1

Ouzounov, D., Bryant, N., Logan, T., Pulinets, S., and Taylor, P. (2006), Satellite thermal IR phenomena associated with some of the major earthquakes in 1999–2003, *Phys. Chem. Earth Parts A/B/C*, *31*(4–9), 154–163, doi:10.1016/j.pce.2006.02.036

Ouzounov, D., Pulinets, S. Tramutoli, V., Liu, T., Hattori, K., Parrot, M., Namgaladze, A., and Solomentsev, D. (2014), Validation of lithosphere–atmosphere–ionosphere coupling concept by geo space observation of natural and anthropogenic processes, paper presented at the *XXXIth International Union of Radio Science General Assembly and Scientific Symposium (URSI GASS)*, 16–23 August.

Oyama, K. I., Kakinami, Y., Liu, J. Y., Abdu, M. A., and Cheng, C. Z. (2011), Latitudinal distribution of anomalous ion density as a precursor of a large earthquake, *J. Geophys. Res. Space Phys.*, *116*(A4), A04319, doi:10.1029/2010ja015948

Parrot, M. (1995), Use of satellites to detect seismo-electromagnetic effects, *Adv. Space Res.*, *15*(11), 27–35.

Parrot, M., Achache, J., Berthelier, J. J., Blanc, E., Deschamps, A., Lefeuvre, F., Menvielle, M., Plantet, J. L., Tarits, P., and Villain, J. P. (1993), High-frequency seismo-electromagnetic effects, *Phys. Earth Planet. Inter.*, *77*(1–2), 65–83.

Parrot, M., Berthelier, J. J., Lebreton, J. P., Sauvaud, J. A., Santolik, O., and Blecki,J. (2006), Examples of unusual ionospheric observations made by the DEMETER satellite over seismic regions, *Phys. Chem. Earth, Parts A/B/C*, *31*(4), 486–495, doi:https://doi.org/10.1016/j.pce.2006.02.011

Pasko, V. P. (2013), *Time-dependent earth surface-ionosphere coupling model: application to earthquake lights and global electric circuit*, Frontiers in Earth System Dynamics Meeting, NCAR, Boulder, CO.

Pulinets, S. A. (2009), Physical mechanism of the vertical electric field generation over active tectonic faults, *Adv. Space Res.*, *44*(6), 767–773.

Pulinets, S., and Boyarchuk, K. (2004), *Ionospheric Precursors of Earthquakes*, Springer, Berlin, 315 pp.

Pulinets, S. A. and Legen'ka, A. D. (2003), Spatial-temporal characteristics of large scale distributions of electron density observed in the ionospheric F-region before strong earthquakes, *Cosmic Res.*, *41*(3), 221–229.

Pulinets, S., and Ouzounov, D. (2011), Lithosphere–atmosphere–ionosphere coupling (LAIC) model—an unified concept for earthquake precursors validation, *J. Southeast Asian Earth Sci.*, *41*(4–5), 371–382, doi:10.1016/j.jseaes.2010.03.005

Pulinets, S. A., Boyarchuk, K. A., Hegai, V. V., Kim, V. P., and Lomonosov, A. M. (2000), Quasielectrostatic model of atmosphere-thermosphere-ionosphere coupling, *Adv. Space Res.*, *26*(8), 1209–1218.

Pulinets, S. A., Legen'ka, A. D., Gaivoronskaya, T. V., and Depuev, V. K. (2003), Main phenomenological features of ionospheric precursors of strong earthquakes, *J. Atmos. Terr. Phys.*, *65*(16–18), 1337–1347, doi:10.1016/j.jastp.2003.07.011.

Pulinets, S. A., Ouzounov, D., Karelin, A. V., Boyarchuk, K. A., and Pokhmelnykh, L. A. (2006), The physical nature of thermal anomalies observed before strong earthquakes, *Phys. Chem. Earth Parts A/B/C*, *31*(4–9), 143–153, doi:10.1016/j.pce.2006.02.042

Richmond, A. D. (1995), Ionospheric electrodynamics using magnetic apex coordinates, *J. Geomagnet. Geoelectri.*, *47*(2), 191–212, doi:10.5636/jgg.47.191

Schekotov, A. Y., Molchanov, O. A., Hayakawa, M., Fedorov, E. N., Chebrov, V. N., Sinitsin, V. I., Gordeev, E. E., Belyaev, G. G., and Yagova, N. V. (2007), ULF/ELF magnetic field variations from atmosphere induced by seismicity, *Radio Sci.*, *42*(6), RS6S90, doi:10.1029/2005RS003441

Shinagawa, H., Iyemori, T., Saito, S., and Maruyama, T. (2007), A numerical simulation of ionospheric and atmospheric variations associated with the Sumatra earthquake on December 26, 2004, *Earth Planets Space*, *59*, 1015–1026.

Shinagawa, H., Tsugawa, T., Matsumura, M., Iyemori, T., Saito, A., Maruyama, T., Jin, H., Nishioka, M., and Otsuka, Y. (2013), Two-dimensional simulation of ionospheric variations in the vicinity of the epicenter of the Tohoku-oki earthquake on 11 March 2011, *Geophys. Res. Lett.*, *40*(19), 5009–5013, doi:10.1002/2013GL057627

Smirnova, N. A. and Hayakawa, M. (2007), Fractal characteristics of the ground-observed ULF emissions in relation to geomagnetic and seismic activities, *J. Atmos. Terr. Phys.*, *69*, 1833–1841, doi:10.1016/j.jastp.2007.08.001

Sorokin, V. and Hayakawa, M. (2013), Generation of seismic-related DC electric fields and lithosphere–atmosphere–ionosphere coupling, *Modern Appl. Sci.*, *7*(6), 1–23, doi: 10.5539/mas.v7n6p1

Sorokin, V., Chmyrev, V., and Hayakawa, M. (2015), *Electrodynamic Coupling of Lithosphere, Atmosphere, Ionosphere of the Earth*. Nova Science Publishers, 1–326.

St-Laurent, F., Derr, J. S., and Freund, F. T. (2006), Earthquake lights and the stress-activation of positive hole charge carriers in rocks, *Phys. Chem. Earth*, *31*, 305–312.

Surkov, V. V. and Hayakawa, M. (2007), ULF electromagnetic noise due to random variations of background atmospheric current and conductivity, *J. Geophys. Res. (Atmos.)*, *112*, D11116, doi:10.1029/2006JD007788.

Thériault, R., St-Laurent, F., Freund, F. T., and Derr, J. S. (2014), Prevalence of earthquake lights associated with rift environments, *Seismol. Res. Lett.*, *85*(1), 159–178, doi:10.1785/0220130059

Uyeda, S., Nagao, T., and Kamogawa, M. (2009), Short-term earthquake prediction: Current status of seismo-electromagnetics, *Tectonophysics*, *470*, 205–213, doi:10.1016/j.tecto.2008.07.019

Varotsos, P. A. (2006), Recent seismic electric signals (SES) activities in Greece, *Acta Geophys.*, *54*, 158–164, doi:10.2478/s11600-006-0019-6

Wilson, C. T. R. (1921), Investigations on lightning discharges and on the electric field of thunderstorms, *Philos. Trans. R.Soc. Lon.*, *221*, 73–115.

Zolotov, O. V., Prokhorov, B. E., Namgaladze, A. A., and Martynenko, O. V. (2011), Variations in the total electron content of the ionosphere during preparation of earthquakes, *Russ. J. Phys. Chem. B*, *5*(3), 435–438, doi:10.1134/s1990793111030146

Zolotov, O. V., Namgaladze, A. A., Zakharenkova, I. E., Martynenko, O. V., and Shagimuratov, I. I. (2012), Physical interpretation and mathematical simulation of ionospheric precursors of earthquakes at midlatitudes, *Geomagn. Aeron.*, *52*(3), 390–397, doi:10.1134/S0016793212030152

Part III: Pre-Earthquake Seismic Phenomena

8

Short-Term Foreshocks and Earthquake Prediction

Gerassimos Papadopoulos, George Minadakis, and Katerina Orfanogiannaki

ABSTRACT

Foreshock recognition before main shocks depends on various factors, e.g. geophysical, catalog completeness, foreshock definition, and spatiotemporal windows. Foreshocks move towards the main shock epicenter as the event approaches, their number increases as time to the event decreases, and their *b*-value decreases. However, only in a very few single foreshock sequences have these three-dimensional patterns been recognized at the same time, e.g. before the 2009 L' Aquila (Italy) earthquake (*Mw* 6.3) and the 2010, 2014 and 2015 major earthquakes (*Mw* 8+) in the Chilean subduction zone. For the first time we found statistically significant three-dimensional foreshock patterns before two small-to-moderate earthquakes occurring on 4 March 2012 (*Mw* 5.2) and 3 July 2013 (*Mw* 4.8) in Athos and Polyphyto, northern Greece. The marked similarity of the patterns found before strong and major earthquakes indicates that the foreshock process is scale invariant in a wide magnitude range. It is likely that the process is independent of the faulting type, at least for dip-slip faulting. There is also a trend of the main shock magnitude to scale with the foreshock area. These findings imply that foreshock activity is likely governed by a universal pattern, which may also reflect universality in the deformation process thus opening new directions for utilization of the foreshock pattern in earthquake prediction.

8.1. INTRODUCTION

Short-term foreshocks increase the probability for the occurrence of a future strong main shock [*Yamashina*, 1981; *Jones*, 1985b; *Agnew and Jones*, 1991; *Console et al.*, 1993; *Savage and DePolo*, 1993; *Dodge et al.*, 1995; *Michael*, 2012]. Therefore, foreshock activity has been considered as one of the most promising precursory phenomena for the prediction of the main shock [e.g. *Wyss*, 1997; *Vidale et al.*, 2001]. For example, the short-term prediction of the large (*M* 7.2) earthquake of 4 February 1975 in Haicheng, China, was based on several precursory phenomena, particularly short-term foreshocks [*Raleigh et al.*, 1977; *Scholz*, 1977; *Jones et al.*, 1982]. Short-term foreshocks are considered in a time frame varying from minutes up to several months before the main shock.

Superiority of foreshock activity as compared to other candidate earthquake precursors is apparent because (a) foreshocks occur exactly in the area that is highly stressed before the main shock and (b) the foreshock three-dimensional distribution in the space, time, and size domains likely follows a certain pattern.

In spite of the progress noted so far as regards the precursory value of foreshocks, there still remain some very important issues of uncertainty. The first regards foreshock occurrence pattern: why are some main shocks preceded by foreshocks while others are not? Of relevance to this question of foreshock incidence concerns whether they are associated with large main shocks only, or also with moderate or even small main shocks. The second

Institute of Geodynamics, National Observatory of Athens, Athens, Greece

Pre-Earthquake Processes: A Multidisciplinary Approach to Earthquake Prediction Studies, Geophysical Monograph 234, First Edition. Edited by Dimitar Ouzounov, Sergey Pulinets, Katsumi Hattori, and Patrick Taylor.

important issue refers to the three-dimensional pattern that governs foreshock activity: do all foreshock sequences have a similar pattern? Is this pattern dependent on the main shock size and faulting type? Importantly, is it possible to use three-dimensional foreshock pattern to discriminate foreshocks from other types of spatiotemporal seismicity clusters, like swarms and aftershocks? Is such a pattern useful for the prediction of the main shock? In this chapter we examine these fundamental issues using seismicity data from several earthquake areas worldwide.

8.2. FORESHOCK RATE

One of the very early reports on the rate of foreshocks preceding main shocks was that of *Imamura* [1937, p. 78], who reported that the number of earthquakes that are ushered in by foreshocks barely reaches 20% of the total number. He added, however, that with the sensitive seismometers now in use this proportion would increase. In the aftermath of the prediction of the 4 February 1975, Haicheng (China) earthquake, increased interest was shown in the predictive value of foreshocks. As a consequence, studies focused on whether foreshocks are a common occurrence. In a global review of large ($M \geq 7$) shallow earthquakes, *Jones and Molnar* [1976] found that 44% in the National Oceanic and Atmospheric Administration list had at least one foreshock. Yet, they proposed that this percentage was too low an estimate since foreshock recognition is strongly influenced by seismograph coverage and event cataloging. In another global review, *Seggern* [1981] found a foreshock rate of <20% for main shocks of magnitude ≥5.8, but again this was dependent on earthquake monitoring capabilities and foreshock definition.

In addition, contradictory results were also found in several seismic regions of the Earth. In California and in the western cordillera of USA, estimations of the foreshock rate varied from 35% [*Jones*, 1985a], to 50% [*Agnew and Jones*, 1991], and to 60% [*Doser*, 1989], depending on the time period and the main shock magnitude threshold considered as well as on the spatiotemporal foreshock definition adopted. In the North Pacific region large interplate earthquakes are preceded by foreshock sequences at a rate of 70% [*Bouchon et al.*, 2013]. On the Chinese mainland, *Chen et al.* [1999] found that only 5% of the 159 main shocks ($M > 1.5$) had foreshocks in the period 1966–1996. However, this result was strongly influenced by the *a priori* foreshock definition adopted by the authors of that paper; foreshock occurrence was considered within 5 days and 20 km of the main shock, with occurrence rate exceeding ten events per day with $M \geq 1.5$. At the other extreme, for the thrust fault system of Chengkung, eastern Taiwan, all six of the larger earthquakes ($M_L \geq 5$) occurring between 1992 and 2003 had foreshocks at a distance of only a few kilometers within just a few days of the main shock [*Lin*, 2004].

Foreshock rate dependence on predefined spatiotemporal windows is also indicated by several studies in various parts of the world, such as in Japan [*Yamashina*, 1981; *Maeda*, 1999; *Imoto*, 2005], *Kamchatka* [*Sobolev*, 2000], New Zealand [*Merrifield et al.*, 2004], Albania [*Peçi et al.*, 1999], Italy and Greece [*Di Luccio et al.*, 1997; *Console et al.*, 1993, *Papadopoulos et al.*, 2000]. A characteristic example on how foreshock recognition depends on the foreshock definition adopted is provided by the case of the 12 November 1999 Düzce earthquake (*Mw* 7.1) of the North Anatolian Fault. *Wu et al.* [2014] investigated for foreshocks preceding that large earthquake within only ~65 h before and within 20 km of the main shock. No foreshock activity was found. To use foreshocks as a potential tool for short-term prediction of the main shock it is of great importance to discriminate between foreshocks and other types of seismicity clusters, e.g. swarms and main-shock–aftershock sequences [e.g. *Ogata et al.*, 1996], but this again depends on the foreshock definition.

The foreshock rate depends also on the completeness of the earthquake catalog [e.g. *Yamaoka et al.*, 1999]; small seismic events, including foreshocks, usually are not included in standard catalogs produced by routine daily seismic analysis; an example is provided by the strong ($Mw = 5.8$) earthquake of 20 October 2005 in the eastern Aegean Sea, Greece, which was reported by *Papadopoulos et al.* [2006]. No foreshocks were recognized in the standard catalog produced by the National Observatory of Athens, Greece, however, analysis of the records at one local station installed close to the epicenter showed that at least 50 foreshocks of small magnitude preceded the main shock.

Incidence or not of foreshocks very likely depends on a variety of geophysical factors, such as the style of faulting, focal depth, and degree of small-scale crustal heterogeneity [e.g. *Abercrombie and Mori*, 1996; *Cheng and Wong*, 2016]. This may explain at least in part why only some main shocks are preceded by short-term foreshocks. On the other hand, catalog incompleteness as well as the adoption of very narrow spatiotemporal limits when searching for foreshocks may lead to foreshock events being missed. Therefore, these important issues should be taken into account in the investigation of foreshock patterns in the space–time–size domains.

8.3. FORESHOCK PATTERNS

The possibility that foreshock sequences perhaps follow recognizable patterns in the space, time, and size domains was explored as early as the 1960s. *Suyehiro et al.* [1964] observed an increase in seismicity before the great Chilean earthquake (*Mw* 9.5) of 22 May 1960, as well as a decrease in *b*-value [*Suyehiro*, 1966]. Increase in precursory signals of crustal failure is also supported by the pioneering laboratory experiments on material fracture conducted by *Mogi* [1963a,b] and *Scholz* [1968]. Subsequently, it was verified that the foreshock increase in time as the earthquake approached follows a pattern of acceleration [e.g. *Papazachos*, 1975; *Ishida and Kanamori*, 1978; *Kagan and Knopoff*, 1978; *Jones and Molnar*, 1979; *Maeda*, 1999; *Yamaoka et al.*, 1999; *Papadopoulos et al.*, 2000, 2010]. Averaging of data over several foreshock sequences revealed that the foreshock acceleration process fits well with a power-law distribution [e.g. *Jones and Molnar*, 1979]; an observation further supported by numerical modeling using spring-blocks [e.g. *Hainzl et al.*, 1999] as well as by modeling of analytical damage mechanics [e.g. *Main*, 2000; *Yamashita and Knopoff*, 2007]. In addition, it has been found that foreshocks tend to move towards the main shock epicenter [e.g. *Engdahl and Kisslinger*, 1978; *Chen et al.*, 1999; *Papadopoulos et al.*, 2010; *Lippiello et al.*, 2012].

In the size domain the critical parameter is *b*, which is the slope of the straight line in the magnitude–frequency (or G–R; Gutenberg–Richter) relation [*Ishimoto and Iida*, 1939; *Gutenberg and Richter*, 1944]

$$\log N = a - bM \qquad (8.1)$$

where N is the cumulative number of events of magnitude equal to or larger than M and a and b are variables determined by the data. Therefore, b is an experimentally observed macroscopic variable measuring the ratio of small-magnitude events relative to large-magnitude events. At the global scale of seismicity, the *b*-value was found to be around unity [e.g. *Frohlich and Davis*, 1993; *Schorlemmer et al.*, 2005]. However, locally or even regionally this parameter varies due to its strong dependence on stress-loading conditions [*Mogi*, 1963a; *Scholz*, 1968] and on crustal heterogeneity [e.g. *Abercrombie and Mori*, 1996]. A low value of b indicates enhancement in higher magnitude events, high stress conditions, and/or increased material heterogeneity. Inversely, a high *b*-value is interpreted as evidence for enhancement of events of smaller magnitudes, low stress, and/or decreased material heterogeneity. Therefore, b could be considered as a stress meter that depends inversely on differential stress [e.g. *Schorlemmer et al.*, 2005]. From simulations on Olami–Feder–Christensen spring-block models, *Avlonitis and Papadopoulos* [2014] showed that the low values of b experimentally observed in foreshock sequences can be modeled by a process of material softening in the seismogenic volume. If foreshock activity is a signature of high stress conditions, then it may be expected that the b value would decrease with respect to background seismicity and aftershocks. This is exactly the finding of several authors, including *Mogi* [1963a,b, 1985], *Suyehiro and Sekiya* [1972], *Papazachos* [1975], *Jones and Molnar* [1979], *Main et al.* [1989], *Molchan et al.* [1999], *Chan et al.* [2012], *Kato et al.* [2012], and *Nanjo et al.* [2012].

Until the occurrence of the strong (*Mw* 6.3) L' Aquila, central Italy, earthquake of 6 April 2009, no single earthquake sequence was known to be characterized by the presence of foreshock patterns in the three domains at the same time. In single foreshock sequences, patterns in only one or two domains were found, thus making questionable their universality. In addition, only in a few cases were the precursory seismicity changes, determined as foreshock activities, tested for their statistical significance. These issues are of crucial importance for organizing a research strategy aiming to investigate three-dimensional foreshock patterns. In this endeavor two alternative approaches may be followed. The first is to examine global or regional earthquake catalogs looking for foreshocks before a set of strong earthquakes, potentially considered as main shocks, with the adoption of a magnitude value cut-off and of pre-selected spatiotemporal limits. Global and regional catalogs are very inhomogeneous, however, particularly in the magnitude domain and are characterized by much variation in their completeness in space and time domains. In addition, as described earlier, foreshock rate depends directly on the definition of foreshock, while preselection of narrow spatiotemporal limits may lead to artificial elimination of foreshocks. Such difficulties may explain the contradictory results obtained about the level of foreshock rate in various seismogenic regions of the Earth.

Another alternative is to investigate good examples of single foreshock sequences in sufficiently complete catalogs of earthquakes using a low cut-off value for magnitude. Such catalogs can be found in areas of dense, continuous monitoring coverage, and are compiled on the basis of standard seismic analysis procedures. This strategy was applied to the L'Aquila 2009 earthquake, which is the first single foreshock sequence where three-dimensional foreshock patterns were recognized to occur at the same time, and which have been tested for statistical significance using the national seismic catalog of Italy's National Institute of Geophysics and Volcanology (INGV) [*Papadopoulos et al.*, 2010; *Daskalaki et al.*, 2016] (Table 8.1).

Another excellent example is the great Iquique, Chile, earthquake (*Mw* 8.1) of 1 April 2014. Several studies supported that foreshock activity very likely controlled the initiation of the great earthquake through gradual unlocking of the plate boundary [*Brodsky and Lay*, 2014; *Kato and Nakagawa*, 2014; *Lay et al.*, 2014; *Ruiz et al.*, 2014; *Schurr et al.*, 2014; *Yagi et al.*, 2014; *Bedford et al.*, 2015; *Meng et al.*, 2015; *Duputel et al.*, 2015]. Different opinions, however, were expressed regarding the initiation time of the activity and the extent

Table 8.1 Parameters of earthquake sequences, E, having significant foreshock patterns in the domains of space, time and magnitude before the earthquake events examined in the present paper.

Event	Date	Time	*Mw*	h (km)	Latitude	Longitude	M_c	M_f	Radius[d] (km)	d_w	d_s
L' Aquila, Italy[a]	6 April 2009	01:32:39	6.3	10	42.42°N	13.39°E	1.3	4.0	15	150	10
Maule, Chile[b]	27 February 2010	06:34:08	8.8	30	36.29°S	73.24°W	3.0	5.6	170	103	-
Athos, Greece[c]	4 March 2012	03:31:07	5.2	23	40.13°N	24.05°E	1.4	5.1	10	53	18
Polyphyto, Greece[c]	3 July 2013	13:28:23	4.8	20	40.14°N	21.87°E	1.1	4.7	10	236	1.1
Iquique, Chile[b]	1 April 2014	23:46:45	8.1	39	19.57°S	70.91°W	2.6	6.7	65	53	17
Illapel, Chile[b]	16 September 2015	22:54:31	8.4	23	31.64°S	71.74°W	3.0	5.3	50	108	-

Mw, moment magnitude; *h*, focal depth; M_c, cut-off magnitude; M_f, maximum foreshock magnitude; d_w, duration in days of weak foreshock stage; d_s, duration in days of strong foreshock stage.

[a]*Papadopoulos et al.* [2010].

[b]*Papadopoulos and Minadakis* [2016].

[c]Present study).

[d]Radius of circular foreshock area considered for the strong foreshock stage; if only weak foreshock stage was detected this was taken into account.

of the foreshock area. For example, *Schurr et al.* [2014] proposed that foreshock activity was developed at least 500 days or so before the main shock, but *Ruiz et al.* [2014] found that the precursory foreshock activity started on 4 January 2014. As for the foreshock area, *Schurr et al.* [2014] suggested that it extended from 19.00°S to 21.00°S (ca. 220 km) along the subduction zone strike, while *Ruiz et al.* [2014] determined a foreshock area that spans ~150 km along the strike of the subduction zone. Such discrepancies can be explained by the realization that interest was focused on investigating how foreshocks contributed to the stress accumulation before the great Iquique 2014 main shock rather than on three-dimensional foreshock patterns and their deviation from randomness through testing for statistical significance; only the activity rate variation was statistically tested by *Schurr et al.* [2014], who found significant changes during the 250 days leading up to the main shock.

Systematic investigation of three-dimensional patterns revealed that the great Iquique 2014 earthquake was preceded by statistically significant patterns quite similar to those found before the L'Aquila 2009 main shock [*Papadopoulos and Minadakis*, 2016] (Table 8.1). Of particular importance is the difference in magnitude and the seismotectonic regimes associated with the two earthquakes: crustal normal faulting for L'Aquila 2009, subduction-related thrust for Iquique 2014. Similar three-dimensional foreshock patterns preceded the great Chilean earthquakes of Maule, 27 February 2010 (*Mw* 8.8) and Illapel, 16 September 2015 (*Mw* 8.4) [*Papadopoulos and Minadakis*, 2016] (Table 8.1). However, in these cases only weak foreshock activity was detected, likely because of catalog incompleteness.

Later in this chapter we examine two additional good examples, those of small-to-moderate earthquakes (*Mw* 5.2 and *Mw* 4.8) that occurred in Greece in 2012 and 2013 (Table 8.1). These cases, along with those of L'Aquila in 2009 and of Chile in 2010, 2014 and 2015, cover a wide spectrum for the main shock magnitude and represent various seismic areas of the world characterized by different seismotectonic regimes.

8.4. METHOD OF ANALYSIS AND COMPUTATIONAL NOTES

Seismicity analysis for the *a posteriori* identification of precursory foreshock patterns was performed using the in-house developed FORMA (FOReshock–Main shock–Aftershock) software that has been explained elsewhere [e.g. *Papadopoulos et al.*, 2010; *Papadopoulos and Minadakis*, 2016]. Therefore, we only briefly describe the method and the computational details that we followed in our analysis.

Foreshock patterns were investigated in the space, time and size (magnitude) domains. In the space domain, the Euclidean distance, D, of earthquake epicenters from the main shock epicenter was used as a metric of the seismicity spatial variation. The seismicity (activity) rate, r, and the b-value of the G–R relation are metrics used to detect changes in the time and size domains, respectively. At the beginning of the investigation procedure we selected sufficiently wide enough spatiotemporal windows so as to avoid missing possible foreshocks. We gradually narrowed the windows until significant space–time seismicity changes, if any, were detected and the target area was determined. Once a seismicity cluster was detected as candidate foreshock activity, we put forward the hypothesis that the entire time interval selected for analysis comprised two different but sequential states of seismicity, the background period of seismicity and the foreshock (testing) period. The statistical significance of changes in D and r were tested with the z-test. Changes in b were checked with the test of *Utsu* [1966, 1992], which measures the statistical significance of the difference in the b-value between two earthquake samples. According to the Utsu test, the probability, P, that two samples may come from the same population is calculated by

$$P \approx \exp\left[-\left(\mathrm{d}A/2\right)-2\right] \tag{8.2}$$

where dA is a function of the number of events and of the b-values in the two samples. The lower the metric P, the higher the significance level.

Our assumption is that detection of an increase in r and decrease in D and b indicate foreshock activity. We considered that the foreshock activity is weak or strong when the three precursory seismicity changes are significant (level more than 90%) or highly significant (level more 95%), respectively. Also, the significant change of only two out of three parameters may indicate weak foreshock activity.

Particular attention is paid to the b-value estimation since this parameter is quite sensitive to the computational techniques used. We adopted the maximum likelihood approximation of *Aki* [1965] for calculation of b, hereafter denoted as b_{ML}. For reasons of comparison we also calculated b_{GR} by applying the weighted least-square method to the G–R magnitude distribution. The parameter b, however, depends on the magnitude cutoff, M_c, and therefore was calculated for various levels of M_c. Also of interest was to examine the time evolution of b in the target area and so a conditional backward windowing technique was applied. An initial value of b_{ML} was calculated starting with a window comprising the last n events just prior to the main shock. The process was repeated for the next n events sliding backwards with steps of 1 event until the whole process reached the last n events from the beginning of the entire time interval selected for analysis. A value of $n = 100$ was selected if enough earthquake events were contained in the catalog; otherwise a lower value of n was selected appropriately. Another condition included in this procedure is that $\Delta = M_{max} - M_{min} > 1.4$, since it was found [*Papazachos*, 1974] that when $M_{max} - M_{min} < 1.4$, the calculation of b becomes quite unstable: where M_{max} and M_{min} are the maximum and minimum magnitudes inserted in the earthquake sample under examination. When the condition $\Delta > 1.4$ was not fulfilled in the backward windowing technique, the calculation was repeated by increasing the window with a step of 1 event at a time until the condition was fulfilled.

Another way of examining b-value variation is to consider the variation of mean magnitude, $<M>$, since it has been shown that in an earthquake sample $<M>$ is a function of the inverse of b [*Utsu*, 1965; *Lomnitz*, 1966; *Hamilton*, 1967]. Therefore, $<M>$ was also calculated with the backward windowing technique.

8.5. FORESHOCKS PRECEDING SMALL-TO-MODERATE GREEK EARTHQUAKES

We have applied our method of seismicity analysis to two seismic sequences recorded in Greece in 2012 and 2013. What the two sequences have in common is that they culminated with main shocks of moderate magnitudes: on 4 March 2012 with *Mw* 5.2 and on 3 July 2013 with *Mw* 4.8. The other common feature is the relatively low magnitude cut-off in the earthquake catalog, which as we explain later was found to be as low as 1.40 and 1.10, respectively, and favors seismicity analysis and testing for statistical significance.

The earthquake catalog used was taken from the Institute of Geodynamics, National Observatory of Athens, Greece (http://www.gein.noa.gr/en/ and http://bbnet.gein.noa.gr/ HL/database; last accessed 17 May 2016), which operates the national seismograph system in the country. In the National Observatory at Athens (NOA), the recorded data are sent to the National Data Center via the SeedLink protocol, where they are processed automatically (SeisComp3 analysis suite) and analyzed manually by operators in a 24/7 mode (Nanometrics processing suite). We used the catalog compiled after manual processing. Since the particular suite for manual processing was established on 1 February 2011, changes were caused in the homogeneity of the earthquake catalog as regards magnitude determination. Therefore, the date 1 February 2011 was considered as the beginning of the background seismicity period in the present analysis. Staff at NOA calculated moment magnitude, *Mw*, and local magnitude, M_L, following standard procedures, the details of which can be found in links in the above website.

8.5.1. Athos, Northern Greece, 4 March 2012, *Mw* 5.2

On 4 March 2012, a moderate earthquake of magnitude *Mw* 5.2/M_L 5.0 occurred offshore Athos Peninsula, northern Greece (Figure 8.1, Table 8.1). This earthquake was preceded by another moderate earthquake occurring on 14 February 2012, 01:34:38 UTC, and measuring *Mw* 5.1/M_L 5.0. Since the two epicenters are nearly identical, hereafter the epicenter of the strongest event is considered as the common epicenter for both events and is called the target point.

Precursory seismicity changes were investigated initially in a circular area of radius 50 km around the target point for the period extending from 1 February 2011 up to the occurrence of the 4 March 2012 event (Figure 8.1). The investigation revealed that from 22 December 2011 the seismicity clustered in a narrow area at a distance of no more than 20 km from the target point (Figure 8.1). Before the *Mw* 5.1 event of 14 February 2012 the activity clustered at a distance of about 10 km from the target point. After a temporary expansion of the activity at a distance ranging from about 10 km to 20 km, the activity moved even closer to the main event target point at distances between only 4 km and 10 km (Figures 8.2 and 8.3). This result is independent of the number of earthquake events used to calculate the Euclidean mean distance, D, of earthquake epicenters from the target point (Figure 8.3).

Based on the spatial clustering of the earthquake activity, seismicity changes were further investigated in the time and size domains in a target area of radius 20 km around the target point. The background period was taken from 1 February 2011 up to 21 December 2011, while the testing period was considered to extend from 22 December 2011 up to the occurrence of the 4 March 2012 main event. Completeness analysis of the earthquake catalog, based on the G–R diagrams, showed that the magnitude cut-off could be taken as low as $M_c = 1.40$ (Figure 8.4, left). For $M_c = 1.40$ the seismicity rate gradually increased from $r_b = 0.17$ events day^{-1} in the period of background seismicity to $r_f = 5.21$ events day^{-1} in the foreshock period (Figure 8.5). This increase was found significant at the level of 100%.

It should not pass unnoticed that there was a pause in seismicity during the first 12 days of February 2011 (Figure 8.5). We checked whether this was an artifact owing to a temporary lack of station records, but this can be rejected since the stations of the national seismograph system in North Greece were operating normally during that period, recording seismic events in a

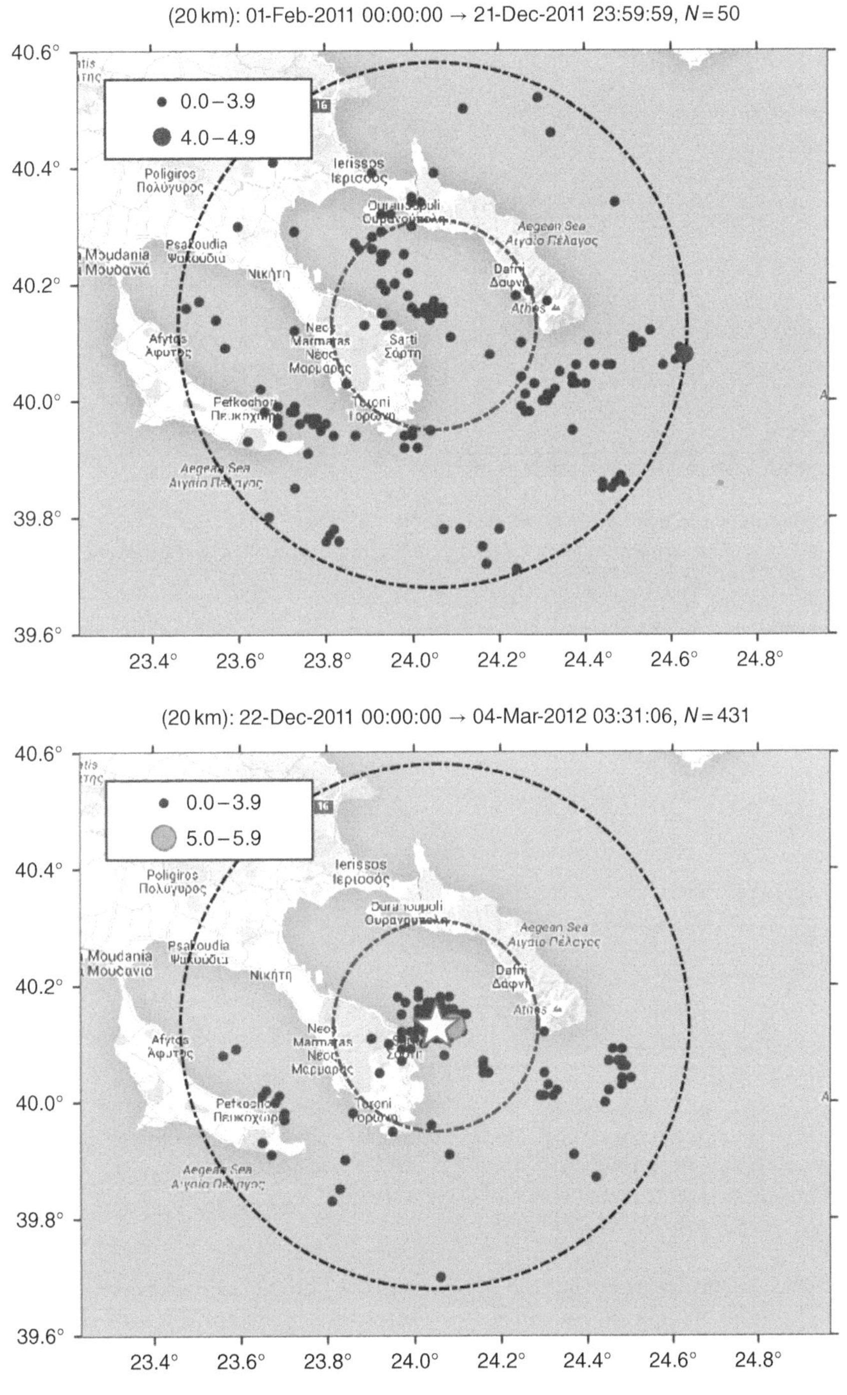

Figure 8.1 Earthquake epicenter plots in the area of Athos, northern Aegean Sea, Greece, for two time periods: (top) from 1 February 2011 to 21 December 2011 and (bottom) from 22 December 2011 to 4 March 2012. Circles are radii of 50 and 20 km from the epicenter (target point) of the 4 March 2012 main event (star). After 21 December 2011 the seismicity was significantly clustered close to the main event epicenter.

wide magnitude range and occurring outside the target area. Therefore, we concluded that the temporary pause in seismicity, which we call foreshock quiescence, was of natural origin.

The G–R diagram (Figure 8.4) clearly illustrates that the value of b_{ML}, equal to 1.22 in the background period, reduced to 0.77 in the foreshock period. This difference is highly significant, as indicated by the Utsu test (equation

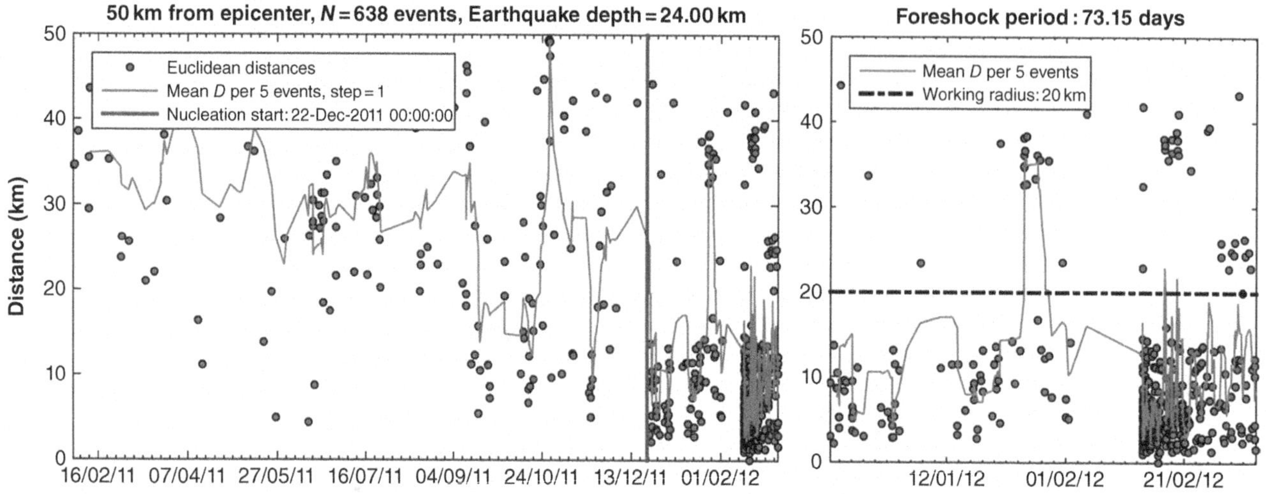

Figure 8.2 Time variation of the average Euclidean distance, D, of epicenters from that of the main event of 4 March 2012 (target point). At each point of time, D was calculated as the average of five events.

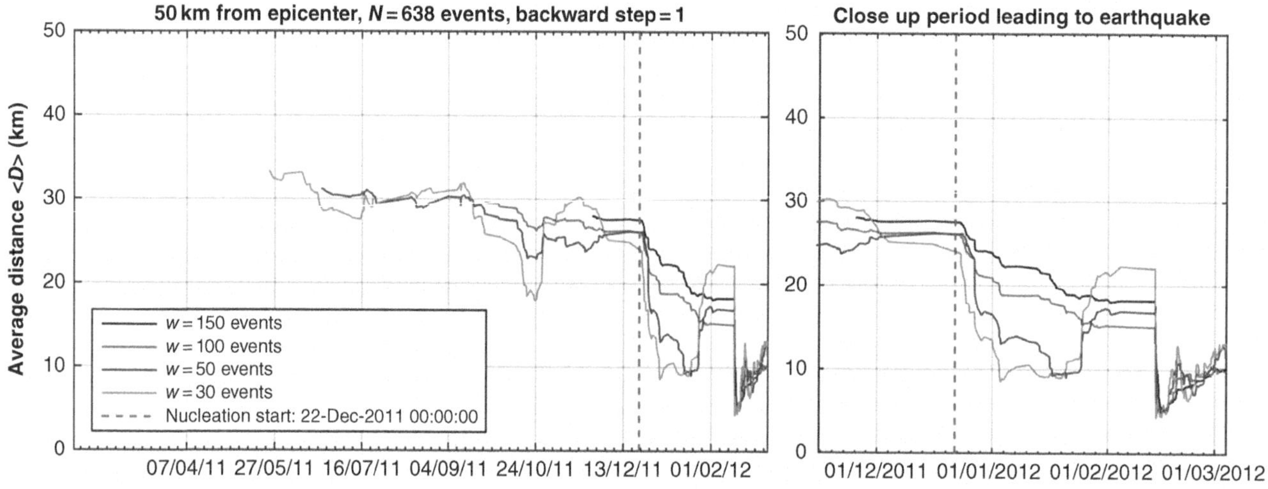

Figure 8.3 Time variation of the average Euclidean distance, D, of epicenters from the target point. At each point of time, D was calculated with the backward technique as the average of 30, 50, 100, or 150 events starting from the origin time of the 4 March 2012 main event and using a step of 1 event.

8.2) which found $P = 0.008$. It is noteworthy that the b_{ML} decrease does not depend on the magnitude cut-off considered (Figure 8.4, bottom). To further check the b_{ML} time variation, a backward testing technique was performed (Figure 8.6). For reasons of comparison the time variation of the mean magnitude, $<M>$, was also tested with the same technique. As might be expected, the two variations are consistent with each other in the sense that they follow inverse patterns. The b_{ML} decrease and the corresponding increase of $< M >$ from the beginning of the testing period up to the occurrence on 14 February 2012 of the strongest event in the testing (foreshock) period, are quite evident. However, after that event the patterns for both b_{ML} and $< M >$ temporarily reversed. At the same time the activity rate increased significantly showing an Omori-like distribution similar to the aftershock time distribution that followed the main event of 4 March 2012 (Figure 8.7). This aftershock-like feature is consistent with the notion that activity expanded spatially, as previously noted. An explanation could be that immediately after the event of 14 February 2012 the earthquake population was enhanced in smaller magnitude events and this temporarily enhancement caused the b-value to increase to around its level during the background seismicity. Soon after, however, both parameters again exhibited changes to be expected in foreshock sequences, i.e., b_{ML} decreased and $< M >$ increased during about the last 2 weeks before the main event occurred (Figure 8.6).

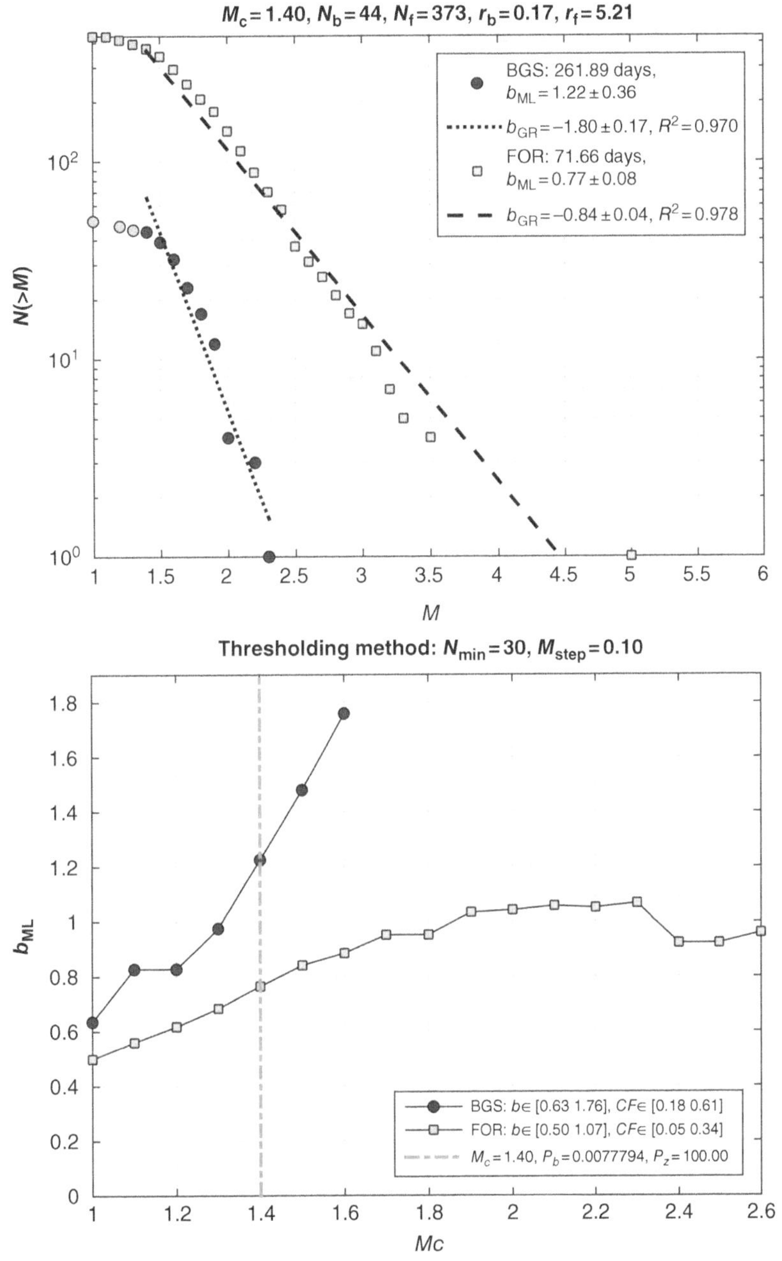

Figure 8.4 (Top) Magnitude–frequency (G–R) distribution of earthquakes occurring in the background seismicity period (BGS) and the foreshock period (FOR) before the main event of 4 March 2012: M_c, cut-off magnitude; N_b, number of events in BGS; N_f, number of events in FOR; r_b, seismicity rate in BGS; r_f, seismicity rate in foreshock period; N_{min}, minimum number of events; b_{ML} and b_{GR} are calculated by the maximum-likelihood and the weighted least-squares methods, respectively. In FOR the b_{ML} remains lower than that in BGS for different M_c levels (bottom). Calculation performed for N_{min} = 30 with a magnitude step of 0.10: *CF*, confidence interval; P_b, Utsu probability test; P_z, z probability-test.

Since the magnitude difference between the main shock and the strongest foreshock is as small as 0.1 and falls within the error limits it may be argued that the later was the main shock and the former was a strong aftershock. It therefore is of interest to check closer the seismicity features from the onset of the suggested foreshock period on 22 December 2011 up to the occurrence of the 14 February 2012 event. We found that the spatial cluster observed before the 14 February 2012 earthquake (Figure 8.7) was associated with an activity rate increase from r_b = 0.17 events day^{-1} in the background period to r_f = 1.59 events day^{-1} in the foreshock period (Figure 8.8).

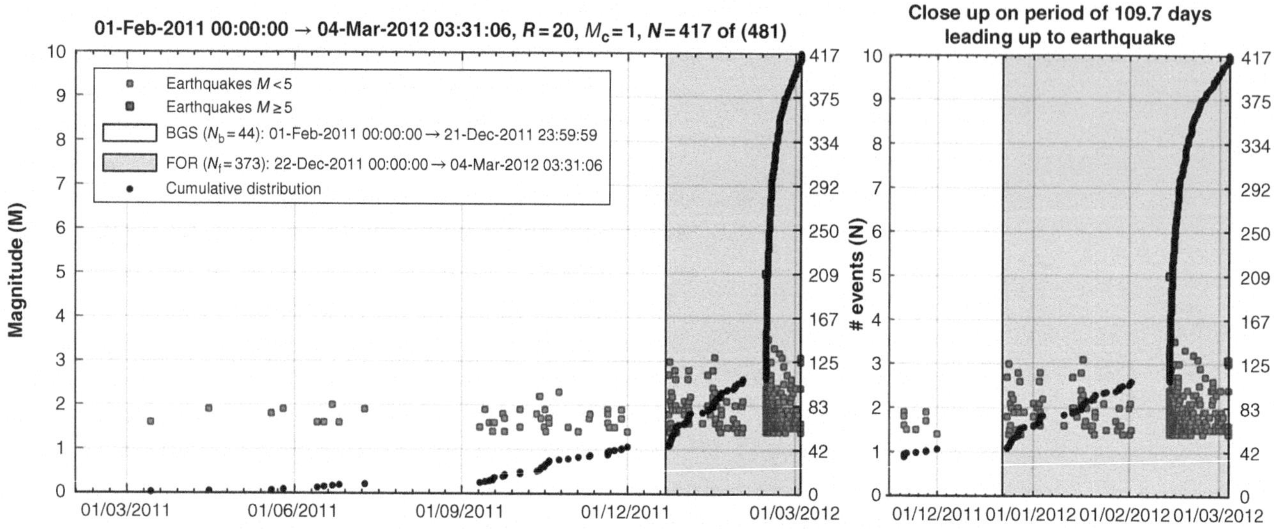

Figure 8.5 Cumulative number of earthquake events of magnitude $M \geq M_c = 1.40$ occurring in the target area ($R = 20$ km) around the epicenter of the 4 March 2012 main event (target point). See Figure 8.4 for notation.

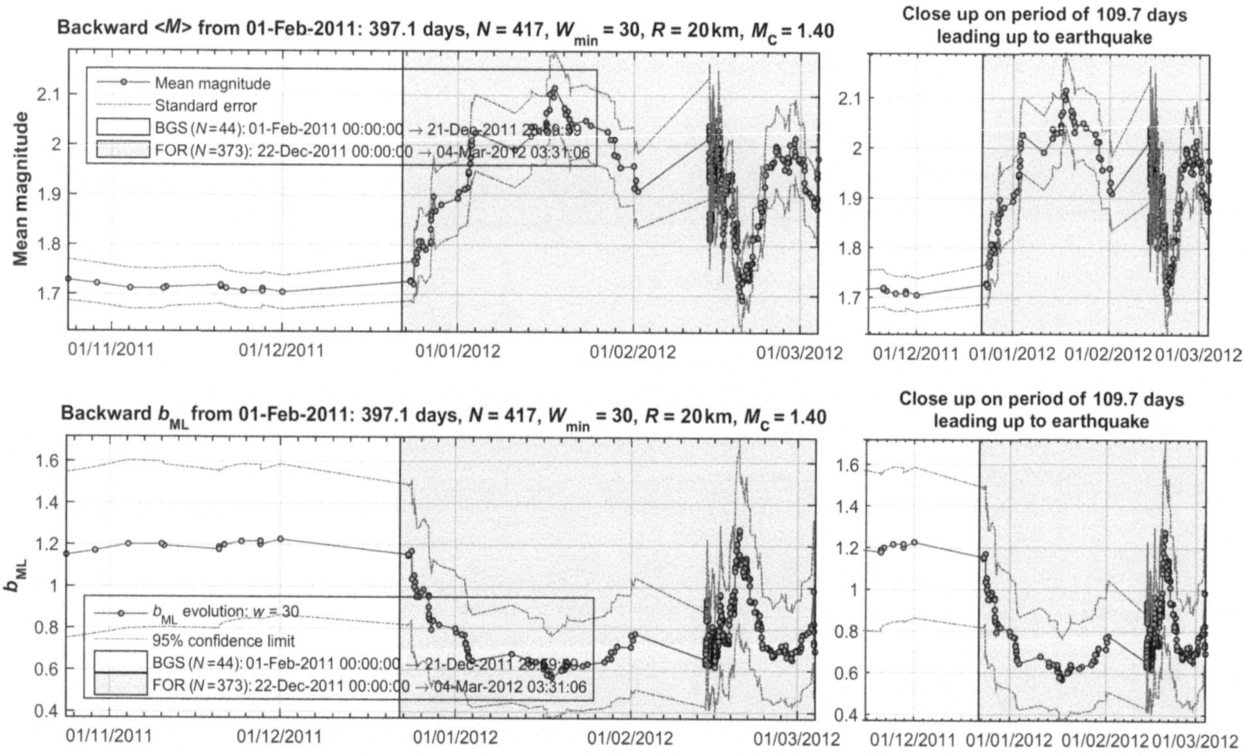

Figure 8.6 Time variation of (upper) mean magnitude, <M>, and (lower) b_{ML} of earthquake events of $M \geq M_c = 1.40$ occurring in the target area ($R = 20$ km) around the target point. Both parameters were calculated with the backward technique by taking window $w = 30$ events starting from the origin time of the 4 March 2012 main event and using a step of 1 event.

This change was significant at the 99.99% level. At the same time, the value of b_{ML} decreased from 1.22 to 0.70, a highly significant change at the $P = 0.008$ level (Figure 8.9, top), and this gradual decrease in b_{ML} was accompanied by a gradual increase of $< M >$ (Figure 8.10). Again the b_{ML} reduction was independent of the magnitude cut-off considered (Figure 8.9, bottom). From these findings it transpires that the 14 February 2012 event was preceded by three-dimensional foreshock patterns similar to those that preceded the 4 March 2012 earthquake. It is

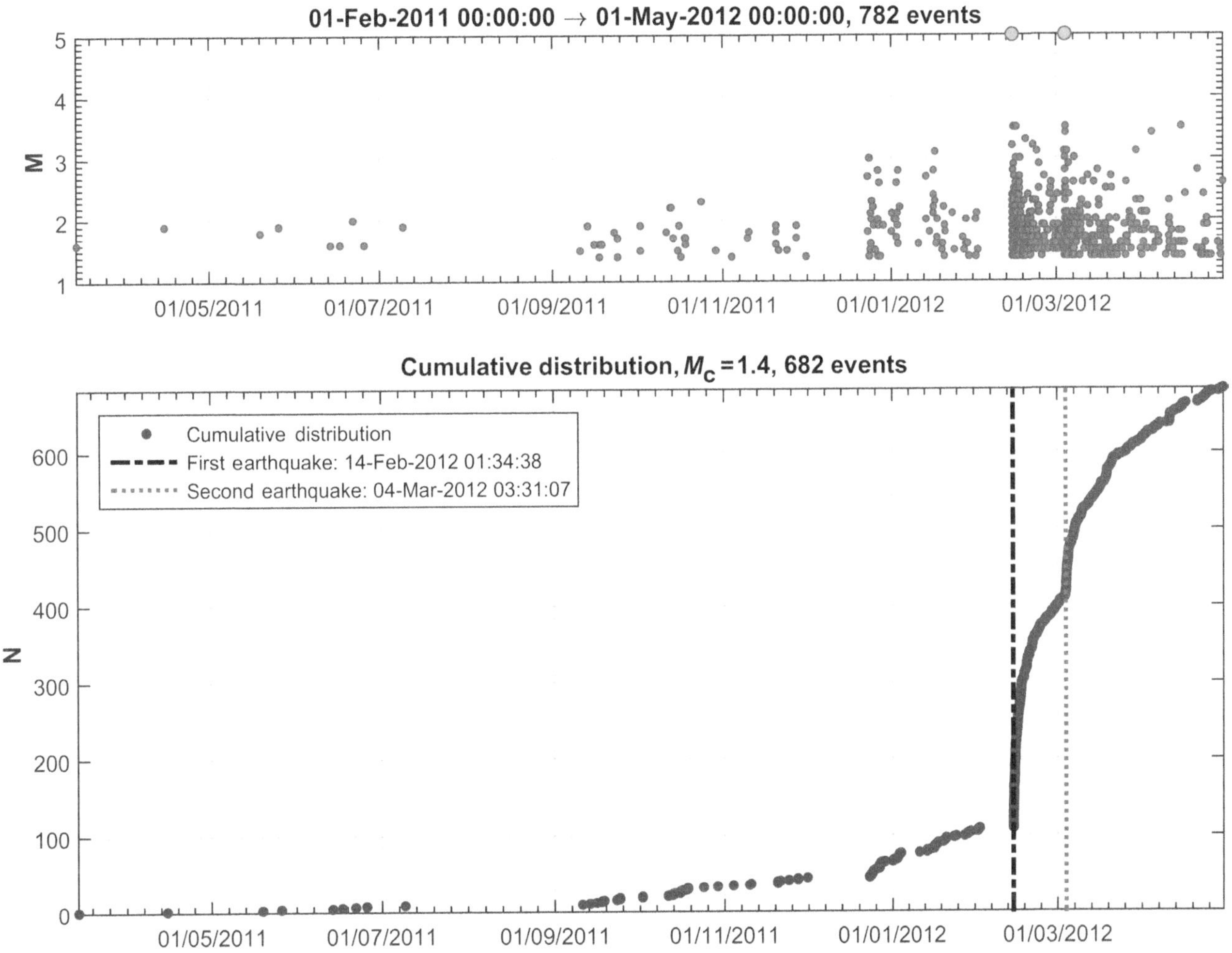

Figure 8.7 Cumulative number, N, of earthquake events of magnitude $M \geq M_c = 1.40$ occurring until the end of April 2012 in the target area ($R = 20$ km) around the target point. Vertical blue and red dotted bars show the origin times of the strong foreshock of 14 February 2012 and of the main shock, respectively. See Figure 8.4 for notation. (*See electronic version for color representation.*)

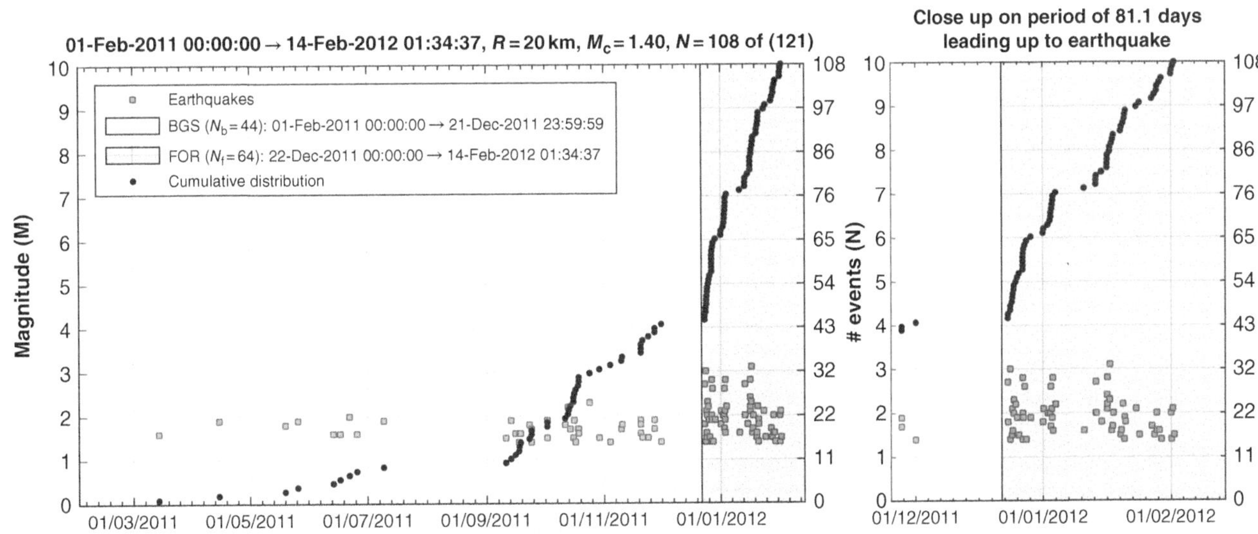

Figure 8.8 Cumulative number of earthquake events of $M \geq M_c = 1.40$ occurring in the target area ($R = 20$ km) around the target point in the BGS and FOR periods; see Figure 8.4 for notation. The 14 February 2012 earthquake coincides with the right-hand vertical axis of both panels.

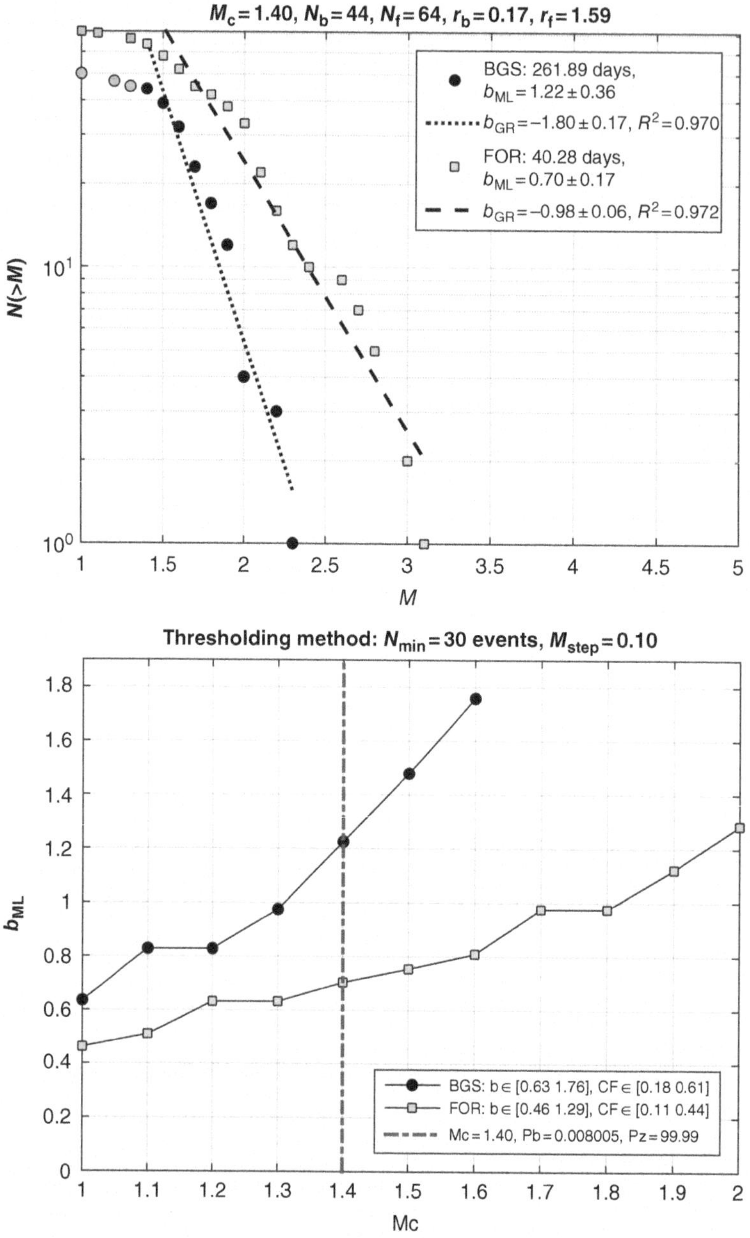

Figure 8.9 (Top) Magnitude–frequency (G–R) distribution of earthquakes occurring in the BGS and FOR periods before the *Mw* 5.1 earthquake of 14 February 2012. See Figure 8.4 for notation. In the FOR period the b_{ML} remained lower than that in the BGS period for different M_c levels (bottom). Calculation performed for N_{min} =30 with a magnitude step of 0.10.

noteworthy that foreshock quiescence also happened before the 14 February 2012 event.

From the seismicity patterns detected before the 4 March 2012 earthquake we concluded that this is another good example of a main shock preceded by foreshock sequence characterized by significant three-dimensional precursory patterns. This case is important for the additional reason that although the main shock was of moderate size its foreshock patterns are quite similar to those observed before larger main shocks, such as the *Mw* 6.3 earthquake of 6 April 2009 in L'Aquila, Italy, and the *Mw* 8.1 earthquake of 1 April 2014 in Iquique, Chile. An additional similarity is the development of the foreshock sequence in two main stages: a weak foreshock stage and a strong foreshock stage. In the Athos case the stage of weak foreshock activity lasted from 22 December 2011 up to the occurrence of the 14 February 2012 (*Mw* 5.1) strong foreshock, which was also preceded by similar

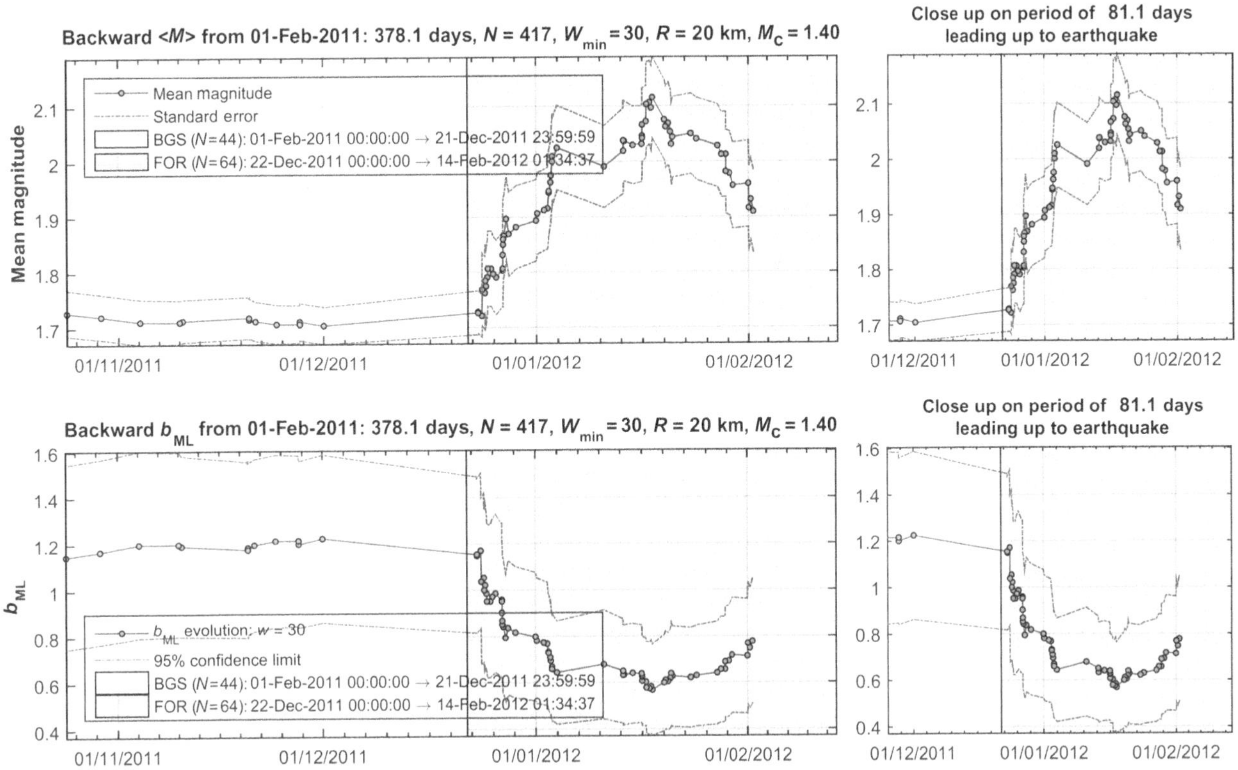

Figure 8.10 Time variation of (top) mean magnitude, <M>, and (bottom) b_{ML} of earthquake events of $M \geq M_c = 1.40$ occurring in the target area ($R=20$ km) around the target point. Both parameters were calculated with the backward technique by considering $w=30$ events starting from the origin time of the *Mw* 5.1 earthquake of 14 February 2012 and using a step of 1 event.

three-dimensional foreshock patterns. With this event the foreshock sequence entered in its strong stage culminating with the event of 4 March 2012, which was the strongest in the sequence. The entire foreshock sequence lasted for about 2.3 months, while the strong foreshock stage lasted for about 18 days. These time intervals are compatible with the duration of foreshock sequences preceding larger events (Table 8.1).

8.5.2. Polyphyto Area, North Greece, 3 July 2013, *Mw* 4.8

On 2 and 3 July 2013 two small-to-moderate earthquakes measuring magnitudes *Mw* 4.7/M_L 4.7 and *Mw* 4.8/M_L 4.7 occurred in the area of Polyphyto city, northern Greece (Table 8.1, Figure 8.11). This event is of particular interest because the area is considered to have a low level of seismic activity, but a very strong (*Mw* 6.6) earthquake struck on 13 May 1995. It is noteworthy that the area is characterized by the presence of the nearby Polyphyto artificial lake. In about the last 30 min before the 1995 main shock some foreshocks were recorded to the south of the source [*Drakatos et al.*, 1998] forcing residents to move outdoors, thus no one was killed [*Pavlides et al.*, 1995] in spite of the extensive building damage caused. The earthquakes of the 2 and 3 July 2013 took place at the southwestern tip of the artificial lake.

In our analysis we again considered an initial area of radius of 50 km around the epicenter of the main event of 3 July 2013. We found that from 8 November 2012 onwards the seismicity significantly clustered around the main event epicenter at a distance of < 25 km (Figure 8.12). This result is independent of the number of events used to calculate the mean distance, *D*, of earthquake epicenters from the main event epicenter (Figure 8.13).

During the last days before the main event a cluster developed very close to main event epicenter due to activity associated with the *Mw* 4.8 earthquake of 2 July. The analysis used here for the investigation of seismicity changes in the time and size domains is the same as that performed in the case of the Athos moderate earthquake event of 4 March 2012. The background and testing periods were considered as extending from 1 February 2011 up to 7 November 2012 and from 8 November 2012 up to the occurrence of the 3 July 2013 main event, which was the largest in the sequence. Seismicity analysis was performed

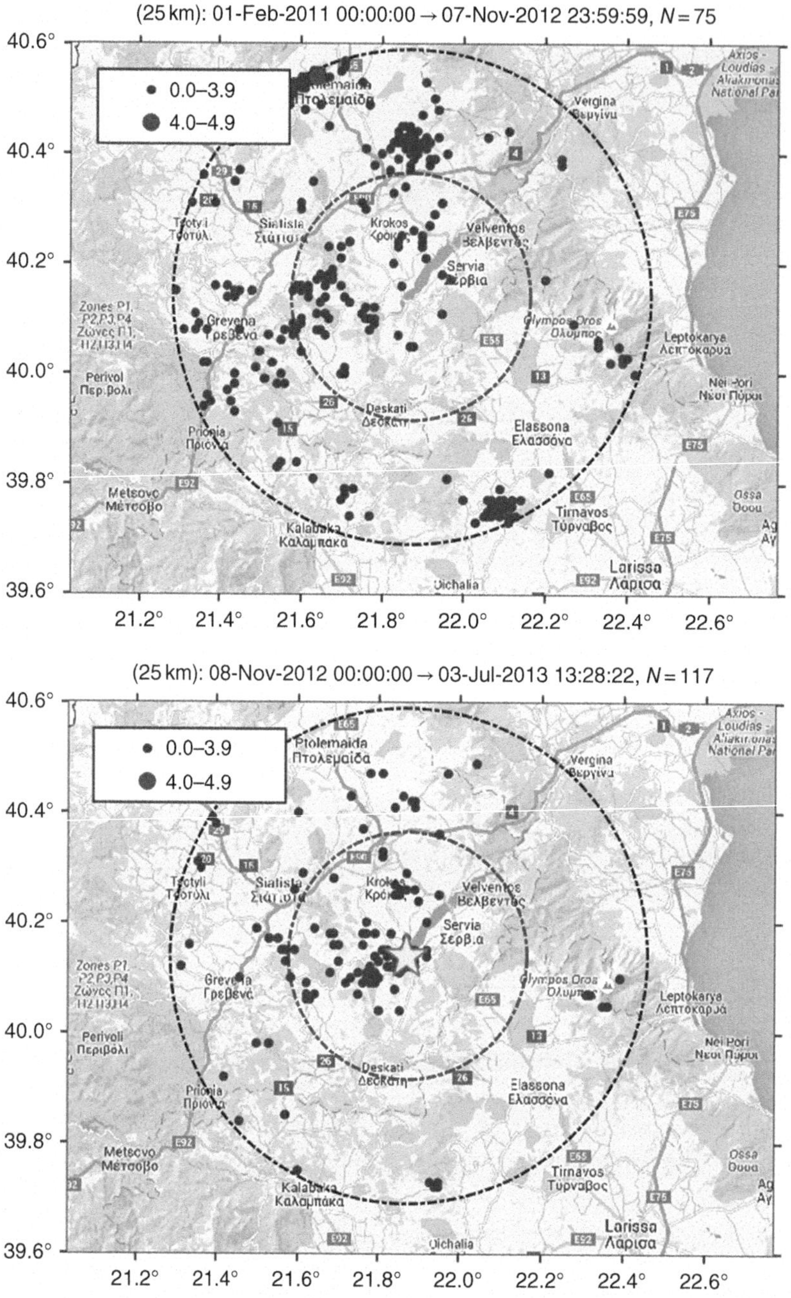

Figure 8.11 Earthquake epicenter plots in the Polyphyto area, northern Greece, for two time periods: (top) from 1 February 2011 to 7 November 2012 and (bottom) from 8 November 2012 to 3 July 2013. Circles are radii of 50 and 20 km from the epicenter of the 4 March 2013 main event (star). After 7 November 2012 the seismicity was clustered close to the main event epicenter at the southwestern tip of the Polyphyto artificial lake.

for magnitude cut-off $M_c = 1.10$ (Figure 8.14, top) in a target area of radius 25 km from the epicenter of the main event.

A significant (level of 98.54%) seismicity rate increase from $r_b = 0.12$ events day^{-1} to $r_f = 0.49$ events day^{-1} in the background and foreshock periods, respectively, was found (Figure 8.15). At the same time b_{ML} decreased from 0.77 to 0.59 (Figure 8.14, top) while the mean magnitude $< M >$ increased (Figure 8.16). The Utsu test showed $P = 0.09$. The b-value decrease is not highly significant due to the relatively small number of events, $N_b = 70$ and $N_f = 115$, contained in the two earthquake samples examined (Figure 8.14, top).

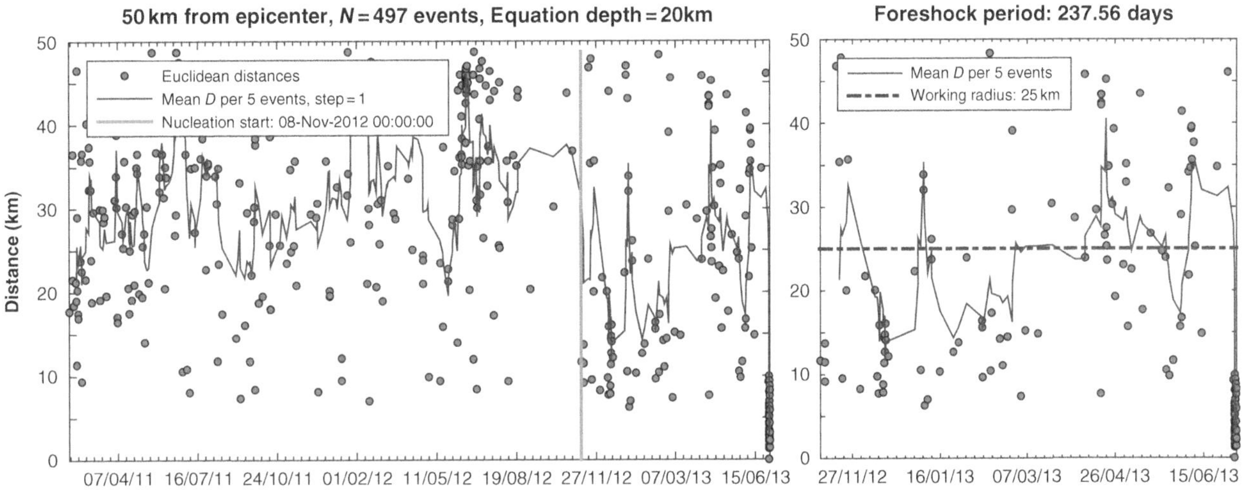

Figure 8.12 Time variation of the average Euclidean distance, *D*, of epicenters from that of the main event of 3 July 2013. At each point of time *D* was calculated as the average of five events.

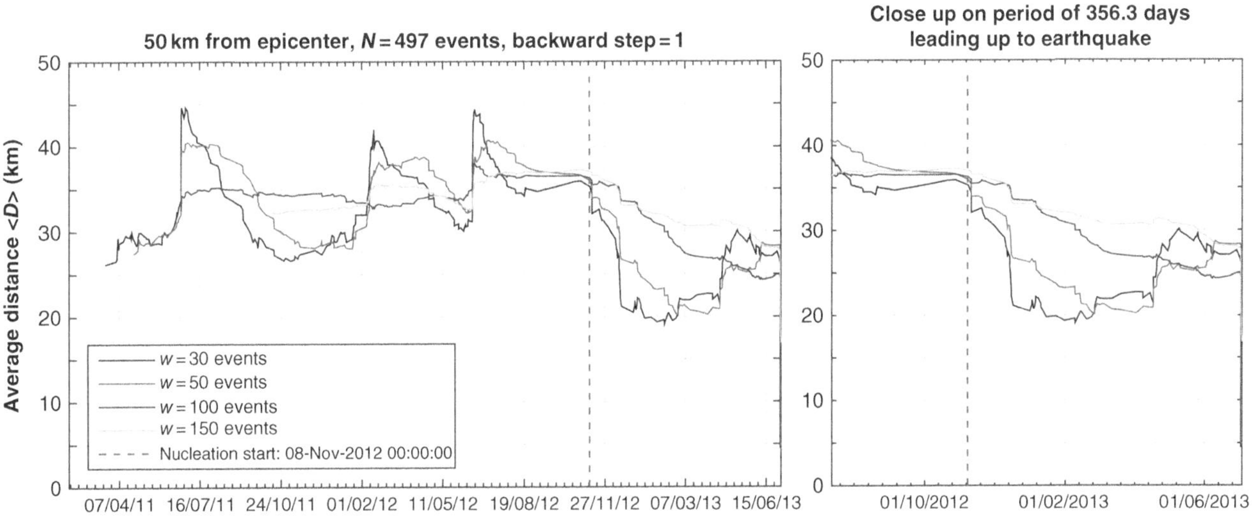

Figure 8.13 Time variation of the average Euclidean distance, *D*, of epicenters from that of the main event of 3 July 2013. At each point of time *D* was calculated with the backward technique as the average of *n* events (*n* was equal to 30, 50, 100, or 150) starting from the origin time of the 3 July 2013 main event and using a step of 1 event.

We note once more that a small-to-moderate event of *Mw* 4.8 was preceded by space, time and size patterns similar to the patterns found in foreshock sequences before larger main shocks. A characteristic feature of the Polyphyto case is the relatively long foreshock sequence, which lasted for nearly 8 months. Considering that the event ($Mw=4.7$) of 2 July 2012 was the strongest foreshock, then the duration of the strong foreshock stage was of about only 27 h. Similarly to the Athos 2013 case, the two stronger events in the Polyphyto earthquake sequence have a very small difference in magnitude. Therefore, we again tested for the case that the first event (2 July 2012) was the main shock and that of 3 July 2012 was a strong aftershock. We found that the foreshock activity that preceded the event of 2 July 2012 was significant at the 91.47% level for the increase in activity rate from 0.13 to 0.25. However, the b_{ML} decrease from 0.77 to 0.63 was of low significance ($P=0.195$). These features imply that the 8-month part of the activity represents the weak stage of the foreshock sequence.

8.6. DISCUSSION AND CONCLUSIONS

A series of critical issues regarding foreshocks and their value for earthquake prediction are still debatable. A review on the foreshock rate showed that contradictory results were obtained by different studies that examined different time periods of seismicity in various regions of

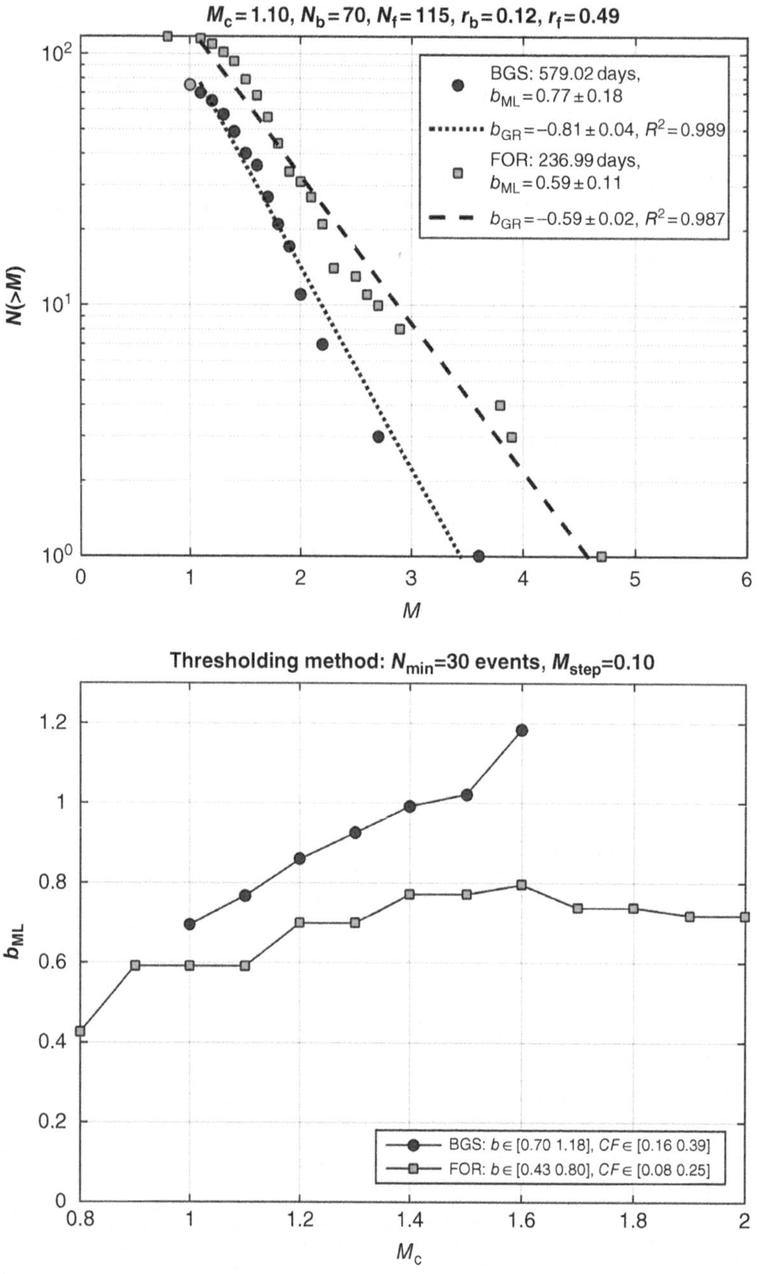

Figure 8.14 (Top) Magnitude–frequency (G–R) distribution of earthquakes occurring in the background seismicity period (BGS) and the foreshock period (FOR) before the main event of 3 July 2013. Notation as in Figure 8.6. In the foreshock period the b_{ML} remains lower than that in the background period for different M_c levels (bottom). Calculation performed for $N_{min}=30$ with a magnitude step of 0.10.

the Earth. It has been shown that the foreshock rate depends not only on crustal properties, such as material heterogeneity, focal depth and fault type, but also on the completeness of the earthquake catalog used and on the foreshock definition adopted. An important lesson learned is that foreshocks may escape recognition if the investigation focus is very narrow in space and time limits and/or the catalog magnitude cut-off is large. Also debatable is the issue of how foreshock incidence scales with the main shock size. So far it has generally been thought that foreshocks may precede only strong and large earthquakes, but very little knowledge is available on foreshock activity documented before moderate or small earthquakes [e.g. *Engdahl and Kisslinger*, 1978].

To our knowledge this is the first time that significant three-dimensional foreshock patterns have been investigated before small-to-moderate main shocks, such as the two Greek cases examined in this paper. This investigation

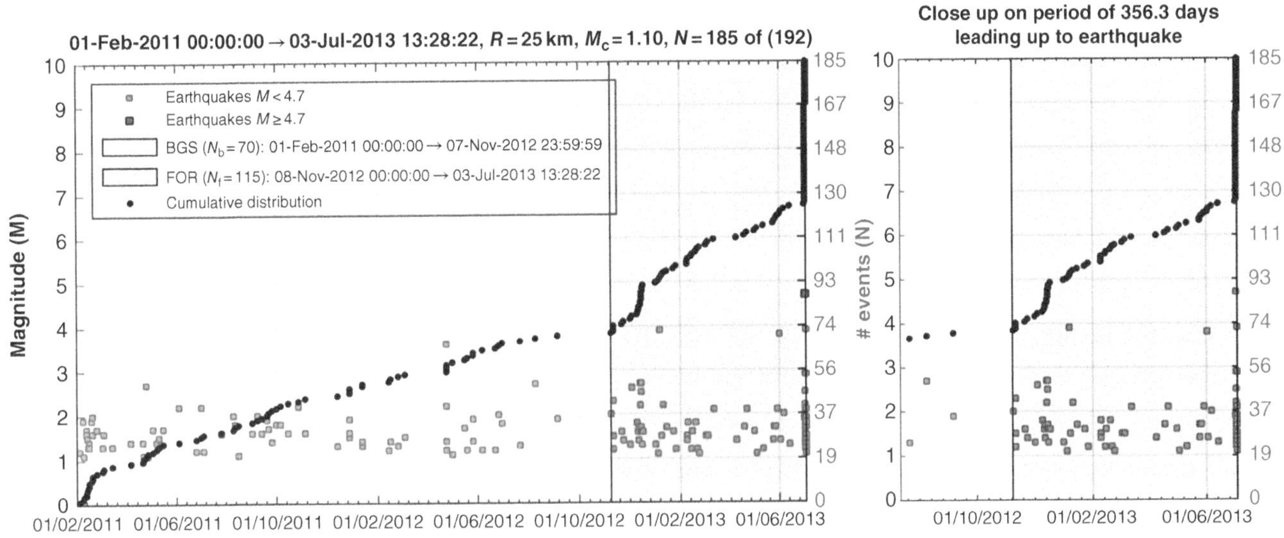

Figure 8.15 Cumulative number of earthquake events of magnitude $M \geq M_c = 1.10$ occurring in the target area ($R = 25$ km) around the epicenter of the main event of 4 March 2012. See Figure 8.4 for notation.

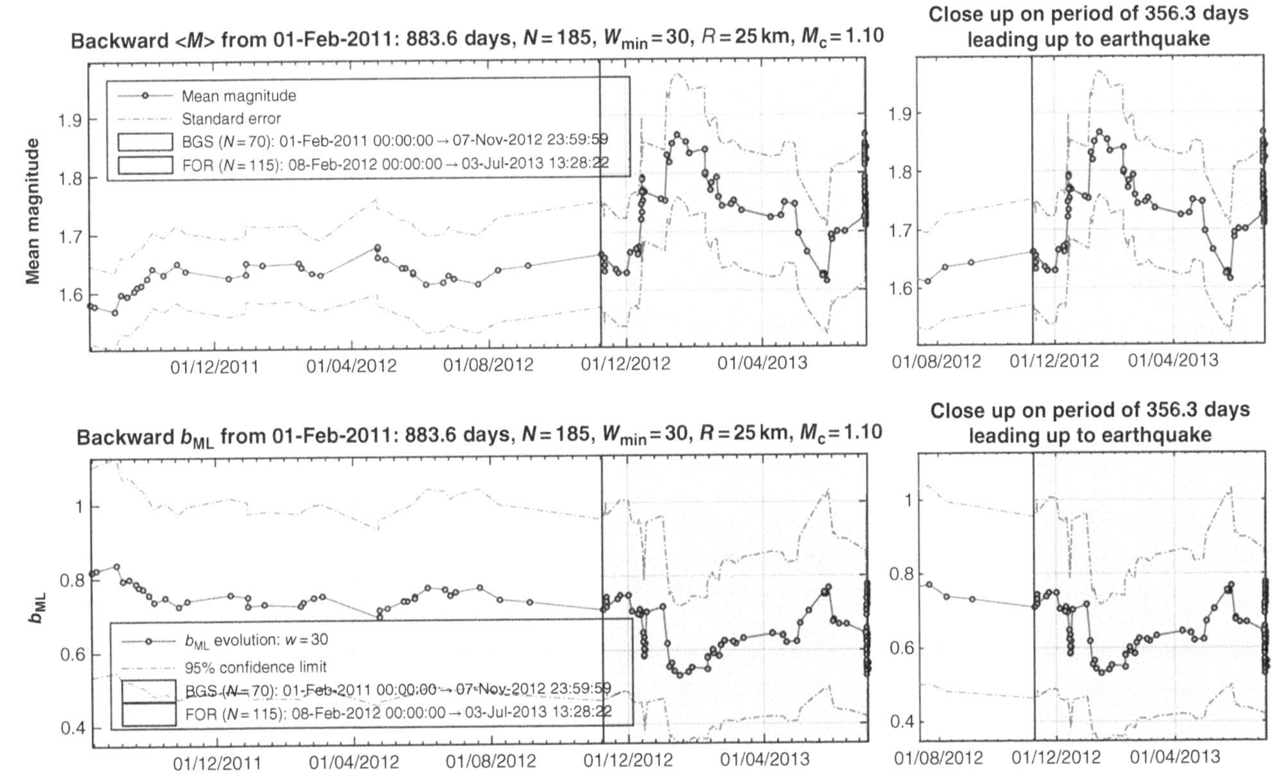

Figure 8.16 Time variation of (top) mean magnitude, <M>, and (bottom) b_{ML} of earthquake events with magnitude $M \geq M_c = 1.10$ occurring in the target area ($R = 25$ km) around the epicenter of the main event of 3 July 2013.

showed that in both cases, 4 March 2012 in Athos and 3 July 2013 in Polyphyto, the main events of *Mw* 5.2 and *Mw* 4.8, respectively, were preceded by foreshock sequences having three-dimensional patterns quite similar to the patterns that preceded strong and large main shocks. Investigation of foreshock activity occurring before earthquakes of moderate or even small magnitude may introduce new possibilities for better understanding the factors that likely control foreshock incidence and three-dimensional foreshock patterns. It is also of great benefit for exploring universality in foreshock processes, i.e., whether patterns recognized before strong and large earthquakes can be also recognized in smaller magnitude main shocks.

Table 8.1 summarizes the main shock and foreshock parameters of the Athos and Polyphyto cases, of the L'Aquila (Italy) strong 2009 earthquake, and of three great earthquakes occurring in Chile in 2010, 2014 and 2015. This list contains only foreshock cases with three-dimensional patterns tested for statistical significance. Of particular importance is to consider the seismic fault types associated with the main shocks listed in Table 8.1. There is a general consensus that the L'Aquila 2009 main shock was associated with a normal fault striking NW–SE [e.g. *Chiarabba et al.*, 2009; *Tinti et al.*, 2014]. Moment tensor solutions produced by NOA (http://bbnet.gein.noa.gr/HL/database) indicated that both the Athos 2012 and Polyphyto 2013 small-to-moderate events in North Greece were also associated with normal faults striking NW–SE in Athos and NE–SW in Polyphyto. In contrast, thrust-type faulting has been recognized for the great Chilean subduction earthquakes of Maule 2010 [*Madariaga et al.*, 2010], Iquique 2014 [*Schurr et al.*, 2014] and Illapel 2015 [*Ye et al.*, 2016].

In four out of six cases the foreshock sequence developed in two sequential stages: first a weak stage followed by a strong stage. In the Chilean cases of 2010 and 2015, however, the strong foreshock stage was lacking, only the weak foreshock stage was detected. This has been attributed to the large magnitude cut-off in the earthquake catalog [*Papadopoulos and Minadakis*, 2016]. A common feature in the four sequences is that the strong foreshock stage lasted < 3 weeks, ranging in duration from about 18 days in Athos (Greece) 2012 and in Iquique (Chile) 2014, to only 1.1 days in Polyphyto (Greece) 2013. The total duration of the six foreshock sequences ranges from nearly 8 months in Polyphyto to about 2.3 months in Athos and Iquique. The long duration of the weak foreshock stage in Polyphyto is likely due to the low rate of tectonic loading in the area, as evident from the low level of background seismicity. We do not rule out, however, that the generation of the Polyphyto sequence might be associated with water loading changes in the nearby dam, a topic that we intend to investigate.

It is quite obvious that the total duration of foreshock sequences, as well as the duration of the strong foreshock stage, do not correlate with the main shock magnitude, M_o, and the maximum foreshock magnitude, M_f. In addition, no correlation is evident between M_o and M_f, a result also found by other studies in the past.

Additionally, it is of interest to examine whether or not the main shock magnitude M_o scales with the surface of the foreshock area measured by radius R in Table 8.1. There is a tendency of M_o to scale with R but the Illapel (Chile, M_o 8.4) 2015 case is a possible deviation. In this case the foreshock area is smaller than that found for the Iquique (Chile, M_o 8.1) 2014 case. This again is possibly due to the higher magnitude cut-off in the Illapel case as compared to the Iquique case. Another problem is the accuracy of epicenter determination, particularly for lower magnitude events such those of Athos and Polyphyto in Greece. Relocation of foreshock sequences would help substantially to improve the estimation of the foreshock area.

Our conclusions are summarized as follows.

1. The estimation of the foreshock rate remains a debatable issue since it is not controlled only by geophysical factors, such as stress conditions, material heterogeneity, focal depth and fault type, but depends also on catalog completeness and the foreshock definition adopted.

2. When present, foreshock sequences likely follow certain three-dimensional patterns in the space, time and size domains: foreshock epicenters move towards the main shock epicenter, the event count increases as the inverse of time towards the main shock origin time, while the b-value decreases, and the mean magnitude $< M > (\sim 1/b)$ increases, with respect to that in background seismicity. However, only very few cases of single foreshock sequences were known so far to follow these three patterns at the same time: the L'Aquila 2009 (central Italy) case which concluded with a strong (*Mw* 6.3) main shock as well as the Chilean cases of Maule 2010, Iquique 2014, and Illapel 2015, which concluded with great earthquakes of Mw 8.8, 8.1, and 8.4, respectively. However, no foreshock sequence before small-to-moderate earthquakes was recognized previously.

3. For the first time in this paper we investigated foreshock sequences preceding small-to-moderate earthquakes. We examined the cases of Athos (*Mw* 5.2) and Polyphyto (*Mw* 4.8), both occurring in northern Greece on 4 March 2012 and 3 July 2013, respectively. It was found that both earthquakes were preceded by statistically significant three-dimensional foreshock patterns quite similar to those found before the L'Aquila strong main shock and the three Chilean great earthquakes.

4. The three-dimensional patterns appear independent of the seismic fault type, given that the Athos, Polyphyto, and L'Aquila main shocks were associated with crustal normal faulting while the three great Chilean earthquakes were typical subduction thrusts.

5. Usually the foreshock sequence developed as a weak stage followed by a strong stage. In the Chilean cases of 2010 and 2015, however, only the weak foreshock stage was detected. The lack of a strong foreshock stage is possibly explained by the large magnitude cut-off in the earthquake catalog. The four sequences comprising a strong foreshock stage lasted between 18 days in Athos (Greece) 2012 and in Iquique (Chile) 2014 and 1.1 days in

Polyphyto (Greece) 2013. The total duration of the six foreshock sequences varies from nearly 8 months in Polyphyto to nearly 2.3 months in Athos and Iquique. Further research is needed to explain the long duration of the weak foreshock stage in Polyphyto.

6. The total foreshock sequence duration and the duration of the strong foreshock stage do not correlate with the main shock magnitude, M_o, and the maximum foreshock magnitude, M_f. Also, the magnitudes M_o and M_f do not correlate.

7. There is a trend, however, for M_o to scale with the foreshock area, although the Illapel (Chile, M_o 8.4) 2015 case is a deviation, possibly due to the relatively high magnitude cut-off in the earthquake catalog of the Illapel area.

Our main finding is that the foreshock patterns in space, time and size domains are likely governed by scale-invariant universality holding in a very wide magnitude spectrum and for at least dip-slip fault types. Should this be a valid result it may indicate that the foreshock incidence is an inherent property of the deformation process leading to the main shock nucleation. Therefore, the results obtained above underline the need to further verify the hypothesis of scale-invariant universality of the foreshock three-dimensional patterns. Towards this aim the investigation of more good examples of statistically significant patterns occurring before earthquakes of various magnitudes and associated with various faulting types becomes a priority in foreshock research.

ACKNOWLEDGMENTS

This is a contribution to the Project "Pilot Study in the Area of Early Warning System for Natural Disasters: ARISTOTLE" funded by the EU DG-ECHO (contract n. ECHO/SER/2015/722144).

REFERENCES

Abercrombie, R. E. and Mori, J. (1996), Occurrence patterns of foreshocks to large earthquakes in the western United States, *Nature*, *381*, 303–307.

Agnew, D. C. and Jones, L. (1991), Prediction probabilities from foreshocks. *J. Geophys. Res.*, *96*(B7), 11959–11971.

Aki, K. (1965), Maximum likelihood estimates of b in the formula $\log N = a - bM$ and its confidence limits, *Bull. Earth. Res. Inst. Univ. Tokyo*, *43*, 237–239.

Avlonitis, M. and Papadopoulos, G. A. (2014), Foreshocks and b value: bridging macroscopic observations to source mechanical considerations, *Pure Appl. Geophys.*, *17S1*(10), 2569–2580, doi: 10.1007/s00024-014-0799-6

Bedford, J., Moreno, M., Schurr, B., Bartsch, M., and Oncken, O. (2015), Investigating the final seismic swarm before the Iquique-Pisagua 2014 *Mw* 8.1 by comparison of continuous GPS and seismic foreshock data. *Geophys. Res. Lett.*, *42*, doi:10.1002/2015GL063953.

Bouchon, M., Durand, V., Marsan, D., Karabulut, H., and Schmittbuhl, J., 2013, The long precursory phase of most large interplate earthquakes, *Nature Geosci.*, *6*(4), 299–302, doi 10.1038/ngeo1770.

Brodsky, E. and Lay, Th. (2014), Recognizing foreshocks from the 1 April 2014 Chile earthquake, *Science*, *344*, 700–702.

Chan, C-H, Wu, Y-M., Tseng, T-L., Lin, T-L., and Chen, C.-C. (2012), Spatial and temporal evolution of b-values before large earthquakes in Taiwan, *Tectonophysics*, *532–535*, 215–222, doi:10.1016/j.tecto.2012.02.004

Chen, Y., Liu, J., and Ge, H., (1999), Pattern characteristics of foreshock sequences, *Pure Appl. Geophys.*, *155*, 2–4, 395–408.

Cheng, Y. and Wong, L. N. Y. (2016), Occurrence of foreshocks in large earthquakes with strike-slip rupturing, *Bull. Seism. Soc. Am.*, *106*, 213–224, doi:10.1785/0120140338.

Chiarabba, C., Amato, A., Anselmi, M., et al., (2009), The 2009 L' Aquila (central Italy) Mw6.3 earthquake: Main shock and aftershocks, *Geophys. Res. Lett.*, *36*, L18308, doi: 10.1029/2009GL039627

Console, R., Murru, M., and Alessandrini, B. (1993), Foreshock statistics and their possible relationship to earthquake prediction in the Italian region. *Bull. Seismol. Soc. Am.*, *83*, 1248–1263.

Daskalaki, E., Spiliotis, K., Siettos, C., Minadakis, G., and Papadopoulos, G. A. (2016), Foreshocks and short-term hazard assessment to large earthquakes using complex networks: the case of the 2009 L'Aquila earthquake, *Nonlin. Processes Geophys.*, *23*, 241–256, https://doi.org/10.5194/npg-23-241-2016

Di Luccio, F., Console, R., Imoto, M., and Murru, M. (1997), Analysis of short time-space range seismicity patterns in Italy, *Ann. di Geofis.*, *XL*, 783–798.

Dodge, D. A., Beroza, G. C., and Ellsworth, W. L. (1995), Foreshock sequence of the 1992 Landers, California, earthquake and its implications for earthquake nucleation. *J. Geophys. Res.*, *100*(B6), 9865–9880.

Doser, D. (1989), Foreshocks and aftershocks of large ($M \geq 5.5$) earthquakes within the western Cordillera of the United States, *Bull. Seismol. Soc. Am.*, *80*, 110–128.

Drakatos, G., Papanastassiou, D., Papadopoulos, G., Skafida, H., and Stavrakakis, G., (1998), Relationship between the 13 May 1995 Kozani-Grevena (NW Greece) earthquake and the Polyphyto artificial lake. *Eng. Geol.*, *51*, 65–74.

Duputel, Z., Jiang, J., Jolivet, R. et al. (2015), The Iquique earthquake sequence of April 2014: Bayesian modeling accounting for prediction uncertainty, *Geophys. Res. Lett.*, *42*, 7949–7957, doi:10.1002/2015GL065402

Engdahl, E. R. and Kisslinger, C. (1978), Seismological precursors to a magnitude 5 earthquake in the Central Aleutian Islands, in C. Kisslinger and Z. Suzuki (eds), *Earthquake Precursors*, pp. 243–250, Center for Academic Publications, Tokyo.

Frohlich, C. and Davis, S. D. (1993), Teleseismic b values; or, much ado about 1.0. *J. Geophys. Res.*, *98*(B1), 631–644.

Gutenberg, B. and Richter, C. (1944), Frequency of earthquakes in California, *Bull. Seismol. Soc. Am.*, *34*, 185–188.

Hainzl, S., Zoller, G., and Kurths, J. (1999), Similar power laws for foreshock and aftershock sequences in a spring-block model for earthquakes, *J. Geophys. Res.*, *104*, 7243–7253.

Hamilton, R. M. (1967), Mean magnitude of an earthquake sequence. *Bull. Seismol. Soc. Am.*, *57*, 1115–1116.

Imamura, A., (1937), *Theoretical and Applied Seismology*, Maruzen, Tokyo, 358 pp.

Imoto, M. (2005), *Use of Potential Foreshocks to Estimate the Short-Term Probability of Large Earthquakes*, Tohoku, Japan.

Ishida, M. and Kanamori, H. (1978), The foreshock activity of the 1971 San Fernando earthquake, California. *Bull. Seismol. Soc. Am.*, *68*, 1265–1279.

Ishimoto, M. and Iida, K. (1939), Observations of earthquakes registered with the microseismograph constructed recently, *Bull. Earthquake Res. Inst. Tokyo Univ.*, *17*, 443–478.

Jones, L. M. (1985a), Foreshocks and time-dependent earthquake hazard assessment in southern California, *Bull. Seismol. Soc. Am.*, *75*(6), 1669–1679.

Jones, L. M. (1985b), Foreshocks (1966–1980) in the San Andreas system California. *Bull. Seismol. Soc. Am.*, *74*, 1361–1380.

Jones, L. M. and Molnar, P. (1976), Frequency of foreshocks, *Nature*, *262*(5570), 677–679.

Jones, L. M. and Molnar, P. (1979), Some characteristics of foreshocks and their possible relationship to earthquake prediction and premonitory slip on faults, *J. Geophys. Res.*, *84*, 3596–3608.

Jones, L. M., Wang, B., Xu, S., and Fitch, Th.-J., (1982), The foreshock sequence of the February 4, 1975, Haicheng earthquake ($M = 7.3$). *J. Geophys. Res.*, *87*(B6), 4575–4584.

Kagan, Y. and Knopoff, L. (1978), Statistical study of the occurrence of shallow earthquakes, *Geophys. J. Roy. Astr. Soc.*, *55*, 67–86.

Kato, A. and Nakagawa, S. (2014), Multiple slow-slip events during a foreshock sequence of the 2014 Iquique, Chile *Mw* 8.1 earthquake, *Geophys. Res. Lett.*, *41*, 5420–5427, doi: 10.1002/2014GL061138

Kato, A., Obara, K., Igarashi, T., Tsuruoka, H., Nakagawa, S., and Hirata N. (2012), Propagation of slow slip leading up to the 2011 *Mw* 9.0 Tohoku-Oki earthquake, *Science*, *335*(6069), 705–708.

Lay, Th., Yue, H., Brodsky, E. E., and An, C., (2014), The 1 April 2014 Iquique, Chile, *Mw* 8.1 earthquake rupture sequence, *Geophys. Res. Lett.*, *41*, doi:10.1002/2014GL060238.

Lin, C.-H., (2004), Repeated foreshock sequences in the thrust faulting environment of eastern Taiwan. *Geophys. Res. Lett.*, *31*, L13601, doi:10.1029/2004GL019883

Lippiello, E., Marzocchi, W., De Arcangelis, L., and Godano, C. (2012), Spatial organization of foreshocks as a tool to forecast large earthquakes, *Sci. Rep.*, *2*, DOI:10.1038/srep00846.

Lomnitz, C. (1966), Magnitude stability in earthquake sequences. *Bull. Seismol. Soc. Am.*, *56*, 247–249.

Madariaga, R., Métois, M., Vigny, Ch., and Campos, J. (2010), Central Chile finally breaks, *Science*, *328*, 181–182, doi: 10.1126/science.1189197

Maeda, K. (1999), Time distribution of immediate foreshocks obtained by a stacking method, *Pure Appl. Geophys.*, *155*(2–4), 381–394.

Main, I. (2000), Apparent breaks in scaling in the earthquake cumulative frequency-magnitude distribution: fact or artifact? *Bull. Seismol. Soc. Am.*, *90*(1), 86–97.

Main, I., Meredith, Ph. G., and Jones, C. (1989), A reinterpretation of the precursory seismic b-value anomaly from fracture mechanics, *Geophys. J. Int.*, *96*, 131–138.

Meng, L., Huang, H., Bürgmann, R., Ampuero, J.-P., and Strader, A. (2015), Dual megathrust slip behaviors of the 2014 Iquique earthquake sequence, *Earth Planet. Sci. Lett.*, *411*, 177–187.

Merrifield, A., Savage, M. K., and Vere-Jones, D., (2004), Geographical distributions of prospective foreshock probabilities in New Zealand, *N. Z. J. Geol. Geophys.*, *47*, 327–339.

MichaeL, A. (2012), Fundamental questions of earthquake statistics, source behavior, and the estimation of earthquake probabilities, *Bull. Seismol. Soc. Am.*, *102*, 2547–2562, doi: 10.1785/0120090184.

Mogi, K. (1963a), The fracture of a semi-infinite body caused by an inner stress origin and its relation to the earthquake phenomena (second paper), *Bull. Earthquake Res. Inst. Univ. Tokyo*, *41*, 595–614.

Mogi, K. (1963b), Some discussion on aftershocks, foreshocks and earthquake swarms – the fracture of a semi-infinite body caused by an inner stress origin and its relation to the earthquake phenomena (third paper), *Bull. Earthquake Res. Inst. Univ. Tokyo*, *41*, 615–658.

Mogi, K. (1985), *Earthquake Prediction*, Academic Press, Tokyo, 355 pp.

Molchan, G. M., Kronrod, T. L., and Nekrasona, A. K. (1999), Immediate foreshocks: time variation of the *b*-value, *Phys. Earth Planet. Int.*, *111*, 229–240.

Nanjo, K. Z., Hirata, N., Obara, K., and Kasahara, K. (2012), Decade-scale decrease in *b* value prior to the *M*9-class 2011 Tohoku and 2004 Sumatra quakes, *Geophys. Res. Lett.*, *39*, L20304, doi:10.1029/2012GL052997

Ogata, Y., Utsu, T., and Katsura, K. (1996), Statistical discrimination of foreshocks from other earthquake clusters, *Geophys. J. Int.*, *127*, 17–30.

Papadopoulos, G. A., Drakatos, G., and Plessa, A. (2000), Foreshock activity as a precursor of strong earthquakes in Corinthos Gulf, Central Greece, *Phys. Chem. Earth*, *25*, 239–245.

Papadopoulos, G. A., Charalampakis, M., Fokaefs, A., and Minadakis, G. (2010), Strong foreshock signal preceding the L'Aquila (Italy) earthquake (*Mw* 6.3) of 6 April 2009. *Nat. Hazards Earth Syst. Sci.*, *10*, 19–24.

Papadopoulos, G. A., Latoussakis, I., Daskalaki, E., Diakogianni, G., Fokaefs, A., Kolligri, M., Liadopoulou, K., Orfanogiannaki, K., and Pirentis, A. (2006), The East Aegean Sea strong earthquake sequence of October–November 2005: lessons learned for earthquake prediction from foreshocks, *Nat. Hazards Earth Syst. Sci.*, *6*, 895–901.

Papadopoulos, G. A. and Minadakis, G. (2016), Foreshock patterns preceding great earthquakes in the subduction zone of Chile, *Pure Appl. Geophys.*, *173*(10–11), 3247–3271.

Papazachos, B. C. (1974), Dependence of the seismic parameter *b* on the magnitude range, *Pure Appl. Geophys.*, *112*, 1059–1065.

Papazachos, B. C. (1975), Foreshocks and earthquake prediction, *Tectonophysics*, *28*, 213–226.

Pavlides, S. B., Zouros, N. C., Chatzipetros, A. A., Kostopoulos, D. S., and Mountrakis, D. M., (1995), The 13 May 1995 western Macedonia, Greece (Kozani-Grevena) earthquake: preliminary results, *Terra Nova*, *7*, 544–549.

Peçi, V., Maeda, K., Matsumura, K., and Irikura, K. (1999). Foreshock activity and its probabilistic relation to earthquake occurrence in Albania and the surrounding area, *Ann. di Geofis.*, *42*, 809–819.

Raleigh, B., Benett, G., Craig, H. et al. (1977), Prediction of the Haicheng earthquake, *EOS (Trans. Am. Geophys. Union)*, *58*, 236–272.

Ruiz, S., Metois, M., Fuenzalida, A. et al. (2014), Intense foreshocks and a slow slip event preceded the 2014 Iquique *Mw* 8.1 earthquake, *Science*, doi:10.1126/science.1256074

Savage, M. K. and DePolo, D. M., (1993), Foreshock probabilities in the western great-basin eastern Sierra Nevada, *Bull. Seismol. Soc. Am.*, *83* (6), 1910–1938.

Scholz, C. H. (1968), Microfractures, aftershocks, and seismicity, *Bull. Seismol. Soc. Am.*, *58*, 1117–1130.

Scholz, C. H. (1977), A physical interpretation of the Haicheng earthquake prediction, *Nature*, *267*, 121–124.

Schorlemmer, D., Wiemer, S., and Wyss, M. (2005), Variations in earthquake-size distribution across different stress regimes, *Nature*, *437*, 539–542.

Schurr, B., Asch, G., Hainzl, S. et al. (2014), Gradual unlocking of plate boundary controlled initiation of the 2014 Iquique earthquake, *Nature*, *512*, 299–302, doi: 10.1038/nature13681

Seggern, D. (1981), Seismicity pattern preceding moderate to major earthquakes, *J. Geophys. Res.*, *86* (B10), 9325–9351.

Sobolev, G. A. (2000), Precursory phases of large Kamchatkan earthquakes, *Volcanol. Seismol.*, *21*, 497–509.

Suyehiro, S. (1966), Difference between aftershocks and foreshocks in the relationship of magnitude to frequency of occurrence for the great Chilean earthquake of 1960, *Bull. Seismol. Soc. Am.*, *56*, 185–200.

Suyehiro, S. and Sekiya, H. (1972), Foreshocks and earthquake prediction, *Tectonophysics*, *14*, 219–225.

Suyehiro, S., Asada, T., and Ohtake, M. (1964), Foreshocks and aftershocks accompanying a perceptible earthquake in central Japan, *Meteorol. Geophys.*, *15*, 71–88.

Tinti, E., Scognamiglio, L., Cirella, A., and Cocco, M. (2014), Up-dip directivity in near-source during the 2009 L' Aquila main shock, *Geophys. J. Int.*, *198*, 1618–163, doi: 10.1093/gji/ggu227

Utsu, T. (1965), A method for determining the value of b in a formula $\log N = a - bM$ showing the magnitude–frequency relation for earthquakes, *Geophys. Bull. Hokkaido Univ.*, *13*, 99–103. (In Japanese.)

Utsu, T. (1966), A statistical test of the difference in b-value between two earthquake groups, *J. Phys. Earth*, *14*, 37–40.

Utsu, T. (1992), Representation and analysis of the earthquake size distribution: A historical review and some new approaches, *Pure Appl. Geophys.*, *155*, 509–535, 1992.

Vidale, J., Mori, J., and Houston, H., (2001), Something wicked this way comes: clues from foreshocks and earthquake nucleation, *EOS (Trans. Am. Geophys. Union)*, *82*, 68.

Wu, C., Meng, X., Peng, Z., and Ben-Zion, Y. (2014), Lack of spatiotemporal localization of foreshocks before the 1999 *Mw* 7.1 Düzce, Turkey, earthquake. *Bull. Seismol. Soc. Am.*, *104*, 560–566.

Wyss, M. (1997), Second round of evaluations of proposed earthquake precursors, *Pure Appl. Geophys.*, *149*, 3–16.

Yagi, Y., Okuwaki, R., Enescu, B. et al., (2014), Rupture process of the 2014 Iquique Chile earthquake in relation with the foreshock activity, *Geophys. Res. Lett.*, *41*, doi: 10.1002/2014GL060274

Yamaoka, K., Ooida T., and Ueda Y. (1999), Detailed distribution of accelerating foreshocks before a M 5.1 earthquake in Japan, *Pure Appl. Geophys.*, *155*(2–4), 335–353.

Yamashina, K. (1981), A method of probability prediction for earthquakes in Japan, *J. Phys. Earth*, *29*, 9–22.

Yamashita, T. and Knopoff, L., (2007), A model of foreshock occurrence, *Geophys. J. Int.*, *96*(3), 389–399.

Ye, L., Lay, Th., Kanamori, H., and Koper, K. D., (2016), Rapidly estimated seismic source parameters for the 16 September 2015 Illapel, Chile *Mw* 8.3 Earthquake. *Pure Appl. Geophys.*, *173*, 321–332, doi: 10.1007/s00024-015-1202-y

9

Recent Developments in the Detection of Seismicity Patterns for the Italian Region

Antonella Peresan

ABSTRACT

Reliable forecasting tools must be formalized and tested, first in retrospect and then in experimental real-time mode. The operational issues of monitoring, validation, and use of forecasts based on seismicity patterns are discussed, reviewing the advances in the application of the CN earthquake prediction algorithm (originally formulated for the California–Nevada region, hence its name) in Italy and surroundings. The CN algorithm is based on pattern recognition concepts and uses a detectable inverse cascade of seismic process to allow for a diagnosis of the intervals of time when a strong event is likely to occur inside a specified region. The results achieved by the CN algorithm in about two decades of rigorous prospective monitoring of seismicity in Italy, considering a regionalization based on the seismotectonic zoning and the main geodynamic features of the area, are tested. The results from real-time experimental testing support the hypothesis that the precursors of a strong shock can be found within homogeneous seismogenic areas associated with a dominating geodynamic process and confirm the feasibility of earthquake forecasting based on the analysis of seismicity patterns at the intermediate-term middle-range scale.

9.1. INTRODUCTION

Earthquake prediction is widely recognized among the most challenging scientific problems, both due to its societal relevance and to the intrinsic complexity of the problem. Can we predict earthquakes with useful precision? Is the information from earthquake forecast/prediction sufficiently reliable to potentially avoid or reduce damage? These and other similar questions are the subject of long-term controversial discussions and debates [e. g., *Geller et al.*, 1997; *Jordan et al.*, 2014; *Wang and Rogers*, 2014; *Kossobokov et al.*, 2015].

By the 1980s the seismically active lithosphere of the Earth had been clearly recognized as a complex, hierarchically (self-) organized, nonlinear dissipative system, with critical phase transitions through larger earthquakes [*Keilis-Borok*, 1990]. Mathematically, this kind of "chaotic" systems is predictable, but with limited precision and after substantial educated averaging. Accordingly, success in forecasting catastrophic earthquakes implies a holistic approach, from the whole to detail. The problem of prediction can then be posed as a successive, step-by-step process, narrowing down the territory and time interval, where and when the incipient earthquake is most likely to happen. Otherwise, an attempt of shooting "a silver bullet" that pin-points the magnitude, source, and time of a future earthquake with an "exact" precision would usually fail, except for a rare case of random coincidence or a rather common case of "shoot first then draw the target" (i.e., model overfitting, from a posteriori analyses).

Significant steps have been made towards assessing earthquake space–time correlations, clustering, and the emergence of seismicity patterns [e. g., *Tiampo and*

National Institute of Oceanography and Experimental Geophysics, OGS-CRS, Udine, Italy

Pre-Earthquake Processes: A Multidisciplinary Approach to Earthquake Prediction Studies, Geophysical Monograph 234, First Edition. Edited by Dimitar Ouzounov, Sergey Pulinets, Katsumi Hattori, and Patrick Taylor.

Shcherbakov, 2012]; accordingly, the observed seismic dynamics before and after many major earthquakes demonstrate common features of predictability and diverse behavior. The holistic approach to the understanding of earthquake occurrence as a complex phenomenon, in particular, has shown the potential for reproducible and testable earthquake forecasting, based on robust information provided by seismicity [*Keilis-Borok and Soloviev*, 2003]. Seismicity, however, is just one possible expression of Earth's complex dynamics that may precede catastrophic earthquakes. Many newly available data collected on a global scale and their well identified patterns, as well as probabilistic models of earthquake occurrence, nowadays provide opportunities for systematic analysis and model testing. A variety of physical observations, ranging from ground-related deformation patterns (detected using global positioning system (GPS), synthetic aperture radar (SAR), etc., data) to pre-earthquake environmental changes (including geochemical, electromagnetic, and thermodynamic processes), may be related to stress variations in the lithosphere prior to a large earthquake and thus may contribute to reducing uncertainties about impending earthquakes, as discussed in the chapters of this volume.

Over the past few decades it has become well established that forecasting methods, particularly those intended for operational use, should be tested against the available data to assess their performances and confidence level [*Allen et al.*, 1976]. Nevetheless, only recently has the focus shifted from searching for precursory phenomena and developing predictive statistical models (mostly based on a posteriori analysis of few well documented cases) to their generalization and testing. Early pioneering efforts aimed at meaningful, systematic analysis of hypotheses concerning precursory phenomena [*Sadovsky*, 1986], eventually motivated the beginning of to experimental forward prediction of the largest earthquakes worldwide [*Healy et al.*, 1992; *Kossobokov et al.*, 1999, 2012]. More recently, the Collaboratory for the Study of Earthquake Predictability (CSEP; http://www.cseptesting.org) has been set up in different regions of the world (including Italy), with the declared general objective of providing a well-controlled environment in which a wide range of scientific experiments on forecast/prediction of earthquakes can be run and evaluated. The CSEP evaluation, however, is based on comparative analysis of a large collection of competing methods, and is made according to the best probability gain achieved in a rather short testing period, which could be very misleading. As pointed out by *Peresan et al.* [2012], the most frequent inaccuracies in a comparative analysis are due to inadequate assessment of the space–time volume of alarms (e. g., areal extent of the affected territory, given a specific probability threshold) associated with probabilistic forecasts, as well as to inadequate or missing criteria to compare alarm-based with probability-based models [*Molchan and Romashkova*, 2011; *Molchan*, 2012].

Remarkably, so far, none of the proposed earthquake precursory signals evaluated by the International Association for Seismology and Physics of the Earth's Interior have shown sufficient evidence to be used as a precursor [*Wyss and Booth*, 1997], and none of the gridded rate-based forecast models passed the rigid CSEP testing [*Jordan*, 2006; *Schorlemmer and Gerstenberger*, 2014].

In Italy, a first successful attempt in forward prediction of earthquakes was carried out by *Caputo* [1983], based on the analysis of sequences of large earthquakes in Southern region; the prediction was confirmed by the occurrence of a moderately large event (*M* 5.4) that took place in May 1984 in the Abruzzo region [*Caputo*, 1988]. Since then, a number forecasting models have been developed and applied to the Italian territory, ranging from long-term nonparametric models of earthquake occurrence [*Rotondi*, 2010], to the very popular short-term models based on the Epidemic Type Aftershock System (ETAS) [e. g., *Console et al.*, 2010]. Different intermediate-term prediction algorithms based on seismicity patterns have also been applied, including CN (named after its development for the California–Nevada region) and M8S algorithms [*Peresan et al.*, 2005, and references therein] and the Pattern Informatics (PI) algorithm [e. g., *Radan et al.*, 2013]. An extensive literature exists, for the Italian territory, in which several forecasting schemes have been proposed, including a number of papers that present models tested within the CSEP – Testing Region Italy (http://cseptesting.org/regions/italy.php) [e. g., *Marzocchi et al.*, 2010]; to my knowledge, no document on the results achieved by this prospective testing has so far been released.

The 2009 L'Aquila earthquake (*M* 6.3), which was preceded by a remarkable swarm and by claims of prediction, eventually motivated the setting up of the International Commission on Earthquake Forecasting (ICEF), with the task to provide an overview of current knowledge and consistent recommendations for operational earthquake forecast/prediction development and use. The report produced by the ICEF Commission [*Jordan et al.*, 2011] was a starting point for critical discussion on different forecast/prediction methodologies. *Peresan et al.* [2012] discussed and complemented the ICEF recommendations, providing information about existing operational practices in Italy, as well as on possible mitigation actions that can be taken in response to formally defined earthquake forecast/predictions. In particular, the authors emphasized that in order to be useful, operational earthquake forecasting (OEF) information must be reliable, tested, confirmed by evidence, and not necessarily probabilistic.

Although a number of possible forecasting methodologies have been proposed and are currently applied both at national and international scales, only few of them, including the CN algorithm considered in this study [*Peresan et al.*, 2005], are defined in a formally testable way and are undergoing rigorous prospective testing over a time span sufficient to validate, in a statistically reliable way, the obtained results [*Peresan et al.*, 2015]. The CN belongs to a family of prediction algorithms that have been developed according to the general scheme of pattern recognition, which allows identification of the space–time volumes most prone to future seismic activation [*Keilis-Borok and Soloviev*, 2003]. These algorithms have provided effective and statistically significant results in rigorous real-time testing, which has been ongoing for more than two decades on a global scale [*Kossobokov*, 2012, 2014; *Kossobokov and Soloviev*, 2015]. This chapter provides an overview of the recent advances and results achieved from using the CN algorithm over about 20 years of rigorous prospective testing in Italy and its surroundings [*Peresan et al.*, 2011, 2015], and addresses general issues such as testing the criteria for selection of the areas of investigation based on the seismotectonic model, as well as operational issues related to the homogeneity of the earthquake catalog used for real-time monitoring of seismicity.

9.2. SEISMICITY PATTERNS AND SEISMOTECTONIC INFORMATION

Recent technological advances, particularly satellite Earth observations (EO), provide steadily increasing amounts of systematic data about a variety of physical phenomena, ranging from ground-related deformation to geochemical, electromagnetic, and thermodynamic properties, which could be related to processes of earthquake preparation [e. g., *Riguzzi et al.*, 2012; *De Santis et al.*, 2015, and references therein]. Consistent with the theory of nonlinear dynamic systems [*Keilis-Borok and Soloviev*, 2003; *De Santis et al.*, 2010], detectable variations in observable phenomena are expected in the build up to a catastrophic event (i.e., critical point, singularity). Nevertheless, the long time-scales associated with lithospheric dynamics, and the sporadic nature in the occurrence of large earthquakes, still limit the possibilities for their systematic analysis and testing. At the moment, catalogs of earthquakes remain the most objective and prolonged record of seismic activity on Earth, which is why most methods presently applied to earthquake forecasting/prediction are based on analysis of these catalogs.

An essential step when analyzing premonitory seismicity patterns consists in the definition of the area where precursors have to be sought, i.e., the area of investigation, which is strictly interrelated with the size of the events to be predicted. Obvious statistical arguments imply considering areas of adequate size, which include a number of earthquakes sufficient for a sound characterization of seismic activity and detection of possible anomalous changes, and which are consistent with the uncertainties of the input data. Moreover, the spatial extent of the earthquake source must be taken into account, since the occurrence of a large event involves a fault with finite dimensions, rather than just a point. Therefore, quite naturally, the size of the study area should increase with increasing rupture size $L = L(M_0)$, where M_0 is the magnitude threshold of the target events. In fact, the preparation of a large earthquake may involve a system of faults rather than a single fault, hence nonlocal precursors, including migration of seismicity, and long-range correlations may reflect some underlying large-scale processes, which cannot be explained by the simple post-seismic stress redistribution in an elastic medium, but which are typical in self-organized critical systems. This fact is supported by independent observations [e.g., *Bowman et al.*, 1998] showing that the size of the region, where accelerated seismicity can be observed, scales with the magnitude and is much greater than that predicted by simple elastodynamic interactions. Similar considerations apply also to other observable phenomena, as pointed out by *Dobrovolsky et al.* [1979], who reported evidence that the region of interest in pre-earthquake processes scales with the magnitude of the impending earthquake.

In particular, for application of the methodology considered in the study detailed in this chapter, it has been found empirically that the linear dimensions of the investigated area must be greater or equal to $5L$–$10L$. This condition is especially relevant in view of the multiscale seismicity (MS) model [*Molchan et al.*, 1997] and of the different generalizations of the Gutenberg–Richter (GR) law [*Gutenberg*, 1956], such as the unified scaling laws for earthquakes occurrence, independently proposed by *Kossobokov and Mazhkenov* [1994] and by *Bak et al.* [2002]. In fact, the GR law was formulated considering seismicity at a global scale; hence it does not provide any explanation as to how the frequency–magnitude relation changes when focusing on a smaller area. According to the MS model, the log-linear relation corresponding to the GR law describes adequately only the ensemble of earthquakes generated by faults with areas that are small with respect to the area of the analyzed region. In agreement with the MS model, when seismicity is analyzed over regions of relatively small area, the frequency–magnitude relation is linear only up to a certain magnitude, while for larger events it usually exhibits an upward bend. Specifically, within the areas delimited for prediction purposes, the number of earthquakes with $M > M_0$ (i.e., the target earthquakes, whose source size is comparable with the region size) usually exceeds the extrapolation of the GR law, and hence these events can be considered abnormally large within the given region. At the same time the MS model guarantees the log-linearity of the frequency–magnitude relation in the magnitude range considered for the analysis of the premonitory patterns ($M < M_0$), namely the self-similarity condition for the small and moderate events. In such a way, as discussed in detail by *Peresan et al.* [2005], the information carried by the many small and moderate earthquakes, statistically obeying the GR law and its generalizations [Kossobokov and Mazhkenov, 1994; Bak et al., 2002; Nekrasova et al., 2011], is used to predict strong earthquakes that do not satisfy the GR law.

Therefore, the seismic precursors of moderate and strong earthquakes (e.g., $M > 5.5$) must be sought within regions having linear dimensions of several hundred kilometers. Even though the shape and position of the areas of investigation have an objective physical significance, there is unavoidable arbitrariness in outlining their boundaries. The selection of the areas of investigation can be carried out following a variety of criteria, ranging from systematic scanning of the territory by geometrically predefined areas (e. g., the circles of investigation (CI) used by *Kossobokov et al.* [2002]), to more specific definition of the boundaries based on the space–time completeness of available data [*Keilis-Borok et al.*, 1990] or accounting for the seismotectonic and geodynamic features of the study region [*Peresan et al.*, 1999].

In this study the CN algorithm [*Keilis-Borok and Rotwain*, 1990] is considered, which belongs to a family of middle range intermediate-term earthquake prediction algorithms based on a formalized, quantitative analysis of the seismic flow. The algorithm is based on a multiple set of premonitory patterns and has been designed following the general concepts of pattern recognition, which automatically imply strict definitions and reproducible prediction results. The CN algorithm allows for a diagnosis of the times of increased probability (TIP) for the occurrence, inside a given region and time window, of earthquakes with magnitude greater than a fixed threshold M_0. Quantification of the seismicity patterns is obtained through a set of empirical functions of seismicity, each representing a reproducible precursor, whose definition has been guided by the theory of complex systems and laboratory experiments on rock fracturing [*Keilis-Borok*, 1996; *Rotwain et al.*, 1997]. Specifically, the functions, which are evaluated on the sequence of the main shocks occurred within the analyzed region, account for increased space–time clustering of moderate size earthquakes, as well as for specific changes in seismic activity, including anomalous activation and quiescence. One of the most relevant properties of the considered precursory patterns is the similarity; after appropriate averaging, the seismicity changes, which can be detected 2–3 years before the occurrence of a strong earthquake, appear to be the same in a wide variety of tectonic settings [*Keilis-Borok and Soloviev*, 2003].

In practice, the seismic activity within the area of investigation is represented, at each time t, by the vector $\mathbf{P}(t) = [p_1(t),\ldots,p_m(t)]$, where the components $p_i(t)$ represent the values of m different functions, evaluated within sliding time windows with common end in t. The problem of diagnosis using TIP is then formulated in term of pattern recognition as follows: knowing $\mathbf{P}(t)$, recognize whether or not the moment t belongs to a TIP (i.e., an alarm period), i.e., if the probability that a strong earthquake (i.e., with magnitude $M \geq M_0$) occurs in the time interval $(t, t+\tau)$ is higher than in normal conditions. This general scheme of analysis is naturally open for inclusion of other traits and different data, not necessarily seismological.

The robust definition of alarm periods as "Times of Increased Probability with respect to normal conditions," is imposed by the fact that any attempt to quantify precisely the probability increase during TIP would require several a priori assumptions (i.e., Poissonian recurrence, independence of TIP, etc.), most of which is poorly constrained by the observations available within each specific region. The quantification that the probability requires an observation time interval of many thousand years. Gross empirical estimates of the increase in probability of strong earthquake occurrence associated with TIP can be obtained based on long-term experimental testing, as discussed by *Peresan et al.* [2015]. Nonetheless, from an operational perspective, providing specific values of probability for the occurrence of a strong earthquake is not necessarily preferable; in fact, the essentially low and very uncertain probability estimates associated with sporadic large events may mistakenly lead to reassuring considerations, even when the increase in probability is high.

Since the early formulation of the algorithm, based on a detailed analysis of seismicity in the California–Nevada region [*Keilis-Borok and Rotwain*, 1990], CN has been applied in a number of regions worldwide, including Italy [*Peresan et al.*, 2005] and, most recently, Iran [*Maybodian et al.*, 2014]. In all the worldwide standard applications the premonitory patterns sought in the seismic flow are those selected for the California–Nevada region, when the prototype of CN [*Gabrielov et al.*, 1986] was developed. Tests carried out on a global scale allowed for a statistical assessment of the forecasting performances of the CN algorithm. Specifically, a preliminary estimate of the significance of the prediction results, obtained for the period 1983–1998 in 22 regions of the world, provided a confidence level around 95% [*Rotwain and Novikova*, 1999]. Such conclusions comply with the statistically significant results obtained by the ongoing rigorous prospective testing of the CN algorithm for the Italian territory since 1998 [*Peresan et al.*, 2015], as well as by its recent extension to the Adriatic region described in this chapter.

9.2.1. CN Regionalization Based on Seismotectonic Zoning

Compared with the California–Nevada region, where the CN algorithm was originally developed, the Italian peninsula and the whole Mediterranean area exhibit a considerable heterogeneity in the tectonic regime, revealed by the coexistence of greatly different, fragmented seismogenic structures [*Meletti et al.*, 2000, 2008; *Cuffaro et al.*, 2011]. The present geodynamic regime of the central Mediterranean region is controlled by Africa–Europe plate interaction and by the subduction of the southwestern margin of the Adria plate, along the Apennines [*Cuffaro et al.*, 2010; *Brandmayr et al.*, 2011; *Doglioni and Panza*, 2015; *Raykova et al.*, 2015]. The Alpine arc, the main structure characterizing Northern region, is generally uplifting [*D'Agostino et al.*, 2005; *Cloentingh*, 2006] with some westerly strike-slip motion, and therefore the majority of focal mechanisms exhibit contractional or transpressive features [*Vannucci and Gasperini*, 2004; *Guidarelli and Panza*, 2006; *Basili et al.*, 2008]. The Apennines form the central part of the Italian peninsula and are characterized by a band of tensional seismotectonic behaviour, with prevailing dip-slip focal mechanisms (normal faulting). The seismotectonic features in the southern part of the Italian peninsula are controlled by the sinking of the Ionic plate under the Calabrian Arc, towards the Thyrrenian Sea [e. g., *Splendore et al.*, 2010; *Brandmayr et al.*, 2011; *Trua et al.*, 2014].

Early experiments in the application of the CN algorithm to the Italian territory showed that defining the areas of investigation according to seismological and tectonic evidence improved its performance, reducing both the percentage of TIP and failures to predict, while increasing their stability. Consequently, *Peresan et al.* [1999] proposed a criterion for the selection of the CN algorithm regionalization based on the seismotectonic model, which takes into account the tectonic complexity of the area, and at the same time decreases the subjectivity in defining the territory of investigation. Given the general rules for the CN algorithm application [*Peresan et al.*, 2005], including considerations about the dimensions of the areas of investigation, regionalization definition required substantial grouping of the seismogenic zones recognized so far for the Italian territory [*Scandone et al.*, 1990; *Meletti et al.*, 2000]. According to the proposed criterion, a defined region for CN algorithm application comprises adjacent zones with homogeneous seismogenic characteristics (e.g., only contractional or only tensional) and by zones with transitional properties. A transitional zone is included in a region if it is between zones of the same kind or if it separates two zones with different regimes and the space distribution of the aftershocks reveals a possible connection. Following these rules and taking into account the main geodynamic features of the area, regionalization covering a large part of the Italian territory has been delineated including the most seismically active areas, as comprehensively described in *Peresan et al.* [1999]. The three regions outlined in this way are shown in Figure 9.1, and have been used for prospective testing of the CN algorithm in Italy since 1998. The satisfactory results obtained so far, described in detail in section 9.3, support the hypothesis that precursors of a strong shock can be found inside seismogenically homogeneous areas associated with a dominating geodynamic process [*Peresan et al.*, 1999].

The idea to consider groups of homogeneous seismogenic zones, with the aim to optimize the selection of the causal system of faults involved in the preparation process, turned out a successful improvement with respect to earlier standard applications of CN in Italy (e.g. *Rotwain and Novikova*, 1999). In fact, it permits to account for the available independent information provided by geology and geodynamics, and not only by seismological observations.

9.2.2. Extending Application of the CN Algorithm Monitoring to the Adriatic Region

The successful results obtained so far by application of the CN algorithm with the regionalization based on the seismotectonic model, which include test on independent data from real-time predictions, warrants its application to the analysis of seismicity in the foreland domain extending along the Adriatic Sea and the Ionian Sea. The goal of applying the CN algorithm to the Adriatic region is twofold: (a) practical, by broadening its present application in the region; (b) theoretical, by exploring the similarity of the earthquake preparation process in seismotectonically different regions. Following the same rules that control the definition of the three regions in the Italian territory, a region suitable for application of the CN algorithm to the Adriatic area has been identified (Figure 9. 2).

Thanks to the normalization of its functions, application of the CN algorithm does not require repetition of the full pattern recognition procedure, requiring only definition of the region and the threshold for selecting the events to be predicted. The magnitude threshold for the normalization of functions and the thresholds for their discretization are then consequentially adjusted on the basis of the analysis of seismicity during the threshold-setting period (TSP). This stage of the analysis is often indicated as the *learning* stage, to distinguish it from the stage of *voting* (i.e., the recognition of the time intervals).

The criterion adopted here for the selection of the area of investigation, which considers zones with homogeneous seismogenic features, naturally complements the general rules for the definition of regions for application of the CN algorithm, namely: (a) the boundaries of the

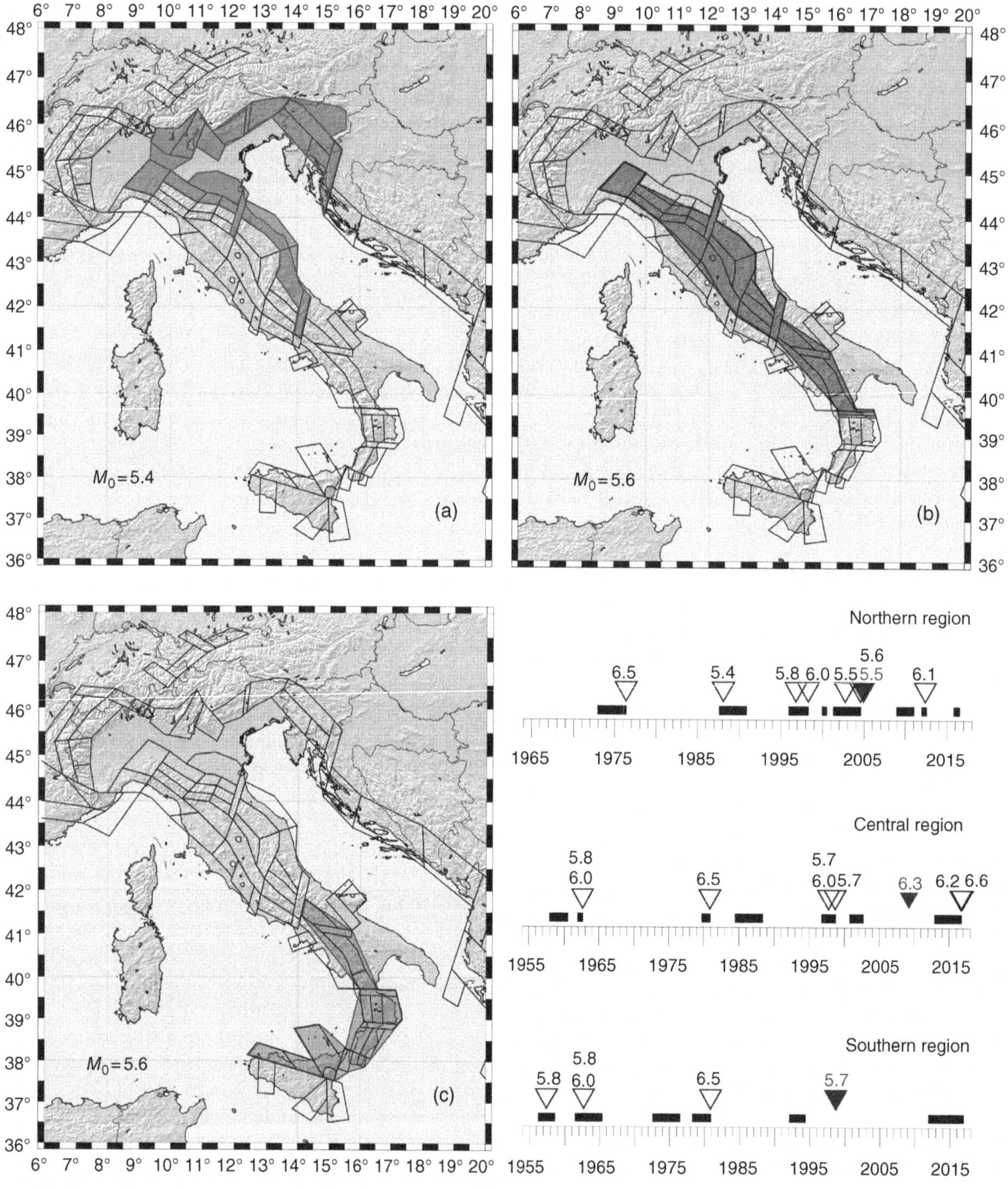

Figure 9.1 Regionalization identified for the application of the CN algorithm in Italy, defined according to the seismotectonic model [*Meletti et al.*, 2000 and references therein]. The magnitude thresholds M_0 for each region are indicated. In the diagrams of the TIP, the black boxes represent periods of alarm, while each triangle indicates the occurrence of a strong event ($M \geq M_0$), together with its magnitude. Two numbers above a triangle indicate the occurrence of two target earthquakes within the same prediction time window (2 months). Full red triangles indicate failure to predict. (*See electronic version for color representation.*)

region must correspond, as much as possible, to minima in seismicity, compatibly with the geometry of the fault system characterizing the area; (b) its linear dimensions must be greater or equal to $5L - 10L$, where L is the length (or linear dimension) of the earthquake source to be predicted; (c) on average at least three events with magnitude larger than the cut-off magnitude, M_c, should occur inside the region each year [*Keilis-Borok*, 1996]. The latter condition indicates that, in addition to the physical complexity of the system, the possibility to

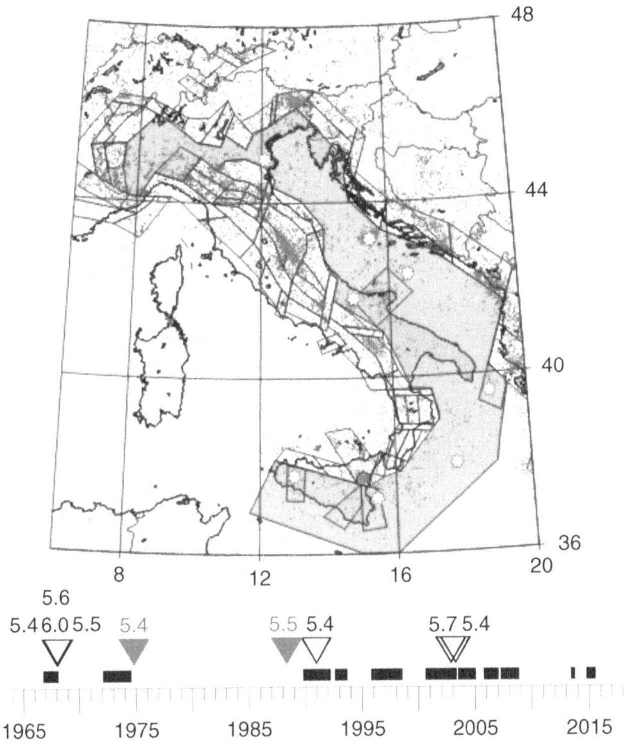

Figure 9.2 Regionalization identified for the application of the CN algorithm to the Adria plate (blue polygon), defined based on the seismotectonic model (thin lines) proposed by *Meletti et al.* [2000]. Red dots mark the epicenters of main shocks with $M \geq 3.0$ reported in the UCI catalog in the time interval 1900–2005. Yellow stars mark the epicenters of target events, i.e., earthquakes with $M \geq M_0 = 5.4$ that have occurred since 1964. The diagram shows the temporal distribution of TIP obtained for the Adriatic region. Key to symbols is the same as in Figure 9.1. (*See electronic version for color representation.*)

reduce the spatial uncertainty of predictions based on the analysis of seismicity patterns is controlled by the completeness level of the available earthquake catalog.

The Adriatic region defined following these criteria (Figure 9.2) comprises foreland areas extending from the Adriatic Sea to Southern Sicily, through the Ionian Sea. Along its northwest boundary the region is adjacent to the Northern region previously defined (Figure 9.1a), while to the east the border coincides with the margin of seismogenic zones delineating the contractional belt along the Dinarides. The northern edge extends up to the contraction belt in the Western Alps, crossing over the Northern region (Figure 9.1a) [*Peresan et al.*, 1999] to include an additional part of the foreland territory along the western Po Plain. The southern boundary of the region, where seismogenic zones are not defined, corresponds to a minimum in the spatial distribution of epicenters, in agreement with the general rules of the CN algorithm application [*Keilis-Borok*, 1996]. The area outlined was partially included in the previous regionalizations considered for Italy [*Keilis-Borok et al.*, 1990] and for *Croatia* [*Herak et al.*, 1999], which were defined mainly based on the spatial completeness of the catalogs available at that time. From a geophysical point of view, the region includes part of the Ionian plate, crosses the Malta escarpment and covers a large part of the Sicily block. Experiments have been carried out taking into account different configurations of the area of investigation, with the aim to reduce the spatial uncertainty of predictions. Splitting the region in two or more parts does not provide satisfactory results. This may indicate a connection between earthquakes that occur within the different geological elements that compose the Adriatic foreland; however, it cannot be excluded that the poor results obtained in each subregion may be the effect of some incompleteness in the catalog, due to the lack of sea-bottom seismographs.

The earthquake catalog used for the CN algorithm analysis in the Adriatic region is the same as that utilized for the routine monitoring of seismicity patterns in the Italian territory, namely the UCI2001 catalog [*Peresan and Panza*, 2002], referred to hereinafter as UCI. The catalog is a compilation of different local Italian catalogs from 1900 to 1986, and since 1986 has been updated using global data from NEIC-USGS (National Earthquake Information Centre, United States Geological Survey) and ISC (International Seismological Centre), as described by *Romashkova and Peresan* [2013]. The UCI catalog is sufficiently complete and homogeneous in time within the outlined region since 1960. As the quantification of the seismic patterns is performed considering the sequence of the main shocks only, the aftershocks are removed by a robust declustering window method [*Keilis-Borok et al.*, 1980]. Research is in progress to assess the robustness of the results with respect to the use of a different declustering method [e. g., *Zaliapin et al.*, 2008], and the favorable results obtained so far have been reported by *Peresan et al.* [2016].

According to the general rules of application for the CN algorithm, once the boundaries of the region are determined, the threshold M_0 for the selection of the events to be predicted is chosen on the basis of the long-term (several tens of years) average occurrence time (not return period) of the earthquakes with $M \geq M_0$ inside the analyzed area, which should be approximately 6–7 years (to guarantee an adequate statistic for the target events). An additional condition requires that the threshold M_0 should be selected as close as possible to a minimum in the histogram of the number of events versus magnitude [*Molchan*, 1990]; this condition, in addition to the above-mentioned physical considerations related to the log-linearity of the frequency–magnitude distribution, guarantees a certain stability of the results, as shown by *Peresan et al.* [2002]. The Adriatic region considered for CN application appears to be characterized by relatively low seismicity, when compared with adjacent areas (Figure 9.2). Nevertheless, based on the frequency–magnitude distribution of earthquakes that have occurred within the region since 1900 (Figure 9.3), a threshold $M_0 = 5.4$ can be defined, which corresponds to an average occurrence time of earthquakes with $M \geq M_0$ of about 7 years (16 main shocks with magnitude greater or equal to 5.4 have occurred since 1900, nine of them since 1960).

Consistent with the completeness of the UCI earthquake catalog, the selected TSP period corresponds to the time interval 1960–1998, the same TSP considered for the CN algorithm analysis in the Northern region [*Peresan et al.*, 2005]. The TIP analysis starts in 1964, since at least 4 years are required for computation of the CN algorithm functions. For the period 1999–2004 the algorithm was applied without any adjustment of parameters, but since 2005 it underwent rigorous prospective testing. The results of the TIP analysis for the Adriatic region are shown in Figure 9.2 and can be summarized as follows: seven out of nine target earthquakes that occurred in the period 1964–2016 are correctly preceded by a TIP, with alarms declared for 33.8% of the considered time span. Remarkably, two earthquakes occurred after the end of TSP and are correctly identified by the algorithm.

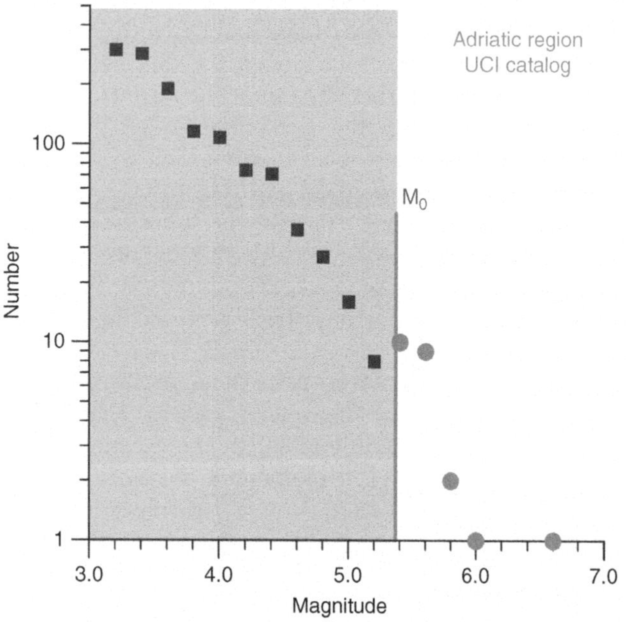

Figure 9.3 Distribution of the number of events versus magnitude for earthquakes that have occurred within the Adriatic region, as reported in the UCI catalog for the time interval 1900–2005. Red dots correspond to the target earthquakes with $M \geq M_0 = 5.4$, whereas the blue squares correspond to the events in the magnitude range considered for the computation of CN functions. (*See electronic version for color representation.*)

The configuration of the Adriatic region and the related parameters, before being used for real-time analysis of seismicity since 2005, underwent a set of control experiments to assess the robustness of the results being provided (*Peresan et al.*, 2006). Two types of errors characterize the results of a prediction, within a fixed region, according to *Molchan* [1990]. The first is the percentage η of failures to predict $\eta = (1 - n/N)$, where n is the number of predicted events and N is the total number of target events. The second is the percentage τ of the total duration of alarms: $\tau = t/T$, where t is the total duration of alarms and T is the length of the analyzed time interval. A measure of the quality of a prediction result is provided by the quantity $\xi = \eta + \tau$, that can be compared with the results of a random guess giving $\xi = 100\%$. Low values of ξ characterize accurate predictions, while as ξ increases toward 100% the quality of results decreases. An additional measure of the prediction quality is the statistic of false alarms: $\kappa = k/K$, where k is the number of false alarms and K is the total number of alarms. The results obtained from application of the CN algorithm in different regions of the world indicate that TIP cover about 30% of the total time and precede about 80% of the strong earthquakes [*Keilis-Borok and Rotwain*, 1990; *Rotwain and Novikova*, 1999]; the statistics $\xi = \eta + \tau$ varies from 20% to 80% and κ is up to 70%.

The control experiments performed show that the prediction results for the Adriatic region are quite stable and satisfactory for values of the threshold M_0 in the range 5.3–5.6. As reported in Table 9.1, the total prediction error $\xi = \eta + \tau$ varies from 28.4% to 57.3%, and κ from 30% to 50%. The average occurrence time for earthquakes with $M \geq 5.3$ and $M \geq 5.6$ is about 4 and 10 years respectively. In addition, the retrospective test referred as "seismic history" [*Gelfand et al.*, 1976] has been applied in order to assess how the variation of TSP time span affects the prediction results. Accordingly, the TSP duration is progressively reduced and the stages of normalization and discretization of functions are repeated for each of the selected TSP, while other parameters of the CN algorithm (e. g., the threshold M_0) are kept fixed. The results of the experiment "seismic history" for $M_0 = 5.4$ are

Table 9.1 Results of retrospective application of the CN algorithm in the Adriatic region, for different values of the magnitude threshold M_0, in the time span 1964–2004.

M_0	n/N	η (%)	τ (%)	$\eta + \tau$ (%)	κ (%)
5.3	7/10	30.0	27.3	57.3	30
5.4	7/9	22.2	31.3	53.5	45
5.5	4/4	0.0	28.4	28.4	33
5.6	2/3	33.3	22.7	56.0	50

N, number of earthquakes to predict; n, number of predicted events; τ, percentage of TIP duration; η, percentage of failures to predict; κ, percentage of false alarms.

Table 9.2 Results of the seismic history experiment for application of the CN algorithm to the Adriatic region, performed for magnitude threshold $M_0 = 5.4$.

End of TSP	n/N	η (%)	τ (%)	$\eta+\tau$ (%)	κ (%)
1 January 2003	7/9	22.2	31.3	53.5	45.0
1 January 1999	7/9	22.2	31.3	53.5	45.0
1 January 1991	7/9	22.2	28.2	50.4	40.0
1 May 1988	7/9	22.2	34.4	56.6	45.0

N, number of earthquakes to predict; n, number of predicted events; τ, percentage of TIP duration; η, percentage of failures to predict; κ, percentage of false alarms.

summarized in Table 9.2 and appear very stable: the total prediction error $\xi=\eta+\tau$ varies from 50.4% to 56.6% and there are no additional failures to predict with respect to the results obtained for TSP up to 1998.

9.3. PROSPECTIVE TESTING OF THE CN ALGORITHM IN ITALY AND SURROUNDING AREAS

It is well recognized since the 1970s that any reliable forecasting tool must be submitted to rigorous testing, first in retrospect, and then in an experimental real-time mode. The US National Research Council, Panel on Earthquake Prediction of the Committee on Seismology came to the indisputable conclusion that: "Only by careful recording and analysis of failures as well as successes can the eventual success of the total effort be evaluated and future directions charted" [*Allen et al.*, 1976]. Still, there are a number of practical issues that need to be addressed, when dealing with the long-term experimental efforts required by rigorous prospective testing of predictions at the intermediate-term middle-range scale. The most common one is eventually related with the necessity to preserve certain homogeneity of the input data (e. g., the earthquake catalog) for a time span of several decades, which is required for a sound statistical formulation and testing of methods targeting strong sporadic earthquakes.

In Italy the effort to set up a systematic and testable analysis of seismicity started in 1990, with the first application of the CN algorithm [*Keilis-Borok et al.*, 1990]. During the initial stages of the experiment the basic problems related to the timely upgrading and to the quality of the input data (i.e., completeness and homogeneity of the earthquake catalog) were identified and solved as follows. When it became clear that the originally available catalog for the Italian territory contained a systematic error in magnitude determinations [*Peresan et al.*, 2000], the revised UCI catalog was compiled [*Peresan and Panza*, 2002]. To allow for prospective testing using the CN algorithm, the UCI has been routinely updated using global data from NEIC since 1998 [*Peresan et al.*, 2005]. In 2009, however, the completeness level of NEIC data over the Italian territory suddenly increased well above the $M_c = 3.0–3.5$ required for application of the CN algorithm. To overcome this problem, starting from January 2009, the ISC data [ISC, 2016] have been merged during the regular update of the UCI catalog with the NEIC data [*Romashkova and Peresan*, 2013]. Among the different magnitudes listed in the UCI catalog, the operating magnitude for application of the CN algorithm, M_{prio}, is selected according to the priority criteria defined in *Peresan et al.* [2005]. A detailed description of the UCI earthquake catalog [*Peresan and Panza*, 2002] used for the routine monitoring of seismicity by the CN algorithm in Italy is available at: http://geologia.units.it/esperimento-di-previsione-dei-terremoti-mt/italian-earthquakes-catalogue

The possibility to test the CN algorithm predictions using the catalog established for CSEP testing over the Italian territory (CSEP-TRI) has also been explored; this catalog, however, is quite heterogeneous and discontinuous and hence not adequate for application of the CN algorithm [*Romashkova and Peresan*, 2013]. The detected magnitude heterogeneity [*Romashkova et al.*, 2009], corroborated by *Gasperini et al.* [2013] independently, poses serious concern for those results obtained based on such input data, particularly those from model testing by CSEP-TRI.

Several experiments were devoted to assess the robustness of the methodology against the unavoidable uncertainties in the input data [*Peresan et al.*, 2000, 2002]. Given the satisfactory results acquired, an ongoing prediction experiment was launched in July 2003, aimed at rigorous real-time prospective testing of the CN algorithm in Italy. The goal of the experiment is to accumulate a collection of right and wrong predictions (the latter include the false alarms and/or failures to predict), in order to assess the predictive capability of the considered methodology. In this framework the CN algorithm predictions have been routinely updated every 2 months (i.e., January, March, May, July, September, November), for the three regions identified in Figure 9.1, following the rules for the real-time application of the CN algorithm in Italy as described in detail by *Peresan et al.* [2005]. Analysis was subsequently expanded to the Adriatic region (Figure 9.2), which was included in the prospective testing since 2005. Any region considered for application of the CN algorithm is characterized by a specific area and magnitude range, which identify the target earthquakes; accordingly, for a target earthquake to be considered successfully predicted, its epicenter must be located within an area alerted for the corresponding magnitude range. When comparing the four maps in Figures 9.1 and 9.2 it is evident that the regions outlined for CN algorithm monitoring in the Italian territory partially overlap; thus, as a rule, if a target earthquake occurs in an area where two regions overlap, it is considered correctly predicted if at least one of the regions is alerted for the corresponding magnitude range.

The routinely updated results and a complete archive of predictions are available online at the website: http://geologia.units.it/esperimento-di-previsione-dei-terremoti-mt/intermediate-term-middle-range-earthquake-prediction). This procedure permits a rigorous validation and independent evaluation of the applied algorithms. The full list of target earthquakes for the applied CN algorithm, both in retrospective and real-time application in Italy and adjacent territory, is provided in Table 9.3.

The results obtained so far by application of the CN algorithm in Italy and adjacent territory are summarized in Table 9.4, considering the following time intervals: (a) the periods of retrospective analysis during TSP (i.e., 1954–1963, Central and Southern regions only; 1964–1997, all four regions); (b) the period of forward analysis 1998–2016; (c) forward prediction obtained in real time only (i.e., including the Adriatic region since January 2005); (d) the overall period of analysis, 1954–2016. All together 22 out of the 26 target earthquakes that occurred within the monitored territory since 1954 have been (correctly) preceded by an alarm (TIP) declared by the applied CN algorithm. About 31% of the overall space–time volume (STV) is occupied by the alarms and the confidence level of such predictions is above 99%. The results obtained in the forward prediction experiment over the extended regionalization were also satisfactory: seven out of nine target earthquakes have been predicted in real-time by the applied CN algorithm, with about 35% STV occupied by the alarms; the probability of obtaining such results by random guess is below 2%. The Emilia earthquake (M 6.1, 20 May 2012), scores amongst the successful predictions by the applied CN algorithm, as it occurred within the Northern region (Figure 9.1a), which had been alerted for an earthquake with a magnitude $M \geq 5.4$ since 1st March 2012. The most recent successful real-time predictions correspond to the Amatrice (M 6.2) and Norcia (M 6.6) earthquakes, which struck central region on 24 August 2016 and 30 October 2016, respectively. The epicenters fall within the Central region (Figure 9.1b), alerted by the applied CN algorithm for an earthquake with magnitude $M \geq 5.6$, starting on 1 November 2012. Quite interestingly, the alarm for the Central region did not end after the Amatrice earthquake (24 August 2016); in fact, at the stage of the CN algorithm September predictions update, based on analysis of seismicity patterns, the TIP was extended up to 1st November 2016, thus resulting in the correct forecast of the Norcia earthquake (30 October 2016).

Since the beginning of the real-time prediction experiment only two events, out of nine target earthquakes, were missed by the applied CN algorithm, namely the Salò earthquake (M 5.5, 24 November 2004) and the L'Aquila earthquake (M 6.3, 6 April 2009). The L'Aquila earthquake scores as a failure to predict because its epicenter was located about 10 km outside the alarm territory identified by the applied CN algorithm for the corresponding time window [*Peresan et al.*, 2011]. The situation with the Salò earthquake, which occurred within the Northern region, was quite different. In fact, three target earthquakes struck the same region in a row, in September 2003, July 2004, and November 2004 (Table 9.3). Both the first and second earthquakes were correctly predicted by the applied CN algorithm, with an alarm starting in May 2001; according to the CN algorithm rules, however, the seismic energy release within the region was sufficiently high to terminate the alarm after the July 2004 earthquake, thus resulting in the failure to predict the later Salò earthquake.

The prediction results obtained so far appear very encouraging, particularly on account of introducing in the experiment of the Adriatic region, an area mostly covered by the sea and crossed by several political borders. These features obviously indicate a less complete and inhomogeneous of the Adriatic region earthquake catalog, compared with that available for the regions defined for the Italian territory. Based on the results obtained for the regionalization shown in Figure 9.1, it has been possible to estimate an increase by about a factor three of the rate of target events during TIP, when compared to their long-term average occurrence rate [*Peresan et al.*, 2015]. The results obtained for the Adriatic region are consistent with such estimates (Figure 9.4). Accordingly, the probability gain associated with CN predictions for the Italian territory and adjacent regions is estimated between 2 and 4, in good agreement with the independent estimates by ICEF [*Jordan et al.*, 2011]. Noticeably, these probability-gain estimates, supported by decades of prospective testing of intermediate-term middle-range forecasts, are very different from the short-term "nominal probability gain factors in regions close to the epicenters of small-magnitude events," which are clearly recognized as "highly uncertain and largely unvalidated" [*Jordan et al.*, 2011]. In fact, due to the fundamental property of the probability gain scoring [*Molchan*, 2003] the maximum is reached when the STV of alarms goes to zero; as a result, a single success obtained by randomly declaring few sporadic alarms may provide very high probability gain.

The STV of alarms given in Table 9.4 is computed accounting for the rate of occurrence of past earthquakes in the area under investigation (relative intensity, RI), following the procedure described by *Kossobokov et al.* [1999]. The STV measure based on the seismicity distribution is more stringent than the measure based on

Table 9.3 List of target earthquakes for retrospective (1954–1997) and prospective (1998–2016) application of the CN algorithm in Italy and adjacent territory. Only the central and southern regions are considered up to 1964. Earthquakes that occurred during the period of real-time monitoring (i.e., since 1998 for the north, centre and south regions and since 2005 for the Adriatic region) are indicated by bold typeface.

Date	Latitude (°N)	Longitude (°E)	Depth (km)	M_{prio}	CN algorithm applied	Region delimited by CN algorithm
20 May1957	38.70	14.10	60	5.8	Yes	Southern
21 August 1962	41.15	15.00	40	6.8	Yes	Central, southern
21 August 1962	41.15	15.00	40	6.0	Yes	Central, southern
17 April1962	42.45	17.37	25	5.5	Yes	Adriatic
30 December 1967	45.00	12.10	35	5.4	Yes	Adriatic
15 January 1968	37.77	13.00	48	5.6	Yes	Adriatic
15 January 1968	37.77	12.98	44	6.0	Yes	Adriatic
12 February 1968	38.00	17.80	40	5.5	Yes	Adriatic
20 October 1974	39.60	18.90	60	5.4	No	Adriatic
6 May 1976	46.23	13.13	12	6.5	Yes	Northern
23 November 1980	40.85	15.28	18	6.5	Yes	Central, southern
1 February 1988	46.31	13.13	8	5.4	Yes	Northern
26 April 1988	42.37	16.60	14	5.4	No	Adriatic
13 December 1990	37.30	15.44	11	5.6	Yes	Adriatic
15 October1996	44.79	10.78	10	5.8	Yes	Northern
26 September 1997	43.08	12.81	10	5.7	Yes	Central
26 September 1997	43.08	12.81	10	6.0	Yes	Central
12 April 1998	**46.24**	**13.65**	**10**	**6.0**	**Yes**	**Northern**
9 September 1998	**40.03**	**15.98**	**10**	**5.7**	**Yes**	**Central, southern**
31 October 2002	41.78	14.87	10	5.7	Yes	Adriatic
29 March 2003	43.26	15.49	33	5.4	Yes	Adriatic
14 September 2003	**44.33**	**11.45**	**10**	**5.6**	**Yes**	**Northern**
12 July 2004	**46.30**	**13.64**	**7**	**5.7**	**Yes**	**Northern**
24 November 2004	**45.63**	**10.56**	**17**	**5.5**	**No**	**Northern**
6 April 2009	**42.33**	**13.33**	**8**	**6.3**	**No**	**Central**
20 May 2012	**44.90**	**11.23**	**6**	**6.1**	**Yes**	**Northern**
24 August 2016	**42.72**	**13.19**	**4**	**6.2**	**Yes**	**Central**
30 October 2016	**42.85**	**13.09**	**10**	**6.6**	**Yes**	**Central**

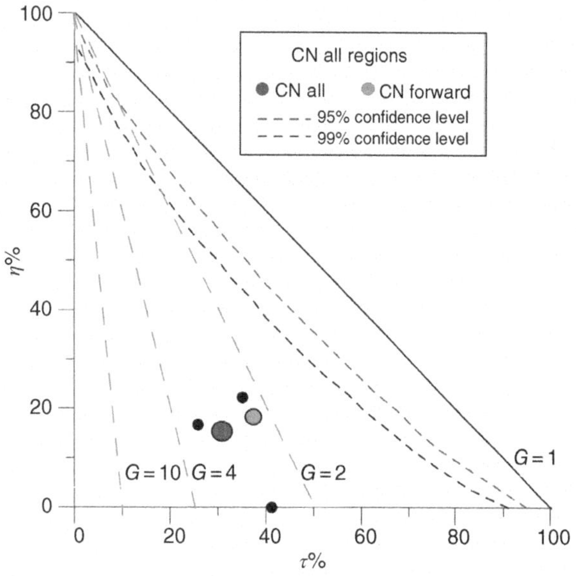

Figure 9.4 Error diagram [*Molchan*, 1990] for the results of the applied CN algorithm in Italy detailed in Table 9.4: η, percentage of failure to predict; τ, total space–time volume occupied by alarms. The diagonal line corresponds to the results of a random guess (gain $G=1$). The blue dots correspond to the results obtained for the three experimental time intervals considered in Table 9.4. The prediction results, obtained considering together all retrospective and real-time testing, are indicated by the red circle, and the curves corresponding to 95% and 99% confidence levels are shown (the CN algorithm score is above 99%, see text). (*See electronic version for color representation.*)

Table 9.4 Space–time volume of alarm indicated by the applied CN algorithm in Italy and its surroundings.

Experiment time interval	Space–time volume of alarm (%)	n/N	Confidence level (%)
1954–1963[a]	41	3/3	93
1964–1997	26	10/12	>99
1998–2016	37	9/11	>99
Forward predictions (1998–2016)[b]	35	7/9	98
All together (1954–2016)	31	22/26	>99

N, total number of target events; *n*, number of predicted earthquakes
[a] Only Central and Southern regions were analyzed.
[b] Began in 2005 for the Adriatic region.

simple spatial extent of alarm areas (in km^2), and allows for the assessment of the significance of the issued predictions with respect to a time-independent and nonclustered seismicity model (i.e., RI), according to which future large earthquakes are considered more likely where higher seismic activity occurred in the past. The problem of using an adequate measure of the alarm volume to properly assess the performances of a forecasting model is discussed in some detail by *Radan et al.* [2013], with specific reference to the PI algorithm; the study demonstrates that to consider the simple spatial extent of the alarm areas may erroneously lead to the impression of high accuracy of the method. Although the predictive capability of the CN algorithm, with respect to a random Poissonian model, has been clearly recognized, it might be interesting in future to compare the CN algorithm predictions against models based on earthquake clustering. However, so far no rigorously validated clustering model for the Italian territory is available to carry out such analysis, even accounting for the model testing within the CSEP—Testing Region Italy (i.e., August 2009; http://cseptesting.org/regions/italy.php); in fact, no document has been released so far on the results achieved by prospective testing.

9.4. REDUCING SPACE-TIME UNCERTAINTY OF PREDICTIONS BY INTEGRATION OF MULTIDISCIPLINARY DATA

The results from application of the CN algorithm to the Italian territory indicate the practical feasibility of earthquake forecasting, although with limited precision, that is with a characteristic alarm-time ranging from a few months to a few years and a linear uncertainty in space of hundreds of kilometers. The reduction of uncertainties about location (where) and time (when) a strong earthquake is to be expected requires the use of additional information, which eventually may be provided by multidisciplinary data (e. g., geodetic, geomorphological, and geochemical), as well as by seismicity data in a lower magnitude range available in high quality local catalogs (e. g., MSc algorithm by *Kossobokov et al.* [1999]). Accordingly, from a practical viewpoint the prediction problem can be posed as a sequential, step-by-step process, narrowing down the territory and time interval, where and when the incipient earthquake is most likely to happen [e. g., *Kossobokov and Shebalin*, 2003]. This multiscale approach may benefit, at any stage, from independent complementary geophysical observations [*Kanamori*, 2003; *Bormann*, 2011; *Panza et al.*, 2013].

The possibilities for an integrated analysis of intermediate-term predictions with different data, spanning from space geodesy to ground-shaking modeling, are exemplified in this section, focusing on physically sound and testable methodologies, which have been developed and successfully applied to the Italian territory.

9.4.1. Integration with Information from Space Geodesy

A highly automated operational system, developed within the framework of the Pilot Project SISMA (https://www.asi.it/it/eventi/workshop/progetto-sisma) and funded by the Italian Space Agency (ASI), demonstrates the power of effective integration of Earth observation (EO) data and advanced methods in geophysical and seismological data analysis [*Crippa et al.*, 2008; *Panza et al.*, 2013]. The synergic use of geodetic EO data and geophysical forward modeling (GFM) deformation maps at the national scale, in particular, complements the real-time analysis of seismicity patterns by the applied CN algorithm and thus reduces the space uncertainty in the localization of the impending strong earthquakes.

The diagnostic elements provided by the SISMA system take advantage from using data and models at different space–time scales, from the regional scale of the entire Italian peninsula down to the local scale of the single seismogenic area. In addition to the validated premonitory seismicity patterns, there are promising indications for pre-earthquake transient phenomena in both GNSS (Global Navigation Satellite System) and SAR (Synthetic Aperture Radar) data and in the modeled stress field. Anomalies in EO data time series, with respect to the linear trend of tectonic strain rate, have been observed during some pre-seismic phases. Accordingly, integration of GNSS and GFM observations at the regional scale can be used to provide the strain rates due to the background tectonic activity that, once identified within a single seismogenic area, define a geodetic anomaly. Transient regional deformation indicators are monitored and modeled at the regional scale [e. g., *Marotta and Sabadini*, 2008; *Splendore et al.*, 2010] with a time step of few months, in agreement with the schedule defined for seismicity monitoring. Whenever an alerted region is identified at the national scale, it is possible to concentrate the efforts on the faults within the alerted area, with a repeat time for EO data analysis of the order of few days, so as to identify possible anomalies related to stress variations, with respect to the linear stationary trend typical of the interseismic phase, in the vicinity of the fault. The displacements fields obtained from EO data provide the input for the geophysical modeling, which eventually indicates whether a specific fault is in a "critical state" [*Panza et al.*, 2013].

At the national scale of Italy, the SISMA system merges the information about alerted areas obtained from the formal analysis of seismicity with the results from GNSS data analysis. The formal integration is carried out according to a Bayesian approach, with posterior probabilities estimated taking into account the geodetic time series available so far and the results collected in about 50 years of retrospective and real time application of the CN algorithm. In fact, as discussed in section 9.3, the probability that a strong earthquake may occur during an alert diagnosed by the applied CN algorithm, is about three times larger than its long-term time independent estimate (i.e., not accounting for prediction results), and about ten to twenty times larger than the probability associated with no-alert diagnosis. Similarly, within each of the geodetic polygons that could be outlined according to the GNSS national network (Figure 9.5a), the geodetic anomalies have been defined with respect to a mean strain field value, by inspecting the variations in the deformation rate over a 6-month moving window; their a posterior probabilities have also been obtained by applying the Bayes theorem. The analysis of regional deformation transients is updated every 2 months, in agreement with the schedule defined for seismicity monitoring. By combining information about seismicity-based alerted areas (Figure 9.5a) and geodetic anomalies (Figure 9.5b), a joint alert map is obtained (Figure 9.5c) that assigns to the different parts of the monitored territory one of the following possible alert levels: (a) geodetic anomaly plus seismic alert; (b) only seismic alert; (c) only geodetic anomaly; (d) neither geodetic anomaly nor seismic alert.

The maps shown in Figure 9.5 exemplify how the combined use of independent information may contribute to reduce uncertainty of predictions. Although the geodetic anomalies and the seismic alerts are defined over areas of comparable linear dimensions, i.e., of the order of tens to hundred kilometres, their intersection leads to definition of smaller areas, characterized by a different level of alarm quantified from the combination of the geodetic anomaly with the seismic alert. Once an alerted region is identified at the national scale, an effort can be made to further reduce the space–time uncertainty of prediction by resorting to local scale models and refined data analysis. Specifically, the EO data processing is performed with a time step of a few days, focusing the analysis on the faults within the alerted area; this may allow identifying possible anomalies with respect to the linear stationary trend typical of the interseismic phase, related to stress variations in the vicinity of the fault.

The integrated forecasting scheme adopted by the SISMA system is quite general and is open for inclusion of different methods for seismic and geodetic data analysis, provided they are formally defined and testable. As an example, the study by *Riguzzi et al.* [2012] suggests that larger earthquakes tend to occur in areas of lower strain rate, namely in areas where faults are locked and elastic energy accumulates. Accordingly, in tectonically active regions, maps of high and low strain rates may provide useful information for identifying those areas more prone to the next earthquakes. However, to be formulated as a testable forecasting

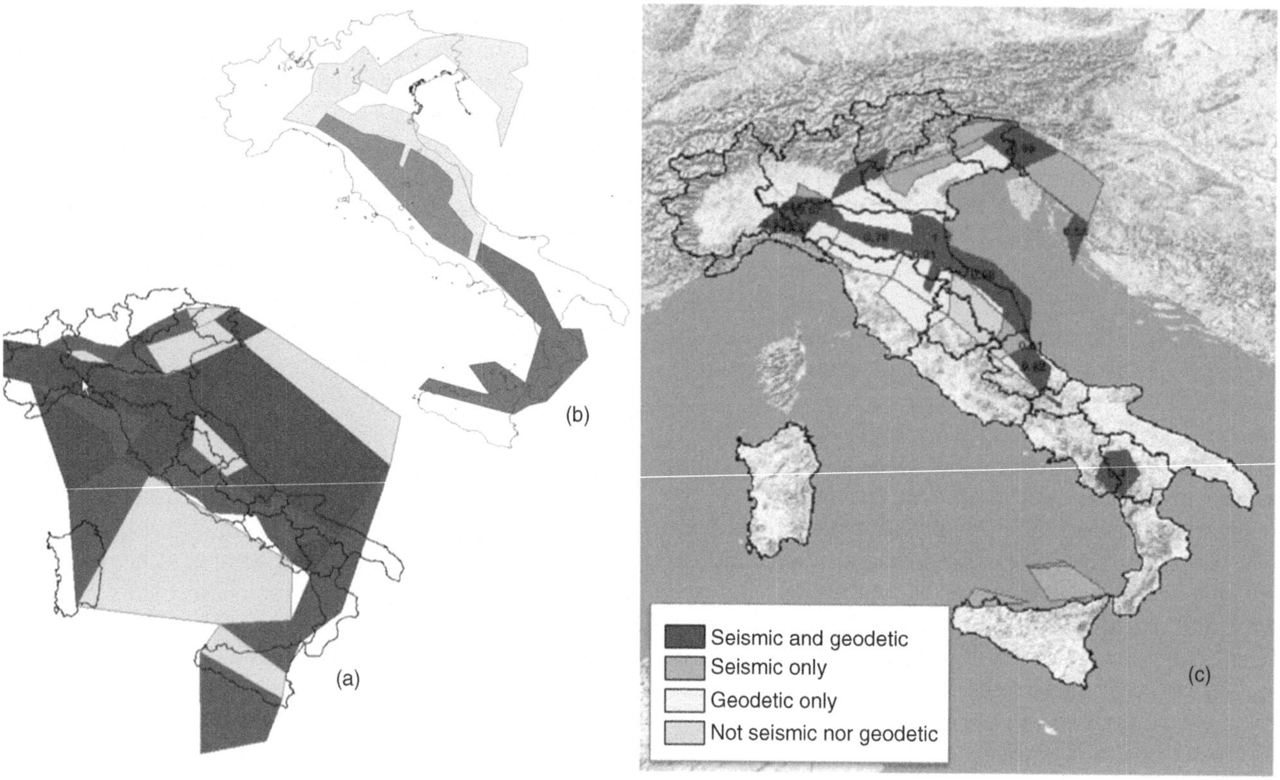

Figure 9.5 Example of an integrated alert map for the Italian territory, obtained from geodetic and seismicity data: (a) GNSS anomalies at a given time (green, polygons with normal behavior with respect to the mean tectonic deformation; red, anomalous polygons; grey, areas where no information is available. (b) Map of territory monitored (grey, yellow and blue areas) and alerted (yellow and blue areas) by the applied CN algorithm, based on the analysis of seismicity patterns (as in Figure 9.1). (c) Joint map of seismic and geodetic anomalies. Different colors correspond to different combinations of seismic and geodetic anomalies. (Source: Reproduced with permission of Springer.) (*See electronic version for color representation.*)

tool, based on strain-rate maps, an unambiguous definition should be provided about when and where such earthquakes may occur, so that the statistical significance of the method can be formally assessed by systematic retrospective and real-time analysis (as described in section 9.3). Given the short time span of geodetic EO data available so far (about 20 years), which may prevent a formal definition of the alerted STVs, coupling with seismic anomalies (e. g., using the CN algorithm) may be a viable approach.

Provided a sufficiently general definition of the geodetic anomalies and assuming similarity of the pre-earthquake processes in different areas worldwide (as done with the seismic anomalies considered in this chapter), large-scale testing could significantly speed up the formalization and verification process of anomalies from EO data, without waiting to collect observations about a full hypothetical seismic cycle within a specific region.

9.4.2. Time-Dependent Seismic Hazard Scenarios: Targeting Areas of Strong Ground Shaking

The space uncertainty typical of the intermediate-term middle-range predictions is intrinsically quite large, on account of the physical and statistical properties of earthquake data. Reducing such uncertainty requires the use of independent, potentially predictive observations, as exemplified in section 9.4.1. Another option is to combine the time-dependent indications given by earthquake forecasting with additional time-independent information available about the earthquake sources and/or their potential impact (deterministic hazard scenarios, e. g., *Panza et al.* [2012]). From an operational perspective, in particular, it may be interesting to focus on the most hazardous areas associated with the alarmed territory. At this stage it is possible to take advantage of the information about the areas most prone to large earthquakes (including identified active faults), as well as about the related ground shaking, as follows.

An example of term-less anticipatory information is provided by the identification of the seismogenic nodes. The pattern-recognition technique can be used to systematically identify the sites capable of generating strong earthquakes, through the combined use of geological, geomorphic, and seismological data and independently from any transient seismic information. The recognition of earthquake prone areas is carried out based on the assumption that large events correlate with the morphostructural nodes, which represent specific structures, formally defined, formed around the intersections of lineaments. Lineaments are identified by the morphostructural zonation (MSZ) method [*Alekseevskaya et al.*, 1977], which delineates a hierarchical block structure of the studied region, using tectonic and geological data, with special care to topography. The boundary zones between blocks are called lineaments and the nodes are defined at the intersections or junctions of two or more lineaments. Among the defined nodes, those prone to earthquakes above a specified (target) magnitude are then identified by pattern-recognition analysis. The quantitative parameters used for the recognition indirectly characterize the intensity of neotectonic movements and fragmentation of the crust around the nodes (e. g., elevation and its variations in mountain belts and watershed areas; orientation and density of linear topographic features; type and density of drainage pattern). For this purpose, the nodes are typically defined as circles, with a radius of a few tens of kilometers, surrounding each point of intersection of lineaments; yet, more refined boundaries of the nodes can be determined, whenever adequately detailed information is available [*Gorshkov et al.*, 2009]. The reliability and statistical significance of the methodology is supported by the number of strong earthquakes that occurred in the previously studied regions: 87% of the events occurred after the pertinent publications, with relevant magnitudes, within the nodes that were recognized in advance as being capable of target earthquakes [*Soloviev et al.*, 2014]. In Italy and surrounding regions, the identification of the sites where strong earthquakes can nucleate has been performed by *Gorshkov et al.* [2002, 2004], considering circles with radius $R=25$ km and two magnitude thresholds, $M_N \geq 6.0$ and $M_N \geq 6.5$. The analysis has been expanded recently to the identification of nodes prone to earthquakes with $M_N \geq 5.0$ in the Po Plain [*Peresan et al.*, 2015].

Integrating the space information provided by the earthquake prone areas with the space–time information provided by CN algorithm predictions, a formal identification of areas most prone to future seismic activation can be performed. The resulting indications can be used to focus the investigation of possible local scale precursors within areas with linear dimensions of a few tens of kilometers, where the probability of a strong earthquake is relatively high, as well as for drawing time-dependent scenarios of ground shaking by the neodeterministic approach to seismic hazard assessment (NDSHA) [*Peresan et al.*, 2015]. Accordingly, a set of scenarios of ground motion at bedrock, which refers to the time interval when a strong event is likely to occur within the alerted area, can be defined by means of full waveform modeling, both at regional and local scales [*Panza et al.*, 2012, and references therein].

In order to define the NDSHA time-dependent scenarios only those sources located within the alarmed areas are considered, following the procedure described in *Peresan et al.* [2011]. Realistic synthetic seismograms are then computed for sites placed at each node of a grid with a step $0.2° \times 0.2°$ that covers the national territory, starting from the available information about seismic sources and regional structural models. From the set of realistic synthetic seismograms, different maps of seismic hazard that describe the maximum ground shaking at the bedrock can be produced, including peak ground displacement (PGD) or velocity (PGV), or design ground acceleration (DGA).

In Italy, CN algorithm predictions, as well as the related time-dependent neodeterministic ground motion scenarios associated with the alarmed areas, have been regularly updated every 2 months since 2006. Figure 9.6 illustrates the scenarios of PGV computed for the four regions monitored by the CN algorithm (see Figures 9.1 and 9.2). The maps show only the sites where ground-shaking values greater than 15 cm s^{-1} are estimated at bedrock, which corresponds to IMCS ≥ IX [*Indirli et al.*, 2011]. Since the velocity of ground motion is well correlated with the damage caused by earthquakes [*Uang and Bertero*, 1990; *Decanini and Mollaioli*, 1998], when an alert is declared by the applied CN algorithm, these maps can focus attention on areas where the risk is rather high, with respect to normal conditions; notably, these areas are significantly narrower than the alerted regions.

The time-dependent ground-shaking scenarios associated with the Northern region (Figure 9.6a), defined for the time interval 1 March 2012 to 1 May 2012, correctly predicted the ground shaking recorded for the Emilia earthquake ($M=6.1$, 20 May 2012), as reported by *Peresan et al.* [2012]. Remarkably, the ground shaking for this earthquake systematically exceeded the values expected at the bedrock in the area according to current Italian seismic regulation, which is based on a classic map from probabilistic seismic hazard assessment (PSHA) [*Gruppo di Lavoro*, 2004]; even accounting for specific soil conditions, i.e., considering the ground shaking estimates for soil type C, the peak ground acceleration (PGA) predicted according to the PSHA map is $0.21g$ [*Gallipoli et al.*, 2014], a value still lower than that observed. Another large earthquake that struck the Italian territory

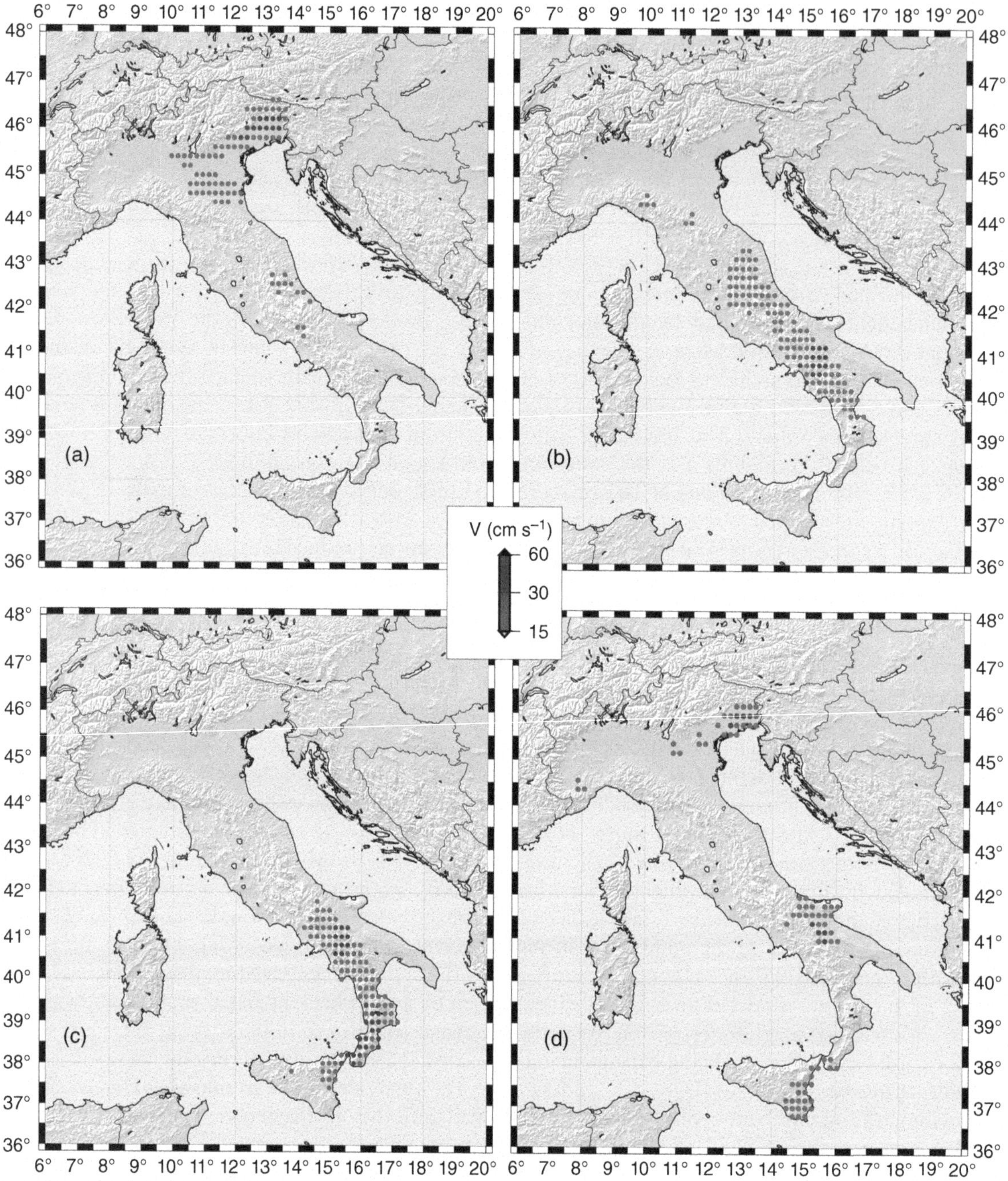

Figure 9.6 Time-dependent scenarios of ground shaking associated with an alarm in each of the regions (shown in Figures 9.1 and 9.2) monitored by the applied CN algorithm: (a) Northern, (b) Central, (c) Southern, (d) Adriatic. Peak ground velocity (PGV) is computed using simultaneously all of the possible sources within the alarmed area and for frequencies up to 10 Hz. Only values of PGV ≥ 15 cm s^{-1} are provided in the maps, which correspond to IMCS ≥ IX [*Indirli et al.*, 2011].

a few years before the Emilia event, was the L'Aquila earthquake ($M=6.3$, 2009). Although the L'Aquila earthquake scores as a failure in the earthquake prediction experiment (see section 9.3), the time-dependent ground-shaking scenario defined for the period 1 March 2009 to 1 May 2009 correctly predicted the macroseismic intensities, as large as IX (MCS), observed for this earthquake.

The latest confirmation of the applied methodology for time-dependent seismic hazard assessment is provided by the recent Amatrice (*M* 6.2) and Norcia (*M* 6.6) earthquakes, which struck the Central region on 24 August 2016 and on 30 October 2016, respectively. (These two earthquakes occurred after manuscript submission, with the latter after the first revision.) Specifically, the events took place within the previously identified seismogenic nodes prone to earthquakes with magnitude $M \geq 6.0$ [*Gorshkov et al.*, 2002], an area that has experienced other destructive earthquakes in the past, including the large *M* 6.9 Valnerina earthquake in 1703. The time-dependent ground-shaking scenario associated with the Central region defined for the period 1 November 2012 to 1 September 2016, was very consistent with the ground shaking recorded for the earthquake [*Peresan et al.*, 2016]. The predicted acceleration was 0.3–0.6 *g* (PGV 15–30 cm s^{-1}, as shown in Figure 9.6b), which was very consistent with the recorded values as high as 0.45 *g* (PGV up to 31 cm s^{-1}) (RAN data: http://ran.protezionecivile.it/IT/dettaglio_evid.php?evid=340867). Similarly, the time-dependent scenario defined for the period 1 September 2016 to 1 November 2016, associated with the continued alert by the applied CN algorithm for the Central region, correctly anticipated the even stronger ground shaking recorded for the larger Norcia earthquake, reaching 0.6 *g* at some sites (PGV up to 56 cm s^{-1}; http://ran.protezionecivile.it/IT/dettaglio_evid.php?evid=345980). Since the NDSHA time-dependent scenarios are regularly computed, from 2006, four large earthquakes have struck the Italian territory, Amatrice and Norcia (2016) along with L'Aquila (2009) and Emilia (2012), and in all cases the integrated method correctly predicted the observed ground motion.

By taking into account the information provided by seismogenic nodes, it has been possible to concentrate the efforts in areas where the largest events are likely to occur within the alerted region. Figure 9.7 depists an example of a scenario corresponding to a node located within Adriatic region delimited by the applied CN algorithm (i.e., node D70, *Gorshkov et al.* [2004]) and identified as prone to events with magnitude larger than 6.0. The node is located near the epicenter of the San Giuliano di Puglia earthquake (*M* 5.7, 2002). Although the magnitude of the event is below the considered threshold, and hence does not score as a success of the method, the occurrence of the earthquake confirms the high hazard of the area anticipated by the MSZ. The ground-shaking velocities observed for this earthquake are well below the values given in Figure 9.7 (http://itaca.mi.ingv.it).

The examples provided of the existing operational practice in predicting seismic ground shaking are well in line with the guidelines and recommendations given in the Report of the International Commission on Earthquake Forecasting [*Jordan et al.*, 2011]. The results acquired in the prospective application of the time-dependent NDSHA approach provide information that can be useful to assign unbiased priorities for timely mitigation actions. As an example, for sites where large ground-shaking values (e. g., PGV > 15 cm s^{-1}) are estimated at bedrock (Figure 9.6), further investigations can be performed taking into account the local soil conditions [e. g., *Vaccari et al.*, 2005, 2009].

9.5. DISCUSSION AND CONCLUSIONS

Operational issues are a pressing concern in earthquake forecasting as well as in seismic hazard assessment, as evidenced by several destructive events, which turn out to be fatal surprises in many countries worldwide, like the Tohoku (Japan, *M* 9.0, 11 March 2011), Haiti (*M* 7.3, 12 January 2010), and the L'Aquila (*M* 6.3, 6 April 2009) and Emilia (*M* 5.9, 20 May 2012) earthquakes in Italy. Any reliable information about possible impending strong earthquakes may permit effective (low-key) mitigation actions to be undertaken and thus to reduce earthquake-related risks, including possible cascading effects on the environment. Conversely, relying on untested information from forecasts or seismic hazard maps can turn out quite expensive: underpredicting earthquake risks can lead to fatalities and significant economic losses [*Kossobokov and Nekrasova*, 2010, 2012; *Wyss et al.*, 2012]; overpredicting risks [e. g., *Nekrasova et al.*, 2015] can, sometimes, lead to investment in unecessary and costly safety measures [*Stein and Stein*, 2013].

Forecasting earthquakes and related ground shaking is not an easy task and requires careful application of statistics to data sets of limited size and different levels of accuracy. Observations and physical models suggest that several processes in the Earth's lithosphere are predictable, but only after substantial averaging and with limited precision. The problem can thus be addressed following a multiscale approach, where earthquake prediction is posed as a step-by-step process, narrowing the magnitude range, territory, and time of expectation, all within the unavoidable limits imposed by physics and data uncertainties.

Uncertainties concerning input data control our capability of describing a system, but the accuracy of the results cannot be better than that of input data. This obvious fact is often overlooked. Quite interestingly, *Zechar et al.* [2016] state that "Researchers hoping to model European seismicity must consider methods for treating imperfect data: catalogs in Europe will not be homogeneous for a very long time, if ever." Actually, methods for operational earthquake forecasting (OEF) already do exist that allow for defective and incomplete data, like the CN and M8 algorithms, and have been applied to global scale catalogs for decades [e. g., *Kossobokov and Nekrasova*, 2012]. Methods for OEF must be robust and

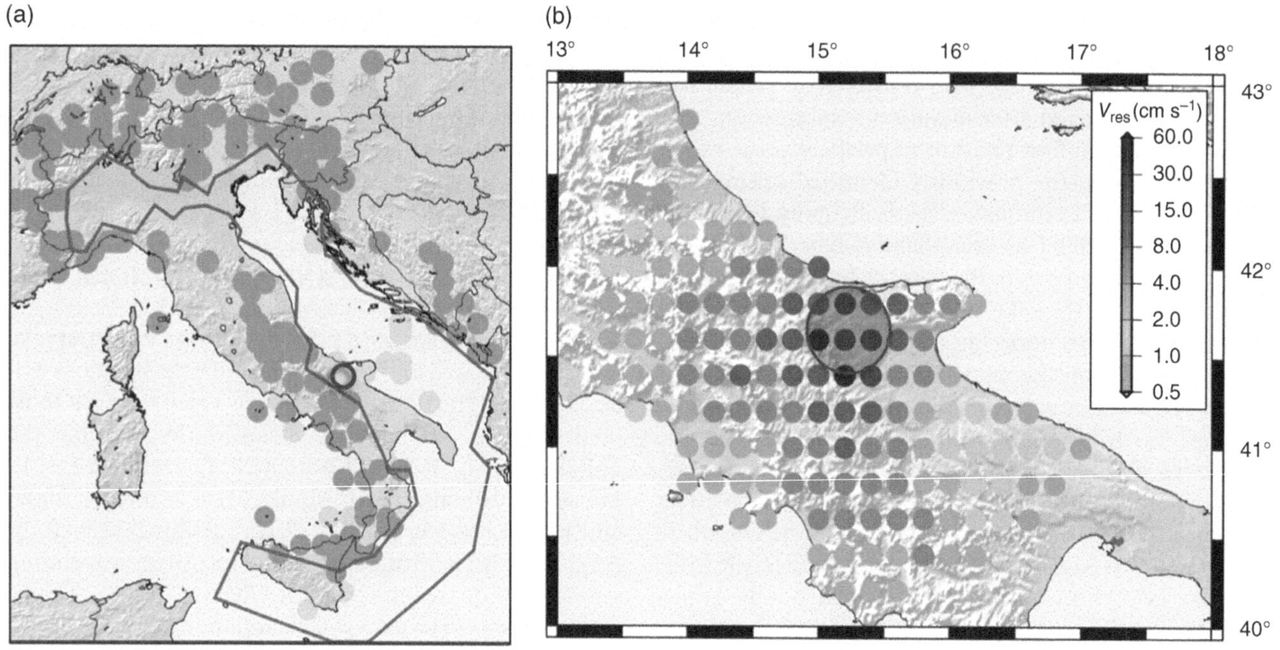

Figure 9.7 Example of ground-shaking scenario associated with a seismogenic node: (a) map of earthquake prone nodes identified for $M_N \geq 6.0$ (light blue) and $M_N \geq 6.5$ (dark blue) according to *Gorshkov et al.* [2002, 2004]. The borders of the Adriatic region (red polygon) and the selected node D70 are outlined in the map. (b) Peak ground velocity (PGV), computed for the potential seismic sources associated with the selected node (large green circle). The maximum observed value is 2.1 cm s^{-1} (http://itaca.mi.ingv.it), well below the values reported in the figure. (*See electronic version for color representation.*)

their robustness should be assessed with respect to uncertainties of randomly defective and incomplete input data [e. g., *Peresan et al.*, 2002]. Any sound method, however, would still fail (or be expected to fail) if the input data provide a systematically inconsistent picture of reality, like in the case reported by *Peresan et al.* [2000]. Given that data from earthquake catalogs are imperfect, as stressed by *Zechar et al.* [2016], the question whether OEF information about impending earthquakes should be given as an alarm (i.e., "prediction") or probability (i.e., "forecast") is not a merely semantic issue. The essential problem is whether detailed probability estimates are meaningful and useful, particularly in view of the unavoidably imperfect data available to compute them. In this framework, probabilities coarsely discretized into a few relative values (e. g., high/medium/low or just high/low probability, as in the case of TIP) appear preferable since they are very consistent with data uncertainties. Making detailed quantitative probabilistic claims may be dramatically misleading and give a deceptive impression of detailed knowledge, which has been proven wrong [e. g., *Kossobokov*, 2005, 2009], particularly for large and sporadic earthquakes. Moreover, decision making does not necessitate a specific value of probability to be assigned, but rather requires reliable information on increased likelihood of incipient disaster [*Guidelines for Earthquake Predictors*, 1983].

As clearly stated by *Kossobokov et al.* [2015] any reliable anticipatory information about impending earthquakes can be (a) effective, (b) complementary to design and construction of seismoresistant infrastructures, and (c) well appreciated by the population as a timely precautious reminder and warning. The practical advantages of OEF, including the possible mitigation actions related to alerts at different space–time scales, have been highlighted by *Peresan et al.* [2012], as well as in earlier publications and reports [e. g., *Panza*, 2006]. It has been remarked by *Kossobokov et al.* [2015] that *Peresan et al.* [2012] never questioned the usefulness of reliable OEF for being "ineffective, distractive, and dangerous," as speciously reported by *Jordan et al.* [2014] and *Field et al.* [2016]. To be effective, however, any forecasting tool must demonstrate its capability in anticipating large earthquake occurrences and related ground shaking, a result that can be attained only through rigorous verification and validation processes. The "usefulness" of a method can then be judged depending on specific mitigation actions of different levels (from low-key to evacuation).

It is indisputable that only falsification tests, by carefully recording failures and successes, can eventually allow to assess the forecasting reliability and effectiveness. Regretably, claims of a high potential and efficiency for a particular methodology are often based on a questionable application of statistics (e. g., nominal probability gain alone, as mentioned in section 9.3), and are hardly suitable for communication to decision makers. The need to apply adequate tools for assessing performances of OEF, e. g., the "error diagram" [*Molchan*, 1990] and the "seismic roulette" null-hypothesis [*Kossobokov et al.*, 1992, 1999], is self-evident. These tools should also be applied to assess the performances of seismic hazard maps, which aim to predict ground shaking from future earthquakes. Such a verification process must be carried out in advance of claiming hazardous areas and/or times. The set of errors, i.e., rate of failures and alerted space–time volume, compared to those obtained in the same number of random guess trials permits evaluation of the effectiveness of the forecasting method and the identification of the optimal choice of the parameters defined by specified cost-benefit functions. The theoretical framework for the optimal choice of disaster preparedness measures, to be undertaken in response to reliable forecast/prediction, has been available since the 1970s [*Kantorovich et al.*, 1974; *Kantorovich and Keilis-Borok*, 1991]. Prudent cost-effective safety measures can be taken when prediction is reliable, but not necessarily precise. The accuracy of reliable intermediate-term middle-range forecasting tools is already sufficient for undertaking earthquake preparedness measures efficiently [e. g., *Davis et al.*, 2012], particularly when used to develop comprehensive earthquake scenarios, as illustrated in section 9.4. Moreover, OEF can be an effective part of a multihazards approach to risk assessment [*Boni et al.*, 2015], where possible chain reactions triggered by the earthquake are considered (e. g., landslides, avalanches, damage of critical facilities such as dams or nuclear power plants). Obviously, the range of practical low-key preparedness options increases in the case of longer than short-term warnings. Accordingly, decision makers are advised to use the full range of a very broad spectrum of possible actions, following a general strategy of response to predictions by escalation or de-escalation of safety measures, depending on expected losses and magnitude–space–time accuracy of reliable forecasting.

This chapter has provided experimental evidence about the feasibility of OEF at the intermediate space–time scale, based on about 20 years of rigorous prospective testing of precursory seismicity patterns in Italy as identified by application of the CN algorithm. Furthermore, the expansion of the monitored territory to the foreland areas located along the Adria microplate, provided indications that are relevant to different aspects of OEF, namely: (a) practical, by broadening the monitored territory; (b) theoretical, by supporting the similarity of earthquake preparation processes in diverse tectonic regions, each characterized by homogeneous seismogenic properties; (c) operational, by establishing satisfactory stability of predictions with respect to the quality of the input catalog, over different time spans (an aspect particularly relevant in this region, due to the presence of several borders and of the sea).

The possibility to reduce the intrinsically large uncertainties associated with intermediate-term middle-range earthquake predictions based on seismicity patterns, has been discussed. It is shown that the coupled use of independent observations, including data from space geodesy and morphostructural analysis, can contribute to better constrain the alerted area, as exemplified in section 9.4. Pattern recognition of earthquake prone areas and identification of areas of anomalous deformation, can help to focus the investigation of possible local scale precursors in the zones where the probability of a strong earthquake is relatively high, as well as for deriving time-dependent scenarios of ground shaking by the NDSHA approach [*Peresan et al.*, 2011]. The procedure for the definition of time-dependent seismic hazard scenarios, based on pattern-recognition techniques and ground motion modeling, has been implemented within the framework of the SISMA project (https://www.asi.it/it/eventi/workshop/progetto-sisma) of the Agenzia Spaziale Italiana, an integrated prototype system for real-time joint processing of seismic and geodetic data streams [*Crippa et al.*, 2008; *Panza et al.*, 2013]. The areas of current TIP were routinely provided to the Protezione Civile della Regione Autonoma Friuli Venezia Giulia (PC FVG) in the form of bimonthly reports since 2005 [*Panza et al.*, 2013, 2014]. The bimonthly reports with maps of ground-shaking expected in the nearest future can be very helpful for operational planning and determining the extent of emergency resources and preparedness.

The results acquired in the prospective application show that the time-dependent NDSHA approach can provide reliable information useful for assigning unbiased priorities for timely mitigation actions. This OEF information, in particular, may enhance preparedness of the population in the alerted areas of TIP, where the increased probabilities during such periods should be used for warning people that low-key mitigation measures must be taken to improve their safety. Moreover, in the areas where the strongest ground shaking is expected, further investigation can be performed, taking into account the local soil conditions, in order to assess the vunerability of relevant structures, including schools [*Panza et al.*, 2015], critical facilities, and historical buildings [e. g., *Vaccari et al.*, 2009].

Hazards from earthquakes have some common elements with those from terror attacks, which are dramatically factored into our daily life. In both cases it is unknown when exactly the attack (the earthquake) will occur, although most but not all of the sensitive areas (the seismogenic zones) are known. As discussed in this chapter, an exact prediction of earthquakes, suitable for declaring a red alert and take highly impacting actions (e. g., evacuation), is not possible at present, if ever will be. Yet, predictions at the intermediate space–time scale are feasible and may permit taking various effective low-key actions. Therefore, though with terrorism the situation is eventually less defined, several strategies developed for protection from terror attacks are suitable, *mutatis mutandis*, for the mitigation of earthquake impacts.

ACKNOWLEDGMENTS

I am grateful to V. Kossobokov, L. Romashkova, and I. Rotwain for their fundamental contribution in setting up and running the intermediate-term middle-range earthquake prediction experiment in Italy. I am also indebted to G. F. Panza for his valuable support and discussions during the preparation of the manuscript. I acknowledge the collaboration of A. Magrin for the computation of the time-dependent scenarios of ground motion associated with alarmed territories. I am especially grateful to Angelo De Santis and four anonymous reviewers for their constructive comments. The research presented in this chapter benefited from funding provided by: PCFVG-ICTP agreement, Protezione Civile della Regione Autonoma Friuli-Venezia Giulia (DGR 1459 dd. 24.6.2009); DPC-INGV Project S3-2012; ASI - Pilot Project "SISMA- Information System for Monitoring and Alert."

REFERENCES

Alekseevskaya, M. A., Gabrielov, A. M., Gvishiani, A. D., Gelfand, I. M., and Ranzman, E. Ya. (1977), Formal morphostructural zoning of mountain territories, *J. Geophys.*, *43*, 227–233.

Allen, C.R., Edwards, W., Hall, W.J., Knopoff, L., Raleigh, C. B., Savit, C. H., Toksoz, M. N. Turner R. H. (1976), *Predicting Earthquakes: A Scientific and Technical Evaluation—with Implications for Society*, Panel on Earthquake Prediction of the Committee on Seismology, Assembly of Mathematical and Physical Sciences, National Research Council, US National Academy of Sciences, Washington, DC.

Bak, P., Christensen, K., Danon, L., and Scanlon, T. (2002), Unified scaling law for earthquakes, *Phys. Rev. Lett.*, *88*, 178501–178504.

Basili, R., Valensise, G., Vannoli, P., Burrato, P., Fracassi, U., Mariano, S., Tiberti, M. M., and Boschi, E. (2008), The Database of Individual Seismogenic Sources (DISS), Version 3: Summarizing 20 years of research on Italy's earthquake geology, *Tectonophysics*, doi:10.1016/j.tecto. 2007.04.014

Boni, G., Mai, M. P., Parodi, A., Peresan, A., Tarolli, P., and Katz, O. (2015), *European Geosciences Union (EGU) Information Briefing: Continued risks of natural disasters in Nepal*, www.egu.eu

Bormann, P. (2011), From earthquake prediction research to time-variable seismic hazard assessment applications, *Pure Appl. Geophys.*, *168*, 329–366, doi:10.1007/s00024-010-0114-0

Bowman, D. D., Oullion, G., Sammis, C. G., Sornette, A., and Sornette, D. (1998), An observational test of the critical earthquake concept, *J. Geophys. Res.*, *103*, 24,359–24,372.

Brandmayr, E., Marson, I., Romanelli, F., and Panza, G. F. (2011), Lithosphere density model in Italy: No hint for slab pull, *Terra Nova*, *23*(5), 292–299.

Caputo, M. (1983), The occurence of large earthquakes in Southern Italy, *Tectonophysics*, *99*, 73–83.

Caputo, M. (1988), The forecast of the magnitude 5.8 May 7th 1984 earthquake in Central Italy, *Rev. Geofis.*, *28*, 101–121.

Cloentingh, S., Tornu, T., Ziedler, P. A., and Beekman, F. (2006), Neotectonics and intraplate topography of the northern Alpine Foreland, *Earth Sci. Rev.*, *74*, 127–196.

Console, R., Murru, M., and Falcone, G. (2010), Probability gains of an epidemic-type aftershock sequence model in retrospective forecasting of $M \geq 5$ earthquakes in Italy, *J. Seismol.*, *14*(1), 9–26.

Crippa, B., Sabadini, R., Chersich, M., Barzaghi, R., and Panza, G. (2008), Coupling geophysical modelling and geodesy to unravel the physics of active faults. Second Workshop on *Use of Remote Sensing Techniques for Monitoring Volcanoes and Seismogenic Areas*, USEReST.

Cuffaro, M., Riguzzi, F., Scrocca, D., Antonioli, F., Carminati, E., Livani, M., and Doglioni, C. (2010), On the geodynamics of the northern Adriatic plate, *Rend. Fis. Acc. Lincei*, doi: 10.1007/s12210-010-0098-9

Cuffaro, M., Riguzzi, F., Scrocca, D., and Doglioni, C. (2011), Coexisting tectonic settings: the example of the southern Tyrrhenian Sea, *Int. J. Earth Sci. Geol Rundsch*, doi: 10.1007/s00531-010-0625-z

D'Agostino, N., Cheloni, D., Mantenuto, S., Selvaggi, G., Michelini, A., and Zuliani, D. (2005). Strain accumulation in the southern Alps (NE Italy) and deformation at the northeastern boundary of Adria observed by CGPS measurements, *Geophys. Res. Lett.*, *32*, L19306, doi:10.1029/2005GL024266.

Davis, C., Keilis-Borok, V., Kossobokov, V., and Soloviev, A. (2012), Advance prediction of the March 11, 2011 Great East Japan Earthquake: A missed opportunity for disaster preparedness, *Int. J. Disaster Risk Red.*, *1*, 17–32.

Decanini, L. D. and Mollaioli, F. (1998), Formulation of elastic earthquake input energy spectra, *Earthquake Eng. Struct. Dyn.*, *27*, 1503–1522.

De Santis, A., Cianchini, G., Qamili, E., and Frepoli, A. (2010), The 2009 L'Aquila (Central Italy) seismic sequence as a chaotic process, *Tectonophysics*, *496*, 44–52.

De Santis, A., De Franceschi, G., Spogli, L., Perrone, L., Alfonsi, L., Qamili, E., Cianchini, G., Di Giovambattista, R., Salvi, S., Filippi, E., Pavón-Carrasco, F. J., Monna, S., Piscini, A., Battiston, R., Vitale, V., Picozza, P. G., Conti, L., Parrot, M., Pinçon, J.-L., Balasis, G., Tavani, M., Argan, A., Piano, G., Rainone, M. L., Liu, W., and Tao, D. (2015). Geospace perturbations induced by the Earth: the state of the art and future trends, *Phys. Chem. Earth*, *85–86*, 17–33.

Dobrovolsky, I. P., Zubkov, S. I., and Miachkin, V. I. (1979). Estimation of the size of earthquake preparation zones, *Pure Appl. Geophys.*, *117*, 1025, doi:10.1007/BF00876083

Doglioni, C. and Panza, G. (2015), Polarized plate tectonics, *Adv. Geophys.*, *56*, 1–167.

Field, E. H., Jordan, T. H., Jones, L. M., Michael, A. J., Blanpied, M. L., and Other Workshop Participants (2016), The potential uses of operational earthquake forecasting, *Sesimol. Res. Lett.*, doi: 10.1785/0220150174

Gabrielov, A. M., Dmitrieva, O. E., Keilis-Borok, V. I., Kossobokov, V. G., Kutznetsov, I. V., Levshina, T. A., Mirzoev, K. M., Molchan, G. M., Negmatullaev, S. Kh., Pisarenko, V. F., Prozorov, A. G., Rinheart, W., Rotwain, I. M., Shelbalin, P. N., Shnirman, M. G., and Schreider, S. Yu (1986), *Algorithms of Long-Term Earthquakes' Prediction*, International School for Research Oriented to Earthquake Prediction-algorithms, Software and Data Handling, Lima.

Gallipoli, M. R., Chiauzzi, L., Stabile, T. A., Mucciarelli, M., Masi, A., Lizza, C., and Vignola, L. (2014), The role of site effects in the comparison between code provisions and the near field strong motion of the Emilia 2012 earthquakes, *Bull. Earthquake Eng.*, *12*(5), 2211–2230.

Gasperini, P., Lolli, B., and Vannucci, G. (2013), Empirical calibration of local magnitude data sets versus moment magnitude in Italy, *Bull., Seism. Soc. Am.*, *103*, 2227–2246, doi: 10.1785/0120120356

Gelfand, I. M., Guberman, Sh. A., Keilis-Borok, V. I., Knopoff, L., Press, F., Ranzman, E. Ya., Rotwain, I. M., and Sadovsky, A. M. (1976), Pattern recognition applied to earthquakes epicenters in California, *Phys. Earth. Planet. Inter.*, *11*, 227-2–283.

Geller, R. J., Jackson, D. D., Kagan, Y. Y., and Mulargia, F. (1997), Earthquakes cannot be predicted, *Science*, *275*, 1616 –1617.

Gorshkov, A. I., Panza, G. F., Soloviev, A. A., and Aoudia, A. (2002), Morphostructural zonation and preliminary recognition of seismogenic nodes around the Adria margin in peninsular Italy and Sicily, *J. of Seismol. Earthquake Eng.*, *4*(1), 1–24.

Gorshkov, A. I., Panza, G. F., Soloviev, A. A., and Aoudia, A. (2004), Identification of seismogenic nodes in the Alps and Dinarides, *Boll. Soc. Geol. Ital.*, *123*, 3–18.

Gorshkov, A., Panza, G. F., Soloviev, A., Aoudia, A., and Peresan, A. (2009), Delineation of the geometry of nodes in the Alps–Dinarides hinge zone and recognition of seismogenic nodes ($M \geq 6$), *Terra Nova*, *21*(4), 257–264, doi 10.1111/j.1365–3121.2009.00879.x

Gruppo di Lavoro (2004), *Redazione della mappa di pericolosità sismica prevista dall'Ordinanza PCM 3274 del 20 marzo 2003*, Rapporto conclusivo per il Dipartimento della Protezione Civile, INGV, Milano-Roma, aprile, 65 pp. + 5 allegati.

Guidarelli, M. and Panza, G. F. (2006), INPAR, CMT and RCMT seismic moment solutions compared for the strongest damaging events (*M*_4.8) occurred in the Italian region in the last decade, *Rend. Accad. Naz. Scienze, Mem. Scienze Fisiche Nat.*, *30*, 81–98.

Guidelines for Earthquake Predictors (1983), *Bull. Seismol. Soc. Am.*, *73*(6), 955–956.

Gutenberg, B. (1956), The energy of earthquakes. *Q. J. Geol. Soc. London*, *112*, 1–14.

Healy, J. H., Kossobokov, V. G., and Dewey, J. W. (1992), A test to evaluate the earthquake prediction algorithm, M8, *US Geol. Surv. Open-File Rep.*, 92-401, 23 pp. + six ppendices.

Herak, D., Herak, M., Panza, G. F., and Costa, G. (1999), Application of the CN intermediate term earthquake prediction algorithm to the area of the Southern External Dinarides, *Pure Appl. Geophys.*, *156*, 689–699.

Indirli, M., Razafindrakoto, H., Romanelli, F., Puglisi, C., Lanzoni, L., Milani, E., Munari, M., and Apablaza, S. (2011), Hazard evaluation in Valparaiso: the MAR VASTO Project, *Pure Appl. Geophys.*, *168*, 543–582.

ISC (2016), International Seismological Centre, Thatcham, United Kingdom, on-line bulletin, http://www.isc.ac.uk

Jordan, T. (2006), Earthquake predictability, brick by brick, *Seismol. Res. Lett.* *77*(1), 3–6.

Jordan, T., Chen, Y., Gasparini, P., Madariaga, R., Main, I., Marzocchi, W., Papadopoulos, G., Sobolev, G., Yamaoka, K., and Zschau, J. (2011), Operational earthquake forecasting. State of knowledge and guidelines for Utilization, *Ann. Geophys.*, *54*(4), doi:10.4401/ag-5350

Jordan, T.H., Marzocchi, W., Michael, A.J., and Gerstenberger, M.C. (2014). Operational earthquake forecasting can enhance earthquake preparedness, *Seismol. Res. Lett.*, *85*(5), 955–959.

Kanamori, H. (2003), Earthquake prediction: An overview, *International Handbook of Earthquake and Engineering Seismology,Int. Assoc. Seismol. Phys. Earth's Int.*, *81B*, 1205–1216.

Kantorovich, L. V. and Keilis-Borok, V. I. (1991), Earthquake prediction and decision-making: social, economic and civil protection aspects, in *Proceedings of the International Conference on Earthquake Prediction: State-of-the-Art*, pp. 586–593, Scientific-Technical Contributions, CSEM-EMSC, Strasbourg, France.

Kantorovich, L. V., Keilis-Borok, V. I., and Molchan, G. M. (1974), Seismic risk and principles of seismic zoning, in *Seismic Design Decision Analysis*, Internal Study Report, Department of Civil Engineering, MIT, 43 pp.

Keilis-Borok, V. I. (1990),The lithosphere of the Earth as a nonlinear system with implications for earthquake prediction, *Rev. Geophys.*, *28*, 19–34.

Keilis-Borok, V. I. (1996), Intermediate term earthquake prediction, *Proc. Natl. Acad Sci. USA*, *93*, 3748–3755.

Keilis-Borok, V. I. and Rotwain, I. M. (1990), Diagnosis of time of increased probability of strong earthquakes in different regions of the world: algorithm CN, *Phys. Earth Planet. Inter.*, *61*, 57–72.

Keilis-Borok, V. I. and Soloviev, A. A. (eds) (2003), *Non-linear Dynamics of the Lithosphere and Earthquake Prediction*, Springer, Heidelberg, 337 pp.

Keilis-Borok, V. I., Knopoff, L., and Rotwain, I. M. (1980), Bursts of aftershocks, long-term precursors of strong earthquakes, *Nature*, *283* (5744), 259–263.

Keilis-Borok, V. I., Kuznetsov, I. V., Panza, G., Rotwain, I. M., and Costa, G. (1990), On intermediate-term earthquake prediction in Central Italy, *Pure Appl. Geophys.*, *134*(1), 79–92.

Kossobokov, V. (2005), Regional earthquake likelihood models: A realm on shaky grounds? *Eos (Trans. Am. Geophys. Union)*, *86*(52), Fall Meeting Supplement, Abstract S41D-08.

Kossobokov, V. (2009), Testing earthquake forecast/prediction methods: "Real-time forecasts of tomorrow's earthquakes in California," in *Some Problems of Geodynamics*, pp. 321–337, KRASAND, Moscow. (In Russian.)

Kossobokov, V. G. (2012), Earthquake prediction: 20 years of global experiment, *Nat. Hazards*, doi 10.1007/s11069-012-0198-1

Kossobokov, V. G. (2014), Times of increased probabilities for occurrence of catastrophic earthquakes: 25 years of hypothesis testing in real time, in M. Wyss and J. Shroder (eds), *Earthquake Hazard, Risk, and Disasters*, pp. 477–504, Elsevier, Oxford.

Kossobokov, V. G. and Mazhkenov, S. A. (1994), On similarity in the spatial distribution of seismicity, in D. K. Chowdhury (ed.), *Computational Seismology and Geodynamics*, pp. 6–15, American Geophysical Union, Washington, DC.

Kossobokov, V. G. and Nekrasova, A. K. (2010), Global seismic hazard assessment program maps are misleading, *Eos (Trans. Am. Geophys. Union)*, *91*(52), Fall Meeting Supplement, Abstract U13A-0020.

Kossobokov, V. G. and Nekrasova, A. K. (2012), Global seismic hazard assessment program maps are erroneous, *Seismic Instr.*, *48*(2), 162–170, doi: 10.3103/S0747923912020065

Kossobokov, V. and Shebalin, P. (2003). Earthquake prediction, in V. I. Keilis-Borok and A. A. Soloviev (eds), *Non-Linear Dynamics of the Lithosphere and Earthquake Prediction*, pp. 141–207, Springer, Heidelberg.

Kossobokov, V. G. and Soloviev A. A. (2015), Evaluating the results of testing algorithms for prediction of earthquakes, *Dokl. Earth Sci.*, *460*(2),192–194, doi: 10.1134/S1028334X15020208

Kossobokov, V. G., Healy, J. H., Keilis-Borok, V. I., Dewey, J. W., and Khokhlov, A. V. (1992), The test of an intermediate-term earthquake prediction algorithm: the design of real-time monitoring and retroactive application, *Dokl. Acad. Nauka*, *325*(1), 46–48. (In Russian.)

Kossobokov, V. G., Romashkova, L. L., Keilis-Borok,V. I., and Healy, J. H. (1999), Testing earthquake prediction algorithms: statistically significant advance prediction of the largest earthquakes in the Circum-Pacific, 1992–1997, *Phys. Earth Planet. Inter.*, *111*. 187–196.

Kossobokov, V. G., Romashkova, L.L., Panza, G. F., and Peresan, A. (2002), Stabilizing intermediate-term medium-range earthquake predictions, *J. Seismol. Earthquake Eng.*, *8*, 11–19.

Kossobokov, V. G., Peresan, A., and Panza, G. F. (2015), On operational earthquake forecast/prediction problems, *Seismol. Res. Lett.*, *86*(2AA), doi: 10.1785/0220140202

Marotta, A. M. and Sabadini, R. (2008), Africa–Eurasia kinematics control of long-wavelength tectonic deformation in the central Mediterranean, *Geophys. J. Int.*, *152*(2), 742–754, doi:10.1111/j.1365-246X.2008.03906.x

Marzocchi, W., Schorlemmer, D., and Wiemer, S. (2010), Preface to the special issue: "An earthquake forecast experiment in Italy," *Ann. Geophys.*, *53*, 3, doi: 10.4401/ag-4851

Maybodian, M., Zare, M., Hamzehloo, H., Peresan, A., Ansari, A., and Panza, G. F. (2014), Analysis of precursory seismicity patterns in Zagros (Iran) by CN algorithm, *Turk. J. Earth Sci.*, *23*(1), 91–99.

Meletti, C., Patacca, E., and Scandone, P. (2000), Construction of a seismotectonic model: the case of Italy, *Pure Appl. Geophys.*, *157*, 11–35.

Meletti, C., Galadini, F., Valensise, G., Stucchi, M., Basili, R., Barba, S., Vannucci, G., and Boschi, E. (2008), A seismic source zone model for the seismic hazard assessment of the Italian territory, *Tectonophysics*, *450*, 85–108, doi:10.1016/j.tecto.2008.01.003

Molchan, G. M. (1990), Strategies in strong earthquake prediction, *Phys. Earth Planet. Inter.*, *61*, 84–98.

Molchan, G. M. (2003), Earthquake prediction strategies: a theoretical analysis, in V. I, Keilis-Borok and A. A. Soloviev (eds), *Nonlinear Dynamics of the Lithosphere and Earthquake Prediction*, pp. 209–237, Springer, Heidelberg.

Molchan, G. (2012), On the testing of seismicity models, *Acta Geophys.*, *60*(3), 624–637, doi: 10.2478/s11600-011-0042-0

Molchan, G. and Romashkova, L. (2011), Gambling score in earthquake prediction analysis, *Geophys. J. Int.*, *184*, 1445–1454, doi:10.1111/j.1365-246X.2011.04930.x

Molchan, G. M., Kronrod, T. L., and Panza, G. F. (1997), Multiscale seismicity model for seismic risk, *Bull. Seismol. Soc. Am.*, *87*(5), 1220–1229.

Nekrasova, A., Kossobokov, V., Peresan, A., Aoudia, A., and Panza, G. F. (2011), A Multiscale application of the unified scaling law for earthquakes in the central Mediterranean area and Alpine region. *Pure Appl. Geophys.*, *168*, 297–327, doi: 10.1007/s00024-010-0163-4.

Nekrasova, A., Peresan,A., Kossobokov, V. G., and Panza, G. F. (2015), A new probabilistic shift away from seismic hazard reality in Italy?, in M. Kouteva-Guentcheva and B. Aneva (eds), *Nonlinear Mathematical Physics and Natural Hazards*, pp. 83–104, Proceedings in Physics, Vol. *163*, Springer.

Panza, G. F. (2006), Integrated deterministic seismic hazard scenarios, International Framework for Development of Disaster Reduction Technology List on Implementation Strategies, "*Disaster Reduction Hyperbase" NIED*, 27–28 February, Tsukuba, Japan.

Panza, G. F., La Mura, C., Peresan, A., Romanelli F., and Vaccari, F. (2012), Seismic hazard scenarios as preventive tools for a disaster resilient society, in R. Dmowska (ed.), *Advances in Geophysics*, 93–165, Elsevier, Oxford.

Panza, G.FG. F., A. Peresan, A., Magrin, A., Vaccari, F., Sabadini, R., Crippa, B., Marotta, A. M., Splendore, R., Barzaghi, R., Borghi, A., Cannizzaro, L., Amodio, A., and Zoffoli, S. (2013), The SISMA prototype system: integrating geophysical modeling and earth observation for time-dependent seismic hazard assessment, *Nat. Hazards*, *69*, 1179–1198, doi: 10.1007/s11069-011-9981-7

Panza, G. F., Peresan, A., and Magrin, A. (2014), *Scenari neo-deterministici di pericolosità sismica per il Friuli Venezia Giulia e le aree circostanti*, Memorie descrittive della Carta Geologica d'Italia, Vol. 94/2014, ISPRA and Italian Geological Service, 104 pp. (In Italian.)

Panza, G. F., Romanelli, F., Vaccari, F., and Altin, G. (2015), *Vademecum for the seismic verification of existing buildings*, http://www.studioaltin.com/wp-content/uploads/2016/04/studioAltin_verifica_sismica_edifici_ProvTS.pdf (accessed August 2015).

Peresan, A. and Gentili, S. (2016), Caratterizzazione statistica di sequenze e sciami sismici nell'Italia Nord Orientale, *35th Assembly of Gruppo Nazionale di Geofisica della Terra Solida*, 22–24 November, Lecce, Italy). (In Italian.)

Peresan, A. and Panza, G. F. (2002), *UCI2001: The Updated Catalogue of Italy*, Internal Report IC/IR/2002/3, The Abdus Salam International Centre for Theoretical Physics, Trieste, Italy.

Peresan, A., Costa, G., and Panza, G. F. (1999), Seismotectonic model and CN earthquake prediction in Italy, *Pure Appl. Geophys.*, *154*, 281–306.

Peresan, A., Panza, G. F., and Costa, G. (2000), CN algorithm and long lasting changes in reported magnitudes: the case of Italy, *Geophys. J. Int.*, *141*, 425–437.

Peresan, A., Rotwain, I., Zaliapin, I., and Panza, G. F., (2002), Stability of intermediate-term earthquake predictions with respect to random errors in magnitude: the case of central Italy, *Phys. Earth Planet. Inter.*, *130*, 117–127.

Peresan, A., Kossobokov, V., Romashkova, L., and Panza, G. F. (2005), Intermediate-term middle-range earthquake predictions in Italy: a review, *Earth Sci. Rev.*, *69*, (1–2), 97–132.

Peresan, A., Zuccolo, E., Vaccari, F., Gorshkov, A., and Panza, G. F. (2011), Neo-deterministic seismic hazard and pattern recognition techniques: time dependent scenarios for north-eastern Italy, *Pure Appl. Geophys.*, *168* (3–4), 583–607, doi 10.1007/s00024-010-0166-1.

Peresan, A., Kossobokov, V. G., and Panza, G. F. (2012), Operational earthquake forecast/prediction, *Rend. Lincei.*, *23*, 131–138, doi 10.1007/s12210-012-0171-7

Peresan, A., Gorshkov, A., Soloviev, A., and Panza, G. F. (2015), The contribution of pattern recognition of seismic and morphostructural data to seismic hazard assessment, *Boll. Geofis. Teor. Appl.*, *56*, 295–328, doi 10.4430/bgta0141

Peresan, A., Kossobokov, V., Romashkova, L., Magrin, A., Soloviev, A., and Panza, G. F. (2016), Time-dependent neo-deterministic seismic hazard scenarios: preliminary report on the *M*6.2 central Italy earthquake, 24th August 2016, Special issue on "The August 2016 Central Italy earthquake," *New Concepts. Global Tect., J.*, *4*(3), 487–493.

Radan, M. Y., Hamzehloo, H., Peresan, A., Zare, M., and Zafarani, H. (2013), Assessing performances of pattern informatics method: a retrospective analysis for Iran and Italy, *Nat Hazards*, *68* (2), 855–881, doi 10.1007/s11069-013-0660-8

Raykova, R. B., Panza, G. F., and Doglioni, C. (2015), Lithosphere–asthenosphere system in the mediterranean region in the framework of polarized plate tectonics, *8th Congress of the Balkan Geophysical Society*.

Riguzzi, F., Crespi, M., Devoti, R., Doglioni, C., Pietrantonio, G., and Pisani, A. R. (2012), Geodetic strain rate and earthquake size: New clues for seismic hazard studies, *Phys. Earth Planet. Inter.*, *206–207*, 67–75.

Romashkova, L. and Peresan, A. (2013), Analysis of Italian earthquake catalogs in the context of intermediate-term prediction problem, *Acta Geophys.*, *61* (3), 583–610, doi: 10.2478/s11600-012-0085-x

Romashkova, L., Peresan, A., and Nekrasova, A. (2009), *Analysis of Earthquake Catalogs for CSEP Testing Region Italy*, Internal Report IC/IR/2009/006, ICTP.

Rotondi, R. (2010), Bayesian nonparametric inference for earthquake recurrence time distributions in different tectonic regimes, *J. Geophys. Res.*, *115*, B01302, doi:10.1029/2008JB006272

Rotwain, I. M. and Novikova, O. (1999), Performance of the earthquake prediction algorithm CN in 22 regions of the world, *Phys. Earth Planet. Inter.*, *111*, 207–213.

Rotwain, I. M., Keilis-Borok, V. I., and Botwina, L. (1997), Premonitory transformation of steel fracturing and seismicity, *Phys. Earth Planet. Inter.*, *101*, 61–71.

Sadovsky, M. A. (ed.) (1986), *Long-Term Earthquake Prediction: Methodological Recommendations*, Academy of Science USSR and Institute of Physics of the Earth, Moscow, 127 pp. (In Russian.)

Scandone, P., Patacca, E., Meletti, C., Bellatalla, M., Perilli, N., and Santini, U. (1990), Struttura geologica, evoluzione cinematica e schema sismotettonico della penisola italiana, *Atti del Convegno GNDT*, Vol. *1*, pp. 119–135. (In Italian.)

Schorlemmer, D. and Gerstenberger, M. C. (2014). Quantifying improvements in earthquake-rupture forecasts through testable models, in J. F. Wyss and M. Shroder (eds), *Earthquake Hazard, Risk and Disasters*, pp. 405—429, Academic Press, doi: http://doi.org/10.1016/B978-0-12-394848-9.00015-8

Soloviev, A. A., Gvishiani, A. D., Gorshkov, A. I., Dobrovolsky, M. N., and Novikova, O. V. (2014), Recognition of earthquake-prone areas: methodology and analysis of the results, *Izv. Phys. Solid Earth*, *50* (2), 151–168, doi: 10.1134/S1069351314020116

Splendore, R., Marotta, A. M., Barzaghi, R., Borghi, A., and Cannizzaro, L. (2010), Block model versus thermo mechanical model: new insights on the present-day regional deformation in the surroundings of the Calabrian Arc, *Geol. Soc. London Spec. Publ.*, *332*, 129–147, doi:10.1144/SP332.9.

Stein, S. and Stein, J. (2013), How good do natural hazard assessments need to be? *GSA Today*, *23*(4/5), 60–61, doi: 10.1130/GSATG167GW.1

Tiampo, K. F. and Shcherbakov, R. (2012), Seismicity-based earthquake forecasting techniques: Ten years of progress, *Tectonophysics*, *522–523*, 89–121.

Trua, T., Marani, M., and Barca, D. (2014), Lower crustal differentiation processes beneath a back-arc spreading ridge (Marsili seamount, Southern Tyrrhenian Sea), *Lithos*, *190–191*, 349–362.

Uang, C. M. and Bertero, V. V. (1990), Evaluation of seismic energy in structures, *Earthquake Eng. Struct. Dyn.*, *19*(1), 77–90.

Vaccari, F., Romanelli, F., and Panza, G. F. (2005), Detailed modelling of strong ground motion in Trieste, *Geol. Tecn. Ambient*, *2*, 7–40.

Vaccari, F., Peresan, A., Zuccolo, E., Romanelli, F., Marson, C., Fiorotto, V., and Panza, G. F. (2009), Neo-deterministic seismic hazard scenarios: application to the engineering analysis of historical buildings, in F. M. Mazzolani (ed.), *Proceedings of PROHITECH 2009—Protection of Historical Buildings*, pp. 1559–1564, Taylor & Francis, London.

Vannucci, G. and Gasperini, P. (2004), The new release of the database of Earthquake Mechanisms of the Mediterranean Area (EMMA Version 2), *Ann. Geophys.*, *47* (Suppl. 1), 307–334.

Wang, K. and Rogers, G. C. (2014), Earthquake preparedness should not fluctuate on a daily or weekly basis, *Seismol. Res. Lett.*, *85*, 569–571.

Wyss, M. and Booth, D. C. (1997), The IASPEI Procedure for the evaluation of earthquake precursors, *Geophys. J. Int.*, *131*, 423–424.

Wyss, M., Nekrasova, A., and Kossobokov, V. (2012) Errors in expected human losses due to incorrect seismic hazard estimates, *Nat. Hazards*, *62*(3), 927–935, doi 10.1007/s11069-012-0125-5

Zaliapin, I., Gabrielov, A., Wong, H., and Keilis-Borok, V. (2008), Clustering analysis of seismicity and aftershock identification, *Phys. Rev. Lett.*, *101*.

Zechar, J. D., Marzocchi, W., and Wiemer, S. (2016), Operational earthquake forecasting in Europe: progress, despite challenges, *Bull. Earthquake Eng*, doi 10.1007/s10518-016-9930-7

10

Combining Probabilistic Seismicity Models with Precursory Information

Peter Shebalin

ABSTRACT

Probabilistic earthquake seismicity models provide just very low local expected rates of large earthquakes. In operational earthquake forecasting this makes it difficult to take preventive measures. Precursory phenomena detected in different geophysical fields may signal periods of significant rise of earthquake probabilities, but usually they may not be expressed in terms of probabilities. However, probabilistic features of the precursory phenomena are expressed in the statistics of their case histories. A differential probability gain approach provides a way of converting precursory phenomena into a probabilistic form and to combine them with seismicity models. Here the approach is described for several examples showing the possibility for obtaining from the resulting models high probabilities of expected earthquakes. The approach is applicable to any precursory phenomena, not only the seismicity patterns considered here.

10.1. INTRODUCTION

The past decades have seen a growing number of reasonably reliable and skillful numerical seismicity forecast models [*Jordan*, 2006; *Gerstenberger et al.*, 2007; *Rhoades and Gerstenberger*, 2009; *Zechar and Jordan*, 2010; *Zechar et al.*, 2010a,b; *Eberhard et al.*, 2012; *Tsuruoka et al.*, 2012; *Taroni et al.*, 2014]. Such models allow calculation of estimated probabilities of earthquakes of a given magnitude within a given time and space range. Assessment of the effectiveness and reliability of a large number of the seismicity forecasting models in prospective tests became possible in the early 2000s due to the activity of the Collaboration for the Study of Earthquake Predictability (CSEP) [*Jordan*, 2006]. Before the creation of the CSEP testing centers, starting in the late 1980s, rigorous tests of prediction algorithms were applied to several alarm-based earthquake prediction algorithms at worldwide and regional scales: M8 and M8S [*Keilis-Borok and Kossobokov*, 1990; *Romashkova et al.*, 1998; *Romashkova and Kossobokov*, 2004], MSc [*Kossobokov et al.*, 1990]; CN [*Keilis-Borok and Rotwain*, 1990; *Peresan et al.*, 1999], and NSE [*Vorobieva*, 1999]. The tests of some methods had already shown statistical significance relative to random forecasts [*Kossobokov et al.*, 1990; *Peresan et al.*, 2005; *Kossobokov*, 2013; *Kossobokov and Soloviev*, 2015].

An important advance in earthquake forecasting/prediction was the development of operational earthquake forecasting [*Jordan and Jones*, 2010; *Peresan et al.*, 2012] through establishing communication between scientists and decision makers with the goal of reducing hazards associated with earthquakes. However, operational earthquake forecasting using probabilistic seismicity models remains a daunting challenge because the probabilities derived from them are still quite small [*Jordan and Jones*, 2010]. The only exception is a period of a few hours (very rarely days) of aftershocks in a fault zone of a large earthquake. Effective operational forecasts

Institute of Earthquake Prediction Theory and Mathematical Geophysics, Russian Academy of Sciences, Moscow, Russia

Pre-Earthquake Processes: A Multidisciplinary Approach to Earthquake Prediction Studies, Geophysical Monograph 234, First Edition. Edited by Dimitar Ouzounov, Sergey Pulinets, Katsumi Hattori, and Patrick Taylor.

obviously require high probability rates in advance of an expected earthquake in order to make substantive decisions (e.g., evacuation or other emergency actions). Part of the problem results from the small time and space regions used in the models. Combining to make larger regions rarely results in probabilities that are sufficiently high to take confident decisions.

In seismicity-based models only high clustering in time and space may cause a multifold growth in the expectation rate of a forthcoming earthquake. Earthquake swarms and bursts of aftershocks are known possible precursors of large earthquakes [*Keilis-Borok et al.*, 1980; *Evison*, 1982; *Gupta and Singh*, 1989; *Yamashita and Knopoff*, 1992]. Derived from earthquake catalogs, data from these precursors may be incorporated in seismicity forecast models by the utilization of branching models like epidemic type aftershock sequence (ETAS) [*Ogata*, 1988]. Other types of precursor and seismicity patterns observed in various geophysical fields also may signal an increasing probability of an earthquake [*Cicerone et al.*, 2009; *Uyeda et al.*, 2009; *Pulinets et al.*, 2015]. There are no direct ways, however, to convert such precursory information into a form suitable for probabilistic seismicity models.

Researchers have suggested a few methods for combining models and/or earthquake precursors. For a set of rate-based models, a weighted average is a natural solution [*Rhoades and Gerstenberger*, 2009; *Marzocchi et al.*, 2012; *Rhoades*, 2013]. In the current implementation of such approaches weighting is not dependent on space, but chosen according to the observed relative performance of the model during a testing period. A direct product of functions describing precursory behavior, however, was used in the register transfer level (RTL) prediction algorithm [*Sobolev et al.*, 1996]. In this case, even if the initial functions are probabilistic, the output is a nonprobabilistic alarm-based model. Recently, *Rhoades et al.* [2015] have suggested a different multiplicative approach for combining models into a likelihood forecast model. Another way to combine models is by using Bayes' formula for conditional probabilities [*Sobolev et al.*, 1991]. However, when using this approach, it is difficult to take into account the interdependence of the combined elements, and resulting estimates are hardly probabilistic.

Starting in the 1970s, the idea of combining several precursors using pattern-recognition methods was intensively investigated by two eminent Russian scientists: M. Gelfand and V. Keilis-Borok. This collaboration yielded a set of earthquake prediction algorithms (EPA) [*Gelfand et al.*, 1972; *Gvishiani and Soloviev*, 1984]: M8 and M8-MSc [*Kossobokov et al.*, 1990], CN [*Keilis-Borok and Rotwain*, 1990], NSE [*Vorobieva*, 1999], SR [*Shebalin and Keilis-Borok*, 1999], ROC and ACCORD [*Keilis-Borok et al.*, 2002], and RTP [*Shebalin et al.*, 2004, 2006]; where CN refers to California–Nevada (the area of its development), NSE refers to next strong earthquake, ROC refers to range of correlation, ACCORD refers to chord (as in music), and RTP to reverse tracing of precursors. However, those algorithms do not provide estimations of the earthquake expectation rates or probabilities. A standard formulation of predictions is: in a given time–space area the probability of an earthquake of a given size is increased. Accordingly, such methods often are called "alarm-based" or deterministic. Unlike the alarm-based methods the seismicity forecasting models provide estimates of the earthquake expectation rate or probability, producing "rate-based", "probability-based," or just "probabilistic" specifications for them.

An obvious advantage of probability-based models is their wider applicability. A probability-based may be easily converted to an alarm-based model by introducing a threshold for the rate or probability of detecting periods and areas of increased likelihood of earthquakes (above the threshold). Alarm-based models, however, may not be converted to probability-based models directly.

Recently, *Shebalin et al.* [2012] have suggested a method for converting alarm-based to rate-based models. Deterministic forecasts have a probabilistic nature expressed in the statistics of the application of the method. Some predictions prove wrong (false alarms), or some target earthquakes occur outside periods of alarms (failure to predict, or missed target). For each method the probabilities of errors of two kinds may be estimated from the results of past predictions. Joint analysis of the two kinds of errors is convenient using error diagrams [*Molchan*, 1991; *Molchan and Keilis-Borok*, 2008]. Usually alarm-based models have one or several control parameters. For simplicity, let us consider a single control parameter, or an "alarm function" [*Zechar and Jordan*, 2008]. Varying the alarm function, a trade-off between false alarms and missed targets can be observed. On the error diagram the joint rates of the errors form a trajectory. The tangent to the error diagram trajectory is called differential probability gain [*Shebalin et al.*, 2012], it depends on the alarm function only. Before explaining the method that can be used to convert a deterministic to a rate-based model, let us consider error diagrams in more detail.

The error diagram compares the effectiveness of a forecast method relative to some seismicity model that reflects current knowledge about the seismic process; let us call this model a "reference model." Initially, the error diagram was used to evaluate models with fixed areas of prediction [*Molchan*, 1990, 1991], and only times of increased probability (TIP) of large earthquakes were detected [*Kossobokov and Shebalin*, 2003]. Accordingly, a uniform reference model was used and the false alarm errors were calculated as the fraction of time occupied by TIPs, usually denoted by τ. The other axis of the error diagram represents the rate of target misses, i.e., the fraction of target earthquakes that occurred outside TIPs, which is usually denoted by ν. The statistics of past predictions by this method plots a point on the error diagram, and the further the distance of this point from the (0, 1:1, 0) diagonal (which corresponds to random, unskilled, prediction methods), the more effective is the prediction method. The distance to the diagonal is measured by various functions, usually called "loss functions," including $\nu + \tau$ and $\min(\max[\nu,\tau])$ [*Molchan*, 1991].

With the development of earthquake prediction algorithms, areas of expected large earthquakes started to be part of predictions in the 1990s. Initially, the spatial reference model for error diagrams was also uniform, and the false alarms rate was counted as a fraction of time–space occupied by alarms [*Zechar and Jordan*, 2008]. An obvious disadvantage of the uniform reference model is the possibility of artificially increasing the area under consideration to aseismic zones, thus decreasing the fraction of time–space occupied by alarms. To avoid this, A. Prozorov suggested measuring the space for the parameter τ in units of the frequency of recorded moderate magnitude earthquakes [*Kossobokov and Shebalin*, 2003; *Molchan and Keilis-Borok*, 2008]. In a more general case, nonuniform distribution for both time and space may be considered as a reference model. If a reference model specifies the expected rates of earthquakes of a given magnitude in each cell of the considered [space × time] volume, then the parameter τ is equal to the ratio of total expected rate in cells that are in a state of alarm to the total rate over the whole space–time considered [*Shebalin et al.*, 2012, 2014].

Returning to differential probability gain, let us look at a schematic error diagram (Figure 10.1). The trajectory on the diagram corresponds to the varying threshold A_0 for the alarm function A specifying time and space of alarms (with high values of the alarm function corresponding to higher probability of an earthquake). The trajectory at the upper left corner is formed by the highest threshold A_0: the time–space volume of alarms τ is zero, but all targets are missed, $\nu = 1$. By contrast, the trajectory at the lower right corner is the lowest threshold of A_0, with alarms occupying the whole time–space, $\tau = 1$, but no misses, $\nu = 0$.

Probability gain, introduced by *Aki* [1981], may be used as a measure of the effectiveness of an earthquake prediction method. In this context, the rate of successful predictions per the expected rate within alarms according to the reference model can be counted. Alarms are set by the condition $A \geq A_0$ and where the total expected rate is equal to the actual number of target earthquakes, the probability gain G is

$$G = \frac{1-\nu}{\tau} \qquad (10.1)$$

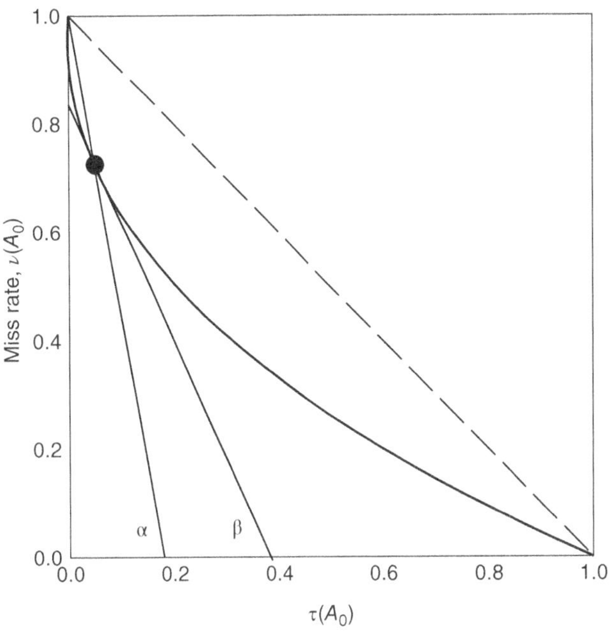

Figure 10.1 Schematic error diagram for a prediction method. The trajectory (thick line) is formed by points corresponding to different values of the alarm function. The diagonal (dashed line) corresponds to arbitrary predictions.

In Figure 10.1 this ratio is given by the slope α: $G = tg(\alpha)$. *Shebalin et al.* [2012, 2014] suggested considering an interval of the alarm function values and accordingly introduced the differential probability gain, denoted here as $g(A)$. For a specified interval (A_1, A_2) differential probability gain counts the ratio of successfully predicted target earthquakes within alarms defined by the condition $(A_1 \leq A < A_2)$ per the number of targets expected by the reference model within those alarms. Where the total expected rate is equal to the actual number of target earthquakes, the differential probability gain g is

$$g = \frac{-\nu(A_2) + \nu(A_1)}{\tau(A_1) - \tau(A_2)}. \qquad (10.2)$$

In a continuous case with infinitely small interval, the differential probability gain is equal to the slope of the tangent to the trajectory, as shown in Figure 10.1, $g = tg(\beta)$.

Differential probability gain can be used to combine two forecast models in the case where at least one of them is rate-based. The rate-based model is used as a reference model. In the case where the second model is also rate-based, the rates given by the model are used directly as an alarm function [*Shebalin et al.*, 2014]. The method of combining two models employs the local differential probability gain values as a multiplier of the reference model rates. If the value is higher than 1 (high slope of the error diagram trajectory), then the initial rate is increased; otherwise it is decreased. In a trivial case of a random alarm-based model (trajectory close to the diagonal (0, 1; 1, 0)) the differential probability gain value is equal to 1 everywhere, and the output and reference models are equivalent.

The differential probability gain combining method introduces a variety of possible combinations. Combining a time-and-space dependent, alarm-based model with a simple time-independent seismicity model actually converts the alarm-based model to a rate-based form [*Shebalin et al.*, 2012]. Combining two models with the same timescale may significantly improve models [*Shebalin et al.*, 2014], but the method can combine models with different timescales. For example, a rate-based model with a time span of 1 month combined with a rate-based model with a time span of 1 day produces as output a rate-based model with a time span of 1 day. A rate-based model may be improved even by using a time-independent second model. This is a possible way to incorporate some geologic information into time-dependent seismicity models. Transient earthquake precursors not well fixed to a specific area, also can improve the forecast models using the differential probability gain method.

In this chapter I introduce examples demonstrating how the differential probability gain combining method works and pay special attention to the task of incorporating earthquake precursors into seismicity forecast models.

10.2. METHODS

10.2.1. Evaluation of Forecast Models

Forecasts are discretized in space and time according to a predefined grid and a given time step. In prospective tests, all forecasts are given for the next time step and a finite magnitude range. Those assumptions correlate with the standards applied in the CSEP. At CSEP testing centers, all the rate-based models are evaluated by using likelihood tests. In contrast, alarm-based models are tested only at the California testing center using error diagrams ("Molchan tests") and the related ROC and area skill score (ASS) tests [*Zechar and Jordan*, 2008; *Zechar*, 2010]. Unfortunately, alarm- and rate-based models are still tested independently. The main reason for this is that the likelihood tests cannot be applied to an alarm-based model with a nonprobabilistic alarm function. To address this problem, *Shebalin et al.* [2012] proposed a method to convert alarm-based models to rate-based forecasts. In addition, two rate-based models can be compared using error diagrams. In practice, the complete rate-based model may be reduced to a single rate value by summing over a given range of magnitude. The reduced model can be treated as a rate-based model and/or as an alarm-based model. Its alarm function is simply composed of the single rate value in each spatial cell.

Likelihood tests and error diagram are complementary and both can be used to estimate the performance of the forecast models. The likelihood tests, however, do not apply to nonprobabilistic alarm-based models.

10.2.1.1. Error Diagram

Error diagrams are used to compare an alarm-based model with a reference model of seismicity defined on the same spatial grid [*Molchan*, 1990]. For any space–time region (x,t), the reference model provides the rate $\lambda(x,t)$ of target earthquakes. The alarm-based model is entirely defined by its alarm function $A(x,t)$. Where this alarm function exceeds a given threshold value A_0, an alarm is issued and a target earthquake is expected to occur. Although it is not necessary, it is usually assumed that A values are ordered from smallest to largest according to the probability of occurrence of a target event. In almost all cases, numerical forecast models can be easily converted to an alarm-based forecast because the information provided by a numerical value assignment on a given space–time grid can be used as an alarm function.

An error diagram compares rates of types I and II errors for varying threshold values A_0 (i.e., a level of alarm). Different A_0 type I errors are measured by

$$\tau\left(A_0\right)=\frac{\Sigma_{A(x,t)\geq A_0}\lambda\left(x,t\right)}{\Sigma\lambda\left(x,t\right)}, \qquad (10.3)$$

where the summation symbols refer to the space–time regions in which the subscript condition is satisfied. The τ value is often interpreted as a fraction of the space–time region occupied by alarms [*Kossobokov and Shebalin*, 2003; *Molchan*, 2010]. In fact, this value measures a probability, according to the reference model, that a target earthquake will occur within an alarm region, at specified conditions. Here, it is important to emphasize that the reference model may depend on time, and summation in equation (10.3) is calculated over both space and time. Type II error rates are the miss rates $\nu(A_0)$, the fraction of target earthquakes that occurred in space–time bins in which $A < A_0$, that is outside the alarm.

In an error diagram, the (τ, ν) curve constructed for all A_0 values forms a trajectory [*Molchan*, 1990; *Zechar and Jordan*, 2008]. Here, we use the term "error diagram" implying the trajectory. This trajectory runs from the point (0, 1) to the point (1, 0) for a decreasing A_0 value. The diagonal connecting these two points corresponds to an unskilled forecast. Below the diagonal, the alarm function may bring additional predictive power [*Shebalin et al.*, 2006] with respect to a given level of significance α [*Zechar and Jordan*, 2008, equations 2 and 3].

The closer the error diagram is to the y axis, the more skill the forecast has. It is often desirable to characterize forecast skills by a scalar value, the so-called loss function. Examples of loss functions are the minimal summary error $\max(1-\tau-\nu)$ [*Molchan*, 1990], the minimax loss function $\inf(\max[\nu, \tau])$ [*Molchan*, 1990], the area above the error diagram [*Zechar and Jordan*, 2008, 2010], and the maximal probability gain $\max((1-\nu)/\tau))$ [*Aki*, 1996]. Many of these expressions have a singularity at $\tau = 0$. This singularity corresponds to 100% rate of misses, so for practical use some reasonable maximal rate of misses may be introduced. Furthermore, a standard situation for choosing the threshold A_0 is a trade-off between the rates of false alarms and failures to predict so that its best value may depend on the goal of the forecast [*Molchan*, 1990]. In all cases a two-dimensional error diagram better visualizes the vector data that are processed, rather than a scalar loss function.

10.2.1.2. Likelihood Tests

Likelihood tests are commonly used to evaluate rate-based models of seismicity. Their forecasts are tested against the number of observed target earthquakes in each bin [*Schorlemmer et al.*, 2007]. For simplicity, individual rates are assumed to follow independent Poisson processes. For all magnitude classes, the complete likelihood counts the Poisson joint log-likelihood of the observed number $\omega(x, t)$ given the forecast $\lambda(x, t)$:

$$L(t) = \Sigma\left(-\lambda(x,t) + \omega(x,t)\log(\lambda(x,t)) - \log \omega(x,t)!\right). \tag{10.4}$$

The closer the joint log-likelihood is to zero, the better the forecast is.

The spatial likelihood is a reduction of the complete likelihood applied to a forecast with rate value normalized to match the total observed number of targets [*Zechar et al.*, 2010a]. In each spatial bin, the single rate values are obtained by summing the expected event rates over the whole range of magnitudes. For the total duration of the experiment, the total sum of log likelihoods over all time steps can be calculated and divided by the total number of observed events to estimate a log likelihood per earthquake.

Likelihood tests applied to forecasts defined on a high-density spatial grid are often criticized because of potential earthquake interactions [*Molchan*, 2012], but the problem of bin independence cannot be solved easily and it is generally thought that the dependence is conditional on earthquake occurrence. For example, we expect many aftershocks after large earthquakes, but a prospective forecast experiment requires any interbin dependence to be provided in advance, before the large earthquake is known [*Zechar*, 2010].

10.2.2. Combining Models Using Differential Probability Gain

The differential probability gain approach uses the probabilistic properties of any forecast, either probabilistic or deterministic, expressed in the statistics of successes and failures. The procedure may be applied successively many times by inserting at each step additional information. In this section I describe one iteration of the combination procedure using a reference model (the initial rate-based forecast model) and an input model (any type of numerical forecast model) to compute a new rate-based forecast model. At subsequent iterations the new rate-based model output may serve as a reference model; multiple application of the approach on an example is described in section 10.4.4.

The entire procedure is based on the error diagram that evaluates retrospectively the performance of the input model with respect to the reference model. For this reason, the input model must be expressed through an alarm function A. For example, the alarm function of a standard CSEP rate-based model may be simply calculated as a sum of the rates over a magnitude range [Kossobokov, 2006].

In an error diagram, we can define the probability gain [Aki, 1981] as

$$G(A_0) = \frac{\Sigma_{A_0 \le A(x,t)}\omega(x,t)}{\Sigma_{A_0 \le A(x,t)}\lambda(x,t)} = \frac{1-\nu}{\tau} \times \frac{\Sigma\omega(x,t)}{\Sigma\lambda(x,t)} \tag{10.5}$$

where A_0 is the threshold value of the alarm function, and the summation symbols refer to the space–time regions in which the subscript condition is satisfied. This G value is a factor that integrates the increase of the rate of the current model within the space–time region in which $A > A_0$ [*Aki*, 1981; *Molchan*, 1991; *Zechar and Jordan*, 2008]. To isolate smaller areas and specific behaviors associated with different ranges of the alarm function, we work with the differential probability gain function that can be defined as the derivative of a continuous error diagram trajectory,

$$g = \frac{\partial \nu}{\partial \tau} \tag{10.6}$$

For each segment and the corresponding range $[A_0;A_0+\delta A_0]$ of alarm function values, we define the differential probability gain as

$$g(A_0)=\frac{\Sigma_{A_0\le A(x,t)<A_0+\delta A_0}\omega(x,t)}{\Sigma_{A_0\le A(x,t)<A_0+\delta A_0}\lambda(x,t)}=\frac{\Delta\nu}{\Delta\tau}\times\frac{\Sigma\omega(x,t)}{\Sigma\lambda(x,t)} \tag{10.7}$$

where the sum is taken over to the space–time regions in which the subscript condition is satisfied. Considering all segments, we can assign a specific $g(A_0)$ value to any A value of the input model. Having done that, we can produce space–time maps of the differential probability gain of the input model and combine them with the reference rate-based model. Then, the output rate-based model is defined as

$$\lambda_{new}(x,t)=g\left(A(x,t)\lambda_{ref}(x,t)\right) \tag{10.8}$$

that is, the initial rates of the reference model increase or decrease according to the local $g(A(x,t))$ value. The $g(A)$ values are always estimated retrospectively over long times in order to be used in future applications. This method of model combination is similar to convolving the reference model with the input model. For this reason, in the following, we denote the procedure by

*new model = reference model * input model*

10.2.2.1. Binary Case

To illustrate the differential probability gain approach let us consider a binary input model. Very often precursors may be expressed just in the form of "present" or "not present", or numerically 1 or 0. In such cases the differential probability gain function for the input relative to a reference model consists of only two values: $g(1)$ and $g(0)$. In order to understand how important knowledge of the presence or absence of this precursor is, we study the joint history of the input information and the reference model using an error diagram (Figure 10.2a). According to the input information the entire time–space is subdivided into areas of alarm, and areas of no alarm. The error diagram consists of two segments with different slopes. Value τ, corresponding to the end of the upper segment, measures the total rate of expected earthquakes (according to the reference model), within alarm areas, divided by the total expected rate. The slope of this segment is equal to the rate of target earthquakes that occurred within alarm areas divided by the value τ.

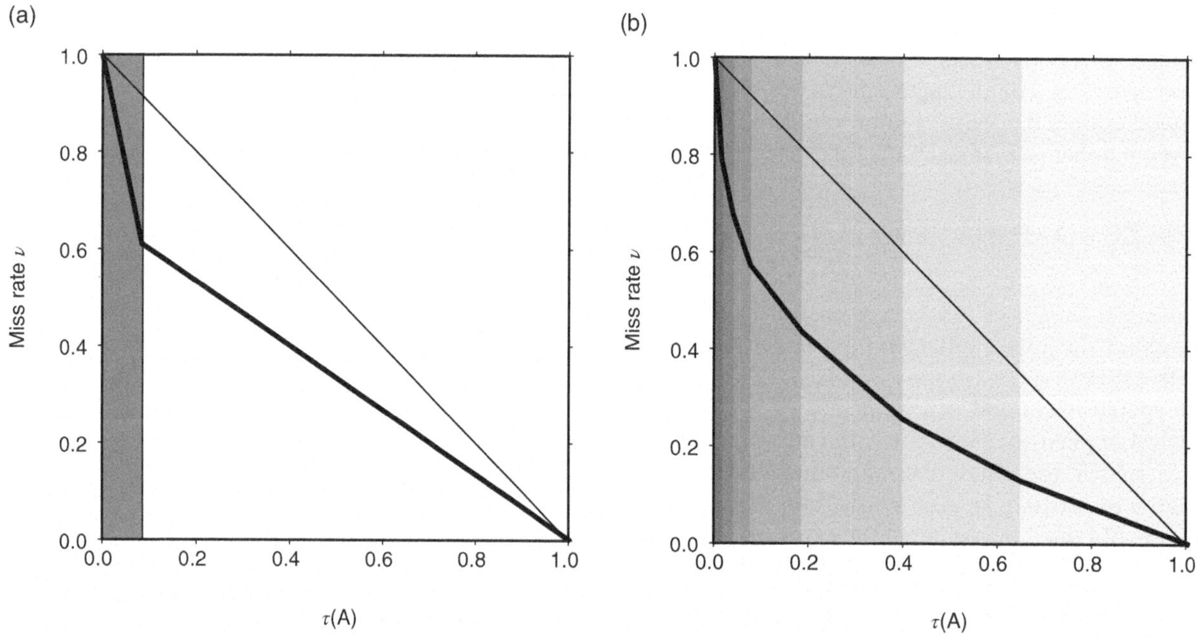

Figure 10.2 Error diagrams for binary and discrete alarm functions. In the binary case (a) the alarm function has only two values ("yes"/"no") and the trajectory has two slopes. In the discrete case (b) several segments with specific slopes correspond to several alarm function values. Differential probability gain $g(A)$ for each discrete value of the alarm function is equal to the slope of the corresponding segment. The continuous alarm function may be discretized using a set of thresholds.

This slope, equal according to equation (10.7) to $g(1)$, describes the probability gain of the alarms defined by the input relative to the reference model: how many times the actual number of target earthquakes within alarm areas was higher than was expected by the reference model. Similarly, the slope of the lower segment in the error diagram is equal to $g(0)$. If its value is < 1, that signifies that the absence of the alarm decreases the probability of target earthquakes (the multiplier $g(0)$ is < 1). The application of the differential probability gain approach in this binary case is extremely simple and clear: the rates expected by the reference model within alarm areas should be multiplied by a factor $g(1)$, and by a factor $g(0)$ outside the alarm areas, as defined by the input. If the error diagram consists of three, four, or more segments (Figure 10.2b) there is no principal difference with the binary case. The alarms my be numerically expressed as 0, 1, 2, etc., and the error diagram for the learning period will consist of several segments with slopes $g(0)$, $g(1)$, $g(2)$, etc., that are then applied as multipliers for the reference model in real-time forecasting.

10.2.2.2. Smoothing Error Diagram Trajectory

Given a finite number N of target earthquakes and the discretization of space, the error diagram trajectory is a step-like function. Examples in section 10.4 show that the step-like shape of the error diagrams is actually often quite irregular. Some steps may occur that are very steep; local values of the differential probability gain (equal to the local slope of the error diagram) for such cases is very high. To avoid overfitting and to make differential probability gain estimates more reliable we smooth the error diagram using a limited number of segments. Here we use an automatic procedure to smooth an error diagram trajectory into N_{seg} segments, where N_{seg} is the parameter of the algorithm. If $N \le N_{seg}$, we consider a segment for each step of the error diagram trajectory, and the vertical coordinates of the segments are i/N, with $I \in \{0,1,\ldots,N\}$. If $N > N_{seg}$, we consider only N_{seg} segments, and the vertical coordinates of the segments are $\lfloor (N_{seg} - i)/N_{seg} \rfloor / N$, with $i \in [0, N]$ and where $\lfloor x \rfloor$ is the largest integer less than or equal to x. At each step where there is a vertical limit of segments, the horizontal coordinate of the segments is the τ value that corresponds to the median of the distribution of the alarm function value for this step. Everywhere in this study, we use $N_{seg} = 20$, but we check the stability of the results for $10 < N_{seg} < 30$. From our experience we also infer that there is not a strong impact of the smoothing method on the result as soon as $N_{seg} > 10$. An example of application of this procedure is shown in Figure 10.5 in section 10.4. For specific cases different smoothing procedures may be developed.

10.2.2.3. Conservation Property of the Differential Probability Gain Combining Method

An important property of the differential probability gain method contributes to its consistency. In the learning period used to estimate the differential probability gain values the total expected rate of earthquakes is equal to that in the reference model. To prove it let us denote N_{seg} the number of segments in the smoothed error diagram (see previous section), and (τ_i, ν_i) and A_i, $i \in \{0, 1,\ldots, N_{seg}\}$, the segment extremity coordinates and the corresponding alarm function values of the input model, respectively. The slope of these segments is the differential probability gain g_i, $i \in \{1, 2,\ldots, N_{seg}\}$. By definition, we have

$$g_i = \frac{\nu_i - \nu_{i-1}}{\tau_i - \tau_{i-1}} \tag{10.9}$$

For the reference and the new model, we may group the expected event rates according to the ranges of the input model alarm function that correspond to the different segments of the error diagram:

$$\lambda_{ref}^i = \sum_{A_i < A(x,t) < A_{i+1}} \lambda_{ref}(x,t), \quad \lambda_{new}^i = \sum_{A_i < A(x,t) < A_{i+1}} \lambda_{new}(x,t), \tag{10.10}$$

where the summation symbols refer to the space–time regions in which the subscript condition is satisfied. We define the total rates of the reference and the new models as

$$\Lambda_{ref} = \sum_{i=1}^{N_{seg}} \lambda_{ref}^i, \quad \Lambda_{new} = \sum_{i=1}^{N_{seg}} \lambda_{new}^i, \tag{10.11}$$

For the period in which the differential probability gain function is estimated, we have

$$\lambda_{ref}^i = \Lambda_{ref}(\tau_i - \tau_{i-1}). \tag{10.12}$$

In that case, using equations (10.9), (10.12), and (10.10) successively we obtain

$$\begin{aligned} \Lambda_{new} &= \sum_{i=1}^{N_{seg}} g_i \lambda_{ref}^i = \Lambda_{ref} \sum_{i=1}^{N_{seg}} g_i (\tau_i - \tau_{i-1}) \\ &= \Lambda_{ref} \sum_{i=1}^{N_{seg}} (\nu_i - \nu_{i-1}) = \Lambda_{ref} \end{aligned} \tag{10.13}$$

For a real-time test or a quasi-prospective test this conservation of the total expected rate is approximate.

10.3. DATA

In this study all seismicity models are based on data from the earthquake catalog utilized. We apply the study to the SCEP California testing area, and the Advanced National Seismic System (ANSS; http://quake.geo.berkeley.edu/cnss/catalog-search.html [last accessed May 2016]) earthquake catalog was searched.

10.4. EXAMPLES OF APPLICATION OF THE DIFFERENTIAL PROBABILITY GAIN MODEL COMBINING METHOD

10.4.1. Conversion of an Alarm-Based Forecast Model to a Rate-Base Form

The first application of the differential probability gain method [*Shebalin et al.*, 2012)] was aimed to convert an alarm-base forecast model EAST (Early Aftershock Statistics) to a rate-based form. Based on the hypothesis that the time delay before the onset of the power-law aftershock decay rate decreases as the level of stress increases, the EAST model is an alarm-based forecast model that uses early aftershock statistics to determine the space–time regions where this time delay is abnormally high [*Shebalin et al.*, 2011]. The model is based on the hypothesis that the time delay before the onset of the power-law aftershock decay rate decreases as the level of stress and the seismogenic potential increase [*Narteau et al.*, 2002, 2005, 2008, 2009]. In contrast to epidemic-type forecast models, the EAST model concentrates only on the temporal properties of large magnitude aftershocks ($1.8 \le M_A$) of small magnitude main shocks ($2.5 \le M_A \le 4.5$) to infer the seismogenic potential of the testing region. The EAST model is being tested in the California CSEP testing center, beginning as of 1 July 2009, with a time step of 3 months and target earthquake magnitude 4.0. The model shows better predictive power than the relative intensity (RI) reference model [*Tiampo et al.*, 2002] at a level of significance of 1%. The RI reference model is commonly used because it is simply obtained by smoothing the location of earthquakes in the past [*Kossobokov and Shebalin*, 2003; *Helmstetter et al.*, 2006; *Molchan and Keilis-Borok*, 2008; *Zechar and Jordan*, 2008].

To isolate space–time regions with a relative high level of stress, the EAST forecast model uses the E_A-value, the ratio between the long-term and short-term estimations of $<t_g>$, the geometric mean of elapsed times between main shocks and early aftershocks within a fixed time window [*Shebalin et al.*, 2011]. However, main-shock–aftershock sequences are not uniformly distributed, and this alarm function cannot be defined everywhere at all time steps. In these cases, the alarm-function value is taken equal to λ_{RI}, the normalized event rate given by the RI reference model [*Shebalin et al.*, 2011]. Finally, the alarm-function of the EAST forecast model is

$$A_{EAST} = \begin{cases} \dfrac{\langle t_g \rangle_{long}}{\langle t_g \rangle_{short} \lambda_g}, & if \langle t_g \rangle_{long} \; and \; \langle t_g \rangle_{short} \; are \; defined \\ \lambda_{RI}, & else, \end{cases} \tag{10.14}$$

where $\langle t_g \rangle_{short}$ and $\langle t_g \rangle_{long}$ are the geometric means of elapsed times between main shocks and the aftershocks during 5 years and 25 years relative to the current time; λ_g is a spatially-smoothed event rate similar to λ_{RI}. λ_{RI} is obtained by uniform smoothing within circles of 12 km radius, while λ_g obtained from a two-dimensional Gaussian function with a standard deviation of 10 km [*Shebalin et al.*, 2011]. This Gaussian filter is chosen to assign a nonzero value to all space–time regions.

Conversion of the EAST model to its rate-based form designated $EAST_R$ was undertaken using equation (10.8) separately for five magnitude intervals of the target earthquakes. The time-independent RI model has been used as a reference model, $\lambda_{ref}(x,t) = \lambda_{RI}(x)$. The differential probability gain values $g_{EAST}^{RI}(A_{EAST})$ for a five magnitude interval, from 4 to 4.5, 4.5 to 5, 5 to 5.5, from 5.5 to 6, and ≥6, have been estimated using the ANSS earthquake catalog (ANSS; http://quake.geo.berkeley.edu/cnss/catalog-search.html), 1984–2008. After smoothing, a table was constructed of the values $g_{EAST}^{RI}(A_{EAST})$ for the five magnitude intervals, and large values up to 10 were found. The table values are independent of both time and space. The conversion of the EAST model to its rate-based form $EAST_R$ is now as simple as finding for each spatial bin at each time step the corresponding element in the $g_{EAST}^{RI}(A_{EAST})$ table (taking into account magnitude), and obtaining the output value by multiplying the λ_{RI} value for the spatial bin by the value $g_{EAST}^{RI}(A_{EAST})$ from the table.

Parameters of the EAST model were fixed before the beginning of the test in July 2009 [*Shebalin et al.*, 2011], and values $g_{EAST}^{RI}(A_{EAST})$ for converting the model into the rate-based form at the end of 2010 [*Shebalin et al.*, 2012]. Thereafter the test of both versions is completely prospective starting 2011. *Shebalin et al.* [2012] have shown for the period from July 2009 to December 2010 that the performance of both models relative to the RI model is very similar in terms of error diagram, and that the $EAST_R$ model significantly outperforms the RI model in terms of likelihood tests. Here I demonstrate that both properties remain valid for the period from January 2011 to March 2016. The area of the test is

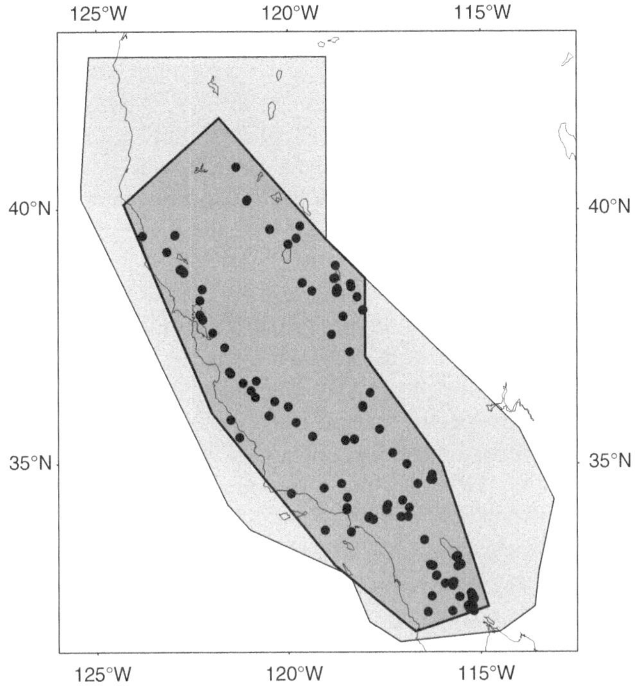

Figure 10.3 The area of the test for the EAST, $EAST_R$, and EAST*EEPAS models. The CSEP California testing region is outlined by the dashed line and the heavier shaded area outlined by the thick line was used for adapting parameters of the EAST model. Dots mark epicenters of $M \geq 4$ events during the period January 2011 to March 2016.

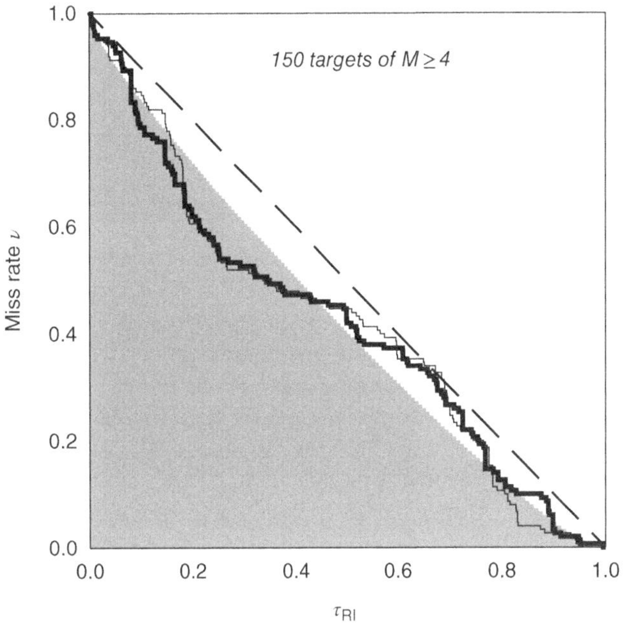

Figure 10.4 Error diagram for EAST and $EAST_R$ forecasts for the period January 2011 to March 2016 with respect to the RI model. The plots for EAST forecasts (thin line) and $EAST_R$ forecasts (thick line) are close to each other. The dashed diagonal line corresponds to unskilled forecasts. The shaded area indicates the zone in which the forecast of the tested models outperforms the forecast of the reference model at a level of significance $\alpha = 1\%$.

shown in Figure 10.3. In the tests I consider a slightly smaller area than CSEP California testing region. It coincides with the area where parameters of the EAST model and the converting factors $g_{\mathrm{EAST}}^{\mathrm{RI}}(A_{\mathrm{EAST}})$ for the $EAST_R$ have been estimated. The complete error diagrams for the EAST and the $EAST_R$ models relative to the RI model for the period 1 January 2011 to 31 March 2016 are shown in Figure 10.4. The diagrams are very similar, which confirms the validity of the differential probability gain method for converting alarm-based models into the rate-based form. The steep section of the diagrams corresponds to the upper part of the alarm function in equation (10.12), and the sloping section corresponds to the part of the alarm function that actually repeats the RI model. Results of the likelihood tests for the $EAST_R$ models and the RI models in the same time and space areas are presented in Table 10.1. In most of the three-month intervals the $EAST_R$ forecasts outperform those of the RI model, only in three intervals does the likelihood test score the RI model slightly better. On average, the log likelihood gain of the $EAST_R$ model relative to the RI model is equal to 0.49 per earthquake.

Table 10.1 Likelihood test results for forecasts of the $EAST_R$ and RI reference models.

Period	$EAST_R$	RI	N_{target}
January–March 2011	**−88.1**	−99.9	11
April–June 2011	**−139.9**	−148.8	14
July–September 2011	**−79.5**	−81.8	9
October–December 2011	**−67.9**	−69.6	7
January–March 2012	**−41.2**	−42.0	5
April–June 2012	**−79.3**	−92.4	10
July–September 2012	**−158.5**	−171.3	19
October–December 2012	**−43.0**	−45.1	5
January–March 2013	**−76.1**	−78.4	8
April–June 2013	**−96.6**	−98.4	9
July–September 2013	−40.6	**−39.6**	4
October–December 2013	**−49.8**	−50.5	5
January–March 2014	**−73.2**	−74.3	8
April–June 2014	−40.0	**−39.4**	4
July–September 2014	−41.7	**−41.3**	4
October–December 2014	**−18.0**	−18.8	2
January–March 2015	**−39.7**	−39.8	4
April–June 2015	**−35.8**	−37.1	4
July–September 2015	**−43.4**	−44.0	5
October–December 2015	**−32.0**	−33.1	3
January–March 2016	**−96.6**	−107.9	10
January 2011–March 2016	−1380.7	−1453.5	150

Bold typeface indicates smaller AIC values for $EAST_R$ relative to RI

10.4.2. Combining Alarm-Based and Rate-Based Forecasting Models

Lets us consider combining these two types of model using the differential probability gain method on the Every Earthquake a Precursor According to Scale (EEPAS) and EAST forecast models in California as an example.

The EEPAS and EAST models are forecast models installed at the CSEP Testing Center in California. They are of special interest for this study because, based upon the joint evaluation using error diagram tests (outside the official CSEP testing process), it was found that they both yield statistically significant better forecasts than the commonly used time-independent RI reference model [*Kossobokov and Shebalin*, 2003; *Helmstetter et al.*, 2006; *Molchan and Keilis-Borok*, 2008; *Zechar and Jordan*, 2008].

The medium-term EEPAS forecast model [*Rhoades and Evison*, 2004, 2007] is based on the precursory phenomenon scale increase and associated predictive scaling relations [*Evison and Rhoades*, 2004]. In this model, every earthquake is a precursor according to scale, with the scale referring to larger earthquakes to follow in the medium to long term. Thus, smaller earthquakes are "witnesses" of the seismogenic process and not "actors" as in the well-known ETAS branching model [*Ogata*, 1989]. Several versions of the EEPAS model have been used to generate forecasts of seismicity for the forthcoming 3 months. Among the five versions of this forecast model, we chose the EEPAS-0F model (henceforth referred to as simply the EEPAS model) because it performs best against the RI reference model.

The EAST model is an alarm-based earthquake forecast model that uses early aftershock statistics [*Shebalin et al.*, 2011], and was described in the previous section. In contrast to the EEPAS model, the EAST model is not a model of seismicity rates. Instead, it possesses a nonprobabilistic alarm function, defined in equation (10.12), that is dedicated to detecting zones with a higher level of stress where an earthquake is more likely to occur. These zones are identified by relatively low values of the geometric mean of elapsed times between main shocks and early aftershocks. The three-month class EAST and EEPAS models have been archived at the California CSEP center since July 2009 and January 2008 respectively.

The EAST and EEPAS models are complementary because they focus on different relevant aspects of seismicity, with each giving an additional amount of predictive information. This complementary nature of the two independent models of seismicity is difficult to detect using likelihood tests. However, we stress the point that it may be an important property for earthquake forecasting and certainly the best case to combine two independent models. Then, the method to combine must preserve the knowledge gain that each model offers. Using the differential probability gain method, we detect additional independent knowledge provided by the rate-based EAST model relative to the reference EEPAS model, and keep the knowledge provided by the EEPAS model.

As shown by *Shebalin et al.* [2012] and in the previous section, there is no significant difference in the predictive power of the EAST and $\mathrm{EAST_R}$ models. Therefore, to avoid potential noise introduced by the RI reference model or the conversion method, it is natural to combine directly the EAST and EEPAS models. To construct the differential probability gain functions $g_{\mathrm{EAST}}^{\mathrm{EEPAS}}$ we considered the retrospective period from January 1984 to June 2009 and three magnitude ranges of 4.95–5.45, 5.45–5.95, and 5.95–∞. Within each of those intervals, the combined model inherits the magnitude distribution of the EEPAS model, which is not constrained to follow the Gutenberg–Richter relation in each cell. Figure 10.5 shows for each interval the error diagrams, their approximation by segments, and the differential probability gain functions of the EAST model with respect to the EEPAS model. We observe that the $g_{\mathrm{EAST}}^{\mathrm{EEPAS}}$ values are >1 for almost two orders of magnitude of the alarm function of the EAST forecast model (Figure 10.5d,e,f).

For the three magnitude intervals the corresponding functions $g_{\mathrm{EAST}}^{\mathrm{EEPAS}}$ and the rates λ_{EEPAS} of the EEPAS model in equation (10.8) have been used to obtain the rates of the new rate-based model EAST*EEPAS. Figure 10.6 shows error diagrams used to evaluate the forecast of the EAST, EEPAS, and EAST*EEPAS models with respect to the RI reference model during the quasi-prospective period (July 2009 to December 2011). Comparison of these trajectories on the diagrams shows that the combined model works better than the two initial models from which it has been derived. This is particularly the case for the smallest $_{\mathrm{RI}}$ value, for which both initial models perform better than the RI reference model at a significance level below $\alpha = 1\%$. Results of the likelihood test (Table 10.2) also indicate a gain of the EAST*EEPAS model with respect to the EEPAS model. Quantitatively, the log-likelihood gain is 0.30 per earthquake. With respect to the $\mathrm{EAST_R}$ model, this gain is 0.34.

Figure 10.7 illustrates how the differential probability gain combining method works using the outputs of the EAST, $\mathrm{EAST_R}$, EEPAS, and EAST*EEPAS models for the forecast period from April to June 2010 along the USA–Mexico border. This space–time region includes the M 7.2 El Mayor-Cucapah earthquake of 4 April 2010. Figure 10.7a shows the map of the EAST alarm function values. For rate-base models (Figure 10.7b–d), the rates are calculated for $M \geq 4.95$ events by summation of all model rates in corresponding magnitude bins. All models exhibit a clear maximum near the epicenter of the El

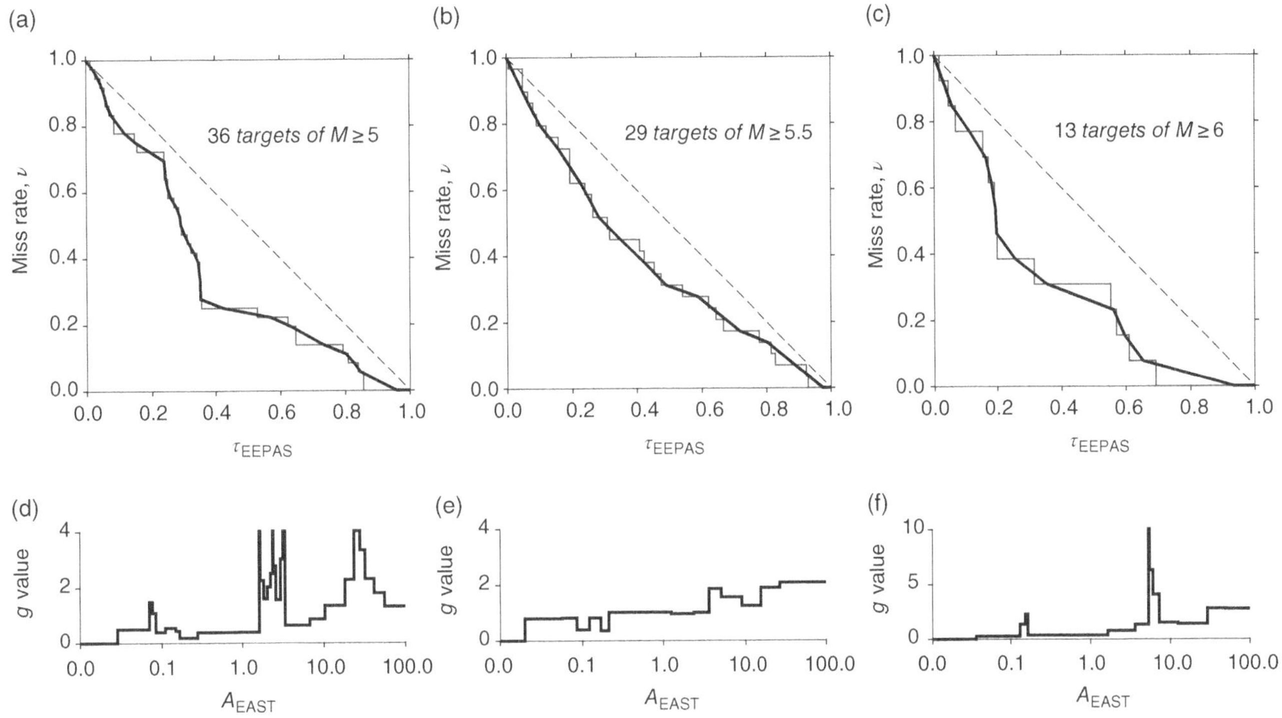

Figure 10.5 Differential probability gain functions for the EAST model with respect to the EEPAS model showing estimation of the differential probability gain functions, $g_{\mathrm{EAST}}^{\mathrm{EEPAS}}$, of the EAST forecast model with respect to the EEPAS model for California from January 1984 to June 2009. (a–c) Error diagrams in which we smooth the trajectory (thin line) by a set of segments (thick lines; see section 10.2.2.2). The $g_{\mathrm{EAST}}^{\mathrm{EEPAS}}$ value is the local slope of these segments. (d–f) Differential probability gain $g_{\mathrm{EAST}}^{\mathrm{EEPAS}}$ as a function of the alarm function A_{EAST} of the EAST models.

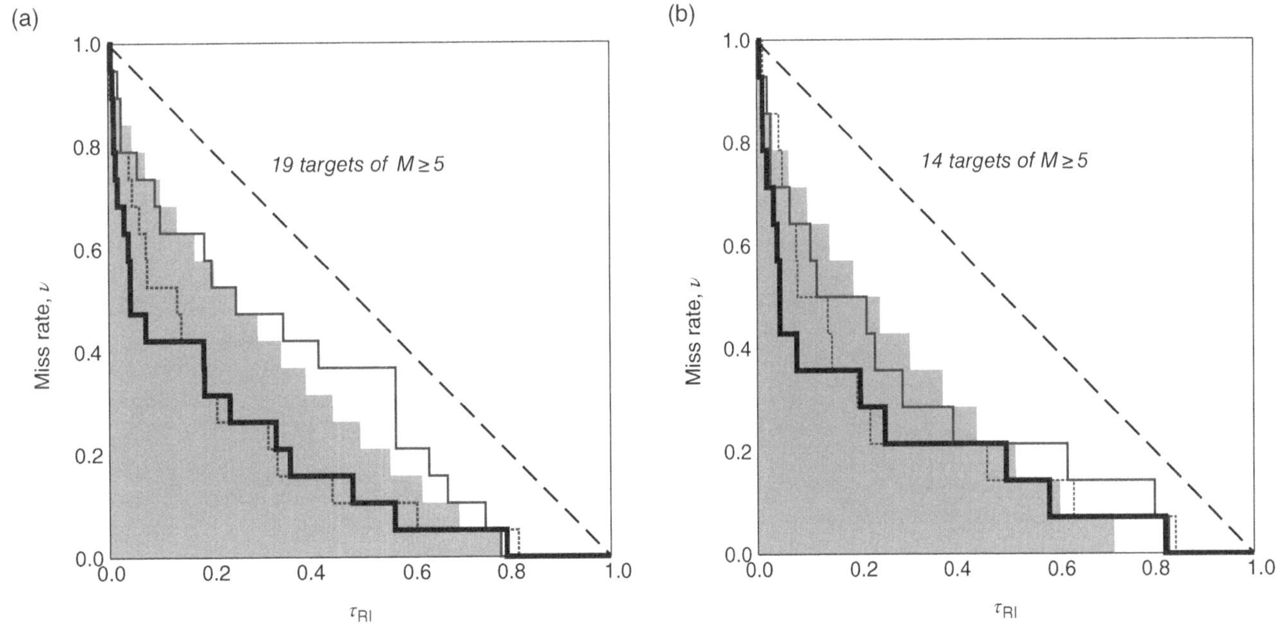

Figure 10.6 Quasi-prospective evaluation of the EAST*EEPAS, EAST_R, and EEPAS models. Tests were carried out in (a) the California CSEP testing region and in (b) the reduced region indicated in Figure 10.3. The forecasts of the EAST*EEPAS (thick lines), EAST_R (thin lines), and EEPAS (thin dashed lines) models are compared with respect to the RI reference model. The dashed diagonal line corresponds to an unskilled forecast. See Figure 10.4 for explanation of shaded area. For all the models we consider single rate values obtained by summing the expected rates of $M \geq 4.95$ target earthquakes.

Table 10.2 Results of the likelihood tests for forecasts of four models.

Period	N_{target}	EAST*EEPAS	EAST+EEPAS	$EAST_R$	EEPAS	RI
July–September 2009	2	−13.62	−15.71	−15.96	−15.58	−18.64
October–December 2009	2	−17.65	−18.82	−19.42	−18.44	−20.93
January–March 2010	2	−24.97	−24.56	−26.76	−23.57	−25.17
April–June 2010	8	−74.79	−76.13	−77.82	−77.35	−81.69
July–September 2010	2	−14.87	−14.25	−17.47	−16.71	−19.65
October–December 2010	0	−2.18	−1.88	−1.36	−2.40	−1.78
January–March 2011	1	−8.22	−7.94	−11.62	−7.70	−11.07
April–June 2011	1	−12.50	−12.23	−12.12	−12.40	−11.55
July–September 2011	0	−1.75	−1.74	−1.36	−2.12	−1.78
October–December 2011	1	−13.48	−12.81	−12.29	−13.41	−11.66
July 2009 – March 2011 (CSEP region)	19	−184.04	−186.07	−196.17	−189.66	−203.91
July 2009 – March 2011 (EAST region)	14	−137.68	−140.08	−138.63	−146.48	−151.46

Mayor-Cucapah earthquake. However, in an area extending northward from this epicenter, the EAST*EEPAS model gives rates that are almost one order of magnitude larger than the rates of the two individual models. This example demonstrates that the combining method sharpens the individual forecasts, providing higher expected earthquake rates in more confined areas. These local increases of the forecast event rates are compensated by decreases in other places so that the total event rate over the whole territory does not significantly change.

We now compare the EAST*EEPPAS model to the $EAST_R$ + EEPAS model, the average of the EAST and EEPAS forecast models. Using the weighted average of two rate-based models is straightforward and therefore is the most common combining method [*Rhoades and Gerstenberger*, 2009; *Marzocchi et al.*, 2012]. In addition, the use of weighted averages may increase the total predictive skill of the resulting model by locally giving more importance to high and low extreme rate values of each model. However, it remains an averaging method. To avoid additional optimizations, we assign equal weights to the EAST and EEPAS models.

Using error diagrams, we then compare the two combined models to the RI reference model. Figure 10.8 shows that the forecast of the EAST*EEPAS model outperforms the forecast of the $EAST_R$ + EEPAS model, especially for the smallest τ_{RI} value, for which both the EAST and EEPAS models perform the best.

The results of the likelihood tests are quite different to those for error diagrams. In fact, if the error diagram trajectory of a linearly combined model is likely to run between the trajectories of the initial models, Table 10.2 shows that the log likelihood for the linearly combined model is closer to zero than for the models from which it is derived. However, the EAST*EEPAS model again exhibits better performance than the $EAST_R$ + EEPAS model. The gain in log likelihood per earthquake is 0.11 for total likelihood and 0.15 for spatial likelihood. In the reduced region, the EAST*EEPPAS performs better than the linearly combined model, which now has a score between that of EEPAS and that of $EAST_R$.

Figure 10.8a shows the cumulative distribution functions of the rates predicted by the EAST*EEPPAS and $EAST_R$ + EEPPAS models. This plot indicates that the EAST*EEPAS model explores a much wider range of values than the $EAST_R$ + EEPAS models. In addition, we observe that more than 50% of target earthquakes occur for only 2% of the highest rates and that, for these events, the rates of the EAST*EEPAS model are about twice those of the $EAST_R$ + EEPAS model. This difference indicates that the combination based on differential probability gain is currently better than a linear combination at increasing the predicted event rates in the limit of high rate values. Obviously, the opposite is true in the limit of low range value. Nevertheless, in this case, both models fail to predict the target earthquake and a possible gain in trying to combine them is worth discussing. Simultaneously, these results confirm that the only restriction for performing a good combination is the need to use two complementary forecast models, i.e., two models that ensure a nontrivial error diagram with respect to one another ($g(A(x,t)) > 1$ in equation (10.6)).

Figure 10.8b and c compares the single rate values of the EAST*EEPAS and $EAST_R$ + EEPAS models in space–time regions where a target earthquake occurred. In the limit of high rates, the rate values of the EAST*EEPAS model are double those of the $EAST_R$ + EEPAS model (Figure 10.6b). This demonstrates that a combination of forecast models based on differential probability gains can increase earthquake probabilities in rate-based forecast models.

10.4.3. Tests of Impact of Noise in Precursory Patterns on the Resulting Forecasts

The differential probability gain approach may be applied to a set of precursory patterns successively. At each step the previous resulting model is used as a reference

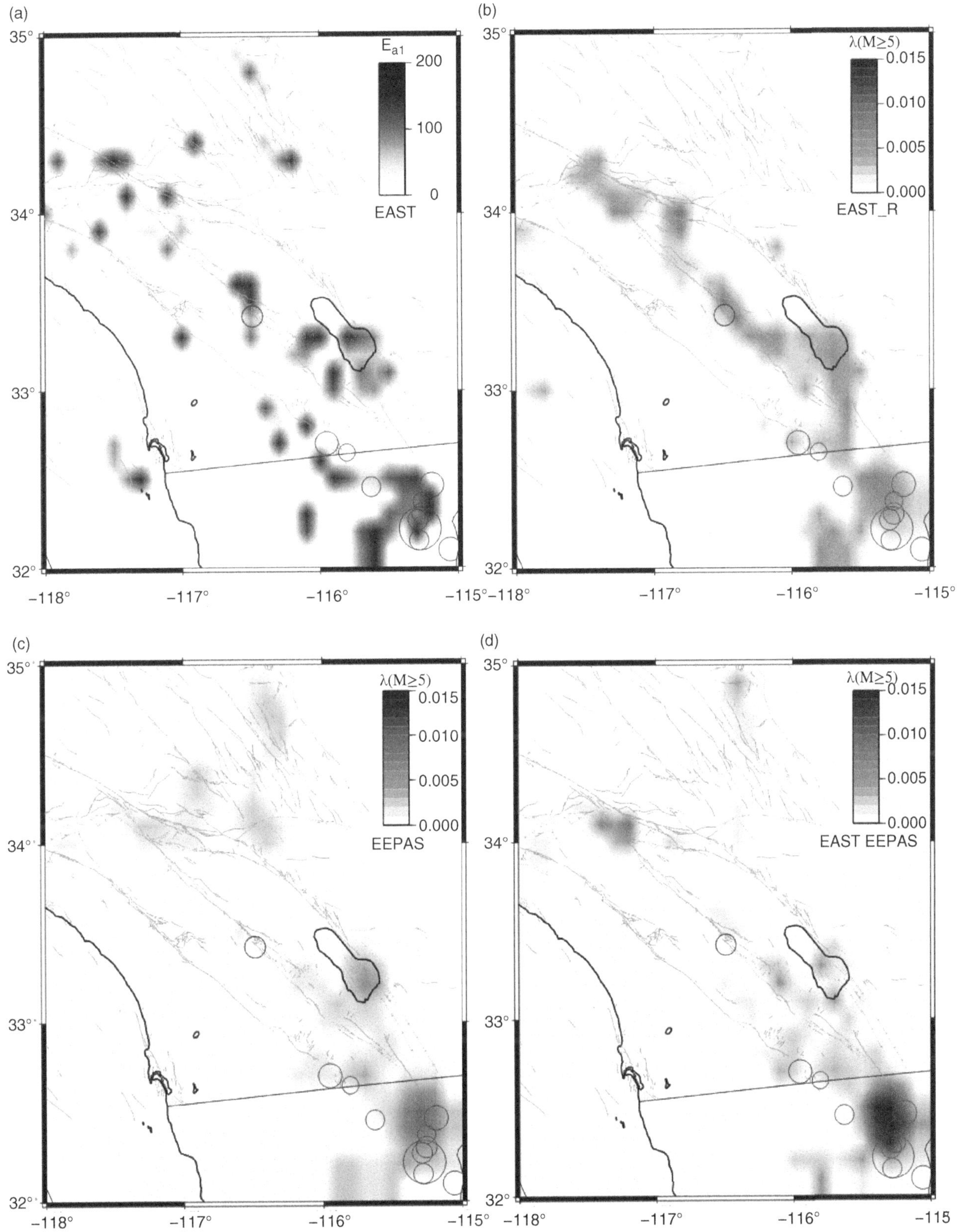

Figure 10.7 Three-month forecasts of the (a) EAST, (b) EASTR, (c) EEPAS, and (d) EAST*EEPAS models for $M \geq 4.95$ earthquakes from April to June 2010 in northern Baja, California along the USA–Mexico border. Circles correspond to $M \geq 4.95$ earthquakes that occurred in this area during this period. For the EAST model the color map varies from zero to the maximum of the alarm function A_{EAST} (equation 10.14). For other models, the same color bar is used to represent the forecast rates of $M \geq 4.95$ earthquakes. Note the higher contrast for the EAST*EEPAS forecasts and the increase in event rate in zones where both the $EAST_R$ and EEPAS models have high event rates. Straight line is the USA–Mexico border. (*See electronic version for color representation.*)

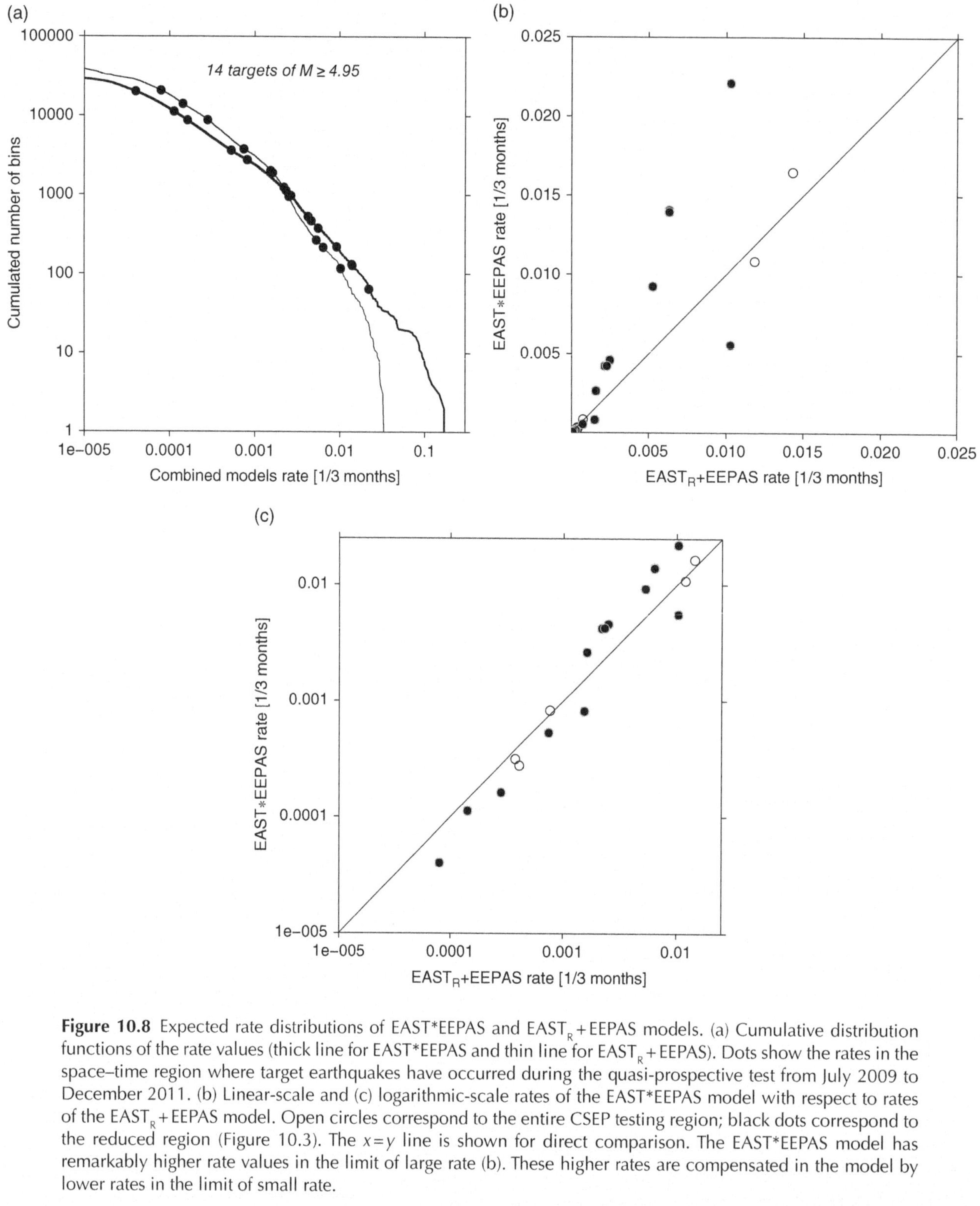

Figure 10.8 Expected rate distributions of EAST*EEPAS and $EAST_R$+EEPAS models. (a) Cumulative distribution functions of the rate values (thick line for EAST*EEPAS and thin line for $EAST_R$+EEPAS). Dots show the rates in the space–time region where target earthquakes have occurred during the quasi-prospective test from July 2009 to December 2011. (b) Linear-scale and (c) logarithmic-scale rates of the EAST*EEPAS model with respect to rates of the $EAST_R$+EEPAS model. Open circles correspond to the entire CSEP testing region; black dots correspond to the reduced region (Figure 10.3). The $x=y$ line is shown for direct comparison. The EAST*EEPAS model has remarkably higher rate values in the limit of large rate (b). These higher rates are compensated in the model by lower rates in the limit of small rate.

model. It may possible, however, that the useful information added by each precursory pattern may be devalued by noise in the patterns. Moreover, we may expect some degradation of the forecasts at each step.

Taking the EAST*EEPAS model as the initial forecast model, we study the impact of a set of alarm-based models containing no actual prediction information on the next-generation forecasts (Figure 10.9a). As before, we

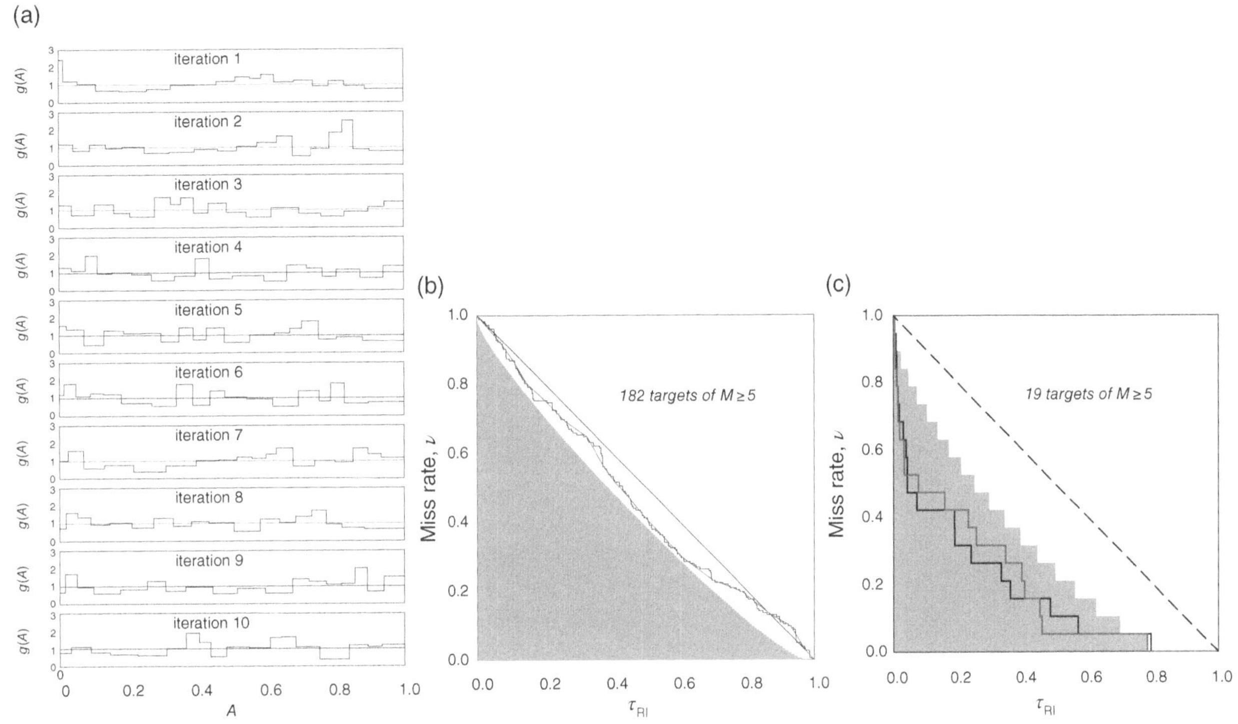

Figure 10.9 Level of noise in combined forecast models using the differential probability gains approach. The EAST*EEPAS model is successively combined with 10 random rate-based models for $M>4.95$ target earthquakes (see text). (a) Differential probability gain $g(A)$ estimated during the learning period from January 1984 to June 2009. (b) For the same period, a comparison of two consecutive generation forecasts before and after the seventh iteration using an error diagram. This iteration exhibits the largest deviation from the diagonal. (c) Evaluation of the initial (thin line) and final (thick line) generation forecasts. In these diagrams, the dashed diagonal line corresponds to an unskilled forecast. See Figure 10.4 for explanation of shaded area.

used the period from January 1984 to June 2009 for learning and the period from July 2009 to December 2011 for testing. We considered a single magnitude interval for target events ($M \geq 4.95$). We simulated ten "alarm-based models" actually consisting of noise only, by drawing random numbers from a uniform distribution between 0 and 1. Then, we iteratively created the next-generation forecasts. As described above, the $g(A(x,t))$ values are estimated during the learning period (Figure 10.9a) and then injected into the current-generation forecasts for both periods. For the learning period, Figure 10.9b shows the error diagram that compares the forecasts before and after the ninth iteration. This iteration exhibits the largest deviation from the diagonal. For the testing period, Figure 10.9c shows the error diagram that evaluates the predictive skills of the initial and the final forecast models with respect to the RI reference model. We find that even after ten iterations with very noisy simulated models, the result remains quite similar to the original one. This result is consistent with all the successive error diagrams, which systematically show that two consecutive forecasts have no specific skills with respect to one another. This analysis reveals that, despite an unavoidable increase of noise, there is no systematic erosion of the forecast skills of a model during the combining procedure.

10.4.4. Multiple Combining Precursors with a Rate-Based Reference Model

Several earthquake precursors or precursory patterns, taken into account together, obviously may improve the quality of forecasting. Where a rate-based model incorporates a set of precursors, it may be hoped that the resulting earthquake probabilities locally reach high values necessary for effective decision making. For now, mostly pattern recognition methods have been used to combine various earthquake precursory patterns [*Gelfand et al.*, 1972; *Gvishiani and Soloviev*, 1984; *Keilis-Borok and Kossobokov*, 1990; *Keilis-Borok and Rotwain*, 1990; *Kossobokov et al.*, 1990; *Peresan et al.*, 1999; *Shebalin and Keilis-Borok*, 1999; *Vorobieva*, 1999; *Keilis-Borok et al.*, 2002; *Romashkova and Kossobokov*, 2004; *Shebalin* et al., 2004; *Shebalin*, 2006]. These methods do not provide estimates of the expected earthquake rates. The differential probability gain combining approach introduces a way to produce rate-based models that include precursory information of various kinds.

Let us consider how to combine n precursors, defined by alarm functions $A_i(x,t)$, I = 1, 2,..., n, and a reference rate-based model $\lambda_{ref}(x,t)$. In a simple case, when all precursors or patterns are independent, the resulting model is obtained using the following equation:

$$\lambda_{new}(x,t)=\prod_{i=1}^{n} g_i(A_i)\lambda_{ref}(x,t), \qquad (10.15)$$

where $g_i(A_i(x,t))$ are found relative the reference rate-base model $\lambda_{ref}(x,t)$

Where the independence of precursors is not evident, the differential probability gain approach should be applied iteratively. At each iteration an intermediate resulting model is obtained, which becomes a reference model for the next iteration:

$$\begin{aligned} &\lambda_1(x,t)=g_{\text{ref}}^1(A_1)\lambda_{\text{ref}}(x,t),\\ &\lambda_2(x,t)=g_1^2(A_2)\lambda_1(x,t),\\ &\dots\\ &\lambda_{\text{new}}(x,t)=g_{n-1}^n(A_n)\lambda_{n-1}(x,t), \end{aligned} \qquad (10.16)$$

where $g_{i-1}^i(A_1)$ are found relative to the intermediate model obtained at the previous iteration.

As shown in section 10.4.3, a multiple application of the differential probability gain approach does not iintroduce significant noise in the resulting model. However, the result may depend on the order in equation (10.16), especially if the data are limited in the learning period used to estimate the differential probability gains. To reduce this ambiguity a recommendation is to split precursors into independent groups. In groups containing two or more not obviously independent precursors an optimal order of combining using equations (10.16) should be found: aggregated differential probability gain factors will be found for such a group. Combining groups is now possible using equation (10.15). A specific example in the next section illustrates this composite approach.

10.4.5. A Composite Earthquake Expectation Rate Model in California Based on Intermediate-Term Precursors

The model considered here is based on four precursory patterns previously used in the RTP earthquake prediction algorithm [*Keilis-Borok et al.*, 2002, 2004; *Shebalin et al.*, 2004, 2006]. For simplicity, as a reference model I use the time-independent RI model, same as described in section 10.4.1. The test of the model covers the CSEP California testing region with a time step of 3 months and spatial grid of 0.1°×0.1° cells. The forecasting targets are earthquakes with magnitude 5 and above (there is no differentiation of magnitudes above the threshold). Intermediate-term precursory patterns, originally considered in the areas of "earthquake chains" [*Shebalin et al.*, 2006, appendix A.3], are considered here in the form of circles. I use the period 1960–1984 to optimize the patterns and to estimate the differential probability gains (learning period). Then the resulting model is retrospectively tested for the period January 1985 to March 2016 (testing period). Note that the learning and the testing periods do not intersect.

10.4.5.1. Definitions

The equations defining the alarm functions for the patterns and parameter values are listed in Table 10.3. In all functions summation is in circles of radius R centered at the centers of the spatial bins (denoted as x), in the interval $t-T\le\tau<t$ (t is the time of the beginning of the

Table 10.3 Intermediate-term precursory patterns for the composite forecasting model.

Name of the precursory pattern and alarm function	R, km	T, years	M_1
"Activity" $A_a(x,t)=N(x,t;R,T,M_1)$	75	5	4
"Gamma" $A_g(x,t)=\dfrac{\Sigma M_i}{N(x,t;R,T,M_1)}$	25	5	4
"Acceleration" $A_c(x,t)=\dfrac{N(x,t;R,T,M_1)}{N(x,t-T;R,T,M_1)}$	25	2	2
"b-micro" $A_b(x,t)=\sum S_i$ where $S_i=\sum_i\sum_{l;t_l-t_i\le\delta t;M_{li}\ge M_2}10^{M_{li}}$, M_{li} are the magnitude of the lth aftershock of the ith main shock, δt=2 days, and M_2=2.0	25	0.5	3

Table 10.4 Correlation coefficients for the pairs of alarm functions.

Alarm functions	Correlation coefficient ρ
A_u and A_g	0.520
A_u and A_c	0.238
A_u and A_b	0.262
A_g and A_c	0.211
A_g and A_b	0.296
A_c and A_b	0.268

forecasting 3-months interval), and for magnitudes $M_i \geq M_1$. $N(x,t;\ R,T,M_1)$ denotes the number of main shocks in those space–time–magnitude volumes. Main shocks are selected using the *Gardner and Knopoff* [1974] algorithm.

The next step appraises whether the alarm functions are correlated, and the simplest way is to estimate the pairwise Pearson correlation coefficients. The CSEP California testing region includes 7682 $0.1° \times 0.1°$ cells. The learning period consists of 100 3-month intervals. Table 10.4 shows the correlation coefficients for the pairs of the four alarm functions considered, calculated using 768,200 pairs of values each.

Only A_u and A_g are slightly correlated, $\rho = 0.52$. Using a subjective choice with threshold $\rho = 0.3$, I consider those two patterns as correlative. For other pairs the correlation is $\rho < 0.3$, and I consider them as uncorrelated. According to the previous section, the construction of the composite forecasting model using the four patterns may consist of three independent groups. Two groups have one pattern each, "Acceleration" and "b-micro." For the group of slightly correlated patterns "Activity" and "Gamma" I apply equations (10.14). Using an error diagram for the "Gamma" pattern relative to the RI reference model (it is automatically smoothed using the method described in section 10.2.2.2), I estimate $g^g_{\mathrm{RI}}(A_g)$ (Figure 10.10a). Next, using equation (6), I construct an intermediate model:

$$\lambda_{g*\mathrm{RI}}(x,t) = g^g_{\mathrm{RI}}(A_g)\lambda_{\mathrm{RI}}(x,t). \qquad (10.17)$$

At the next step, for the pattern "Gamma," the smoothed error diagram is constructed relative to this intermediate model, then the differential probability gain factor $g^g_{g*\mathrm{RI}}(A_u)$ is determined (Figure 10.10b). For the remaining two patterns "Acceleration" and "b-micro," according to equation (10.17) it is sufficient to estimate the differential probability gains relative to the RI model, $g^c_{\mathrm{RI}}(A_c)$ and $g^b_{\mathrm{RI}}(A_b)$ respectively (Figure 10.10 c and d). Finally, the output composite model is defined by the equation:

$$\lambda_{g*u*c*b*\mathrm{RI}}(x,t) = g^g_{\mathrm{RI}}(A_g)g^g_{g*\mathrm{RI}}(A_u)g^c_{\mathrm{RI}}(A_c)$$
$$g^b_{\mathrm{RI}}(A_b)\lambda_{\mathrm{RI}}(x,t). \qquad (10.18)$$

10.4.5.2. Retrospective Tests

I performed both error diagram (section 10.2.1.1) and likelihood tests (section 10.2.1.2) for the learning period, 1960–1984 (Figure 10.11a), and the testing period, 1985-2016 (Figure 10.11b). Note that the data from the testing period have not been used for the construction of the model. The likelihood tests using equation 10.4 are summarized in Table 10.5.

As we can see from Figure 10.11 and from the analysis of the likelihood tests, the forecasts for the testing period even outperform those for the learning period.

The composite model produces much more concentrated space and time areas for expected earthquake rates. Figure 10.12 illustrates such a redistribution of the rates. The two panels show maps for the forecasting interval 1 April 2010 to 30 June 2010: the composite model on the left, RI model on the right. Colored cells are highest rate cells in which 50% of the earthquakes during the interval are expected. For the composite model those 50% target earthquakes are expected in only 35 spatial $0.1° \times 0.1°$ cells, whereas for the RI model they are expected in 624 cells. For the composite model this median rate is equal to 0.0132 (earthquakes of magnitude $M \geq 5$ every 3 months in a $0.1° \times 0.1°$ cell). For the RI model the median rate is equal to 0.00067. Thus the effective rise of the earthquake probabilities in the composite model is about 20 times relative to the RI model.

10.4.5.3. Dependence of the Results on the Order of Applying the Differential Probability Gain Approach

A hypothetical disadvantage of the differential probability gain approach is that the result theoretically may depend on the sequence in which precursory patterns are added to the model, in case the patterns are not independent. In the considered application only the "Activity" and "Gamma" patterns have been found slightly correlated. First I constructed the intermediate model (equation 10.17) and then found the differential probability gains $g^u_{g\times\mathrm{RI}}(A_u)$ of the pattern "Activity" relative to the intermediate model. A different order was also possible: first construct an intermediate model $\lambda_{u\times\mathrm{RI}}(x,t)$ and then find the factors $g^g_{u\times\mathrm{RI}}(A_g)$. A direct comparison of the rates obtained using those two different orders shows that the difference is small. The correlation coefficient for the forecasts in the period 1985 to 2016 is equal to 0.997. The maximum difference of the rates between the two versions is 18%. The difference in log likelihood is as small as 0.04 per earthquake.

10.5. CONCLUSIONS

There are different types of precursory phenomena observed prior to large earthquakes. This kind of information may be useful to improve forecast models. High

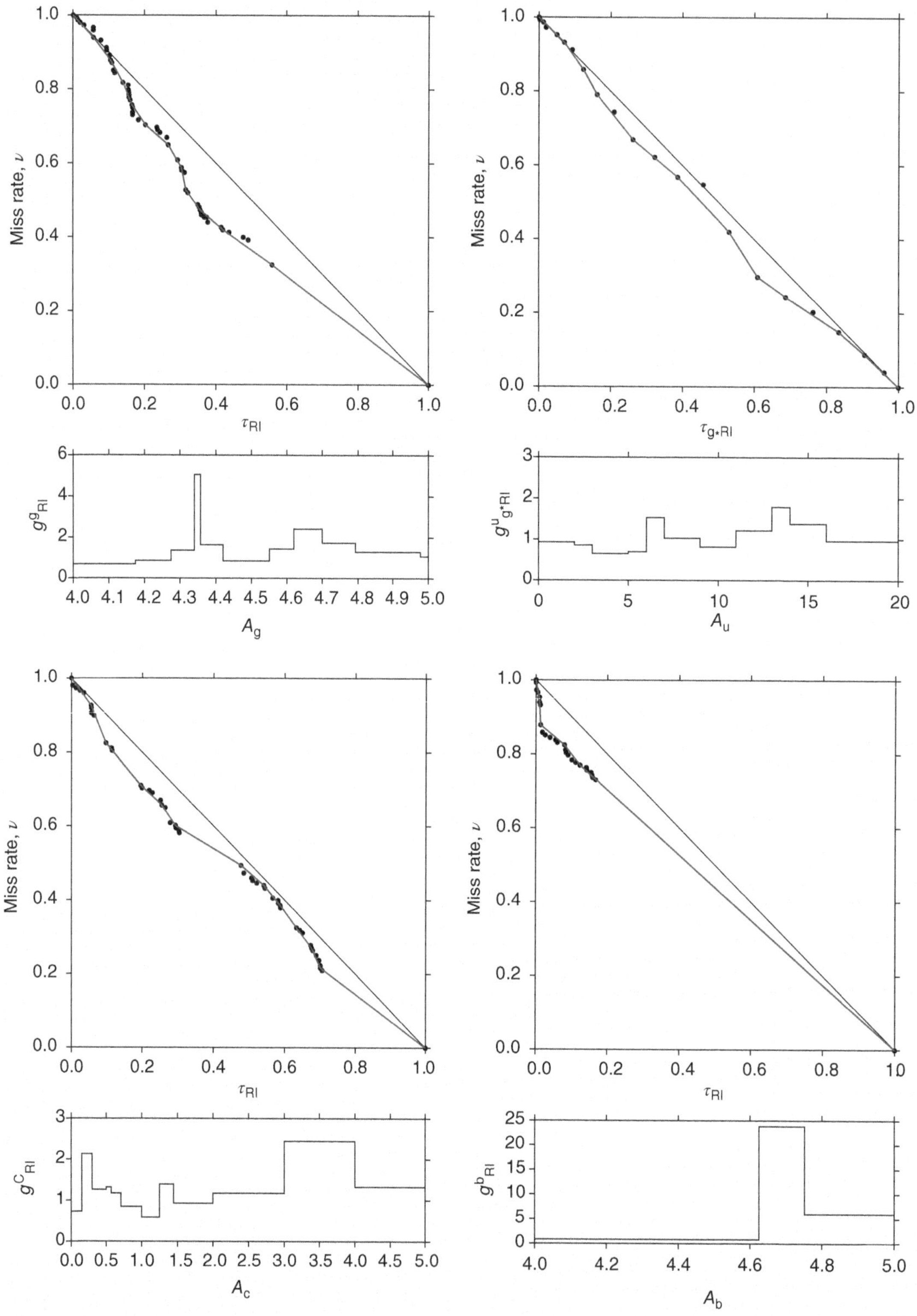

Figure 10.10 Estimation of the differential probability gains for four seismicity precursory patterns. (a) The "Gamma" pattern relative to the RI model; (b) the "Activity" pattern relative the intermediate model, $g \times$ RI, obtained at the first step; (c) the "Acceleration" pattern relative to the RI model; (d) the "b-macro" pattern relative to the RI model. Panels at the top show error diagrams for the patterns relative the reference models (RI or $g \times$ RI). Panels at the bottom show the differential probability gain values as a function of the corresponding alarm function. Calculations were made for the CSEP California testing region using the ANSS earthquake catalog for 1960–1984.

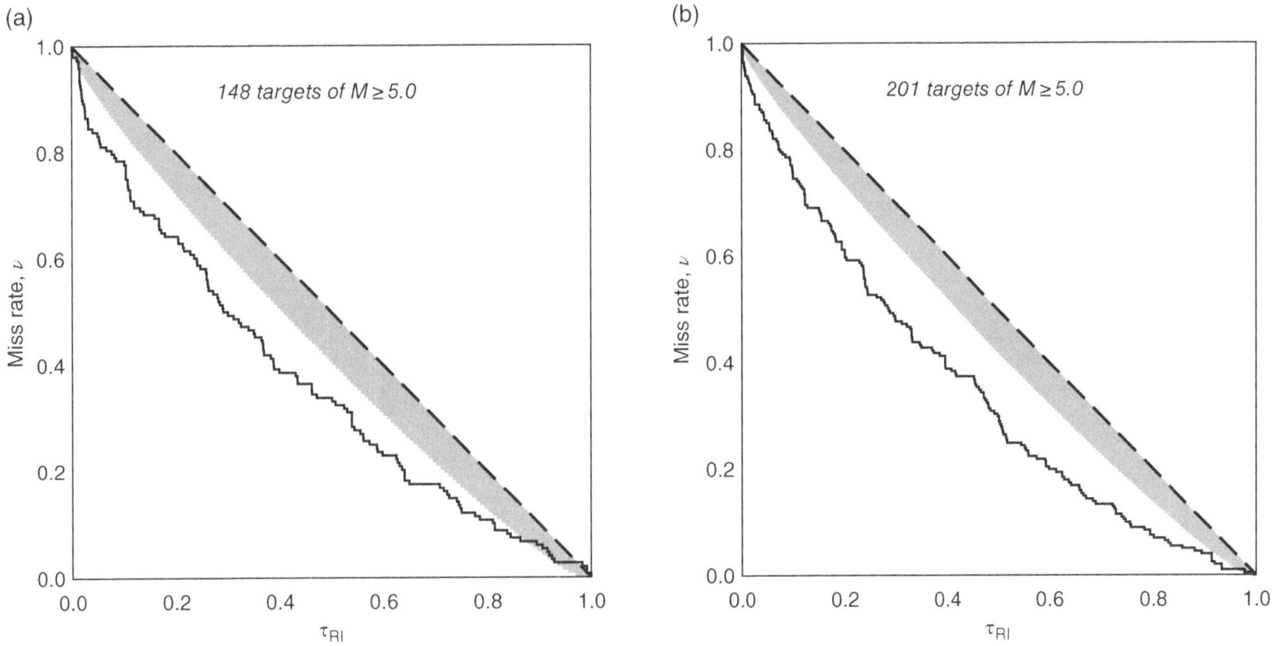

Figure 10.11 Error diagrams of the composite earthquake expectation rates model based on four intermediate-term precursors. Error diagrams for the composite model relative to the RI model: (a) learning period 1960–1984; (b) testing period 1985–2016. The shaded area indicates the zone in which the difference between forecast of the forecasts of the tested model and underperforms the forecast of the reference model at a level of significance $\alpha = 1\%$ are not statistically significant; successful forecasts within this zone should be considered as random.

Table 10.5 Comparison of likelihood test (equation 10.4) results for the composite and RI models.

	Learning period, 1960–1984	Testing period, 1985–2016
Composite model	−842.97	−1103.05
RI model	−1229.82	−1663.07
Number of target earthquakes	148	201
Log likelihood gain per target	2.55	2.79

local expected earthquake rates may be achieved by incorporating their information into seismicity models known to forecast just very low rates of large earthquakes. An important advantage of the method is the possibility of combining individual forecasts of different classes of statistical models, and models constructed at different spatial and time scales.

The output of the application of the method is always the model forecasting rates of earthquakes, independently of whether the input precursory information is probabilistic or deterministic. The probabilistic properties of the precursors expressed in the statistics of their manifestation are the basis of the differential probability gain approach that makes it possible to combine seismicity models with precursors of any types.

This method can be described as an operational device to identify and combine independent sources of information related to earthquake phenomena. The quality of the combined model does not have to depend on causal relationships between the different models being combined. Actually, the procedure applies to any of the forecast models that have additional forecast skill. Nevertheless, since we cannot formulate the model in terms of traditional classification problems without a loss of generality, the overall performance of a combined forecast model can be only established on purely empirical grounds.

In contrast to linear methods, the differential probability gain procedure does not average the local expected rates of different models. Instead, it redistributes in space the rates of the reference model according to the additional knowledge carried by the input model. Then, as shown by Figure 10.7 for a combined EAST*EEPPAS model, the combined model may cover a larger range of rate values, especially in the limit of high rates that are critical for operational forecast. An essential property of this redistribution process is to keep constant the total expected rate of the reference rate-based model. Nevertheless, if this conservation property is verified for the retrospective period during which the differential probability functions have been determined, it may be approximate only during the quasi-prospective and real-time tests.

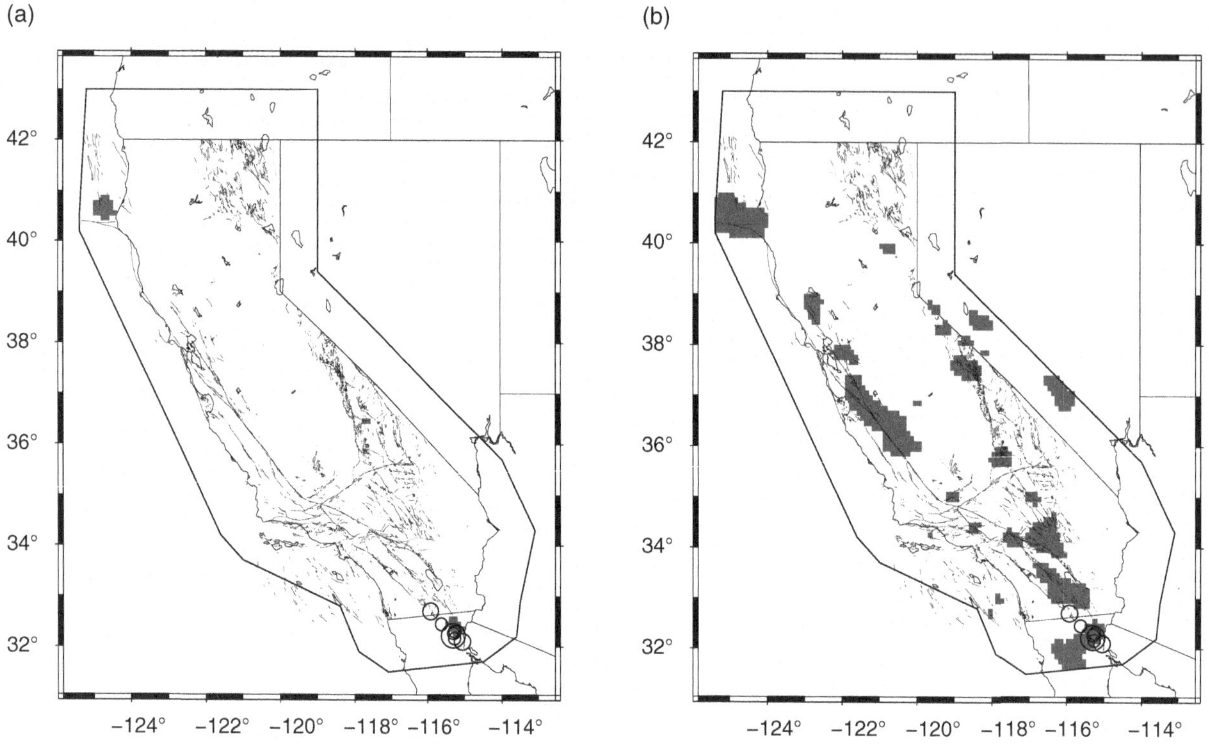

Figure 10.12 Composite model and RI forecasts for the interval 1 April 2010 to 30 June 2010. Red indicates highest rate cells in which 50% of target earthquakes during the 3-month forecasting interval can be expected. For (a) the composite model 50% of target earthquakes are expected in 35 cells (out of 7682), for (b) RI 50% of target earthquakes are expected in 624 cells. Solid line shows the SCEP California testing region. Circles represent epicenters of target earthquakes with magnitude $M \geq 5$ that occurred during the forecasting interval. The largest circle corresponds to the El Mayor-Cucapah M 7.3 earthquake of 4 April 2010. (*See electronic version for color representation.*)

For the composite model with four precursory patterns this redistribution effect with high concentration of expected rates is even more clearly defined. For example, as Figure 10.12 shows, for the interval from 1 April to 30 June 2010, 50% of expected earthquake rates were concentrated in 0.46% of the testing area (including epicenter of El Mayor – Cucapah $M = 7.3$ earthquake of April 4, 2010), while for the reference RI model 50% of the total rate are in 8.2% of highest rate cells. Thus the effective rise of the earthquake probabilities in this case is about 20 times.

The differential probability gain approach can be used to combine successively different forecast models. However, each model brings not only additional information but also some noise. Therefore, changes in the level of noise must be estimated when many models are combined together. In real-time testing, it is practically impossible to quantify the level of noise in individual models because the available case histories are always quite short. Accordingly, making theoretical estimates of the overall noise is not yet possible. Instead, numerical experiments have been performed. The test shows that even after 10 iterations with highly noisy simulated models, the result remains quite similar to the original one.

The differential probability gain combining method is not commutative. To study this, I analyzed two version of the composite model with four precursory patterns obtained with different order of input patterns. Although the two versions are different, they are quite similar. The correlation coefficient for the forecasts in the testing period 1985–2016 is equal to 0.997, the maximum difference of the rates is 18%, the difference in log likelihood is 0.04 per earthquake.

The major difference between error diagrams and likelihood tests resides in the way the rate variable is weighted. Error diagrams are based on the sum of the expected event rates, whereas likelihood tests are based on the sum of the logarithm of these rates. The comparison of Figure 10.8b and c illustrates the difference between these two tests. In error diagrams, the relative performance of two models may be measured by the probability gain. For two rate-based models, this probability gain may also be estimated as the slope of the best-fit line in a cross-distribution of the rates (i.e., the diagram in which the rates of one model are plotted with respect to the other where target earthquakes have occurred). In Figure 10.8b, this slope is close to 2. In addition, we can graphically verify that events at high rates have more weight than events at small rates in determining this slope. In likelihood tests, the relative performance of two models is expressed as the difference of their log likelihoods per earthquake (Equation 2). If the total expected rates are the same in both models, this is equivalent to the average vertical distance to the diagonal in Figure 10.8c. In contrast to Aki's probability gain (Figure 10.8b), this averaging gives the same weight to all rates. As a consequence, positive distances at high rates and negative distances at low rates cancel each other out.

The combining method based on differential probability gains shows promising results compared to individual earthquake forecast models and other linear combination techniques. Overall, this procedure opens new opportunities for operational forecasting by substantially increasing the forecast event rates. It is also possible to imagine applying an iterative application of the method to combine several forecast models of different types using the differential probability gain method. A composite model using four seismic precursory patterns has demonstrated promising results. The combination will be best if the input models and patterns are constructed from different concepts, data, or seismic precursors.

ACKNOWLEDGMENTS

The work was supported by Russian Foundation of Basic Research, Project No. 13-05-00541. I am grateful to two anonymous reviewers for valuable comments.

REFERENCES

Aki, K. (1981), A probabilistic synthesis of precursory phenomena, in D. Simpson and P. Richards (eds), *Earthquake Prediction: An International Review*, pp. 566–574, American Geophysical Union, Washington, DC.

Aki, K. (1996), Scale dependence in earthquake phenomena and its relevance to earthquake prediction, *Proc. Natl. Acad. Sci. USA*, *93*, 3740–3747.

Cicerone, R. D., Ebel, J. E., and Britton, J. (2009), A systematic compilation of earthquake precursors, *Tectonophysics*, *476*(3–4), 371–396.

Eberhard, D. A., Zechar, J. D., and Wiemer, S. (2012), A prospective earthquake forecast experiment in the western Pacific, *Geophys. J. Int.*, *190*(3), 1579–1592.

Evison, F. F. (1982), Generalised precursory swarm hypothesis, *J. Phys. Earth*, *3*, 155–170.

Evison, F. F. and Rhoades, D. A. (2004), Demarcation and scaling of long-term seismogenesis. *Pure Appl. Geophys.*, *161*, 21–45, doi:10.1007/s00024-003-2435-8

Gardner, J. and Knopoff, L. (1974), Is the sequence of earthquakes in S. California with aftershocks removed Poissonian? *Bull. Seismol. Soc. Am.*, *64*(5), 1363–1367.

Gelfand, I. M., Guberman, Sh., Izvekova, M. L., Keilis-Borok, V. I., and Ranzman, E. Ya. (1972), Criteria of high seismicity determined by pattern recognition, *Tectonophysics*, *13*(1–4), 415–422.

Gerstenberger, M. C., Jones, L. M., and Wiemer, S. (2007), Short-term aftershock probabilities: case studies in California, *Seismol. Res. Lett.*, *78*, 66–77.

Gupta, H. K. and Singh, H. N. (1989), Earthquake swarms precursory to moderate to great earthquakes in the northeast India region, *Tectonophysics*, *167*(2–4), 285–298.

Gvishiani, A. D. and Soloviev, A. A. (1984), Recognition of places on the Pacific coast of the South America where strong earthquakes may occur, *Earthquake Predict. Res.*, *2*, 237–243.

Helmstetter, A., Kagan, Y. Y., and Jackson, D. D. (2006), Comparison of short-term and time-independent earthquake forecast models for southern California, *Bull. Seismol. Soc. Am.*, *96*, 90–106.

Jordan, T.H. (2006), Earthquake predictability, brick by brick, *Seismol. Res. Lett.*, *77*, 3–6.

Jordan, T. H. and Jones, L. M. (2010), Operational earthquake forecasting: some thoughts on why and how, *Seismol. Res. Lett.*, *81*, 571–574.

Keilis-Borok, V. I. and Kossobokov, V. G. (1990), Premonitory activation of earthquake flow: algorithm M8, *Phys. Earth Planet. Inter.*, *61*, 73–83.

Keilis-Borok, V. I. and Rotwain, I. M. (1990), Diagnosis of time of increased probability of strong earthquakes in different regions of the world: algorithm CN, *Phys. Earth Planet. Inter.*, *61*, 57–72.

Keilis-Borok, V. I., Knopoff, L., and Rotvain, I. M. (1980), Bursts of aftershocks, long-term precursors of strong earthquakes, *Nature*, *283* (5744), 259–263, DOI: 10.1038/283259a0.

Keilis-Borok, V. I., Shebalin, P. N., and Zaliapin, I. V. (2002), Premonitory patterns of seismicity months before a large earthquake: five case histories in Southern California, *Proc. Natl. Acad. Sci.*, *99*, 16562–16567.

Kossobokov, V. (2006), Testing earthquake prediction methods: the west Pacific short-term forecast of earthquakes with magnitude $M_w \geq 5.8$, *Tectonophysics*, *413*, 25–31.

Kossobokov, V. (2013), Earthquake prediction: 20 years of global experiment, *Nat. Hazards*, *69*(2), 1155–1177.

Kossobokov, V. and Shebalin, P. (2003), Earthquake prediction, in V. I. Keilis-Borok and A. A. Soloviev (eds), *Nonlinear Dynamics of the Lithosphere and Earthquake Prediction*, pp. 141–205, Springer, Berlin–Heidelberg.

Kossobokov, V. and Soloviev, A. (2015), Evaluating the results of testing algorithms for prediction of earthquakes, *Dokl. Earth Sci.*, *460*(2), 192–194.

Kossobokov, V. G., Keilis-Borok, V. I., and Smith S. W. (1990), Reduction of territorial uncertainty of earthquake forecasting, *Phys. Earth Planet. Inter.*, *61*(1–2), R1–R4, doi:10.1016/0031-9201(90)90101-3

Marzocchi, W., Zechar, J. D., and Jordan, T. H. (2012), Bayesian forecast evaluation and ensemble earthquake forecasting, *Bull. Seismol. Soc. Am.*, *102*, 2574–2584, doi:10.1785/0120110327

Molchan, G. (1990), Strategies in strong earthquake prediction, *Phys. Earth Planet. Inter.*, *61*, 84–98.

Molchan, G. (1991), Structure of optimal strategies in earthquake prediction, *Tectonophysics*, *193*, 267–276.

Molchan, G. (2010), Space-time earthquake prediction: the error diagrams, *Pure Appl. Geophys.* *167*(8–9), 907–917, doi:10.1007/s00024-010-0087-z

Molchan, G. (2012), On the testing of seismicity models, *Acta Geophysica*, *60*(3), 624–637, doi:10.2478/s11600-011-0042-0

Molchan, G. and Keilis-Borok, V. (2008), Earthquake prediction: probabilistic aspect, *Geophys. J. Int.*, *173*, 1012–1017.

Narteau, C., Shebalin, P., and Holschneider, M. (2002), Temporal limits of the power law aftershock decay rate, *J. Geophys. Res.*, *107*, doi:10.1029/2002JB001868

Narteau, C., Shebalin, P., and Holschneider, M. (2005), Onset of power law aftershock decay rates in Southern California, *Geophys. Res. Lett.*, *32*, doi:10.1029/2005GL023951

Narteau, C., Shebalin, P., and Holschneider, M. (2008), Loading rates in California inferred from aftershocks, *Nonlin. Proc. Geophys.*, *15*, 245–263.

Narteau, C., Byrdina, S., Shebalin, P., and Schorlemmer, D. (2009), Common dependence on stress for the two fundamental laws of statistical seismology, *Nature*, *462*, 642–645, doi:10.1038/nature08553

Ogata, Y. (1988), Statistical models for earthquake occurrences and residual analysis for point processes, *J. Am. Stat. Assoc.*, *83*, 9–27.

Ogata, Y. (1989), Statistical models for standard seismicity and detection of anomalies by residual analysis, *Tectonophysics*, *169*, 159–174.

Peresan, A., Costa, G., and Panza, G. (1999), Seismotectonic model and CN earthquake prediction in Italy, *Pure Appl. Geophys.*, *154*, 281–306.

Peresan, A., Kossobokov, V. G., Romashkova, L., and Panza, G. F. (2005), Intermediate-term middle-range earthquake predictions in Italy: a review, *Earth Sci. Rev.*, *69*(1–2), 97–132.

Peresan, A., Kossobokov, V. G., and Panza, G. F. (2012), Operational earthquake forecast/prediction, *Rend. Lin. Sci. Fis. Nat.*, *23*(2), 131–138.

Pulinets, S. A., Ouzounov, D. P., Karelin, A. V., and Davidenko, D. V. (2015), Physical bases of the generation of short-term earthquake precursors: A complex model of ionization-induced geophysical processes in the lithosphere-atmosphere-ionosphere-magnetosphere system, *Geomag. Aeron.*, *55*(4), 521–538, doi: 10.1134/S0016793215040131

Rhoades, D. A. (2013), Mixture models for improved earthquake forecasting with short-to-medium time horizons, *Bull. Seimol. Soc. Am.*, *103*, 2203–2215, doi:10.1785/0120120233

Rhoades, D. A. and Evison, F. F. (2004), Long-range earthquake forecasting with every earthquake a precursor according to scale, *Pure Appl. Geophys.*, *161*, 47–72, doi:10.1007/s00024-003-2434-9.

Rhoades, D. A. and Evison, F. F. (2007), Application of the EEPAS model to forecasting earthquakes of moderate magnitude in southern California, *Seismol. Res. Lett.*, *78*, 110–115, doi:10.1785/gssrl.78.1.110

Rhoades, D. A. and Gerstenberger, M. C. (2009), Mixture models for improved short-term earthquake forecasting, *Bull. Seismol. Soc. Am.*, *99*, 636–646, doi:10.1785/0120080063

Rhoades, D. A., Christophersen, A., and Gerstenberger, M. C. (2015), Multiplicative earthquake likelihood models based on fault and earthquake data, *Bull. Seismol. Soc. Am.*,*105* (6), 2955–2968, doi: 10.1785/0120150080

Romashkova, L. L. and Kossobokov, V. G. (2004), Intermediate-term earthquake prediction based on spatially stable clusters of alarms, *Dokl, Earth. Sci.*, *398*, 947–949.

Romashkova, L. L., Kossobokov, V. G., Panza, G. F., and Costa, G. (1998), Intermediate-term predictions of earthquakes in Italy: Algorithm M8, *Pure Appl. Geophys.*, *152*(1), 37–55.

Schorlemmer, D., Gerstenberger, M., Wiemer, S., Jackson, D. D., and Rhoades, D. A. (2007), Earthquake likelihood model testing, *Seismol. Res. Lett.*, *78*, 17–29.

Shebalin, P. (2006), Increased correlation range of seismicity before large events manifested by earthquake chains, *Tectonophysics*, *424*, 335–349.

Shebalin, P. N. and Keilis-Borok, V. I. (1999), Phenomenon of local "seismic reversal" before strong earthquakes, *Phys. Earth Planet. Inter.*, *111*(3), 215–227.

Shebalin, P., Keilis-Borok, V., Zaliapin, I., Uyeda, S., Nagao, T., and Tsybin, N. (2004), Advance short-term prediction of the large Tokachi-oki earthquake, September 25, 2003, $M = 8.1$—A Case History, *Earth Planets Space*, *56*, 715–724.

Shebalin, P., Kellis-Borok, V., Gabrielov, A., Zaliapin, I., and Turcotte, D. (2006), Short-term earthquake prediction by reverse analysis of lithosphere dynamics, *Tectonophysics*, *413*, 63–75.

Shebalin, P., Narteau, C., Holschneider, M., and Schorlemmer, D. (2011), Short-term earthquake forecasting using early aftershock statistics, *Bull. Seimol. Soc. Am.*, *101*, 297–312, doi:10.1785/0120100119

Shebalin, P., Narteau, C., and Holschneider, M. (2012), From alarm-based to rate-based earthquake forecast models, *Bull. Seismol. Soc. Am*, *102*(1), 64–72, doi:10.1785/0120110126

Shebalin, P., Narteau, C., Zechar, J. D., and Holschneider, M. (2014), Combining earthquake forecasts using differential probability gains, *Earth Planets Space*, *66*, 37, doi:10.1186/1880-5981-66-37

Sobolev, G. A., Chelidze, T. L., Zavyalov, A. D., Slavina, L. B., and Nikoladze, V. E. (1991), Maps of expected earthquakes based on a combination of parameters, *Tectonophysics*, *193*, 255–265, doi:10.1016/0040-1951(91)90335-P

Sobolev, G. A., Tyupkin, Y. S., and Smirnov, V. B. (1996), Method of intermediate term earthquake prediction, *Dokl. Akad. Nauk*, *347*, 405–407.

Taroni, M., Zechar, J., and Marzocchi, W. (2014), Assessing annual global *M*6+ seismicity forecasts, *Geophys. J. Int.*, *196*(1), 422–431.

Tiampo, K. F., Rundle, J. B., McGinnis, S., Gross, S. J., and Klein, W. (2002), Mean-field threshold systems and phase dynamics: an application to earthquake fault systems, *Europhys. Lett.*, *60*(3), 481–488.

Tsuruoka, H., Hirata, N., Schorlemmer, D., Euchner, F., Nanjo, K. Z., and Jordan, T. H. (2012), CSEP testing center and the first results of the earthquake forecast testing experiment in Japan, *Earth Planets Space*, *64*(8), 661–671.

Uyeda, S., Nagao, T., and Kamogawa, M. (2009), Short-term earthquake prediction: Current status of seismo-electromagnetics, *Tectonophysics*, *470*(3–4), 205–213.

Vorobieva, I. A. (1999), Prediction of a subsequent large earthquake, *Phys. Earth Planet. Inter.*, *111*, 197–206.

Yamashita, T. and Knopoff, L. (1992), Model for intermediate-term precursory clustering of earthquakes, *J. Geophys. Res.*, *97*(B13), 19873–19879, doi:10.1029/92JB01216

Zechar, J. D. (2010), Evaluating earthquake predictions and earthquake forecasts: a guide for students and new researchers, *Community Online Resource for Statistical Seismicity Analysis*, 1–26, doi:10.5078/corssa-77337879

Zechar, J. and Jordan, T. (2008), Testing alarm-based earthquake predictions, *Geophys. J. Int. 172*, 715–724, doi 10.1111/j.1365-246X.2007.03676.x

Zechar, J. D. and Jordan, T. H. (2010), The area skill score statistic for evaluating earthquake predictability experiments, *Pure Appl. Geophys.*, *167*, 893–906.

Zechar, J. D., Gerstenberger, M. C., and Rhoades, D. A. (2010a), Likelihood-based tests for evaluating space--rate–magnitude earthquake forecasts, *Bull. Seism. Soc. Am.*, *100*(3), 1184–1195, doi:10.1785/0120090192

Zechar, J. D., Schorlemmer, D., Liukis, M., Yu, J., Euchner, F., Maechling, P. J., and Jordan, T. H. (2010b), The collaboratory for the study of earthquake predictability perspective on computational earthquake science, *Concurr. Comput. Pract. Exp.*, *22*, 1836–1847.

Part IV: Ground Geochemical and Electromagnetic Observations of Major Seismicity

11

Continuous Monitoring of Fluid and Gas Geochemistry for Seismic Study in Taiwan

Ching-Chou Fu and Lou-Chuang Lee

ABSTRACT

The purpose of this chapter is to present the relationship between geochemical anomalies and major seismicity in order to improve the possibility of forecasting an impending earthquake. The establishment and development of geochemical monitoring in Taiwan are introduced. From a dataset covering several years of continuously monitored soil gas, some anomalously high Rn concentrations at specific times were identified using a statistical technique. We found that many Rn anomalies were observed in soil gas prior to the occurrence of impending earthquakes. Each station is sensitive to only the seismic events in its specific surrounding area. Comparing the preseismic events at different monitoring stations, we can further identify the sensitivity zone of each monitoring system and then may be able to propose the potential area of the epicenter of the impending earthquake. The relative peaks of soil-gas anomalies, at two or more stations at their respective occurrence times, could eventually enable satisfactorily precise predictions of the location, time, and magnitude of an impending earthquake. The long-term monitoring of gas or fluid composition at appropriate sites or near fault zones is helpful for understanding the mechanism and process of fault and earthquake activity.

11.1. INTRODUCTION

Earthquake precursors are difficult to recognize because of the complexity of the Earth's crust, the various types of earthquake mechanisms, and the lack of extensive geophysical and geochemical monitoring in most areas. On the other hand, both the precursory anomalies and the earthquakes may result from the same broad-scale but small incremental stress changes in an inhomogeneous crust. Thus, systematic characterization of all possible precursors may help us to better understand the mechanisms and processes of an earthquake [*Biagi et al.*, 2000]. Geochemical and hydrological signals preceding significant earthquakes have been reported since the 1960s and have been used for earthquake prediction, particularly in China, Japan, Russia, eastern Europe, and the United States [*King*, 1986]. Changes in gas compositions, including He/Ar, CH_4/Ar, N_2/Ar, and $^3He/^4He$ ratios, have also been proposed as potential precursors [*Sugisaki*, 1978; *Kawabe*, 1985; *Sano et al.*, 1998; *Italiano and Martinelli*, 2001]. Comparing the precursory anomalies associated with the 1995 Kobe and the 1978 Izo-Oshima earthquakes, *Silver and Wakita* [1996] found that co-varying preseismic signals in ground-water level, Rn and Cl^- concentrations, and strain were observed at many monitoring stations. They concluded that many preseismic anomalies were true precursors, although the anomalies may exhibit a heterogeneous response.

Bolt [1999] identified five evolutionary stages in the development of an earthquake-prone region. These are (a) the build up of elastic strain; (b) dilatancy and development of microfractures; (c) influx of water and unstable deformation in the fault zone; (d) the earthquake; and (e) a sudden drop in stress followed by aftershocks.

Institute of Earth Sciences, Academia Sinica, Taipei, Taiwan

Pre-Earthquake Processes: A Multidisciplinary Approach to Earthquake Prediction Studies, Geophysical Monograph 234, First Edition. Edited by Dimitar Ouzounov, Sergey Pulinets, Katsumi Hattori, and Patrick Taylor.

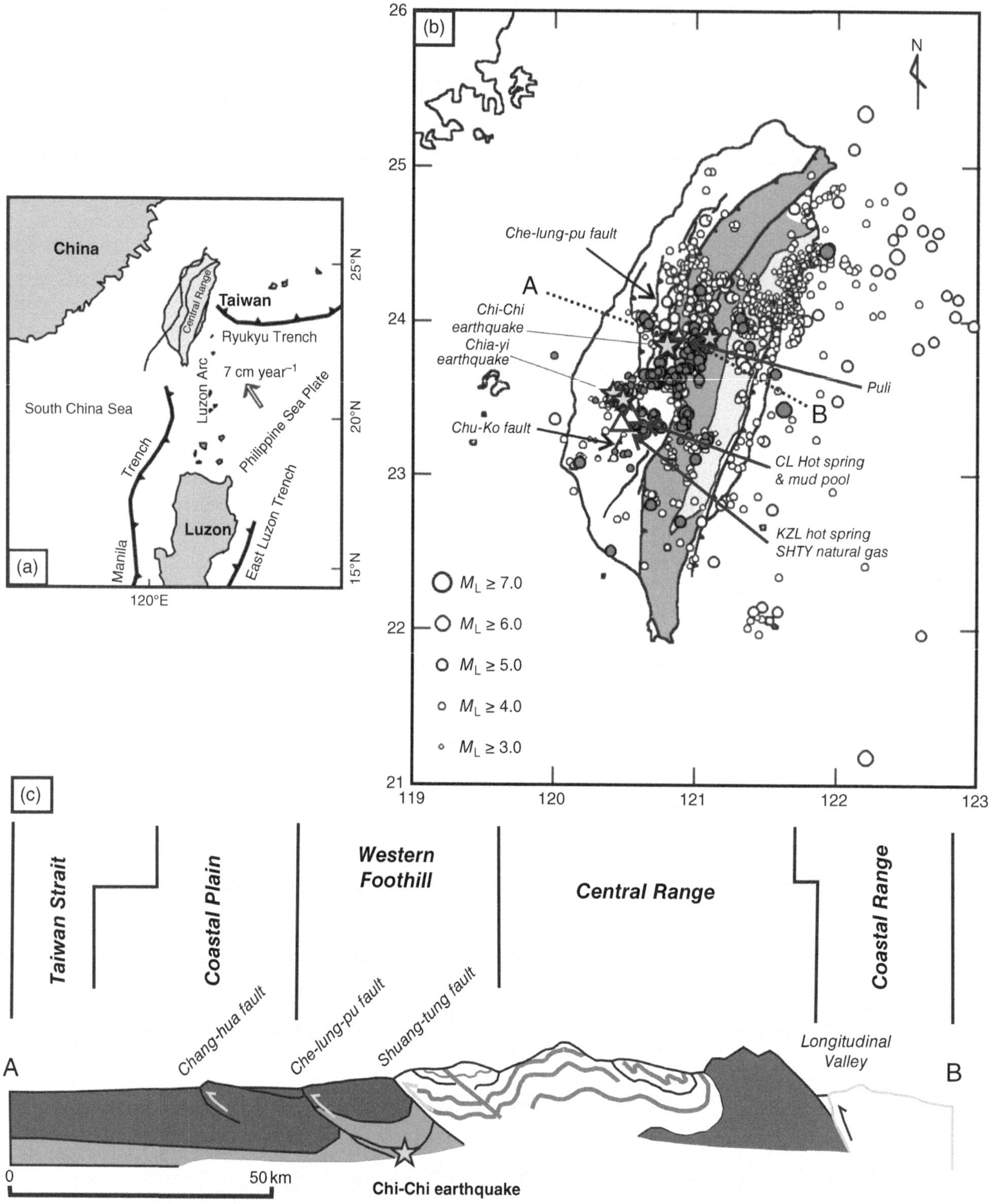

Figure 11.1 Tectonic settings and the earthquakes that have occurred around Taiwan from 1 September to 31 December 1999. (a) Simplified tectonic features around Taiwan. (b) Distributions of the epicenters of M_L 3.0 earthquakes recorded around Taiwan: red stars indicate the epicenters of the main shocks discussed that occurred during this period; shaded circles indicate earthquakes that may have shaken the monitoring stations; earthquake data are taken from the website of Central Weather Bureau, R.O.C. (http://www.cwb.gov.tw); CL, Chung-lun; KZL, Kuan-tze-ling; SHTY, Suei-ho-tong-yuan. (c) Geological profile along the A-B cross-section shown in Figure 11.1b. (*See electronic version for color representation.*)

If the various precursory anomalies mentioned above are triggered to occur at specific stages, one would expect to be able to correlate the observed precursory signals with the different evolutionary stages of a major earthquake, such as the 1999 Chi-Chi *M* 7.3 earthquake in Taiwan.

Systematic Rn monitoring as a potential tool for earthquake prediction had been performed continuously since 1980 in northern and northeastern Taiwan [*Liu et al.*, 1985]. However, the monitoring project was terminated in 1990 due to shortage of funds. After the Chi-Chi earthquake, a new phase of geochemical monitoring in southwestern Taiwan started in late 1999. Many hot springs and mud volcanoes were chosen for regular collection of fluid samples and were considered suitable for monitoring the activity of faults or earthquakes in this area (Figure 11.1). Among them, the Chung-lun (CL) and Kuan-tze-ling (KZL) hot springs are located near the Chu-Ko fault zone, which is considered to be one of the faults with the greatest potential for reactivation in southwestern Taiwan.

11.2. GEOCHEMICAL ANOMALIES AS PRECURSORS TO SEISMIC EVENTS

11.2.1. Hydrochemical Precursors

Utilizing knowledge gained from the 1995 Kobe earthquake [*Igarashi et al.*, 1995; *Tsunogai and Wakita*, 1995] and the 1996 Pyrenean earthquake [*Toutain et al.*, 1997], *Song et al.* [2003] collected commercially bottled groundwater from Puli, which is located approximately 20 km to the northeast of the 1999 Chi-Chi earthquake's epicenter. The water was collected immediately after the earthquake in order to investigate possible changes in groundwater composition. Beginning in mid-April, the concentrations of SO_4^{2-} and NO_3^- increased rapidly, fluctuated, and peaked at the end of June, 3 months before the Chi-Chi earthquake. The highest measured concentrations were approximately 77% to 106% higher than the background values. Subsequently, they gradually decreased with some fluctuations and reached lowest levels, approximately 44% and 32% lower than the background value in mid-October, about 1 month after the earthquake. Comparing these results with the rainfall data of Puli clearly indicated that the variations in NO_3^- and SO_4^{2-} concentration could not be explained by variations in precipitation (Figure 11.2). Hence, those variations are proposed to reflect pre-earthquake changes in the groundwater composition in the Puli area. Similarly, *Grant et al.* [2011] reported chemical anomalies in spring water prior to the 2011 Van *M* 7.1 earthquake. Hydrochemical changes observed before an earthquake can be attributed to preseismic stress and strain changes. Due to the earthquake, increasing stress might cause collapse of rock pores, the opening of cracks, or a change in the local rock permeability. This leads to an expulsion of pore fluids with high concentrations of SO_4^{2-} and NO_3^- into the circulating groundwater system, resulting in an increase in SO_4^{2-} and NO_3^- concentrations in groundwater. Subsequent strain changes may cause different aquifers to mix [*Italiano and Martinelli*, 2004], resulting in variable concentrations of SO_4^{2-} and NO_3^-. Alternatively, it has been suggested that stress-activated electronic charge carriers flowing out of the hypocenter cross the rock-to-water interface and produce verifiable electrochemical reactions [e.g., *Balk et al.*, 2009; *Freund*, 2011; *Grant et al.*, 2011]. With flowing electronic charge carriers, the rock is subjected to stress and releases major cations and anions into the water.

Similar hydrochemical variations were found at the Kuan-tze-ling hot spring [*Song et al.*, 2006]. Cl^- concentrations increased a few days before the Chi-Chi earthquake, although SO_4^{2-} concentrations did not change significantly. These anomalous chemical events are closely correlated with the earthquakes (Figure 11.3) recorded by the seismic station near the monitoring site that detected the intensity of each earthquake.

11.2.2. Geochemical Precursors of Hot Spring Gases

From the compiled available data, at least four gas components can be identified in this area (Figures 11.4 and 11.5):

1. mantle-derived CO_2 with high helium isotopic ratios (~7 R_a);
2. crustal CO_2 with low helium isotopic ratios (< 0.2 R_a);
3. crustal CH_4 with low helium isotopic ratios (< 0.2 R_a) and high helium contents (> 50 ppm);
4. dissolved air in saturated groundwater with $^3He/^4He \equiv 1$ R_a, and very low helium contents (< 1 ppm).

Most gas compositions of samples collected from this area can be explained by mixing the above components in different proportions.

Before the Chi-Chi earthquake, bubbling gases of the CL hot spring contained 42% CO_2 and 27.8% CH_4 with unexpectedly high $^3He/^4He$ ratios ($^3He/^4He = 4.0–5.9$ R_a). The high ratios indicate that a mantle source component may play an important role as a gas source in this area [*Yang et al.*, 2005a]. This in turn implies that the Chu-ko fault (Figure 11.1) may be a very deep fault, which may serve as a pathway to the surface for gases with mantle signatures. In contrast, the gases of the KZL hot spring exhibit much lower $^3He/^4He$ ratios (0.34–0.57 R_a), indicating that atmospheric air and crustal gases are major source components (Figure 11.4). Although both hot springs are close to the Cho-ko fault, they obviously belong to different source reservoirs.

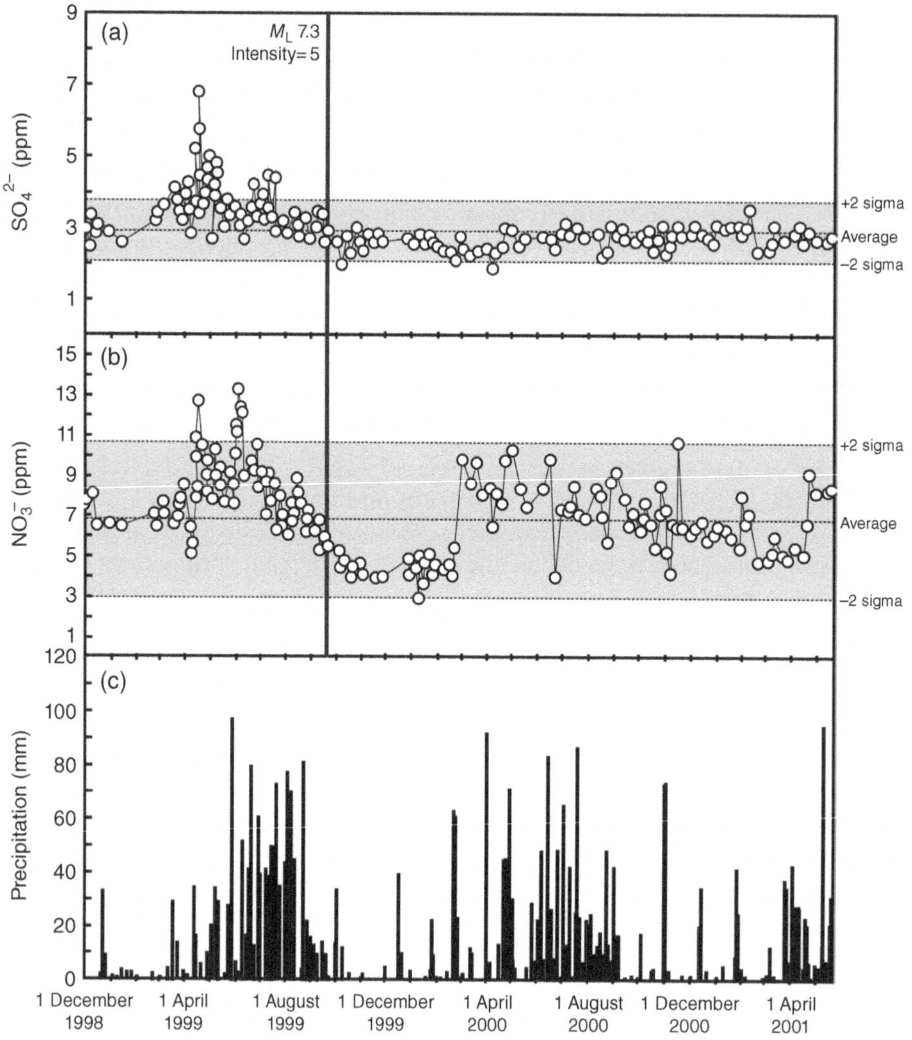

Figure 11.2 Variation of (a) SO_4^{2-} and (b) NO_3^- of the Puli groundwater before and after the 1999 Chi-Chi earthquake. (c) Precipitation is the daily amount obtained in the Puli area [*Song et al.*, 2003]. (Source: Reproduced with permission of Terrestrial, Atmospheric and Oceanic Sciences.)

The helium isotopic compositions of the CL hot spring gases decreased significantly (from 4.2 to 2.9 R_a) after the Chi-Chi main shock (Figure 11.5). Similar anomalous variations of $^3He/^4He$ ratio were also observed at the KZL hot spring (Figure 11.5). Thus, the gas was obviously mixed with a crustal component possessing a high CO_2 concentration with a very low $^3He/^4He$ ratio (< 0.1 Ra) or with an atmospheric air component with a low $^3He/^4He$ ratio (1 R_a) due to the earthquake.

Another earthquake at Chia-Yi (M_L 6.4) with a hypocenter depth of approximately 12 km (Figure 11.1) struck southwestern Taiwan on 22 October 1999 at 10:19 local time. Its epicenter was closer to both hot springs. Hence, it was likely that these hot springs would display significant variations in gas compositions. The $^3He/^4He$ ratio of CL gas dropped dramatically from 3.7 to 0.7 R_a 1 week before the event. Subsequently, the ratio fell to 0.13 R_a indicating that it was clearly dominated by crustal components. These variations in gas composition are closely related to the earthquakes. Even though continuously collected gas composition data prior to the Chi-Chi earthquake is lacking, clear changes in gas composition observed a few days before the Chia-Yi earthquake and its aftershock demonstrated that a change in gas compositions is likely a precursor. Similarly, *Yen et al.* [2004] reported that geomagnetic fluctuations were found at a distance of 42 km from the southern end of the Chelungpu fault and 30 km from the Chia-Yi earthquake. These geomagnetic disruptions with highly anomalous amplitudes were associated with the Chi-Chi and Chia-Yi earthquakes.

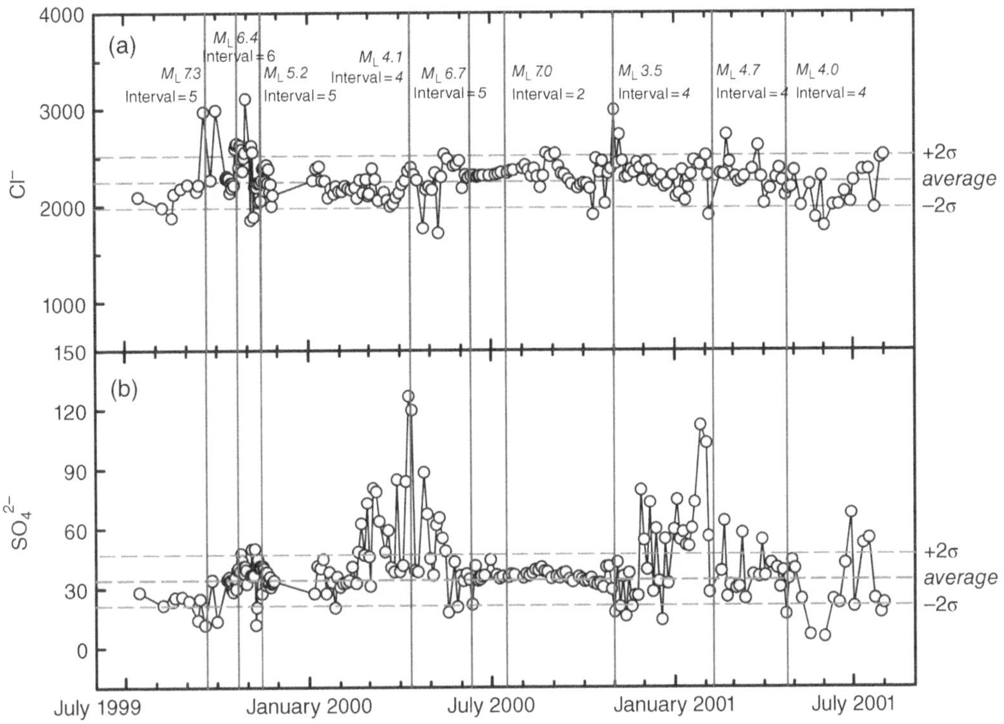

Figure 11.3 Variation of (a) Cl^- and (b) SO_4^{2-} of the Kuan-tze-ling hot spring before and after the 1999 Chi-Chi earthquake. The vertical lines indicate the major earthquakes that occurred during this period [*Song et al.*, 2006]. (Source: *Song et al.* [2006]. Reproduced with the permission of Springer.)

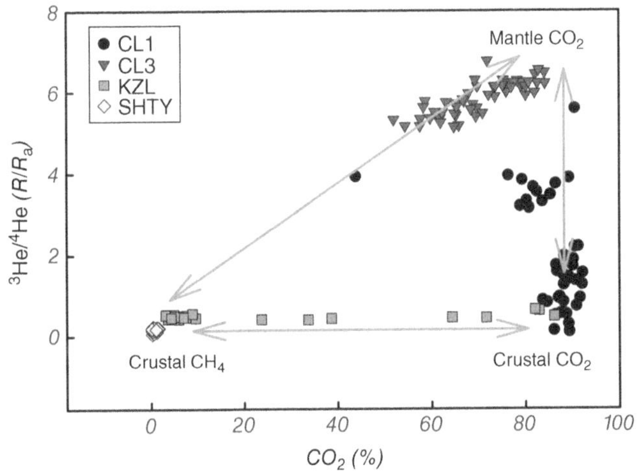

Figure 11.4 The three components, mantle CO_2, crustal CO_2, and crustal CH_4 can be identified in the studied area. Clearly, CO_2 originated from two different sources, namely the mantle and crust. However, the CH_4 from natural gases in southwestern Taiwan mainly originates in the crust. Hence, all gas compositions are clearly explained by mixing the different components. CL1, Chung-lun hot spring; CL3, Chung-lun mud pool; KZL, Kuan-tze-ling hot; SHTY, Suei-ho-tong-yuan fire and water spring.

Except for samples from the CL and KZL areas, CH_4-dominated samples usually showed relatively constant gas compositions. This indicates that they mainly originated from the crustal CH_4 reservoir and were not disturbed by other components, even major earthquake events that occurred. In contrast, bubbling gases from CL and KZL hot springs have larger variations indicating that they were mixed from different gas components. The CL hot spring gases, especially, showed extreme variations and were clearly associated with earthquake events during the studied period. This suggests that natural gases originated from multiple sources in the suitable tectonic environment, and the variations in gas compositions may be sensitive to earthquake events. Hence, the CL area has been considered as one of the best locations in southwestern Taiwan for monitoring gas variations related to earthquakes [*Yang et al.*, 2003, 2005b]. Consequently, a continuous gas monitoring station, mainly equipped with a quadrupole mass spectrometer (QMS), has been set up for this purpose [*Yang et al.*, 2005b, 2006].

Considering the variation in gas composition and its sensitivity to earthquakes, the CO_2/CH_4 ratio can amplify the individual CO_2 and CH_4 variations and minimize the influence of potential air contamination (both occur in trace amounts in atmospheric air). With reference to the background

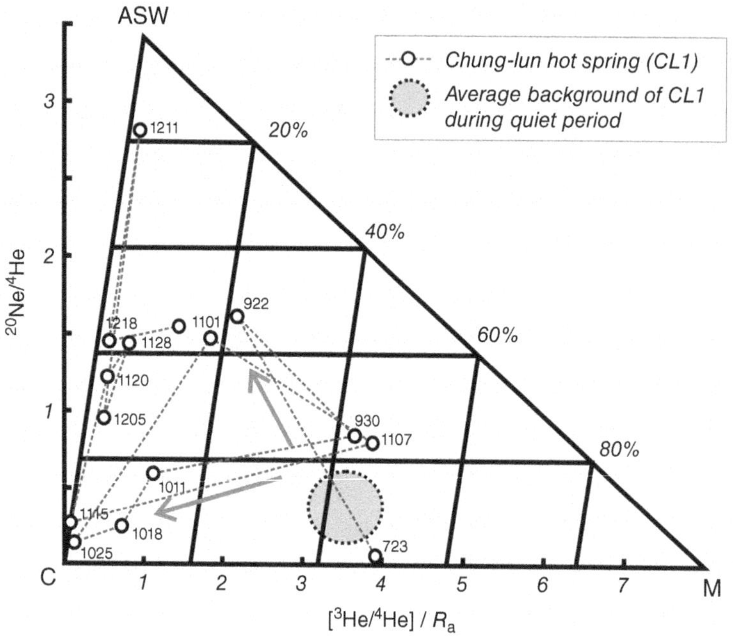

Figure 11.5 A three-component plot of helium isotopes of the gas compositions in southwestern Taiwan: ASW, air saturated water; C, crust; M, mantle. The gas compositions of the Chung-lun hot spring show the most diverse variations. At least three components are necessary to account for the gas compositions. It is worth noting that the gas composition shows significant variations before the major earthquakes, which occurred on 21 September (*M* 7.3), 22 October (*M* 6.4), and 15 November (*M* 5.1).

values, 23 anomalous variation peaks were identified for further correlation with relevant earthquakes, as shown in Figure 11.6 [*Yang et al.*, 2006]. However, anomalously high O_2 concentration occasionally occurred, which was probably due to the production of H_2O_2 by an electrochemical interface reaction [*Balk et al.*, 2009]. Interestingly, this phenomenon is always accompanied by heavy rainfall. In contrast to the decreasing CO_2 concentration, CH_4, C_2H_6, N_2, and Ar concentrations increased suddenly in association with the O_2 anomalies. This implies that local heavy rain may have disturbed the mud-pool. Therefore, the bubbling gas may be contaminated by atmospheric air or dissolved air (O_2, N_2, and Ar), and by released organic gases (CH_4 and C_2H_6) through the disturbance of the rainwater, from the organic-rich sediments at the bottom of the mud-pool. Meanwhile CO_2 concentration decreased because of the dilution effect [*Yang et al.*, 2006].

11.3. MONITORING NETWORK OF SOIL-GAS GEOCHEMISTRY FOR EARTHQUAKE IN TAIWAN

11.3.1 Principle of the Soil-Gas Monitoring Method

The method of selecting sites for continuous soil-gas monitoring is important. Active fault zones usually have higher permeability than surrounding strata and therefore can provide conduits for gases originating from the deep crust or mantle to migrate upward to the surface. In addition, it is a common phenomenon that the mantle-derived gases exhibit higher concentration and higher flux rate due to greater strains [*Trique et al.*, 1999; *Pulinets and Dunajecka*, 2007]. General overviews of the geochemical, structural, and seismic features in tectonically active areas have shown some evidence of correlation between soil-gas anomalies and tectonic activities [*Toutain et al.*, 1992; *Ciotoli et al.*, 1998; *Toutain and Baubron*, 1999; *Fu et al.*, 2008; *Yuce et al.*, 2017]. Soil gases useful for earthquake prediction include the noble gases radon and helium, which play important roles in fault delineation and earthquake precursory studies [*Claesson et al.*, 2004; *Chyi et al.*, 2005; *Walia et al.*, 2005a,b; *Yang et al.*, 2006; *Kumar et al.*, 2009a; *Reddy and Nagabhushanam*, 2011]. Hence, soil-gas samples were systematically collected from a 200–300 m grid in the studied area [*Fu et al.*, 2008, 2009]. After the soil-gas survey, a sensitive site on the active fault zone, identified by high concentrations in both major and trace gases, was chosen for long-term monitoring (Figure 11.7). Based on the soil-gas data, several automatic soil-gas monitoring stations were constructed at appropriate sites near the fault zones around Taiwan, including the CL, Tapingti (TPT), Gukeng (GK), Pingtung (PT), and Chihshang (CS) stations.

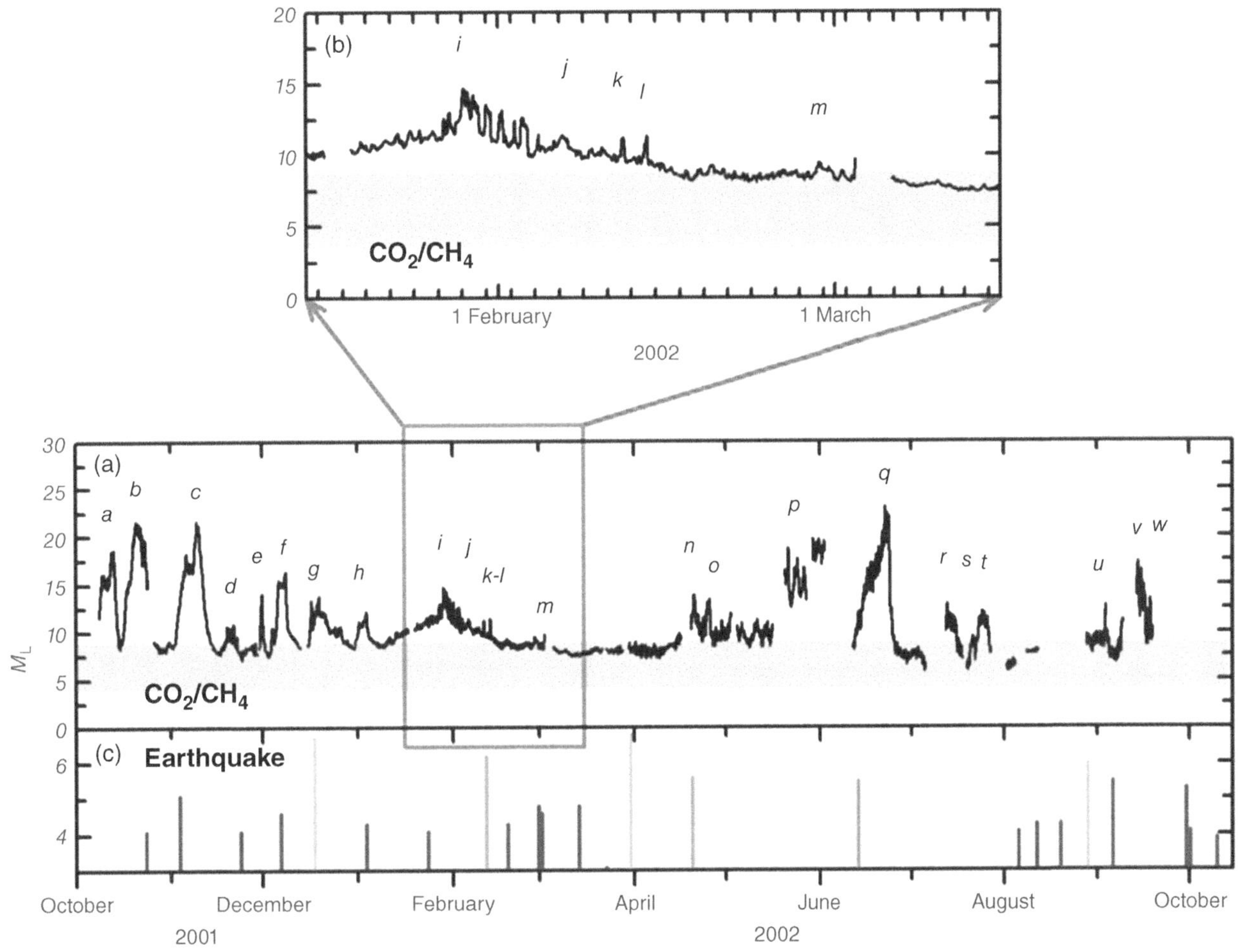

Figure 11.6 Variations of CO_2/CH_4 ratios of bubbling gases from the Chung-lun mud-pool and associated related earthquake events. Twenty-three anomalous peaks (marked with letters) can be recognized [*Yang et al.*, 2006]. (Source: *Yang et al.* [2006]. Reproduced with the permission of Springer.)

11.3.2. Continuous Measurement Setup

For the above mentioned reasons, an automatic CL station for soil-gas monitoring was set up on an active Chuko fault zone near the CL mud-pool in order to avoid possible meteorological effects on the bubbling gas compositions. In addition to the Rn and a QMS, the station is also equipped with a strong motion accelerograph and a thermohygrometer–ombrometer in order to detect the intensity of earthquakes and collect local meteorological data over time. The main exhaust pipe for soil gases, 2.3 m long and 10 cm in diameter, is made of polyvinyl chloride (PVC) with its lower end merging with an artificial ditch. The ditch is lined with a layer of gravel 50 cm in thickness, which facilitates additional collection of soil gas in our self-designed system and draining the water away on rainy days. The water trap and electric gas cooler are used to maintain a relatively low humidity when gases are internally pumped into the detector. Soil gases and meteorological data are recorded once every 15 min and transmitted to the laboratory. This method is similar in design to that described in detail by *Walia et al.* [2009] and *Fu et al.* [2017a, 2017b].

11.4. ANALYSIS OF DATA

Rn anomalies are usually recognized by using the $\pm 2\sigma$ method in which the anomaly thresholds are calculated using the mean value and two standard deviations [*Fu et al.*, 2005, 2008; *Ghosh et al.*, 2007; *Walia et al.*, 2009]. In this study, we use a residual signal processing technique in order to diminish the effect of recording regular and frequent data. For our analysis, a residual signal of concentration, dA, was determined as the difference between the observed signal intensity (soil Rn concentration) and the average of days preceding the observation day:

$$dA(t) = A(t) - RA(t) \tag{11.1}$$

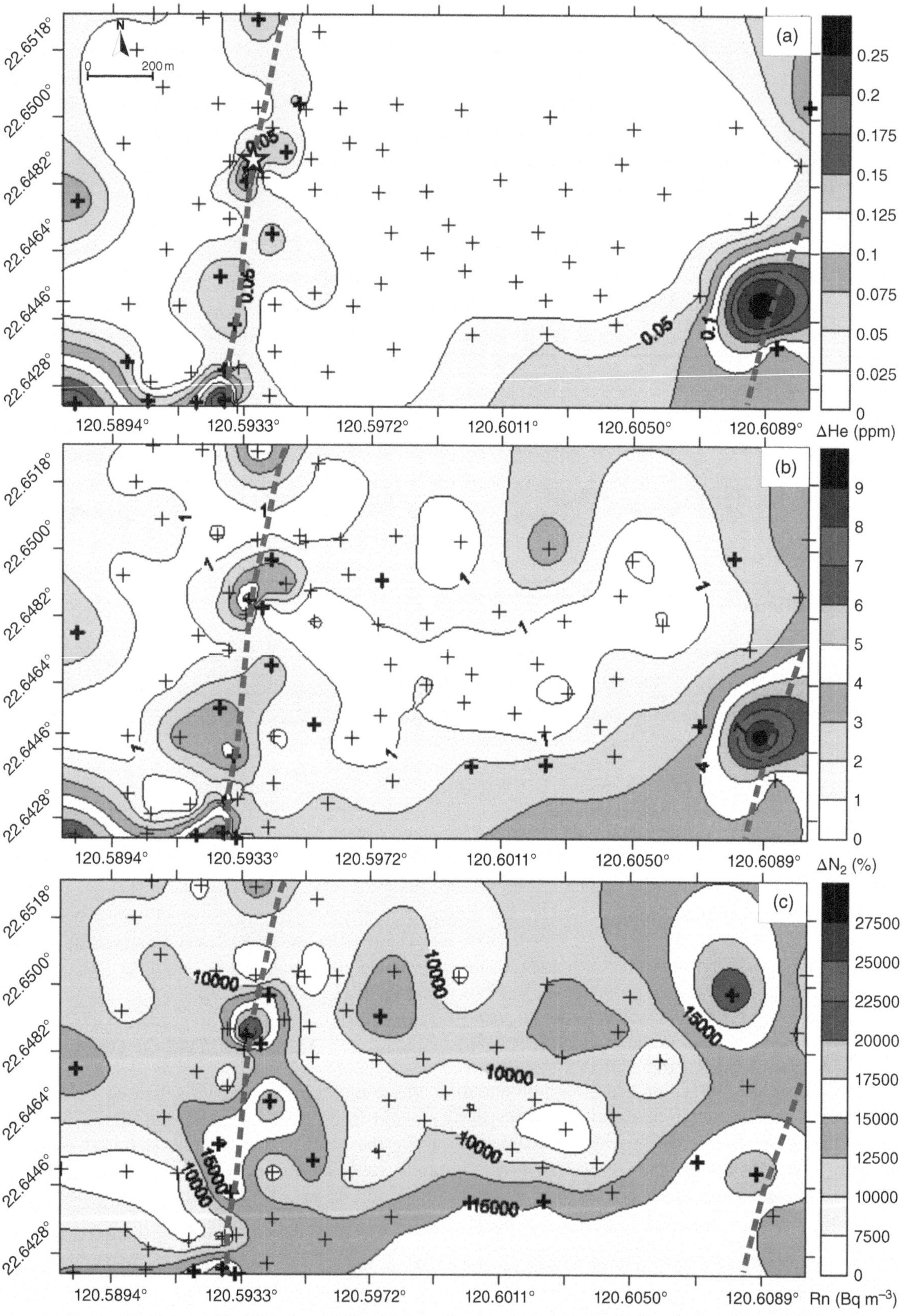

Figure 11.7 Distribution of soil-gas anomaly sites in the study area for (a) He, (b) N_2, and (c) Rn. Cross symbols indicate the sampling sites for the soil-gas survey in this study; bold cross symbols indicate the sites that exhibit anomalous gas concentration; dashed lines indicate the fault distribution; star indicates the location of the station that was chosen based on high concentrations of both major and trace gases [*Fu et al.*, 2008]. (Source: *Fu et al.* [2006]. Reproduced with the permission of Elsevier.)

where $A(t)$ is the trend of soil Rn concentration at a time, t, for an observed day and $RA(t)$ is the rolling average at the same time, t, for −14 days (14 days before the observed day).

For example, the time series data of soil Rn from 1 December 2012 to 1 December 2013 at Chihshang station are shown later in Figure 11.11. The standard deviation of the residual results between the trend and the rolling average of soil Rn concentration was determined (Figure 11.8). An anomalous threshold value was then defined using a similar figure when the standard deviation of $dA(t)$ exceeded 1.5, which was used to identify impending large seismic events ($M_L \geq 5.5$). Except for the final earthquake event, each large earthquake (Figure 11.8 black arrows) was associated with one precursory anomalous peak (Figure 11.8 grey arrows) occurring 2–3 weeks previously. It is noted that some small earthquakes occurring close to the monitoring site may be effective despite the fact that their magnitudes were smaller than 5.0.

11.4.1. The Monitoring Network and its Working Hypothesis

During the studied period, potential precursory signals were recorded at each monitoring station in Taiwan for some earthquakes that occurred in the specific region. Moreover, a sensitive zone for probable earthquakes of approximately 120 km radius was established based on the soil-gas anomalies from each station (as shown in Figure 11.9, and also later in Figure 11.13) [*Yang et al.*, 2005b, 2011]. Similar to concepts previously proposed [*Hartmann and Levy*, 2006; *Chaudhuri et al.*, 2011; *Walia et al.*, 2013], an earthquake preparation area, which occurs within the potential intersection area from individual circles of each monitoring station, may produce similar anomalous variations during the same period. Nevertheless, if an anomaly appears only at one of the stations, an impending local earthquake with a small magnitude (approximately $M_L < 4.5$) is likely to occur

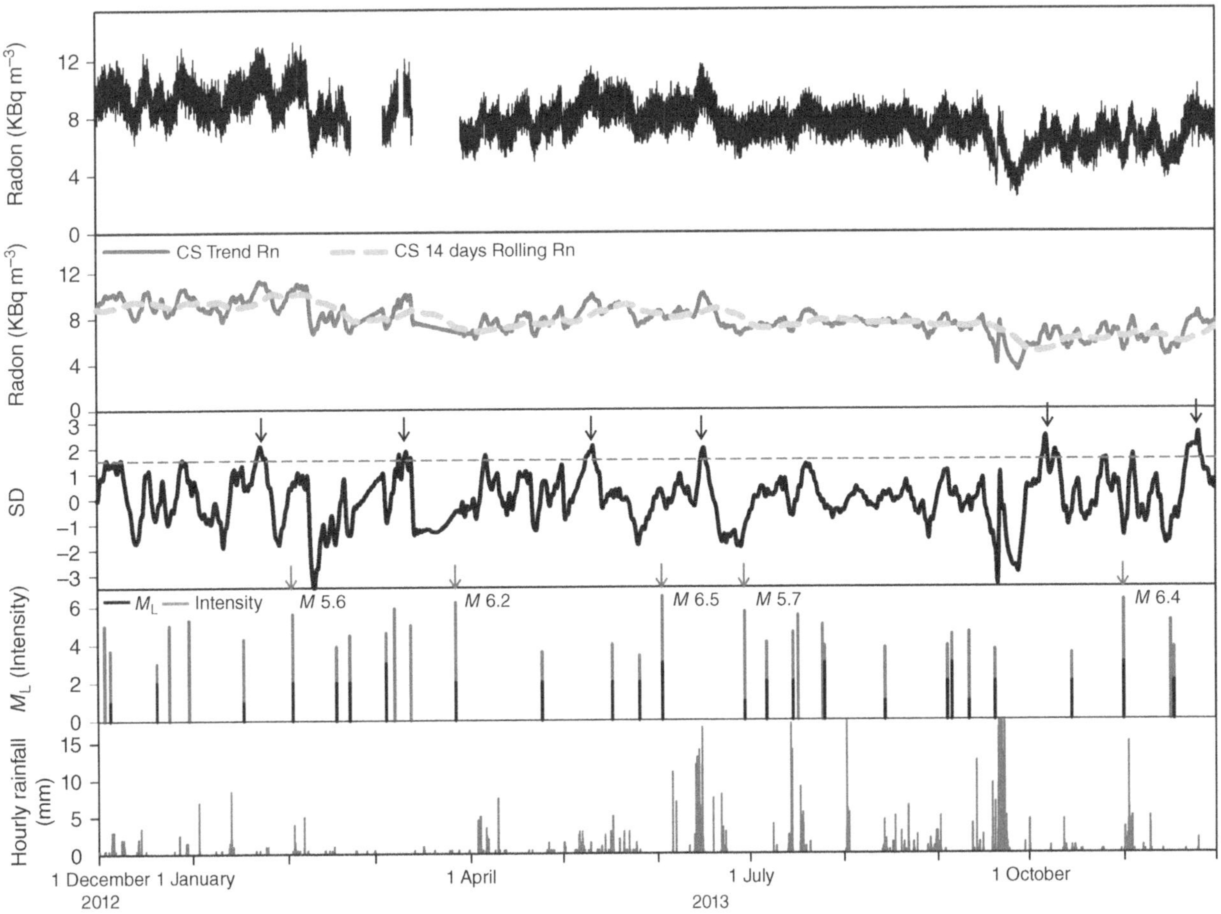

Figure 11.8 Variation of the soil-gas data, signal trend and rolling average of soil Rn concentration, threshold values, seismic events, and rainfall at Chihshang station (CS) from 1 December 2012 to 1 December 2013. The horizontal dashed line indicates the threshold value (+1.5σ); black arrows indicate the highest anomalous points; grey arrows indicate the $M_L \geq 5.5$ seismicity events.

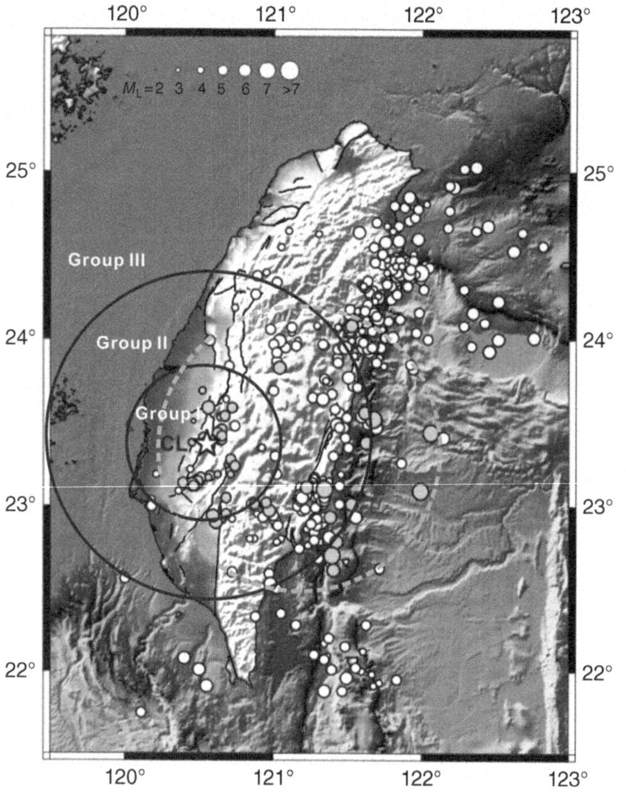

Figure 11.9 Seismicity distributions and Group I earthquakes with epicenter distances of ≤ 30 km; Group II earthquakes with epicenter distances between 60 and 120 km; and Group III earthquakes with epicenter distances ≥ 120 km. The dashed circle shows an observed precursory radon anomaly that indicates with high probability an impending earthquake is likely to be located in the area. It could be considered a sensitive zone for probable earthquakes of the CL station.

near the monitoring station. According to this working hypothesis, variations of soil Rn occurring at more than two stations almost simultaneously may indicate the approximate location of the impending earthquake with high confidence.

11.5. SOME CLASSIC SOIL-GAS PRECURSORS TO EARTHQUAKES IN TAIWAN

Significant precursory variations of soil-gas concentrations are recorded approximately 7 days before local earthquakes, and then soil-gas concentrations return to background values (see Figure 11.10 for a good example). Furthermore, 488 earthquakes were recorded around Taiwan from 1 January 2003 to 31 July 2004 (Figure 11.9). Thirty earthquake events (M_L > 4.5 and intensity > 2) were considered meaningful with respect to the CL monitoring site. The earthquakes were divided into three groups based on distance between epicenters and the monitoring site: 14 "effective" earthquakes were placed in Group I (epicenters ≤ 60 km and M_L ≥ 4.5), 11 into Group II (60–120 km and M_L ≥ 5.0), and five into Group III (≥ 120 km and M_L ≥ 5.5). This indicates that only an earthquake occurring close to the monitoring site was effective unless its magnitude was > 5.0. Except for four earthquakes, each earthquake was correlated with one precursory anomalous peak in the 1–20 day period before the earthquake [*Yang et al.*, 2005b]. In other words, a corresponding precursory Rn–Tn anomaly occurred a few days before all the earthquakes classified as Group I, but only 80% of earthquakes in Groups II and III had associated precursors.

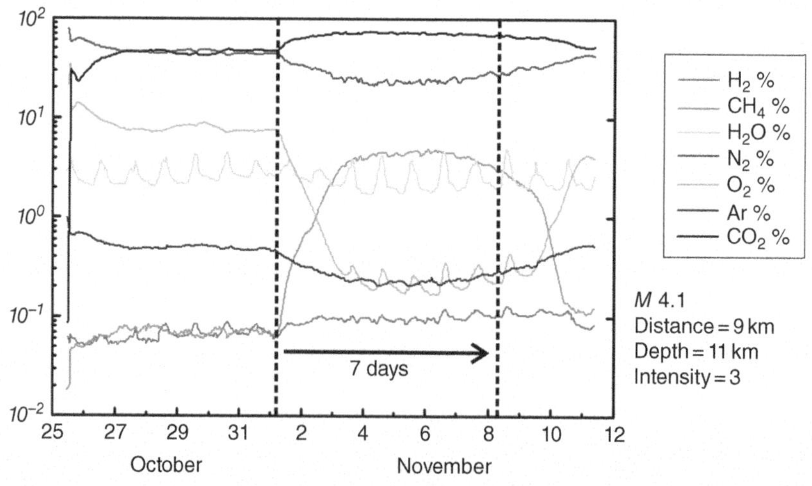

Figure 11.10 Variations in concentrations of soil gases at the Chung-lun station from 25 October to 17 November 2002, showing a precursory anomaly before the local earthquake on 8 November 2002. Each colored line represents a different soil-gas component measured. (*See electronic version for color representation.*)

In recent years, there have been an increasing number of studies focusing on the utilization of gas measurements in Taiwan. For example, Rn anomalies have been observed in soil gas prior to an earthquake and hence their potential as an earthquake precursor has been recognized [*Chyi et al.*, 2001, 2007, 2010, 2011; *Fu et al.*, 2005, 2008, 2009, 2015, 2017a, 2017b; *Kumar et al.*, 2009a, 2009b; *Walia et al.*, 2009, 2010, 2013; *Yang et al.*, 2005b, 2006]. Geochemical monitoring for earthquake studies at the established soil-gas stations along fault zones around Taiwan has been carried out continuously since 2002.

Precursory soil-gas anomalies have been recognized by previous studies that suggest that the precursory signals are distinguished as real fluctuations from background values (derived from annual or seasonal averages) before an earthquake [*Fu et al.*, 2009; *Walia et al.*, 2009]. For example, *Walia et al.* [2009] found that soil-gas variations showed close correlations with the impending offshore twin earthquakes of 26 December 2006. These earthquakes occurred at a depth of approximately 44 and 52 km and had magnitudes of 7 with local intensities of 4 (Figure 11.11), which is the level of the largest earthquakes experienced in southern Taiwan. The Rn concentration varied around 40 kBq m^{-3}, deviating approximately 15 kBq m^{-3} in normal conditions. Rn concentrations steadily increased after 21 November 2006, reached a maximum of approximately 80 kBq m^{-3} on 3 December 2006 (more than 2σ the values for the observation period; shown by the horizontal solid line in Figure 11.11), and subsequently decreased to normal values of approximately 50 kBq m^{-3}. These trends of increasing, peaking, and declining to normal values continued until the final anomaly on 16–17 December 2006, approximately

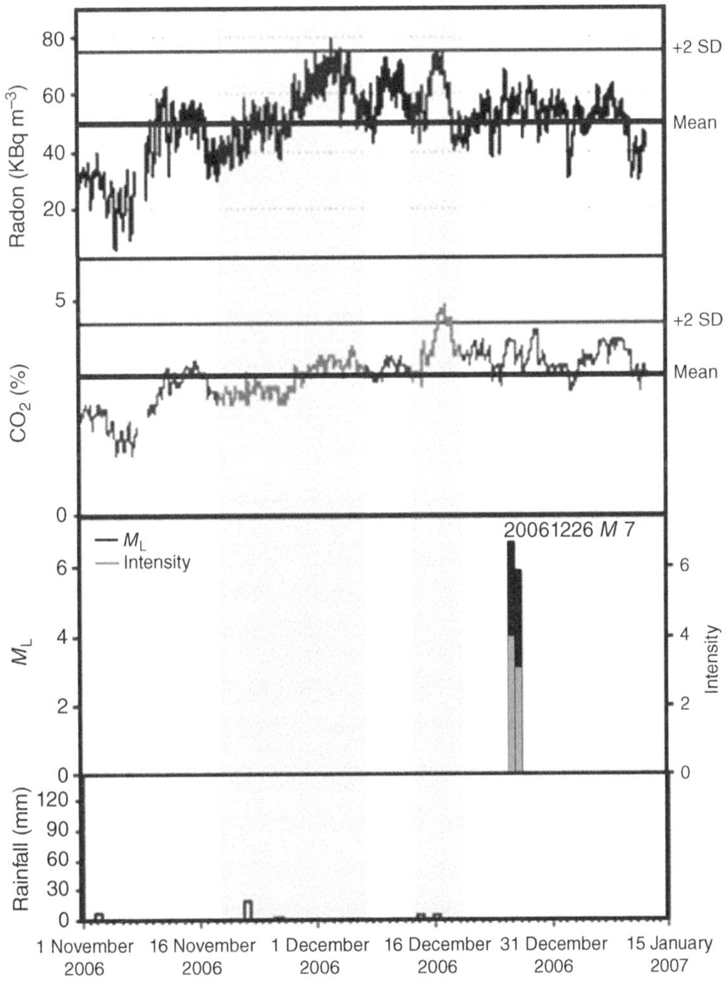

Figure 11.11 Variations of Rn, CO_2, and rainfall at the Hsinhua monitoring station and their association with twin earthquakes on 26 December 2006. The vertical yellow bands indicate the increasing trend in gas concentrations during the observation period [*Walia et al.*, 2009]. (Source: *Walia et al.* [2009]. Reproduced with the permission of Elsevier.) (*See electronic version for color representation.*)

10 days before the earthquake. The CO_2 carrier gas for Rn in the region also showed anomalous values on 16–17 December 2006 (Figure 11.11). These high values of Rn and CO_2 observed at the monitoring station before the 26 December 2006 Pingtung M 7.0 earthquake were precursory signals that developed due to the build up of stress in the region before the main event.

A gamma-ray station has been established recently along the Longitudinal Valley in eastern Taiwan, which is a north-northeast-striking suture zone on the boundary between the Eurasia plate and the Philippine Sea plate. Stresses arise in the crust because of crustal deformation resulting from plate motions and geodynamic processes. Gamma-ray emission associated with the radioactive decay of Rn, which is discharged to the surface from deep source areas, is continuously recorded at the station [*Giuliani et al.*, 2013].

The temporal distribution of earthquakes from 10 March to 27 May 2014 is shown in Figure 11.12. *Fu et al.* [2015] found that in eastern Taiwan anomalous variations in the gamma-ray counting rate may appear as precursors a few days before earthquake swarms, a large earthquake, or both. It was noted that there were two significant increases in the gamma-ray counting rate during the periods from 20 to 26 March and from 22 April to 1 May 2014. These increases were observed approximately 1 week before two earthquake swarms that took place from 26 to 27 March and from 3 to 5 May 2014. Furthermore, a progressive anomaly with a continuous increase (~ 8%) in gamma-ray count was recorded 2 weeks before the Fanglin earthquake (M_L 5.9, 21 May 2014) occurring in the Longitudinal Valley area. The occurrence of earthquakes shifting to the south includes the Fanglin earthquake. The focal mechanism with strike-slip components and orientation corresponds to a NW–SE shortening, caused by movement of the Philippine Sea plate in the northwest direction. Another series of earthquakes occurred afterward, along the southern segment of the Longitudinal Valley. Three different stages of stress transfer from north to south along the Longitudinal Valley system were identified during this period. These deformation stages may indicate that the pore pressure is dominated by intermittent fault propagation.

11.5.1. Soil Radon Anomalies Related to the 4 March 2010 Jiasian *M* 6.4 Earthquake

The M_L 6.4 Jiasian earthquake with a focal depth of 22.6 km occurred on 4 March 2010 in southwestern Taiwan, caused injury and damage [*Huang et al.*, 2011]. There are five soil-gas monitoring stations, and four soil-radon anomalies were found before the 4 March 2010 Jiasian Earthquake in southern Taiwan (Figures 11.13–11.17). The variations of soil-gas records, seismic events, and rainfall at the TPT station from 1 February to 20 March 2010 are shown in Figure 11.14. The annual average of soil Rn concentration is about 9 kBq m^{-3} at this station. From 7 February 2010, the Rn concentration started increasing progressively until 12 February from 11 kBq m^{-3} to 17 kBq m^{-3} before the rainfall and there was a sudden drop to 1 kBq m^{-3} on 12 February. The variations of the calculated standard deviation (SD) show a positive anomaly higher than the defined threshold value (1.5σ) before a M_L 5.earthquake with that occurred on 12 February 2010. After the middle of February, the radon concentrations of soil gas displayed a stable trend with no significant fluctuations, either before or after the Jiasian Earthquake.

The variations of soil gas records for the GK station from 1 February to 20 March 2010 are shown in Figure 11.15. The annual average of soil Rn concentrations is approximately 3.8 kBq m^{-3} at this station. The Rn concentration increased from 2 kBq m^{-3} to 6 kBq m^{-3} from 12 to 14 February 2010, and was followed by a local earthquake that occurred on 15 February approximately 70 km northeast of the GK station. From 16 February 2010, the Rn concentration increased continuously until 21 February, from 4 kBq m^{-3} to 9 kBq m^{-3}, and thereafter, from 21 to 27 February, it showed a progressive decrease to 1 kBq m^{-3}. After 27 February the Rn concentration remained stable for the 5 days before the Jiasian earthquake. The variation of standard deviation approached the threshold value on 19 February, before the Jiasian earthquake. However, failure of power supply caused a loss of monitoring data after the Jiasian earthquake from 4 to 20 March.

Figure 11.16 shows the variations of soil-gas records, seismic events, and rainfall for the CL station. The annual average of soil Rn concentrations is approximately 45 kBq m^{-3}. From 17 to 26 February 2010, the Rn concentrations increased from 45 kBq m^{-3} to 65 kBq m^{-3}, and then, from 26 to 27 February, Rn concentrations progressively decreased to 30 kBq m^{-3}. After 27 February the Rn concentrations remained stable for the 5 days before the Jiasian earthquake. These recordings were similar to the observations at the GK station. The variation of SD approached the threshold value on 21 February as an earthquake precursor.

Figure 11.17 presents the variations of monitoring results for the CS station discussed below. The annual average of soil Rn concentrations is approximately 6.6 kBq m^{-3}. From 2 to 12 February, the Rn concentrations increased progressively from 7 kBq m^{-3} to 11 kBq m^{-3} and then decreased to 6.6 kBq m^{-3}. This may be related to the earthquake (M_L 5.2) that occurred on 12 February with a hypocenter approximately 87 km from the CS station. From 19 to 26 February 2010, the Rn concentrations increased from 6.6 kBq m^{-3} to 11 kBq m^{-3}, and then

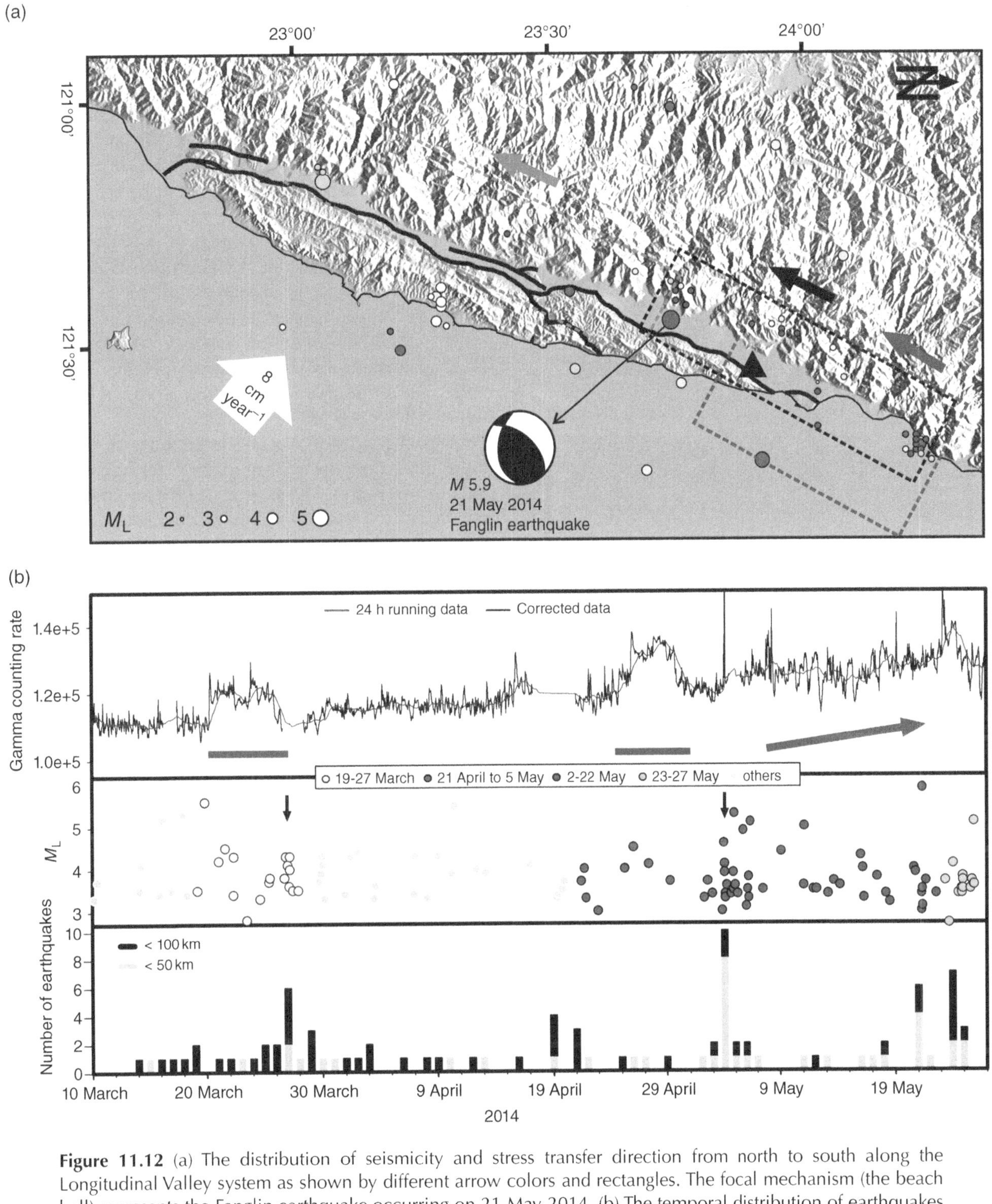

Figure 11.12 (a) The distribution of seismicity and stress transfer direction from north to south along the Longitudinal Valley system as shown by different arrow colors and rectangles. The focal mechanism (the beach ball) represents the Fanglin earthquake occurring on 21 May 2014. (b) The temporal distribution of earthquakes and gamma-ray variations from 10 March to 27 May 2014 at Donghua station in eastern Taiwan. The red arrow and lines indicate the duration of gamma-ray anomalies. The black arrows indicate the earthquake swarms. The accumulated number of earthquakes at different distances from the hypocenter is shown as a histogram [*Fu et al.*, 2015]. (Source: *Fu et al.* [2015]. Reproduced with the permission of Elsevier.) (*See electronic version for color representation.*)

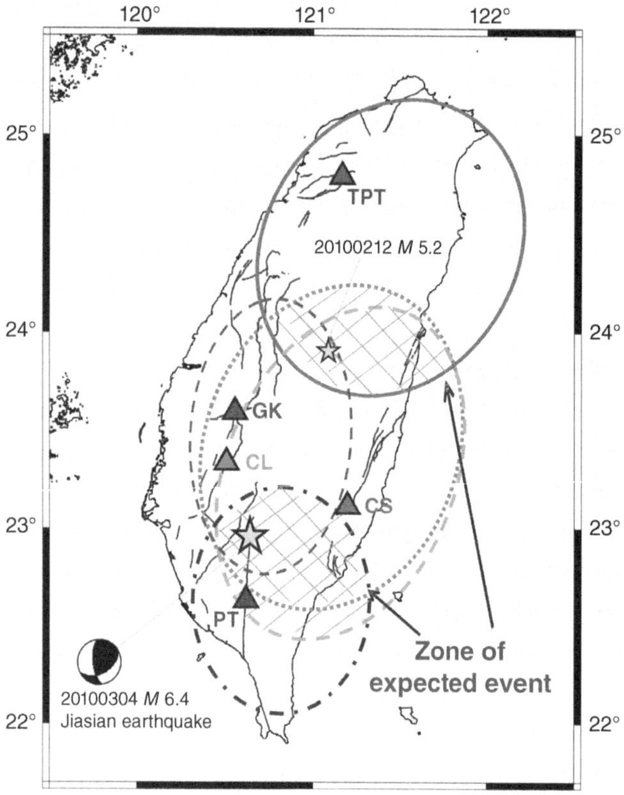

Figure 11.13 A simplified tectonic map of Taiwan and distribution of earthquake detectable zones at each soil-gas station shown by different circles. The triangles represent the location of the soil-gas stations: TPT, Tapingti; GK, Gukeng; CL, Chunglun; PT, Pingtung; CS, Chihshang. The focal mechanism (the beach ball) represents the Jiasian earthquake occurring on 4 March 2010.

progressively decreased to 8 kBq m-3 from 26 February to 4 March 2010, when the Jiasian earthquake occurred. The increase in Rn concentrations from 8 kBq m^{-3} to 10 kBq m^{-3} from 4 to 7 March 2010 was possibly related to an earthquake (M_L 4.9), which occurred on 8 March. From 11 to 16 March 2010, the Rn concentrations increased further from 6 kBq m^{-3} to 10 kBq m^{-3} and then suddenly decreased to the annual average value from 16 to 17 March 2010. This was related to an earthquake with magnitude M_L 4.0, which occurred on 18 March along the Longitudinal Valley fault and had a hypocentral distance of approximately 100 km, or surface creeping of the Chihshan Fault [*Fu et al.*, 2009]. The variation of SD approached the threshold value on 22 February and it may be a precursor related to the Jiasian earthquake.

Figure 11.18 displays the variations of soil-gas records, seismic events, and rainfall for the PT station from 1 February to 20 March 2010. The annual average of soil Rn concentrations is approximately 15 kBq m^{-3}. From 15 to 20 February 2010, the Rn concentrations increased from 14 kBq m^{-3} to 27 kBq m^{-3} and then progressively decreased to 8 kBq m^{-3} from 20 to 27 February. The Rn concentrations then remained stable for 5 days, when the Jiasian earthquake occurred. These observations were also similar to those from the GK and CL stations. From 9 to 11 March there was a sudden rise from 5 kBq m^{-3} to 15 kBq m^{-3}, and then the Rn concentration progressively decreased before the impending local M_L 3.7 earthquake on 13 March. The distance between the hypocenter and the PT station was approximately 46 km. The threshold value was approached on 18 February as a significant precursor related to the Jiasian earthquake.

The above results clearly show that precursory signals independently or simultaneously may be used to predict the approximate location of an impending earthquake with high confidence. The location of an impending seismic event is expected to be within the intersecting areas between two earthquake detectable zones of the monitoring stations (as shown for the Jiasian earthquake denoted by a star in Figure 11.13). The soil Rn anomalies are distributed in the region at an epicentral distance of 48–120 km. One station showing no obvious anomaly is located in northern Taiwan 300 km away. The earthquake preparation area of the Jiasian earthquake was estimated as 565 km according to the equation [*Dobrovolsky et al.*, 1979; *Hauksson and Goddard*, 1981]

$$R = 10^{0.43M} \tag{11.2}$$

where M is the magnitude of the earthquake and R is the radius in km. In this case, the anomalies should be observed at all stations in the preparation area prior to the Jiasian earthquake. However, the crust is not homogeneous as it is made up of different materials, tectonic structures, etc., which may affect the stress transfer and strain accumulation. This indicates that although all monitoring stations are located inside the probable preparation zone of the earthquake, the Rn precursory anomalies were not observed at all the stations because the crust was not homogeneous.

11.6. SUMMARY

Several precursory anomalies have been observed for the 1999 Chi-Chi earthquake. Anomalous variations of hydrochemical characteristics and gas composition occurred a few days to a few months before the earthquake and are likely to be associated with the dilatancy and microfracturing stages of earthquake development. These results suggest that many preseismic anomalies are true precursors of the earthquake and demonstrate that extensive, multidisciplinary, and real-time monitoring is effective for alerting the earthquake alarm system.

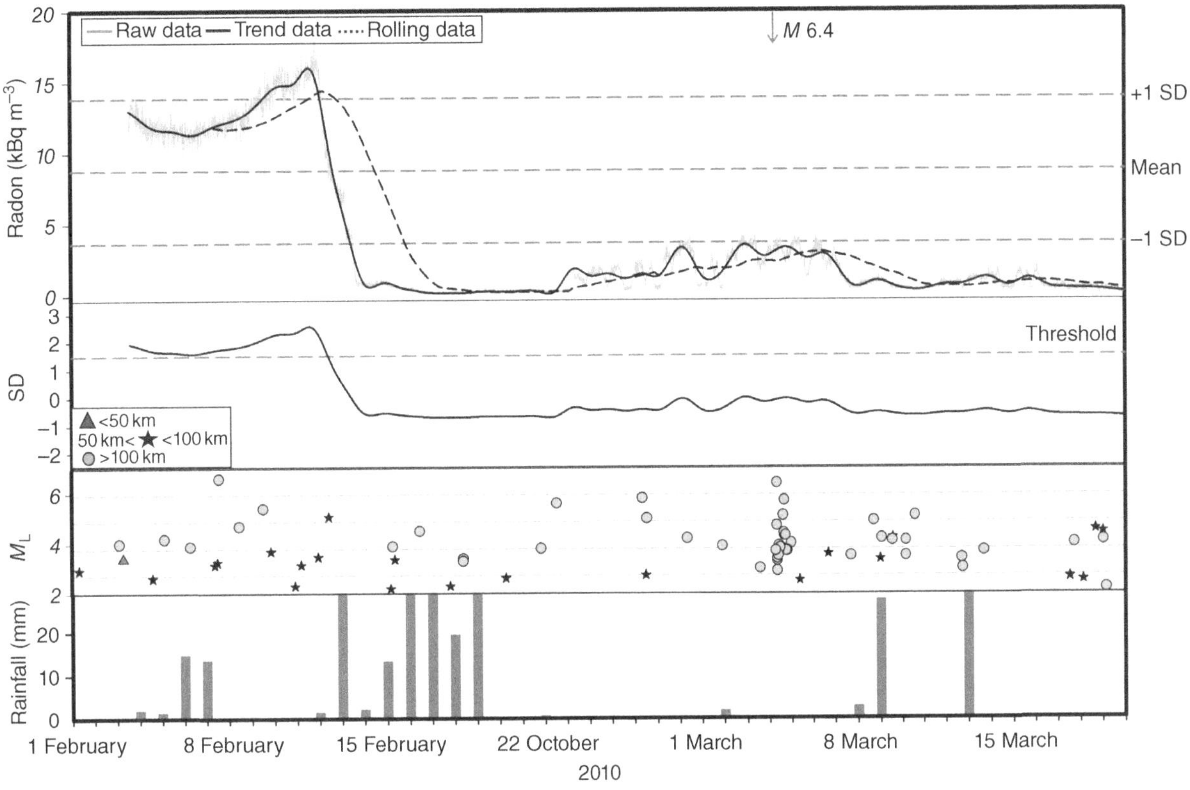

Figure 11.14 Variation of the soil-gas data, seismic events, and rainfall at Tapingti station from 1 February to 20 March 2010. Horizontal dashed lines indicate the annual average of soil gas and threshold value (+1.5σ). Rainfall and earthquake occurrence and magnitude data are also plotted for comparison. The triangles, stars, and circles represent different distance zones from the earthquake hypocenters.

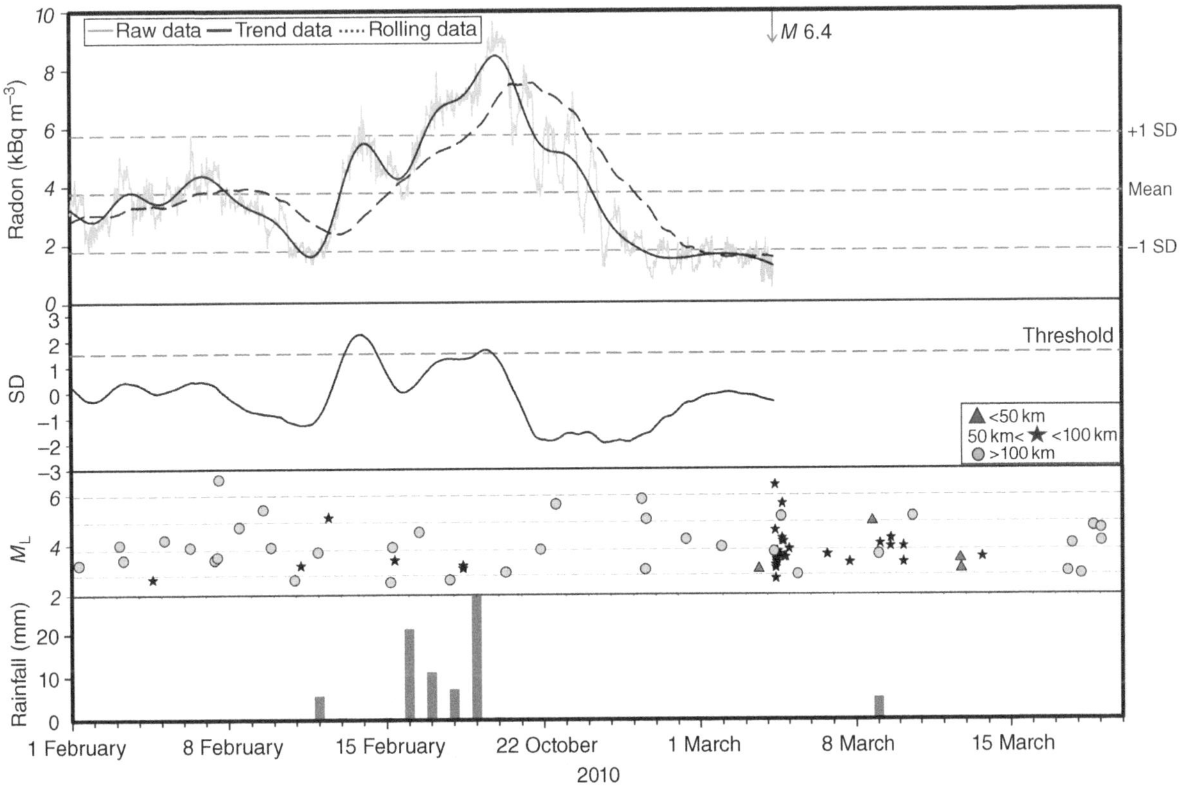

Figure 11.15 Variation of the soil gas data, seismic events, and rainfall at Gukeng station from 1 February to 20 March 2010. Horizontal dashed lines indicate the annual average of soil gas and threshold value (+1.5σ). Rainfall and earthquake occurrence and magnitude data are also plotted for comparison. The triangles, stars, and circles represent different distance zones from the earthquake hypocenters.

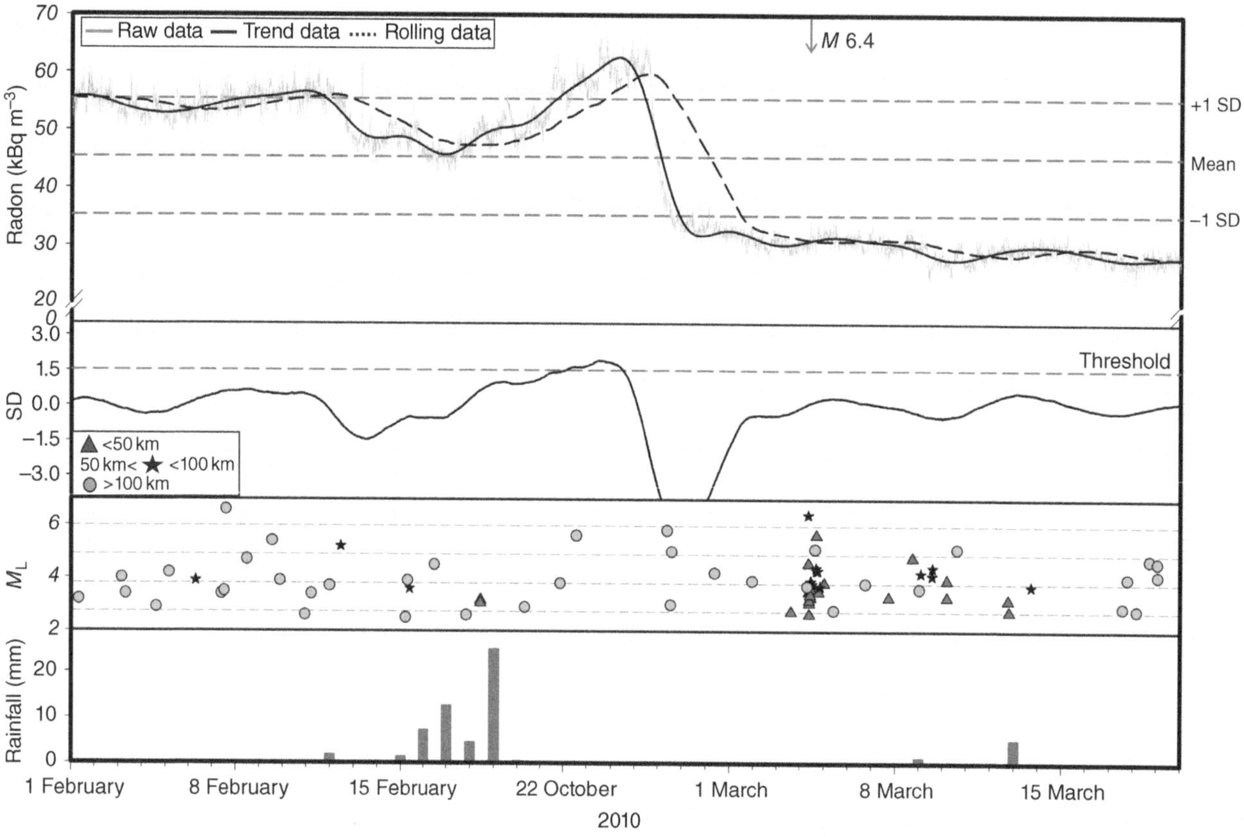

Figure 11.16 Variation of the soil gas data, seismic events, and rainfall at Chung-lun station from 1 February to 20 March 2010. Horizontal dashed lines indicate the annual average of soil gas and threshold value (+1.5σ). Rainfall and earthquake data are also plotted for comparison. The triangles, stars, and circles represent different distance zones from the earthquake hypocenters.

Continuous monitoring of the soil-gas data indicates that Rn concentrations exhibit some precursory anomalies a few days before local and small earthquakes. Significant changes in soil Rn at four stations originated from a stress accumulation period and were recorded 2 weeks before the major Jiasian earthquake in southwestern Taiwan.

Statistical analyses of soil-gas records were performed that can help us to identify anomalies and improve the reliability of soil Rn variations under automatic detection. Furthermore, according to our working hypothesis, simultaneous variations of soil Rn at different stations show precursory anomalies that can aid in predicting the approximate location of the impending seismic event based on an earthquake detectable zone with high confidence (Figure 11.13). Furthermore, the increasing gamma counting rate periods may also be related to induced stress changes, which may be related to the impending earthquake swarms, a large earthquake, or both. Hence, the geochemical measurements provide a useful tool for exploring earthquake precursors in Taiwan. The main observations of the presentation can be summarized in Table 11.1.

ACKNOWLEDGMENTS

This research is dedicated to Professor Tsanyao Frank Yang (TFY), who passed away on 12th March 2015. Born in 1961, he worked for many years at the Department of Geosciences, National Taiwan University (NTU), Taipei in Taiwan. His passion for research set a new standard for his students and coworkers. He will be greatly missed in the Earth science community. TFY set up the geochemical monitoring stations to understand pre-earthquake signals and proposed a potential model for their future use in Taiwan. We deeply appreciate all his contributions. This work is financially supported by grant 104-5226904000-03-01 through the Central Geological Survey of Taiwan, and by grant MOST104-2119-M-001-004. The Taiwan Earthquake Research Center (TEC) is funded through the Ministry of Science and Technology (MOST). The TEC contribution number for this article is 00134.

Table 11.1 Characteristic parameters of the geochemical precursors in Taiwan.

Observation site	Type	Parameters	Anomalies	Reference
Ilan	Hot and cold groundwater	Rn concentrations	Increase	*Liu et al.* [1985]
Puli	Groundwater	SO_4^{2-} and NO_3^- concentrations	Increase	*Song et al.* [2003]
Kuan-tze-ling	Hot spring	Cl^- and SO_4^{2-} concentrations	Increase	*Song et al.* [2006]
Chung-lun	Hot spring	Helium isotopes ($^3He/^4He$)	Decrease	This study
Chung-lun mud pool	Bubbling gas	CO_2/CH_4 ratios of bubbling gases	Increase	*Yang et al.* [2006]
Chihshang	Soil gas	Rn concentrations	Increase	*Walia et al.* [2009]
Chung-lun	Soil gas	^{222}Rn and ^{220}Rn concentrations	Increase	*Yang et al.* [2005b]
Chung-lun	Soil gas	CO_2 and CH_4	Increase	This study
Hsin-hua	Soil gas	Rn and CO_2 concentrations	Increase	*Walia et al.* [2009]
Dong-hua	Gamma rays	Gamma rays	Increase	*Fu et al.* [2015]
Gukeng Chung-lun Pingtung Chihshang.	Soil gas	Rn concentrations	Increase	This study

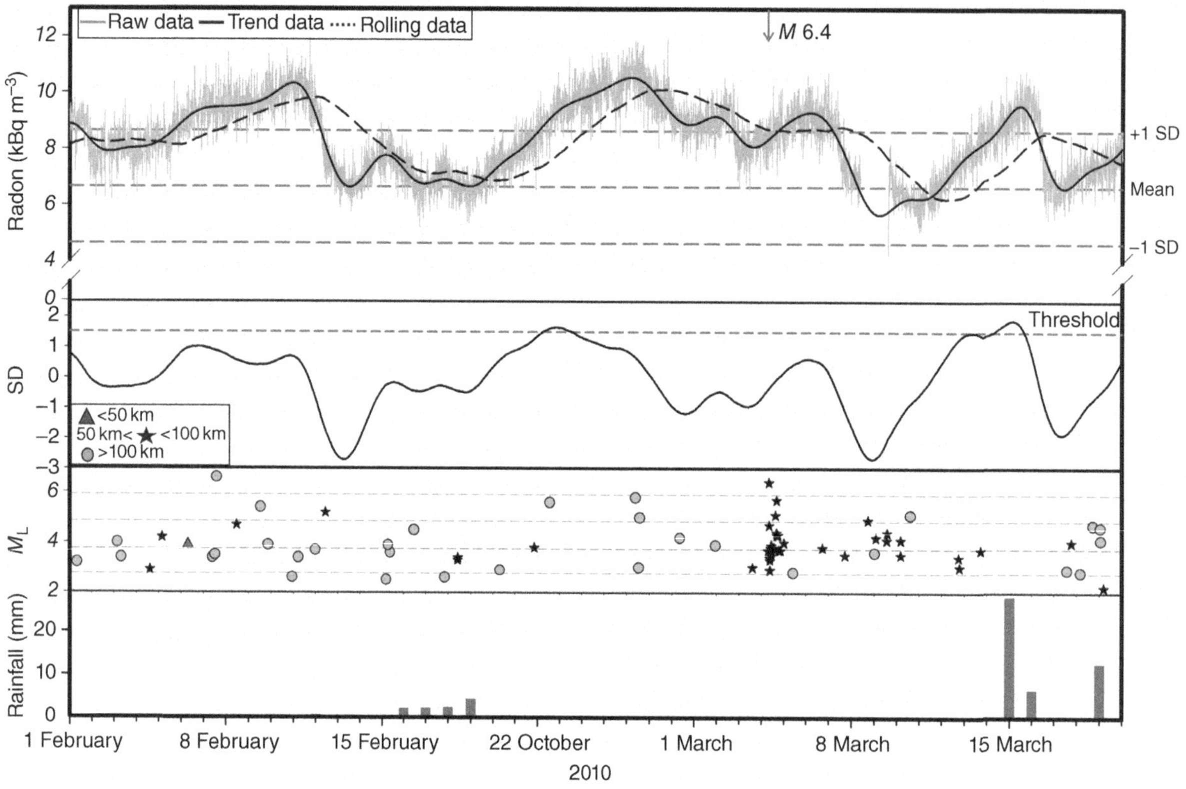

Figure 11.17 Variation of the soil gas data, seismic events, and rainfall at Chihshang station from 1 February to 20 March 2010. Horizontal dashed lines indicate the annual average of soil gas and threshold value (+1.5σ). Rainfall and earthquake data are also plotted for comparison. The triangles, stars, and circles represent different distance zones from the earthquake hypocenters.

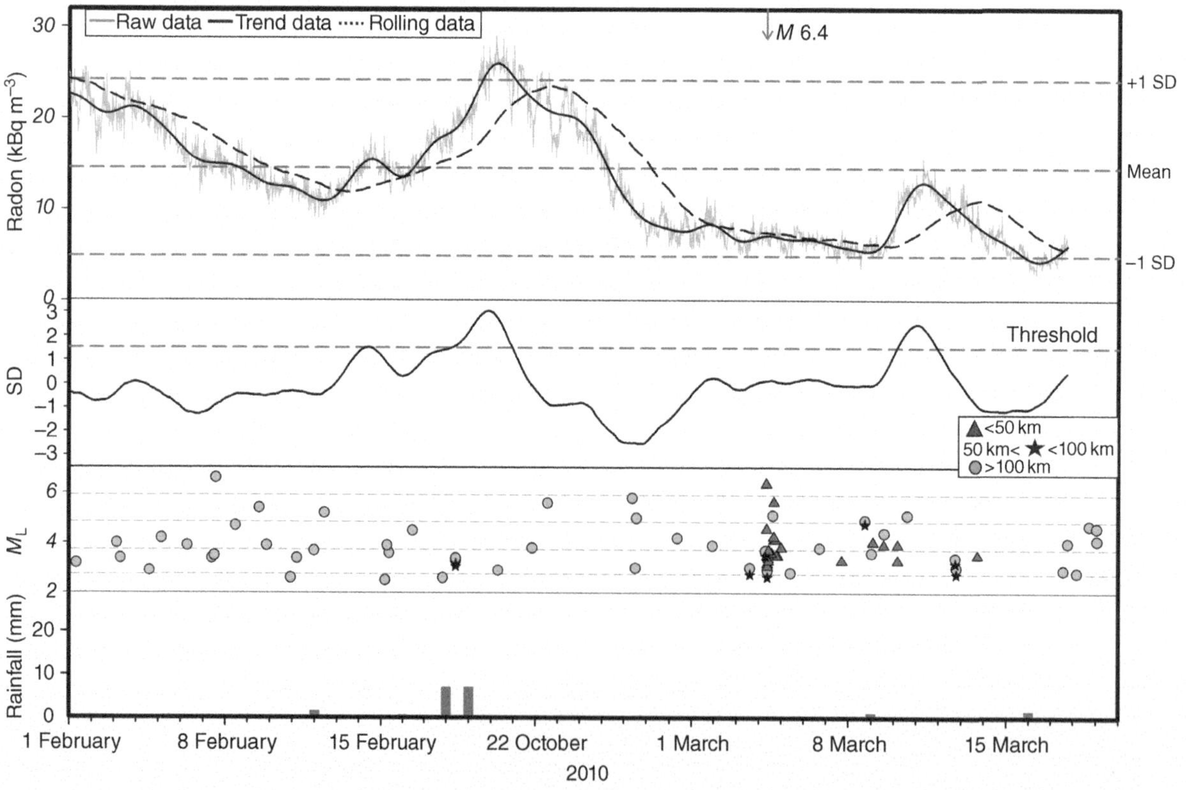

Figure 11.18 Variation of the soil gas data, seismic events, and rainfall at Pingtung station from 1 February to 20 March 2010. Horizontal dashed lines indicate the annual average of soil gas and threshold value (+1.5σ). Rainfall and earthquake data are also plotted for comparison. The triangles, stars, and circles represent different distance zones from the earthquake hypocenters.

REFERENCES

Balk, M., Bose, M., Gözen, E., Rogoff, D. A., Rothschild, L. J., and Freund, F. T. (2009), Oxidation of water to hydrogen peroxide at the rock–water interface due to stress-activated electric currents in rocks, *Earth Planet. Sci. Lett.*, *283*(1–4), 87–92.

Biagi, P. F.. Kingsley, S. P., and Vallianatos, F. (2000), Earthquake precursors, *Phys. Chem. Earth Part A.*, *25*(3), 225–225.

Bolt, B. A. (1999), *Earthquakes*, W.H. Freeman, New York.

Chaudhuri, H., Bari, W., Iqbal, N., Bhandari, R., Ghose, D.,Sen, P., and Sinha, B. (2011), Long range gas-geochemical anomalies of a remote earthquake recorded simultaneously at distant monitoring stations in India, *Geochem. J.*, *45*, 137–156.

Chyi, L. L., Chou, C. Y., Yang, T. F., and Chen, C.-H. (2001), Continuous radon measurements in faults and earthquake precursor pattern recognition, *Western Pacific Earth Sci.*, *1*(2),43–72.

Chyi, L. L., Quick, T. J., Yang, F. T., and. Chen, C.-H. (2005), Soil gas radon spectra and earthquakes, *Terr. Atmos. Ocean. Sci.*, *16*, 763–774.

Chyi, L. L., Quick, T. J., Yang, F. T., and Chen, C.-H. (2007), Soil radon time series and earthquake forecast, in P. Senand N. K. Das (eds), *Proceedings of the International Brainstorming Session on Geochemical Precursors for Earthquakes*, pp. 38–41, Saha Institute of Nuclear Physics, MacMillan.

Chyi, L. L., Quick, T. J., Yang, T. F., and Chen, C.-H. (2010), The experimental investigation of soil gas radon migration mechanisms and its implication in earthquake forecast, *Geofluids*, *10*, 556–563.

Chyi, L. L., Quick, T. J., Yang, F. T., and Chen, C.-H. (2011), The origin and detection of spike-like anomalies in soil gas radon time series, *Geochem. J.*, *45*, 431–438.

Ciotoli, G., Guerra, M., Lombardi, E., and Vittori, E. (1998), Soil gas survey for tracing seismogenic faults: A case study in the Fucino basin, Central Italy, *J. Geophys. Res.*, *103*, 23781–23794.

Claesson, L., Skelton, A., Graham, C., Dietl, C., Morth, M., Torssander, P., and Kockum, I. (2004), Hydrogeochemical changes before and after major Earthquake, *Geology*, *32*, 641–644.

Dobrovolsky, I. P., Zubkov, S. I., and Miachkin, V. I. (1979), Estimation of the size of earthquake preparation zone, *Pure Appl. Geophys.*, *117*, 1025–1044.

Freund, F. T. (2011), Pre-earthquake signals: Underlying physical processes, *J. Asian Earth Sci.*, *41*, 383–400.

Fu, C. C., Yang, T. F., Walia, V., and Chen, C.-H. (2005), Reconnaissance of soil gas composition over the buried fault and fracture zone in southern Taiwan, *Geochem. J.*, *39*, 427–439.

Fu, C. C., Yang, T. F., Du, J., Walia, V., Liu, T. K., Chen, Y. G., and Chen, C.-H. (2008), Variations of helium and radon concentrations in soil gases from an active fault zone in southern Taiwan, *Radiat. Meas.*, *43*, 348–352.

Fu, C. C., Yang, T. F., Walia, V., Liu, T. K., Lin, S. J., Chen, C.-H., and Hou, C. S. (2009), Variations of soil-gas composition around the active Chihshang Fault in a plate suture zone, eastern Taiwan, *Radiat. Meas.*, *44*, 940–944.

Fu, C. C., Wang, P. K., Lee, L. C., Lin, C. H., Chang, W. Y., Giuliani, G., and Ouzounov, D. (2015), Temporal variation of gamma rays as a possible precursor of earthquake in the Longitudinal Valley of eastern Taiwan, *J. Asian Earth Sci.*, *114*(2), 362–372.

Fu, C. C., Yang, T. F., Chen, C.-H., Lee, L. C., Wu, Y. M., Liu, T. K., Walia, V., Kumar, A., and Lai, T. H. (2017a), Spatial and temporal anomalies of soil gas in northern Taiwan and its tectonic and seismic implications, *J. Asian Earth Sci.*, *149*, 64–77, doi.org/10.1016/j.jseaes.2017.02.032

Fu, C. C., Yang, T. F., Tsai, M. C., Lee, L. C., Liu, T. K., Walia, V., Chen, C.-H.,Chang, W. Y., Kumar, A., and Lai, T. H. (2017b), Exploring the relationship between soil degassing and seismic activity by continuous radon monitoring in the Longitudinal Valley of eastern Taiwan, *Chem. Geol.*, *469*, 163–175, doi.org/10.1016/j.chemgeo.2016.12.042

Ghosh, D., Deb, A., Sengupta, R., Bera, S., and Patra, K. K. (2007), Pronounced soil radon anomaly—precursor of recent earthquakes in India, *Radiat. Meas.*, *42*, 466–471.

Giuliani, G., Attanasio, A., and Fioravanti, G. (2013), Gamma detectors for continuous monitoring of radon, *J. Int. Environ. Appl. Sci.*, *8*(4), 541–550.

Grant, R. A., Halliday, T., Balderer, W. P., Leuenberger, K., Newcomer, M., Cyr, G., and Freund, F. T. (2011), Ground water chemistry changes before major earthquakes and possible effects on animals, *Int. J. Environ. Res. Public Health*, *8*, 1936–1956.

Hartmann, J. and Levy, J. K. (2006), The influence of seismotectonics on precursory changes in groundwater composition for the 1995 Kobe earthquake, *Hydrogeol. J.*, *14*, 1307–1318.

Hauksson, E. and Goddard, J. G. (1981), Radon earthquake precursor studies in Iceland, *J. Geophys. Res.*, *86*(B8), 7037–7054.

Huang, H. H., Wu, Y. M., Lin, T. L., Chao, W. A., Shyu, J. B. H., Chan, C. H., and Chang, C. H. 2011), The Preliminary Study of the 4 March 2010 *Mw*6.3 Jiasian, Taiwan, Earthquake Sequence, *Terr. Atmos. Oceanic Sci.*, *22*(3), 283–290.

Igarashi, G. and Wakita, H. (1995), Geochemical and hydrological observations for earthquake prediction in Japan, *J. Phys. Earth*, *43*(5), 585–598.

Igarashi, G., Saeki, S., Takahata, N., Sumikawa, K., Tasaka, Y., Sasaki, Y., Takahashi, M., and Sano, Y. (1995), Groundwater radon anomaly before the Kobe earthquake in Japan, *Science*, *269*, 60–61.

Inan, S., Balderer, W. P., Leuenberger-West, F., Yakan, H., Uozvan, A., and Freund, F. T. (2012), Springwater chemical anomalies prior to the *Mw*=7.2 Van Earthquake (Turkey), *Geochem. J.*, *46*, 11–16.

Italiano, F. and Martinelli, G. (2001), Anomalies of mantle-derived helium during the 1997–1998 seismic swarm of Umbria-Marche, Italy, *Geophys. Res. Lett.*, *28*(5), 839–842.

Italiano, F. and Martinelli, G. (2004), Geochemical evidence of seismogenic-induced anomalies in the dissolved gases of thermal waters: A case study of Umbria (Central Apennines, Italy) both during and after the 1997–1998 seismic swarm, *Geochem. Geophys. Geosys.*, *5*, Q11001, doi:10.1029/2004GC000720

Kawabe, I. (1985), Anomalous changes of CH_4/Ar ratio in subsurface gas bubbles as seismo-geochemical precursors at Matsuyama, Japan, *Pure Appl. Geophys.*, *122*, 194–214.

King, C. Y. (1986), Gas geochemistry applied to earthquake prediction: an overview, *J. Geophys. Res.*, *91*(B12), 12269–12281.

Kumar, A., Singh, S., Mahajan, S., Bajwa, B. S., Kalia, R., and Dhar, S. (2009a), Earthquake precursory studies in Kangra valley of North West Himalayas, India, with special emphasis on radon emission, *Appl. Radiat. Isot.*, *67*, 1904–1911.

Kumar, A., Walia, V., Singh, S., Bajwa, B. S., Mahajan, S., Dhar, S., and Yang, T. F. (2009b), Earthquake precursory studies at Amritsar Punjab, India using radon measurement techniques, *Int. J. Phys. Sci.*, *7*(42), 5669–5677.

Liu, K. K., Yui, T. F., Yeh, Y. H., Tsai, Y. B., and Teng, T. (1985), Variations of radon content in ground waters and possible correlation with seismic activities in northern Taiwan, *Pure Appl. Geophys.*, *122*, 231–244.

Pulinets, S. A. and Dunajecka, M. A. (2007), Specific variations of air temperature and relative humidity around the time of Michoacan earthquake *M*8.1 Sept. 19, 1985 as a possible indicator of interaction between tectonic plates, *Tectonophysics*, *431*, 221–230.

Reddy, D. V. and Nagabhushanam, P. (2011), Groundwater electrical conductivity and soil radon gas monitoring for earthquake precursory studies in Koyna, India, *Appl. Geochem.*, *26*, 731–737.

Sano, Y., Takahata, N., Igarashi, G., Koizumi, N., and Sturchio, N. C. (1998), Helium degassing related to the Kobe earthquake, *Chem. Geol.*, *150*(1–2), 171–179.

Silver, P. G. and Wakita, H. (1996), A search for earthquake precursors, *Science*, *273*, 77–78.

Song, S. R., Ku, W. Y., Chen, Y. L., Lin, Y. C., Liu, C. M., Kuo, W., Yang, T. F., and Lo, H. J. (2003), Groundwater chemical anomaly before and after the Chi-Chi earthquake in Taiwan, *Terr. Atmos. Oceanic Sci.*, *14*(3), 311–320.

Song, S. R., Ku, W. Y., Chen, Y. L., Liu, C. M., Chan, P. S., Chen, Y. G., Yang, T. F., Chen, C.-H., Liu, T. K., and Lee, M. (2006), Hydrogeochemical anomalies in the springs of the Chiayi area in west-central Taiwan as likely precursors to earthquakes, *Pure Appl. Geophys.*, *163*, 675–691.

Sugisaki, R. (1978), Changing He/Ar and N_2/Ar ratios of fault air may be earthquake precursors, *Nature*, *275*, 209–211.

Toutain, J. P. and Baubron, J. C. (1999), Gas geochemistry and seismotectonics: a review, *Tectonophysics*, *304*, 1–27.

Toutain, J. P., Baubron, J. C., Le Bronec, J., Allard, P., Briole, P., Marty, B., Miele, G., Tedesco, D., and Luongo, G. (1992), Continuous monitoring of distal gas emanations at Vulcano, southern Italy, *Bull. Volcanol.*, *54*, 147–155.

Toutain, J. P., Munoz, M., Poitrasson, F., and Lienard, A. C. (1997), Springwater chloride ion anomaly prior to a $M_L = 5.2$ Pyrenean earthquake, *Earth Planet. Sci. Lett.*, *149*, 113–119.

Trique, M., Richon, P., Avouac, J. P., and Sabroux, J. C. (1999), Radon emanation and electric potential variations associated with transient deformation near reservoir lakes, *Nature*, *399*, 137–141.

Tsunogai, U. and Wakita, H. (1995), Precursory chemical changes in ground water: Kobe earthquake, Japan, *Science*, *269*, 61–63.

Walia, V., Su, T. C., Fu, C. C., and Yang, T. F. (2005a), Spatial variations of radon and helium concentration in soil gas across Shan-Chiao fault, Northern Taiwan, *Radiat. Meas.*, *40*, 513–516.

Walia, V., Virk, H. S., Yang, T. F., Mahajah, S., Walia, M., and Bajwa, B. S. (2005b), Earthquake prediction studies using radon as a precursor in N-W Himalayas, India: A case study, *Terr. Atmos. Oceanic Sci.*, *16*, 775–804.

Walia, V., Lin, S. J., Hong, W. L., Fu, C. C., Yang, T. F., Wen, K. L., and Chen, C.-H. (2009), Continuous temporal soil-gas composition variation for earthquake precursory studies along Hsincheng and Hsinhua faults in Taiwan, *Radiat. Meas.*, *44*, 934–939.

Walia, V., Lin, S. J., Fu, C. C., Yang, T. F., Hong, W. L., Wen, K. L., and Chen, C.-H. (2010), Soil-gas monitoring: a tool for fault delineation studies along Hsinhua Fault (Tainan), southern Taiwan, *Appl. Geochem.*, *25*, 602–607.

Walia, V., Yang, T. F., Lin, S. J., Kumar, A., Fu, C. C., Chiu, J. M., Chang, H. H., Wen, K. L., and Chen, C.-H. (2013), Temporal variation of soil gas compositions for earthquake surveillance in Taiwan, *Radiat. Meas.*, *50*, 154–159.

Yang, T. F., Chou, C. Y., Chen, C.-H., Chyi, L. L., and Jiang, J. H. (2003), Exhalation of radon and its carrier gases in SW Taiwan, *Radiat. Meas.*, *36*, 425–429.

Yang, T. F., Lan, T. F., Lee, H. F., Fu, C. C., Chuang, P. C., Lo, C. H., Chen, C.-H., Chen, C. T. A., and Lee, C. S. (2005a), Gas compositions and helium isotopic ratios of fluid samples around Kueishantao, NE offshore Taiwan and its tectonic implications, *Geochem. J.*, *39*, 469–480.

Yang, T. F., Walia, V., Chyi, L. L., Fu, C. C., Chen, C.-H., Liu, T. K., Song, S. R., Lee, C. Y., and Lee, M. (2005b), Variations of soil radon and thoron concentrations in a fault zone and prospective earthquakes in SW Taiwan, *Radiat. Meas.*, *40*, 496–502.

Yang, T. F., Fu, C. C., Walia, V., Chen, C.-H., Chyi, L. L., Liu, T. K., Song, S. R., Lee, M., Lin, C. W., and Lin, C. C. (2006), Seismo-geochemical variations in SW Taiwan: multi-parameter automatic gas monitoring results, *Pure Appl. Geophys.*, *163*(4), 693–709.

Yang, T. F., Liu, T. K., and Kuo, M. C. T. (2011), Observation and research on active faults by geochemical method (II), *US Geol. Surv. Open-File Rep.*, *100-08*, 378 pp.

Yen, H. Y., Chen, C.-H., Yeh, Y. H., Liu, J. Y., Lin, C. J., and Tsai, Y. B. (2004), Geomagnetic fluctuations during the Chi-Chi, Taiwan earthquake, *Earth Planets Space*, *56*, 39–45.

Yuce, G., Fu, C. C., D'Alessandro, W., Gulbay, A. H., Lai, C. W., Bellomo, S., Yang, T. F., Italiano, F., and Walia, V. (2017), Geochemical characteristics of soil radon and carbon dioxide within the Dead Sea Fault and Karasu Fault in the Amik Basin (Hatay), Turkey, *Chem. Geol.*, doi.org/10.1016/j.chemgeo.2017.01.003.

12

Geochemical and Fluid-Related Precursors of Earthquakes: Previous and Ongoing Research Trends

Giovanni Martinelli[1] and Andrea Dadomo[2]

ABSTRACT

Hydrogeologic and geochemical analyses are carried out in groundwaters and in gaseous emissions on order to contribute to earthquake prediction research. At presently the main test sites are in Japan, China, and Taiwan. Catalogs of presumed precursory episodes have been compiled over the years and provided preliminary conclusions about site-selection techniques. Controlled experimental sites have recently given the opportunity to better investigate the physical mechanisms originating the recorded preseismic anomalies. The main characteristics and limitations of hydrogeologic and geochemical parameters are discussed. An in depth review of results obtained from the most relevant test site areas identifies future research trends and projectections of future instrumental networks, including geophysical parameters and remote sensing techniques oriented to hazard reduction policies.

12.1. INTRODUCTION

Hydrogeologic and geochemical parameters have been considered considered over past 60 years in research oriented toward earthquake prediction. In particular, radon was widely utilized in the period 1950–2000, while water level and helium have attracted the attention of researchers and of government agencies in more recent times, together with satellite-based techniques able to detect possible anomalous concentrations of optically active gases like CO_2. A relatively large amount of point data have been collected by manual techniques in various geological contexts, while a much lower amount of automatically recorded data has been collected in some experimental areas such as Japan, USA, former USSR, China, Turkey, and Iceland. Considered parameters include oil-flow rate, water level, water temperature, water dissolved ions, CO_2, CH_4, radon, helium, and mercury. Possible geochemical and hydrogeologic precursors have been observed hours to months before some strong earthquakes in "sensitive" monitoring sites among many insensitive sites [*Thomas*, 1988; *King and Igarashi*, 2002]. Sensitive sites are usually found along active faults, in thermal springs or in deep wells reaching confined deep reservoirs [*Arieh and Merzer*, 1974; *Sugisaki*, 1981] able to act as natural strainmeters [*Bodvarsson*, 1970; *Wang and Manga*, 2010] (Figures 12.1 and 12.2). Data collected worldwide have been reviewed in previous research [*Hauksson*, 1981; *Friedmann*, 1985; *Roeloffs*, 1988; *Kissin and Grinevsky*, 1990; *Toutain and Baubron*, 1999; *Hartmann and Levy*, 2005; *Cicerone et al.*, 2009; *Wang and Manga*, 2010; *Matsumoto and Koizumi*, 2011; *Woith*, 2015] in attempts to find general laws on the behavior of deep fluids before earthquakes in order to better constrain possible time and location of a forthcoming seismic event. This chapter reviews actual state of the art of the topic with the purpose to identify well-assessed findings in non-volcanic areas in view of possible experimental application and to highlight past and recent research trends.

[1] *ARPAE Emilia-Romagna, Reggio Emilia, Italy*
[2] *GEOINVEST Srl, Piacenza, Italy*

Pre-Earthquake Processes: A Multidisciplinary Approach to Earthquake Prediction Studies, Geophysical Monograph 234, First Edition. Edited by Dimitar Ouzounov, Sergey Pulinets, Katsumi Hattori, and Patrick Taylor.

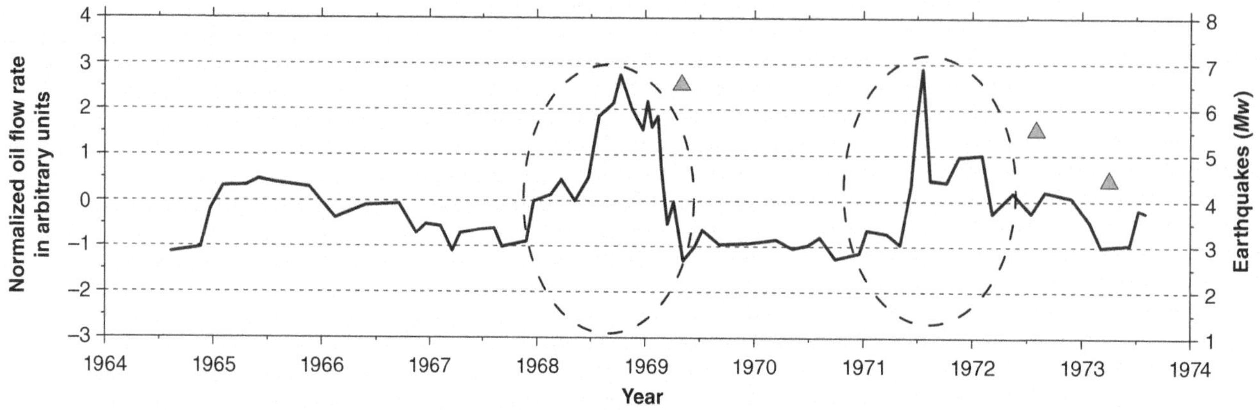

Figure 12.1 Oil-flow-rate variations in a 10-years time series recorded in the Sinai area in a well 3 km deep before two *Mw*>5.5 earthquakes occurred within a radius of 150 km. Crustal deformation preceding shocks has induced an increase in spontaneous oil flow from a deep hydrocarbon well. Similar phenomena may affect groundwaters and thermal waters. Confined aquifers may act as natural strainmeters. (Source: Reproduced with the permission of Nature Publishing Group.)

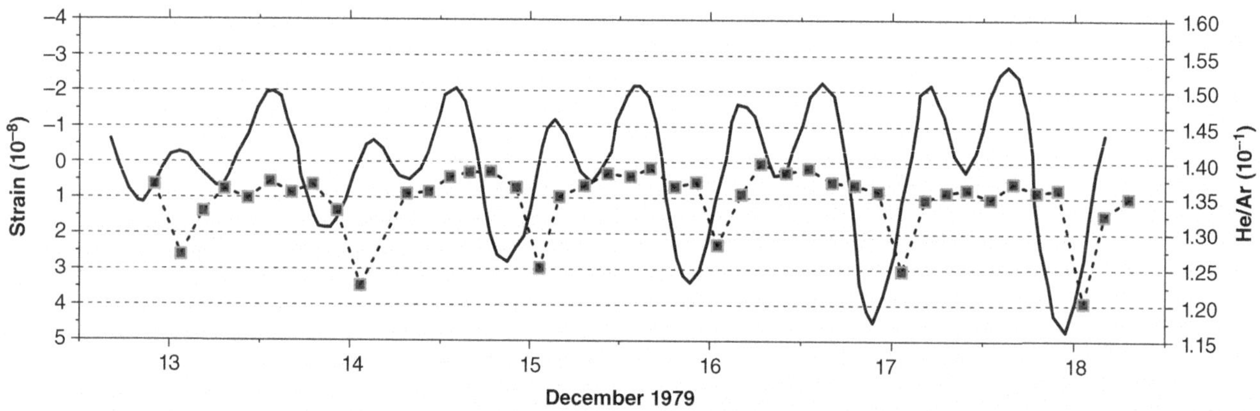

Figure 12.2 Helium/argon concentration in a thermal spring in Japan following crustal deformation induced by earth tides. Crustal deformation may induce variations in deep fluid flow rate or emissions. This phenomenon is utilized in monitoring activities in earthquake precursors research. (Source: Reproduced with the permission of The American Association for the Advancement of Science.)

12.2. THE ORIGIN OF FLUID-RELATED PRECURSORS

Groundwaters, hydrocarbons, and gases are pervasive fluids that fill deep-seated geological strata where porosity values control accumulation or circulation of fluids [*Fyfe et al.*, 1978]. Usually hydrocarbons fill porous sedimentary anticlines, while groundwaters can be of meteorologic or of marine origin.

Marine originated waters (formation waters) fill sedimentary geological strata often in association with hydrocarbons, while meteorologic originated groundwaters can fill all kinds of rocks. Formation waters are very old and characterized by ages similar to the hydrocarbons with which they are associated, while meteoric water ages are in the range of 1–100,000 years. Young groundwaters (1–100 years old) are usually hosted in phreatic aquifers and linked to present hydrologic cycles, while confined aquifers and geothermal circuits host relatively old groundwaters (100–100,000 years old) characterized by low circulation velocities and scarce or no link with present hydrologic cycles. Geothermal systems can be fault controlled and generate thermal springs and carbon-dioxide-dominated gas emissions. Carbon dioxide is chiefly generated by thermometamorphic reactions in the crust [*Frezzotti et al.*, 2009] or by mantle degassing in volcanic systems. Further gas emissions, represented in particular, by methane may affect fault-controlled accumulations of hydrocarbons.

Geochemical anomalies observed before earthquakes include stable isotopes of water [*Skelton et al.*, 2014], dissolved ions, dissolved gases, and soil gases [*Thomas*, 1988; *Segovia et al.*, 1989; *King and Igarashi*, 2002]. The main gases involved in Earth's degassing activity are CH_4 and CO_2, which are considered responsible for water–gas–rock interaction processes able to induce chemical variations in groundwater composition. Geochemical variations in the chemical composition of groundwaters for the most part are attributed to aquifer mixing processes [*Thomas*, 1988], in particular when accompanied by temperature variations of fluids. Precursory variations observed in radon, helium, and hydrogen in soil gases have been accompanied by carrier gas CO_2 or CH_4 flow rate variations. Generation mechanisms of hydrologic and geochemical precursors have been reviewed by *Roeloffs* [1988], *Thomas* [1988], *King and Igarashi* [2002], and *Matsumoto and Koizumi* [2011], and nearly all geochemical anomalies observed before earthquakes have been directly or indirectly attributable to pressure variations in deep fluids induced by crustal deformation processes, since fluid pressure is proportional to stress and volumetric strain. The stress–strain relationship for an isotropic, linearly elastic porous medium was studied, among others, by *Rice and Cleary* [1976, and references therein] and by *Roeloffs* [1996]. In particular the stress tensor $\boldsymbol{\sigma}_{kk}$, the volumetric strain $\boldsymbol{\varepsilon}_{kk}$, and the fluid pressure p for undrained conditions can be described as:

$$p = -B\boldsymbol{\sigma}_{kk}/3 \qquad (12.1)$$

$$p = -2GB(1+\nu_u)\varepsilon_{kk}/3(1+\nu_u) \qquad (12.2)$$

where G is the shear modulus, B is the Skempton coefficient, and ν_u is the Poisson's ratio for undrained conditions. As a consequence the fluid pressure is proportional to stress and volumetric strain. Thus groundwaters can be, in principle, utilized as natural strainmeters, being as water is incompressible, to monitor large-scale networks. To better constrain the candidate area eventually identified by anomalous signals detected from large-scale networks, attempts have been made to identify areas affected by water-level maximum amplitude signals [*Popov and Vartanyan*, 1990; *Vartanyan et al.*, 1992; *Chelidze et al.*, 2010; *Chen et al.*, 2015].

12.3. EXPERIMENTS UNDERTAKEN THROUGHOUT THE WORLD

Geochemical prospections and hydrologic studies have been widely utilized in past and present research in order to locate faults and to characterize tectonically active areas. Identified areas were partly subjected to monitoring over time utilizing one or more parameters at one or more sites. Most relevant results were obtained in places were manual or instrumental networks monitored wells or springs for periods longer than 5 years, in particular in the former USSR, China, USA, and Japan.

Observational networks of wells (depth range 0.2–2.9 km) or of thermal spring sources comprising about 100 sampling points (1 sampling site every 10,000 km^2, but at a higher density in the so-called experimental "polygons") were utilized in the period 1970–1990 in Tadjikistan, Uzbekistan, Kazakhstan, Caucasus, and Kamchatka. *Sidorenko et al.* [1984], *Asimov et al.* [1984] and *Sidorin* [2003] reported that some local strong seismic events were preceded some hours or days before by precursory water-level variations, although with a bias in epicentral localization that could reach 100 km. Further networks for groundwater monitoring were also set up in China, where water level and gaseous content have been analyzed in over 600 thermal springs or wells (depth range 0.1–2.0 km) (1 sampling site every 3000 km^2, but at a higher density in experimental "polygons"). Precursory variations in water level were observed before some strong earthquakes [*Huang et al.*, 2004; *Shi et al.*, 2013], while tectonic related geochemical variations required wider study in order to better understand possible fluid phenomena related to geodynamic processes [*Zheng et al.*, 2013; *Zhou et al.*, 2015]. Most of findings in the former USSR and China were published in local languages and complete time series of water recorded data were scarcely published, *Thomas* [1988], *Roeloffs* [1988] and *King* and Igarashi (2002) retained reliable the precursory character of observed anomalies.

In the period 1984–1989 groundwater level (six monitoring stations) and radon (five monitoring stations) were recorded together with other geophysical parameters in Turkey [*Zschau and Ergunay*, 1989]. The lack of strong earthquakes during the observation period hampered definitive conclusions but significant methodological improvements were reached in groundwater monitoring techniques and in data interpretation [*Woith et al.*, 2013]. A network of 12 wells was set up in California and water level data recorded and compared to other geophysical parameters [*Bakun*, 1988]. A precursory variation in groundwater level was detected in 1985 [*Roeloffs et al.*, 1997] but the fluid monitoring network was partially dismantled in the following years due to successive missed anomalies before local earthquakes.

In Japan, many research groups have recorded hydrologic and geochemical data in single spring sources or in small networks of wells [*Matsumoto and Koizumi*, 2011], and some precursory anomalies were detected although with no reliable signals [*King et al.*, 2000]. Recently a network of 14 deep wells (depth 0.6 km) started to monitor activity in the area of Nankai, Tonankai and Tokai [*Koizumi et al.*, 2008] in view of a possible expected strong earthquake in the locality. Further observations characterized by relatively shorter time series (<5 years) have been carried out in France, Italy, Spain, Greece, India, Germany, Iceland, Bulgaria, Afghanistan, Iran, Israel, Mexico, and Taiwan, and all available data were considered in reviews by *Hauksson* [1981], *Friedmann* [1985], *Kissin and Grinevsky* [1990], *Toutain and Baubron* [1999], *Hartmann and Levy* [2005], *Cicerone et al.* [2009], *Ghosh et al.* [2009], *Wang and Manga* [2010], *Petraki et al.* [2015] and *Woith* [2015].

12.4. MECHANISMS RESPONSIBLE FOR FLUID-RELATED ANOMALIES: CRUSTAL DEFORMATION

Nine review papers on geochemical and fluid parameters were considered (Table 12.1). All review publications considered directly or indirectly conclude that a deformation process is responsible for observed anomalies in fluids. Most scrutinized works are related to "a posteriori" identified precursory signals, although official alarms were issued before the 1975 Haicheng M 7.3 [*Wang et al.*,

Table 12.1 Most relevant review papers on possible precursory geofluids published in past 25 years.

Hauksson [1981]	Reviewed all available radon precursory anomalies that occurred the worldwide for the period 1971–1981 and argued that time duration of anomalies increases with magnitude. He observed that radon anomalies reflect small changes in the local stress intensity factor, which controls the velocity of crack growth
Friedmann [1985]	Reviewed all available radon precursory anomalies published in the period 1969–1982 and reached the conclusion that faster spike-like anomalies are more frequent close to the epicentral area, while bay-shaped anomalies are more frequent far from the epicenter
Toutain and Baubron [1999]	Reviewed gaseous anomalies (chiefly radon) published in the period 1980–1995 and observed that amplitudes of anomalies are independent of magnitudes and epicentral distances of related earthquakes, while time and duration of anomalies increase with magnitude and epicentral distance
Hartmann and Levy [2005]	Reviewed gaseous and water-related anomalies (level and geochemistry) published in the period 1978–1997, reaching the conclusion that signal duration is proportional to the magnitude of the seismic event
Kissin and Grinevsky [1990]	Reviewed available anomalies in water-level data that occurred in the period 1948–1980 finding that precursory time and signal amplitude are proportional to magnitude, in particular most groundwater variations occurred within 50 km from the epicenter and in areas characterized by intense deformation processes
Cicerone et al. [2009]	Reviewed geochemical, hydrologic, and geophysical anomalies that occurred in the period 1948–2001, finding that signal amplitude is proportional to earthquake magnitude and that all observed precursory anomalies have been found in regions of Earth where deformation is largest largest
Ghosh et al. [2009]	Reviewed radon data recorded in many places of the world in the period 1983–2002 and discussed relations among radon anomalies, magnitude, and time of occurrence of seismic events
Wang and Manga [2010]	Reviewed nine well-documented earthquakes that occurred in the period 1975–2003 and were preceded by geochemical or water-level variations in groundwaters, finding that combined deformation and water level measurements can be useful to understand relationships between hydrogeological changes and crustal deformation
Petraki et al. [2015]	Reviewed radon data recorded worldwide in the period 1966–2014 and discussed the debate of the time regarding the physical meaning of radon anomalies, recognizing that stress and strain could be responsible for the observed geochemical anomalies
Woith [2015]	Reviewed radon and further geochemical data recorded worldwide in the period 1967-2014, paying attention to data quality and to the length of the time series. He discussed the possible link of radon and of other geochemical anomalies with deformation processes and discovered that the number of published preseismic anomalies is proportional to the inverse of the length of time series and concluded that nonseismotectonics signals may strongly affect geochemical time series

2006] and the 1978 Pamir M 7.2 earthquakes [*Asimov et al.*, 1984] the exact time and locations of the forthcoming events were not appropriately constrained. *Roeloffs* [2006] reviewed data about deformation processes before earthquakes and found that in at least ten seismic events for which data were available an aseismic deformation change characterized by time durations in the range 10 min to 15 years occurred. Not all earthquakes seem to be preceeded by detectable crustal strain changes in the epicentral area and this could explain the lack of fluid - related precursors observed in many cases. Nevertheless, because water is incompressible, water level can be considered as the signal of a naturally sensitive (10^{-7}–10^{-8}) strainmeter waiting to record a deformation process in the candidate area of an earthquake. Furthermore anticipated signal characteristics of a forthcoming anomaly could be calculated in advance, as in the last generation network of groundwater monitoring set up in the Tokai area [*Matsumoto et al.*, 2007].

This incompressible character is lacking in gases, thus radon data compared with strainmeter data, when available, do not produce unambiguous results [*Roeloffs*, 1999], although progress in time series analysis has been achieved [*Finkelstein et al.*, 1998]. The calculation of a stress tensor from radon data is still an unsolved problem, but radon or helium data could still be utilized in future to monitor semiquantitative signals of tectonic reactivation in faulted areas. In particular, radon originates in the upper crust and may change due to velocity variations of carrier gases like CO_2 or CH_4 and to eventual increase in permeability of rock formations. Helium chiefly originates in the crust but an isotope of it, ^{3}He, originates in the mantle and could provide extremely interesting information about very deep geophysical processes. To this purpose *Barry et al.* [2009] experimented with new equipment that enabled a high sampling rate and produced a long ^{3}He time series. The International Association for Seismology and Physics of the Earth's Interior Sub-Commission Earthquake Prediction reviewed available precursory anomalies and added radon and water level to a list of possibly significant precursors [*Wiss*, 1991; *Wiss and Booth*, 1997].

12.5. POSSIBLE EXPERIMENTAL APPLICATIONS

Epicenters of earthquakes often migrate toward in a particular direction without this being the result of a large-scale crustal deformation processes [*Kasahara*, 1979]. Indirect and direct experimental confirmation of the role of regional strain on possible precursory fluid anomalies were collected by *Vartanyan et al.* [1992], *Silver et al.* [1993], *Facchini et al.* [1993], *Albarello and Martinelli* [1994], *Buntebarth and Chelidze* [2005], and *Chen et al.* [2015]. Migration of a deformation fronts may trigger earthquakes and is characterized by velocities in the range 10–100 km year^{-1}. These phenomena are linked to rheologic differentiation of the lithosphere [*Rice*, 1980] and may play a fundamental role in medium- and long-term earthquake prediction [*Mantovani et al.*, 2010]. Groundwater monitoring of well networks may contribute to a better understanding of these processes for research oriented toward earthquake forecasting.

Most relevant recent networks on water level or fluid monitoring include China, Taiwan, Turkey, Japan, Russia, and USA, while no systematic research appears to have been carried out in Europe. Most seismically active areas of Italy, France, Germany, Czech Republic, Greece, Slovenia, and of the Balkan area have hosted short experimental monitoring research in the past, but no conclusive results were obtained due to too short a monitoring time over too small areas during nonrelevant seismic periods. Hydro-geodeformation fronts have been monitored in various areas of the former USSR [*Vartanyan et al.*, 1992; *Chelidze et al.*, 2010] and of Italy [*Albarello and Martinelli*, 1994]. The availability of a significant number of thermal springs, of gaseous emissions, and of wells in Italy, Greece, Spain, Albania, and in the former Yugoslavia in areas characterized by a relatively high frequency of strong earthquakes (see also *Martinelli and Albarello* [1997]) provide potential to set up monitoring networks in the near future. The most suitable technologies include sensors for water level, temperature, electric conductivity, and gas components in selected wells or thermal springs.

At the same time the continuous precise measurement of tectonic motions by strainmeters and tiltmeters is not easily implementable. *Agnew* [1986] and *Kumpel* [1992] reviewed data summarized by *Roeloffs* [1988] and by *Kissin and Grinevsky* [1990] and underlined the fact that the weakness of conventional wells as stress sensors results from the fact that the process of fluid flow often lacks a sufficient control on possible inhomogeneities of the reservoir and on fluid circulation modalities. To better constrain experimental conditions and reduce problems due to natural geological strata, *Swolfs and Walsh* [1990] experimented using an "artificial confined aquifer," which entailed emplacing a liquid-filled pressurized cell in a cavity drilled in rock. A precursor of a local shock was detected in the experiment, confirming the need for advanced networks able to continuously monitor crustal deformation processes. The advantage of a natural aquifer for such monitoring, if appropriately chosen, is the size-dependent sensitivity not attained by artificial devices.

Recent advantages in monitoring techniques have increased accompanying advances in electronics and informatics. Automatic monitoring was utilized by *Woith et al.* [2006], which observed possible nonrandom phenomena in confined aquifers of Turkey linked to crustal deformation processes. Their monitoring activity demonstrated that transient phenomena in confined aquifers can be characterized by relatively short durations (hours or days). *Roeloffs et al.* [1997] reported the possible existence of short-duration transient phenomena recorded during monitoring in the Parkfield area, and feature also recorded in other studies [e.g. *Sidorenko et al.*, 1984] in various geological contexts.

Short-term possible precursory phenomena in groundwater or in gas emissions can be observed only by automatic monitoring equipment. In the past 15 years short-term possible fluid-related precursory signals in southern Europe (Greece, Italy, Spain, and Slovenia), Turkey, and Israel have been recorded by automatic monitoring stations, as reported by *Contadakis and Asteriadis* [2001], *Lapenna et al.* [2004], *Colangelo et al.* [2005], *Riggio and Sancin* [2005], *Colangelo et al.* [2007], *Cioni et al.* [2007], *Richon et al.* [2007], *Perez et al.* [2008], *Heinicke et al.* [2010], *Torkar et al.* [2010], *Yuce et al.* [2010], *Steinitz et al.* [2003]. and in central Europe by *Heinicke et al.* [1995] and *Stejskal et al.* [2009]. Further automatic monitoring studies have been carried out in Taiwan [*Walia et al.*, 2013; *Chen et al.*, 2015], Kamchatka [*Firstov et al.*, 2007], and Mexico [*Taran et al.*, 2005]. In spite of a relatively high sampling rate and encouraging data, completely definitive results were not obtained since the number of false alarms or of missed alarms was relatively high for practical purposes, probably due to inadequate array density, inadequate site selection, or simply beyond the capability of the task, e.g., pore pressure change at seismogenic depth < 0.1 MPa [*Johnston and Linde*, 2002].

Since experimental data are lacking, it is unclear whether aseismic precursory crustal deformation processes always accompany earthquakes. Nevertheless, lateral migration of the stress field after dislocation events has been recognized as being responsible for generation of further earthquakes. Tectonic pumping phenomena, able to generate groundwater and gas anomalies, could be easily explained in this view. A great number of the recorded geochemical and hydrologic anomalies reported in the cited review papers were detected in Tadjikistan, Uzbekistan, Kazakhstan, Caucasus, Kamchatka, Yunnan Province, Hebei Province, Iceland, and Japan. These areas are characterized by anomalous (>60 mW m^{-2}) heat-flow values [*IHFC*, 2011], a feature that implies the existence of hydrothermal circuits where earthquakes similar to volcanic seismic events can occur. In this sense the observed short-term precursory anomalies could have been generated by fluid-kinetics-related processes, which are typical of volcanic areas. In volcanic areas fluid kinetics is the driver of seismicity, while fluids have a secondary role in plate-tectonic earthquake generation. In geothermal areas the brittle–ductile transition can be shifted to shallower depths [*Zencher et al.*, 2006], enhancing the effects of crustal deformation due to a more effective deformability of ductile layers induced by temperatures > 300 °C [e.g. *Ranalli*, 1995]. For this reason, dilatancy processes are scarcely invoked to explain earthquake triggering in volcanic areas. In volcanic areas precursory geochemical and hydrological phenomena often have been observed, and fluid monitoring is presently part of civil defense routine control for volcanic eruption risk reduction. Thus observed geochemical and hydrological short-term earthquake precursors could have been generated in hydrothermal areas whose seismicity could be similar, in principle, to that observed in volcanic area. These features could explain the apparently encouraging results obtained in Central Asia, China, and Japan and the low scores obtained in low-heat-flow seismically active areas. In this sense volcanic fluid monitoring is at the scientific maturity stage while earthquake fluid precursor monitoring is still in its infancy.

Zhang et al. [2013] recently reported encouraging results obtained in the area of Beijing. In particular, over ten local seismic events were apparently preceded by 1–2 years of anomalous signals related to geophysical and geochemical parameters. The monitoring system considers about 19 different parameters. In this way a more multidisciplinary approach is adopted and a comprehensive super index is evaluated every 3 months in an attempt to forecast earthquakes at least in the medium term. the study has not reached a true alarm system but the authors believe that a multiparameter approach could be more suitable than monitoring a single parameter, at least in medium-term forecasting. No definitive results were obtained from short-term earthquake precursors monitoring since possible precursors are not detectable for all earthquakes, and a relatively high number of missed alarms were reported. Hydrogeologic and geochemical monitoring could, however, better contribute to constraining the time and location of occurrence of shocks forecasted by other methods [e.g. *Peresan et al.*, 2005; *Mantovani et al.*, 2010], since the existence of possible short-term precursory signals could, in principle, not be ruled out. In particular, where the monitored aquifer, as evidenced in the work of previous reviews, is directly fed by a deep geothermal system. This being the case, thermal springs and of deep wells should be considered as priority monitoring sites. Data obtained so far, although incomplete, look encouraging for future more in-depth research oriented to hazard reduction.

12.6. INDIRECT GEOCHEMICAL MONITORING BY SATELLITE TECHNIQUES

Several studies on earthquake precursors published in the past 30 years were carried out by means of satellite techniques [e.g. *Gorny et al.*, 1988; *Tronin*, 1996; *Qiang et al.*, 1997; *Pulinets and Ouzounov*, 2011]. In particular some studies detected anomalous space–time transients in the thermal infrared (TIR) radiation emitted by Earth and possibly due to the expulsion of underground gases induced by crustal deformation during the earthquake preparatory stage. *Pulinets et al.* [2006] reported that anomalies could be caused by ionization processes due to radon expulsion from the ground. *Surkov* [2015] and *Martinelli et al.* [2015] reported that radon contribution to eventual anomalies is negligible, while *Tramutoli et al.* [2013] reported that the expulsion of greenhouse gases like CO_2 and CH_4 could be effective in anomaly generation. *Ouzounov and Freund* [2004] reported that the enhanced infrared emission could be also due to motions of electric charges as a result of rocks being under stress. Crustal deformation may be accompanied by fluid expulsion when available, thus the introduction of a new research field of fluid monitoring. Ground-based and satellite-based joint measurements will probably better explain the observed phenomena.

12.7. THE NEED TO IMPROVE SEISMIC AND CRUSTAL DEFORMATIONS MONITORING

In the initial decades, geochemical and fluid monitoring was characterized by an empirical approach. The immediate correlation between a possible fluid-related anomaly and a forthcoming seismic event was followed in many cases by further research. Experiences, new technologies, and more accurate data-processing procedures indicated great difficulties in constraining time and location of a possible forthcoming seismic event. All experimental data obtained in fluid monitoring to date do not clearly indicate that the observed anomaly is generated by the hypocentral volume of a forthcoming earthquake. Crustal deformation, when detectable, results from regional large-scale processes. Thus eventual geochemical or fluid-related anomalies can be attributed, in general, to a large-scale process capable of generating strong seismic events or a variety of low-magnitude events. Recent data indicate that fluid-related anomalies could be interpreted as an indicator of stress-field evolution over time rather than as signals generated by a hypothetic focal volume of a forthcoming seismic event. A large part of CO_2, CH_4, and water reservoirs are located in the upper 2-3 km of the crust, whereas most seismic events occur at a depth of 10–30 km. Thus large-scale deformation processes, if able to reach Earth's surface, may easily affect fluid reservoirs, although it is difficult to believe in an active role of a single forthcoming seismic source. Fluid reservoirs and seismic sources are usually not directly linked and data analysis of further geophysical parameters could better constrain information of ongoing tectonic processes. Analysis of the evolution of minor seismicity, GPS stations, and satellite-based techniques including interferometric sensors could contribute to future research trends in terrestrial fluid monitoring.

12.8. CONCLUSIONS

The existence of migration processes associated with crustal deformation fronts requires monitoring crustal topography and crustal volumes subjected to stress-induced strain. Thermal springs and deep wells in confined aquifers could be monitored as high-resolution natural strainmeters. Automatic devices able to monitor water level, temperature, and electric conductivity in most seismically active areas of the world could be used effectively in short- and medium-term earthquake forecasting experiments. Gaseous component monitoring of carrier and noble gases could be jointly carried out, in particular in selected degassing areas characterized by relatively high heat-flow values. Instrumental arrays should be improved with respect to past pioneering experiences and characterized by at least one observational point (piezometer or thermal spring or gas emission) every 500 km^2, as in the best-monitored areas of Japan, China, and Taiwan. Geofluid monitoring networks should ultimately be embedded in existing GPS, tiltmeter, and strainmeter networks and in satellite-based monitoring activities. Data interpretation should be comprehensive and take into account all geophysical and geochemical monitored parameters.

DECLARATION

The authors declare that there are no conflicts of interest, that all data utilized in this chapter are obtainable from the references cited or deducible from figures, and that no funding was received to write the chapter.

REFERENCES

Agnew, D. C. (1986), Strainmeters and tiltmeters, *Rev. Geophys.*, *24*, 579–624, doi: 10.1029/RG024i003p00579

Albarello, D. and Martinelli, G. (1994), Piezometric levels as possible geodynamic indicators: analysis of the data from a regional deep waters monitoring network in Northern Italy, *Geophys. Res. Lett.*, *21*, 1955–1958, doi: 10.1029/94GL01598

Arieh, E. and Merzer, A. M. (1974), Fluctuations in oil flow before and after earthquakes, *Nature*, *247*, 534–535, doi:10.1038/247534a0

Asimov, M. S., Mavlyanov, F. A., Erzhanov, J. S., Kalmurzaev, K. E., Kurbanov, M. K., Kashin, L. N., Negmatulaev, S. H., and Nersesov, I. L. (1984), On the state of research concerning earthquake prediction in the Soviet republics of central Asia, in *Proceedings of the International Symposium on Eartquake Prediction*, pp. 565–573. Terra Scientific Publishing Company, Paris.

Bakun, W. H. (1988), Parkfield: The prediction and the promise, *Earthquakes Volcan.*, *20*(2), 40–91.

Barry, P. H., Hilton, D. R., Tryon, M. D., Brown, K. M., and Juologonski, J. T. (2009), A new syringe pump apparatus for the retrivial and temporal analysis of helium in groundwaters and geothermal fluids, *Geochem. Geophys. Geosyst.*, *10*(5), doi: 10.1029/2009GC002422

Bodvarsson, G. (1970), Confined fluids as strain meters, *J. Geophys. Res.*, *75*(14), 2711–2718, doi: 10.1029/JB075i014p02711

Buntebarth, G. and Chelidze, T. (eds) (2005), *Time-Dependent Microtemperature and Hydraulic Signals Associated with Tectonic/Seismic Activity*, Nodia Institute of Geophysics, Georgian Academy of Sciences, Tbilisi, 244 pp.

Chelidze, T., Matcharashvili, T., and Melikadze, G. (2010), Earthquakes' signatures in dynamics of water level variations in boreholes, in V., de Rubeis, Z. Czechowski and R. Teysseyre (eds), *Synchronization and Triggering: from Fracture to Earthquake Processes*, Springer-Verlag, Berlin–Heidelberg, pp. 287–303, doi: 10.1007/978-3-642-12300-9_17

Chen, C. H., Tang, C. C., Cheng, K. C., Wang, C. H., Wen, S., Lin, C. H., Wen, Y. Y., Meng, G., Yeh, T. K., Jan, J. C., and Yen, H. Y. (2015), Groundwater–strain coupling before the 1999 *Mw* 7.6 Taiwan Chi-Chi earthquake, *J. Hydrol.*, *524*, 378–384, doi: 10.1016/j.jhydrol.2015.03.006

Cicerone, R. D., Ebel, J. E., and Britton, J. (2009), A systematic compilation of earthquake precursors, *Tectonophysics*, *476*(3–4), 371–396, doi: 10.1016/j.tecto.2009.06.008

Cioni, R., Guidi, M., Pierotti, L., and Scozzari, A. (2007), An automatic monitoring network installed in Tuscany (Italy) for studying possible geochemical precursory phenomena, *Nat. Hazards Earth Syst. Sci.*, *7*(3), 405–416, doi: 10.5194/nhess-7-405–2007

Colangelo, G., Heinicke, J., Koch, U., Lapenna, V., Martinelli, G., and Telesca, L. (2005), Results of gas flux records in the seismically active area of Val d'Agri (Southern Italy), *Ann. Geophys.*, *48*(1), 55–63, doi: 10.4401/ag-3179

Colangelo, G., Heinicke, J., Lapenna, V., Martinelli, G., Mucciarelli, M., and Telesca, L. (2007), Investigating correlations of local seismicity with anomalous geoelectrical, hydrogeological and geochemical signals jointly recorded in Basilicata Region (Southern Italy), *Ann. Geophys.*, *50*(4), 527–538, doi: 10.4401/ag-3066

Contadakis, M. E. and Asteriadis, G. (2001), Recent results of the research for preseismic phenomena on the underground water and temperature in Pieria, northern Greece, *Nat. Hazards Earth Syst. Sci.*, *1*, 165–170, doi:10.5194/nhess-1-165–2001.

Facchini, U., Garavaglia, M., Magnoni, S., Rinaldi, F., Sordelli, C., and Delcourt-Honorez, M. (1993), Radon levels in a deep geothermal well in the Po plain, in V. Spagna and E. Schiavon (eds), *Proceedings of the Scientific Meeting on the Seismic Protection "Migration of Fluids in the Subsoil and Seismic Events: Compared Experiences"*, pp. 220–226, Venice.

Finkelstein, M., Brenner, S., Eppelbaum, L., and Ne'eman, E. (1998), Identification of anomalous radon concentrations due to geodynamic processes by elimination of Rn variations caused by other factors, *Geophys. J. Int.*, *133*(2), 407–412, doi: 10.1046/j.1365-246X.1998.00502.x

Firstov, P. P., Yakovleva, V., Shirokov, V. A., Rulenko, O. P., Filippov, Y. A., and Malysheva, O. P. (2007), The nexus of soil radon and hydrogen dynamics and seismicity of the northern flank of the Kuril-Kamchatka subduction zone, *Ann. Geophys.*, *50*(4), 547–556, doi: 10.4401/ag-3068

Frezzotti, M. L., Peccerillo, A., and Panza, G. (2009), Carbonate metasomatism and CO_2 lithosphere-astenosphere degassing beneath the Western Mediterranean: An integrated model arising from petrological and geophysical data, *Chem. Geol.*, *262*(1–2), 108–120, doi: 10.1016/j.chemgeo.2009.02.015

Friedmann, H. (1985), Anomalies in the Radon Content of Spring Water as Earthquake Precursor Phenomena, *Earthquake Prediction Research*, *1*, 179–189.

Fyfe, W. S., Price, N. J., and Thompson, A. B. (1978), *Fluids in the Earth's Crust*, Elsevier Scientific Publishing Company, Amsterdam, 383 pp.

Ghosh, D., Deb, A., and Sengupta, R. (2009), Anomalous radon emission as precursor of earthquake, *J. Appl. Geophys.*, *69*(2), 67–81, doi: 10.1016/j.jappgeo.2009.06.001

Gorny, V. I., Salman, A. G., Tronin, A. A., and Shilin, B. B. (1988), The Earth outgoing IR radiation as an indicator of seismic activity, *Proc. Acad. Sci. USSR*, *301*, 67–69.

Hartmann, J. and Levy, J. K. (2005), Hydrogeological and gas-geochemical earthquake precursors—a review for applications, *Nat. Hazards*, *34*(3), 279–304, doi: 10.1007/s11069-004-2072-2

Hauksson, E. (1981), Radon content of groundwater as an earthquake precursor: Evaluation of worldwide data and physical basis. *J. Geophys. Res. Solid Earth*, *86*(B10), 9397–9410, doi: 10.1029/JB086iB10p09397

Heinicke, J., Koch, U., and Martinelli, G. (1995), CO_2 and radon measurements in the Vogtland Area (Germany). A contribution to earthquake prediction research, *Geophys. Res. Lett.*, *22*(7), 771–774, doi: 10.1029/94GL03074

Heinicke, J., Italiano, F., Koch, U., Martinelli, G., and Telesca, L. (2010), Anomalous fluid emission of a deep borehole in a seismically active area of Northern Apennines (Italy), *Appl. Geochem.*, *25*(4), 555–571, doi:10.1016/j.apgeochem.2010.01.012

Huang, F., Jian, C., Tang, Y., Xu, G., Deng, Z., Chi, G-C., and Ferrar, C. D. (2004), Response changes of some wells in the mainland subsurface fluid monitoring network of China, due to the September 21, 1999, *Ms* 7.6 Chi-Chi earthquake, *Tectonophysics*, *390*, 217–234.

IHFC (2011), *International Heat Flow Commission* http://www.heatflow.und.edu/index2.html (last update January 12, 2011).

Johnston, M. J. S. and Linde, A. T. (2002), Implications of crustal strain during conventional, slow, and silent earthquakes, *Int. Handb. Earthquake Eng. Seismol.*, *81A*, 589–605.

Kasahara, K. (1979), Migration of crustal deformation, *Tectonophysics*, *52*(1–4), 329–341, doi:10.1016/0040-1951(79)90240-3.

King, C.-Y. and Igarashi, G. (2002), Earthquake-related hydrologic and geochemical changes, *Int. Handb. Earthquake Eng. Seismol.*, *81A*, 637–645.

King, C.-Y., Azuma, S., Ohno, M., Asai, Y., He, P., Kitagawa, Y., Igarashi, G., and Wakita, H. (2000), In search of earthquake precursors in the water-level data of 16 closely clustered wells at Tono, Japan, *Geophys. J. Int.*, *143*(2), 469–477, doi: 10.1046/j.1365-246X.2000.01272.x

Kissin, I. G. and Grinevsky, A. O. (1990), Main features of hydrogeodynamic earthquake precursors, *Tectonophysics*, *178*(2–4), 277–286, doi:10.1016/0040-1951(90)90154-Z.

Koizumi, N., Matsumoto, N., Itaba, S., and Ohtani, R. (2008), Integrated observation well network of geological survey of Japan, AIST for study of Nankai and Tonankai earthquakes, in C.L. Shieh K. Koizumi, and N. Matsumoto (eds), *Proceedings of 6th Taiwan–Japan International Workshop on Hydrological and Geochemical Research for Earthquake Prediction*, GSJ Open File Report 484, Geological Survey of Japan.

Kumpel, H.-J. (1992), About the potential of wells to reflect stress variations within inhomogeneous crust. *Tectonophysics*, *211*(1–4), 317–336, doi:10.1016/0040-1951(92)90068-H

Lapenna, V., Martinelli, G., and Telesca, L. (2004), Long-range correlation analysis of earthquake-related geochemical variations recorded in Central Italy, *Chaos, Solitons Fractals*, *21*(2), 491–500, doi:10.1016/j.chaos.2003.12.008

Mantovani, E., Viti, M., Babbucci, D., Albarello, D., Cenni, N., and Vannucchi, A. (2010), Long-term earthquake triggering in the Southern and Northern Apennines, *J. Seismol.*, *14*, 53–65, doi: 10.1007/s10950-008-9141-z

Martinelli, G. and Albarello, D. (1997), Main constraints for siting monitoring networks devoted to the study of earthquake related hydrogeochemical phenomena in Italy, *Ann. Geophys.*, *40*(6), 1505–1525, doi: 10.4401/ag-3827

Martinelli, G., Solecki, A. T., Tchorz-Trzeciakiewicz, D. E., Piekarz, M., and Grudzinska, K. K. (2015), Laboratory measurements on radon exposure effects on local environmental temperature: Implications for satellite TIR measurements, *Phys. Chem. Earth*, *85–86*, 114–118, doi:10.1016/j.pce.2015.03.007

Matsumoto, N. and Koizumi, N. (2011), Recent hydrological and geochemical research for earthquake prediction in Japan, *Nat. Hazards*, *69*(2), 1247–1260, doi: 10.1007/s11069-011-9980-8

Matsumoto, N., Kitagawa, Y., and Koizumi, N. (2007), Groundwater-level anomalies associated with a hypothetical preslip prior to the anticipated Tokai earthquake: Detectability using the groundwater observation network of the Geological Survey of Japan, *AIST, Pure Appl. Geophys.*, *164*, 2377–2396, doi: 10.1007/978-3-7643-8720-4_2

Ouzounov, D. and Freund, F. (2004), Mid-infrared emission prior to strong earthquakes analyzed by remote sensing data, *Adv. Space Res.*, *33*(3), 268–273, doi: 10.1016/S0273-1177(03)00486-1

Peresan, A., Kossobokov, V., Romashkova, L., and Panza, G. (2005), Intermediate-term middle-range earthquake predictions in Italy: a review. *Earth Sci. Rev.*, *69*(1–2), 97–132, doi:10.1016/j.earscirev.2004.07.005

Perez, N., Hernandez, P. A., Igarashi, G., Trujillo, I., Nakai, S., Sumino, H., and Wakita, H. (2008), Searching and detecting earthquake geochemical precursors in CO_2-rich groundwaters from Galicia, Spain, *Geochem. J.*, *42*, 75–83, doi: 10.2343/geochemj.42.75

Petraki, E., Nikolopoulos, D., Panagiotaras, D., Cantzos, D., Yannakopoulos, P., Nomicos, C., and Stonham, J. (2015), Radon-222: a potential short-term earthquake precursor, *J. Earth Sci. Clim. Change*, *6*, 282, doi: 10.4172/2157-7617.1000282

Popov, E. and Vartanyan, G. S. (1990), Investigation of geodynamic processes in connection with strong earthquakes prediction, *Proceedings of the ECE/UN Seminar on Prediction of Earthquakes*, pp. 73–85, Lisbon.

Pulinets, S. A. and Ouzounov, D. (2011), Lithosphere–atmosphere–ionosphere coupling (LAIC) model—a unified concept for earthquake precursors validation, *J. Asian Earth Sci.*, *41*(4–5), 371–382, doi:10.1016/j.jseaes.2010.03.005.

Pulinets, S. A., Ouzounov, D., Karelin, A. V., Boyarchuk, K. A., and Pokhmelnykh, L. A. (2006), The physical nature of thermal anomalies observed before strong earthquakes, *Phys. Chem. Earth*, *31* (4–9), 143–153, doi:10.1016/j.pce.2006.02.042

Qiang, Z. J., Xu, X. D., and Dian, C. D. (1997), Thermal infrared anomaly precursor of impending earthquakes, *Pure Appl. Geophys.*, *149*(1), 159–171, doi: 10.1007/BF00945166

Ranalli, G. (1995), *Rheology of the Earth*, Chapman and Hall, Padstow, 413 pp.

Rice, J. R. (1980), The mechanics of earthquake rupture, in A. Dziewonski and E. Boschi (eds), *Physics of the Earth's Interiors*, pp. 555–644, North Holland, Amsterdam.

Rice, J. R. and Cleary, M. P. (1976), Some basic stress diffusion solutions for fluid-saturated elastic porous media with compressible constituents, *Rev. Geophys. Space Phys.*, *14*(2), 227–241, doi: 10.1029/RG014i002p00227

Richon, P., Bernard, P., Labed, V., Sabroux, J.-C., Beneito, A., Lucius, D., Abbad, S., and Robe, M.-C. (2007), Results of monitoring ^{222}Rn in soil gas of the Gulf of Corinth region, Greece, *Radiat. Meas.*, *42*(1), 87–93, doi:10.1016/j.radmeas.2006.06.013

Riggio, A. and Sancin, S. (2005), Radon measurements in Friuli (N.E. Italy) and earthquakes: first results, *Boll. Geofis. Teor. Appl.*, *46*(1), 47–58.

Roeloffs, E. (1988), Hydrologic precursors to earthquakes: A review, *Pure Appl. Geophys.*, *126*(2), 177–209, doi: 10.1007/BF00878996

Roeloffs, E. (1996), Poroelastic techiques in the study of earthquake related-related hydrologic phenomena, *Adv. Geophys.*, *37*, 135–195, doi: 10.1016/S0065-2687(08)60270-8

Roeloffs, E. (1999), Radon and rock deformation, *Nature*, *399*, 104–105, doi:10.1038/20072

Roeloffs, E. (2006), Evidence for aseismic deformation rate changes prior to earthquakes, *Ann. Rev. Earth Planet. Sci.*, *34*, 591–627, doi: 10.1146/annurev.earth.34.031405.124947

Roeloffs, E., Quilty, E., and Scholtz, C. H. (1997), Water level and strain changes preceding and following the August 4, 1985 Kettleman Hills, California, earthquake, *Pure Appl. Geophys.*, *149*(1), 21–60, doi: 10.1007/BF00945160

Segovia, N., De La Cruz-Reyna, S., Mena, M., Ramos, E., Monnin, M., and Seidel, J. L. (1989), Radon in soil anomaly observed at Los Azufres geothermal field, Michoacan: A possible precursor of the 1985 Mexico earthquake (Ms=8.1), *Nat. Hazards*, *1*(4), 319–329, doi: 10.1007/BF00134830

Shi, Z., Wang, G., and Liu, C. (2013), Advances in research on earthquake fluids hydrogeology in China: a review, *Earthquake Sci.*, *26*(6), 415–425, doi: 10.1007/s11589-014-0060-5

Sidorenko, A. V., Sadovsky, M. A., Nersesov, I. L., Popov, E. A., and Soloviev, S. L. (1984), Soviet experience in earthquake prediction in the USSR and the propspects for its development, *Proceedings of the Interantional Symposium on*

Eartquake Prediction, pp. 565–573, Terra Scientific Publishing Company, Paris.

Sidorin, A. Ya. (2003), Search for earthquake precursors in multidisciplinary data monitoring of geophysical and biological parameters, *Nat. Haz. Earth Syst. Sci.*, *3*, 153–158.

Silver, P. G., Valette-Silver, N. G., and Kolbek, O. (1993), Detection of hydrothermal precursors to large Northern California earthquakes, in M.J.S. Johnston (ed.), *The Loma Prieta, California, Earthquake of October 17, 1989: Earthquake Occurrence. Preseismic Observations, US Geol. Surv. Prof. Pap.*, *1550-C*, c73–c80.

Skelton, A., Andrén, M., Kristmannsdòttir, H., Stockmann, G., Morth, C.-M., Sveinbjornsdottir, A., Jonsson, S., Sturkell, E., Gudrunardottir, H. R., Hjartarson, H., Siegmund, H., and Kockum, I. (2014), Changes in groundwater chemistry before two consecutive earthquakes in Iceland, *Nature Geosci.*, *7*, 752–756, doi:10.1038/ngeo2250

Steinitz, G., Begin, Z. B., and Gazit-Yaari, N. (2003), Statistically significant relation between radon flux and weak earthquakes in the Dead Sea rift valley, *Geology*, *31*(6), 505–508, doi: 10.1130/0091-7613(2003)031<0505:SSRBRF>2.0.CO;2

Stejskal, V., Kasparek, L., Kopilova, G. N., Lyubushin, A. A., and Skalsky, L. (2009), Precursory groundwater level changes in the period of activation of the weak intraplate seismic activity on the NE margin of the Bohemian Massif (central Europe) in 2005, *Stud. Geophys. Geod.*, *53*(2), 215–238, doi: 10.1007/s11200-009-0014-x

Sugisaki, R. (1981), Deep-seated gas emission induced by the Earth tide: A basic observation for geochemical earthquake prediction, *Science*, *212*, 1264–1266, doi: 10.1126/science.212.4500.1264

Surkov, V. V. (2015), Pre-seismic variation of atmospheric radon activity as a possible reason for abnormal atmospheric effects. *Ann. Geophys.*, *58*(5), doi: 10.4401/ag-6808

Swolfs, H. S. and Walsh, J. B. (1990), The theory and prototype development of a stress-monitoring system, *Bull. Seismol. Soc. Am.*, *80* (1), 197–208.

Taran, Y. A., Ramirez-Guzman, A., Bernard, R., Cienfuegos, E., and Morales, P. (2005), Seismic-related variations in the chemical and isotopic composition of thermal springs near Acapulco, Guerrero, Mexico. *Geophys. Res. Lett.*, *32*(14), L14317, doi: 10:1029/2005GL022726

Thomas, D. (1988), Geochemical precursors to seismic activity, *Pure Appl. Geophys.*, *126*(2), 241–266, doi: 10.1007/BF00878998

Torkar, D., Zmazek, B., Vaupotic, J., and Kobal, I. (2010), Application of artificial neural networks in simulating radon levels in soil gas, *Chem. Geol.*, *270*(1–4), 1–8, doi:10.1016/j.chemgeo.2009.09.017

Toutain, J.-P. and Baubron, J.-C. (1999), Gas geochemistry and seismotectonics: a review, *Tectonophysics*, *304*(1–2), 1–27, doi:10.1016/S0040-1951(98)00295-9

Tramutoli, V., Aliano, C., Corrado, R., Filizzola, C., Genoano, N., Lisi, M., Martinelli, G., and Pergola, N. (2013), On the possibile origin of thermal infrared radiation (TIR) anomalies in earthquake-prone areas observed using robust satellite techniques (RST), *Chem. Geol.*, *339*, 157–168, doi:10.1016/j.chemgeo.2012.10.042

Tronin, A. A. (1996), Satellite thermal survey-a new tool for the study of seismoactive regions, *Int. J. Remote Sens.*, *17*(8), 1439–1455, doi: 10.1080/01431169608948716.

Vartanyan, G. S. (2014), Fast deformation cycles in the lithosphere and catastrophic earthquakes: was it possible to prevent the Fukushima tragedy? *Izvest. Atmos. Ocean. Phys.*, *50*(8), 805–823, doi: 10.1134/S0001433814080088

Vartanyan, G. S., Bredehoeft, J. D., and Roeloffs, E. (1992), Hydrogeological methods for studying tectonic stress, *Sovetsk. Geol.*, *9*, 3–12. (In Russian.)

Walia, V., Yang, T. F., Lin, S.-J., Kumar, A., Fu, C.-C., Chin, J.-M., Chang, H.-H., Wen, K.-L., and Chen, C.-H. (2013), Temporal variation of soil gas compositions for earthquake surveillance in Taiwan, *Radiat. Meas.*, *50*, 154–159, doi:10.1016/j.radmeas.2012.11.007

Wang, C.-Y. and Manga, M. (2010), *Earthquakes and Water*, Springer-Verlag, Berlin–Heidelberg, 225 pp, doi: 10.1007/978-3-642-00810-8

Wang, K., Chen, Q.-F., Sun, S., and Wang, A. (2006), Predicting the 1975 Haicheng earthquake, *Bull. Seismol. Soc. Am.*, *96*(3), 757–795, doi: 10.1785/0120050191

Wiss, M. (1991), *Evaluation of Proposed Earthquake Precursors*. American Geophysical Union, Washington, DC, 94 pp, doi: 10.1029/90EO10300

Wiss, M. and Booth, D. C. (1997), The IASPEI procedure for the evaluation of earthquake precursors, *Geophys. J. Int.*, *131* (3), 423–424, doi: 10.1111/j.1365-246X.1997.tb06587.x

Woith, H. (2015), Radon earthquake precursor: A short review, *Eur. Phys. J. Spec. Topics*, *224*(4), 611–627, doi: 10.1140/epjst/e2015-02395-9

Woith, H., Venedikov, A. P., Milkereit, C., Parlaktuna, M., and Pekdeger, A. (2006), Observation of crustal deformation by means of wellhead pressure monitoring, *Bull. Inform. Mar. Terres.*, *141*, 11277–11285.

Woith, H., Wang, R., Maiwald, U., Pekdeger, A., and Zschau, J. (2013), On the origin of geochemical anomalies in groundwaters induced by the Adana 1998 earthquake, *Chem. Geol.*, *339*, 177–186, doi:10.1016/j.chemgeo.2012.10.012

Yuce, G., Ugurluoglu, D. Y., Adar, N., Yalcin, T., Yaltirak, C., Streil, T., and Oeser, V. (2010), Monitoring of earthquake precursors by multi-parameter stations in Eskisehir region (Turkey). *Appl. Geochem.*, *25*(4), 572–579, doi: 10.1016/j.apgeochem.2010.01.013

Zencher, F., Bonafede, M., and Stefansson, R. (2006), Near-lithostatic pore pressure at seismogenic depths: a thermoporoelastic model, *Geophys. J. Int.*, *166*(3), 1318–1334, doi: 10.1111/j.1365-246X.2006.03069.x

Zhang, Y., Gao, F., Ping, J., and Zhang, X. (2013), A synthetic method for earthquake prediction by multidisciplinary data, *Nat. Hazards*, *69*(2), 1199–1209, doi: 10.1007/s11069-011-9961-y

Zheng, G., Xu, S., Liang, S., Shi, P., and Zhao, J. (2013), Gas emission from the Qingzhu River after the 2008 Wenchuan Earthquake, Southwest China, *Chem. Geol.*, *339*,187–193, doi:10.1016/j.chemgeo.2012.10.032

Zhou, X., Chen, Z., and Cui, Y. (2015), Environmental impact of CO_2, Rn, Hg, degassing from the rupture zones produced by Wenchuan M_s 8.0 earthquake in western Sichuan, China, *Environ. Geochem. Health*, *38*(5), 1067–1082, doi: 10.1007/s10653-015-9773-1

Zschau, J. and Ergunay, O. (eds) (1989), *Turkish–German Earthquake Research Project*, Ankara, 211 pp., doi:10.1029/2007GL029222

13

Statistical Analysis and Assessment of Ultralow Frequency Magnetic Signals in Japan As Potential Earthquake Precursors

Katsumi Hattori[1] and Peng Han[2]

ABSTRACT

Electromagnetic measurements from the space to the ground have documented a number of signals associated with large earthquakes over the past few decades. Although there is still active debate on the seismo-electromagnetic phenomena, a large number of statistical investigations have shown correlation between electromagnetic anomalies and large earthquakes. In this chapter, we first briefly review results of previous studies, and then we discuss the potential precursory information in the seismo-electromagnetic data. We use Molchan's error diagram to evaluate the potential earthquake precursory information in the ultralow-frequency (ULF) magnetic data recorded in the Kanto region of Japan during 2000–2010. The results show that the earthquake detections based on magnetic anomalies are clearly better than a random guess, indicating that the magnetic data contain potentially useful information on earthquake forecasts.

13.1. INTRODUCTION

The detection of electromagnetic perturbations before fault ruptures has been proposed as a useful way to monitor activities of Earth's crust. So far, there have been many reports on electromagnetic field changes associated with earthquakes in a very wide frequency range from megahertz to quasi-DC [e.g., *Hayakawa and Fujinawa*, 1994; *Hayakawa*, 1999; *Hayakawa and Molchanov*, 2002; *Uyeda et al.*, 2009a]. These seismo-electromagnetic phenomena have been intensively studied worldwide for some decades [e.g., *Park*, 1996; *Johnston*, 1997; *Pulinets et al.*, 2003; *Pulinets and Boyarchuk*, 2004; *Hattori*, 2004; *Sarkar et al.*, 2007; *Huang*, 2011a]. Generally, seismo-electromagnetic phenomena can be classified into two types: (a) electric and magnetic changes in the lithosphere and (b) perturbations in the atmosphere and the ionosphere.

[1] *Graduate School of Science, Chiba University, Chiba, Japan*
[2] *Southern University of Science and Technology, Shenzhen, China*

The seismo-electromagnetic phenomena in the lithosphere are commonly recorded by passive ground-based observations of the electric field and/or the magnetic field [*Hayakawa and Molchanov*, 2002]. One of the most promising candidates is the measurement of ultralow-frequency (ULF) electromagnetic phenomena, because of their deeper skin depths [e.g., *Park*, 1996; *Johnston*, 1997; *Hattori*, 2004]. A large number of ULF electromagnetic phenomena associated with earthquakes has been reported [*Varotsos and Lazaridou*, 1991; *Hayakawa et al.*, 1996; *Ismaguilov et al.*, 2001; *Nagao et al.*, 2002; *Uyeda et al.*, 2002, 2009b; *Hattori et al.*, 2004a,b, 2013a; *Serita et al.*, 2005; *Zhuang et al.*, 2005; *Telesca and Hattori*, 2007; *Telesca et al.*, 2008; *Bleier et al.*, 2009; *Han et al.*, 2009, 2015, 2016; *Chavez et al.*, 2010; *Chen et al.*, 2010, 2013; *Dunson et al.*, 2011; *Hirano and Hattori*, 2011; *Huang*, 2011a,b, 2015; *Prattes et al.*, 2011; *Wen et al.*, 2012; *Xu et al.*, 2013; Schekotov and Hayakawa, 2015].

Although various electromagnetic anomalies have been proposed, they were not detected before all large earthquakes, nor were they always followed by them. The electromagnetic anomaly seems to be neither a necessary

Pre-Earthquake Processes: A Multidisciplinary Approach to Earthquake Prediction Studies, Geophysical Monograph 234, First Edition. Edited by Dimitar Ouzounov, Sergey Pulinets, Katsumi Hattori, and Patrick Taylor.

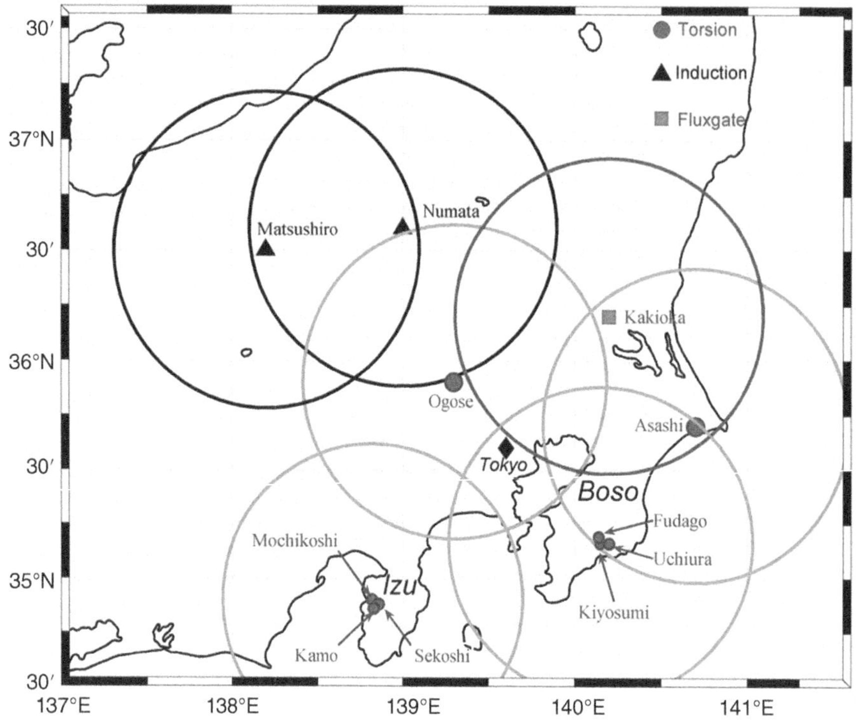

Figure 13.1 The ULF geomagnetic observation network in the Kanto–Tokai region of Japan. The radius of the circles is 80 km.

nor a sufficient condition for a large earthquake. These observations have led to active debates on the feasibility of seismo-electromagnetic study for earthquake prediction [*Kappler et al.*, 2010; *Varotsos et al.*, 2013]. To address this problem, some statistical studies have been performed based on long-term continuous observations [e.g., *Zhuang et al.*, 2005; *Schekotov et al.*, 2007; *Hattori et al.*, 2013a; *Han et al.*, 2014]. The statistical results suggest a significant increase of probability of electromagnetic anomalies before large earthquakes, implying that there is a relation between the appearance of anomalies and the occurrence of earthquakes. However, whether these anomalies contain precursory information and how they can improve the forecasting of sizable earthquakes have not been determined yet. In this study, we present the statistical analysis of siesmo-electromagnetic phenomena in the Kanto Region of Japan, and assess the potential precursory information for earthquakes.

13.2. JAPANESE ULF GEOMAGNETIC NETWORK OBSERVATION SYSTEM

According to *Hattori et al.* [2004a], ULF emissions could be observed within about 60 km from the source region for an earthquake with $M \geq 6.0$, and within about 100 km in the case of an earthquake with $M \geq 7.0$. Based on this result, a sensitive geomagnetic network of torsion magnetometers [*Hattori et al.*, 2004a] with interstation distances about 100 km was installed in the Kanto region of Japan. In this study, we focus on the earthquake-related ULF geomagnetic signature observed in the Izu and Boso Peninsulas where seismic activities are intensive.

Figure 13.1 displays the distribution of ULF geomagnetic stations in Izu Peninsula (Seikoshi, Mochikoshi, and Kamo) and Boso Peninsula (Kiyosumi, Fudago, and Uchiura), and the Kakioka station. The details of these stations are given in *Hattori et al.* [2004a].

The original sampling rate is 50 Hz. The data are resampled down to 1 Hz [*Hattori et al.*, 2004a] and transferred to a database server at Chiba University every day via internet or public telephone line. The observation system sometimes stopped recording or registered error data due to power failure, thunderstorm, or instrument malfunction. Because the Seikoshi (SKS) and Kiyosumi (KYS) stations have the most complete data in Izu and Boso [*Hattori et al.*, 2013a], respectively, we mainly focus on these two stations for these two areas. We also later focus on data from the Kakioka (KAK) station, which is operated by the Japan Meteorological Agency (JMA), with high data quality.

13.3. DATA PROCESSING

Generally, ground-based ULF geomagnetic data are a superposition of several signals: global magnetic perturbations, artificial noises, and magnetic signals possibly

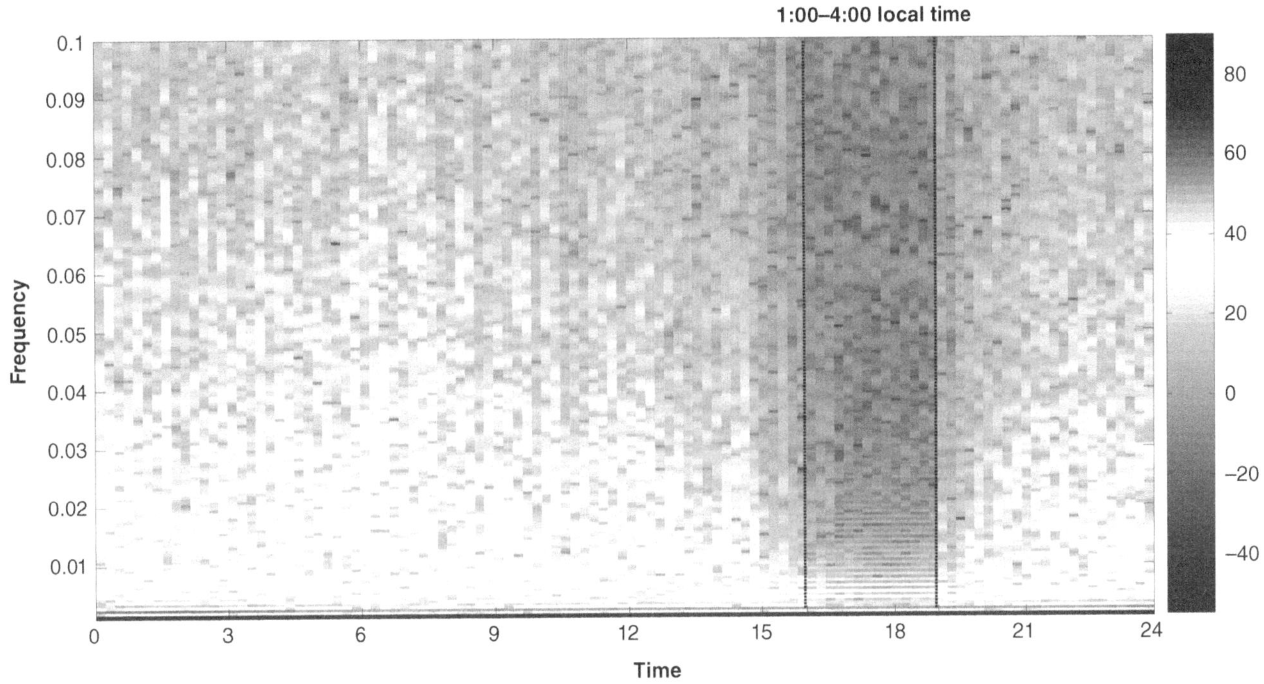

Figure 13.2 A typical spectrum of geomagnetic field (*Z* component) observed at the KYS station (Boso) on 3 February 2001. (*See electronic version for color representation.*)

due to underground activities. Therefore, the key point in seismomagnetic studies is to distinguish the earthquake-related signals from the others. Figure 13.2 shows a typical spectrogram of the *Z* component of geomagnetic data at the KYS station on 3 February 2001, with a clear time zone of lower background noise from midnight to early morning in local time (LT) when the train system is shut down. Thus, to minimize the influence of such artificial noises, we only utilize the data during 01:00–04:00 LT for the KYS.

The spectrum of magnetic signals at the SKS (Izu) and KAK stations are also investigated. It is found that there are lower background noises during 01:00–04:00 LT at the SKS station and 02:30–04:00 LT at the KAK station. We utilize the data observed during the relevant period at each station.

In previous studies, one of the most frequently reported periods of seismo-electromagnetic phenomena was around 100 s [*Fraser-Smith et al.*, 1990; *Hayakawa et al.*, 1996; *Uyeda et al.*, 2002; *Hattori*, 2004; *Hattori et al.*, 2004a,b, 2013a,b; *Han et al.*, 2011]. In this study we apply the wavelet transform to the 1-Hz geomagnetic data and extract the signals at the frequency around 0.01 Hz. After comparison among several types of wavelets, we adopt Daubechies 5 (db5) as the mother wavelet, following *Jach et al.* [2006]. Details of such wavelet analysis are found in *Han et al.* [2011] and *Hattori et al.* [2013a,b].

Early studies have indicated that there are possible energy enhancements of ULF geomagnetic signals in the *Z* component prior to large earthquakes [*Hattori*, 2004; *Hattori et al.*, 2004a, 2013a; *Han et al.*, 2011]). Therefore, we focus on the geomagnetic energy variation in the *Z* component and discuss the relationship between the energy enhancements and earthquake activities. First, we apply a six-level discrete wavelet transform using the db5 mother wavelet to the daily 1-Hz nighttime geomagnetic data. Next, we extract the signals in the sixth level where the central frequency is 0.01 Hz, and finally we compute the daily average energy of the obtained 0.01 Hz signals (see *Hattori et al.* [2013a] for more details).

13.4. STATISTICAL ANALYSIS

Although there have been a number of reports on ULF seismomagnetic phenomena worldwide, the underlying physics of these observed results are still unclear. The earthquake process is actually rather complex. There could be some geomagnetic anomalies originating from underground activities that do not lead to large earthquakes, and some large earthquakes that do not create geomagnetic anomalies. Therefore, we should not expect a one-to-one correspondence between anomalies and earthquakes. This makes the study of seismomagnetic phenomena rather complex and difficult. Statistical tests are required to verify the correlation between geomagnetic anomalies and local seismicity. In this section, statistical studies based on superposed epoch analysis (SEA) [e.g. *Hattori et al.*, 2013a] have been performed at the KYS and SKS stations during 2000–2010, and the KAK station during 2001–2010.

13.4.1. SEA in Izu and Boso

To undertake a statistical study, an important consideration is how to select earthquakes. There is a threshold for a detectable ULF seismomagnetic signature according to *Hattori et al.* [2004a]. In this study, we use the E_s parameter, which considers the magnitude and distance of an earthquake simultaneously to select statistical samples. The E_s index is the daily sum of the local earthquake energy obtained by the following formula [*Hattori et al.*, 2006; *Hirano and Hattori*, 2011]

$$Es = \sum_{1\text{day}} \frac{10^{4.8+1.5M}}{r^2} \tag{13.1}$$

where M and r indicate the magnitude of earthquake and hypocenter distance from a geomagnetic station. We define an event day when the E_s parameter exceeds 10^8. For statistical analysis in Izu, we select earthquakes within 100 km from SKS station; for Boso, within 150 km from KYS station. As a result, there are in total 60 event days in Izu and 92 event days in Boso, satisfying the criterion, respectively.

Another important parameter for statistical studies is the criterion for geomagnetic anomaly definition. According to case studies in *Hattori et al.* [2013a], the energy of the Z component seems to increase prior to intensive underground activities. Thus, we define an anomaly when the energy of the Z component exceeds the median + $1.5IQR$, where IQR is the interquartile range. Considering that ionospheric disturbances may also lead to enhancement in the Z component, we exclude the anomaly when the energy of the H component in the reference station exceeds median + $3IQR$.

SEA is a statistical method to highlight a weak signal from noisy data. This technique can reveal typical characteristics, periodicities, and consequences of a special event [*Adams et al.*, 2003; *Hocke*, 2008], and has been introduced in the studies of earthquake related GPS–TEC (total electron content) anomalies [*Kon et al.*, 2011] and ULF geomagnetic anomalies [*Hattori et al.*, 2013a]. In this study, we apply the SEA method to test ULF seismomagnetic phenomena in the Kanto region of Japan. First, for each earthquake event, we create a dataset of the geomagnetic parameters for 45 days before and after its day of occurrence. The time span of each dataset is 91 days and centered on the event day. Next we investigate the predefined anomalous values. If the anomaly exists, we count 1 for the corresponding day. We repeat this procedure for all the 60 (92) selected earthquake events in Izu (Boso) and we add up the counts for all datasets. We then obtain the SEA result in terms of the seismomagnetic anomaly. To evaluate the statistical significance, we randomly pick 60 (92) days from the entire analyzed period for SKS (KYS) station instead of 60 (92) earthquake days, and then perform the same procedure described above, to obtain the random SEA for a ULF anomaly. We repeat such random SEA tests for 100,000 times to compute the mean (hereafter random_mean) and the corresponding standard deviation (σ).

Figure 13.3 summarizes the results of statistically calculating ULF magnetic anomalies by SEA in Izu and Boso during 2000-2010 (modified from *Hattori et al.* [2013a]). For comparison, the 5-day counts were normalized using the corresponding random_mean + 2σ values, which were then calculated via 10,000 times random SEA tests. Thus, if the 5-day counts are more than the corresponding random_mean + 2σ values, the normalized value will be > 1; otherwise it will be < 1. Statistically, if the times of magnetic anomalies distribute randomly in relation to the earthquake events, the counts of anomalies probably do not exceed random_mean + 2σ values. Conversely, if the counts exceed random_mean + 2σ values, the time of magnetic anomalies and that of earthquake events probably have some correlation. In Figure 13.3, the counts of several intervals before the earthquake are clearly above the random_mean + 2σ value, suggesting a possible correlation between magnetic anomalies and earthquake events.

13.4.2. SEA at the KAK Station

We adopt the same criterion used in Izu and Boso: an earthquake event occurs when the E_s index in that day exceeds 10^8. Here the region within 100 km from the KAK station is named region A. There are 50 earthquake events. Furthermore, to investigate whether there is any distance dependence of seismomagnetic anomalies, we choose another 50 samples with $E_s > 10^8$ in a more distant region (region B) surrounding region A (Figure 13.4). The earthquake events in regions A and B are shown in Figure 13.4, and detailed information is reported in *Han et al.* [2014]. Here we choose 216 km as the outer boundary for region B, such that there are equal numbers of earthquake events in regions A and B during 2001–2010.

We focus on the Z component as it might record possible naturally-enhanced energy signals prior to a large earthquake. We estimate the global disturbances at the KAK station by applying a linear least-square method to the data from a reference point in Kanoya (KNY station), based on their high correlation of 10-year daily energy. Finally, we define a new parameter, the κ value,

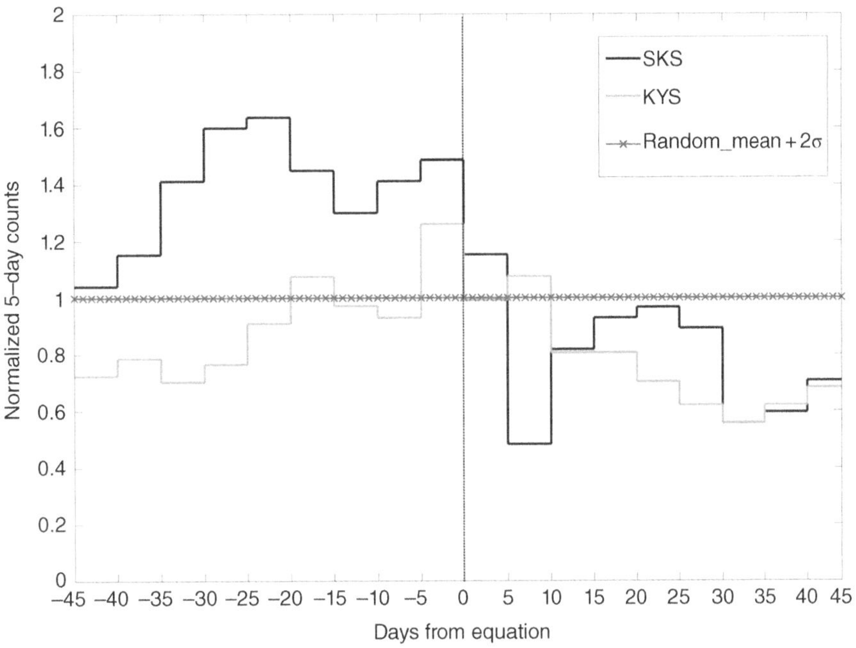

Figure 13.3 Statistical results of ULF magnetic anomalies by superposed epoch analysis (SEA) at the Izu (SKS) and Boso (KYS) stations during 2000–2010. (*See electronic version for color representation.*)

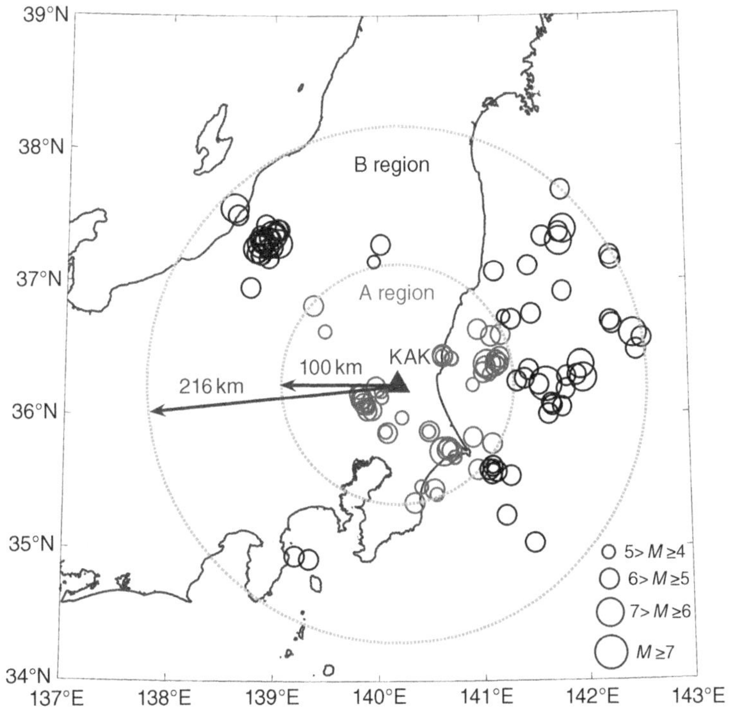

Figure 13.4 Spatial distributions of major earthquakes with $E_s > 10^8$ around the KAK station during 2001–2010. The blue triangle indicates the location of the KAK station. The red and black open circles indicate earthquakes in regions A and B, respectively. (*See electronic version for color representation.*)

which is the ratio between observed daily energy data Z_{KAK} and its model Z^*_{KAK}:

$$\kappa = \frac{Z_{KAK}}{Z^*_{KAK}} \quad (13.2)$$

Details of this process are described in *Han et al.* [2014]. The value could effectively reflect the local energy enhancements at the KAK station and therefore contain precursory information for forecasting local sizable earthquakes [*Han et al.*, 2014]. We define an anomaly when the value exceeds median + 1.5*IQR*.

We apply SEA to examine the correlation between the defined geomagnetic anomalies at the KAK station and the earthquake events in region A. Furthermore, to investigate whether the occurrence of the seismomagnetic anomalies depends on the epicenter distance, similar SEA has been applied in region B. Here, the number of earthquake events is also 50, and the number of geomagnetic anomalies is 324, which are the same as those used in region A. Considering that the results of 5-day counts are more stable and can provide more general characteristics of the data, hereafter we mainly discuss the results of 5-day counts.

For comparison, Figure 13.5 plots the 5-day counts of regions A and B simultaneously. The blue and green lines with cross markers show the random_mean and random_mean + 2σ, respectively. These two values are identical for regions A and B, because there are equal numbers of earthquake events and geomagnetic anomalies. In Figure 13.5, significant anomalies are found within 6–15 days before the earthquake events in region A, while there are no such clear anomalies in region B. These results suggest that the ULF geomagnetic anomalies observed at the KAK station are more likely to be related to closer earthquake events in region A rather than more distant events in region B. In other words, the ULF seismomagnetic anomalies are more sensitive to closer earthquakes, and are dependent on the epicenter distance.

13.5. ASSESSING THE PRECURSORY INFORMATION

In the following, we investigate whether the geomagnetic anomalies contain precursory earthquake information through Molchan's error diagram.

13.5.1. Molchan's Error Diagram

Assuming that there is a time lag between an earthquake event and the onset of a precursory anomaly, the lead time (Δ) and the alarm window (L) are introduced to

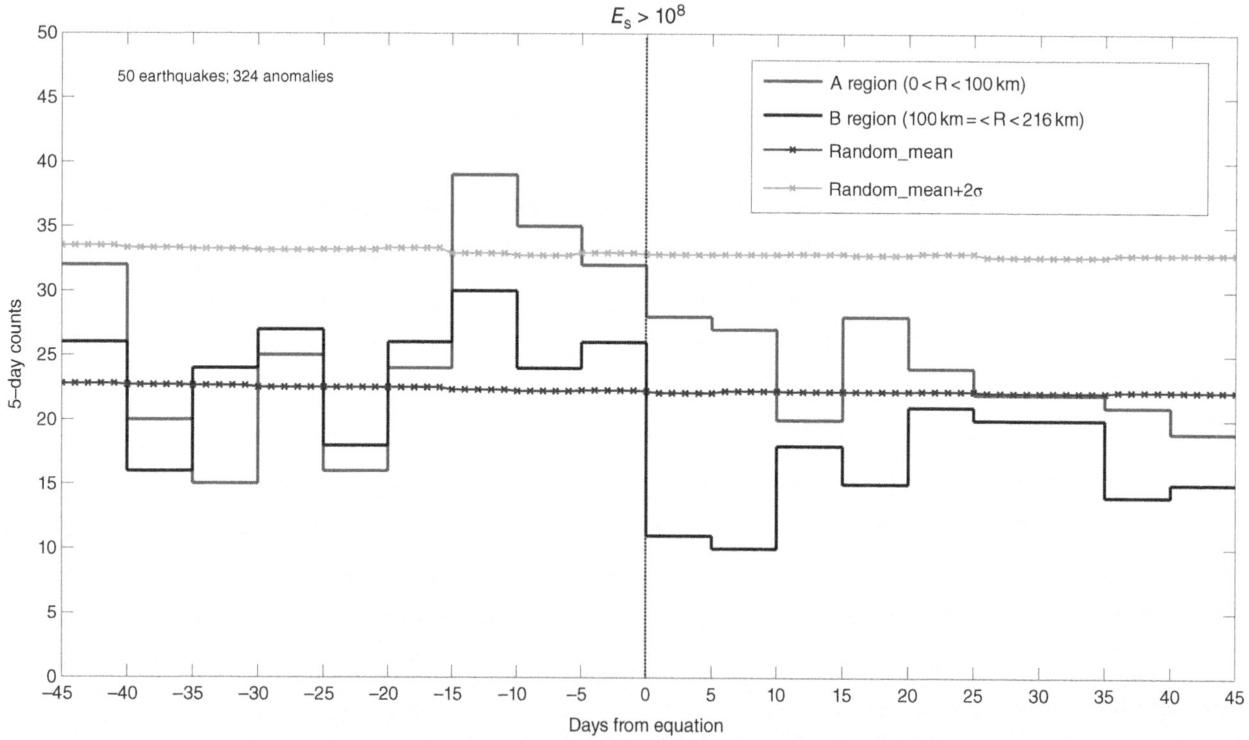

Figure 13.5 Statistical results of ULF magnetic anomalies by superposed epoch analysis (SEA) at the KAK station during 2001–2010. The red and the black lines demonstrate the results of 5-day counts for regions A and B, respectively. The blue and green lines with cross markers show random_mean and random_mean + 2σ, respectively. (*See electronic version for color representation.*)

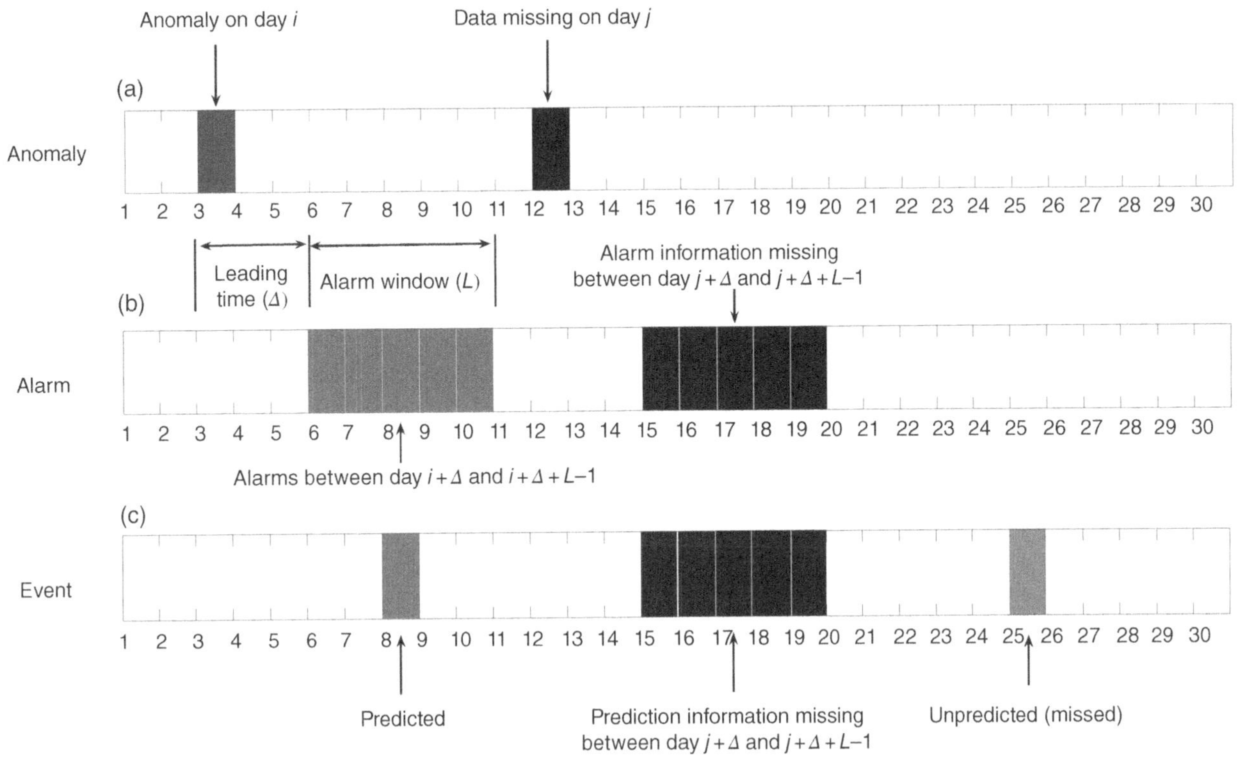

Figure 13.6 A schematic illustration of the alarm implementation scheme. Figures from the top to the bottom (a–c) show a series of anomaly, alarm, and earthquake events, respectively. The units of the horizontal axis are days.

specify the alarm function, where Δ is the time length between a detected anomaly and the following alarm to issue, and L is the number of days for which the issued alarm lasts. The time unit is 1 day for both Δ and L. First, we selected a value for threshold P to define the magnetic anomalies. Next, we issued an alarm for the occurrence of an "earthquake event" with $E_s > 10^8$ in the alarm interval between Day $i+\Delta$ and Day $i+\Delta+L-1$ if the magnetic energy on Day i (or the ith day) showed an anomaly, as demonstrated in Figure 13.6. Based on this alarm algorithm, each day in the alarm series was determined by the anomaly information in the L-day-long window between $\Delta+L-1$ day before and Δ day before. Subsequently, we took account of missing data in the anomaly series. As shown in Figure 13.6, if there are data missing on Day j, the alarm information between Day $j+\Delta$ and Day $j+\Delta+L-1$ will be affected, as we cannot tell whether Day j is an anomaly day. In such a case, to ensure that each day has equal information, we assumed the affected day in the alarm series as missing alarm information, and regarded the corresponding day in the event series as missing prediction information. We excluded these information-missing days in further analysis.

To evaluate the performance of the prediction algorithm, we define the detection rate (ν) and alarm rate (τ) as the functions of alarm threshold P. First, we count an "event day" as "predicted" if it falls within an alarm interval, or "missed" if it falls outside an alarm interval. Second, we set N = the total number of days during the entire period analyzed, n = the total number of event days during the period, N_1 = the number of alarmed days during the period, and n_1 = the number of predicted event days during the period, respectively. The "detection rate," the proportion of predicted earthquake events, can be given as $\nu(P) = n_1/n$. The "alarm rate," the proportion of alarmed days, can be given as $\tau(P) = N_1/N$ [*Molchan*, 1991; *Wang et al.*, 2013]. For a practical purpose, P is set to change from the maximum to the minimum of the energy data. Correspondingly, τ and ν range from 0 to 1.

Molchan's error diagram used here plots detection rate ν against alarm rate τ. It is modified from its original version, which plotted unpredicted rate versus alarm rate [*Molchan*, 1991]. Generally, the detection rate increases with the alarm rate. In the Poisson model, if the alarms are made randomly, the detection rate will be almost equal to the alarm rate [*Molchan*, 1991; *Zechar and Jordan*, 2008]. Therefore, the diagonal line indicates the prediction by a random guess (a la Poisson model). Any prediction with the τ–ν curve above this diagonal line indicates that the proportion of predicted earthquakes is greater than that of alarmed days, i.e., the prediction is better than a random guess. If the prediction is below the diagonal line, it is worse than a random guess. Note that, at a given alarm rate τ, the probability of n_1 "detections" using a random guess follows a binomial distribution:

$$B(n_1 \mid n,\tau) = \binom{n}{n_1}(\tau)^{n_1}(1-\tau)^{n-n_1} \qquad (13.3)$$

As a result, one can compute the 95% confidence interval of 'detection' at each given alarm rate. If the prediction curve is above the upper threshold of confidence intervals, the prediction is significantly better than a random guess. In other words, the anomalies here contain precursory information and can improve the forecasting of earthquake events.

13.5.2. Efficiency of Magnetic Anomalies for Earthquake Forecasting

In Figure 13.3 there are statistical significances in the −5 day to −1 day window at both the SKS and KYS stations. In this study the lead time (Δ) is set at 1 day and the alarm window (L) at 5 days, so that an earthquake event can be alarmed if there is any magnetic anomaly 1–5 days before it. Figures 13.7a and 13.7b show the modified Molchan's error diagram using magnetic anomalies as precursors at the SKS station (Izu) and the KYS station (Boso), respectively. The solid red lines are prediction curves using observed magnetic data. The black lines show predictions by a random guess and the blue lines represent the 95% confidence intervals. The prediction curves of both the SKS and KYS stations are clearly better than a random guess, and most of them exceed the upper 95% confidence threshold, which suggests the effectiveness of the magnetic data for short-term earthquake forecasting.

For practical application to short-term earthquake forecasting, we need to choose a certain value of P threshold for issuing alarms. Consequently, the actual prediction result should be only a single point on the curve. In general, a higher detection rate causes a large number of false alarms, whereas a lower alarm rate causes a large number of missed earthquakes. Thus, we cannot simply choose one single parameter (ν or τ) for better prediction. The value of the P threshold must be determined according to the operator's priority: whether gain (ν) or cost (τ) is preferred. One useful index could be probability gain (PG), which is the ratio of detection rate ν to alarm rate τ [*Molchan*, 1991]. Another useful index could be the difference (D) between ν and τ. $PG=1$ or $D=0$ indicates completely random prediction. The maximum PG and D values suggesting the optimal P threshold are indicated in Figure 13.7.

In Figure 13.5 the most prominent result was 11–15 days prior to sizeable earthquakes at the KAK station.

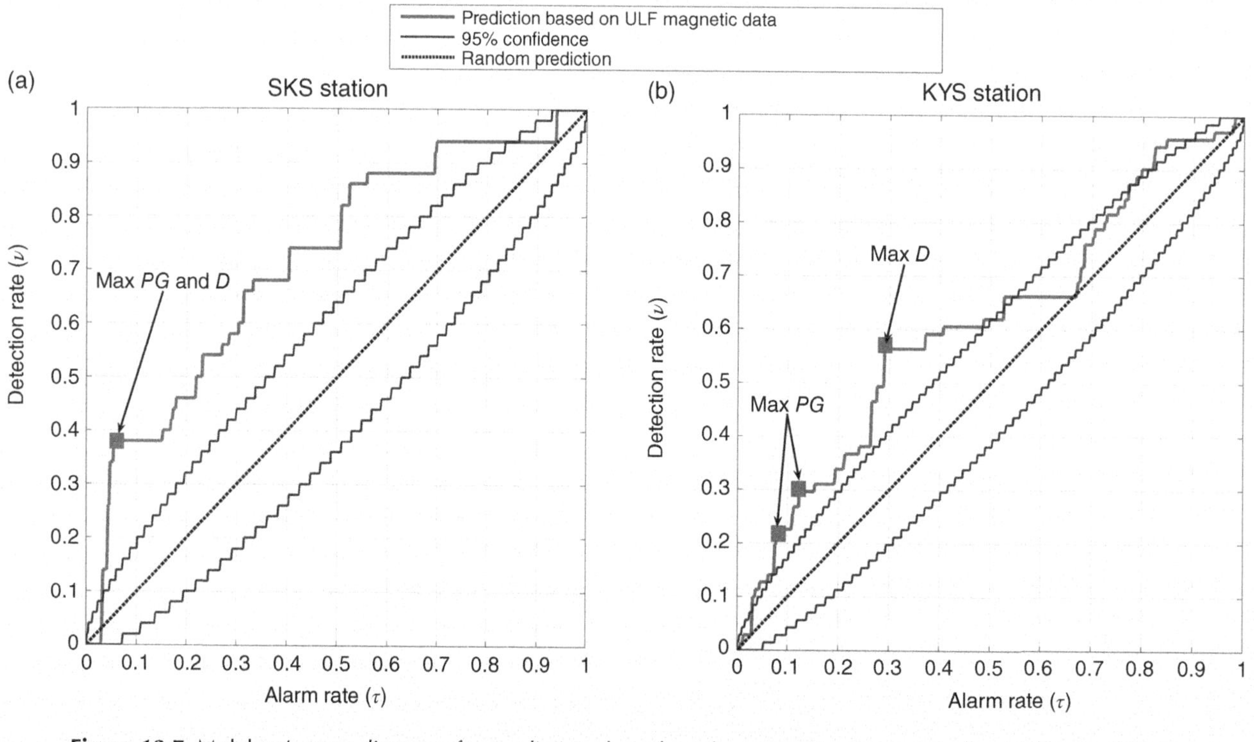

Figure 13.7 Molchan's error diagrams for predictions based on the magnetic anomalies observed at the SKS (a) and KYS (b) stations. Δ is set to 1 day and L to 5 days; $R=100$ km and $E_s=10^8$ for the SKS station, and $R=150$ km and $E_s=10^8$ for the KYS station. The arrows indicate the locations of the maximum PG and D values, respectively. (*See electronic version for color representation.*)

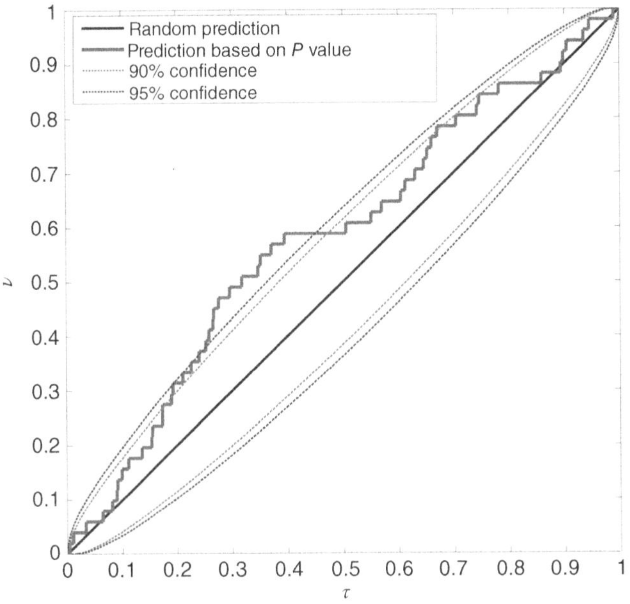

Figure 13.8 Molchan's error diagram for predictions based on the magnetic anomalies observed at the KAK station. Δ is set to 11 days and L to 5 days; $R = 100$ km and $E_s = 10^8$. (*See electronic version for color representation.*)

Based on these results, the leading time (Δ) is first set to 11 days and the alarm window (L) to 5 days. The red solid line in Figure 13.8 shows the prediction based on magnetic data, and the light and dark blue lines represent the 90% and 95% confidence intervals for the random prediction, respectively. The prediction shown by the red line is clearly better than a random guess, and exceeds the 95% confidence threshold when τ is 0.25–0.45.

13.5.3. Dependence of Forecasting Performance on Δ and *L*

The prediction curve could change significantly by using different Δ and L values. Next, we will discuss the dependence of forecasting performance using the KAK data. To find out the most effective values, we must quantify the precursory information of each prediction curve. To date, several parameters have been proposed as a measure of prediction efficiency. They include the minimax strategy and the sum of errors [*Molchan*, 1997; *Molchan and Keilis-Borok*, 2008]. The present study, however, adopts the area skill score (Zechar and Jordan, 2008) with some modifications.

The modified area skill score $S(\Delta, L)$ measures the area between the actual prediction curve and the random prediction line, where:

$$S_{(\Delta,L)} = \int_0^1 \left(\nu(p) - \tau(p)\right) d\tau(p) \qquad (13.4)$$

In this method, if the area is located above the random prediction line, it is counted as positive; otherwise, it is counted as negative. The area skill score $S(\Delta, L)$ is the algebraic sum of all the areas, and the score ranges (−0.5, 0.5). A score of 0.5 indicates perfect skill and a score of −0.5 indicates complete nonskill. The expected score for a random prediction is 0. The scores for the upper boundaries of 90% and 95% confidence intervals in Figure 13.7 are 0.09 and 0.11, respectively.

Figure 13.9a shows the area skill score of prediction in color gradation from blue (low score) to red (high score), based on using different Δ and L values. There are mainly two high score regions (higher than 0.11): one with a leading time Δ around 1 week and an alarm window $L < 4$ days, and the other with $\Delta = 13$–14 days and $L < 1$ week. The maximum S, which suggests the optimal prediction strategy, has a value of 0.13, and is located at $\Delta = 8$ days and $L = 1$ day.

Figure 13.9b demonstrates the Molchan's error diagram of the optimal prediction strategy using $\Delta = 8$ days and $L = 1$ day. The prediction curve here clearly exceeds the 95% confidence interval when τ is 0.10–0.57. It is the best prediction that contains the largest amount of precursory information of the sizeable earthquakes in the KAK region.

13.6. SUMMARY AND CONCLUSION

In order to clarify the ULF seismomagnetic phenomena, a sensitive geomagnetic network has been installed in the Kanto region of Japan. In this study, we have analyzed geomagnetic data observed during the past decade in Izu and Boso Peninsulas, and at the Kakioka station. Energy of ULF geomagnetic signals at the frequency around 0.01 Hz has been investigated by wavelet transform analysis. Statistical studies by SEA have indicated that before a sizeable earthquake there are clearly higher probabilities of ULF anomalies than after the earthquake. Meanwhile, we have used Molchan's error diagram to evaluate the potential earthquake precursory information in the ULF magnetic data recorded in the Kanto region of Japan during 2000–2010. The results show that the earthquake detections based on magnetic anomalies are clearly better than a random guess, indicating that the magnetic data contain potentially useful information on earthquake forecasts. We have also demonstrated the influences of Δ and L on short-term earthquake forecasting. To find out the best prediction parameters, a modified area skill score was introduced to assess the efficiency of different prediction strategies. The results indicate that the ULF magnetic data of the KAK station contain higher precursory information when (a) Δ is around 1 week and L is <4 days, or (b) Δ is 13–14 days and L is <1 week. The optimal strategy for

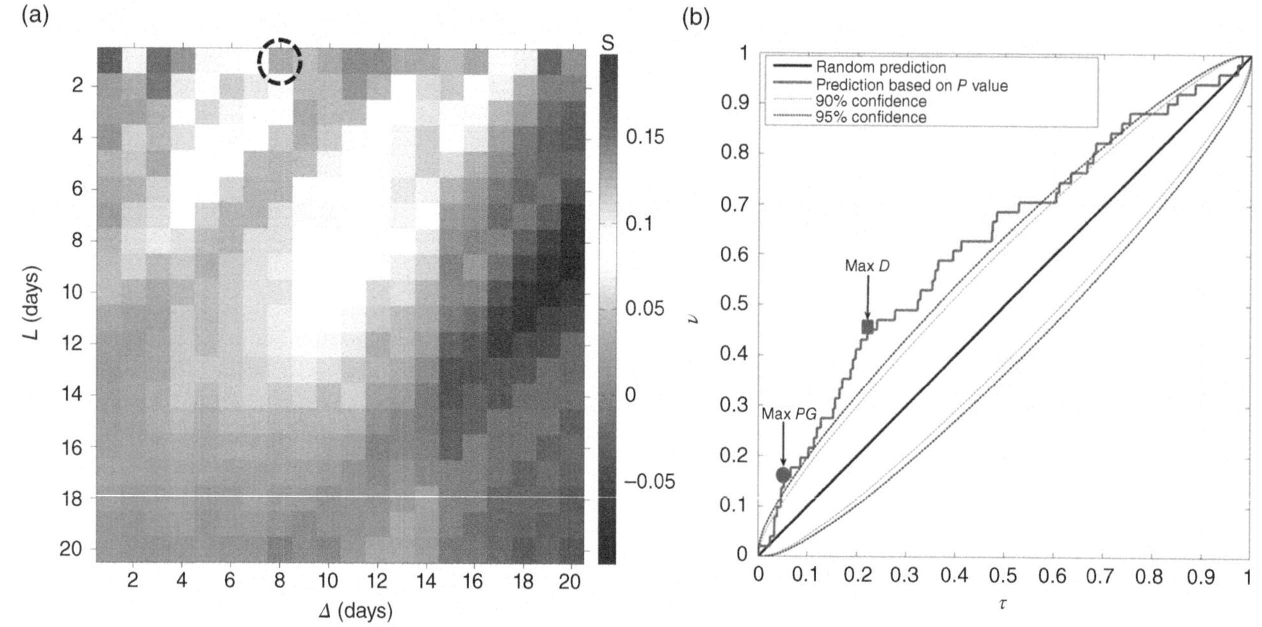

Figure 13.9 (a) Area skill scores for predictions with different Δ and L and (b) Molchan's error diagram of the optimal prediction strategy ($\Delta = 8$ days and $L = 1$ day). The arrows indicate the locations of the maximum PG and D values, respectively. (*See electronic version for color representation.*)

short-term forecasts is $\Delta = 8$ days and $L = 1$ day. The methodology proposed in this study could help evaluate the prediction policy and find out the optimal solution of different measurements for short-term earthquake forecasting.

ACKNOWLEDGMENTS

The authors would like to thank the Japan Meteorological Agency for providing geomagnetic data and earthquake catalogs. This research is partly supported by Grand-in-Aids for Scientific Research of Japan Society for Promotion of Science (19403002, 26249060 and 26240004) and National Institute of Information and Communication Technology (R & D promotion funding international joint research).

REFERENCES

Adams, J. B., Mann, M. E., and Ammann, C. M. (2003), Proxy evidence for an El Niño-like response to volcanic forcing, *Nature*, *426*, 274–278.

Bleier, T., Dunson, C., Maniscalco, M., Bryant, N., Bambery, R., and Freund, F. (2009), Investigation of ULF magnetic pulsations, air conductivity changes, and infrared signatures associated with the 30 October 2007 Alum Rock *M*5.4 earthquake, *Nat. Hazards Earth Syst. Sci.*, *9*, 585–603.

Chavez, O., Millan-Almaraz, J. P., Pérez-Enríquez, R., Arzate-Flores, J. A., Kotsarenko, A., Cruz-Abeyro, J. A., and Rojas, E. (2010), Detection of ULF geomagnetic signals associated with seismic events in Central Mexico using discrete wavelet transform, *Nat. Hazards Earth Syst. Sci.*, *10*, 2557–2564 doi:10.5194/nhess-10-2557-2010

Chen, C. H., Liu, J. Y., Lin, P. Y., Yen, H. Y., Hattori, K., Liang, W. T., Chen, Y. I., Yeh, Y. H., and Zeng, X. (2010), Pre-seismic geomagnetic anomaly and earthquake location, *Tectonophysics*, *489*, 240–247, doi:10.1016/j.tecto.2010.04.018

Chen, C. H., Hsu, H. L., Wen, S., Yeh, T. K., Chang, F. Y., Wang, C. H., Liu, J. Y., Sun, Y. Y., Hattori, K., Yen, H. Y., and Han, P. (2013), Evaluation of seismo-electric anomalies using magnetic data in Taiwan, *Nat. Hazards Earth Syst. Sci.*, *13*, 597–604, doi:10.5194/nhess-13-597-2013

Dunson, J. C., Bleier, T. E., Roth, S., Heraud, J., Alvarez, C. H., and Lira, A. (2011). The pulse azimuth effect as seen in induction coil magnetometers located in California and Peru 2007–2010, and its possible association with earthquakes. *Nat. Haz. Earth Syst. Sci.*, *11*(7), 2085.

Fraser-Smith, A. C., Bernardi, A., McGill, P. R., Ladd, M., Helliwell, R. A., and Villard, O. G. (1990). Low-frequency magnetic field measurements near the epicenter of the *Ms* 7.1 Loma Prieta earthquake, *Geophys. Res. Lett.*, *17*(9), 1465–1468.

Han, P., Huang, Q. H., and Xiu, J. G. (2009), Principle component analysis of geomagnetic diurnal variation associated with earthquakes: Case study of the *M*6.1 Iwateken Nairiku Hokubu earthquake, *Chin. J Geophys.*, *52* (6), 1556–1563, doi: 10.3969/j.issn.0001-5733.2009.06.017

Han, P., Hattori, K., Huang, Q., Hirano, T., Ishiguro, Y., Yoshino, C., and Febriani, F. (2011), Evaluation of ULF electromagnetic phenomena associated with the 2000 Izu Islands earthquake swarm by wavelet transform analysis, *Nat. Hazards Earth Syst. Sci.*, *11*, 965–970, doi:10.5194/nhess-11-965-2011

Han, P., Hattori, K., Hirokawa, M., Zhuang, J., Chen, C., Febriani, F., Yamaguchi, H., Yoshino, C., Liu, J., and Yoshida, S. (2014), Statistical analysis of ULF seismomagnetic phenomena at Kakioka, Japan, during 2001–2010, *J. Geophys. Res., Space Phys.*, *119*, doi: 10.1002/2014JA019789

Han, P., Hattori, K., Xu, G., Ashida, R., Chen, C.-H., Febriani, F., and Yamaguchi, H. (2015), Further investigations of geomagnetic diurnal variations associated with the 2011 off the Pacific coast of Tohoku earthquake (*Mw*9.0), *J. Asian Earth Sci.*, doi:10.1016/j.jseaes.2015.02.022

Han, P., Hattori, K., Huang, Q., Hirooka, S., and Yoshino, C. (2016), Spatiotemporal characteristics of the geomagnetic diurnal variation anomalies prior to the 2011 Tohoku earthquake (*Mw* 9.0) and the possible coupling of multiple pre-earthquake phenomena. *J. Asian Earth Sci.*, *129*, 13–21.

Hattori, K. (2004), ULF geomagnetic changes associated with large earthquakes, *Terr. Atmos. Ocean Sci.*, *15*, 329–360.

Hattori, K., Takahashi, I., Yoshino, C., Isezaki, N., Iwasaki, H., Harada, M., Kawabata, K., Kopytenko, E., Kopytenko, Y., Maltsev, P., Korepanov, V., Molchanov, O., Hayakawa, M., Noda, Y., Nagao, T., and Uyeda, S. (2004a) ULF geomagnetic field measurements in Japan and some recent results associated with Iwateken Nairiku Hokubu earthquakes in 1998, *Phys. Chem. Earth*, *29*, 481–494, doi:10.1016/j.pce.2003.09.019

Hattori, K., Serita, A., Gotoh, K., Yoshino, C., Harada, M., Isezaki, N., and Hayakawa, M. (2004b) ULF geomagnetic anomaly associated with 2000 Izu islands earthquake swarm, Japan, *Phys. Chem. Earth*, *29*, 425–436, doi:10.1016/j.pce.2003.11.014

Hattori, K., Serita, A., Yoshino, C., Hayakawa, M., and Isezaki, N. (2006), Singular spectral analysis and principal component analysis for signal discrimination of ULF geomagnetic data associated with 2000 Izu Island Earthquake Swarm. *Phys. Chem. Earth*, *31*, 281–291, doi:10.1016/j.pce.2006.02.034

Hattori, K., Han, P., Yoshino, C., Febriani, F., Yamaguchi, H., and Chen, C. (2013a), Investigation of ULF seismo-magnetic phenomena in Kanto, Japan during 2000–2010: Case studies and statistical studies, *Surv. Geophys.*, *34*,293–316.

Hattori, K., Han, P., and Huang, Q. (2013b), Global variation of ULF geomagnetic fields and detection of anomalous changes at a certain observatory using reference data, *Elec. Eng. Jpn*, *182*, 9–18, doi: 10.1002/eej.22299

Hayakawa, M. (ed.) (1999), *Atmospheric and Ionospheric Electromagnetic Phenomena Associated with Earthquakes*, Terra Scientific, Tokyo.

Hayakawa, M. and Fujinawa, Y. (eds) (1994) *Electromagnetic Phenomena Related to Earthquake Prediction*, Terra Scientific, Tokyo, 677 pp.

Hayakawa, M. and Molchanov, O. A. (eds) (2002), *Seismo Electromagnetics Lithosphere–Atmosphere–Ionosphere Coupling*, Terra Scientific, Tokyo.

Hayakawa, M., Kawate, R., Molchanov, O. A., and Yumoto, K. (1996), Results of ultra-low-frequency magnetic field measurements during the Guam earthquake of 8 August 1993, *Geophys. Res. Lett.*, *23*, 241–244.

Hirano, T. and Hattori, K. (2011), ULF geomagnetic changes possibly associated with the 2008 Iwate–Miyagi Nairiku earthquake, *J. Asian Earth Sci.* *41*, 442–449, doi:10.1016/j.jseaes.2010.04.038

Hocke, K. (2008), Oscillations of global mean TEC, *J. Geophys. Res.*, *113*, A04302, doi:10.1029/2007JA012798

Huang, Q. H. (2011a) Rethinking earthquake-related DC-ULF electromagnetic phenomena: towards a physics-based approach, *Nat. Hazards Earth Syst. Sci.*, *11*, 2941–2949, doi:10.5194/nhess-11-2941-2011

Huang, Q. H. (2011b) Retrospective investigation of geophysical data possibly associated with the *Ms*8.0 Wenchuan earthquake in Sichuan, China, *J. Asian Earth Sci.* *41*, 421–427, doi:10.1016/j.jseaes.2010.05.014

Huang, Q. H. (2015), Forecasting the epicenter of a future major earthquake. *Proc. Nat. Acad. USA*, *112*(4), 944–945, doi: 10.1073/pnas.1423684112

Ismaguilov, V. S., Kopytenko, Y. A., Hattori, K., Voronov, P. M., Molchanov, O. A., and Hayakawa, M. (2001), ULF magnetic emissions connected with under sea bottom earthquakes, *Nat. Haz. Earth Syst. Sci.*, *1*(1/2), 23–31.

Jach, A., Kokoszka, P., Sojka, J., and Zhu, L. (2006), Wavelet-based index of magnetic storm activity, *J. Geophys. Res. Space Phys.*, *111*, doi:10.1029/2006JA011635

Johnston, M. (1997), Review of electric and magnetic fields accompanying seismic and volcanic activity, *Surv. Geophys.*, *18*, 441–476, doi:10.1023/A:1006500408086

Kappler, K. N., Morrison, H. F., and Egbert, G. D. (2010), Long-term monitoring of ULF electromagnetic fields at Parkfield, California, *J. Geophys. Res.*, *115*, B04406, doi:10.1029/2009JB006421

Kon, S., Nishihashi, M., and Hattori, K. (2011), Ionospheric anomalies possibly associated with *MP* 6.0 earthquakes in the Japan area during 1998–2010: Case studies and statistical study, *J. Asian Earth Sci.* *41*, 410–420, doi:10.1016/j.jseaes.2010.10.005

Molchan, G. M. (1991), Structure of optimal strategies of earthquake prediction, *Tectonophysics*, *193*, 267–276.

Molchan, G. M. (1997), Earthquake prediction as a decision making problem, *Pure Appl. Geophys.*, *149*, 233–247.

Molchan, G. M. and Keilis-Borok, V. (2008), Earthquake prediction: Probabilistic aspect, *Geophys. J. Int.*, *173*, 1012–1017.

Nagao, T., Enomoto, Y., Fujinawa, Y., Hata, M., Hayakawa, M., Huang, Q., Izutsu, J., Kushida, Y., Maeda, K., Oike, K., Uyeda, S., and Yoshino, T. (2002), Electromagnetic anomalies associated with 1995 Kobe earthquake, *J. Geodyn.*, *33*, 401–411.

Park, S. K. (1996), Precursors to earthquakes: seismoelectromagnetic signals, *Surv. Geophys.*, *17*(4), 493–516.

Prattes, G., Schwingenschuh, K., Eichelberger, H. U., Magnes, W., Boudjada, M., Stachel, M., Vellante, M., Villante, U., Wesztergom, V., and Nenovski, P. (2011), Ultra low frequency (ULF) European multi station magnetic field analysis before and during the 2009 earthquake at L'Aquila regarding regional geotechnical information, *Nat. Haz. Earth Syst. Sci.*, *11*, 1959–1968, doi:10.5194/nhess-11-1959-2011

Pulinets, S. and Boyarchuk, K. (2004), *Ionospheric Precursors of Earthquakes*, Springer Science and Business Media.

Pulinets, S. A., Legen'Ka, A. D., Gaivoronskaya, T. V., and Depuev, V. K. (2003), Main phenomenological features of ionospheric precursors of strong earthquakes, *J. Atmos. Solar Terr. Phys.*, *65*(16), 1337–1347.

Sarkar, S., Gwal, A. K., and Parrot, M. (2007). Ionospheric variations observed by the DEMETER satellite in the mid-latitude region during strong earthquakes. *J. Atmos. Solar Terr. Phys.*, *69*(13), 1524–1540.

Schekotov, A. and Hayakawa, M. (2015). Seismo-meteo-electromagnetic phenomena observed during a 5-year interval around the 2011 Tohoku earthquake, *Phys. Chem. Earth, Parts A/B/C*, *85–86*, 167–173.

Schekotov, A. Y., Molchanov, O. A., Hayakawa, M., Fedorov, E. N., Chebrov, V. N., Sinitsin, V. I., and Yagova, N. V. (2007), ULF/ELF magnetic field variations from atmosphere induced by seismicity, *Radio Sci.*, *42*(6).

Serita, A., Hattori, K., Yoshino, C., Hayakawa, M., and Isezaki, N. (2005), Principal component analysis and singular spectrum analysis of ULF geomagnetic data associated with earthquakes, *Nat. Haz. Earth Syst. Sci.*, *5*(5), 685–689.

Telesca, L. and Hattori, K. (2007), Non-uniform scaling behavior in ultra low frequency (ULF) earthquake-related geomagnetic signals, *Physica A*, *384*, 522–528.

Telesca, L., Lapenna, V., Macchiato, M., and Hattori, K. (2008), Investigating non-uniform scaling behavior in ultra low frequency (ULF) earthquake-related geomagnetic signals, *Earth Planet. Sci. Lett.*, *268*, 219–224, doi:10.1016/j.epsl.2008.01.033

Uyeda, S., Hayakawa, M., Nagao, T., Molchanov, O., Hattori, K., Orihara, Y., Gotoh, K., Akinaga, Y., and Tanaka, H. (2002), Electric and magnetic phenomena observed before the volcano-seismic activity in 2000 in the Izu Island Region, Japan, *Proc. Nat. Acad. Sci. USA*, *99*(11), 7352–7355

Uyeda, S., Nagao, T., and Kamogawa, M. (2009a) Short-term earthquake prediction: Current status of seismo-electromagnetics. *Tectonophysics*, *470*(3), 205–213.

Uyeda, S., Kamogawa, M., and Tanaka, H. (2009b) Analysis of electrical activity and seismicity in the natural time domain for the volcanic-seismic swarm activity in 2000 in the Izu Island region, Japan, *J. Geophys. Res.*, *114*, B02310.

Varotsos, P. and Lazaridou, M. (1991), Latest aspects of earthquake prediction in Greece based on seismic electric signals, *Tectonophysics*, *188*, 321–347.

Varotsos, P., Sarlis, N., Skordas, E., and Lazaridou, M. (2013), Seismic electric signals: an additional fact showing their physical interconnection with seismicity. *Tectonophysics*, *589*, 116–125.

Wang, T., Zhuang, J., Kato, T., and Bebbington, M. (2013), Assessing the potential improvement in short-term earthquake forecasts from incorporation of GPS data, *Geophys. Res. Lett.*, *40*, doi:10.1002/grl.50554.

Wen, S., Chen, C.-H., Yen, H. Y., Yeh, T. K., Liu, J. Y., Hattori, K., Han, P., Wang, C. H., and Shin, T. C. (2012), Magnetic storm free ULF analysis in relation with earthquakes in Taiwan, *Nat. Hazards Earth Syst. Sci.*, *12*, 1747–1754, doi:10.5194/nhess-12-1747-2012

Xu, G., Han, P., Huang, Q., Hattori, K., Febriani, F., and Yamaguchi, H. (2013), Anomalous behaviors of geomagnetic diurnal variations prior to 2011 off the Pacific coast of Tohoku earthquake (*Mw*9.0), *J. Asian Earth Sci.*, *77*, 59–65, doi: 10.1016/j.jseaes.2013.08.011

Zechar, J. D. and Jordan, T. H. (2008), Testing alarm-based earthquake predictions, *Geophys. J. Int.*, *172*, 715–724.

Zhuang, J., Vere-Jones, D., Guan, H., Ogata, Y., and Ma, L. (2005), Preliminary analysis of observations on the ultra-low frequency electric field in a region around Beijing, *Pure Appl. Geophys.*, *162*, 1367–1396, doi:10.1007/s00024-004-2674-3

M11	*Panda et al.*, 2007	EOS/MODIS	$\Delta T(x,y,t) = T(x,y,t) - \mu_T(x,y,t)$	5–10 K	111,000	7 days before	$Mw = 7.6$	V
M12	*Halle et al.*, 2008	NOAA/ AVHRR	As M6 with $V(x,y,t) = LST(x,y,t)$	$\otimes(x,y,t) > 2–3$	2.600–5,000	2–10 days before 4–7 days after	6.4 ÷ 7.8	V and C
M13	*Eneva et al.*, 2008	EOS/MODIS	As M6 with $V(x,y,t) = LST(x,y,t)$ and with $\mu_{\Delta V}(x,y)$ and $\sigma_{\Delta V}(x,y)$ computed on 31 days before t	$\otimes(x,y,t) > 2.5 \div 3.5$	–	20 days before 20 days after	4.5 ÷ 6.6	V and C
M14	*Huang et al.*, 2008	EOS/MODIS	Visual inspection	3–5 K	–	1 day before	$M_s = 8$	V
M15	*Ouzounov et al.*, 2006; *Bleier et al.*, 2009	EOS/MODIS GOES/IMAGER	$T(x,y,t_i) = T_0 + at_i$ (6 pm < t_i < 6 am)	$a > 0$	–	1–13 days before	$Mw = 7.7$ $M = 5.4$	V
M16	*Piroddi* 2011; *Piroddi and Ranieri*, 2012; *Piroddi et al.*, 2014	MSG/SEVIRI	$<T(x,y,t_i)> = T_0 + at_i$ (6 pm < t_i < 4 am)	$a > 0$	10,000	7 days before	$Mw = 6.3$	V and C
M17	*Chen et al.*, 2010; *Ma et al.*, 2010; *Saradjian and Akhoondzadeh*, 2011	NOAA/AVHRR EOS/MODIS	Wavelet transform	4–5 K	–	15 days after	$M > 7$	V
M18	*Yang and Guo*, 2010	MTSAT	$\Delta T_{year}(x,y,d) = [T_{year}(x,y,d) - T_{year\text{-}n}(x,y,d)] - T_{year}(x,y,d-1)$	4–5 K	30,000	1-14 days before	$M_s = 6.2$	V
M19	*Zhang et al.*, 2010, 2011; *Xie et al.*, 2013	FY-2C FY-2E	Wavelet transform	4–10 K	10,000–600,000	Several days to 2 months before	M_s 7.2÷9	V
M20	*Saradjian and Akhoondzadeh*, 2011	EOS/MODIS	Interquartile, wavelet transform and Kalman filter method	1–4 K	–	1–20 days before	Mw 6.1 ÷ 6.6	V
M21	*Zoran*, 2012;*Zoran et al.*, 2016	EOS/MODIS	$\Delta LST(x,y,t) = (LST(x,y,t) - <LST>(t))/LST(x,y,t)$	10 K	30,000	15 days before	$Mw = 9$	V
M22	*Xiong et al.*, 2013, 2015a,b	AATSR	As M6 using $\otimes_{\Delta V}(x,y,t) = [\Delta V(x,y,t) - \mu_{\Delta V}(x,y)]/\sigma_{\Delta V}(x,y)$	$\otimes(x,y,t) > 4$	130,000	33 days before 1 day after	$Mw = 6.3$	V and C

$T(x,y,t)$ = TIR signal measured in correspondence of the geographical coordinates (x,y) at the time t; $LST(x,y,t)$ = LST products computed in correspondence of the geographical coordinates (x,y) at the time t; $\mu T(t,D)$ = spatial average over a seismically unperturbed zone (D) on the same image; $\mu T(t,A)$ = spatial average over the same area (A) of punctual air temperature data (from meteorological stations and other sources). $\mu T(t,H)$ = spatial average over a selected restrict area (H) on the same image (cloud-free, seismically unperturbed); $\otimes\Delta V(x,y,m) = \mu\otimes(x,y)$ = monthly average of daily RETIRA index $\otimes\Delta V(x,y,t)$; $<T(x,y,t_i)>$ = average of $T(x,y,t_i)$ on 10 days before; LSTRMS(t_i) = square root of the mean value of the quantity LST2(x,y,t_i) computed in an area of $M \times N$ km^2 (in the considered case 100 × 100 km^2) centered on the epicenter; LSTRMS = temporal average of LSTRMS(t_i) computed on the preceding 60 or 90 days; $T(x,y,t) = T_0(x,y) + a(x,y)\bullet t$ is the linear regression function computed on the basis of 41 MSG-SEVIRI TIR values (41, 15-min slots between 6 pm and 4 am) $T(x,y,t_i)$ corresponding to the averages on the previous 9 days where $a(x,y)$ is the coefficient of the linear regression and $<T(x,y,t)>$ is the average of $T(x,y,t_i)$ computed for each; LST$y(d)$ = spatial average of the LST(x,y,t) image collected on the nighttime of day d of the year y over an area of $M \times N$ km^2 (in the considered case 100 × 100 km^2) centered on the epicenter. d, Julian day; V, validation performed to verify the presence of TIR anomalies in space–time relationship with earthquakes; C, confutation performed to verify the absence of TIR anomalies in absence of earthquakes in the same area and period of the year; $a > 0$ refers to the intensity of anomalies—the authors consider in their work all the positive value of anomalies $T(x,y,ti) = T_0 + at_i$, $<T(x,y,ti)> = T_0 + at_i$; MFG/MVIRI, Meteosat First Generation/Meteosat Visible and InfraRed Imager; NOAA/AVHRR, National Oceanic and Atmospheric Administration/Advanced Very High Resolution Radiometer; GMS, Geostationary Meteorological Satellite; GOES, Geostationary Operational Environmental Satellite/IMAGER; MSG/SEVIRI, Meteosat Second Generation/Spinning Enhanced Visible and Infrared Imager; EOS/MODIS, Earth Observing System/Moderate Resolution Imaging Spectroradiometer; MTSAT, Multifunctional Transport Satellites; FY-2C, Fengyan 2C.

concentration of CO_2 (from three to four times its normal value) and of other greenhouse gases (such as H_2 and H_2O) in the atmosphere was reported by *Qiang et al.* [1991] in association with such TIR anomalies, with an abrupt reduction just 1 day before the event.

Using a similar approach *Tronin* [1996, 2000] and *Tronin et al.* [2002, 2004], identified the presence of positive preseismic TIR anomalies corresponding with large linear structures and fault systems within Earth's crust in Central Asia (e.g., *M* ~7.3 Gazli earthquake on 19 March 1984) and Japan (e.g., *M* ~6.9 Kobe earthquake on 16 January 1995). These authors report positive, 50 km wide, TIR anomalies (from 3 to 6 K), appearing from 6 to 24 days before, and even 1 week after, $M > 4.7$ earthquakes and up to distances of 200–1000 km from the epicenters.

Saraf and Choudhury [2004, 2005a,b,c], *Choudhury et al.* [2006], *Rawat et al.* [2011] and *Saraf et al.* [2008, 2009, 2012], using visual inspection of TIR data from NOAA/AVHRR, report anomalies of the TIR signal (about 5–7 K) close to the epicentral zone, from 1–10 days before and within a few days after the seismic events considered.

Similarly, using the Japanese Geostationary Meteorological Satellite (GMS), *Xu et al.* [2000] noted that about 66% of *M*s ~6 earthquakes in eastern China from 1988 were preceded by infrared anomalies. Similar results are reported by: *Yoshioka et al.* [2005] who, using standard AVHRR-LST (Land Surface Temperature) observations over Japan, reported a relative rise of 2–10 K just 2–3 days before the event studied; *Lixin et al.* [2006] and *Liu et al.* [2007] on the basis of a visual interpretation of NOAA/AVHRR TIR images; *Huang et al.* [2008] who, on the basis of MODIS (Moderate Resolution Imaging Spectroradiometer) TIR data, reported signal excess up to 5 K just 1 day before the M_s ~8 Sichuan (China) earthquake of 12 May 2008.

14.1.2. Methodologies Based on a Multitemporal Approach

Lü et al. [2000], in order to identify possible preseismic TIR anomalies, used the difference between the AVHRR image at hand and a reference image acquired in the preceding days, chosen to be representative of "normal" thermal condition (i.e., low or no seismic activity, no meteorological disturbances, etc.) over the area considered. They applied such a methodology to study the Ms ~6.2 Zhangbei earthquake (China, 10 January 1998), reporting TIR anomalies (*BT* differences up to 8 K), a few days before the earthquake occurrence. For the same seismic event, *Yang and Guo* [2010] applied the same approach to data provided by the Japanese geostationary satellite MTSAT-1R (Multifunctional Transport Satellite), identifying TIR anomalies on the basis of the differences between a TIR image collected before the earthquake and another one collected in the same period of a precedent year. In addition, air temperature data of 700 weather stations from all over China were considered to take into account occasional warming/cooling due to meteorological factors. They report TIR anomalies within 2 weeks before the main shock, at < 250 km from the epicenter. *Ouzounov and Freund* [2004] adopted a similar approach, defining TIR anomalies in relation to the difference of measured *BT* on the same day in two different years.

The LST product can be used, instead of the simple TIR signal usually collected around 11 μm, for those sensors (like AVHRR and MODIS) having a second splitwindow TIR channel around 12 μm. Different to the simple TIR radiances whose value depends on surface emissivity (highly variable with vegetation cover and soil moisture content as well as with atmospheric water vapor and aerosols content), LST is expected (with errors, however, which are higher than 3 K) to give an estimate of the land surface temperature corrected for the effects of atmospheric water vapor content and surface emissivity.

Ouzounov et al. [2006], on the basis of MODIS data, computed the difference $\Delta\text{LST}(t_i)$ between the daily LST root mean square, $LST_{RMS}(t_i)$ (i.e., the square root of the mean value of the quantity $\text{LST}(x,y,t_i)$ computed over an area of $\text{M} \times \text{N km}^2$ (in that case $100 \times 100\,\text{km}^2$) centred on the epicenter) and its temporal average LST_{RMS}, computed for the preceding 60 or 90 days. In order to identify possible preseismic anomalies, the quantity $\Delta\text{LST}(t_i) = LST_{RMS}(t_i) - LST_{RMS}$ is compared day by day with the same quantity computed for the same days (t_i, $i = 1\ldots60$) in the preceding year(s). They report ΔLST(ti) values up to 4 K, 5–6 days before the events considered. Like *Ouzounov et al.* [2006], *Panda et al.* [2007] used LST MODIS images, but they computed the reference LST image by averaging the LST images collected on the same days in the previous 5 years. To identify thermal anomalies in some connection with the studied seismic event, ΔLST is computed by subtracting from the LST images collected during the days preceding the earthquake the corresponding reference LST images. Applying such a methodology to the *M*w ~7.6 Kashmir earthquake (8 October 2005), they report preseismic anomalies affecting a quite large area (about 111,000 km²) a week before the event.

Zoran [2012] used a similar approach, computing the quantity (LST − <LST>)/LST, with LST being the spatial average of LST over the area considered and <LST> its multiyear mean value. The author reports thermal anomalies (LST − <LST> values up to 10 K by using MODIS observations, up to 5 K using AVHRR) around the epicentral area of the *Mw* 9.0 Tohoku/Sendai earthquake (Japan, 11 March 2011), starting from 2 weeks before the main event as well as after the main shock. *Zoran et al.* [2016] used the same methodology to study the *Mw* 5.9 Vrancea region earthquake (Romania, 27 October 2004), reporting MODIS LST variations (5–10 °C of LST increase) around the epicentral area 4–7 days or more before the seismic event.

Qu et al. [2006], using nighttime cloud-free NOAA/AVHRR data, computed a spatial average of TIR values over preselected ($130 \times 10\,km^2$) regions of the fault area of the events studied (Dayao, Yunnan Province, China, on 21 July 2003, M ~6.2 and on 16 October 2003, M ~6.1 respectively). Annual variations of such averaged values were compared considering the year of the earthquake studied and years without strong seismic events ($M > 6$) in the selected area. The authors report a persistent hot belt along the fault as well as higher TIR averaged values along the hot belt in the year of earthquake occurrence.

Huang and Luo [1992] used a similar approach to identify TIR anomalies on AVHRR images, but they compared TIR images with a decadal temperature map obtained by interpolating 10 days averaged (using data acquired by meteorological stations) temperature measurements.

Saraf and Choudhury [2005c] computed weekly averages of the microwave (MW) signal measured by SSMI (Special Sensor Microwave Imager aboard DMSP—Defense Meteorological Satellite Program) and, to identify anomalous values, they used a simple visual comparison with the mean climatological temperature values computed for the same week in the previous 14 years. Applying such an approach, they report preseismic thermal anomalies (2–10 K of excesses in the measured signal) in the same week of the M_S 7.8 Izmit (17 August 1999) and the *Mw* 6.1 Hindukush (25 March 2002) earthquakes, and the week before the *Mw* 6.1 Kalat event (Pakistan, 4 March 1990).

Chen et al. [2006], *Ma et al.* [2010], *Zhang et al.* [2010, 2011], *Saradjian and Akhoondzadeh* [2011], and *Xie et al.* [2013] performed a wavelet analysis to isolate possible preseismic anomalies in Earth's thermal emission field, from those (normal) variations due to the annual and daily solar cycle, and other meteorological factors (including rain clouds, cold–heat air currents). They used time-series of TIR and/or LST images of different satellite sensors (both polar such as AVHRR and MODIS and geostationary such as FY-2C and FY-2E), reporting:

- the possibility that anomalies in the low-frequency band (period > 1 year) of AVHRR LST data could offer a certain indicator for judging the tectonic activity of earthquakes [*Chen et al.*, 2006];
- evidence of some correlation between tectonic activity and long-term variations of MODIS-LST residuals [*Ma et al.*, 2010];
- preseismic FY-2C TIR anomalies recorded over a wide area (more than $10{,}000\,km^2$) and an extended time range (up to 35 days before three strong earthquakes occurred in China in 2008 [*Zhang et al.*, 2010], and even up to 6 months before the M_s 7.2 Myanmar earthquake, 24 March 2011 [*Zhang et al.*, 2011]);
- anomalies in both time and frequency domains apparent from analyzing daily infrared data collected by the Chinese geostationary meteorological satellite FY-2E [*Xie et al.*, 2013].

Starting from *N. Bryant*'s idea [personal communication, 2007], some studies [*Ouzounov et al.*, 2006; *Bleier et al.*, 2009; *Piroddi and Ranieri*, 2012; *Piroddi et al.*, 2014] exploited the high temporal repetition of geostationary TIR sensors to investigate possible "nocturnal heating" before strong earthquakes. In order to characterize the behavior of the nocturnal cooling expected in normal conditions, *Bleier et al.* [2009] used 3 years of GOES-W IR data, while *Piroddi and Ranieri* [2012], and later *Piroddi et al.* [2014], considered MSG-SEVIRI (Meteosat Second Generation—Spinning Enhanced Visible and Infrared Imager) LST products, on the previous 9 days, at the same time (slot) of the day. On the basis of the expected behavior, positive (i.e. anomalous) slopes of the nocturnal TIR signal are reported 3 days before the M 5.4 Alumni Rock earthquake (31 October 2007) by *Bleier et al.* [2009], just the night before the *Mw* 7.7 Gujarat earthquake (26 January 2001,) by *Ouzounov et al.* [2006], and during all 8 days preceding the *Mw* 6.3 Abruzzo earthquake (6 April 2009) by *Piroddi and Ranieri* [2012] and *Piroddi et al.* [2014].

14.2. ROBUST SATELLITE TECHNIQUE APPROACH AND ROBUST ESTIMATOR OF TIR ANOMALIES INDEX

Tramutoli et al. [2001, 2005] start their work moving from a critical review of the previously quoted methods, the limitations of which justified the caution shown by the scientific community in accepting their results.

In particular, with reference to the methods reported in Table 14.1 (hereafter referred to by the reference in column 1), with the exception of M6 and its variations used in the critical studies M12, M13 and M22, no attempt was made to attribute any statistical significance to the excesses of the thermal signal that were reported as preseismic anomalies. In these studies, variations $\Delta T(x,y,t)$ of the measured TIR signal $T(x,y,t)$ of only a few degrees were reported as anomalies just because they were higher than some reference value $\mu_T(t,D)$ (e.g., a spatial or temporal average of $T(x,y,t)$) without comparing them with the 'normal' (i.e. not related to seismic events) variability of the signal observed in the past in similar observational conditions. $\mu_T(t,D)$ is usually computed on a 'seismically unperturbed' portion D:

- of the same scene (like in methods M1, M2, M3, M4, M8, M9, M10, M14);

Table 14.2 Main factors contributing to TIR signal variability.

Main factors contributing to TIR signal variability	Description
Surface spectral emissivity	Quite constant (~0.98) over oceans. Over land it is highly variable, taking values within 0.90 and 0.98 depending mainly on fraction of soil and vegetation
Atmospheric spectral transmittance	Depends mainly on atmospheric temperature and humidity vertical profiles
Surface temperature (temporal variations)	Related to the regular daily and yearly solar cycles but sensitive also to meteorological (and climatological) factors
Surface temperature (spatial variations)	Depend on local geographical (altitude above sea level, solar exposition, geographic latitude) factors
Observational conditions (spatial variations)	Variations across the same scene of satellite zenithal angles introduce spatial variations of the registered signal not related to actual fluctuations of near-surface thermal emission
Observational conditions (temporal variations of satellite view angle)[a]	The same location is observed, at each revisiting time, at a different satellite zenithal angle: this introduces a spurious temporal variation of the measured signal due simply to the change in observational conditions (e.g. air mass)
Observational conditions (temporal variations of ground resolution cells)[a]	The change of satellite view angle also determines a sensible change in the size of the ground resolution cell. Spurious temporal variations of the measured signal have to be expected then because of the change in size of the ground-resolution cell
Observational conditions (variations of the time of the satellite pass)[a]	Satellite pass occurs each day at different times falling in a time-slot up to 3 h around the nominal time of pass. Significant [e.g. *Peck*, 1996] variations of the measured signal could be associated to such (time) variability of observation condition

[a] Only for instrumental packages aboard polar satellite (not applicable to geostationary platforms).

• of a previous interval of time in the past (like in methods M5, M7, M11, M15, M16, M17, M18, M19, M20, M21).

Moreover, in only a few cases [*Saraf and Choudhury*, 2004, 2005a,c; *Yoshioka et al.*, 2005; *Ouzounov et al.*, 2006; *Piroddi*, 2011; *Piroddi and Ranieri*, 2012; *Piroddi et al.*, 2014] was a confutation attempt made to verify the absence of similar TIR anomalies in the absence of seismic events. In all the other cases, the absence of a confutation attempt, does not exclude other possible origins (e.g. meteorological) of their appearance that are not related to the preparation of an impending earthquake.

Significant progress in this direction has been achieved by the application of the general Robust Satellite Technique (RST) [*Tramutoli*, 1998, 2005, 2007] approach to this kind of study. The present chapter is devoted to this method, which has been evaluated previously by independent studies funded by the National Space Agencies of Italy (ASI), Germany (DLR), and United States (NASA).

The RST proposed by *Tramutoli et al.* [2001, 2005] is based on the general Robust AVHRR Technique (RAT) approach [*Tramutoli*, 1998]. It is based solely on satellite data at hand (not requiring ancillary data whatsoever) so that it:

• can be completely automated for operational real-time monitoring purposes;

• is intrinsically exportable to different geographic areas;

• is immediately exportable to different satellite packages, which is the reason why RAT became RST [*Tramutoli*, 2005, 2007].

The RST approach was applied by *Tramutoli et al.* [2001, 2005] for defining and discriminating possible pre-seismic TIR anomalies from all those signal variations normally related to known (see Table 14.2) or also unknown, natural/observational factors that can be responsible for proliferation of "false alarms." To meet this objective the Robust Estimator of TIR Anomalies (RETIRA) index $\otimes(\mathbf{r},t')$ was introduced [*Filizzola et al.*, 2004; *Tramutoli et al.*, 2005], which can be computed as follows:

$$\otimes(\mathbf{r},t') \equiv \frac{\left[\Delta T(\mathbf{r},t') - \mu_{\Delta T}(\mathbf{r})\right]}{\sigma_{\Delta T}(\mathbf{r})} \tag{14.1}$$

where $\mathbf{r} \equiv (x,y)$ represents the geographic coordinates at the center of the ground-resolution cell from which the TIR radiance $T(\mathbf{r},t')$ has been collected; t' is the time of acquisition of the measured TIR radiance with $t' \in \{\Im\}$, in which $\{\Im\}$ defines the homogeneous domain of satellite imagery collected in the same time-slot (hour) of the day and period (e.g. in the same month) of the year.

The TIR signal strongly depends on surface temperature T_S and spectral emissivity ε_{TIR}. Surface temperature mostly depends on the daily and annual solar cycle, whereas ε_{TIR} is quite constant (≈0.98) over the sea and slowly variable over land (mostly in relation to the fraction of vegetation cover within the ground-resolution cell) during the year. However, significant temporal variation of the measured TIR signal can be expected as a consequence of the emissivity variations due, for instance, to seasonal changes of vegetation cover (from 1 K to 3 K for only $\Delta\varepsilon_{TIR} \approx 0.01$, e.g., *Tramutoli et al.*, [2001] and references therein), which can range from $\varepsilon_{TIR} \approx 0.90$ to $\varepsilon_{TIR} \approx 0.98$ moving from bare to fully vegetated soils. The choice to select only images collected at the same time of the day reduces the effects of the daily T_S cycle on the measured TIR signal. The choice to consider only images collected during the same month reduces both the effects of the seasonal T_S cycle and that of emissivity changes due to the variation in the fraction of vegetation cover, which can be expected to be no more rapid than few weeks. Other more conservative choices, however, are possible (like the 31-day moving window, from day t'_{-31} up to day t'_{-1}, used in Eneva et al., [2008], or even better by using the t'_{-15} up to t'_{+15} temporal window) in order to reduce observational noise [e.g. *Tramutoli*, 1998, 2005, 2007; *Tramutoli et al.*, 2001, 2005], even if requiring a greater computation effort (365 instead of only 12 reference fields to be computed).

Returning to equation (14.1): $\Delta T(\mathbf{r},t') = T(\mathbf{r},t') - T(t')$ is the difference between the current $(t = t')$ TIR signal, $T(\mathbf{r},t')$, measured at location $\mathbf{r}$, and its spatial average $T(t')$, computed in place on the image at hand, discarding cloudy pixels and considering only sea pixels, if $\mathbf{r}$ is located on the sea, only land pixels, if $\mathbf{r}$ is located over the land; $\mu_{\Delta T}(\mathbf{r})$ and $\sigma_{\Delta T}(\mathbf{r})$ are the time average and standard deviation values of $\Delta T(\mathbf{r},t')$, at location $\mathbf{r}$, computed on cloud-free satellite records belonging to the selected homogeneous data set $(t' \in \{\mathfrak{I}\})$. Note that the choice of differential variable $\Delta T(\mathbf{r},t')$ instead of $T(\mathbf{r},t')$ is expected to reduce possible contributions (e.g. occasional warming) due to day-to-day and/or year-to-year climatological change and/or seasonal time-drifts.

The $\otimes(\mathbf{r},t')$ index gives the *llocal* (the double *l* was introduced by *Tramutoli* [1998], and is used hereafter, to highlight a dependence not only on a specific place $\mathbf{r}$ but also on a specific time t') excess of the current $\Delta T(\mathbf{r},t')$ signal compared with its historical mean value and weighted by its historical variability at the considered location. Both $\mu_{\Delta T}(\mathbf{r})$ and $\sigma_{\Delta T}(\mathbf{r})$ are computed for each location $\mathbf{r}$, processing several years of historical satellite records acquired in similar observational conditions $(t' \in \{\mathfrak{I}\})$. The excess $\Delta T(\mathbf{r},t') - \mu_{\Delta T}(\mathbf{r})$ then represents the *Signal* (*S*) that is to be investigated for its possible relation with earthquake occurrence in both space and time. It is always evaluated by comparison with the corresponding natural/observational *Noise* (*N*), represented by $\sigma_{\Delta T}(\mathrm{r})$, which describes the overall (*llocal*) variability of *S*, including all (natural and observational, known and unknown) sources of its variability as historically observed at the same site in similar observational conditions. In this approach the intensity of anomalous TIR transients can be evaluated in terms of the *S*/*N* ratio by the RETIRA index $\otimes(\mathbf{r},t')$. As the RETIRA index is computed with reference to measurements collected in similar observational conditions (same location, time of the day, period of the year, etc.) it is expected to be less affected by all the known (Table 14.2) and unknown sources of natural/observational noise [*Tramutoli et al.*, 2001, 2005].

Since its first application to the *M* ~6.9 Irpinia-Basilicata earthquake (23 November 1980) [Tramutoli et al., 2001], the RST approach has demonstrated that the use of the RETIRA index can reduce the dependence on site properties like topography, vegetation cover, etc. *Di Bello et al.* [2004] demonstrated that a doubling of the *S*/*N* ratio can be achieved by using AVHRR-based LST products (which take into account the atmospheric water vapor variability) instead of simple TIR radiances, as in *Tramutoli et al.* [2001]. The study of *Filizzola et al.* [2004] demonstrated, in the case of the M_S ~5.9 Athens's earthquake (7 September, 1999), the possibility to reach *S*/*N* ratios up to 1.5 by using daily (rather than monthly averaged) RETIRA indexes, $\otimes_{\Delta LST}(\mathbf{r},t')$.

In this chapter, space–time persistence of TIR Anomalies (TA) is also introduced as a further critical requirement in order to discriminate Significant Thermal Anomalies (STA) from residual spurious effects due to simple outliers, geolocation errors, or nighttime warm cloud shadows (see also *Aliano* et al. [2008a]). Moreover, *Filizzola et al.* [2004] demonstrated for the first time the advantages to be expected by the use of TIR sensors aboard geostationary, rather than polar orbiting, satellite platforms: in comparison with results previously achieved by using the more refined LST product based on AVHRR observations from the NOAA polar satellite, *Filizzola et al.* [2004] practically doubled the *S*/*N* ratio $(\otimes_{\Delta T}(\mathbf{r},t') > 3)$ of TIR anomalies observed from the MFG (Meteosat First Generation) geostationary satellite corresponding to the same (Athens, 7 September, 1999) event. This was surely a not trivial result, one that could determine abandonment of the use of polar satellites (in favour of the geostationary ones) in the applications of the RST approach to monitoring active seismic areas. This result was (as reported in *Tramutoli* [2013]) ignored by *Blackett et al.* [2011a,b], who used results obtained by the application of RST to polar satellites in order to negate the existence of thermal anomalies observed, for the same test case analysed by *Genzano et al.* [2007] using geostationary data.

Such a strong *S*/*N* increase is attributable only to the drastic reduction of observational noise, represented by the denominator $\sigma_{\Delta T}$ in the definition of RETIRA, and is mostly due to the fact that the last three elements of

Table 14.3 Reported cases of preseismic thermal anomalies identified by using the RST approach.

Event (date and magnitude)	RST techniques	Reference data set (sensor, month, years, hour)	K
Irpinia-Basilicata, Italy 23 November 1980 M_s 6.9	$\otimes_{\Delta T}(x,y,m)$ monthly average [*Tramutoli et al.*, 2001]	NOAA-AVHRR November (1994–1998) 17:00 19:00	0.6
	$\otimes_{\Delta LST}(x,y,m)$ monthly average [*Di Bello et al.*, 2004]		1
Athens, Greece 7 September 1999 M_s 5.9	$\otimes_{\Delta LST}(x,y,t)$ daily analysis [*Filizzola et al.*, 2004]	NOAA-AVHRR August and September (1995–1998) 01:00 04:00	1.5
		METEOSAT August and September (1995–1998) 24:00 GMT	3
Zhangbei, China 10 January 1998 M_s 6.2	$\otimes_{\Delta LST}(x,y,t)$ daily analysis [*Li et al.*, 2007]	NOAA-AVHRR December–January (1996–1999) 19:00	0
Kocaeli-Izmit, Turkey 17 August 1999 M_s 7.8	$\otimes_{\Delta T}(x,y,t)$ daily analysis [*Tramutoli et al.*, 2005]	METEOSAT August (1992–1998, 2000) 24:00 GMT	3.5
	$\otimes_{T}(x,y,t)$ daily analysis [*Aliano et al.*, 2008a]	METEOSAT August (1995–2000) 24:00 GMT	2
	$\otimes_{\Delta SST}(x,y,t)$ daily analysis [*Halle et al.*, 2008]	AVHRR 1997–2004 Daytime	2–3
	$\otimes_{\Delta LST}(x,y,t)$ daily analysis [*Halle et al.*, 2008]	AVHRR 1998–2004 Daytime–nighttime	–
Patras, Greece 28 May 1995 M_b 4.7	$\otimes_{\Delta T}(x,y,t)$ daily analysis [*Corrado et al.*, 2005]	METEOSAT May and June (1992–1999) 24:00 GMT	3
Cyprus 29 May 1995 M_b 5.3			3
Crete, Greece 3 June 1995 M_b 4.2			3
Crete, Greece 18 June 1995 M_b 4.3			3
Erzurum, Turkey 4 May 1996 M_b 4.3			3
Ionian Sea, Greece 13 June 1996 M_b 4.2			3
Patras, Greece 16 June 1996 M_b 4.3			3
Crete, Greece 17 June 1996 M_b 4.0			3
Isparta, Turkey 29 June 1996 M_b 5.1			3

Table 14.3 (Continued)

Event (date and magnitude)	RST techniques	Reference data set (sensor, month, years, hour)	K
Boumerdes, Algeria 21 May 2003 M_s 6.9	$\otimes_{\Delta T}(x,y,t)$ daily analysis [*Aliano et al.*, 2007, 2009]	METEOSAT April and May (1992–1999) 24:00 GMT	3
Gujarat, India 26 January 2001 M_s 7.9	$\otimes_{\Delta T}(x,y,t)$ daily analysis [*Genzano et al.*, 2007]	METEOSAT January and February (1999–2004) 24:00 GMT	3
Umbria-Marche, Italy 26 September 1997 M_s 5.9 to 6.4	$\otimes_{\Delta T}(x,y,t)$ daily analysis [*Aliano et al.*, 2008b]	METEOSAT September (1992–2000) 24:00GMT	2
Hector Mine, California 16 October 1999 M_s 7.4	$\otimes_{\Delta T}(x,y,t))$ daily analysis [*Aliano et al.*, 2008a]	GOES (7-9-10) October (1996–1999) 24:00 LT	2.5
Mestia Tianeti, Georgia 23 October 1992 *M* 6.3	$\otimes_{\Delta T}(x,y,t)$ daily analysis [*Genzano et al.*, 2009a]	METEOSAT 7 October (1992–1999) 24:00 GMT	3
83 Eq Southwestern USA February 2000 and December 2006 *M* 4.5 ÷ 6,6	$\otimes_{\Delta LST}(x,y,t)$ statistical correlation analysis [*Eneva et al.*, 2008]	EOS-MODIS February 2000 to December 2006 2442 daytime images	2.5
		EOS_MODIS July 2002 to December 2006 1625 nighttime images	2.5
Abruzzo, Italy 6 April 2009 *Mw* 6.3	$\otimes_{\Delta T}(x,y,t)$ daily analysis [*Genzano et al.*, 2009b]	MSG-SEVIRI March and April (2005–2009) 24:00 GMT	4
	$\otimes_{\Delta T}(x,y,t)$ daily analysis [*Pergola et al.*, 2010]	EOS-MODIS March and April (2000–2009) 24:00 GMT	3.5
	$\otimes_{\Delta T}(x,y,t)$ daily analysis [*Lisi et al.*, 2010]	NOAA-AVHRR March and April (1995–2009) 24:00 GMT	3.5
	$\otimes_{\Delta LST}(x,y,t)$ daily analysis [*Lisi et al.*, 2015]	LST product from MSG-SEVIRI March and April (2005–2009) 24:00 GMT	3.5

variability of TIR signal in Table 14.2 do not apply to sensors aboard geostationary satellites. In fact, in contrast to what happens with polar satellites, the signal measured from a geostationary platform in correspondence to a specific location (x,y) is always "observed" under the same satellite zenithal angle and is always originated by the same ground-resolution cell of constant size. Moreover, their high temporal repetition allows a very homogeneous dataset to be built composed of images collected at exactly the same time of day.

Tramutoli et al. [2005] report quite high ($\otimes_{\Delta T}(\mathbf{r},t') > 3.5$) TIR anomalies a few days before and immediately after the M_S 7.8 Kocaeli-Izmit earthquake (Turkey, 17 August 1999), in apparent agreement with the 'dilatancy model' of *Scholz et al.* [1973]. In subsequent RST applications in four different continents and seismic regimes, space–time correlations between thermal anomalies and earthquakes were also confirmed for earthquakes with magnitude as low as *M* 4 [*Corrado et al.*, 2005].

Table 14.3 summarizes all seismic events that have been studied (and published) by applying the RST approach, up to the *Mw* 6.3 L'Aquila earthquake (6 April 2009), which was analysed by independent RST studies on data from different satellite systems (namely MSG/SEVIRI, NOAA/AVHRR, and EOS/MODIS). In this case significant and simultaneous TIR anomalies were found by *Genzano et al.* [2009b], *Lisi et al.* [2010], and *Pergola et al.* [2010] in the epicentral area starting just 1 week before the main shock and a few hours before its strongest foreshock (M_L ~4.1on 30 March 2009 at 13:38 UTC).

Confutation analysis was performed in all the previously quoted cases by considering the same period of time (month) in a different, seismically less affected, year. In the confutation phase the presence of space–time

Table 14.4 Long-term correlation analyses among SSTAs and earthquake occurrences.

Test Area	Study period	Number of processed TIR images	Number of SSTAs	SSTAs in possible connection with earthquakes $M>4$		False/ positive (%)	
				Number	(%)	Number	(%)
Italy[a] [*Tramutoli et al.*, 2015c]	June 2004 to December 2014 (01:00 LT)	2861	28	17	61	11	39
Greece [*Eleftheriou et al.*, 2016]	May 2004 to December 2013 (02:00 LT)	3151	62	58	93	4	7
Southwest USA [*Tramutoli et al.*, 2014b]	June–August 2006–2011 (00:00 LT)	402	17	11	65	6	35
Taiwan [*Genzano et al.*, 2015]	September 1995 –2002 (21:00 LT)	240	18	18	100	0	0
Turkey[b] [*Lisi et al.*,2016]	May 2004 to October 2015 (02:00 LT)	3831	155	115	74	40	26
Japan[c] [*Genzano et al.*, 2016]	June 2005 to December 2015 (00:30 LT)	3447	60	37	62	23	38
Totals	**57 years**	**13932**	**340**	**256**	**75**	**84**	**25**

SSTAs, significant sequences of TIR anomalies.

[a] Results achieved by applying RST to spatially degraded images obtained by averaging TIR signal within boxes of 3×3 full resolution pixels discarding boxes having less than five full resolution pixels clouds-free.

[b] As for Italy.

[c] As for Italy. For Japan only earthquakes with $M \geq 5$ have been considered.

persistent TIR anomalies of similar intensity was rarely observed and practically always in connection with minor (but often with $M>4$) seismic events.

To justify such a relation several physical models have been proposed that refer to different (eventually concurrent) processes. Among them should be noted:

• that described in *Tramutoli et al.* [2013], which attributes occurrence of thermal anomalies to the local greenhouse effect produced by an increased degassing activity in the preparatory phases of an earthquake (mostly optically active gases like CO_2 and CH_4) [*Qiang et al.*, 1991; *Tronin*, 1996; *Tramutoli et al.*, 2001, 2005, 2013; *Singh et al.*, 2010; *Zhang et al.*, 2010, and references therein] together with deep-water rise and convective heat flow towards the surface [*Tronin et al.*, 2002; *Surkov et al.*, 2006] with increasing soil moisture and surface emissivity [e.g. *Qin et al.*, 2013];

• the activation of positive-hole pairs in rocks under stress [*Ouzounov and Freund*, 2004; *Freund et al.*, 2006, 2007; *Freund*, 2007a,b];

• the release of condensation latent heath due to the abrupt aggregation of water vapour molecules around the nucleus of atmospheric molecules ionized by increased radon (Rn) emissions over active tectonic faults and tectonic-plate boundaries [*Yasuoka et al.*, 2006; *Pulinets*, 2009; *Pulinets and Ouzounov*, 2011; *Pulinets et al.*, 2015]. It should be noted that while earthquakes themselves (foreshocks, main shocks, aftershocks) are generally expected (as in the model of *Scholz et al.* [1973]) to increase fluid emissions, the role of deformation processes in modulating such emissions depends on local tectonic (see for instance *Doglioni et al.* [2014] and references therein) and geochemical settings, which makes generalizations difficult.

14.3. LONG-TERM ANALYSES FOR THE ASSESSMENT OF RST APPROACH POTENTIAL IN IDENTIFYING PRESEISMIC TIR ANOMALIES

Recently evaluations of the potential contribution of TIR anomalies identified by the RST approach to a multiparametric system for a time Dependent Assessment of Seismic Hazard (t-DASH; *Tramutoli et al.* [2014a]) have been undertaken by *Tramutoli et al.* [2014c, 2015c] for Italy, *Genzano et al.* [2015] for Taiwan, *Eleftheriou et al.* [2016] for Greece; *Lisi et al.* [2016] for Turkey, and *Genzano et al.* [2016] for Japan. In these studies a formal definition of individual TAs was given on the base of the RETIRA index (equation 14.1) value, taking into consideration those pixels in the scene having

$$\otimes(r,t') > K$$

with K being a value (usually > 2) that qualifies the relative intensity of the anomaly. Moreover, following *Filizzola et al.* [2004] and *Tramutoli et al.* [2005], in order to discard signal outliers (characterized by a transitory character both in the temporal and spatial domains) related to particular natural [see *Aliano et al.*, 2008a; *Genzano et al.*, 2009b] or observational [see *Filizzola et al.*, 2004; *Aliano et al.*, 2008b] conditions, only spatially and temporally persistent TAs are retained for further analyses.

Other well-known [see *Filizzola et al.*, 2004; *Aliano et al.*, 2008a; *Genzano et al.*, 2009a] cases of spurious (even spacetime persistent) TAs have been identified in relation with overcast images or in the presence of an asymmetrical distribution of clouds across the scene, producing the *cold spatial average effect* reported by *Aliano et al.* [2008a] and *Genzano et al.* [2009b]. Other TAs discarded, in the above quoted studies, from further analyses are those [see *Aliano et al.*, 2008a; *Genzano et al.*, 2009b]:

1. generated by the nighttime passage of clouds producing a local warming of the underlying Earth surface;
2. due to errors in the image navigation/colocation process [see also *Filizzola et al.*, 2004];
3. associated with extreme (i.e. very rare in the time domain considered $\{\Im\}$) events producing persistent space–time increase of surface temperature (e.g. extremely extended forest fires) or emissivity (e.g. extremely extended floods).

On this basis an *operational definition of STA* is proposed by considering a location (x,y) affected at time by a STA with these conditions:

1. relative intensity—$\otimes_{\Delta T}(x,y,t') \geq K$ (with $K \geq 3$);
2. control on spurious effects—absence of known sources of spurious TAs (see above);
3. spatial persistence—it is not an isolated phenomenon but part of a group of TAs covering at least 150 km^2 within an area of 1° × 1° (longitude and latitude);
4. temporal persistence—previous conditions (i.e. the existence of a group of TAs covering at least 150 km^2 within an area of 1° × 1° around x,y) are satisfied at least once more in the 7 days preceding/following t.

The above mentioned rules were applied to identify *Significant Sequences of Thermal Anomalies* (SSTAs) over about a total of 57 years of satellite observations over Italy [by *Tramutoli et al.*, 2014c, 2015c], California [by *Tramutoli et al.*, 2014b], Taiwan [by *Genzano et al.*, 2015], Greece [by *Eleftheriou et al.*, 2016], Turkey [by *Lisi et al.*, 2016], and Japan [by *Genzano et al.*, 2016].

An example of SSTAs is shown in Figure 14.1, where several STAs appear near tectonic lineaments from 9 to 23 July 2008 over the Peloponnesus region. Table 14.4 reports the results of the correlation analysis that were performed in order to establish the extent to which the STAs' occurrence could be related to location, time, and magnitude of earthquakes with M > 4.

With the objective of establishing SSTAs, specific validation rules, mostly based on the long-term results achieved by applying the RST approach in four different continents, in different tectonic settings, for hundreds of earthquakes with magnitudes ranging from 4.0 to 7.9 (see Table 14.3), have been defined to drive the retrospective correlation analysis process. According to these rules the occurrence of a single STA, observed at time t at location (x,y) is considered possibly related to a seismic event if:

1. it belongs to a previously identified SSTA; and
2. an earthquake of $M \geq 4$ occurs:
   - 30 days after its appearance or within 15 days before (*temporal window*)—for models that indicate the occurrence of similar anomalies also immediately after the quake, see for instance *Scholz et al.* (1973] and, with reference to TIR anomalies, *Tramutoli et al.* [2005, 2013];
   - within a distance D, from the considered STA, so that $150\,\text{km} \leq D \leq R_D$, with $R_D = 10^{0.43M}$ being the Dobrovolsky [*Dobrovolsky et al.*, 1979] distance (*spatial window*) expressed in kilometers.

The results obtained in terms of reliability analyses are reported in Table 14.4. In order to obtain them a total of 57 years of (more than 13,000) daily TIR images were processed as follows: a total of 33 years of MSG/SEVIRI for Italy, Greece and Turkey, 6 years of GOES/IMAGER for Southwest USA, 8 years of Visible and Infrared Spin-Scan Radiometer (VISSR) aboard Geostationary Meteorological Satellite-5 (GMS-5) for Taiwan, and 10.5 years of MTSAT observations for Japan.

The analysis shows that on average more than 78% (from 100% in Taiwan to 61% of Italy; note that for Taiwan only 1 month (for eight different years) was considered for a total of 256 image processed) of the 340 identified SSTAs occur in the prefixed space–time window around the occurrence time and location of earthquakes ($M \geq 4$), with a false positive rate < 22% (from 0% for Taiwan up to 39% for Italy). It is important to note that error diagram tests [*Molchan*, 1990, 1991, 1997; *Molchan and Kagan*, 1992] performed in the most significant cases of Italy [*Tramutoli et al.*, 2014c], Greece [*Eleftheriou et al.*, 2016], Turkey [*Lisi et al.*, 2016], and Japan [*Genzano et al.*, 2016] have demonstrated that the results achieved are not casual, with significant probability gains (in comparison with a random guess) in all the above mentioned cases.

When also considering the results achieved by other independent studies (see Table 14.3) it is possible to report that SSTAs (identified using the RST approach and RETIRA index) have been observed, as foreseen by models like that of *Scholz et al.* [1973], not only before (few days/weeks) an earthquake, but also in the coseismic and postseismic phases (up to few days/weeks after the

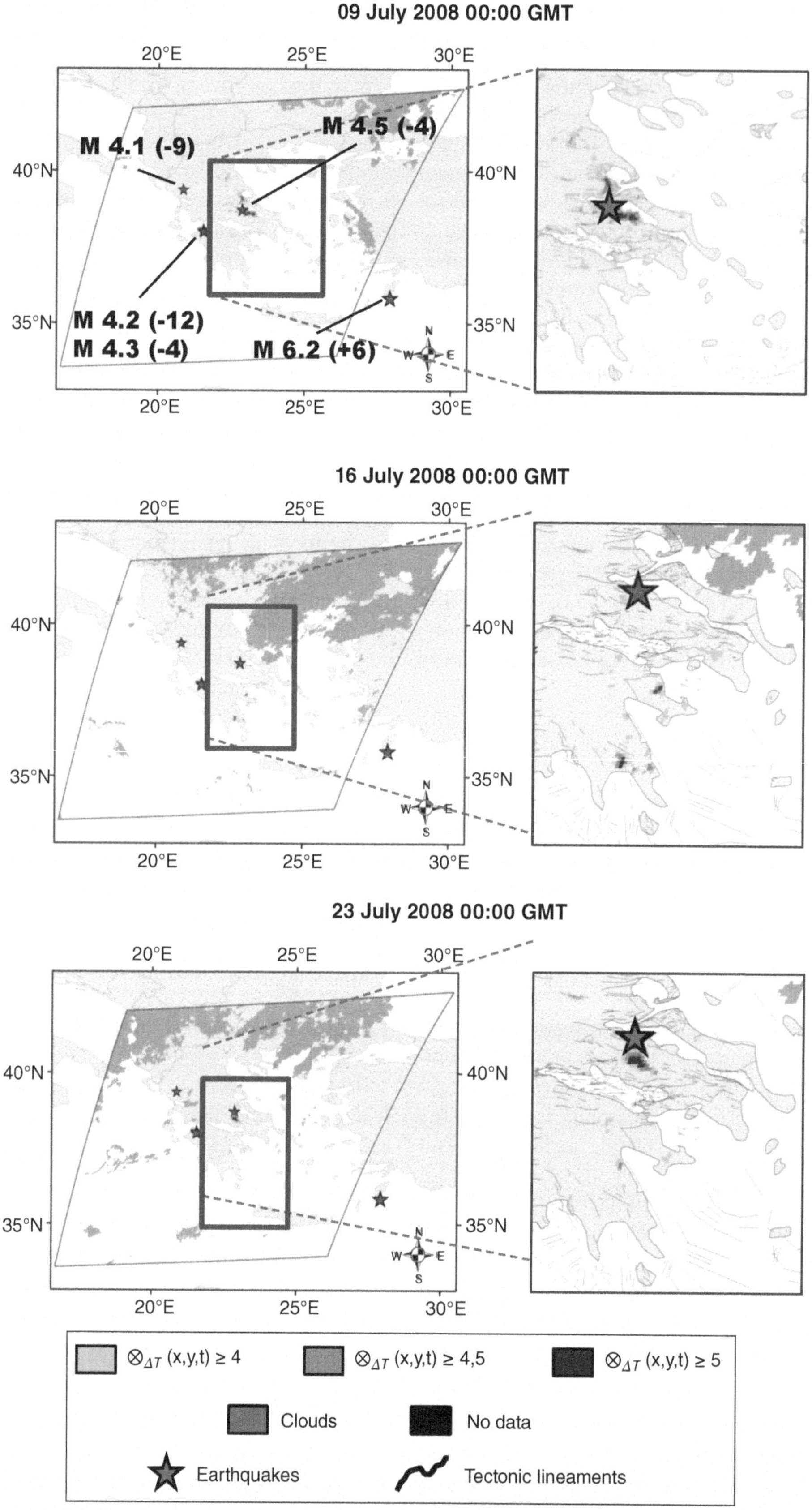

Figure 14.1 Examples of SSTAs identified in the period 09–23 July 2008. Significant Thermal Anomalies (STAs) with $\otimes_{\Delta T}(\boldsymbol{r},t')\geq 4$ (depicted in different colors according with corresponding RETIRA values) appear in the Peloponnesus area before and after $M\geq 4$ seismic events. In addition to the magnitude of earthquakes, the temporal gap from the first appearance of STAs is also indicated by a number (N) in parentheses (+/– N means that the earthquake occurred N days after/before the first appearance of STAs). The red contoured box indicates the limits of analyzed SEVIRI TIR scenes (thermal anomaly map area). (*See electronic version for color representation.*)

earthquake). Moreover their appearance is generally very rare, particularly in seismically quiet periods.

The above-mentioned studies do not reveal particular dependence of TIR anomalies intensities on earthquake magnitudes (but see also the attempts to establish such a correlation made by *Li et al.* [2007]), although possible relationships among the spatial extent of the area affected by TIR anomalies and the magnitude of the impending earthquake have been suggested in the past (e.g. *Pulinets et al.* [2007] and reference therein).

ACKNOWLEDGMENTS

The research leading to these results has been supported by the INGV-DPC S3 project "Short Term Earthquake prediction and preparation." It also has been supported by the European Commission through the FP7 Programme PRE-EARTHQUAKES project *Processing Russian and European EARTH observations for earthQUAKE precursors Studies* (Grant Agreement No. 263502) and by the FSE Basilicata (2007/2013) Project SESAMO (*Sviluppo E Sperimentazione di tecnologie integrate Avanzate per il MOnitoraggio della pericolosità sismica*). The authors would like to also thank the support given by the International Space Science Institute (ISSI) in the framework of the Project *Multi-instrument Space-Borne Observations and Validation of the Physical Model of the Lithosphere-Atmosphere-Ionosphere-Magnetosphere Coupling*. The content of this document is the sole responsibility of the authors and it does not represent the opinion of the INGV-DPC, ISSI, and European Commission who are not responsible for any use that might be made of the information contained herein.

REFERENCES

Aliano, C., Corrado, R., Filizzola, C., Pergola, N., and Tramutoli, V. (2007), Robust satellite techniques (RST) for seismically active areas monitoring: the case of 21st May, 2003 Boumerdes/Thenia (Algeria) earthquake, *International Workshop on the Analysis of Multi-temporal Remote Sensing Images*, pp. 1–6, IEEE.

Aliano, C., Corrado, R., Filizzola, C., Genzano, N., Pergola, N., and Tramutoli, V. (2008a), Robust TIR satellite techniques for monitoring earthquake active regions: limits, main achievements and perspectives, *Ann. Geophys.*, *51*, 303–317.

Aliano, C., Corrado, R., Filizzola, C., Pergola, N., and Tramutoli, V. (2008b), Robust satellite techniques (RST) for the thermal monitoring of earthquake prone areas: the case of Umbria-Marche October, 1997 seismic events, *Ann. Geophys.*, *51*, 451–459.

Aliano, C., Corrado, R., Filizzola, C., Genzano, N., Lanorte, V., Mazzeo, G., Pergola, N., and Tramutoli, V. (2009), Robust satellite techniques (RST) for monitoring thermal anomalies in seismically active areas, in *IEEE International Geoscience and Remote Sensing Symposium*, pp. III-65–III-68, IEEE.

Blackett, M., Wooster, M., and Malamud, B. D. (2011a), Exploring land surface temperature earthquake precursors: A focus on the Gujarat (India) earthquake of 2001, *Geophys. Res. Lett.*, *38*, L15303.

Blackett, M., Wooster, M., and Malamud, B. D. (2011b), Correction to "Exploring land surface temperature earthquake precursors: A focus on the Gujarat (India) earthquake of 2001," *Geophys. Res. Lett.*, *38*, L18307, doi:10.1029/2011GL049428

Bleier, T., Dunson, C., Maniscalco, M., Bryant, N., Bambery, R., and Freund, F. (2009), Investigation of ULF magnetic pulsations, air conductivity changes, and infra red signatures associated with the 30 October Alum Rock *M*5.4 earthquake, *Nat. Hazards Earth Syst. Sci.*, *9*, 585–603.

Chen, S. Y., Liu, P. X., and Liu, L. Q. (2006), Wavelet analysis of thermal infrared radiation of land surface and its implication in the study of current tectonic activities, *Chin. J. Geophys.*, *49*, 824–830.

Chen, Y., Shen, X., Jing, F., and Xiong, P. (2010), Application of outgoing longwave radiation data for earthquake research, in *Proceedings—2010 IEEE International Conference on Intelligent Computing and Intelligent Systems*, Vol. *2*, pp. 46–48, Xiamen.

Choudhury, S., Dasgupta, S., Saraf, A. K., and Panda, S. (2006), Remote sensing observations of pre-earthquake thermal anomalies in Iran, *Int. J. Remote Sens.* *27*, 4381–4396.

Corrado, R., Caputo, R., Filizzola, C., Pergola, N., Pietrapertosa, C., and Tramutoli, V. (2005), Seismically active area monitoring by robust TIR satellite techniques: a sensitivity analysis on low magnitude earthquakes in Greece and Turkey, *Nat. Haz. Earth Syst. Sci.*, *5*, 101–108.

Di Bello, G., Filizzola, C., Lacava, T., Marchese, F., Pergola, N., Pietrapertosa, C., Piscitelli, S., Scaffidi, I., and Tramutoli, V. (2004), Robust satellite techniques for volcanic and seismic hazards monitoring, *Ann. Geophys.*, *47*, 49–64.

Dobrovolsky, I. P., Zubkov S. I., and Miachkin V. I. (1979), Estimation of the size of earthquake preparation zones, *Pure Appl. Geophys.*, *117*, 1025–1044, doi:10.1007/BF00876083

Doglioni, C., Barba, S., Carminati, E., and Riguzzi, F. (2014), Fault on–off versus coseismic fluids reaction, *Geosci. Front.*, *5*, 767–780.

Eleftheriou, A., Filizzola, C., Genzano, N., Lacava, T., Lisi, M., Paciello, R., Pergola, N., Vallianatos, F., and Tramutoli, V. (2016), Long term (2004–2013) RST analysis of anomalous TIR sequences in relation with earthquake occurrence in Greece, *Pure Appl. Geophys.*, *1*(173), 285–303, doi: 10.1007/s00024-015-1116-8

Eneva, M. D., Adams, N., Wechsler, Y., Ben-Zion, Y., and Dor, O. (2008), *Thermal Properties of Faults in Southern California from Remote Sensing Data*, Report sponsored by NASA under contract to SAIC No. NNH05CC13C, 70 pp.

Filizzola, C., Pergola, N., Pietrapertosa, C., and Tramutoli, V. (2004), Robust satellite techniques for seismically active areas monitoring: a sensitivity analysis on September 7, 1999 Athens's earthquake, *Phys. Chem. Earth*, *29*, 517–527.

Freund, F. T. (2007a), Pre-earthquake signals—Part I: Deviatoric stresses turn rocks into a source of electric currents, *Nat. Hazards Earth Syst. Sci.*, *7*, 535–541.

Freund, F. T. (2007b), Pre-earthquake signals—Part II: Flow of battery currents in the crust, *Nat. Hazards Earth Syst. Sci.*, *7*, 1–6.

Freund, F. T., Takeuchi, A., and Lau, B. W. S. (2006), Electric currents streaming out of stressed igneous rocks—a step towards understanding pre-earthquake low frequency EM emissions, *Phys. Chem. Earth, Parts A/B/C*, *31*, 389–396.

Freund, F. T., Takeuchi, A., Lau, B. W. S., Al-Manaseer, A., Fu, C. C., Bryant, N., and Ouzounov, D. (2007), Stimulated infrared emission from rocks: assessing a stress indicator, *eEarth Disc.*, *1*(2), 97–121

Genzano, N., Aliano, C., Filizzola, C., Pergola, N., and Tramutoli, V. (2007), Robust satellite technique for monitoring seismically active areas: The case of Bhuj-Gujarat earthquake, *Tectonophysics*, *431*, 197–210.

Genzano, N., Aliano, C., Corrado, R., Filizzola, C., Lisi, M., Paciello, R., Pergola, N., Tsamalashvili, T., and Tramutoli, V. (2009a), Assessing of the robust satellite techniques (RST) in areas with moderate seismicity, *Proceedings of Multitemp 2009*, pp. 307–314, Mistic, Connecticut, USA, 28–30 July 2009.

Genzano, N., Aliano, C., Corrado, R., Filizzola, C., Lisi, M., Mazzeo, G., Paciello, R., Pergola, N., and Tramutoli, V. (2009b), RST analysis of MSG-SEVIRI TIR radiances at the time of the Abruzzo 6 April 2009 earthquake, *Nat. Haz. Earth Syst. Sci.*, *9*, 2073–2084.

Genzano, N., Filizzola, C., Paciello, R., Pergola, N. and Tramutoli, V. (2015), Robust satellite techniques (RST) for monitoring earthquake prone areas by satellite TIR observations: the case of 1999 Chi-Chi earthquake (Taiwan), *J. Asian Earth Sci.*, *114*, 289–298, doi: org/10.1016/j.jseaes.2015.02.010

Genzano, N., Filizzola, C., Hattori, K., Lisi, M., Paciello, R., Pergola, N., and Tramutoli, V. (2016), Robust satellite techniques for monitoring earth emitted radiation in the Japanese seismic area by using MTSAT observations in the TIR spectral range, *European Geosciences Union (EGU)*, 17–22 April, Vienna.

Gorny, V. I., Salman, A. G., Tronin, A. and Shilin, B. B. (1988), The Earth outgoing IR radiation as an indicator of seismic activity, *Proc. Acad. Sci. USSR*, *301*, 67–69.

Halle, W., Oertel, D., Schlotzhauer, G., and Zhukov, B. (2008), *Early Warning Of Earthquakes By Space-Borne Infrared Sensors* [Erdbebenfrüherkennung mit InfraRot Sensoren aus dem Weltraum], pp. 1–106.

Huang, G. and Luo, Z. L. (1992), Monitoring and predicting strong earthquakes using N.O.A.A data, *ISPRS'92 Congress X.V.I.I part B*, pp. 499–503, Washington, DC, 2–14 August.

Huang, J., Mao, F., Zhou, W., and Zhu, X. (2008), Satellite thermal IR associated with Wenchuan earthquake in China using MODIS data, *Proceeding of the 14th World Conference on Earthquake Engineering*, pp. 12–17.

Li, J., Wu, L., Dong, Y., Liu, S., and Yang, X. (2007), An quantitative model for tectonic activity analysis and earthquake maginitude predication based on thermal infrared anomaly, *International Geoscience and Remote Sensing Symposium (IGARSS)*, pp. 3039–3042, Barcelona.

Lisi, M., Filizzola, C., Genzano, N., Grimaldi, C. S. L., Lacava, T., Marchese, F., Mazzeo, G., Pergola, N., and Tramutoli, V. (2010), A study on the Abruzzo 6 April 2009 earthquake by applying the RST approach to 15 years of AVHRR TIR observations, *Nat. Haz. Earth Syst. Sci.*, *10*, 395–406.

Lisi, M., Filizzola, C., Genzano, N., Paciello, R., Pergola, N., and Tramutoli, V. (2015), Reducing atmospheric noise in RST analysis of TIR satellite radiances for earthquakes prone areas satellite monitoring, *Phys. Chem. Earth Sci. 85–86*, 87–97, doi: 10.1016/j.pce.2015.07.013

Lisi, M., Corrado, A., Filizzola, C., Genzano, N., Paciello, R., Pergola, N., and Tramutoli, V. (2016), Long-term RST analysis of anomalous TIR sequences in relation with earthquakes occurred in Turkey in the period 2004–2015, *EGU General Assembly, Vienna*, *18*, EGU2016-4660.

Liu, S., Yang, D., Ma, B., Wu, L., Li, J., and Dong, Y. (2007), On the features and mechanism of satellite infrared anomaly before earthquakes in Taiwan region, *International Geoscience and Remote Sensing Symposium (IGARSS)*, pp. 3719–3722, Barcelona.

Lixin, W., Shanjun, L., and Yuhua, W. (2006), The experiment evidences for tectonic earthquake forecasting based on anomaly analysis on satellite infrared image, *International Geoscience and Remote Sensing Symposium (IGARSS)*, pp. 2158–2162, IEEE Conference.

Lü, Q. Q., Ding, J. H., and Cui, C. Y. (2000), Probable satellite thermal infrared anomaly before the Zhangbei M_s 6.2 earthquake on January 10, 1998, *Acta Seismologica Sinica*, *13*, 203–209.

Ma, J., Chen, S., Hu, X., Liu, P., and Liu, L. (2010), Spatial-temporal variation of the land surface temperature field and present-day tectonic activity, *Geosci. Front.*, *1*, 57–67.

Molchan, G. M. (1990), Strategies in strong earthquake prediction, *Phys Earth Planet Inter*, *61*, 84–98.

Molchan, G. M. (1991), Structure of optimal strategies in earthquake prediction. *Tectonophysics 193*:267–276, doi: 10.1016/0040-1951(91)90336-Q

Molchan, G. M. (1997), Earthquake prediction as a decision-making problem, *Pure Appl. Geophys.*, *149*, 233–247, doi: 10.1007/BF00945169

Molchan, G. M. and Kagan, Y. Y. (1992), Earthquake prediction and its optimization. *J. Geophys. Res.*, *97*, 4823, doi: 10.1029/91JB03095

Ouzounov, D. and Freund, F. (2004), Mid-infrared emission prior to strong earthquakes analyzed by remote sensing data, *Adv. Space Res. 33*, 268–273.

Ouzounov, D., Bryant, N., Logan, T., Pulinets, S. A., and Taylor, P. (2006), Satellite thermal IR phenomena associated with some of the major earthquakes in 1999–2003, *Phys. Chem. Earth, Parts A/B/C*, *31*, 154–163.

Panda, S. K., Choudhury, S., Saraf, A. K., and Das, J. D. (2007), MODIS land surface temperature data detects thermal anomaly preceding 8 October 2005 Kashmir earthquake, *Int. J. Remote Sens.*, *28*, 4587–4596.

Peck, L. (1996), Temporal and spatial fluctuations in ground cover surface temperature at a Northern New England site, *Atmospheric Research 41*, 131–160.

Pergola, N., Aliano, C., Coviello, I., Filizzola, C., Genzano, N., Lacava, T., Lisi, M., Mazzeo, G., and Tramutoli, V. (2010), Using RST approach and EOS-MODIS radiances for monitoring seismically active regions: a study on the 6 April 2009 Abruzzo earthquake, *Nat. Haz. Earth Syst. Sci.*, *10*, 239–249.

Piroddi, L. (2011), *Sistemi Di Telerilevamento Termico per il Monitoraggio e la Prevenzione dei Rischi Naturali: il Caso Sismico*, PhD Thesis, University of Cagliari.

Piroddi, L. and Ranieri, G. (2012), Night thermal gradient: A new potential tool for earthquake precursors studies. An application to the seismic area of L'Aquila (Central Italy), *IEEE J. Select. Topics Appl. Earth Observ. Remote Sens.*, *5*(1), 307–312.

Piroddi, L., Ranieri, G., Freund, F., and Trogu, A. (2014), Geology, tectonics and topography underlined by L'Aquila earthquake TIR precursors, *Geophys. J. Int.*, *197*, 1532–1536.

Pulinets, S. A. (2009), Lithosphere–atmosphere–ionosphere coupling (LAIC) model, in M. Hayakawa (ed.), *Electromagnetic Phenomena Associated with Earthquakes*, pp. 235–254, Transworld Research Network, Kerala.

Pulinets, S. and Ouzounov, D. (2011), Lithosphere–atmosphere–ionosphere coupling (LAIC) model—a unified concept for earthquake precursors validation, *J. Asian Earth Sci.*, *41*(4–5), 371–382.

Pulinets, S. A., Biagi, P., Tramutoli, V., Legenka, A. D., and Depuev, V. K. H. (2007), Irpinia Earthquake 23 November 1980—Lesson from Nature reviled by joint data analysis, *Ann. Geophys.*, *50*, 61–78.

Pulinets, S., Ouzounov, D., Karelin, A. V. and Davidenko, D. V. (2015), Physical bases of the generation of short-term earthquake precursors: A complex model of ionization-induced geophysical processes in the lithosphere-atmosphere-ionosphere-magnetosphere system, *Geomag. Aeron.* *55*(4), 521–538.

Qiang, Z. J. and Dian, C. G. (1992), Satellite thermal infrared impending temperature increase precursor of Gonghe earthquake of magnitude 7.0, Qinghai Province, *Geoscience*, *6*, 297–300.

Qiang, Z. J., Xu, X. D., and Dian, C. G. (1991), Thermal infrared anomaly precursor of impending earthquakes, *Chin. Sci. Bull.*, *36*, 319–323.

Qiang, Z. J., Dian, C. G., Wang, X. J., and Hu, S. Y. (1992), Satellite thermal infrared anomalous tempeature increase and impending earthquake precursor, *Chin. Sci. Bull.*, *37*(19), 1642–1646.

Qiang, Z. J., Xu, X., and Dian, C. (1997), Thermal infrared anomaly precursor of impending earthquakes, *Pure Appl. Geophys*, *149*, 159–171.

Qin, K., Wu, L., and Zheng, S. (2013), A deviation-time-space-thermal (DTS-T) method for global Earth observation system of systems (GEOSS)-based earthquake anomaly recognition: criterions and quantify indices, *Remote Sens.*, *5*, 5143–5151, doi:10.3390/rs5105143

Qu, C. Y., Ma, J., and Shan, X.-J. (2006), Counterevidence for an earthquake precursor of satellite thermal infrared anomalies, *Chin. J. Geophys. (Acta Geophys. Sin.)*, *49*, 490–495.

Rawat, V., Saraf, A. K., Das, J., Sharma, K., and Shujat, Y. (2011), Anomalous land surface temperature and outgoing long-wave radiation observations prior to earthquakes in India and Romania, *Nat. Hazards*, *59*, 33–46.

Saradjian, M. R. and Akhoondzadeh, M. (2011), Thermal anomalies detection before strong earthquakes ($M>6.0$) using interquartile, wavelet and Kalman filter methods, *Nat. Haz. Earth Syst. Sci.*, *11*, 1099–1108.

Saraf, A. K. and Choudhury, S. (2004), Satellite detects pre-earthquake thermal anomalies associated with past major earthquakes, *Proceedings of Map Asia Conference*, p. 40.

Saraf, A. K. and Choudhury, S. (2005a), Satellite detects surface thermal anomalies associated with the Algerian earthquakes of May 2003, *Int. J. Remote Sens.*, *26*, 2705–2713.

Saraf, A. K. and Choudhury, S. (2005b), Cover: NOAA-AVHRR detects thermal anomaly associated with the 26 January 2001 Bhuj earthquake, Gujarat, India, *Int. J. Remote Sens. 26*, 1065–1073.

Saraf, A. K. and Choudhury, S. (2005c), Thermal remote sensing technique in the study of pre-earthquake thermal anomalies, *J. Ind. Geophys. Union*, *9*, 197–207.

Saraf, A. K., Rawat, V., Banerjee, P., Choudhury, S., Panda, S., Dasgupta, S., and Das, J. (2008), Satellite detection of earthquake thermal infrared precursors in Iran, *Nat. Hazards*, *47*, 119–135.

Saraf, A. K., Rawat, V., Choudhury, S., Dasgupta, S., and Das, J. (2009), Advances in understanding of the mechanism for generation of earthquake thermal precursors detected by satellites, *Int. J. Appl. Earth Observ. Geoinform.*, *11*, 373–379.

Saraf, A. K., Rawat, V., Das, J., Zia, M., and Sharma, K. (2012), Satellite detection of thermal precursors of Yamnotri, Ravar and Dalbandin earthquakes, *Nat. Hazards*, *61*, 861–872.

Scholz, C. H., Sykes, L. R., and Aggarwal, Y. P. (1973), Earthquake prediction: A physical basis, *Science*, *181*, 803–810.

Singh, R. P., Kumar, S., Zlotnicki, J., and Kafatos, M. (2010), Satellite detection of carbon monoxide emission prior to the Gujarat earthquake of 26 January 2001, *Appl. Geochem.*, *25*(4), 580–585, doi:10.1016/j.apgeochem.2010.01.014

Surkov, V. V., Pokhotelov, O. A., Parrot, M., and Hayakawa, M. (2006), On the origin of stable IR anomalies detected by satellites above seismo-active regions, *Phys. Chem. Earth*, *31*, 164–171.

Tramutoli, V. (1998), Robust AVHRR techniques (RAT) for environmental monitoring: theory and applications, In G. Cecchi and E. Zilioli (eds), *Earth Surface Remote Sensing II*, pp. 101–113, SPIE, Proceedings Vol. *3496*.

Tramutoli, V. (2005), Robust satellite techniques (RST) for natural and environmental hazards monitoring and mitigation: ten years of successful applications, *Proceeding of the 9th International Symposium on Physical Measurements and Signatures*, pp. 792–795, Remote Sensing Vol. XXXVI,7/W.

Tramutoli, V. (2007), Robust satellite techniques (RST) for natural and environmental hazards monitoring and mitigation: theory and applications, *Proceeding of 2007 International Workshop on the Analysis of Multi-Temporal Remote Sensing Images* pp. 1–6, Vol.*1*, doi: 10.1109/Multitemp.2007.4293057

Tramutoli, V. (2013), *Comments on: "Exploring land surface temperature earthquake precursors: A focus on the Gujarat earthquake of 2001" paper by Blackett et al., 2011*, Technical Report, doi: 10.13140/2.1.1684.3041

Tramutoli, V., Di Bello, G., Pergola, N., and Piscitelli, S. (2001), Robust satellite techniques for remote sensing of seismically active areas, *Ann. Geofis.*, *44*, 295–312.

Tramutoli, V., Cuomo, V., Filizzola, C., Pergola, N., and Pietrapertosa, C. (2005), Assessing the potential of thermal infrared satellite surveys for monitoring seismically active areas. The case of Kocaeli (Izmit) earthquake, August 17th, 1999, *Remote Sens. Environ.*, *96*(3–4), 409–426.

Tramutoli, V., Aliano, C., Corrado, R., Filizzola, C., Genzano, N., Lisi, M., Martinelli, G. and Pergola, N. (2013), On the possible origin of thermal infrared radiation (TIR) anomalies in earthquake-prone areas observed using robust satellite techniques (RST), *Chem. Geol.*, *339*, 157–168, doi: 10.1016/j.chemgeo.2012.10.042

Tramutoli, V., Jakowski, N., Pulinets, S., Romanov, A., Filizzola, C., Shagimuratov, I., Pergola, N., Ouzounov, D., Papadopulos, G., Genzano, N., Lisi, M., Alparslan, E., Wilken, V., Romanov, A., Zakharenkova, I., Paciello, R., Coviello, R., Romano, G., Tsybulia, K., Inan, S., and Parrot, M. (2014a), From PRE-EARTHQUAKES to EQUOS: how to exploit multi-parametric observations within a novel system for time-dependent assessment of seismic hazard (T-DASH) in a pre-operational civil protection context, *Second European Conference on Earthquake Engineering and Seismology (2ECEES)*, Turkey, 24–29 August.

Tramutoli, V., Armandi, B., Filizzola, C., Genzano, N., Lisi, M., Paciello, R., and Pergola, N. (2014b), RST (robust satellite techniques) analysis for monitoring earth emitted radiation in seismically active area of California (US): a long term (2006–2011) analysis of GOES-W/IMAGER thermal data, *American Geophysical Union (AGU) Fall Meeting*, 15–19 December, San Francisco.

Tramutoli, V., Armandi, B., Coviello, I., Eleftheriou, A., Filizzola, C., Genzano, N., Lacava, T., Lisi, M., Paciello, R., Pergola, N., Satriano, V., and Vallianatos, F. (2014c). Long term RST analyses of TIR satellite radiances in different geotectonic contexts: results and implications for a time dependent assessment of seismic hazard (t-DASH), *American Geophysical Union (AGU) Fall Meeting*, 15–19 December, San Francisco.

Tramutoli, V., Corrado, R., Filizzola, C., Genzano, N., Lisi, M., and Pergola, N. (2015a), From visual comparison to robust satellite techniques: 30 years of thermal infrared satellite data analyses for the study of earthquake preparation phases, *Boll. Geof. Teor. Appl.*, *56*(2), 167–202,

Tramutoli, V., Corrado, R., Filizzola, C., Genzano, N., Lisi, M., Paciello, R., and Pergola, N. (2015b), One year of RST based satellite thermal monitoring over two Italian seismic areas, *Boll. Geof. Teor. Appl.*, *56*(2), 275–294.

Tramutoli, V., Filizzola, C., Genzano, N., Lisi, M., Paciello, R., and Pergola, N. (2015c), A retropective long-term (2004–2014) correlation analysis of TIR anomalies and earthquakes ($M > 4$) Occurrence Over Italy, In *34° Convegno Nazionale, Gruppo nazionale di Geofisica della Terra Solida (GNGTS)*, 17–19 Novembre, Trieste.

Tronin, A. A. (1996), Satellite thermal survey—a new tool for the study of seismoactive regions, *Int. J. Remote Sens.*, *17*, 1439–1455.

Tronin, A. A. (2000), Thermal IR satellite sensor data application for earthquake research in China, *Int. J. Remote Sens.*, *21*, 3169–3177.

Tronin, A. A., Hayakawa, M., and Molchanov, O. A. (2002), Thermal IR satellite data application for earthquake research in Japan and China, *J. Geodyn.*, *33*, 519–534.

Tronin, A. A., Biagi, P. F., Molchanov, O. A., Khatkevich, Y. M., and Gordeev, E. I. (2004), Temperature variations related to earthquakes from simultaneous observation at the ground stations and by satellites in Kamchatka area, *Phys. Chem. Earth, Parts A/B/C*, *29*, 501–506.

Wang, L. and Zhu, C. (1984), Anomalous variations of ground temperature before the Tangsan and Haiheng earthquakes, *J. Seismol. Res.*, *7*, 649–656.

Xie, T., Kang, C. L., and Ma, W. Y. (2013), Thermal infrared brightness temperature anomalies associated with the Yushu (China) $M_s = 7.1$ earthquake on 14 April 2010, *Nat. Hazards Earth Syst. Sci.*, *13*, 1105–1111.

Xiong, P., Gu, X. F., Bi, Y. X., Shen, X. H., Meng, Q. Y., Zhao, L. M., Kang, C. L., Chen, L. Z., Jing, F., Yao, N., Zhao, Y. H., Li, X. M., Li, Y., and Dong, J. T. (2013), Detecting seismic IR anomalies in bi-angular Advanced Along-Track Scanning Radiometer data, *Nat. Hazards Earth Syst. Sci.*, *13*, 2065–2074.

Xiong, P., Shen, X., Gu, X., Meng, Q., Zhao, L., Zhao, Y., Li, Y., and Dong, J. (2015a), Seismic infrared anomalies detection in the case of the Wenchuan earthquake using bi-angular advanced along-track scanning radiometer data, *Ann. Geophys.*, *58*(2), S0217, doi:10.4401/ag-6706

Xiong, P., Shen, X., Gu, X., Meng, Q., Bi, Y., Zhao, L., Zhao, Y., Li, Y., and Dong, J. (2015b), Satellite detection of IR precursors using bi-angular advanced along-track scanning radiometer data: a case study of Yushu earthquake, *Earthquake Sci.*, *28*(1), 25–36, doi: 10.1007/s11589-015-0111-6

Xu, X. D., Xu, X. M., and Wang, Y. (2000), Satellite infrared anomaly before the Nantou $M_s = 7.6$ earthquake in Taiwan, China, *Acta Seismol. Sin.*, *13*, 710–713.

Yang, Y. and Guo, G. (2010), Studying the thermal anomaly before the Zhangbei earthquake with MTSAT and meteorological data, *J. Remote Sens.*, *31*, 2783–2791, doi:10.1080/01431160903095478

Yasuoka, Y., Igarashi, G., Ishikawa, T., Tokonami, S., and Shinogi, M. (2006), Evidence of precursor phenomena in the Kobe earthquake obtained from atmospheric radon concentration, *Appl. Geochem.*, *21*, 1064–1072.

Yoshioka, R., Yamano, Y., Tanaka, Y., and Gotoh, K. (2005), Analysis of ground surface temperature before occurrence of earthquake by satellite thermal infrared data, *International Geoscience and Remote Sensing Symposium (IGARSS)*, Vol. *3*, pp. 1822–1825, Seoul.

Zhang, Y. S., Guo, X., Zhong, M. J., Shen, W. R., Li, W., and He, B. (2010), Wenchuan earthquake: Brightness temperature changes from satellite infrared information, *Chin. Sci. Bull.*, *55*, 1917–1924.

Zhang, Y.-S., Guo, X., Wei, C. X., Shen, W. R., and Hui, S. X. (2011), The characteristics of seismic thermal radiation of Japan Ms9.0 and Myanmar Ms7.2 earthquake, *Chin. J. Geophys. (Acta Geophys. Sin.)*, *54*, 2575–2580.

Zoran, M. (2012), MODIS and NOAA-AVHRR l and surface temperature data detect a thermal anomaly preceding the 11 March 2011 Tohoku earthquake, *Int. J. Remote Sens.*, *33*(21), 6805–6817, doi: 10.1080/01431161.2012.692833

Zoran, M., Savastru, D., and Mateciuc, D. (2016), Earthquake precursors assessment in Vrancea Region through satellite and in situ monitoring data, in R. Vacareanu and C. Ionescu (eds), *Proceedings of the Symposium Commemorating 75 Years from November 10, 1940 Vrancea Earthquake*, pp. 305–314, Springer Natural Hazards.

15

Thermal Radiation Anomalies Associated with Major Earthquakes

Dimitar Ouzounov[1], Sergey Pulinets[2], Menas C. Kafatos[1], and Patrick Taylor[3]

ABSTRACT

Recent developments of remote sensing methods for Earth satellite data analysis contributes to our understanding of earthquake-related thermal anomalies. It was realized that the thermal heat fluxes over areas of earthquake preparation are a result of air ionization by radon (and other gases) and consequent water vapor condensation on newly formed ions. Latent heat (LH) is released as a result of this process and leads to the formation of local thermal radiation anomalies (TRA) known as outgoing longwave radiation. We compare the LH energy, obtained by integrating surface latent heat flux over the area and time with released energies associated with these events. Extended studies of the TRA using the data from the most recent major earthquakes allowed us to establish the main morphological features. It was also established that the TRA are part of a more complex chain of the short-term pre-earthquake generation, which is explained within the framework of a lithosphere–atmosphere coupling processes.

15.1. INTRODUCTION

For many years a search for physically based preseismic signals has been conducted [e.g., *Martinelli*, 1998]. Multiple observations of earthquake precursory signals have been published previously [*Hayakawa*, 1999; *Hayakawa and Molchanov*, 2002; *Pulinets and Boyarchuk*, 2004). Recent analysis of data from multi-instrument spaceborne and ground observations has provided evidence for the existence of pre-earthquake atmospheric signals [*Liu et al.*, 2000, 2010; *Kon et al.*, 2010; *Ouzounov et al.*, 2011b; *Hayakawa*, 2012; *Han et al.*, 2014, 2016; *Pulinets and Davidenko*, 2014; *Tramutoli et al.*, 2015a,b]. These studies have contributed to our understanding of the physics of earthquakes and the phenomena that precede their energy release. Recent progress in earth observing space technology has also helped to advance the scientific understanding of the nature of pre-earthquake phenomena in the atmosphere. We are searching for preseismic observations that might give warning of a major earthquake. Our investigation is based on a search for possible connection between satellite observations (latest National Polar-Orbiting Operational Environmental Satellite System (NPOESS) and NASA Earth Observing System (EOS)) of anomalous atmospheric thermal transient signals and subsequent major earthquakes.

Figure 15.1, the Earth's energy budget (https://earthobservatory.nasa.gov/), shows that more than 70% of the solar radiation is reflected to the atmosphere. While the Earth surface absorbs incident shortwave (SW) solar energy, nearly one-third is reflected back into space. The heat absorbed at the surface does not reach any significant depth because the low thermal conductivity of the surface. During the night, the absorbed energy instead is re-emitted as long wavelength radiation (LWR). The energy directly released into the atmosphere is known as sensible heat (SH) and is transferred as radiant thermal energy or heat. Latent heat (LH) is the energy released or

[1] *Center of Excellence in Earth Systems, Modeling and Observations (CEESMO), Chapman University, Orange, CA, USA*
[2] *Space Research Institute Russian Academy of Sciences, Moscow, Russia*
[3] *NASA GSFC, Greenbelt, MD, USA*

Pre-Earthquake Processes: A Multidisciplinary Approach to Earthquake Prediction Studies, Geophysical Monograph 234, First Edition. Edited by Dimitar Ouzounov, Sergey Pulinets, Katsumi Hattori, and Patrick Taylor.

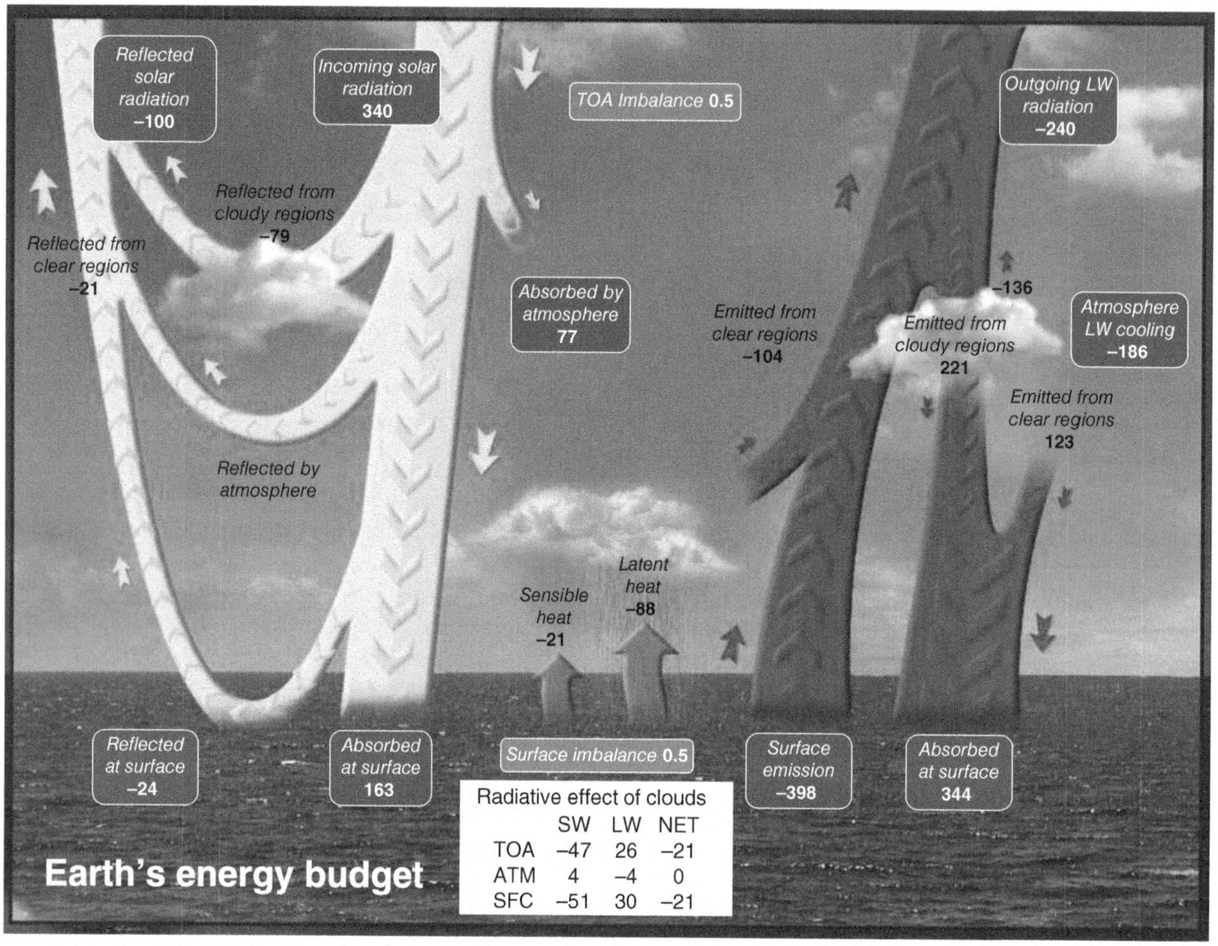

Figure 15.1 Earth atmospheric energies. (Source: Reproduced with the permission of NASA [2012]: https://earthobservatory.nasa.gov/)

absorbed by a substance during a phase change. A balance must exist between incoming and radiated solar energy or else the surface of Earth would permanently heat up. This process of radiative exchange is complex. Generally the LWR emitted by the Earth into the atmosphere gets absorbed and maintains a thermal equilibrium with the atmosphere. When the sky is clear the LWR escapes into space since the atmosphere is "transparent" to the radiation in the 8.0–12.0-μm spectral wavelengths. Sometimes this atmospheric window can be partially blocked by clouds or pollution and outgoing LWR could be reabsorbed by the atmosphere and re-emitted in all directions.

This outgoing LWR at, the top of the atmosphere (TOA), has been recorded by satellite sensors as outgoing longwave radiation (OLR). The net result is that radiative losses from the atmosphere are generally balanced by the energy re-emitted as heat from Earth's surface. Later, mainly during the formation of clouds, the water vapor condensates and the LH are released into the atmosphere (estimated by surface latent heat flux (SLHF) from the National Centers for Environmental Prediction (NCEP)). This leads to heat transfer from the surface into the atmosphere, which is one of the main drivers of the general atmospheric circulation [*Goosse et al.*, 2016].

15.2. METHODS FOR RECOGNITION OF THERMAL RADIATION ANOMALIES

For more than 2000 years scientists have been searching for signals preceding earthquakes. According to ancient Greek philosopher Aristotle, "pneuma" (wind/gas) are involved before earthquakes, producing strange atmospheric effects [*MacArthur*, 1980]. The fundamental meaning of "pneuma" is air in motion or electricity in the airs. Fogs and clouds have been recognized as observational evidence for activities prior to major seismicity since the days of Aristotle and Pliny (Roman Empire), and by many researchers in ancient China [*Tributsch*, 1978]. John Milne, an English seismologist and geologist who invented the first modern seismograph, published the first quantitative analysis of atmospheric signals associated with seismicity [*Milne*, 1913]. He found that for 387 earthquakes in northern Japan, the mean monthly temperatures were generally a little ahead of the crest of the waves, indicating the earthquake arrival [*Milne*, 1913].

Just recently a comprehensive catalog of atmospheric pre-earthquake signals based on historical and instrumental records for more than 1500 earthquakes between 550 BC until AD 2000 has been published [*Tronin*, 2011]. Several atmospheric parameters (atmospheric heat, drought, cooling, clouds, air pressure variations, winds, fogs, etc.) have been found before the occurrence of seismic shocks related to about 700 earthquakes (~50% from all cases) and only a temperature increase has been seen in 10% of the cases, which is still a good indicator, bearing in mind that the thermometer was invented only in the 16th century. In addition to the atmospheric phenomena, a variety of hydrological changes has been reported, usually followed by gas release (bad smell) and foggy atmosphere [*Tronin*, 2011].

Recent advances in remote-sensing instrumentation have helped to advance the scientific understanding of atmospheric earthquake signals. Satellite thermal imaging data reveal stationary (long-lived) thermal anomalies associated with large linear structures and fault systems in Earth's crust [*Carreno et al.*, 2001], and also transient (short-lived) anomalies prior to major earthquakes [*Quing et al.*, 1991; *Salman et al.*, 1992]. Their spatial extent and temporal evolution may be dependent on local geology and tectonics, nature of the focal mechanism, meteorological conditions, and other factors. Studies on the relationship between satellite thermal infrared (TIR) data and earthquake precursors have been based on data from both single and multi-instruments. *Gorny et al.* [1988], *Tronin et al.* [2002, 2004], *Dey et al.* [2004], and *Saraf and Choudhury* [2005] have used imagery recorded by the Advanced Very-High-Resolution Radiometer (AVHRR), analysis methods based on a comparisons between before-and-after images over an earthquake epicenter. Newer techniques have been proposed, using subpixel level coregistration and georeferenced data from both polar-orbiting and geosynchronous satellites (Geostationary Operational Environmental Satellite (NOAA), Meteosat, AVHRR, and Landsat [*Bryant et al.*, 2003; *Di Bello et al.*, 2004]. One of the problems in detecting TIR anomalies is defining abnormal TIR fluctuations from a normal baseline. To address this problem, an approach was developed using a time series of TIR data over earthquake-prone regions. Using pixel-level thermal radiation variance from established base lines, it was possible to identify anomalous TIR signals [*Tramutoli et al.*, 2001; *Filizzola et al.*, 2004; *Cervone et al.*, 2006; *Ouzounov et al.*, 2007]. After the launch of the EOS satellites (Terra in 1999 and Aqua in 2002), a new approach for detecting pre-earthquake anomalies was developed, based on land surface temperature (LST) derived from the 11-μm data [*Ouzounov and Freund*, 2004]. Observations by NPOESS and Aqua's Atmospheric Infrared Sounder (AIRS) of atmospheric environmental parameters have revealed an increase in radiation and an intermediate change in OLR in the 8–12 μm range [*Ouzounov et al.*, 2007]. The OLR transitional changes recorded at the TOA over seismically active regions have been proposed as being related to thermodynamic processes within Earth's crust that leads to earthquakes [*Ouzounov et al.*, 2011a; *Pulinets and Ouzounov*, 2011].

15.2.1. Robust Satellite Technique

Space–time anomalies of Earth's emitted radiation in the TIR spectral range measured from satellite data months to weeks before earthquakes, have been interpreted, by several authors, as pre-earthquake signals [Gorny et al., 1988; *Quing et al.*, 1991; *Tronin*, 1996, 2006; *Tramutoli et al.*, 2005].

The proposed connection between TIR emission and seismic activity has been considered, for a long time, with some caution by the scientific community mainly because of the insufficiency of validated data sets and the occurrence of other phenomenon (e.g. meteorological and geomagnetic) rather than seismic activity, which could be responsible for the observed TIR signal fluctuations (for a review see *Tramutoli et al.* [2015a,b] and references herein). The robust satellite technique (RST) is based on a preliminary multitemporal analysis of several years of homogeneous historical data of satellite TIR records, which are devoted to interpreting the TIR signal for each pixel of the satellite observations. Anomalous TIR patterns are identified using a specific index, RETIRA (robust estimator of TIR anomalies) [*Filizzola et al.*, 2004; *Tramutoli et al.*, 2005].

Quality of the TIR data analyses and of the validation processes devoted to assess the possible correlation among TIR anomalies and seismic activity have been improved in recent years [*Tramutoli et al.*, 2015b], increasing the possibility of including them among the observable phenomena that are expected to positively contribute to an advanced multiparametric system for a time-dependent assessment of seismic hazard [*Tramutoli et al.*, 2014a,b; *Eleftheriou et al.*, 2015]. In particular, the RST proposed in 1998 by *Tramutoli* [see also *Genzano et al.*, 2007] has been applied successfully in order to identify pre-earthquake space–time TIR anomalies, in both very variable (satellite view angle, land topography and coverage, etc.) and normal (e.g. meteorological) observational conditions. In subsequent RST applications to different seismically active areas, the space–time correlation between thermal anomalies and earthquakes occurrence was confirmed, for instance, by using both polar (see *Lisi et al.* [2010] and *Pergola et al.* [2010] for NOAA/AVHRR and EOS/MODIS, respectively) and geostationary (MSG/SEVIRI; see *Genzano et al.* [2009]) sensors, in the case of the M_w 6.3 L'Aquila earthquake (6 April 2009) (Figure 15.2).

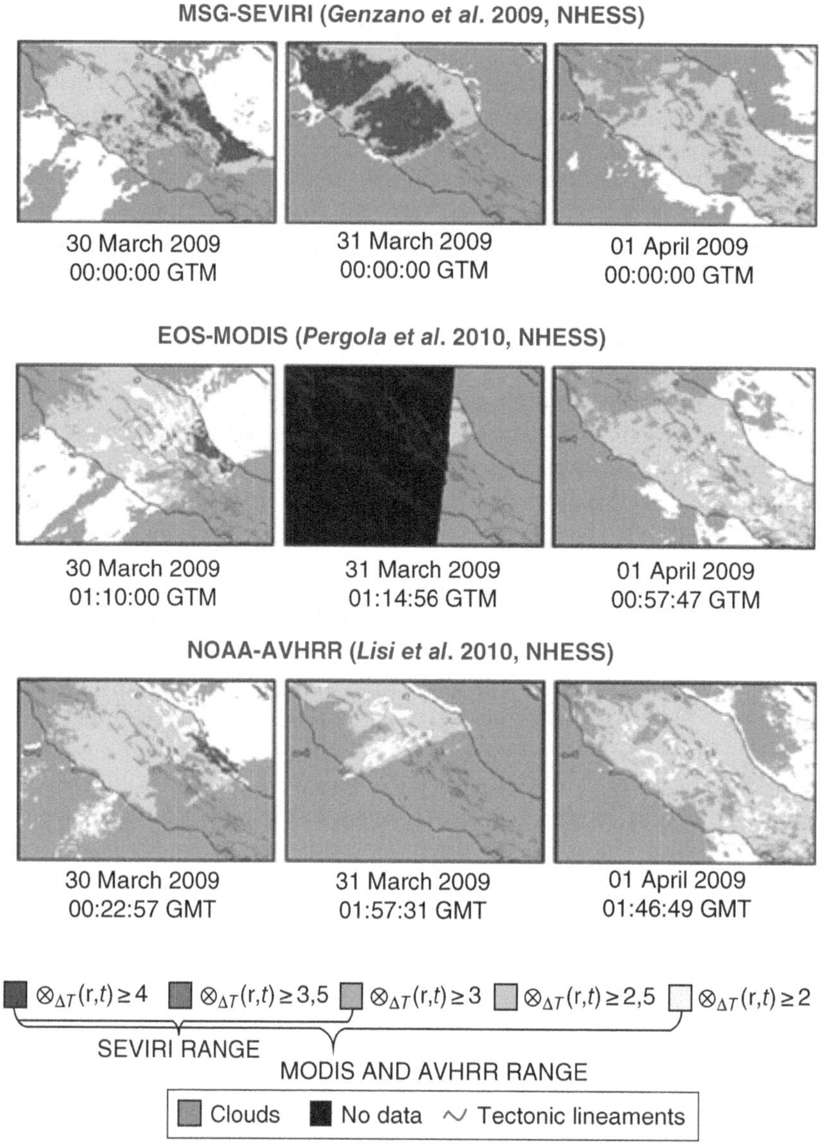

Figure 15.2 Application of the robust satellite technique to seismically active areas shows space–time correlation between thermal anomalies and earthquakes occurrence. In the case of the *Mw* 6.3 L'Aquila earthquake (6 April 2009, (*Lisi et al.* [2010] and *Pergola et al.* [2010] for NOAA/AVHRR and EOS/MODIS) and geostationary (MSG/SEVIRI; *Genzano et al.* [2009]) sensors.

15.2.2. Outgoing Longwave Radiation Analysis

One of the main parameters used to characterize Earth's radiation environment is OLR from Earth [*Liebmann and Smith*, 1996]. The OLR, measured at the TOA, has been associated with integration of emissions from the ground, lower atmosphere, and clouds [*Ohring and Gruber*, 1982]. This was used primarily to study Earth's radiative budget and for climate studies [*Gruber and Krueger*, 1984; *Mehta and Susskind*, 1999]. Daily OLR data have been used to study its variability in the zone of earthquake activity [*Liu and Kang*, 1999; *Ouzounov et al.*, 2007, 2011; *Xiong at al.*, 2010]. An increase in radiation and a transient change in OLR were proposed as being related to thermodynamic processes in the atmosphere over seismically active regions, and were described as thermal radiation anomalies (TRA). Because of the present variety in OLR algorithms and data products, such as RBUD—NOAA/CLASS, CERES LWR—TOA fluxes, NOAA/AVHRR—OLR, and EOS Aqua—CLROLR fluxes, we introduce the TRA as a common name for the types of anomaly identified using different data for computing the transitional LWR. The characteristic of TRA was suggested by *Ouzounov et al.* [2007] as a

Table 15.1 List of earthquakes.

Name	Date (mm/dd/yyyy)	Geographic coordinates	Time (UTC)	M	Depth (km)
Van, Turkey	October 25 2011	38.62°N/43.48°E	16:41:00	7.2	7.1
Aegean Sea, Greece	May 24 2014	40.3°N/25.45°E	09:25:00	6.9	10
NapaValley, California, USA	August 24 2014	38.21°N/122.31°W	10:24:44	6.0	11.11

Source: US Geological Survey.

statistical maximum change in the rate of OLR for a specific spatial location and predefined time and was constructed corresponding to the anomalous thermal field [*Tramutoli et al.*, 1999, 2013]. The anomaly represents the different amplitude for a specific spatial location and predefined times:

$$\text{anomaly}(x,y,t) = \frac{S(x,y,t) - \bar{S}(x,y,t)}{\tau(x,y,t)} \tag{15.1}$$

$$\bar{S}(x,y,t) = \frac{1}{N}\sum_{i-1}^{N} S(x_i,y_i,t_i) \tag{15.2}$$

$$\tau(x,y,z) = \sqrt{\frac{\Sigma\left(S(x,y,t) - \bar{S}(x,y,t)\right)^2}{N}} \tag{15.3}$$

where $S(x,y,t)$ is the current OLR value, $\bar{S}(\text{x},y,t)$ is the computed mean of the background field, defined as the daily y mean value of OLR $S(x_i,y_i,t_j)$ over an area of longitude x and latitude y on the tth day of N years, and $\tau(x,y,z)$ is the standard deviation over the same location (x,y) and local time (t) as the OLR value. This rapid enhancement of radiation could be explained by an anomalous flux of the LH over the area of increased tectonic activity. The input data have been processed with a resolution of (2.5° × 2.5°) and the output maps have been processed using a two step procedure: (a) grid resolution enhancement (1° × 1°) with additional preprocessing to avoid aliasing of short wavelengths; and (b) regridding with spatial filtering and grid points recomputing based on a "minimum curvature" algorithm used by standard computation packages [*Fisher et al.*, 2008].

In our studies we used OLR data from the AVHRR of the NCEP/NOAA. A daily mean global database, with a spatial resolution of 2.5° by 2.5°, was used to study the OLR activity and variability in the region of three recent major earthquakes: (a) M 7.1 Van, Turkey, on 24 October 2011; (b) M 6.9 Aegean Sea, Greece, on 24 May 2014; and (c) M 6.0 Napa, California, on 24 August 2014 (see Table 15.1).

Our method includes recording the OLR at the TOA several days before the onset of an earthquake and characterizing the state of the atmosphere before the earthquake using an established baseline. Calculated at the TOA, the OLR from NOAA NPOESS were used to study the Earth's radiation budget, because they represent emissions from the Earth's surface, lower atmosphere, and clouds, and they are sensitive to near-surface and cloud temperatures. Observations from the NOAA NPOESS were based on the longwave flux estimation of *Ellingson et al.* [1989]. Daily mean OLR values were calculated from these raw data, using algorithms based on equations (15.1)–(15.3), and customized for each region. The NOAA Climate Prediction Center (http://www.cdc.noaa.gov/) provides daily and monthly OLR data and the OLR algorithm for analyzing the AVHRR data. A daily mean, covering a significant area of Earth (90°N–90°S, 0°–357.5°E) and with a spatial resolution of 2.5° × 2.5° recorded.

The "daily values," "normal state," and "TRA anomalies" in relation to pre-earthquake OLR signals were introduced initially by *Ouzounov et al.* [2007] and *Xiong et al.* [2010]. The "TRA anomaly" has been calculated as a deviation of daily values from the normal state and normalized by the multiyear standard deviation for the same pixel (equation 15.1) and shown on Figures 15.3–15.5 in red. The "daily values" are raw values of OLR for the same pixel and same local time observed by the polar orbit satellite (Figures 15.3–15.5; gray curve). The "normal state" for OLR was estimated by a multiyear average (from 2004 to the present) for each pixel (equation 15.2) (Figures 15.3–15.5; black solid curve). The OLR "anomalies" represent the maximum change in the daily variations of the OLR in comparison to the "normal state."

The continuous satellite monitoring of OLR data over Turkey shows a rapid increase of emitted radiation during the mid-September to October 2011 time frame (Figure 15.3). An anomaly at TOA was detected at 19:00 LT on 19 October 2011, southeast of the epicenter area. The anomaly was spatially extended and temporally persistent for 8 h and was the largest TRA anomaly over the entire European continent at that time [*Ouzounov et al.*, 2014].

Continuous satellite monitoring of OLR data over Greece obtained from the NPOESS of NOAA showed rapid increase during the middle of April 2015 and they indicated a probable large earthquake preparation process in the Aegean Sea (Figure 15.4). A strong anomaly in the

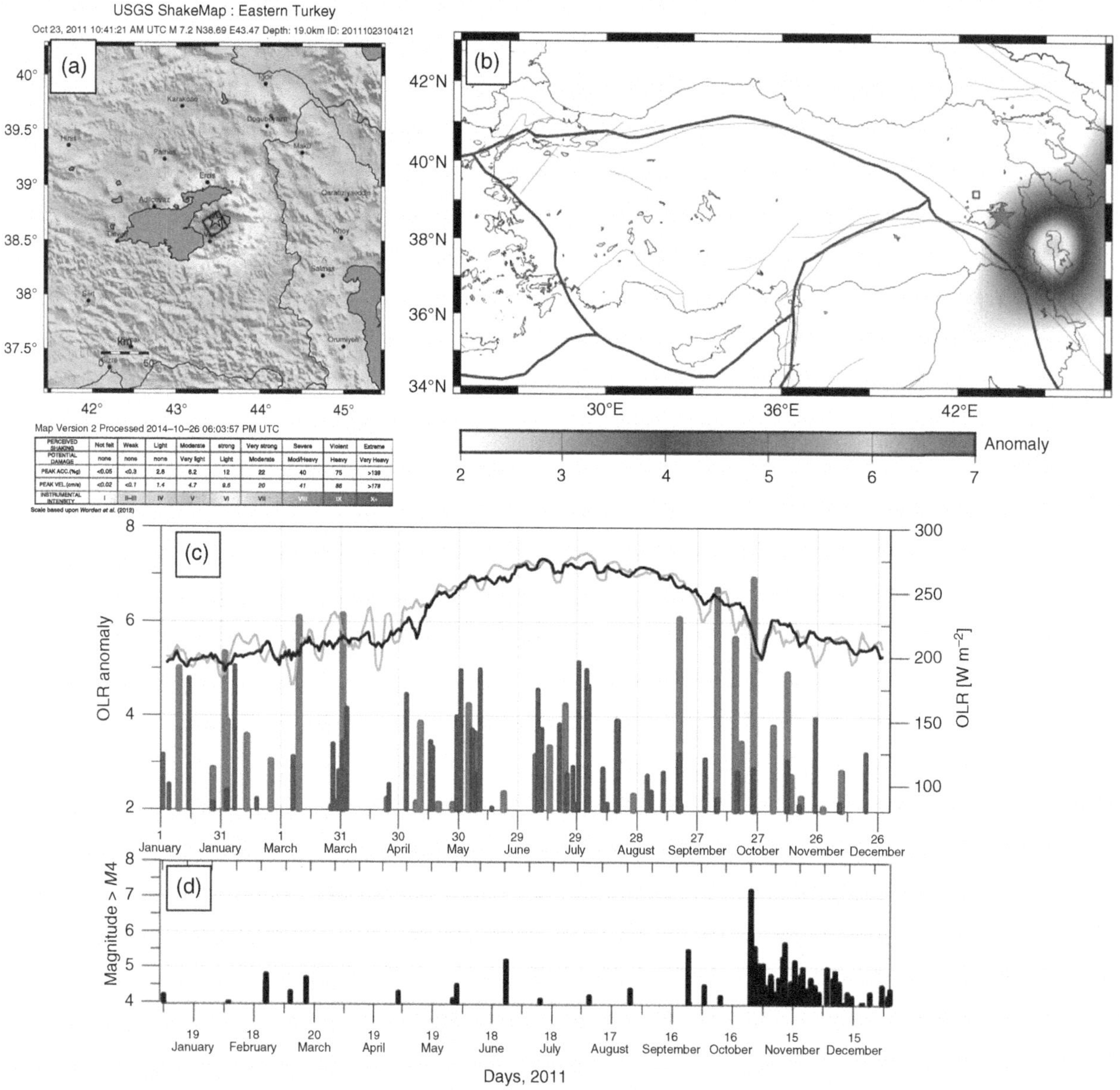

Figure 15.3 Thermal radiation anomalies (TRA) associated with M 7.3 Van, Turkey, eartquake of 23 October 2011. (a) US Geological Survey shake map. (b) Nighttime TRA maps observed on 19 October 2011, 4 days in advance. Epicenter is marked with red star, tectonic plate boundaries by heavy red line, and major faults by feint brown lines. (c) Yearly time series of nighttime OLR over the epicentral area: (red) anomalous values; (gray) 2011 mean value OLR; (black) 2006–2011 mean value; (blue) 2010 anomalous trend with no major seismicity for comparison. (d) $M>4$ seismicity (European-Mediterranean Seismological Centre catalog) near the epicentral area, 2011. (*See electronic version for color representation.*)

atmosphere was detected (prospectively) at 19:00 LT on 14 May 2015 with 2.5 σ significance over 25 years of data analysis [*Ouzounov et al.*, 2015b]. On 24 May 2015, after the anomaly was detected a M 6.9 earthquake occurred in the area of the observed atmospheric anomaly.

At the end of July and into August 2014 we detected (imminent) a large TRA transient field over northern California (Figure 15.5). The location was shifted northwest by about 100 km from the M 6.0 epicentral area of 24 August. For this particular case we used a hybrid OLR product between NOAA's AVHHR and EOS's AIRS observations in order to increase the temporal coverage of the San Francisco area. In the southern Napa Valley, California, earthquake, the value of the OLR compared to the reference field of August 2004 to 2014 indicated a rapid change in the anomalous thermal flux rate at 7:00 AM PDT

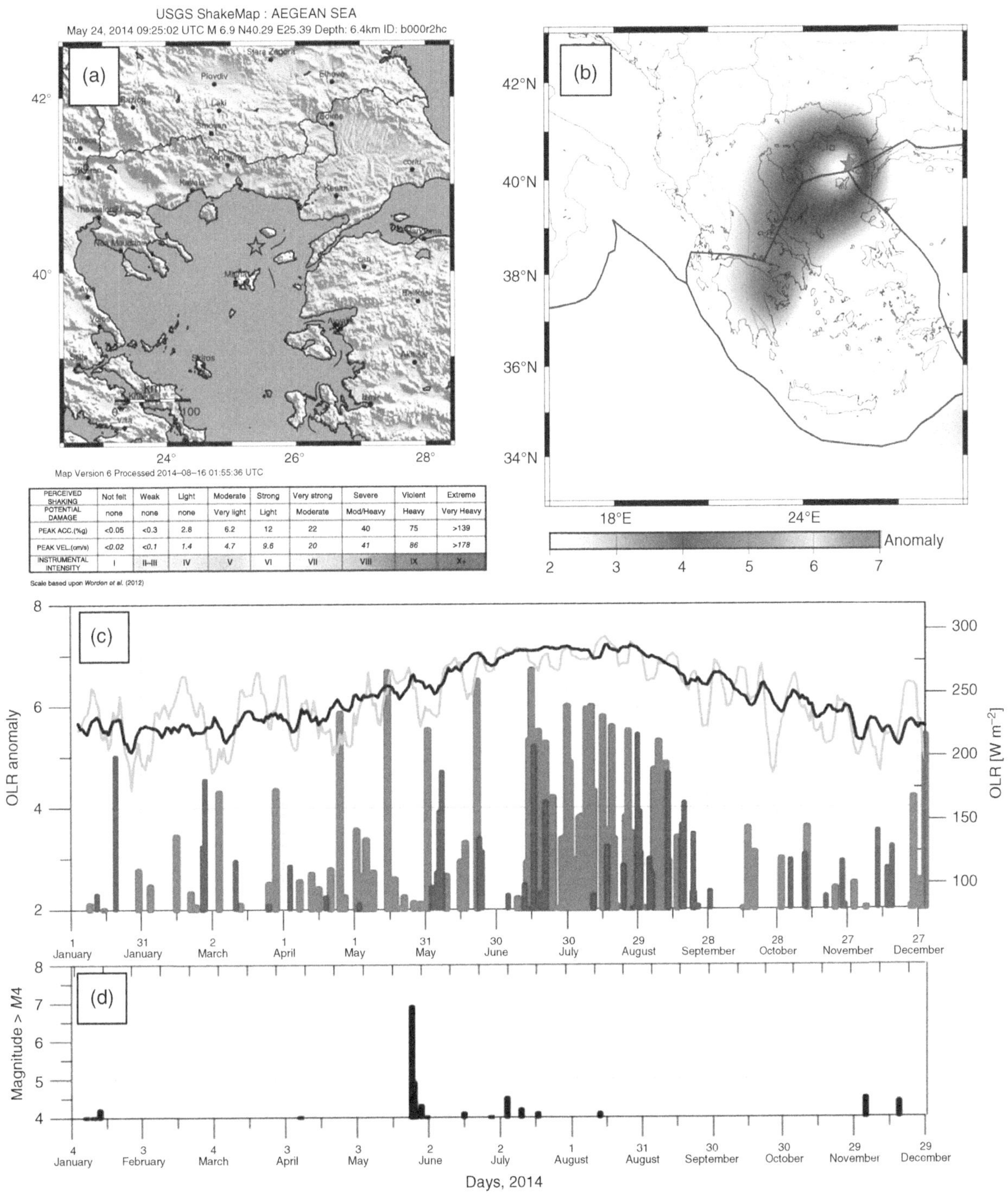

Figure 15.4 Thermal radiation anomalies (TRA) associated with *M* 6.9 Aegean Sea, Greece, earthquake of 24 May, 2014. (a) US Geological Survey shake map. (b) Nighttime TRA map observed on 14 May 2015, 10 days in advance. Epicenter is marked with red star, tectonic plate boundaries by heavy red line, and major faults by feint brown lines. (c) Yearly time series of nighttime OLR over the epicentral area: (red) anomalous values; (gray) 2014 mean value OLR; (black) 2006–2014 mean value; (blue)2013 anomalous trend with no major seismicity for comparison. (d) *M*>4 seismicity (European-Mediterranean Seismological Centre catalog) near the epicentral area, 2014. *(See electronic version for color representation.)*

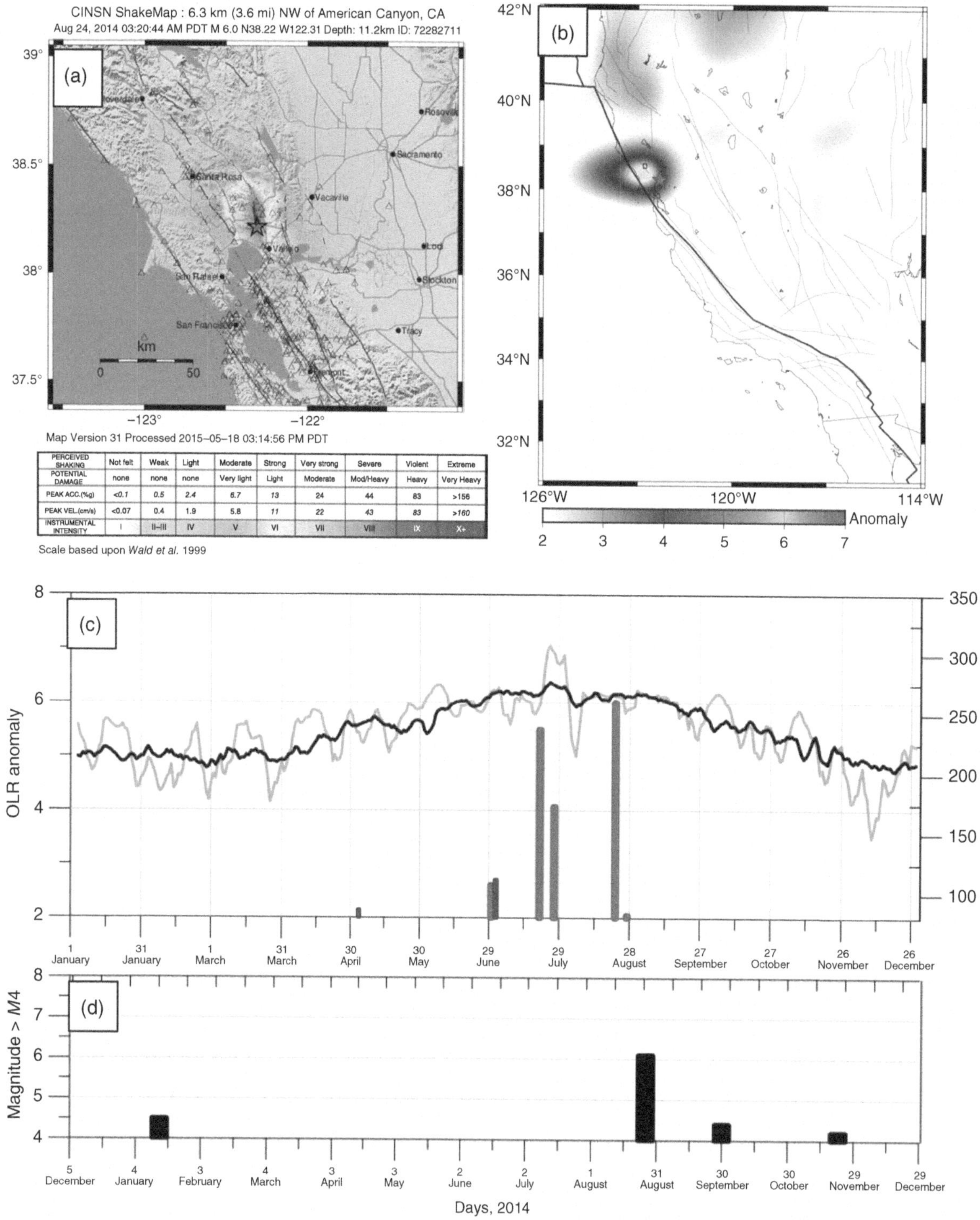

Figure 15.5 Thermal radiation anomalies (TRA) associated with the *M* 6 Napa Valley, California, earthquake of 24 August 2014. (a) US Geological Survey shake map. (b) Nighttime TRA anomaly map observed on 22 August 2014; 2 days before. Epicenter is marked with red star, tectonic plate boundaries by heavy red line, and major faults by feint brown lines. (c) Yearly time series of nighttime OLR over the epicentral area: (red) anomalous values; (gray) 2014 mean value OLR; (black) 2006–2014 mean value; (blue) 2013 anomalous trend with no major seismicity for comparison; and (d) *M*>4 seismicity (European-Mediterranean Seismological Centre catalog) near the epicentral area, 2014. (*See electronic version for color representation.*)

on 23 August. The registered anomalous pattern was the largest anomaly energy flux over California at this time [*Ouzounov et al.*, 2014].

The daily OLR variations (OLR anomalous values ≤ 2) are caused by the daily environmental variations in the vertical atmospheric circulations in the middle atmosphere. Significant deviation from the normal state for the Van, Turkey, case started on 15 September, reached a maximum value (OLR, 7) on 22 October and went back to normal 15 December 2011 (Figure 15.3). For the 2014 Aegean Sea event the breakout started at the beginning of February, the anomaly reached a maximum around 14 May and came back to normal on 15 October 2014 (Figure 15.4). For South Napa the break started on 20July with a systematic increase reaching the maximum level (OLR, ~ 7) in OLR acceleration on 22 August and returning to normal in mid-September 2014 (Figure 15.5).

At the ending of the event the anomalous pattern coincided with the same period as the ending of the major aftershock activities in all of the cases. To test the significance of the results for confutation, i.e. absence of anomalous OLR signals in the absence of major seismic events [*Tramutoli et al.*, 2005], we calculated the OLR anomalies for the same location for one full year before the earthquake. The yearly time distribution of OLR shows significant lower level (blue columns in Figures 15.3–15.5) with a randomly distributed pattern. Thus it is not reproducing the strength and temporal evolution of the OLR anomalous signals generated during the year (red columns Figures 15.3–15.5). The pre-earthquake OLR anomalies for the *M* 7.0 Van (Figure 15.3) and *M* 6.9 Greece (Figure 15.4) earthquakes are not overlapping in time with the maxima of the seasonal variations of OLR (black curve in Figures 15.3 and 15.4)—usually higher during June–August in the Northern Hemisphere—which is an indication for the nonweather related origin of OLR anomalous values and their connection primarily with the geodynamics. The *M* 6.0 South Napa earthquake occurred in the middle of the summer, 24 August 2014, and shows that the computational approach used can rival an anomalous pattern even during a period of natural enhancement of the background radiation field (Figure 15.5). Analogous findings were observed within a few days prior to the most recent major earthquakes in Japan (*M* 9, Tohoku, 2011), China (*M* 7.9, Wenchuan, 2008), Italy (*M* 6.3, L'Aquila, 2009), Samoa (*M* 7, 2009), Haiti (*M* 7, 2010) and Chile (*M* 8.8, 2015) [*Ouzounov et al.*, 2015a,b; *Pulinets et al.*, 2015].

15.3. THERMAL ENERGY ASSOCIATED WITH SOME LARGE EARTHQUAKES

We examined the different energy components of recent strong earthquakes ($M > 8$) using observations and theory. Our motivation is to explore the physics of the associated thermal phenomena, i.e., to compare the energy of the earthquakes, as well as the LH released prior to the events. We have examined the energy budget of earthquakes via theory and observations for the three megaquakes of Sumatra (*M* 9.1, 26 December 2004; *M* 8.7, 28 March 2005; *M* 9.0, Tohoku, 2011) and evaluated mechanical and thermal energies. In other words, we evaluated the thermal energy required to create the energy budget associated with these large earthquakes [*Pulinets et al.*, 2006a,b; *Kafatos et al.*, 2010]. Space-based observations and model outputs can be used to monitor preseismic energy. We expect that our analysis will shed light on the underlying physics (such as of our proposed pre-earthquake lithosphere–atmosphere interaction) associated with some of the largest earthquakes and potentially would be useful for warning of future events. It is large events that provide information about the total energetics and, therefore, they are very useful for yielding information about the underlying physics.

For the seismic moment, M_0, the rupture length, rupture width and slip can e used to obtain values of 8.9×10^{22} J, 1×10^{22} J and 3.2×10^{22} J for the 26 December 2004, 28 March 2005, and 11 March 2011 events, respectively. We calculated that 2000 km^3 of land mass was displaced in the 26 December 2004 Sumatra earthquake, while 1300 km^3 of land mass was displaced in the Japan earthquake. The seismic moment can be considered to be the total energy in the pre-earthquake stressed configuration, and is essentially an estimate of the total available energy budget. It is also close to an estimate of the change in gravitational potential energy mgh, where m is the mass of the plate, g is gravitational acceleration, and h is the vertical distance the plate is displaced. The original Richter energy was supposed to refer to local damage near the epicenter, but the available energies are much higher than the energies that create havoc at the surface. We can estimate the energies of M_0, LH, and moving of landmasses. It is not surprising then that *M* 9 megaquakes seem to affect the entire Earth. We can estimate the change in the rotational energy of Earth for these large events as, for example, the rotation period of the Earth decreased 2.68 ms while oblateness decreased by 1 part in 1010 for the main Sumatra event [*Cook-Anderson and Beasley*, 2005].

We compared the LH energy, obtained by integrating SLHF over the area and the time of the earthquake, accounting for the overall thermal and mechanical associated energies. The SLHF data used are taken from the NCEP/NCAR data set maintained at the International Research Institute (IRI) on Climate Prediction (http://iri.columbia.edu). Thermal OLR data can also be used.

We compared the magnitudes of the energies associated with the 9.3, 8.7, and 9.0 events and their accompanying phenomena. Using the usual conversion of magnitudes to energy [*Båth*, 1966] associated with surface waves, $E_Q \sim 4.5 \times 10^{18}$ J, $E_Q \sim 5.5 \times 10^{17}$ J, and $E_Q \sim 1.5 \times 10^{18}$ J respectively. The total energy release of the entire series of earthquakes, including the many aftershocks, is dominated by these three events. The LH energy released prior to the earthquake, E_{LH}, can be estimated by:

$$\Delta SLHF(\mathbf{r}) = \sum_{i=1}^{N-1} \frac{SLHF(\mathbf{r})}{N-1} \tag{15.4}$$

where **r** is taken in practice by summing pixel by pixel (**r**) and integrating over time. The result is insensitive because of the strongly peaked nature of the signal occurring around 7 December 2004, 19 days in advance of the 26 December 2004 Sumatra earthquake [*Cervone et al.*, 2004]; and similarly on 3–10 March, associated with the Tohoku earthquake [*Ouzounov et al.*, 2011b]. The total LH associated with the three main events are $E_{LH} \sim 8.0 \times 10^{18}$ J, $E_{LH} \sim 3.1 \times 10^{19}$ J, and $E_{LH} \sim 1.9 \times 10^{19}$ J associated with the 28 March 2005, 26 December 2004, and 11 March 2011 earthquakes, respectively (see Table 15.2).

The above energy values give us the main energy channels involved. It would be important to research the interconnection between the mechanical processes and their energies, including LH and associated thermal energies (see below), which might be observed as surrogates for the total energies involved in earthquakes. It may very well turn out that the dominant term in the estimate of the global energy release is the LH energy (~3×10^{19} J), as we are observing LH released through the ocean, while the mechanical/kinetic energies are an order of magnitude less

Our analysis shows that LH has been detected several days prior to an earthquake. On 15 December 2004, 11 days before the *M* 9.3 Sumatra earthquake a strong anomaly was observed along the rupture zone north of the site of the future epicenter, with a value>4 σ (Figure 15.6a). Similar phenomena with lower intensity

Table 15.2 Energy release comparison for the major earthquakes in Sumatra 2004 and 2005, and Tohoku, Japan, 2011.

Location	Date	*Mw*	E_Q Richter × 10^{17} J	Rupture length, km	LH energy × 10^{18} J	SLHF, W m^{-2}	SLHF duration, days
Sumatra	26 December 2004	9.1–9.3	420	~1500	~310	100	10
Sumatra	28 March 2005,	8.7	5.7	~300	~8	80	5
Japan	3 November 2011,	9.0	150	~500	~190	90	8

LH, latent heat; SLHF, surface latent heat flux.

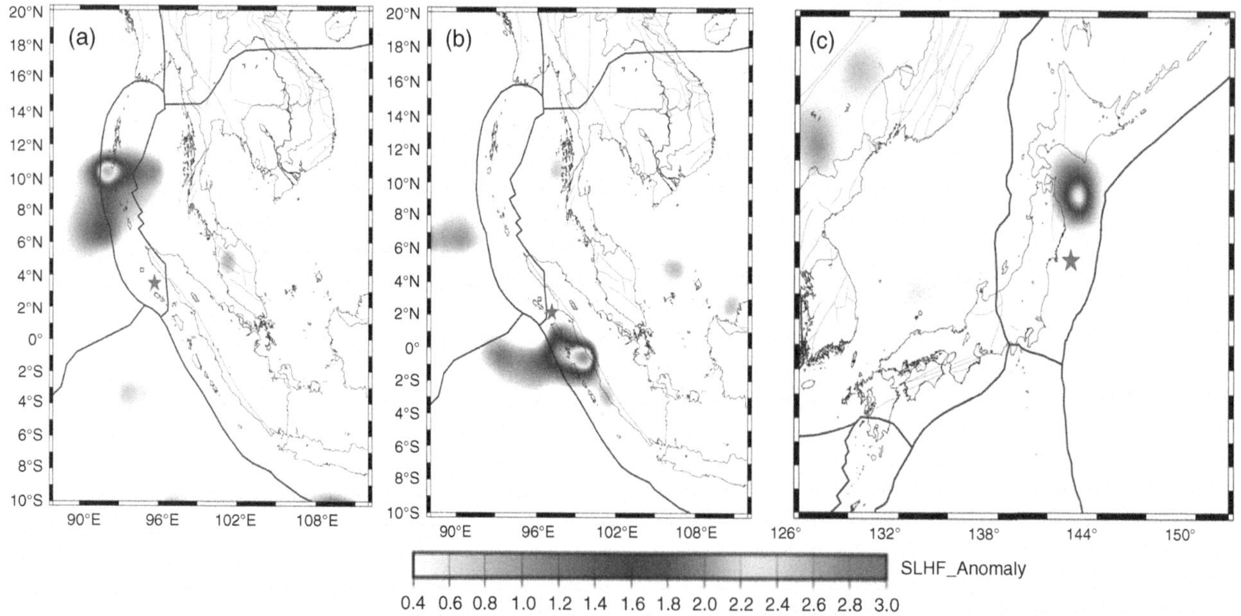

Figure 15.6 Anomalous daily maps of surface latent heat flux. (a) 15 December 21004, 11 days before *M* 9.3 Sumatra earthquake of 26 December 2004. (b) 6 March 2005, 20 days before *M* 8.7 Sumatra earthquake of 28 March 2005. (c) 5 March 2011, 6 days before *M* 9 Tohoku, Japan, earthquake of 3 November 2011. Epicenters are marked with red star, tectonic plate boundaries by heavy red line, and major faults by feint brown lines. (*See electronic version for color representation.*)

and spatial extension were observed on 6 March 2005, 20 days before the *M* 8.6 earthquake of 26 March 2005 (Figure 15.6b). For the 2011 Tohoku earthquake a similar pattern occurred, on 5 March, only 6 days before the main shock, a strong SLHF anomaly was observed north of the epicentral area, along the future rupture zone (Figure 15.6c). As such, the overall earthquake phenomenon apparently causes LH release, suggesting that terrestrial internal energy is used to produce evaporation. The internal energy changes on a short timescale, such as the *M* 9.3 earthquake of December 2004 and the *M* 8.6 of March 2005 in Sumatra, and the *M* 9.0 Tohoku earthquake in March 2011. We found that the ratios of the seismic moment, and the change of mechanical energy, to latent heat, are also a function of the size of the rupture, indicating that LH is a direct surrogate of the total geophysical energies involved.

15.4. THERMAL ANOMALIES AND THE LITHOSPHERE-ATMOSPHERE COUPLING

Several processes have been considered as possible contributors to the transient short-lived "thermal anomalies": (a) rising fluids that would lead to the emanation of warm gases [*Gorny et al.*, 1988; *Salman et al.*, 1992]; (b) rising well-water levels and CO_2 spreading laterally and causing a "local greenhouse" effect [*Quing et al.*, 1991; *Tronin et al.*, 2002; *Tramutoli et al.*, 2005]; (c) activating positive-hole pairs during rock deformation [*Freund*, 2002]; (d) frictional heat around the active fault [*Tagami et al.*, 2008]; and (e) air ionization by radon and LH change due to change of air humidity [*Pulinets and Ouzounov*, 2011].

The TIR anomalies observed from satellites and associated with earthquake processes are less likely to be the result of: (a) Earth crust heat flow (because of short-lived phenomena); (b) convective transport as a result of friction processes associated with active faulting (because of their rapid build up); or (c) meteorological origin (because of the long persistence of the TIR over the same region). The most probable cause of the physical mechanism of thermal anomaly generation could be the morphology and release of thermal energy emanations. The most important consideration is that thermal anomalies are observed over both the land and the ocean. Only strong gas discharges (including underwater gas discharges), as described by *Khilyuk et al.* [2000], could be a potential candidate for this possible mechanism.

Although their detailed underlying mechanisms are still a matter of debate [e.g. *Freund*, 2011; *Pulinets and Ouzounov*, 2011], in the case of earthquakes, the general process that could affect the atmosphere and ionosphere "from below" might be the generation of TRA and the penetration into the ionosphere of anomalous electric fields originating close to the surface [*Kuo et al.*, 2011]. It has been documented instrumentally that the latest stage of the earthquake cycle [*Dobrovolsky et al.*, 1979] is characterized by mechanical changes in Earth's crust accompanied by geochemical and electromagnetic anomalies [*Scholz et al.*, 1973; *Kasahara*, 1981]. The link between crustal transformation under loading and asperities on faults has been studied [*Schorlemmer et al.*, 2004]. The formation of asperities, increasing crustal porosity, leads to changes in gas migration, evidently by an increase of gas release into the boundary layer of the atmosphere.

At the last stage of earthquake preparation we observe the activation of faults within the area of the earthquake, which leads to increased emanation of gases such as carbon dioxide, methane, hydrogen, helium, and radon (Step 1, Figure 15.7). It has been established that together with diffusion within the crust, radon it is actively released in the atmosphere by carrier gases, which usually are carbon dioxide and methane. A correlation has been established between crust deformation and radon release [*Aumento*, 2002]. Radon, due to its radioactivity, produces air ionization [*Harrison et al.*, 2010] (Step 2, Figure 15.7).

Air ionization accelerates, drastically, the formation of cluster ions due to newly formed ion hydration. This process is called ion-induced nucleation (IIN) [*Laakso et al.*, 2002] (Step 3, Figure 15.7). This process leads to formation of large charged ion clusters consisting of a charged core ion with an envelope of water molecules, which attach to the ions due to their high dipole moment [*Sekimoto and Takayama*, 2007]. The IIN process is essentially nonlinear because it is a catalytic exothermic process with two aggregation phase states of water simultaneously coexisting: condensed and vapor.

We now discuss the key question in the relationship between the generation of TRA in the atmosphere and the pre-earthquake process: What are the drivers producing a strong modification of the atmosphere? There is a paradox between the catalytic process, and the energy release, but the source of the energy is different. In the case of TRA it is the LH of the water vapor in the air. Here are some estimates.

The LH constant is $Q = 40.683$ kJ mol^{-1}, which means that the heat released per one molecule is $U_0 = Q/N_A = 0.422$ eV, where $N_A = 6.022 \cdot 10^{23}$ L mol^{-1} (Avogadro's number). Ionization effectiveness depends on the number of water molecules attached to the ion. If the particle grows to some value of critical mass, m_{max}, the estimated energy released will be, $w = m_{max}\ U_0$. If the ion production rate is dN/dt, the heat released in the atmosphere could be expressed as $Pa = w \cdot dN/dt$. Clusters of the order of 1000 nm were found experimentally [*Pulinets and Ouzounov*, 2011]. To ionize any air molecules we need energy of 10^{-15} eV. One particle of 1000 nm size contains $0.4 \cdot 10^{12}$ water

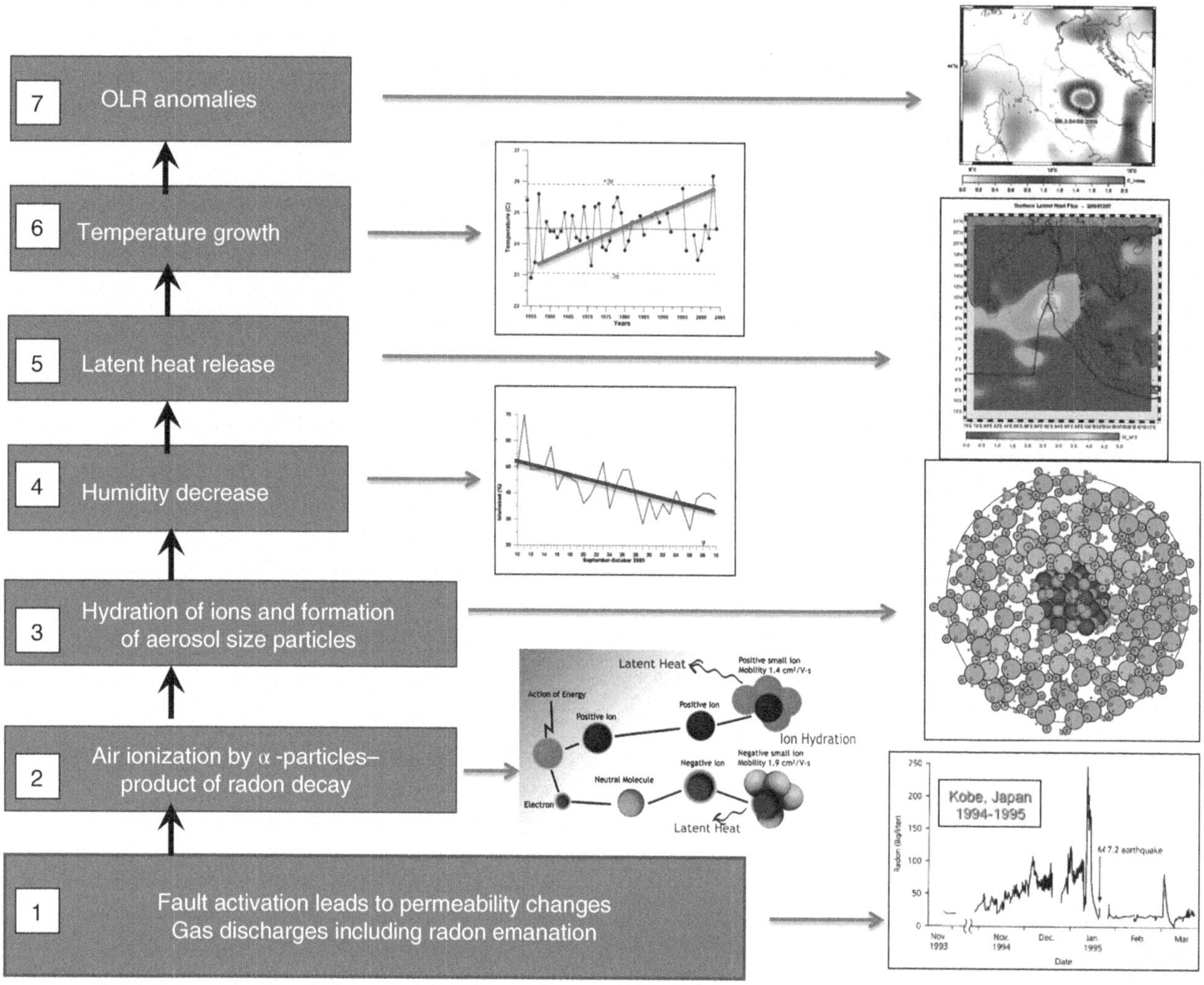

Figure 15.7 Principle diagram for generation of pre-earthquake thermal radiation anomalies according to the lithosphere–atmosphere–ionosphere coupling concept [*Pulinets and Ouzounov*, 2011, for more details see Chapter 6]. (Source: Reproduced with the permission of Elsevier.) (1) Anomalous variation in radon measured in the vicinity of Kobe (Japan) days, weeks before the M7.2 of Jan 1995 [*Igarashi et al.*, 1995]. (2) Principle scheme for negative/positive ion production, ion's hydration and latent heat release (3) Example for hydrated ion cluster formation (4) Abnormal variation of daily minimal relative humidity at Islamabad (Pakistan), days before M7.6 Kashmir October 8th, 2005. (5) Surface latent flux anomaly of Dec 7, 2004, days before the M9.0 Sumatra 2004 of Dec 28 [*Sing et al.*, 2007] (6) Anomaly (50 years baseline) in the mean monthly air temperature for January, before M7.6 Colima of Jan 22, 2003 (Mexico) (7) TRA anomaly over Abruzzo -Italy on April 2, 2009, days before the M6.3 L'Aquila (Italy).

molecules, which produces a heat release of 0.422 eV × $0.4 \cdot 10^{12} = 1.7 \cdot 10^{11}$ eV. So it can be easily seen that the energy gain due to LH release is $1.7 \cdot 10^{11}/15 = \sim 10^{10}$. This is the essential value—radon and ionization do not produce any energy, the energy is instead released due to condensation of the water vapor on ions, and consequent release of the LH contained in the air water vapor.

We can now explain the rectangular boxes in Figure 15.7: Step 4, relative air humidity drops because of water vapor condensation on ions; Step 5, LH release; and Step 6, increase in air temperature. It is well known from meteorology that sharp decreases in humidity are accompanied by an increase in air temperature. Because of the general atmospheric circulation the additional heat flux release due to the LH release could be registered in the longwave part of the infrared emission within the transparency window of the atmosphere 8–12 μm observed by satellites (OLR; Step 7, Figure 15.7). What is the amount of outgoing energy flux that we should expect at the TOA as OLR flux as shown in Figure 15.1? Each α-particle emitted by ^{222}Rn with an average energy of $E\alpha = 5.46$ Mega eV can produce $\sim 3 \cdot 10^5$ electron-ion pairs.

From actual observation of radon activity, before an earthquake the level could be ~2000 Bq m^{-3} [*Inan et al.*, 2008]. The ion production rate of Rn is $\sim 6 \cdot 10^8$ s^{-1}. From our above estimation, we know that a 1000 nm size particle contains $0.4 \cdot 10^{12}$ water molecules. During the water vapor condensation process the LH release is $U_0 \sim 40.68 \cdot 10^3$ J mol^{-1} (1 mol $= 6.022 \cdot 10^{23}$). This means that the given radon activity with formation of particles of 1000 nm size produces a thermal energy output of 16 W m^{-2}. That is exactly the range of ~20 W m^{-2} we were able to observe as TRA anomalies at the TOA by using NOAA/AVHRR data [*Ouzounov et al.*, 2007].

There are two main consequences of the IIN process for the electromagnetic coupling of the atmosphere and ionosphere:

1. LH release due to water vapor condensation on ions;
2. changes in air conductivity leading to local changes in the Global Electric Circuit.

The chain of these physical processes has been proposed as an essential part of the lithosphere–atmosphere–ionosphere–magnetosphere coupling concept [*Pulinets and Ouzounov*, 2011; *Pulinets et al.*, 2015]. Using a multiparameter approach [*Ouzounov et al.*, 2011a] enables the presence of TIR anomalies at different levels (ground level, troposphere, tropopause) and space plasma anomalies including GPS/TEC (total electron content) to be registered from space.

15.5. SUMMARY AND CONCLUSIONS

Our results show that several days before some large earthquakes (Figures 15.3–15.5) infrared signals related to earthquake processes were observed near the epicentral areas by the NOAA NPOESS satellite as OLR hotspots. The OLR hotspots appeared quickly, stayed over the same regions from several hours to up to a few days, and then disappear rapidly. The time lag for the *M* 7.3 earthquake in Van, Turkey, was 4 days; for the *M* 6.9 earthquakes in the Aegean Sea, Greece, it was 10 days, and for the *M* 6.0 earthquake in California it was 2 days. This enhancement of OLR could be explained as a result of water vapor condensation on ions, with a large amount of LH being released. The initial process involves an ionization of the near-ground layer due to an increased concentration of gases (including radon) emitted from active tectonic faults [*Pulinets and Ouzounov*, 2011]. The transient nature in radiative emission preceding large earthquakes follows a general temporal–spatial evolution pattern, which is similar to other large earthquakes worldwide [*Kuo et al.*, 2011; *Ouzounov et al.*, 2016]. The transitional OLR anomalous data usually varied between 16 and 21 W m^{-2}. These are residuals derived from the daily mean OLR compared and with the background field. The latter was derived from multiple years of observations, over the same location and local time, and normalized by the standard deviation [*Ouzounov et al.*, 2007, 2011b]. From spaceborne observations of atmospheric conditions, we have shown the consistent occurrence of TRA at the TOA, over the region of maximum stress associated with, and preceding, large earthquakes. Because of their relatively long duration, these anomalies do not appear to be meteorological. We evaluated the mechanical and thermal energies of three major earthquakes to explore the physics of associated thermal phenomena. Our estimates show that the LH released prior to the events is larger than the seismic energy released during the quake. We show that in large earthquakes the associated phenomena may stand out energetically with measurements above variance that arises from other geophysical processes. As we have a greater energy budget for these large events, we emphasize that it can be used to establish a more general earthquake energy phenomenology. Our analysis of transitional thermal fields of recent major earthquakes (*M* 7.3 Van, Turkey, 2011; *M* 6.9 Aegean Sea, Greece, 2014; and *M* 6.0 Napa Valley, California, 2014) has demonstrated the presence of correlated variations of TRA in the atmosphere, implying their connection with pre-earthquake processes. Our results suggest the existence of an atmospheric response triggered by the coupling processes between the lithosphere and atmosphere.

ACKNOWLEDGMENTS

The authors thank NOAA's Climate Prediction Center (USA), NASA Goddard Earth Science Center (GES DAAC) for OLR data and the International Research Institute for Climate and Society for SLHF data. Special thanks go to the US Geological Survey and European-Mediterranean Seismological Centre for providing earthquake information services and data. Dimitar Ouzounov thanks all his graduate students from NASA DEVELOP National Program at Goddard SFC and Chapman University for helping in the processing of satellite data. Dimitar Ouzounov and Sergey Pulinets thank the International Space Science Institute (Bern and Beijing) for support of the team "Validation of Lithosphere–Atmosphere–Ionosphere–Magnetosphere Coupling (LAIMC) as a concept for geospheres interaction by utilizing space-borne multi-instrument observations." Menas Kafatos thanks Kent S. Wood for useful input in the energetics involved.

REFERENCES

Aumento, F. (2002), Radon tides on an active volcanic island: Terceira, Azores, *Geofis. Int.*, *41*(4), 499–505.

Båth, M. (1966), Body wave energy from seismograms, in L. H. Ahrens, F. Press, and S. K. Runcorn (eds), *Physics and Chemistry of the Earth: Progress Series*, Vol. 7, pp. 133–137, Pergamon Press, London.

Bryant, N., Zobrist, A., and Logan, T. (2003), Automatic co-registration of space-based sensors for precision change detection and analysis, *IGARSS 2003 Transactions*, Vol. *7*, pp. 21–26, France.

Carreno, E., Capote, R., and Yague, A. (2001), Observations of thermal anomaly associated to seismic activity from remote sensing, *General Assembly of European Seismology Commission*, pp. 265–269, Portugal, 10–15 September.

Cervone, G., Kafatos, M., Napoletani, D., and Singh, R. P. (2004), Wavelet maxima curves of surface latent heat flux associated with two recent Greek earthquakes, *Nat. Haz. Earth Syst. Sci.*, *4*(3), 359–374.

Cervone, G., Maekawa, S., Singh, R. P., Hayakawa, M., Kafatos, M., and Shvets, A. (2006), Surface latent heat flux and nighttime LF anomalies prior to the Mw=8.3 Tokachi-Oki earthquake, *Nat. Hazards Earth Syst. Sci.*, *6*, 109–114.

Cook-Anderson, G. and Beasley, D. (2005), *NASA Details Earthquake Effects on the Earth*, Press Release, 10 January, NASA.

Dey, S., Sarkar, S., and Singh, R. P. (2004), Anomalous changes in column water vapor after Gujarat earthquake, *Adv. Space Res.*, *33*(3), 274–278.

Di Bello, G., Filizzola, C., Lacava, T., Marchese, F., Pergola, N., Pietrapertosa, C., Piscitelli, S., Scaffidi, I., and Tramutoli, V. (2004), Robust satellite techniques for volcanic and seismic hazards monitoring, *Ann. Geophys.*, *47*(1), 49–64.

Dobrovolsky, I. R., Zubkov, S. I., and Myachkin, V. I. (1979), Estimation of the size of earthquake preparation zones, *Pure Appl. Geophys.*, *117*, 1025–1044.

Eleftheriou, A., Filizzola, C., Genzano, N., Lacava, T., Lisi, M., Paciello, R., Pergola, N., Vallianatos, F., and Tramutoli, V. (2015), Long-Term RST Analysis of Anomalous TIR Sequences in Relation with Earthquakes Occurred in Greece in the Period 2004–2013, *Pure Appl. Geophys.*, *173*, 285–303.

Ellingson, R. G., Yanuk D. J., Lee, H.-T., and Gruber, A. (1989), A technique for estimating outgoing longwave radiation from HIRS radiance observations, *J. Atmos. Ocean. Tech.*, *6*, 706–711.

Filizzola, C., Pergola, N., Pietrapertosa, C., and Tramutoli, V. (2004), Robust satellite techniques for seismically active areas monitoring: a sensitivity analysis on September 7th 1999 Athens's earthquake, *Phys. Chem. Earth*, *29*, 517–527.

Fisher, S., Sulia, K., Ouzounov, D., Policelli, F., and Ferrucci, M. (2008), Data exploration of atmospheric thermal signals over regions of tectonic faulting and earthquake processes, *AGU Fall Meeting Abstracts*, 13 December.

Freund, F. (2002), Charge generation and propagation in rocks, *J. Geodyn.*, *33*, 545–572.

Freund, F. (2011), Pre-earthquake signals: Underlying physical processes, *J. Asian Earth Sci.*, *41*, 383–400.

Genzano, N., Aliano, C., Filizzola, C., Pergola, N., and Tramutoli, V. (2007), A robust satellite technique for monitoring seismically active areas: the case of Bhuj - Gujarat earthquake, *Tectonophysics*, *431*, 197–210.

Genzano, N., Aliano, C., Corrado, R., Filizzola, C., Lisi, M., Mazzeo, G., Paciello, R., Pergola, N., and Tramutoli, V. (2009), RST analysis of MSG-SEVIRI TIR radiances at the time of the Abruzzo April 6th 2009 earthquake, *Nat. Haz. Earth Syst. Sci. 9*, 2073–2084.

Goosse, H., Barriat, P. Y., Lefebvre, W., Loutre, M. F., and Zunz, V. (2016), *Introduction to Climate Dynamics and Climate Modeling*, Online textbook available at http://www.elic.ucl.ac.be/textbook

Gorny, V. I., Salman, A. G., Tronin, A. A., and Shilin, B. B. (1988), The Earth outgoing IR radiation as an indicator of seismic activity, *Proc. Acad. Sci. USSR*, *301*, 67.

Gruber, A. and Krueger, A. (1984), The status of the NOAA outgoing longwave radiation dataset. *Bull. Am. Meteorol. Soc.*, *65*, 958–962

Han, P., Hattori, K., Hirokawa, M., Zhuang, J., Chen, C.-H., Febriani, F., Yamaguchi, H., Yoshino, C., Liu, J. Y., and Yoshida, S. (2014), Statistical analysis of ULF seismomagnetic phenomena at Kakioka, Japan, during 2001–2010, *J. Geophys. Res. Space Phys.*, *119*, 4998–5011.

Han, P., Hattori, K., Huang, Q., Hirooka, S., and Yoshino, C. (2016), Spatiotemporal characteristics of the geomagnetic diurnal variation anomalies prior to the 2011 Tohoku earthquake (*Mw* 9.0) and the possible coupling of multiple pre-earthquake phenomena, *J. Asian Earth Sci. 129*, 13–21.

Harrison, R. G., Aplin, K. L., and Rycroft, M. J. (2010), Atmospheric electricity coupling between earthquake regions and the ionosphere, *J. Atmos. Solar Terr. Phys.*, *72*, 376–381, 2010.

Hayakawa, M. (1999), *Atmospheric and Ionospheric Electromagnetic Phenomena with Earthquakes*, Terra Scientific Publishing, Tokyo.

Hayakawa, M. (ed.) (2012), *Frontier of Earthquake Short-Term Prediction Study*, Nihon-Senmontosho, Japan, 794 pp.

Hayakawa, M. and Molchanov, O. A. (2002), *Seismo Electromagnetics, Lithospheric–Atmospheric–Ionospheric coupling*, Terra Scientific Publishing, Tokyo.

Inan, S., Akgu, T., Seyis, C., Saatc, R., Baykut, S., Ergintav, S., and Bas, M. (2008). Geochemical monitoring in the Marmara region (NW Turkey): A search for precursors of seismic activity, *J. Geophys. Res.*, *113*, B03401, doi:10.1029/2007JB005206

Kafatos, M., Ouzounov, D., Pulinets, S., Hattori, K., Liu, J., Parrot, M., and Taylor, P. (2010), Multi sensor approach of validating atmospheric signals associated with major earthquakes, *EGU General Assembly Conference Abstracts*, Vol. *12*, p. 1418.

Kasahara, K. (1981), *Earthquake Mechanics*, Cambridge University Press, Cambridge, pp. 1–248.

Khilyuk, L. F., Chillingar, G. V., Robertson, J. O. Jr., and Endres, B. (2000), *Gas Migration. Events Preceding Earthquakes*, Gulf Publishing Company, Houston, TX, 390 pp.

Kon, S., Nishihashi, M., and Hattori, K. (2010), Ionospheric anomalies possibly associated with M ≥ 6 earthquakes in Japan during 1998–2010: Case studies and statistical study, *J. Asian Earth Sci.*, *41*(4), 410–420, doi: 10.1016/j.jseaes.2010.10.005

Kuo, C. L., Huba, J. D., Joyce, G., and Lee, L. C. (2011), Ionosphere plasma bubbles and density variations induced by pre-earthquake rock currents and associated surface charges, *J. Geophys. Res.*, *116*, 1029–1039.

Laakso, L., Mäkelä, J. M., Pirjola, L., and Kulmala, M. (2002), Model studies on ion-induced nucleation in the atmosphere. *J. Geophys. Res.*, *107*(D20), 4427, doi:10.1029/2002JD002140

Liebmann, B. and Smith, C. A. (1996), Description of a complete (interpolated) outgoing longwave radiation dataset, *Bull. Am. Meteorol. Soc.*, *77*, 1275–1277.

16

Very-Low- to Low-Frequency Sounding of Ionospheric Perturbations and Possible Association with Earthquakes

Masashi Hayakawa[1,2], Tomokazu Asano[1], Alexander Rozhnoi[3], and Maria Solovieva[3]

ABSTRACT

The purpose of this chapter is to review the use of subionospheric very-low- to low-frequency (VLF–LF) signals to detect ionospheric perturbations in possible association with earthquakes. This VLF–LF method is described first, and then the results are presented based on the observations with the Far East VLF–LF network consisting of several stations in Japan and a few stations in Russia. The primary purpose is to detect pre-earthquake ionospheric perturbations, but we have to know the background of VLF–LF signals and also any disturbances induced by other agents from above and below as interferences to the seismogenic effects. Initially we have investigate the conventional solar–terrestrial effects such as geomagnetic storms and atmospheric disturbances including typhoons. Then, we move on to our main topic of pre-earthquake ionospheric perturbations. We start with clear signatures of seismogenic ionospheric perturbations for the 1995 Kobe earthquake, and followed with results for the huge 2011 Tohoku earthquake. A statistical correlation is presented between such VLF–LF perturbations and earthquakes. Finally, we comment on the mechanism of why and how the ionosphere is perturbed prior to an earthquake.

16.1. INTRODUCTION

The recent sequence of highly destructive earthquakes around the world, including the latest 2011 Tohoku earthquake, has heightened awareness of earthquakes, and the inability of seismology as a discipline to derive information of increasing earthquake hazards in the weeks and days before major seismic events. In order to mitigate the earthquake disaster, especially human lives, it would be immensely meaningful to forecast the approach of a large earthquake on a timescale of hours, days, and week (this is called "short-term" earthquake prediction) [e.g., *Hayakawa*, 2015]. The requirement for a short-term earthquake prediction seems to have drastically changed in Japan during the past two decades, since the Kobe earthquake in 1995, because the conventional earthquake prediction based on the measurement of crustal movements (so-called medium-term earthquake prediction), was concluded to be not so useful for short-term earthquake prediction. In that period we have had a new wave of measurements of electromagnetic effects, and accumulated substantial evidence that electromagnetic phenomena do take place prior to earthquakes [e.g., *Hayakawa*, 1999, 2009a, 2012, 2013; *Hayakawa and Molchanov*, 2002; *Pulinets and Boyarchuk*, 2004; *Molchanov and Hayakawa*, 2008]. For this reason these electromagnetic effects have attracted a lot of attention from scientists as a promising candidate of short-term earthquake prediction.

The electromagnetic method for earthquake prediction principally can be classified into two categories: (a) detection of radio emissions from the earthquake hypocenter

[1] *Hayakawa Institute of Seismo Electromagnetics, Co. Ltd, University of Electro-Communications (UEC), Incubation Center, Chofu, Tokyo, Japan*

[2] *University of Electro-Communications, Chofu, Tokyo, Japan*

[3] *Institute of Physics of the Earth, Russian Academy of Sciences, Moscow, Russia*

Pre-Earthquake Processes: A Multidisciplinary Approach to Earthquake Prediction Studies, Geophysical Monograph 234, First Edition. Edited by Dimitar Ouzounov, Sergey Pulinets, Katsumi Hattori, and Patrick Taylor.

(or epicenter), and (b) detection of indirect effects of earthquakes taking place in the atmosphere and ionosphere by means of the pre-existing radio transmitters (known as "radio sounding").

This chapter deals with our method of very low frequency (VLF; 3–30 kHz) to low frequency (LF; 30–300 kHz) sounding of seismo-ionospheric perturbations within category (b) above. We first review the physical agents affecting ionospheric perturbations because the lower ionosphere is sensitive to different physical parameters (not only from above, but also from below), and then we move on to the perturbations being in possible association with earthquakes. Specific earthquakes are considered, including the 1995 Kobe and the 2011 Tohoku earthquakes. We then establish a statistical correlation between ionospheric perturbations and earthquakes on the basis of our long-term observations. Finally, we present the mechanism of how the ionosphere is perturbed prior to an earthquake.

16.2. VERY-LOW- TO LOW-FREQUENCY SOUNDING OF THE LOWER IONOSPHERIC PERTURBATIONS

A number of nations currently operate large very-low-to low-frequency (VLF–LF) transmitters primarily for navigation, radio watches, and communication with military submarines. To radiate electromagnetic waves efficiently at these lower frequencies requires an antenna with dimension on the order of a wavelength of the radiation, which suggests that VLF–LF transmitter antennas must be very large, typically many hundreds of meters high [*Watt*, 1967; *Hayakawa*, 2015].

Our method of VLF–LF sounding is a well-known radiophysical technique [*Hayakawa*, 2007, 2009b; *Rozhnoi et al.*, 2013], in which the amplitude and phase of radio signals from navigational transmitters propagating inside the Earth–ionosphere waveguide are monitored. If the transmitter frequency and receiver distance are fixed, then the observed VLF signal parameters are determined mainly by the position of the reflection height, which depends on the value and gradients of electron density near the atmosphere–ionosphere boundary. It is typically 80–85 km in daytime and about 90 km at night. These altitudes are too high for balloons and too low for satellites, making measurements in situ extremely rare. The only possible means for probing the D and E regions of the lower ionosphere is by monitoring VLF–LF subionospheric radio signals. The D and E regions are very sensitive to any kind of agent of disturbance, not only from above but also from below, and also the dynamic range of lower ionospheric density change is extremely large as compared with that in the upper F region, providing an important advantage for our VLF–LF method.

Disturbances from above (space) constitute the well known solar–terrestrial effects, for which the VLF method has become a standard for recording short-time electron density variations in the lower ionosphere and upper atmosphere connected with solar flares (e.g. Roentgen flares), cosmic rays (Forbush effect), magnetic storms [*Belrose and Thomas*, 1968; *Potemra and Rosenberg*, 1973; *Kikuchi*, 1981; *Sauer et al.*, 1987], and precipitation of energetic particles due to wave (whistler) and particle interaction in the magnetosphere [*Inan et al.*, 1985; *Dowden and Adams*, 1988]. In this chapter we present our own results on the effect of geomagnetic activity with a duration of at least of a few days on the lower ionosphere being the most serious challenge to the detecting and distinguishing of seismo-ionospheric perturbations.

The effects from below (troposphere) can be associated with (a) nuclear weapons testing and (ii) lightning-induced ionization (and/or heating) [e.g., *Nickolaenko and Hayakawa*, 1995; *Craig and McCormik*, 2006]. Ionospheric perturbation due to lightning discharge is very short-lived, as is the precipitation effect of energetic particles from the magnetosphere associated with a solar flare. Accordingly these effects are not treated in this chapter because such short-lived effects are not important in the detection and distinction of seismo-ionospheric perturbations. Instead, we will present our recent results on meteorological effects, such as typhoons, as being the more long-lasting source of ionospheric perturbations.

16.3. VLF–LF PERTURBATIONS DUE TO SOLAR-TERRESTRIAL EFFECTS AND METEOROLOGICAL EFFECTS

16.3.1. VLF–LF Anomalies Associated with Geomagnetic Solar Effects

We present our latest results on VLF–LF anomalies associated with solar–terrestrial effects, the details of which can be found in *Rozhnoi et al.* [2012a, 2013]. Thus, the essential points of this chapter are described here.

Our OmniPAL receiver was installed at Petropavlovsk-Kamchatsky (PTK) (geographic coordinates: 53.09°N, 158.55°E) in June 2000 within the framework of the Japanese–Russian ISTC project (see Figure 16.1). The receiver in Kamchatka is part of the Far East network formed after the installation of seven Japanese stations. All stations receive the amplitude and phase of signals from the four transmitters: JJY (40 kHz, Fukushima) and JJI (22.2 kHz, Miyazaki) in Japan; NWC transmitter in Australia (19.8 kHz); and NPM transmitter in Hawaii (21.4 kHz). Among them the signal from the JJY transmitter is most interesting for our analysis because of the fortunate position of the PTK station relative to the transmitter. The area of sensitivity along the wavepath including the

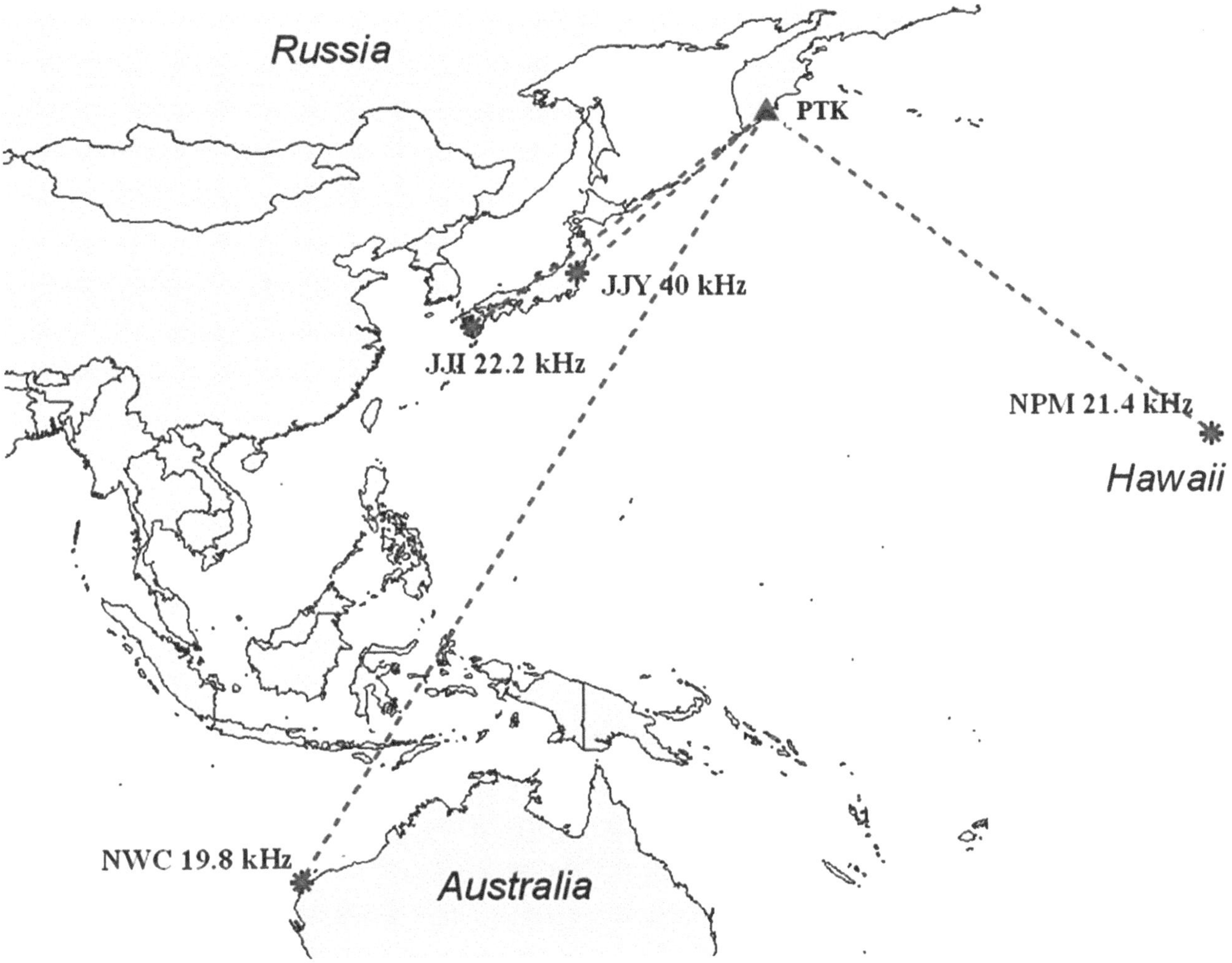

Figure 16.1 Positions of the VLF receiver in Petropavlovsk-Kamchatsky (PTK) and the radio transmitters JJY, JJI, NWC (Australia), and NPM (Hawaii).

3rd region of the Fresnel zone is seen to cover almost completely the highly seismically active Izu-Bonin and Kurile-Kamchatka arcs. The length of the wavepath JJY–PTK is about 2300 km, and the sampling frequency is 20 s. The semiwidth (y) of the Fresnel zone 1st region is $y \approx (\lambda^2/4 + \lambda x(1 - x/D))0.5$ where λ is the wavelength, x is the coordinate along the path, and D is the transmitter-receiver distance. The minor semiaxis is $b \approx 70$ km.

The main feature of the VLF–LF signal is daily (diurnal) variation, which depends on the frequency, wavepath length, illumination conditions along the path, and direction of the signal propagating relative to the magnetic field of Earth. Each trace has its own diurnal variation, and diurnal runs of the amplitude and phase of any VLF–LF signal significantly change from month to month. Therefore for our analysis we use a residual signal of the phase dP or amplitude dA defined as the difference between the observed signal ($A(t)$, $P(t)$) and the average during a few quiet days immediately preceding or following the current day ($<A>$,$<P>$): $dA(t) = A(t) - <A>$, $dP(t) = P(t) - <P>$ [*Rozhnoi et al.*, 2004; *Hayakawa et al.*, 2010]. We define anomaly signals as dP or dA exceeding its corresponding 2σ (σ, standard deviation).

VLF–LF waves are reflected from the lowest region of the ionosphere (the D region during daylight and the lower E region at night), and apart from the sunrise and sunset periods, they exhibit propagation characteristics that are very stable in both amplitude and phase. Therefore for data collection the night- and daytime periods are chosen such that they exclude the sunrise and sunset periods. Because the daytime variations of VLF–LF signal are smaller than those at night and the VLF–LF signal is strongly influenced by sudden ionospheric disturbances (SID) caused by X-rays emitted during a solar flare on the dayside of the Earth, we have chosen "night conditions" for our analysis.

When using the method of VLF–LF signal for the purpose of searching for earthquake precursors, we must know the background variations in the signal caused by planetary waves, solar activity (proton and X-ray flares and electron fluxes), magnetic storms, and atmosphere circulation [*Rozhnoi et al.*, 2003, 2006; *Kleimenova et al.*, 2004]. We take into consideration the following parameters: K_p and D_{st} indices of magnetic field activity; X-rays and energetic electron (and proton) fluxes derived from the geostationary satellite GOES; data from geomagnetic mid-latitude observatories of Moshiri, Japan (MSR, geomagnetic latitude = 37.4°) and Magadan, Russia (MGD, 53.6°), which are located nearly in the same magnetic meridian as our wavepaths; magnitudes of earthquakes from the US Geological Survey catalog within the sensitivity zone of our wavepaths, the amplitude and phase of the signal, and their residual values.

First of all, we study the sensitivity of the VLF–LF signal amplitude and phase to ionospheric perturbations associated with the variation in magnetic and solar activities on the basis of 9-year observation. For the analysis we use an average residual signal of the amplitude and phase in nighttime (averaged over the night period) and the data series on the D_{st} index and high-pitch angle particle fluxes (electrons and protons) registered on the geostationary satellite GOES-10. For proton analysis we choose channel P1 (0.6–4.2 MeV) and channel E1 (>2 MeV) for electron analysis. Every analyzable parameter (D_{st}, electron and proton fluxes) is divided into certain intervals and the number of days with the chosen parameters in every interval was calculated (N). Then in every interval among the chosen days we selected days for which the average dP or dA or their dispersions exceed 2σ (N_i). The ratio N_i/N is regarded as the sensitivity of LF signal amplitude or phase to an analyzed parameter.

Results of analysis of geomagnetic and solar factors are shown in Figure 16.2. The correlations of phase and amplitude variations of the LF signal at JJY with the D_{st} index, proton and electron fluxes are obvious. It is significant that the nature of this relation is the same for all receiving stations (PTK, MSR, and Kochi (KCH)) and especially the amplitude of the LF signal verifies its higher stability in comparison with the phase of the signal. Similar results were obtained also for dispersions of amplitude and phase of the LF signal.

The VLF–LF phase anomalies are typical for the main and recovery phases of magnetic storms. It was found that "bay-like" phase and amplitude anomalies during bursts of Pi3 geomagnetic pulsations at nighttime are observed (sometimes with a delay of a few tens of minutes) [*Rozhnoi et al.*, 2003; *Kleimenova et al.*, 2004]. At the same time the intensity of LF signal variations is not dependent linearly on the strength of a magnetic storm

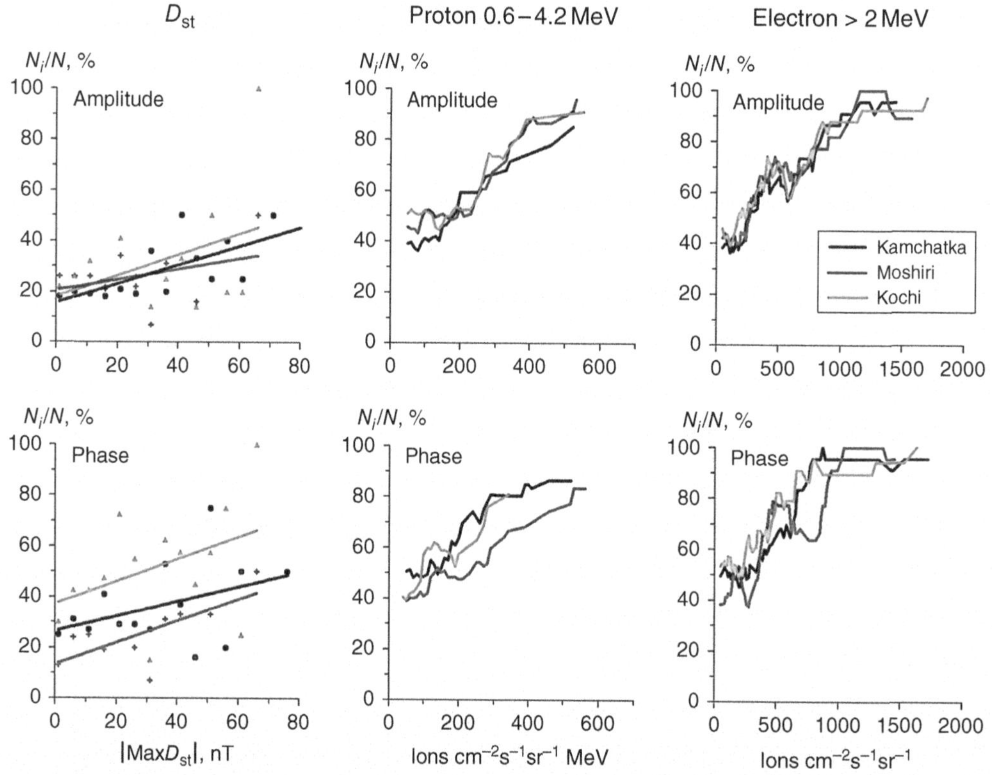

Figure 16.2 Results of statistical analysis for three receiving stations (Kochi, Moshiri, and Petropavlovsk-Kamchatsky). N is the number of days with the chosen parameters for D_{st}, electron and proton fluxes; N_i is the number of days in the same interval with the data exceeding the corresponding 2σ criterion.

or substorm. For example, a strong magnetic storm on 31 March 2001 (D_{st} = 360 nT) was accompanied by a small positive phase anomaly, but a rather modest geomagnetic disturbance on 21 January 2001 (D_{st} = 40 nT) caused a large negative phase anomaly.

Our main goal is to distinguish local VLF–LF perturbations connected with earthquakes from large-scale or global anomalies related to magnetic storms, solar activity, and atmospheric circulations. As our experience shows, the picture is very complicated and the separation of seismic anomalies against the background of all other perturbations (especially when two possibilities coexist) is a difficult problem because the morphology of different types is similar and effects are approximately equal. The only chance to distinguish one anomaly type from another is a precise correlation with any factor. Anomalies in the LF signal can be considered as earthquake precursors only when all other known reasons are discarded. Reliability in the definition of the nature of observed anomalies increases using the data from a network of stations.

16.3.2. VLF–LF Anomalies in Association with Meteorological Conditions (Typhoons)

We present the results of our latest work on VLF–LF propagation anomalies in possible association with tropical typhoons [*Rozhnoi et al.*, 2014]. The details are presented in this chapter, and so some essential points are described here.

Meteorological effects in the ionosphere have been studied for a long time. Since the work by *Bauer* [1958] there have been many studies on the response in the F region of the ionosphere (foF2) in terms of the total electron content (TEC) to the passage of tropical typhoons [*Bishop and Straus*, 2006; *Liu et al.*, 2006; *Xu et al.*, 2008; *Polyakova and Perevalova*, 2011; *Vanina-Dart et al.*, 2011; *Lin*, 2012; *Zakharov and Kunitsyn*, 2012]. Further studies on the effects of meteorological phenomena on the upper ionosphere have been performed with HF Doppler shift data [*Huang et al.*, 1985; *Xiao et al.*, 2007] and also by satellite observations [*Mikhailova et al.*, 2001; *Isaev et al.*, 2002; *Sorokin et al.*, 2015].

In contrast, there are very few publications on the effects of tropical typhoons on the lower ionosphere. The only anomaly in subionospheric VLF–LF signals which was likely to be associated with a strong tropical typhoon was reported by *Hayakawa et al.* [2008] in analyzing the ionospheric perturbations associated with the 2007 Niigata Chuetsu-oki earthquake.

The VLF–LF data observed in PTK, Yuzhno-Sakhalinsk (YSH; geographic coordinates 46.95°N, 142.75°E), and Yuzhno-Kurilsk (YUK; 44.03°N, 145.86°E) are used for this analysis, and we have studied several typhoons here. The period selected for our analysis is August 2012, when four tropical cyclones of different intensity were formed in the South Asia region and moved across the paths under consideration. The receiver in PTK was temporally out of operation during this period, so we analyzed the signals at the other two stations (YSH and YUK). Figure 16.3 illustrates the results.

The interval of analysis was from 23 July to 31 August. During this period three strong earthquakes occurred, with the epicenters inside the sensitivity zones of our paths, on 18 August (*M* 6.3), 26 August (*M* 6.6), and 31 August (*M* 7.6). The first tropical cyclone Damrey (1210) in the interval of analysis was formed as a tropical depression west of Minamitorishima Island on 27 July 2012, and was upgraded to TS (tropical storm) intensity the next day. Damrey was upgraded to STS (severe tropical storm) intensity on 30 July and upgraded to TY (typhoon) intensity over the Yellow Sea the next day; it reached its peak intensity with maximum sustained winds of 70 kt (36 m s^{-1}) and a central pressure of 965 hPa. This tropical cyclone entered the sensitivity zone of the path NWC-YUK on 30 July and moved across the paths under consideration until 1 August. Following Damrey, another tropical cyclone Haikui (1211) was formed on 1 August. It was upgraded to TS intensity on 3 August, upgraded to STS intensity northwest of Okinawa Island on 5 August, and subsequently upgraded to TY intensity, reaching peak intensity with maximum sustained winds of 65 kt (33 m s^{-1}) and a central pressure of 965 hPa on 7 August. Haikui moved across the NWC-YUK and NWC-YSH sensitivity paths from 3 to 5 August. Tropical cyclone 1210 apparently caused a decrease in the signal on the path NWC-YSH observed on 1 August. The decrease in the signal on the path NWC-YUK on 2 August is most likely to be under the influence of tropical cyclone 1211.

Tropical cyclone Kai-tak (1213) was generated as a tropical depression east of the Philippines, which is inside sensitivity zones of the paths under analysis, on 12 August. It was upgraded to TS intensity on 13 August and it left the sensitivity zones on 14 August before reaching its peak intensity. The decrease of the signals on both propagation paths on 14 August can be due to the tropical cyclone influence or it can be a preseismic effect before the earthquake on 18 August. The next very strong tropical cyclone Bolaven (1215) was formed as a tropical depression west of the Mariana Islands on 19 August and was upgraded to TS intensity 24 h later. It was then upgraded to TY intensity on 21 August. Bolaven reached its peak intensity with maximum sustained winds of 100 kt (51 m s^{-1}) and with central pressure of 910 hPa on 25 August. It passed around Okinawa Island with TY intensity the next day. The typhoon moved across the sensitivity zones of our wavepaths from 22 to 26 August in its peak intensity. Nevertheless, it seems that this tropical

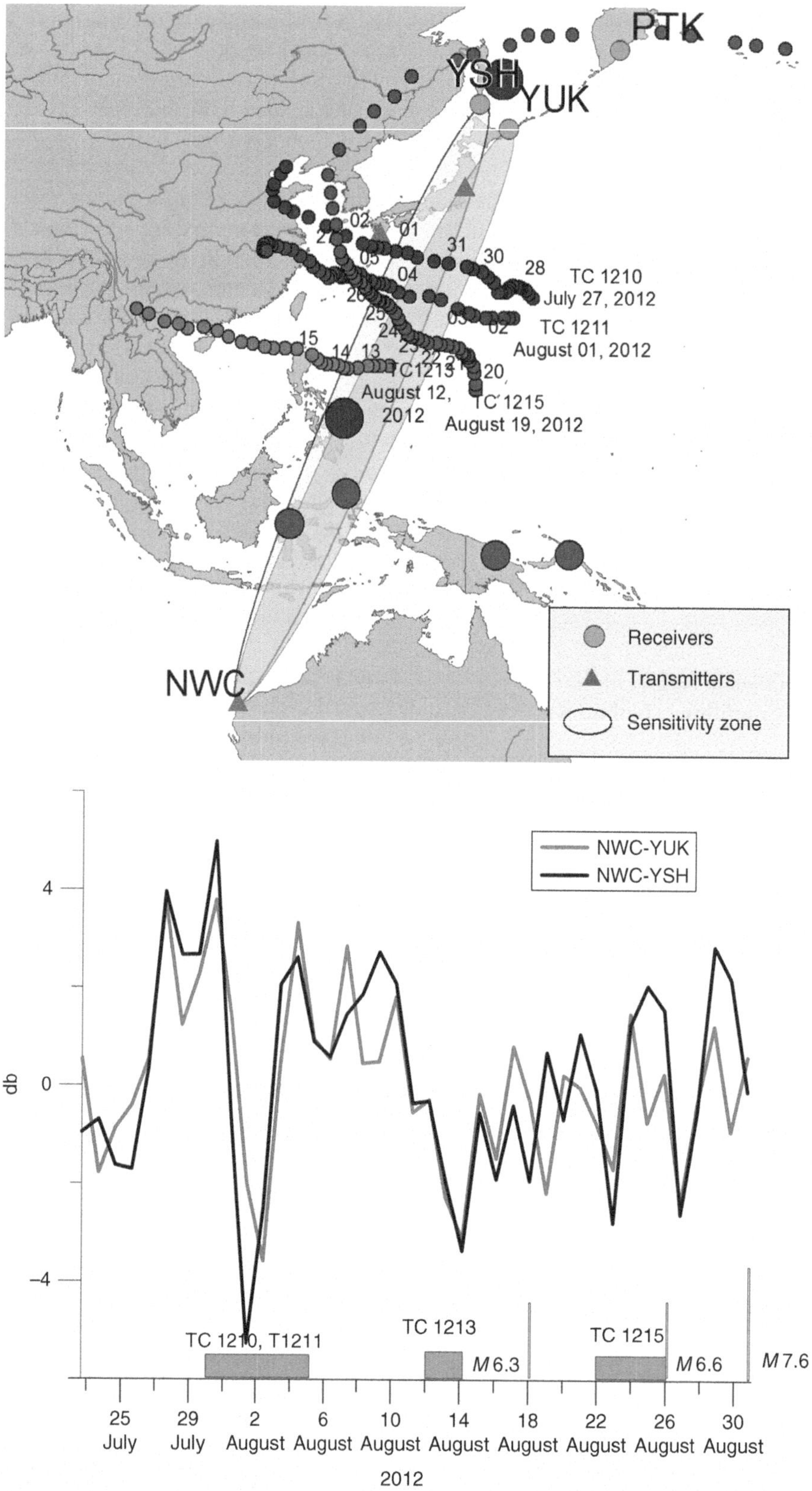

Figure 16.3 The anomalies in the NWC signal recorded at the YSH and YUK stations during the passage of several tropical cyclones in August 2012. In the bottom panel horizontal grey bars on the abscissa show the periods when the tropical cyclones crossed the sensitivity zones of the paths under consideration, and blue vertical bars refer to times of earthquake occurrence. The epicenters of earthquakes with $M>6$ that took place in the Pacific region during the period of analysis are shown in the top panel by large solid brown circles. (*See electronic version for color representation.*)

cyclone did not have significant influence, if any, on the VLF signal. Some decrease in the signal on both paths can be seen on 22 August and 26 August, but we are not sure that this decrease is unambiguously related to this tropical cyclone because of the occurrence of two earthquakes on 26 and 31 August.

In summary the results demonstrate the sensitivity of VLF–LF signals to the variations of atmospheric pressure, humidity, wind velocity, and temperature, and the VLF–LF record at the station of YUK is found to be most sensitive to those variations of atmospheric parameters. The region under consideration is characterized by high winter cyclonic activity in mid-latitudes and strong summer and autumn typhoon activity in low latitudes. The VLF–LF signal variations during eight tropical cyclones with different intensity are considered. Negative nighttime anomalies in the signal amplitude that are most probably caused by tropical cyclone activity are found for six events. Those anomalies are observed over 1–2 days when tropical cyclones move inside the sensitivity zones of the subionospheric paths. Perturbations of the VLF signal observed during two tropical cyclones can be caused by both the tropical cyclone influence and seismic activity, but no correlation between tropical cyclone intensity and magnitude of the signal anomalies is found.

Here we comment briefly on the mechanism of cyclone-related ionospheric perturbations, because this topic is closely related to the question of why the ionosphere is perturbed prior to an earthquake, which will be discussed extensively later. A few hypotheses have been proposed, but the most likely seems to be upward propagation of internal gravity waves excited by typhoons. An alternative theory of atmosphere–ionosphere coupling is based on the influence of the electric fields induced by tropical typhoons.

16.4. VLF–LF ANOMALIES IN ASSOCIATION WITH EARTHQUAKES

Russian scientists first suggested using this VLF–LF method in earthquake research about 20 years ago [*Gokhberg et al.*, 1989; *Gufeld et al.*, 1992]. Nighttime "baylike" anomalies of the phase and amplitude from "Omega" VLF signals were detected before a number of strong earthquakes in the 3rd region of the Fresnel zone for longwave paths. Another data processing method (the so-called terminator time (TT) method) was developed in Japan during an analysis of the famous M 7.1 Kobe earthquake of 17 January 1995 [*Hayakawa et al.*, 1996; *Molchanov and Hayakawa*, 1998; *Hayakawa*, 2007], which is described in detail in the following section. This VLF–LF method is becoming very popular for earthquake prediction [*Hayakawa*, 2007, 2011; *Chakrabarti*, 2010; *Hayakawa and Hobara*, 2010; *Rozhnoi et al.*, 2013].

16.4.1. VLF Ionospheric Perturbation for the 1995 Kobe Earthquake

The most convincing evidence on the seismo-ionospheric perturbations using VLF sounding was obtained by *Hayakawa et al.* [1996] for the famous Kobe earthquake on 17 January 1995. Some important peculiarities observed in that study are summarized as follows: (a) the propagation distance from Tsushima Omega (34.37°N, 129.27°E) to Inubo observatory (35.42°N, 140.52°E) is relatively short (~1 Mm (1000 km) at VLF, as compared with 5–9 Mm used in previous Russian studies [*Gokhberg et al.*, 1989; *Gufeld et al.*, 1992]; and (b) they found that the nighttime fluctuation method as used previously, was not so effective for the short-propagation path, so they developed another, the TT method. The TT is defined as the time when the diurnal phase (or amplitude) variation exhibits a minimum around sunrise and sunset (which we call morning (t_m) and evening (t_e) TTs). We found a significant shift in TTs before the earthquake; that is, t_m shifts to earlier hours and t_e to later hours; see details of the Kobe results in *Hayakawa et al.* [1996], *Molchanov et al.* [1998], and *Hayakawa* [2007, 2009b].

A further extensive study by Molchanov and Hayakawa [1998] was based on many more events over 13 years (11 events with $M > 6.0$ and within the first region of the Fresnel zone) for the same propagation path from the VLF Omega, Tsushima to Inubo, and they came to the following conclusions.

1. For shallow (depth < 30 km) earthquakes, four from five exhibited the same TT anomaly as for the Kobe earthquake with the same 2σ criterion and lead time of less than 1 week.
2. For medium-depth earthquakes (range 30–100 km), there two events were observed, one exhibited the same TT anomaly, while the other indicated a different type of anomaly.
3. Deep (depth > 100 km) earthquakes (four events) did not accompany any anomaly; two had an extremely large magnitude (> 7.0), but had no propagation anomaly.

In summary this might indicate a relatively high probability of the propagation anomaly (in the form of a TT anomaly) on the order of 70–80% for larger ($M > 6.0$) and shallow earthquakes located relatively close to the great-circle path (e.g., first region of the Fresnel zone).

Another important finding is that when we have such a TT propagation anomaly (ionospheric perturbations), a harmonic analysis on the data of the TTs exhibits an enhanced modulation with periodicities of 5 days or 9–11 days (these periods are those of planetary waves). This implies that atmospheric oscillations with those periodicities may play a significant role in the coupling of the lithosphere to the ionosphere. Recently we proposed atmospheric gravity waves

(AGWs) (internal gravity waves) as the carrier because of their stronger tendency of upward propagation in the lithosphere–ionosphere coupling, with the planetary wave as the modulating signal [*Hayakawa*, 2009b]. Based on the study of fluctuation spectra of our observational data (on amplitude and phase), we found an enhanced occurrence of fluctuation power in the frequency range (10 min to 2 h) of AGWs, probably associated with earthquakes [*Molchanov et al.*, 2001]. These findings may provide a fundamental basis for the study of lithosphere–atmosphere–ionosphere coupling (LAIC), as will be discussed later.

Hayakawa et al. [1996] and *Molchanov et al.* [1998] suggested an explanation for the change in the lower ionosphere by means of the full-wave theory of subionospheric VLF propagation over a short distance (~1 Mm), for which there exist several modes of propagation (i.e., TT is the consequence of wave interference of the ground and those sky waves). On the basis of the comparison of theoretical estimations with the experimental data, we concluded that the lower ionosphere might have been lowered by a few kilometers. Here we present a comprehensive view on the importance of TT shift in the subionospheric VLF–LF diurnal variation and its use in inferring the lower ionospheric changes associated with earthquakes. Unlike the studies mentioned above based on full-wave computations [*Hayakawa et al.*, 1996], *Yoshida et al.* [2008] made full use of the wave-hop method (theory) to interpret the TT changes in terms of the wave interference between the ground and sky waves, and indicated how to estimate the change (normally decrease) in ionospheric height by means of the observed shift of TTs.

16.4.2. VLF Ionospheric Perturbations for the 2011 Tohoku Earthquake

The details of the VLF ionospheric perturbations for the recent huge 2011 Tohoku earthquake have already been published [*Hayakawa et al.*, 2012a, 2013a, b], so in this section we describe only some of the important findings.

An extremely huge earthquake (M 9.0) occurred beneath the seabed in the Pacific Ocean off the Tohoku area of Japan. This earthquake took place at 14:46:18 LT on 11 March 2011 with its epicenter at the geographic coordinates 36°6.2'N, 142°51.6'E, as shown in Figure 16.4 by a red star with its date, at a depth of ~20 km. This was a very typical oceanic earthquake of the plate type around Japan, and is very different from the extensively studied fault-type earthquakes such as that at Kobe earthquake [Hayakawa et al., 1996] and the Niigata-chuetsu [*Hayakawa et al.*, 2006; *Yamauchi et al.*, 2007].

We established our Japanese and Pacific network for subionospheric VLF–LF propagation just after the 1995 Kobe earthquake within the framework of a frontier project of the former National Space Development Agency (NASDA) of Japan [*Hayakawa et al.*, 2004]. This observation network has been in continuous operation to date. The main observatories within Japan are MSR in Hokkaido, Chofu (CHF) in Tokyo, Kasugai (KSG) near Nagoya, and KCH on Shikoku island, as shown by black dots in Figure 16.4. At each receiving station we normally detect simultaneously the signals from two Japanese transmitters of JJY (in Fukushima, 40 kHz) as shown by a blue diamond and JJI (in Miyazaki, Kyusyu, 22.2 kHz; not shown) and also a few foreign transmitters (i.e., NWC (North West Cape, Australia), NPM (Hawaii) (these two are not shown in the figure) and NLK (Jim Creek, USA)). Details of this University of Electro-Communications (UEC) network and corresponding VLF receiving system can be found in *Hayakawa et al.* [2004, 2010] and *Hayakawa* [2009b].

This subionospheric VLF–LF network has been extended to cover a wider area of the Pacific ocean, including one station in Russia, Petropavlovsk-Kamchatsky (PTK) shown as a green dot in Figure 16.5 [*Uyeda et al.*, 2002; *Molchanov and Hayakawa*, 2008]. Observations at PTK have been performed regularly resulting in significant scientific output [*Rozhnoi et al.*, 2004, 2006, 2007, 2012a,b]. One more station, Yuzhno-Sakhalinsk (YSH shown as a green dot in Figure 16.5), has recently been established. These two stations are equipped with the same type of VLF–LF receiving system used at the Japanese stations.

Figure 16.4 illustrates one path from JJY to MSR (and its corresponding 5th region Fresnel zone as the wave sensitive area (thin black ellipse)) and the three paths from NLK (Jim Creek, USA) to Japanese VLF–LF observatories (CHF, KSG and KCH). Furthermore, the 5th region Fresnel zones (thin black ellipses) for the propagation paths from NLK to CHF, KSG and KCH are the wave sensitive areas for these paths and are much larger than that for the path from JJY to MSR because of the much greater distance of the NLK-CHF path.

As for the analysis method, we do not follow the TT method as initially developed for the Kobe earthquake [*Hayakawa et al.*, 1996], but apply an alternative "the nighttime fluctuation method" as previously described [*Rozhnoi et al.*, 2004; *Maekawa et al.*, 2006; *Kasahara et al.*, 2008; *Hayakawa et al.*, 2010]. We first read the temporal evolution of amplitude $A(t)$ at a current time t during the local nighttime on a particular day, while $<A(t)>$ is estimated as the average amplitude at the same time t during the period from 1 day to 30 days before the current day. Then, we can estimate the residue $dA(t) = A(t) - <A(t)>$. Using this residue, we can estimate the most important parameter, "trend" as the nighttime average amplitude (mean value of $dA(t)$ over local time). The second parameter

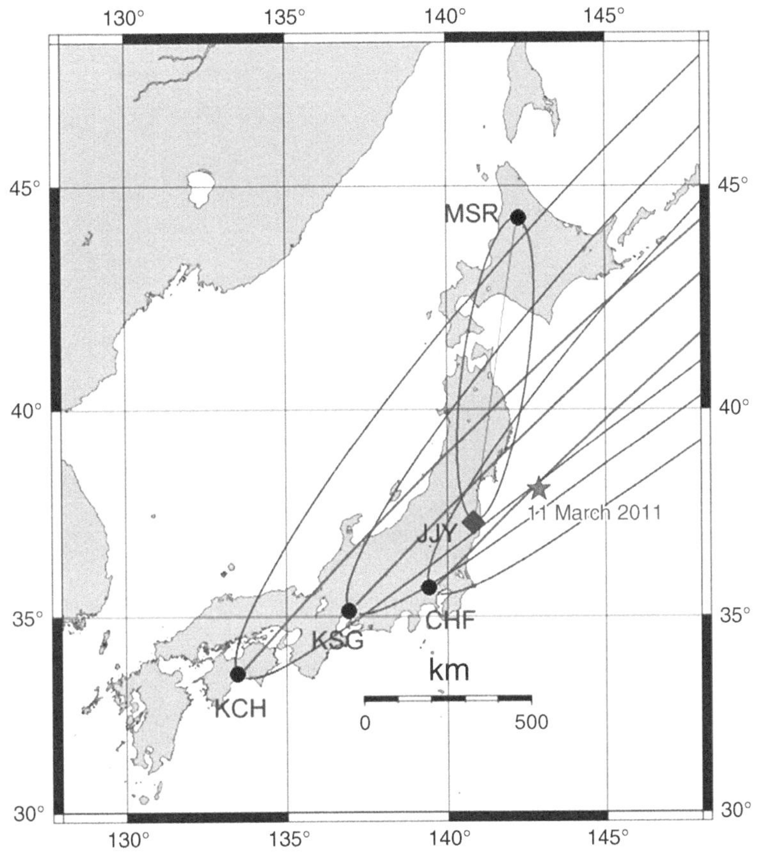

Figure 16.4 The relative locations of two Japanese VLF–LF transmitters (with call signs of JJY (Fukushima) indicated by a blue diamond and JJI (Miyazaki) (not shown) and VLF–LF receiving stations (Moshiri (MSR), Chofu (CHF), Kasugai (KSG) and Kochi (KCH) shown with black dots). The wave-sensitive area defined by the Fresnel zone for the propagation path of JJY–MSR and also that for the propagation path of NLK (Jim Creek, USA) to CHF are plotted (thin black ellipses). Further, the great-circle paths (thin red lines) and the corresponding wave-sensitive areas (thin black lines) are indicated for the paths NLK–KSG and NLK–KCH. The epicenters of the main shock is indicated by the red star. (*See electronic version for color representation.*)

is dispersion, which is characterized by how much the amplitude fluctuates around the average. These two parameters are independent variables. All of these parameters are normalized by their corresponding standard deviations (σ) over 30 to 1 day before the current day of the event. Further details of this nighttime fluctuation method can be found in *Rozhnoi et al.* [2004], *Kasahara et al.* [2008], and *Hayakawa et al.* [2010].

As for the definition of nighttime period, we take the UT period of UT = 11 − 19 h for the propagation path from JJY to MSR because the LT in Japan = UT + 9 h. While the definition of nighttime is considerably complicated for the E–W long-distance propagation from NLK to Japanese stations (such as CHF) (distance = 7–8 Mm). By considering the sunrise and sunset both at the transmitter and receiving observatory (that is, terminator times [*Hayakawa et al.*, 1996]) and also checking the real diurnal variations for the relevant NLK–CHF path, we have taken UT = 10 to 12 h for the nighttime for the NLK–CHF path (i.e., only during this period is the propagation path completely in the dark).

Figure 16.5 illustrates the relative location of the Japanese VLF–LF transmitters (JJY in Fukushima and JJI in Miyazaki) and two Russian observatories, PTK and YSH. The wave-sensitive areas for all combinations of transmitter–receiver are also shown (i.e., JJY–YSH, JJY–PTK, JJI–YSH, and JJI–PTK), together with the locations of the main shock and aftershocks.

Next we have to discuss the nighttime interval for the Russian data because we use the same nighttime fluctuation method. The night in February is UT = 10:30–18:40 and UT = 11:00–16:30 for May. Correspondingly, the nighttime for March and April is within this interval; UT = 10:30–11:00 for sunset and 16:30–18:40 for sunrise. The data analysis for Russian data is exactly the same as the analysis for Japanese data detailed above. The analysis period is taken from 1 January to 22 May 2011, including our target earthquake on 11 March.

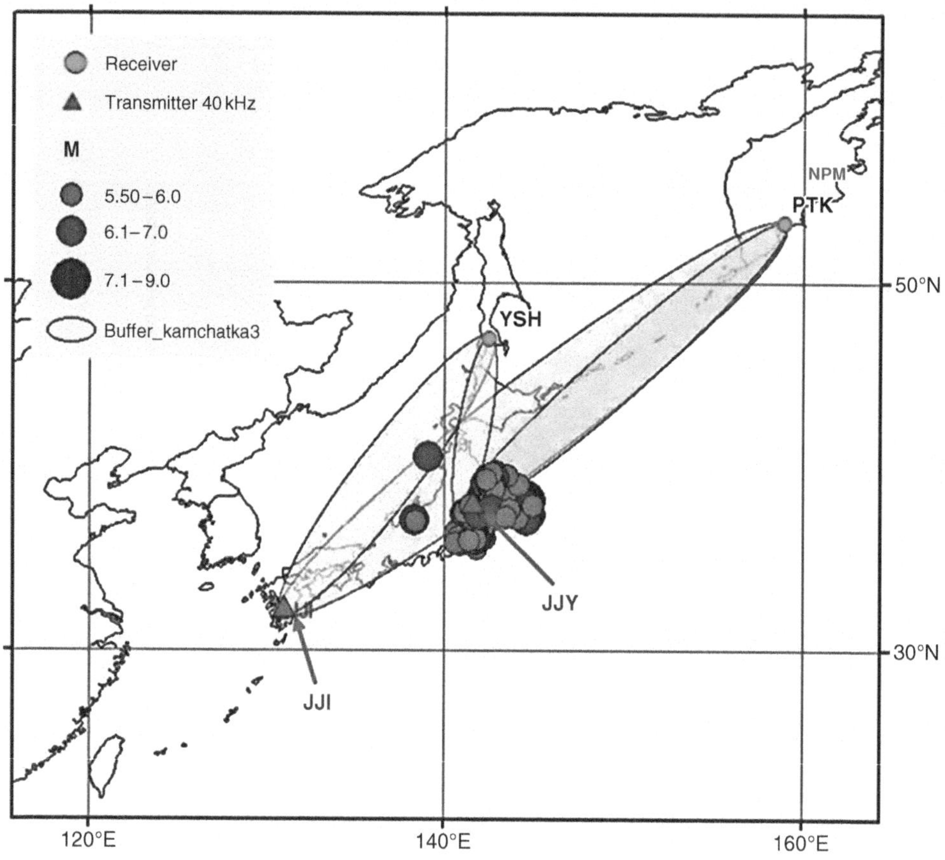

Figure 16.5 Locations of the two Japanese VLF–LF transmitters (JJY and JJI, red triangles) and two observing stations (Petropavlovsk-Kamchatsky (PTK) and Yuzhno- Sakhalinsk (YSH), green dots). The wave-sensitive areas (elliptic zones) for the propagation paths of JJY–YSH, JJY–PTK, JJI–YSH, and JJI–PTK are plotted. Further, earthquake main shocks and aftershocks are plotted, with sizes being proportional to magnitude. (*See electronic version for color representation.*)

Unlike the 2005 Miyagi-oki earthquake [*Muto et al.*, 2009a], the epicenter of this 11 March earthquake was found to be located considerably distant from the wave-sensitive area of the JJY–MSR path, because this earthquake occurred ~150 km away from the coastline of the Tohoku area [*Hayakawa et al.*, 2012a]. Although not shown as a figure in this chapter, we found in our previous study [*Hayakawa et al.*, 2012a] based on preliminary analysis that there is definitely no time interval from 1 March to 9 March before the earthquake on the JJY–MSR path in Figure 16.4 in which the trend shows a notable decrease together with the simultaneous increases in the dispersion, as is the case in a tremendous number of land earthquakes.

We next analyzed the propagation paths of JJY–YSH and JJI–YSH. The path of JJY–YSH is relatively close to the previous path of JJY–MSR as seen in Figure 16.4, because the JJY–YSH path is likely to be just an extension of the JJY–MSR path. We have analyzed the paths of JJY–YSH and JJI–YSH, but found no definite significant effects (no propagation anomalies), although not presented here as figures.

Figure 16.6 suggests that the propagation paths from Japanese receiving stations (CHF, KSG, and KCH) to the American transmitter NLK (Jim Creek, USA) are favorably located with respect to the epicenter of this oceanic earthquake and this is especially so for the NLK–CHF path passing just above the earthquake epicenter; the corresponding wave-sensitive area for this NLK–CHF path is plotted as a thin line in Figure 16.4. Two other propagation paths from NLK to KSG and from NLK to KCH are also favorable for us to detect any corresponding ionospheric perturbations. In response to these theoretical expectations, Figure 16.6 illustrates the real temporal evolutions of propagation characteristics for these relevant paths: Figure 16.6a refers to the NLK–CHF path, Figure 16.6b to the NLK–KSG path, and Figure 16.6c to the NLK–KCH path. In Figure 16.6 we have illustrated, from top to the bottom, the trend and dispersion, with these parameters being all normalized by their corresponding standard deviations (σ). Let us look at the top panel (trend) of Figure 16.6a for the most important propagation path from NLK to CHF during the period from 1 January. We find that over the entire period the

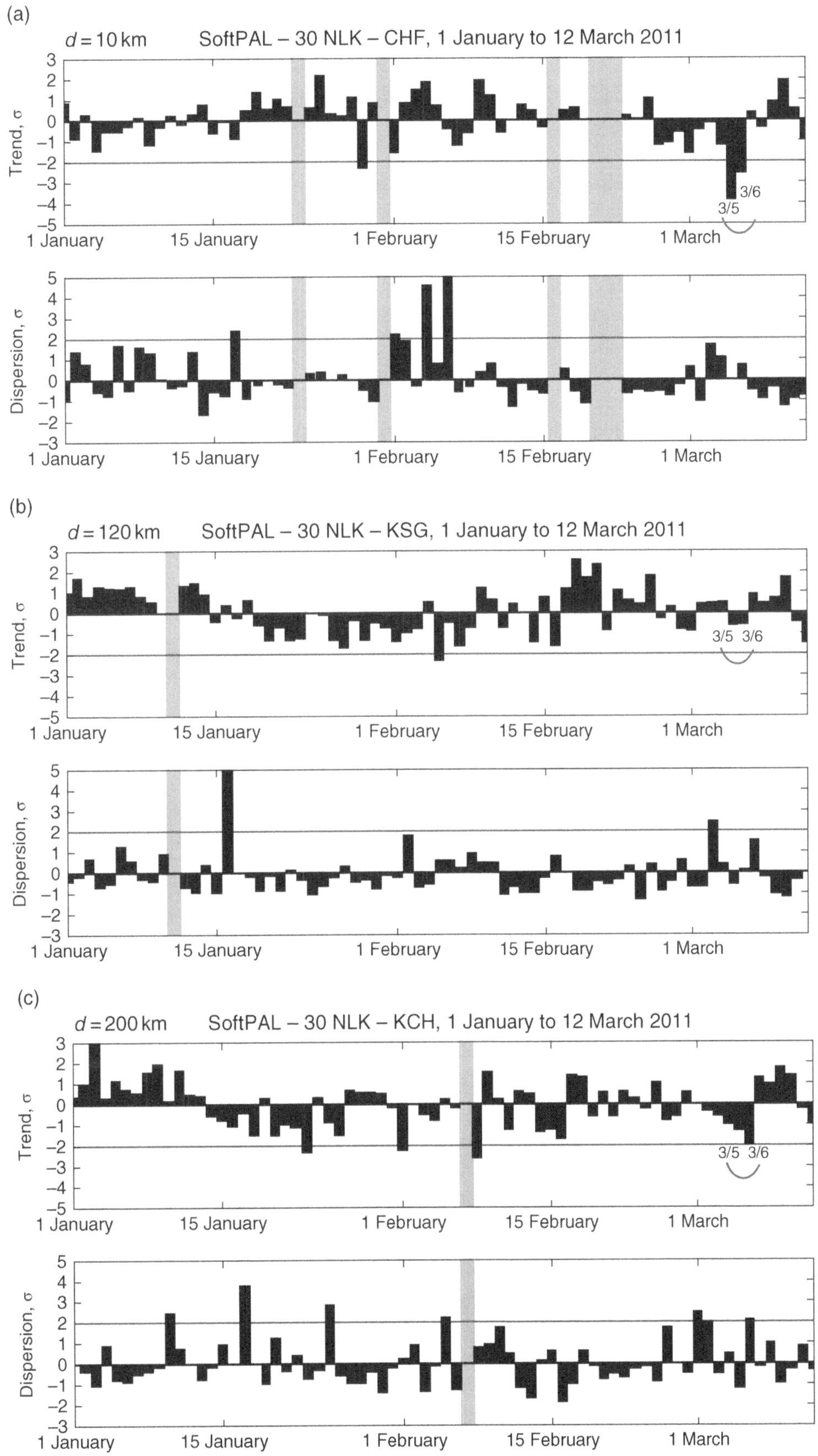

Figure 16.6 Temporal evolution of the propagation characteristics for the three propagation paths: (a) NLK–CHF, (b) NLK–KSG, and (c) NLK–KCH. In each figure, the top panel refers to the average nighttime amplitude (trend), and the bottom panel to the dispersion. All of these values are normalized by their corresponding standard deviations (σ). A clear anomaly is seen on 5 and 6 March. The distance (d) of the earthquake epicenter to each propagation path is indicated on the top left in each figure. (Source: Reproduced with permission of Terrestrial, Atmospheric and Oceanic Sciences.)

trend does not decrease to a -2σ level, except on 29 January and on the two days of 5 and 6 March indicating an extremely significant propagation anomaly is present. The propagation anomaly on 5 March is characterized by a remarkable decrease in trend (exceeding -3σ and even approaching -4σ), together with nearly simultaneous increases in dispersion (approaching $+2\sigma$). The corresponding anomaly is also recognized in Figure 16.6c for the propagation path from NLK to KCH. The anomaly for this path is evident in such a way that the most important parameter, trend, exhibited a significant decrease reaching -2σ level. In contrast, the anomaly for the path of NLK to KSG in Figure 16.6b is less enhanced on the same days of 5 and 6 March, but it is important to note that the response to this earthquake is still very evident.

Here we should comment on other propagation anomalies seen in Figure 16.6. In *Hayakawa et al.* [2010] we have tentatively chosen an earthquake magnitude threshold of *M* 6 (only rather strong earthquakes) and we have obtained a quite significant correlation exceeding well beyond the 2σ criterion. However, even if we lower the magnitude threshold down to *M* 5.5, we know that the correlation between VLF–LF anomalies and earthquakes is still significant, just around 2σ level [*Rozhnoi et al.*, 2004; *Maekawa et al.*, 2006]. Therefore, we try to associate other decreases in trend in Figure 16.6 to earthquakes in the relevant region. First, we comment on the anomaly of 29 January in Figure 16.6a. Probably in possible association with this anomaly, two earthquakes have occurred off the coast of Iwate (on 3 February) and Fukushima (on 10 February, *M* 5.3). Further comments are required on other decreases in trend in Figure 16.6c. The decrease on 23 January is likely to be related to an earthquake off the coast of Chiba on 25 January (*M* 5.1) and those on 1 and 8 February (exceeding -2σ level) are likely to be related to other earthquakes, in Chiba-oki on 5 February (*M* 5.2) and in Miyagi-oki on 15 February (*M* 5.5), respectively. Finally, the decrease on 5 February in Figure 16.6b is likely to be a precursor to an earthquake in Fukushima-oki (*M* 5.3).

Among the three Russian propagation paths, we have found a conspicuous effect only on the propagation path from JJI (Miyazaki, Kyushu) to PTK (Kamchatka). Figure 16.7 illustrates the temporal evolution of the nighttime average amplitude (trend) (top panel). The second panel refers to the conventional dispersion and the bottom indicates the evolution of the earthquakes with $M > 5.5$. In the top panel, the horizontal dotted line indicates the -2σ level. In the middle panel of dispersion, the $+2\sigma$ line is plotted as a horizontal dotted line. Figure 16.7 shows that a significant and prolonged decrease in nighttime amplitude takes place during a rather long period from 28 February to 6 March on the path from JJI to PTK with a maximum decrease on 3 and 4 March. Corresponding increases in dispersion are observed during the same prolonged period, with the maximum on 3 and 4 March. The dates with the VLF–LF propagation anomaly on the Russian path are shifted somewhat compared with that for the NLK–CHF path in Figure 16.6, but the anomaly on this propagation path is considered to be the same as that of NLK–CHF because we know that some inhomogeneity exists in the time and space of the ionospheric perturbation [e.g., *Yamauchi et al.*, 2007].

Finally, we comment on the Russian path JJY–PTK. The wave-sensitive area for this propagation path is seen from Figure 16.5 to be completely within the wave-sensitive area of the above-mentioned JJI–PTK path with significant anomalies. Though not shown as a figure, it is found that the trend shows a significant decrease on 4 March, not exceeding -2σ but approximately -1.5σ. Finally, an anomaly is also observed for this path on 4 March, but the nature of this is indicative of the highly heterogeneous property of the ionospheric perturbation.

By making full use of the Japanese–Russian subionospheric VLF–LF network, the following observations have emerged in possible relation to the 11 March 2011 Japanese earthquake.

1. No definite anomaly has been detected for the three propagation paths JJY–MSR, JJY–YSH, and JJI–YSH.

2. On the other hand, clear and significant propagation anomalies have been observed for the two propagation paths of the NLK to Japanese stations (CHF, KSG and KCH) and JJI–PTK. The propagation anomaly for the path NLK–CHF takes place on 5 and 6 March, which is characterized by a significant decrease in trend (nighttime average amplitude) well exceeding the -3σ level, together with the simultaneous increases in dispersion. While the anomaly on the path of JJI–PTK shows a broad decrease from 28 February to 6 March, with maximum decreases on 3 and 4 March, which also is characterized by a significant decrease in trend and an increase in dispersion. So, the remarkable ionospheric perturbation is likely to be persistent, at least for 4 days (3–6 March). Finally, as shown in 16.3.1, because a small geomagnetic storm occurred on 1 March the effect of geomagnetic activity, which might influence ionospheric perturbations, should not be neglected; *Hayakawa et al.* [2013b] have discussed the effect of this storm.

16.4.3. Statistical Correlation of VLF Ionospheric Perturbations with Earthquakes

A statistical study on the correlation of VLF ionospheric perturbations with earthquakes has been presented by *Hayakawa et al.* [2010] based on long-term observations in Japan and Kamchatka. We have used data for a total 7 years from 1 January 2001 to 31 December 2007. Based on the previous statistical studies [*Rozhnoi et al.*, 2004;

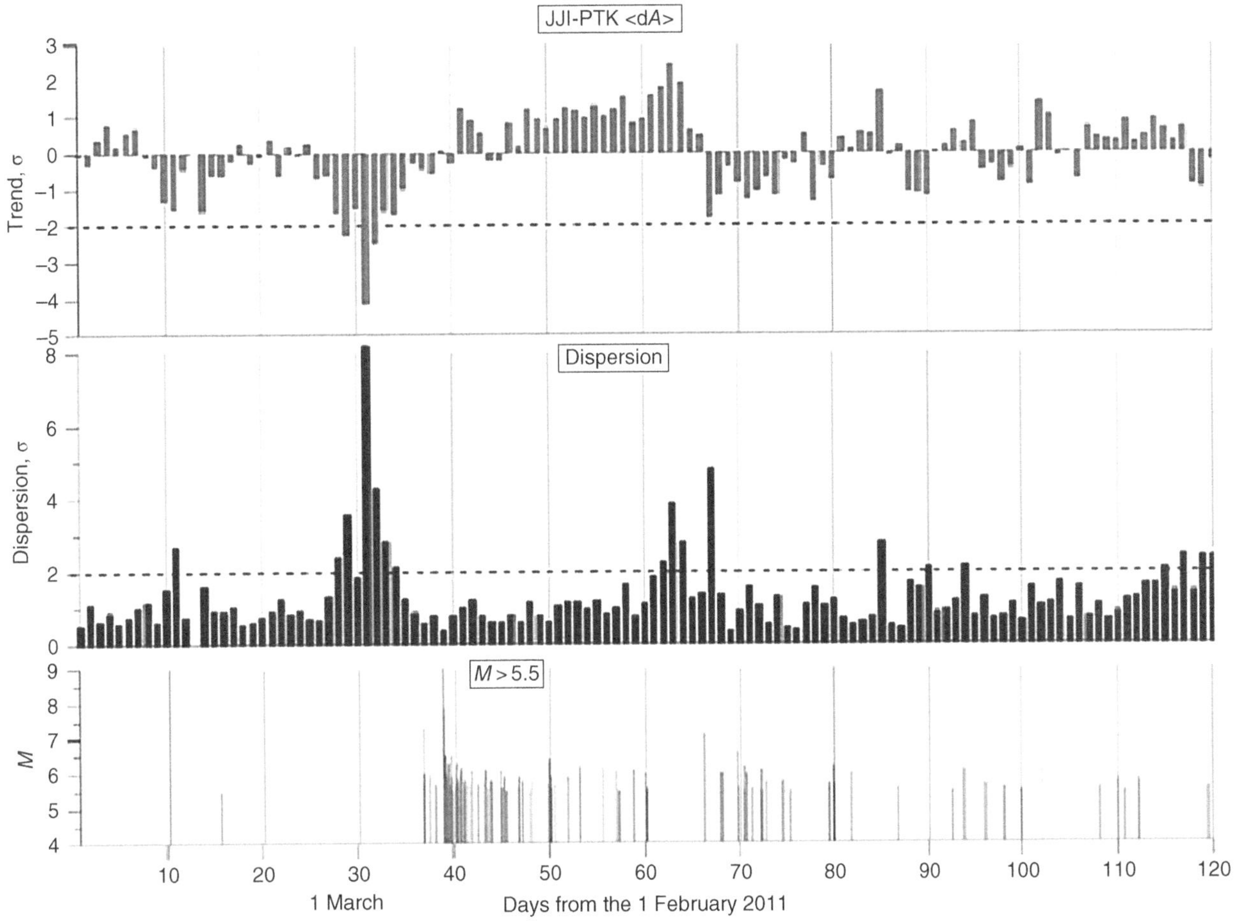

Figure 16.7 Temporal evolution of the propagation characteristics for the wavepath of JJI–PTK. The top panel refers to the average nighttime amplitude (trend; horizontal broken line indicates -2σ level); middle panel plots the dispersion (horizontal broken line, $+2\sigma$ level). Both trend and dispersion are normalized by their standard deviations (σ). The bottom panel indicates the temporal evolution of the seismic activity.

Maekawa et al., 2006; *Kasahara et al.*, 2008], a M 5.5 limit is found to be just at the border to obtain any significant correlation with 2σ criterion between the VLF–LF propagation anomalies and earthquakes, we therefore chose M 6.0 here as a rather severe criterion of selecting earthquakes. By imposing this condition, we found 37 earthquakes taking place within the wave-sensitive areas defined by the fifth-region of the Fresnel zone of the great-circle paths of different propagation paths, as in Table 16.1. For our analysis we divide the earthquake depth into two regions: shallower or deeper than 40 km in order to find the dependence on earthquake depth. As seen from the configuration of the propagation paths in Figure 16.5, we can imagine that some earthquakes are common on a few propagation paths. In the case of shallow earthquakes (depth < 40 km), three are common for three paths and 13 are common for two paths. Similarly, for deep earthquakes (depth ≥ 40 km), four are seen for three paths and 16 are common on two paths. Therefore, the total number of propagation paths that cover the earthquakes with depth < 40 km is 35, while the corresponding number is 38 for those with depth ≥ 40 km. We treat the data for each propagation path as independent events, so that the total number of events is 73. This independent treatment seems to be validated by our previous work. *Yamauchi et al.* [2007] examined a few propagation

Table 16.1 Number of earthquakes identified below and above 40 km depth along each wavepath.

Path	d < 40 km	d ≥ 40 km
JJY–MSR	6	4
JJY–KCK	15	15
JJI–TYM	1	1
JJI–MSR	1	3
JJI–KCK	10	15
JJY–KCH	2	0
Total	35	38

paths for the 2004 Mid-Niigata earthquake by means of the TT method, and found that the anomaly in TT does not always happen on the same day. This could suggest that the seismo-ionospheric perturbation is very inhomogeneous both in space and in time, leading to the effect of very dissimilar variations on different propagation paths even for the same earthquake. Here we use the nighttime fluctuation method, in which we estimate the two physical quantities [*Rozhnoi et al.*, 2004; *Maekawa et al.*, 2006]: (a) trend (as the average of nighttime amplitude) (in dB), and (b) dispersion (D) (in the following we use its square root, i.e., standard deviation, but we use the terminology of dispersion in order to avoid the confusion of using standard deviation too often). *Maekawa et al.* [2006] and *Kasahara et al.* [2008] have indicated that any seismogenic effect is characterized by the almost simultaneous significant decrease in trend and significant increase in dispersion.

It is necessary to mention how data on different propagation paths should be treated, because the variability in VLF–LF amplitude data is very different from one path to another. Thus it is a requirement to treat homogeneously the VLF–LF data when we analyze different propagation paths. We have proposed a so-called "standardization" in the following way.

That is, when taking one particular path, we deal with the two physical quantities of amplitude (trend) and dispersion (D) and we estimate the normalized trend (trend*) and normalized dispersion (D*). When taking an earthquake with a particular date, we estimate the trend on this day and then calculate the average <trend> over ±15 days around that date. Then trend* is defined as (trend − <trend>)/σ_T (σ_T is standard deviation over ±15 days around the current date). The same principle is applied to the dispersion in order to obtain D*.

By using trend* and D* we make full use of a superimposed epoch analysis [e.g., *Rozhnoi et al.*, 2004; *Maekawa et al.*, 2006], which is of extreme importance in enhancing the signal-to-noise ratio by stacking the data around the earthquake day as a reference day. Although we have chosen earthquakes with $M \geq 6.0$ we pay more attention to the effect of earthquake depth here because this point is poorly studied, even though *Maekawa et al.* [2006] have suggested this point qualitatively.

Figures 16.8a and 16.8b are the final results on trend* and D* on the basis of superimposed epoch analysis, which can be summarized as follows.

1. Trend* is found to show a significant decrease (exceeding the $2\sigma_T$ criterion) before the shallow earthquake (with depth < 40 km; red line Figure 16.8a). This anomaly takes place 5 days before the earthquake as a conspicuous peak. With greater earthquake depth, > 40 km, a similar tendency can be observed (blue line in

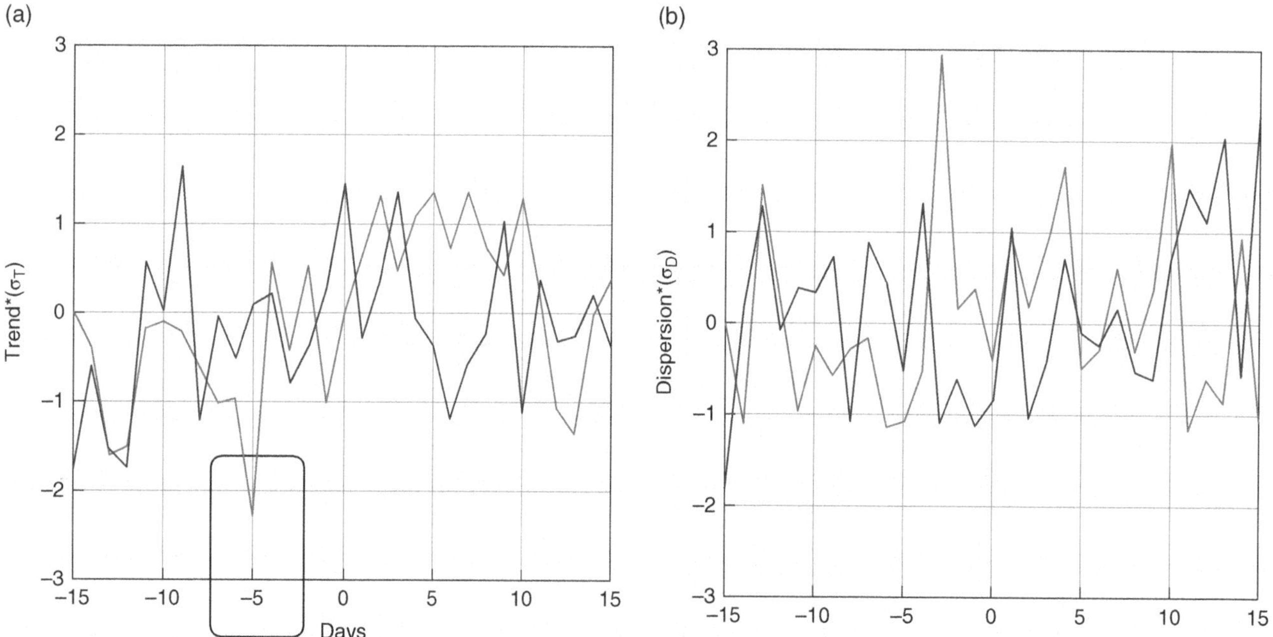

Figure 16.8 Superimposed epoch analysis for (a) the normalized trend (trend*), and (b) the normalized dispersion (D) (dispersion*). The red line refers to shallow earthquakes (depth < 40 km) and the blue line refers to earthquakes with depth > 40 km. The abscissa indicates the day with respect to the earthquake day (0), with minus indicating days before plus days after. (Source: Reproduced with permission of John Wiley and Sons.) (*See electronic version for color representation.*)

Figure 16.8a), such that trend* approaches (but does not exceed) the $2\sigma_T$ criterion 12 days before the earthquake.

2. D^* for earthquake depths < 40 km (red line in Figure 16.8b) is found to exhibit a significant increase 3 days before the earthquake (exceeding the $2\sigma_D$ criterion and even approaching the $3\sigma_D$ level). With earthquake depth > 40 km (blue line in Figure 16.8b), there is no clear precursory signal evident.

We now describe some other possible interference effects on VLF–LF perturbations, as mentioned 16.3.1, and the most confounding effect may be geomagnetic storms. When constructing Figure 16.8 we paid no attention to geomagnetic activity. Here we comment on the effect of geomagnetic storms on ionospheric perturbations because *Kleimenova et al.* [2004] have presented an example of the effect of geomagnetic storms on subionospheric VLF propagation. As evident from Figure 16.8, seismo-ionospheric disturbances are found to take place before an earthquake (on the order from day 0 to about 10 days before the earthquake), so we have examined geomagnetic activity and geomagnetic storm incidence for all earthquakes identified in the previous analysis. For geomagnetic activity we estimated the mean 3 h K_p index for 10 days prior to each earthquake. The average 3 h K_p index during these 10 days is less than 2.0 for 32% of earthquakes, 2.0–3.0 for 46%, 3.0–4.0 for 19%, and 4.0–5.0 for 3%. So, we can say for all earthquakes that geomagnetic activity was not significant.

However, it is more meaningful to analyze geomagnetic storms, which are non-stationary and would influence the lower ionosphere. We have checked whether there were geomagnetic storms during 10 days prior to all earthquakes. We found that one earthquake accompanied two relatively large geomagnetic storms ($|D_{st}| = 359$ and 460 nT), and there were several earthquakes with small $|D_{st}|$ (< 200 nT). So it could be possible that only one particular earthquake was disturbed by geomagnetic activity, but it is reasonable to suggest that all of the ionospheric perturbations are the consequence of earthquakes (i.e., those results in Figure 16.8).

16.5. DISCUSSION OF THE GENERATION MECHANISM OF SEISMO-IONOSPHERIC PERTURBATIONS

As confirmed extensively by means of statistical and case studies, and by study of the modulation effects, etc., it seems highly likely that the ionosphere is disturbed before an earthquake; how the ionosphere is perturbed by the precursory seismic activity in the lithosphere remains poorly understood, however. *Hayakawa et al.* [2004] have already proposed a few possible hypotheses on the mechanism of coupling between lithospheric activity and ionosphere: (a) chemical channels, (b) atmospheric oscillation (or acoustic) channels, and (3) electromagnetic channels: Figure 16.9 illustrates these three coupling channels [Hayakawa et al., 2004] schematically. In the chemical channel, radon emanation induces perturbation in the conductivity of the atmosphere, change in the atmospheric electric field, leading to the ionospheric modification through the atmospheric electric field [e.g., *Pulinets and Boyarchuk*, 2004; *Sorokin et al.*, 2006]. In the acoustic channel atmospheric oscillations, acoustic waves (AW) or atmospheric gravity waves (AGW; or internal gravity waves), play a key role in LAIC, and perturbations at Earth's surface (such as temperature, pressure) in a seismologically active region excites the atmospheric oscillations traveling up to the ionosphere, thereby inducing ionospheric density perturbations [*Molchanov et al.*, 2001; *Miyaki et al.*, 2002; *Shvets et al.*, 2004; *Korepanov et al.*, 2009]. In the electromagnetic channel, radio emissions (in any frequency range) generated in the lithosphere propagate up to the ionosphere, thereby modifying the ionosphere through heating and/or ionization processes. This mechanism, however, is found to be insufficient because of the weak intensity of lithospheric radio emissions [*Molcchanov et al.*, 1993]. So at the moment chemical and acoustic mechanisms are likely more plausible candidates for this coupling [*Molchanov and Hayakawa*, 2008].

Pulinets and Boyarchuk [2004] insisted that the chemical channel is the most promising candidate for the ionospheric perturbations associated with earthquakes. That is, the emanation of radon is suggested as the main player in seismo-ionospheric perturbation [*Pulinets and Ouzounov*, 2011], but there seems to be little experimental (observational) evidence in support of their hypothesis. Of course, we know that radon emanation itself has been reported [e.g., *Molchanov and Hayakawa*, 2008] as a precursor to an earthquake, but it is poorly understood whether the radon emanation might result in ionospheric perturbation. If it does, how it is realized is not well understood. There have been very few studies on the correlation between phenomena at Earth's surface (such as surface latent-heat flux) and ionospheric perturbation as seen from VLF–LF subionospheric perturbation [*Cervone et al.*, 2006].

This role of the chemical channel as a major player has been criticized recently by *Sorokin and Hayakawa* [2013] and *Sorokin et al.* [2015], who have further suggested an alternative mechanism based on the electromotive force (EMF) due to the injection of charged aerosols from the ground to the atmosphere. They have found that this new hypothesis enables explanation of both experimental factors observed prior to an earthquake: (a) no significant change of atmospheric electric field on the ground surface, and (b) significant enhancement of the electric field in the ionosphere.

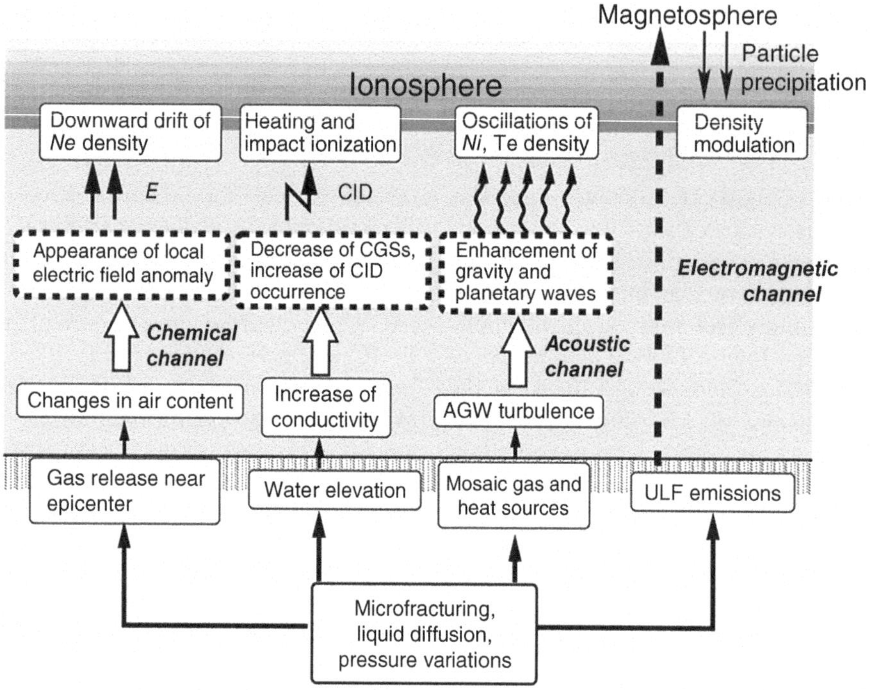

Figure 16.9 Schematic illustration of lithosphere–atmosphere–ionosphere coupling and three channels: chemical (+electric field), acoustic, and electromagnetic. (After *Hayakawa et al.* [2004] and *Hayakawa* [2009b, 2011].)

Although not shown in Figure 16.9, a new hypothesis of electrostatic channel has been proposed by *Freund* [2009] on the basis of the discovery of positive hole charge carriers in crustal rocks, alongside electrons. Normally, these charge carriers lie dormant in the crystal structures of the constituent minerals. When deviatoric stresses are applied, they wake up, turning the stressed rock volume into a battery, from which electric currents can flow out. When the positive holes arrive at the Earth's surface, they can cause a variety of effects, including ionization of air at the air–ground interface, ionospheric perturbations, etc.

Compared with studies of the chemical channel, much evidence has accumulated on the importance of the acoustic channel (due to atmospheric oscillations), mainly through analysing VLF–LF subionospheric data. Below we indicate several observations in support of the acoustic channel (see details in *Hayakawa et al.* [2011]).

1. Observation of AGW modulations in subionospheric VLF–LF data. *Molchanov et al.* [2001] and *Miyaki et al.* [2002] made the first attempt to identify AGW modulation in the subionospheric VLF–LF data during (or before) an earthquake, and found that such AGW modulation is clearly enhanced before earthquakes. Since then, there further evidence has accumulated on those AGW modulations in the VLF data [*Rozhnoi et al.*, 2004, 2007; *Shvets et al.*, 2004; *Muto et al.*, 2009b; *Kasahara et al.*, 2010].

Horie et al. [2007] have studied the renowned Sumatra earthquake of 26 December 2004 on the propagation paths from the Australian transmitter NWC to several stations in Japan. The VLF amplitudes at Japanese stations indicated a depression in nighttime amplitude and also an enhancement in amplitude fluctuation as precursors to an earthquake. An additional important point is that the nighttime fluctuation is composed of wave-like structures, and by performing wavelet and cross-correlation analyses they found a significant enhancement in the fluctuation spectra in the period of 20–30 min to ~100 min (the frequency range of AGWs). The cross-correlation between two propagation paths indicates that the wave-like structures tend to propagate horizontally from the NWC–Kochi path to the NWC–Chiba path with a delay of ~2 h, corresponding to a propagation speed of ~20 m s^{-1}.

2. A statistical study on AGW modulation. *Kasahara et al.* [2010] carried out a statistical analysis on the correlation between AGW modulation and $M>6$ earthquakes by means of superimposed epoch analysis. They conclused that there is a significant correlation between the two.

3. Modulation of VLF–LF data by planetary waves. As previously shown, harmonic analysis of TTs has indicated the presence of modulation in subionospheric VLF–LF data (amplitude, phase) with periods of the planetary waves (2, 5 days, 1011 days or so) [Molchanov and Hayakawa, 1998].

4. VLF–LF Doppler-shift observation. We have established new equipment for observation of Doppler shifts of the LF transmitter signal of JJY (Fukushima, Japan) [*Asai et al.*, 2011]. *Hayakawa et al.* [2012b] found that Doppler shifts are readily observed in the frequency range of AGW and AW before an earthquake when ionospheric perturbations are present. This is direct evidence of the presence of AGW (and AW) being involved in seismo-ionospheric perturbation. Further, the observed Doppler shift enabled us to estimate the vertical velocity of AGW as ~10 m s^{-1}, which is in good agreement with theoretical estimates.

5. Ground-satellite coordination. *Korepanov et al.* [2009] studied the correlation of ground effects using satellite observation, with special reference to meteorological disturbances, and concluded that that AGW is the main agent of LAIC. Following this work, *Nakamura et al.* [2013] attempted to correlate pre-earthquake effects with ionospheric perturbations as detected in VLF–LF data. The scenario is as follows. Pre-earthquake effects appear on Earth's surface, leading to change in atmospheric pressure and excitement of atmospheric oscillations. The atmospheric waves propagate upward and modify the dynamo region (ionospheric perturbations). The change in the dynamo region might result in change in the ULF magnetic field on the ground. Three physical parameters are compared extensively with the use of wavelet analysis. This study is very challenging, but the results are not so clear. It seems that there existed a time delay of a few hours between the VLF–LF fluctuation and atmospheric pressure fluctuation and nearly no delay between the VLF fluctuation and ULF magnetic field variation for one particular earthquake.

The above studies are based on the assumption that some kind of effect is present on the ground before an earthquake, which disturbs the atmosphere. This assumption has long been criticized because there are very few reports on those surface changes before an earthquake. Fortunately, surface deformation before an earthquake has been found recently using sophistical signal processing of GPS data [*Chen et al.*, 2011; *Kamiyama et al.*, 2014], indicating short-term precursors. Also, *Kamiyama et al.* [2013] extensively compared ground deformation with electromagnetic precursors for the 2011 Tohoku earthquake, suggesting a close correlation between the two.

Sun et al. [2011] have tried to understand what is happening between the F region and the ground surface when there is a seismo-ionospheric perturbation. With the use of information on T_n (neutral particle temperature) collected by the SABER (Sounding of the Atmosphere using Broadband Emission Radiometry) instrument on board the TIMED satellite, *Sun et al.* [2011] have concluded the importance of seismogenic internal gravity waves in the dynamo region, providing support for the role played by the acoustic channel in LAIC.

Extensive further studies, both observational and theoretical, are required before we come to the conclusion on which channel is most relevant. We comment briefly here on some computer simulations on LAIC. *Kuo et al.* [2011] performed a theoretical simulation of LAIC on the assumption of a seismogenic source in the lithosphere. *Klimenko et al.* [2012] simulated the effects of AGWs and the penetration of a vertical electric field, and found that the simulation results with the AGW hypothesis are in good agreement with the actual GPS TEC data, in support of the acoustic channel. Even though it is clear that the acoustic channel is more probable at the moment, we need to carry out further work on different channels before a definite conclusion can be drawn.

16.6. RESULTS FROM A NEW VLF–LF NETWORK IN JAPAN

Research results presented above are based on data collected using the UEC VLF–LF network established within the framework of NASDA's frontier project, when one of the authors (MH) was a professor in UEC [*Hayakawa et al.*, 2004]. Unfortunately the network was mainly dedicated to scientific research, and so was not suitable for regular prospective earthquake prediction because sometimes there were long data gaps. These arose because of malfunction of the receiver at a certain station, for example, which could not be repaired immediately, resulting in no data during several days. Consequently, in 2010 we established a venture company, "Earthquake Analysis Laboratory," intending to perform very continuous VLF–LF observation and to release the earthquake prediction information to the public, mainly on smart-phones.

The initial objective of this private venture was to establish a new VLF–LF network in Japan, with the main principle being exactly the same as that of the former VLF–LF system. Figure 16.10(a) illustrates the new VLF–LF network composed of eight observation sites: from the north, NSB (Nakashibetsu), STU (Suttsu), AKT (Akita), KTU (Katsuura) in Chiba prefecture, KMK (Kamakura) in Kanagawa prefecture, IMZ (Imizu) in Toyama prefecture, TYH (Toyohashi) in Aichi prefecture, and ANA (Anan) in Tokushima prefecture. All of those stations are equipped with identical receivers that register simultaneously the amplitude and phase of both MSK (minimum shift keying) and ASK (amplitude shift keying) narrow-band modulated signals in the frequency range of 10–50 kHz from several transmitters. At each station, we receive the signals from two Japanese transmitters with call signs of JJY (40 kHz) (in ASK) and JJI (22.2 kHz) (no information on modulation) as in Figure 16.10a. Additionally, signals from three American transmitters (NWC (19.8 kHz), NPM (21.4 kHz) and NLK (24.8 kHz); all in MSK) are received at each station as shown in Figure 16.10b. The reception at each station is made with an electric rod antenna, and we measure the vertical electric field of the signal. The receiver can record signals with time resolution ranging from 50 ms to 60 s, but for our purpose we use a sampling frequency of 1 s. We started installing the new VLF receiver in October 2013, and completed the network in February 2014. We have continuous records since then; if something happens at any station, the longest data gap is made < 1 day.

Here we present some interesting results from this network in order to demonstrate how our new VLF–LF data are usefully utilized for earthquake prediction in Japan. Figure 16.11 illustrates the distribution of earthquakes in and around Japan during 2014. Different colors refer to earthquake magnitude, with red circles corresponding to earthquakes with $M > 6.3$. Among the $M > 6.3$ earthquakes, we pay a particular attention to a huge M 6.7 earthquake (Nagano-ken hokubu, west of Nagano prefecture) on 22 November 2014 with a depth of 5 km and epicenter located at 36.41°N, 137.53°E. This earthquake

(a)

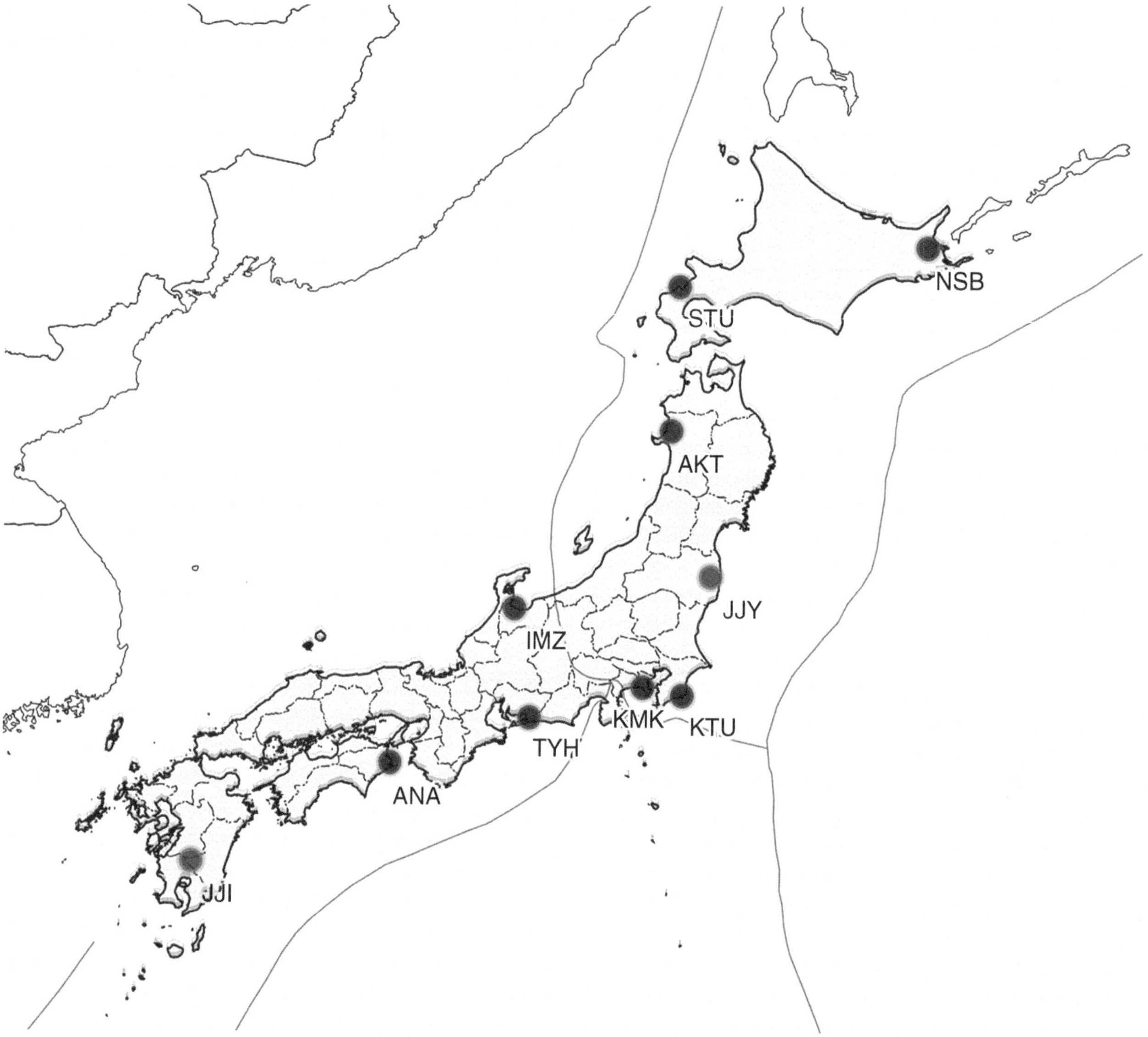

Figure 16.10 (a) The new VLF–LF network in Japan comprising eight stations: from north, NBC (Nakashibetsu), AKT (Akita), KTU (Katsuura), KMK (Kamakura), IMZ (Imizu), TYH (Toyohashi), and ANA (anan). The sites of two Japanese transmitters are also shown: JJY (40 kHz, Fukushima) and JJI (22.2 kHz, Miyazaki). (b) Signals from five transmitters (JJY and JJI, and NWC (19.8 kHz, Australia), NPM (21.4 kHz, Hawaii), and NLK (24.8 kHz, Jim Creek, USA) are received at each station, and the corresponding propagation paths are indicated.

(b)

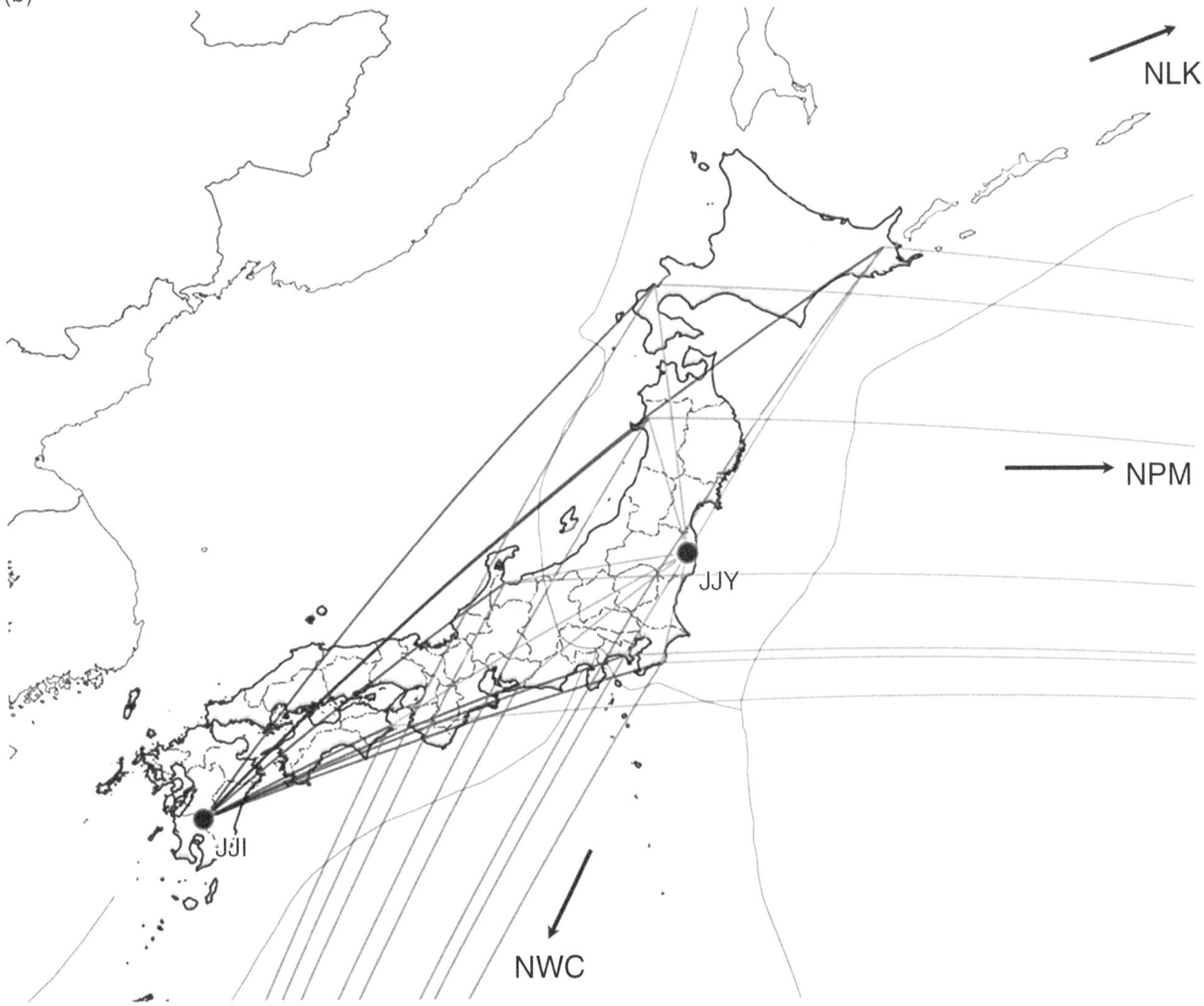

Figure 16.10 (Continued)

is inland and of the fault-type, similar to the 1995 Kobe earthquake, so that we can show the records of different propagation paths that may be useful for the detection of a precursor to this particular earthquake, as in Figure 16.12. That is, we have dealt with four propagation paths (JJY–IMZ, JJY–TYH, JJI–AKT, and JJI–STU) as shown in Figure 16.13. Each propagation path has two plots: the upper plot refers to the nighttime average amplitude (trend (TR) as compared with that during the previous 30 days) in units of standard deviation (σ), and the second, the fluctuation around the average value (again in units of σ). The period shown in Figure 16.13 is just 1 month, including the earthquake date toward the end of period, shown as a downward arrow at the top. A propagation anomaly is characterized by a decrease in trend (e.g., exceeding -2σ level) and an enhancement in dispersion [*Hayakawa et al.*, 2010]. Thus, we can detect significant decreases in trend to take place on 12 and 13 November for all of these propagation paths, as shown with circles in each TR plot. Those minimum values exceed -3σ (and sometimes -4σ), which are definitely a precursor to the earthquake on 22 November, indicating a lead time of about 10 days. Figure 16.13 is based on the average value during local nighttime, but Figure 16.14 illustrates the sequential plots of dP (phase) and dA (amplitude) (both residual from the monthly mean) of daily (diurnal) variation (time in UT) (nighttime is defined by an interval between two green vertical lines, and the average value during this nighttime is used in Figure 16.13) for propagation path JJY–IMZ: left panels refer to phase, and right, amplitude. The period is 9–22 November. Clear nighttime anomalies in amplitude are noticeable starting on 11 November, and culminating on 12 and 14 November, as shown in Figure 16.13. In

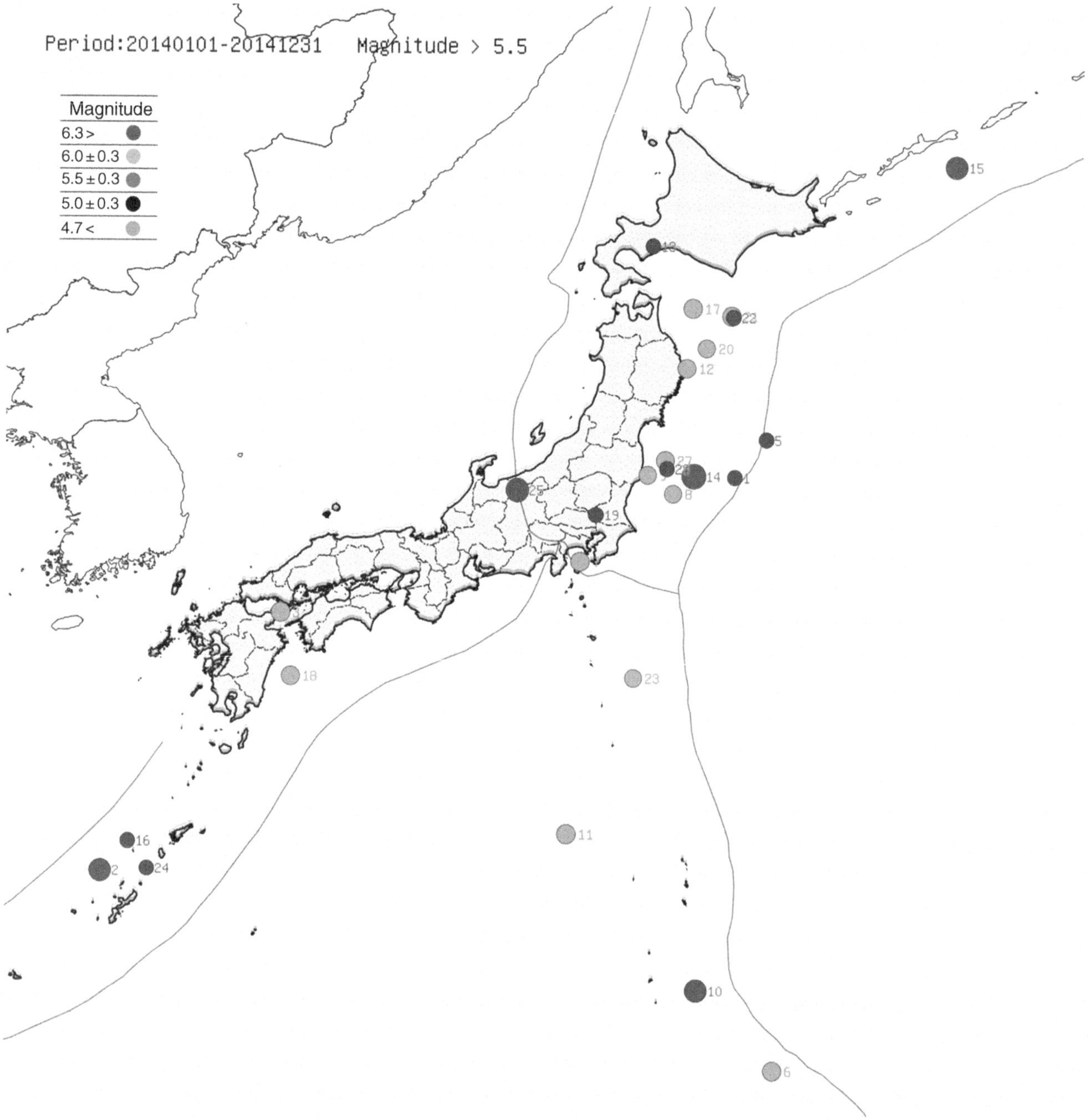

Figure 16.11 Distribution of earthquakes in and around Japan in 2014. Different colors refer to the earthquake magnitude, where red indicates earthquakes with $M > 6.3$. (*See electronic version for color representation.*)

contrast, anomalies in phase (dP) in the left panels seem to be long lasting, from 11 to 18 November, and to be not so clear as compared to the amplitude information.

Our earthquake prediction business normally provides a possible area for an impending earthquake. The success rate during the previous 4 years of operation is 65–70% with an accuracy of ±2–3 days in time, and maximum deviation of 50 km in position and of ±0.5 in magnitude.

16.7. OUTLOOK OF VLF–LF METHOD

The main goal of this chapter was to distinguish local VLF–LF perturbations related to earthquakes from large-scale or global anomalies connected with atmospheric circulations, and magnetic and solar activities. Therefore, we investigated the sensitivity of the amplitude and phase of VLF–LF transmitter signals to the background variations in the signal caused by magnetic storms, proton flares, etc.

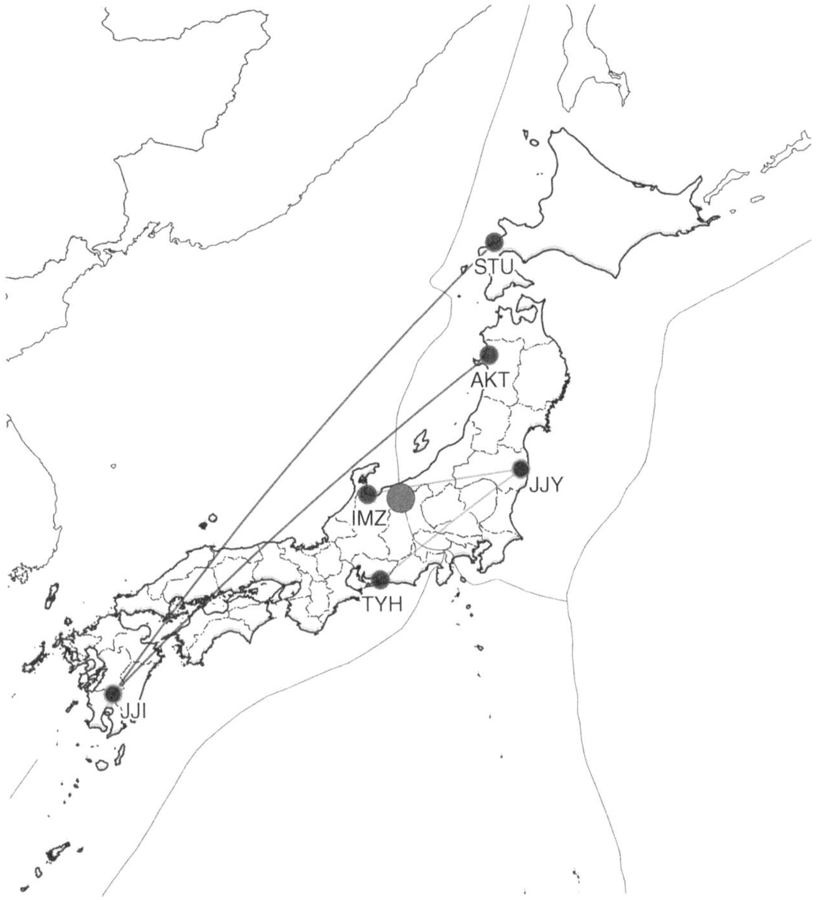

Figure 16.12 Propagation paths used for *M* 6.7 the Nagano-ku Hokubu earthquake (location indicated by red circle): two paths for JJY (JJY–TYH and JJY–IMZ) and two paths for JJI (JJI–AKT and JJI–STU). (*See electronic version for color representation.*)

Correlation of VLF–LF propagation anomalies with the D_{st} index was established, and it was found that VLF–LF anomalies are typical for the main phase of a magnetic storm if its sudden commencement happens in the nighttime. The next major disturbance to seismogenic effects seems to be caused by meteorological disturbances such as cyclones or typhoons. We found negative nighttime anomalies in the signal amplitude caused by tropical cyclones. Thus when analyzing VLF–LF data to search for seismogenic effects, we should be careful about the anomalies corresponding with geomagnetic and meteorological effects, which are the major challenge for seismogenic studies.

Concerning VLF–LF anomalies associated with earthquakes, we initially presented the first convincing evidence of subionospheric propagation anomalies for the 1995 Kobe earthquake, on the basis of a new analysis method of TTs. The TT method has been extended a great deal by *Sasmal and Chakrabarti* [2009], *Sasmal et al.* [2010], and *Ray et al.* [2010], who have obtained further statistical verification on the correlation of TT anomalies with earthquakes. On the basis of long-term observation in and around Japan, we have presented a significant statistical correlation between VLF–LF propagation anomalies (nighttime average amplitude) and shallow (depth < 40 km) earthquakes. Namely, an anomaly is characterized by a decrease in nighttime amplitude and an enhancement in amplitude fluctuation, about 1 week before an earthquake. Finally, we have described a newly established network in Japan, and presented a case study for the *M* 6.7 Nagano-ken hokubu earthquake of 22 November 2014.

Concerning VLF–LF network, the authors are please to note that this kind of observation network for monitoring subionospheric signals from VLF–LF transmitters is becoming a world standard for short-term earthquake prediction, as testified by worldwide establishment of VLF networks, including in Europe [*Biagi et al.*, 2004], India [*Singh et al.*, 2004; *Chakrabarti et al.*, 2009], Russia [*Rozhnoi et al.*, 2007], and Brazil [*Raulin et al.*, 2009].

VLF–LF radio waves can monitor the bottomside of ionosphere, but what about the situation in the upper ionosphere, such as the F layer? The correlation of subionospheric VLF–LF anomalies with earthquakes has been

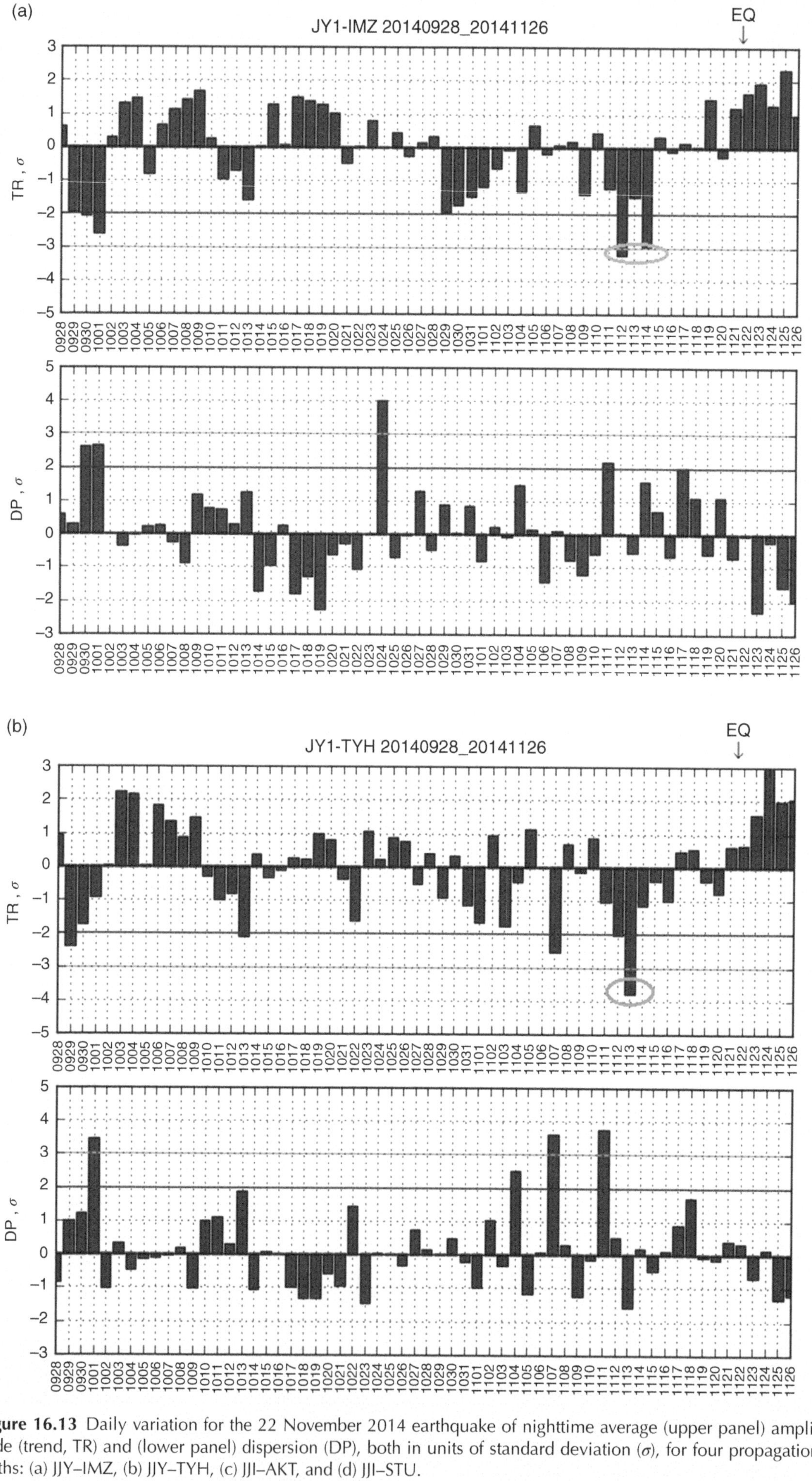

Figure 16.13 Daily variation for the 22 November 2014 earthquake of nighttime average (upper panel) amplitude (trend, TR) and (lower panel) dispersion (DP), both in units of standard deviation (σ), for four propagation paths: (a) JJY–IMZ, (b) JJY–TYH, (c) JJI–AKT, and (d) JJI–STU.

(c)

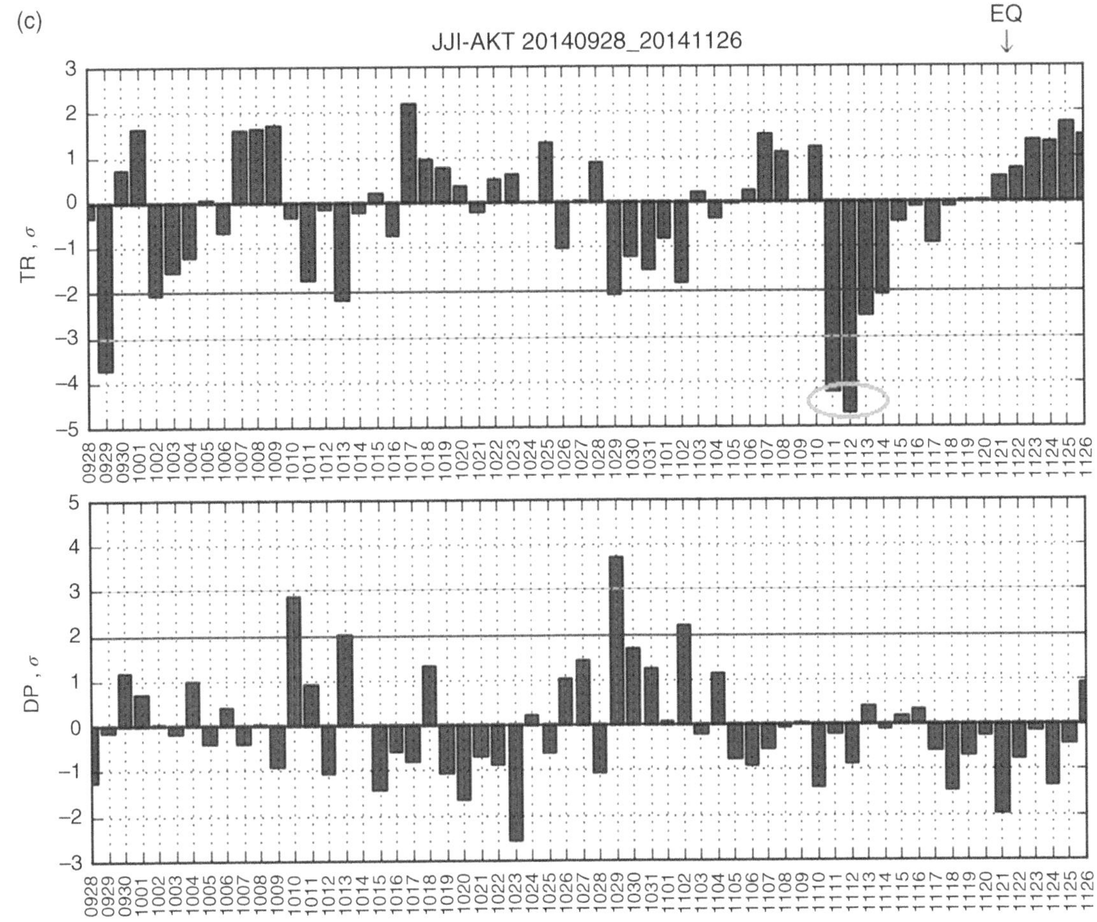

(d)

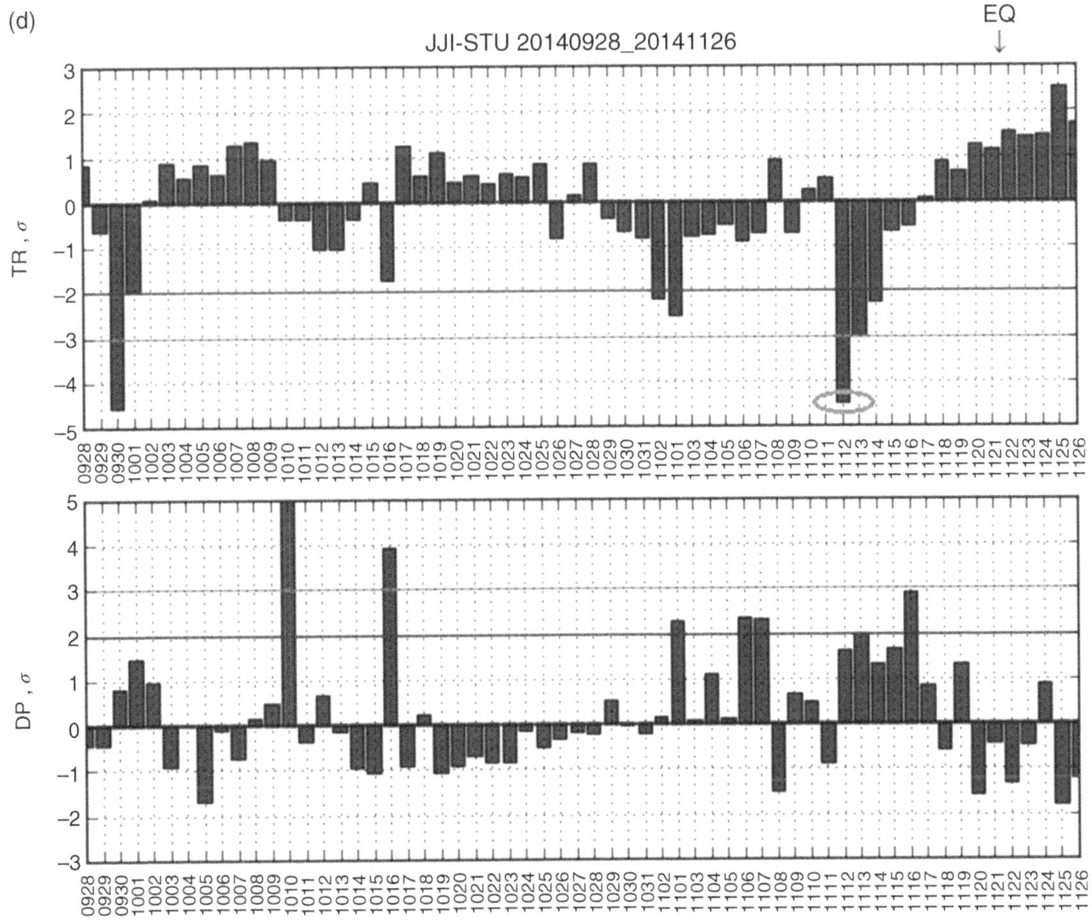

Figure 16.13 (Continued)

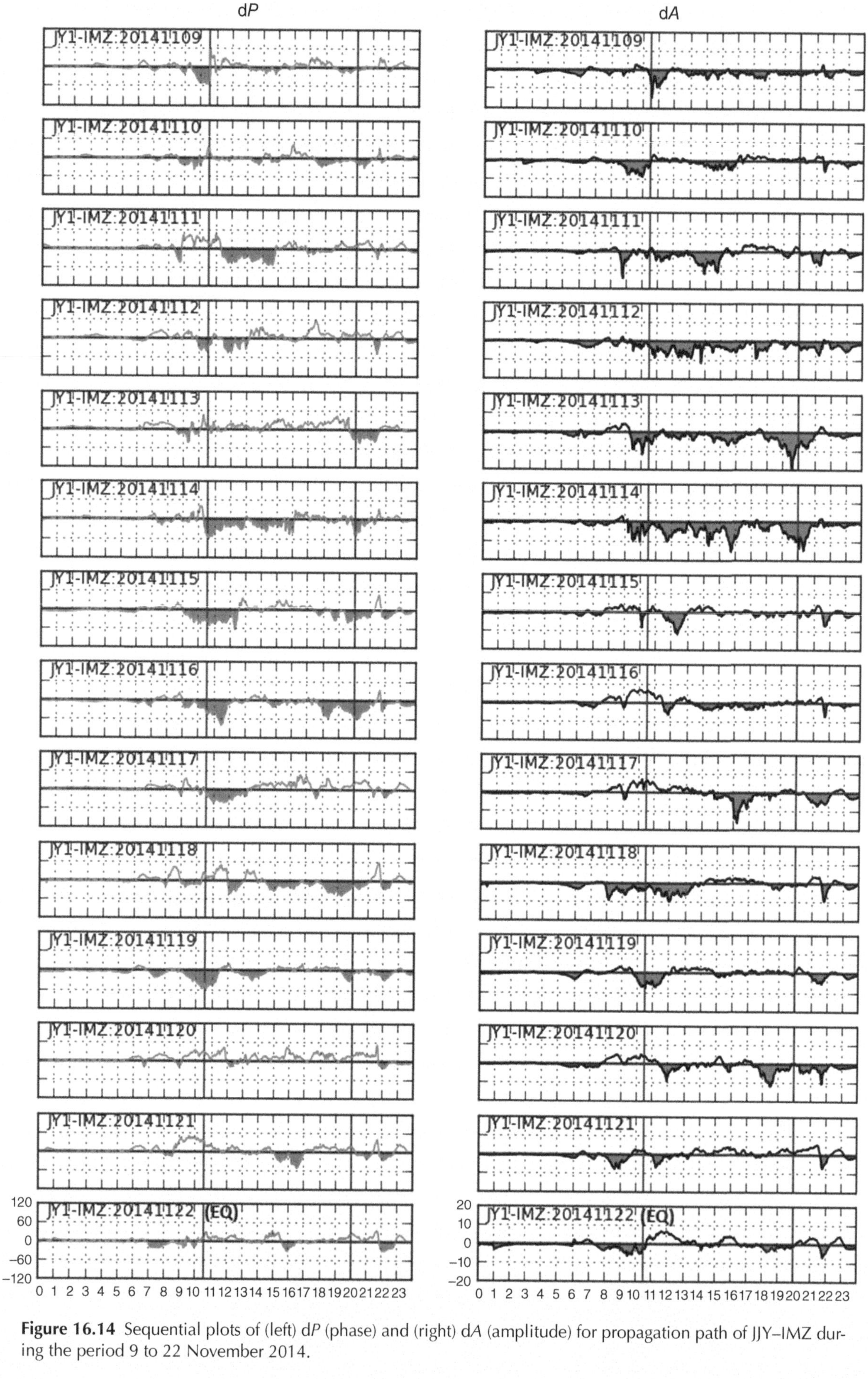

Figure 16.14 Sequential plots of (left) d*P* (phase) and (right) d*A* (amplitude) for propagation path of JJY–IMZ during the period 9 to 22 November 2014.

established by many studies, as described in this chapter, and nearly the same situation is indicated in the upper F layer. In addition to various case studies, *Liu* [2009] and *Liu et al.* [2013] have presented a clear correlation of F-layer density anomalies with earthquakes on the basis of bottomside sounding (foF2 measurement) and also TEC measurement. In this sense, a general consensus is forming that the ionosphere, not only in the lowest part, but also in the upper F layer, is perturbed about 1 week before an earthquake, even though the reasons why and how such ionospheric perturbations are formed prior to an earthquake are not well understood.

We have indicated what is required in the future if we want to understand the reasons why and how ionospheric perturbations are formed prior to an earthquake [*Pulinets and Boyarchuk*, 2004; *Molchanov and Hayakawa*, 2008; *Sorokin et al.*, 2015]. In the analysis of subionospheric VLF–LF data we have been engaged in the analysis of nighttime amplitude and phase as an average. Of course, we sometimes analyzed the fluctuations in amplitude and phase, which has yielded an important finding on the role of AGWs in the frequency range of a few tens of minutes to a few hours before an earthquake [*Molchanov and Hayakawa*, 2008; *Rozhnoi et al.*, 2013]. The next step we have to take is the detailed "temporal and spatial evolution of seismo-ionospheric perturbations" by making full use of our new VLF–LF network data (not only amplitude, but also phase) in Japan. The multiple stations of the network enable us to locate an ionospheric perturbation, and also to estimate its dimension and degree of perturbation. We further trace its movement, if it moves. Of course, we would like to utilize the information on phase effectively in this process. Furthermore, simultaneous information on the upper F layer would be of great importance in understanding whether the effect of lower ionospheric perturbation propogates upwards toward the F region or not. We plan to use ionosonde sounding data and oblique reflection data from the VHF transmitters in Japan. When interpreting the spatiotemporal evolution of the observed seismo-ionospheric perturbation, a comparison will be added with the theoretical modeling with the use of different numerical methods such as FDTD (finite difference time domain), wave-hop method, etc.

Further interdisciplinary coordination will be of great importance for the better understanding of the mechanism of LAIC. Information on the ionosphere (both in the D/E region and upper F region) can be obtained with different kinds of observations, but little information has been available on the atmosphere from the ground surface to the D/E region, although *Sun et al.*'s [2011] study was the first attempt in this direction. Thus it is highly desirable to coordinate observations of the atmosphere and ground-surface motion [e.g., *Chen et al.*, 2011; *Kamiyama et al.*, 2014], together with chemical data such as radon, ions, etc.

ACKNOWLEDGMENTS

Some of the work in this chapter have been made possible within the framework of the former NASDA's Frontier Project, to which we are grateful.

REFERENCES

Asai, S., Yamamoto, S., Kasahara, Y., Hobara, Y., Inaba, T., and Hayakawa, M. (2011), Measurement of Doppler shifts of short-distance subionospheric LF transmitter signals and seismic effects, *J. Geophys. Res.*, doi: 10.1029/2010JA016055

Belrose, J. S. and Thomas, L. (1968), Ionization in the middle latitude D-region associated with geomagnetic storms, *J. Atmos. Terr. Phys.*, *30*, 1397–1413.

Biagi, P. F., Piccolo, R., Castellana, L., Maggipinto, T., Ermini, A., Martellucci, S., Bellecci, C., Perna, G., Capozzi, V., Molchanov, O. A., Hayakawa, M., and Ohta, K. (2004), VLF/LF radio signals collected at Bari (South Italy): a preliminary analysis on signal anomalies associated with earthquakes, *Nat. Hazards Earth Syst. Sci.*, *4*, 685–689.

Bishop, R. L. and Straus, P. R. (2006), Characterizing ionospheric variations in the vicinity of hurricanes and typhoons using GPS occultation measurements, *AGU Fall Meeting*, San Francisco, 11–15 December, *Eos (Trans. Am. Geophys. Union)*, *87*(Suppl.), SA33B-0276.

Cervone, G., Maekawa, S., Singh, R. P., Hayakawa, M., Kafatos, M., and Shvets, A. (2006), Surface latent heat flux and nighttime LF anomalies prior to the *Mw*=8.3 Tokachi-Oki earthquake, *Nat. Hazards Earth Syst. Sci.*, *6*, 109–114.

Chakrabarti, S. K. (Ed.) (2010), *Propagation Effects of Very Low Frequency Radio Waves, Conference Proceedings*, Vol. 1286, 362 pp, American Institute of Physics.

Chakrabarti, S. K., Sasmal, S., and Chakrabarti, S. (2009), Ionospheric anomaly due to seismic activities—Part 2: Evidence from D-layer preparation and disappearance times, *Nat. Hazards Earth Syst. Sci.*, *10*, 1751–1757.

Chen, C. H., Yeh, T. K., Liu, J. Y., Wang, C. H., Wen, S., Yen, H. Y., and Chang, S. H. (2011), Surface deformation and seismic rebound: implications and applications, *Surv. Geophys.*, *32*(3), 291–313, doi: 10.1007/s10712-011-9117-3

Craig, J. R. and McCormick, R. J. (2006), Remote sensing of the upper atmosphere by VLF, in, M. Fullkrung, E. A. Mareev, and M. J. Rycroft (eds), *Elves and Intense Lightning Discharges*, pp. 167–190, NATO Science Series.

Dowden, R. L. and Adams, C. D. D. (1988), Phase and amplitude echoes from lightning-induced electron precipitation, *J. Geophys. Res.*, *93*(11), 543–550.

Freund, F. T. (2009), Stress-activated positive hole charge carriers in rocks and the generation of pre-earthquake signals, in M. Hayakawa (ed.), *Electromagnetic Phenomena Associated with Earthquakes*, pp. 41–96, Transworld Research Network, Trivandrum, India.

Gokhberg, M. B., Gufeld, I. L., Rozhnoi, A. A., Marenko, V. F., Yampolsky, V. S., and Ponomarev, E. A. (1989), Study of seismic influence on the ionosphere by super long wave probing of the Earth-ionosphere waveguide, *Phys. Earth Planet. Inter.*, *57*, 64–67.

Gufeld, I. L., Rozhnoi, A. A., Tyumensev, S. N., Sherstuk, S. V., and Yampolsky, V. S. (1992), Radiowave disturbances in period to Rudber and Rachinsk earthquakes, *Phys. Solid Earth*, *28*(3), 267–270.

Hayakawa, M. (ed.) (1999), *Atmospheric and Ionospheric Electromagnetic Phenomena Associated with Earthquakes*, Terra Scientific Publishing, Tokyo, 996 pp.

Hayakawa, M. (2007), VLF/LF radio sounding of ionospheric perturbations associated with earthquakes, *Sensors*, *7*, 1141–1158.

Hayakawa, M. (ed.) (2009a), *Electromagnetic Phenomena Associated with Earthquakes*, Transworld Research Network, Trivandrum, India, 279 pp.

Hayakawa, M. (2009b), Lower ionospheric perturbations associated with earthquakes, as detected by subionospheric VLF/LF radio waves, in M. Hayakawa (ed.), *Electromagnetic Phenomena Associated with Earthquakes*, pp. 137–185, Transworld Research Network, Trivandrum, India.

Hayakawa, M. (2011), Probing the lower ionospheric perturbations associated with earthquakes by means of subionospheric VLF/LF propagation, *Earthquake Sci.*, *24*(6), 609–637.

Hayakawa, M. (ed.) (2012), *The Frontier of Earthquake Prediction Studies*, Nihon- senmontosho-Shuppan, Tokyo, 794 pp.

Hayakawa, M. (ed.) (2013), *Earthquake Prediction Studies: Seismo Electromagnetics*, Terra Scientific Publishing, Tokyo, 168 pp.

Hayakawa, M. (2015), *Earthquake Prediction with Radio Techniques*, John Wiley & Sons, Singapore, 294 pp.

Hayakawa, M. and Hobara, Y. (2010), Current status of seismo-electromagnetics for short-term earthquake prediction, *Geomatics Nat. Hazards Risk*, *1*(2), 115–155.

Hayakawa, M. and Molchanov, O. A. (eds) (2002), *Seismo-Electromagnetics: Lithosphere-Atmosphere-Ionosphere Coupling*, Terra Scientific Publishing, Tokyo, 477 pp.

Hayakawa, M., Molchanov, O. A., Ondoh, T., and Kawai, E. (1996), The precursory signature effect of the Kobe earthquake on VLF subionospheric signals, *J. Comm. Res. Lab., Tokyo*, *43*, 169–180.

Hayakawa, M., Molchanov, O. A., and NASDA/UEC team (2004), Summary report of NASDA's earthquake remote sensing frontier project, *Phys. Chem. Earth*, *29*, 617–625.

Hayakawa, M., Ohta, K., Maekawa, S., Yamauchi, T., Ida, Y., Gotoh, T., Yonaiguchi, N., Sasaki, H., and Nakamura, T. (2006), Electromagnetic precursors to the 2004 Mid Niigata Prefecture earthquake, *Phys. Chem. Earth*, *31*, 356–364.

Hayakawa, M., Horie, T., Yoshida, M., Kasahara, Y., Muto, F., Ohta, K., and Nakamura, T. (2008), On the ionospheric perturbation associated with the 2007 Niigata Chuetsu-oki earthquake, as seen from subionospheric VLF/LF network observations, *Nat. Hazards Earth Syst. Sci.*, *8*, 573–576, doi: 10.5194/nhess-8-573-2008

Hayakawa, M., Kasahara, Y., Nakamura, T., Muto, F., Horie, T., Maekawa, S., Hobara, Y., Rozhnoi, A. A., Solovieva, M., and Molchanov, O. A. (2010), A statistical study on the correlation between lower ionospheric perturbations as seen by subionospheric VLF/LF propagation and earthquakes, *J. Geophys. Res.*, *115*, A09305, doi: 10. 1029/2009JA015143

Hayakawa, M., Kasahara, Y., Nakamura, T., Hobara, Y., Rozhnoi, A., Solovieva, M., Molchanov, O. A., and Korepanov, V. (2011), Atmospheric gravity waves as a possible candidate for seismo-ionospheric perturbations, *J. Atmos. Electr.*, *31*(2), 129–140.

Hayakawa, M., Hobara, Y., Yasuda, Y., Yamaguchi, H., Ohta, K., Izutsu, J., and Nakamura, T. (2012a), Possible precursor to the March 11, 2011, Japan earthquake: ionospheric perturbations as seen by subionospheric very low frequency/low frequency propagation, *Ann. Geophys.*, *55*(1), 95–99, doi: 10.4401/ag-5357

Hayakawa, M., Kasahara, Y., Endoh, T., Hobara, Y., and Asai, S. (2012b), The observation of Doppler shifts of subionospheric LF signal in possible association with earthquakes, *J. Geophys. Res.*, *117*, A09304, doi: 10.1029/2012JA017752

Hayakawa, M., Rozhnoi, A., Solovieva, M., Hobara, Y., Ohta, K., Schekotov, A., and Fedorov, E. (2013a), The lower ionospheric perturbation as a precursor to the 11 March 2011 Japan earthquake, *Geomatics Nat. Hazards Risk*, *4*(3), 275–287, doi: org/10.1080/19475705.2012.751938

Hayakawa, M., Hobara, Y., Rozhnoi, A., Solovieva, M., Ohta, K., Izutsu, J., Nakamura, T., and Kasahara, Y. (2013b), The ionospheric precursor to the 2011 March 11 earthquake based upon observations obtained from the Japan-Pacific subionospheric VLF/LF network, *Terr. Atmos. Ocean. Sci.*, *24*(3), 393–408, doi: 10.3319/TAO.2012.12.14.01(AA)

Horie, T., Yamauchi, T., Yoshida, M., and Hayakawa, M. (2007), The wave-like structures of ionospheric perturbation associated with Sumatra earthquake of 26 December 2004, as revealed from VLF observation in Japan of NWC signals, *J. Atmos. Solar Terr. Phys.*, *69*, 1021–1028.

Huang, Y. N., Cheng, K., and Chen, S. W. (1985), On the detection of acoustic-gravity waves generated by typhoon by use of real time HF Doppler frequency shift sounding system, *Radio Sci.*, *20*, 897–906, doi: 10.1029/RS020i004p00897

Inan, U. S., Carpenter, D. L., Helliwell, R. A., and Katsufrakis, J. P. (1985), Subionospheric VLF/LF phase perturbations produced by lightning whistler induced particle precipitation, *J. Geophys. Res.*, *90*, 7457–7469.

Isaev, N. V., Sorokin, V. M., Chmyrev, V. M., and Serebryakova, O. N. (2002), Ionospheric electric fields related to sea storms and typhoons, *Geomagn. Aeron.*, *42*, 638–643.

Kamiyama, M., Sugito, M., Kuse, M., Schekotov, A., and Hayakawa, M. (2014), On the precursors to the 2011 Tohoku earthquake: crustal movements and electromagnetic signatures, *Geomatics Nat. Hazards Risk*, doi: 10.1080/19475705.2014.937773

Kasahara, Y., Muto, F., Horie, T., Yoshida, M., Hayakawa, M., Ohta, K., Rozhnoi, A., Solovieva, M., and Molchanov, O. A. (2008), On the statistical correlation between the ionospheric perturbations as detected by subionospheric VLF/LF propagation anomalies and earthquakes, *Nat. Hazards Earth Syst. Sci.*, *8*, 653–656.

Kasahara, Y., Nakamura, T., Hobara, Y., Hayakawa, M., Rozhnoi, A., Solovieva, M., and Molchanov, O. A. (2010), A statistical study on the AGW modulation in subionospheric VLF/LF propagation data and consideration of the generation mechanism of seismo-ionospheric perturbations, *J. Atmos. Electr.*, *30*(2), 103–112.

Kikuchi, T. (1981), VLF phase anomalies associated with substorm, Mem. *Nat. Inst. Polar Res., Tokyo, Spec. Iss.*, *18*, 3–23.

Kleimenova, N. G., Kozyreva, O. V., Rozhnoi, A. A., and Solovieva, M. S. (2004), Variations in the VLF signal parameters on the Australia-Kamchatka radio path during magnetic storms, *Geomagn. Aeron.*, *44*, 385–393.

Klimenko, M. V., Klimenko, V. V., Zakharenkova, I. E., and Karpov, I. V. (2012), Modeling of local disturbance formation in the ionosphere electron concentration before strong earthquakes, *Earth Planets Space*, *64*, 441–450.

Korepanov, V., Hayakawa, M., Yampolski, Y., and Lizunov, G. (2009), AGW as a seismo-ionospheric coupling responsible agent, *Phys. Chem. Earth, Parts A/B/C*, *34*(6–7), 485–495.

Kuo, C. L., Huba, J. D., Joyce, G., and Lee, L. C. (2011), Ionosphere plasma bubbles and density variations induced by pre-earthquake rock currents and associated surface charges, *J. Geophys. Res.*, *116*, doi: 10.1029/2011JA016628.

Lin, J. W. (2012), Study of ionospheric anomalies due to impact of typhoon using principal component analysis and image processing, *J. Earth Syst. Sci.*, *121*, 1001–1010.

Liu, J. Y. (2009), Earthquake precursors observed in the ionospheric F-region, in M. Hayakawa (ed.), *Electromagnetic Phenomena Associated with Earthquakes*, pp. 187–204, Transworld Research Network, Trivandrum, India.

Liu, J. Y., Chen, C. H., and Tsai, H. F. (2013), A statistical study on seismo-ionospheric precursors of the total electron content associated with 146 $M \geq 6.0$ earthquakes in Japan during 1998–2011, in M. Hayakawa (ed.), *Earthquake Prediction Studies: Seismo Electromagnetics*, pp. 17–30, Terra Scientific Publishing, Tokyo.

Liu, Y. M, Wang, J. S., and Suo, Y. C. (2006), Effects of typhoon on the ionosphere, *Adv. Geosci.*, *29*, 351–360, http://www.adv-geosci.net/29/351/2006/

Maekawa, S., Horie, T., Yamauchi, T., Sawaya, T., Ishikawa, M., Hayakawa, M., and Sasaki, H. (2006), A statistical study on the effect of earthquakes on the ionosphere, based on the subionospheric LF propagation data in Japan, *Ann. Geophys.*, *24*, 2219–2225.

Mikhailova, G. A., Mikhailov, Y. M., and Kapustina, O. V. (2000), ULF/VLF electric fields in the external ionosphere over powerful typhoons in Pacific Ocean, *Geomagn. Aeron.*, *2*, 153–158.

Miyaki, K., Hayakawa, M., and Molchanov, O. A. (2002), The role of gravity waves in the lithosphere-ionosphere coupling, as revealed from the subionospheric LF propagation data, in M. Hayakawa and O. A. Molchanov (eds), *Seismo-Electromagnetics (Lithosphere–Atmosphere–Ionosphere Coupling)*, pp. 229–232. Terra Scientific Publishing, Tokyo.

Molchanov, O. A. and Hayakawa, M. (1998), Subionospheric VLF signal perturbations possibly related to earthquakes, *J. Geophys. Res.*, *103*, 17489–17504.

Molchanov, O. A. and Hayakawa, M. (2008), *Seismo Electromagnetics and Related Phenomena: History and Latest Results*, Terra Scientific Publishing, Tokyo, 189 pp.

Molchanov, O. A., Mazhaeva, O. A., Goliavin, A. N., and Hayakawa, M. (1993), Observations by the intercosmos-24 satellite of ELF-VLF electromagnetic emissions associated with earthquakes, *Ann. Geophys.*, *11*, 431–440.

Molchanov, O. A., Hayakawa, M., Ondoh, T., and Kawai, E. (1998), Precursory effects in the subionospheric VLF signals for the Kobe earthquake, *Phys. Earth Planet. Inter.*, *105*, 239–248.

Molchanov, O. A., Hayakawa, M., and Miyaki, K. (2001), VLF/LF sounding of the lower ionosphere to study the role of atmospheric oscillations in the lithosphere-ionosphere coupling, *Adv. Polar Upper Atmos. Res., Tokyo*, *15*, 146–158.

Muto, F., Horie, T., Yoshida, M., Hayakawa, M., Rozhnoi, A., Solovieva, M., and Molchanov, O. A. (2009a), Ionospheric perturbations related to the Miyagi-oki earthquake on 16 August 2005, as seen from Japanese VLF/LF subionospheric propagation network, *Phys. Chem. Earth, Parts A/B/C*, *34*(6–7), 449–455.

Muto, F., Kasahara, Y., Hobara, Y., Hayakawa, M., Rozhnoi, A., Solovieva, M., and Molchanov, O. A. (2009b), Further study on the role of atmospheric gravity waves on the seismo-ionospheric perturbations as detected by subionospheric VLF/LF propagation. *Nat. Hazards Earth Syst. Sci.*, *9*, 1111–1118, doi: 10.5194/nhess-9-1111-2009

Nakamura, T., Korepanov, V., Kasahara, Y., Hobara, Y., and Hayakawa, M. (2013), An evidence on the lithosphere-ionosphere coupling in terms of atmospheric gravity waves on the basis of a combined analysis of surface pressure, ionospheric perturbations and ground-based ULF variations, *J. Atmos. Electr.*, *33*(1), 53–68.

Nickolaenko, A. P. and Hayakawa, M. (1995), Heating of the lower ionosphere electrons by electromagnetic radiation of lightning discharges, *Geophys. Res. Lett.*, *22*, 3015–3018.

Polyakova, A. S. and Perevalova, N. P. (2011), Investigation into impact of tropical cyclones on the ionosphere using GPS sounding and NCEP/NCAR Reanalysis data, *Adv. Space Res.*, *48*, 1196–1210.

Potemra, T. A. and Rosenberg, T. J. (1973), VLF propagation disturbances and electron precipitation at mid-latitudes, *J. Geophys. Res.*, *78*, 1572–1580.

Pulinets, S. A. and Boyarchuk, K. (2004), *Ionospheric Precursors of Earthquakes*, Springer, Berlin, 315 pp.

Pulinets, S. A. and Ouzounov, D. (2011), Lithosphere–atmosphere–ionosphere coupling (LAIC) model—a unified concept for earthquake precursors validation, *J. Asian Earth Sci.*, *41*, 371–382.

Raulin, J. P., David, P., Hadano, R., Saraiva, A. C. V., Correia, E., and Kaufman, P. (2009), The South America VLF NETwork (SAVNET), *Earth, Moon Planets*, *104*, 247–261.

Ray, S., Chakrabarti, S. K., Sasmal, S., and Choudhury, A. K. (2010), Correlations between the anomalous behavior of the ionosphere and the seismic events for VTX-MALDA VLF propagation, in S. Chakrabarti (ed.), *Propagation Effects of Very Low Frequency Radio Waves*, Conference Proceedings Vol. *1286*, 298–308, American Institute of Physics.

Rozhnoi, A. A., Kleimenova, N. G., Kozyreva, O. V., Molchanov, O. A., and Solovieva, M. S. (2003), Nighttime midlatitude variations in the LF (40 kHz) signal parameters and Pi3 geomagnetic pulsations, *Geomagn. Aeron.*, *43*(4), 518–525.

Rozhnoi, A., Solovieva, M. S., Molchanov, O. A., and Hayakawa, M. (2004), Middle latitude LF (40 kHz) phase variations associated with earthquakes for quiet and disturbed geomagnetic conditions. *Phys. Chem. Earth*, *29*, 589–598, doi: 10.1016/j.pce.2003.08.061

Rozhnoi, A. A., Solovieva, M. S., Molchanov, O. A., Hayakawa, M., Maekawa, S., and Biagi, P. F. (2006), Sensitivity of LF signal to global ionosphere and atmosphere perturbations in the network of stations, *Phys. Chem. Earth*, *31*, 409–415.

Rozhnoi, A., Solovieva, M., Molchanov, O., Biagi, P. F., and Hayakawa, M. (2007), Observation evidences of atmospheric gravity waves induced by seismic activity from analysis of subionospheric LF signal spectra, *Nat. Hazards Earth Syst. Sci.*, *7*, 625–628, doi: 10.5194/nhess-7-625-2007

Rozhnoi, A., Solovieva, M., and Hayakawa, M. (2012a), Search for electromagnetic earthquake precursors by means of sounding of upper atmosphere-lower ionosphere boundary by VLF/LF signals, in M. Hayakawa (ed.), *The Frontier of Earthquake Prediction Studies*, pp. 652–677, Nihon-senmon-tosho-Shuppan, Tokyo.

Rozhnoi, A., Solovieva, M., Parrot, M., Hayakawa, M., Biagi, P. F., and Schwingenschuh, K. (2012b), Ionospheric turbulence from ground-based and satellite VLF/LF transmitter signal observations for the Simushir earthquake (November 15, 2006), *Ann. Geophys.*, *55*, 187–192, doi: 10.4401/ag-5190

Rozhnoi, A., Solovieva, M., and Hayakawa, M. (2013), VLF/LF signals method for searching of electromagnetic earthquake precursors, in M. Hayakawa (ed.), *Earthquake Prediction Studies: Seismo Electromagnetics*, pp. 31–48, Terra Scientific Publishing, Tokyo.

Rozhnoi, A., Solovieva, M., Levin, B., Hayakawa, M., and Fedun, V. (2014), Meteorological effects in the lower ionosphere as based on VLF/LF signal observations, *Nat. Hazards Earth Syst. Sci.*, *14*, 2671–2679, doi: 10.5194/nhess-14-2671-2014

Sasmal, S. and Chakrabarti, S. K. (2009), Ionospheric anomaly due to seismic activities, Part 1: Calibration of the VLF signal of VTX 18.2 kHz station from Kolkata and deviation during seismic events, *Nat. Hazards Earth Syst. Sci.*, *9*, 1403–1408.

Sasmal, S., Chakrabarti, S. K., and Chakrabarti, S. (2010), Studies of the correlation between ionospheric anomalies and seismic activities in the Indian subcontinent, in S. Chakrabarti (ed.), *Propagation Effects of Very Low Frequency Radio Waves*, Conference Proceedings Vol. *1286*, pp. 270–290, American Institute of Physics.

Sauer, H. H., Spjeldvik, W. N., and Steel, F. K. (1987), Relationship between longterm phase advances in high-latitude VLF wave propagation and solar energetic fluxes, *Radio Sci.*, *22*, 405–424.

Shvets, A. V., Hayakawa, M., Molchanov, O. A., and Ando, Y. (2004), A study of ionospheric response to regional seismic activity by VLF radio sounding, *Phys. Chem. Earth*, *29*, 627–637.

Singh, V., Singh, B., Hayakawa, M., Kumar, M., Kushwah, V., and Singh, O. P. (2004), Nighttime amplitude decrease in 19.8 kHz NWC signals observed at Agra possibly caused by moderate seismic activities along the propagation path, *J. Atmos. Electr.*, *24*, 1–15.

Sorokin, V. and Hayakawa, M. (2013), Generation of seismic-related DC electric fields and lithosphere-atmosphere-ionosphere coupling, *Modern Appl. Sci.*, *7*(6), 1–25, doi: 10.5539/mas.v7n6p1

Sorokin, V. M., Yaschenko, A. K., Chmyrev, V. M., and Hayakawa, M. (2006), DC electric field amplification in the mid-latitude ionosphere over seismically active faults, *Phys. Chem. Earth*, *31*, 447–453.

Sorokin, V., Chemyrev, V., and Hayakawa, M. (2015), *Electrodynamic Coupling of Lithosphere-Atmosphere-Ionosphere of the Earth*, Nova Science Publishing. 326 pp.

Sun, Y. Y., Oyama, K. I., Liu, J. Y., Jhuang, H. K., and Cheng, C. Z. (2011), The neutral temperature in the ionospheric dynamo region and the ionospheric F region density during Wenchuan and Pingtung Double earthquakes, *Nat. Hazards Earth Syst. Sci.*, *11*, 1759–1768, doi: 10.5194/nhess-11-1759-2011.

Uyeda, S., Nagao, T., Hattori, K., Noda, Y., Hayakawa, M., Miyaki, K., Molchanov, O., Gladychev, V., Baransky, L., Schekotov, A., Belyaev, G., Fedorov, E., Pokhotelov, O., Andreevsky, S., Rozhnoi, A., Khabazin, Y., Gorbatikov, A., Gordeev, E., Chebrov, V., Lutikov, A., Yunga, S., Kosarev, G., and Surkov, V. (2002), Russian–Japanese complex geophysical observatory in Kamchatka for monitoring of phenomena connected with seismic activity, in M. Hayakawa and O. A. Molchanov (eds), *Seismo Electromagnetics (Lithosphere - Atmosphere - Ionosphere Coupling)*, pp. 413–419, Terra Scientific Publishing, Tokyo.

Vanina-Dart, L. B., Romanov, A. A., and Sharkov, E. A. (2011), Influence of a tropical cyclone on the upper ionosphere according to tomography sounding data over Sakhalin Island in November 2007, *Geomagn. Aeron.*, *51*, 774–782.

Watt, A. D. (1967), *VLF Radio Engineering*, Pergamon Press, Oxford, 703 pp.

Xiao, Z., Xiao, S. G., Hao, Y. Q., and Zhang, D. H. (2007), Morphological features of ionospheric response to typhoon, *J. Geophys. Res. Space Phys.*, *112*, A04304, doi: 10.1029/2006JA011671

Xu, G., Wan, W., She, C., and Du, L. (2008), The relationship between ionospheric total electron content (TEC) over East Asia and the tropospheric circulation around the Qinghai–Tibet Plateau obtained with a partial correlation method, *Adv. Space Res.*, *42*, 219–223.

Yamauchi, T., Maekawa, S., Horie, T., Hayakawa, M., and Soloviev, O. (2007), Subionospheric VLF/LF monitoring of ionospheric perturbations for the 2004 Mid-Niigata earthquake and their structure and dynamics, *J. Atmos. Solar Terr. Phys.*, *69*, 793–802.

Yoshida, M., Yamauchi, T., Horie, T., and Hayakawa, M. (2008), On the generation mechanism of terminator times in subionospheric VLF/LF propagation and its possible application to seismogenic effects, *Nat. Hazards Earth Syst. Sci.*, *8*, 129–134.

Zakharov, V. I. and Kunitsyn, V. E. (2012), Regional features of atmospheric manifestations of tropical cyclones according to groundbased GPS network data, *Geomagn. Aeron.*, *52*, 533–545, doi: 10.1134/S0016793212040160

17

Application of Total Electron Content Derived from the Global Navigation Satellite System for Detecting Earthquake Precursors

Jann-Yenq (Tiger) Liu[1], Katsumi Hattori[2], and Yuh-Ing Chen[3]

ABSTRACT

The time series of the total electron content (TEC) at a certain location derived by the local network of ground-based Global Navigation Satellite System (GNSS) receivers or extracted from the global ionosphere map (GIM) is useful for detecting seismo-ionospheric anomalies at the regional level. When the detected anomalies are similar to those repeatedly appearing before large earthquakes in the same region, it might be perceived as a temporal seismo-ionospheric precursor (SIP). To discriminate the possible SIPs from global effects (such as solar disturbances, magnetic storms, etc.), a global search for anomalies using GIM TEC data is an ideal approach. Spatial analysis simultaneously detects anomalies similar to the temporal SIP at each lattice and indicates the global distribution or pattern of the detected anomalies. When the detected anomalies specifically and continuously appear within the monitoring region, we can observe spatial SIPs of the GIM TEC. To further study the fine structure and dynamics of the observed SIPs, a dense network of ground-based GNSS receivers is required. By applying the residual minimization training neural network (RMTNN) tomographic approach to the TEC between the GNSS satellite and the network receivers, the three-dimensional fine structure of the ionospheric electron density can be obtained.

17.1. INTRODUCTION

Anomalous changes in the electron density and/or electromagnetic emissions in the ionosphere before large earthquakes have been studied intensively [*Hayakawa and Fujinawa*, 1994; *Hayakawa*, 1999, 2000; *Liu et al.*, 2000; *Hayakawa and Molchanov*, 2002; *Pulinets and Boyarchuk*, 2004]. *Liu et al.* [2001] pioneered applying the total electron content (TEC) derived from measurements of a local network of 13 ground-based global positioning system (GPS) receivers in Taiwan to detect the temporal variation and spatial distribution of anomalies associated with the *Mw* 7.6 Chi-Chi earthquake on 21 September 1999. Here, the TEC is defined as the integration of the electron density between a ground-based receiver and its associated GPS satellite. Usually, for the full constellation, a ground-based receiver can simultaneously record signals from more than ten GPS satellites [*Liu et al.*, 1996]. Thus, *Liu et al.* [2001] produced an image over the network area constructed by about 130 TEC points every 30 s. The GPS TEC at a monitoring point in central Taiwan (24°N, 121°E), which is zero dimensional in space, extracted from a time series of the images significantly decreased in the afternoon period on 17, 18 and 20 September 1999, i.e., 4, 3, and 1 days before the earthquake.

Many statistical analyses [*Chen et al.*, 2004; *Liu et al.*, 2004a,b, 2006] show an anomalous and significant decrease in ionospheric GPS TEC in the afternoon/evening period 1–6 days before $M \geq 5.0$ earthquakes in Taiwan. Therefore, in Taiwan, significant and persistent decreases

[1] *Institute of Space Science, National Central University, Taoyuan, Taiwan*

[2] *Graduate School of Science, Chiba University, Chiba, Japan*

[3] *Graduate Institute of Statistics, National Central University, Taoyuan, Taiwan*

Pre-Earthquake Processes: A Multidisciplinary Approach to Earthquake Prediction Studies, Geophysical Monograph 234, First Edition. Edited by Dimitar Ouzounov, Sergey Pulinets, Katsumi Hattori, and Patrick Taylor.

of the GPS TEC appearing in the afternoon period could be considered temporal seismo-ionospheric precursors (SIPs). Note that SIPs may vary from place to place in terms of their polarity characteristics (i.e., decrease or increase; negative or positive), local appearance time, lead day, duration, etc. Thus, to detect SIPs related to the TEC at a certain place, statistical analyses are required to first determine its associated characteristics. Since SIPs often expand widely in longitudinal direction, plots of the TEC along the monitoring point (i.e., latitude–time–TEC (LTT), one dimensional in space) are also constructed in order to confirm the temporal SIP detection. With the LTT plot, *Liu et al.* [2001] observed that the crest of the equatorial ionization anomaly (EIA) in the Northern Hemisphere significantly decreased on days 4, 3, and 1 before the Chi-Chi earthquake. Finally, the instant two-dimensional images of LLT maps revealed that the GPS TEC indeed significantly and specifically decreased over the epicenter on those 3 days the SIP was observed. Alternatively, many global phenomena, e.g., solar flares, magnetic storms, solar/lunar tides, etc., could also affect the ionospheric electron density and/or TEC [*Liu et al.*, 1996, 1999, 2004b, 2013; *Lin et al.*, 2003]. Thus, even when the anomalies detected meet the SIP characteristics very well [*Chen et al.*, 2015], they could still coincidently result from one of the global phenomena.

To discriminate SIPs from the effects of global features, the TEC in the global ionosphere map (GIM) derived from the global navigation satellite system (GNSS: GP, GLONAS, Galileo or Beidou systems) are ideally utilized. Currently, the GIM is routinely published for a 1 h or 2 h time interval with 1–2 days delay. The spatial resolutions of the GIM for ±87.5°N latitude and ±180°E longitude are 2.5° and 5°, respectively. Therefore, each map consists of 5183 (=71 × 73) grid points (lattices). Following *Liu et al.* [2001], the temporal and spatial SIPs from TEC in the GIM associated with the *M* 9.3 Sumatra-Andaman earthquake on 26 December 2004 [*Liu et al.*, 2010], the *Mw* 8.0 Wenchuan earthquake on 12 May 2008 [*Liu et al.*, 2009], and the *M* 7.0 Haiti earthquake on 12 January 2010 [*Liu et al.*, 2011] are reported. The approach of these studies is to select a lattice in the monitoring region within which to detect anomalies. If the anomalies detected at a certain point (zero-dimensional) meet the characteristics reached by statistical studies in the region, then temporal SIPs are being observed. Following a temporal study, plots of the LTT along the monitoring point (one-dimensional) as well as a global two-dimensional search on the distribution, the occurrence frequency, and the persistence of anomalies similar to the detected temporal SIPs would be conducted. Note that any global effect would simultaneously affect 5183 lattices, and by contrast SIPs should appear only at a specific small area, but be persistent. Therefore, if SIP-type anomalies are persistent and spatially specific within the monitoring region, this could indicate that a special SIP has been detected.

A massive *M* 9.0 earthquake occurred on Friday 11 March 2011, off the Pacific coast of the northeastern part of the Japanese mainland (Tohoku Region), causing devastating damage (hereafter "the 2011 Tohoku earthquake"). Here, we take the 2011 Tohoku earthquake as an example to demonstrate how to detect a temporal (zero-dimensional) SIP as well as one- and two-dimensional spatial SIPs by means of TEC of the GIM. Leading to a high-resolution three-dimensional structure of the ionospheric electron density, with the TEC derived from a dense ground-based GNSS receiver network in Japan [*Hirooka et al.*, 2011].

17.2. SEISMO-IONOSPHERIC PRECURSOR DETERMINATION FROM THE GIM TEC DATA

The 2011 Tohoku earthquake occurred at 14:46 JST (05:46 UTC) at 38°6.2′N, 142°51.6′E and with depth at 24 km. The planetary magnetic, K_p, index was < 2 and the disturbance storm time, D_{st}, index around −20 nT. Except for a moderate geomagnetic storm with a maximum depression −88 nT occurring on 1 March (http://wdc.kugi.kyoto-u.ac.jp), the geomagnetic condition was generally quiet 1–10 days before the earthquake. However, the 10.7 cm solar radio flux, the F10.7 index, monotonically increased from 105 sfu (solar flux unit, 1 sfu = 10^{-22} $Wm^{-2}Hz^{-1}$) on 1 March and reached a maximum of 160 sfu on 7 March. Figure 17.1 reveals a GIM image of the TEC at 04:00–0600 UT on 11 March 2011, the day of the *M* 9.0 Tohoku earthquake.

To observe temporal SIPs, the TECs at 38°N, 142°E are extracted from a sequence of GIM images for February and March of 2011. To detect anomalous signals of the GIM TEC variations, a quartile-based process is performed. At each time point, we compute the median $\tilde{M}$ of every successive 15 days of the TEC as well as find the deviation between that observed on the 16th day and the computed $\tilde{M}$. To provide the information about the deviation, we also calculate the first (or lower) and the third (or upper) quartiles, denoted by *LQ* and *UQ*, respectively. Note that assuming a normal distribution with mean m and standard deviation σ for the TEC, the expected values of *LQ* and *UQ* should be $m - 1.34\sigma$ and $m + 1.34\sigma$, respectively [*Klotz and Johnson*, 1983]. To make the criterion stringent we set the lower bound as $\mathrm{LB} = \tilde{M} - 1.5(\tilde{M} - \mathrm{LQ})$ and upper bound as $\mathrm{UB} = \tilde{M} + 1.5(\mathrm{UQ} - \tilde{M})$.

Therefore, the probability of a new TEC in the interval (LB, UB) is approximately 68%. The median together with the associated LB and UB then provide references for the TEC variations on the 16th day. When an observed TEC on the 16th day is not in the associated (LB–UB),

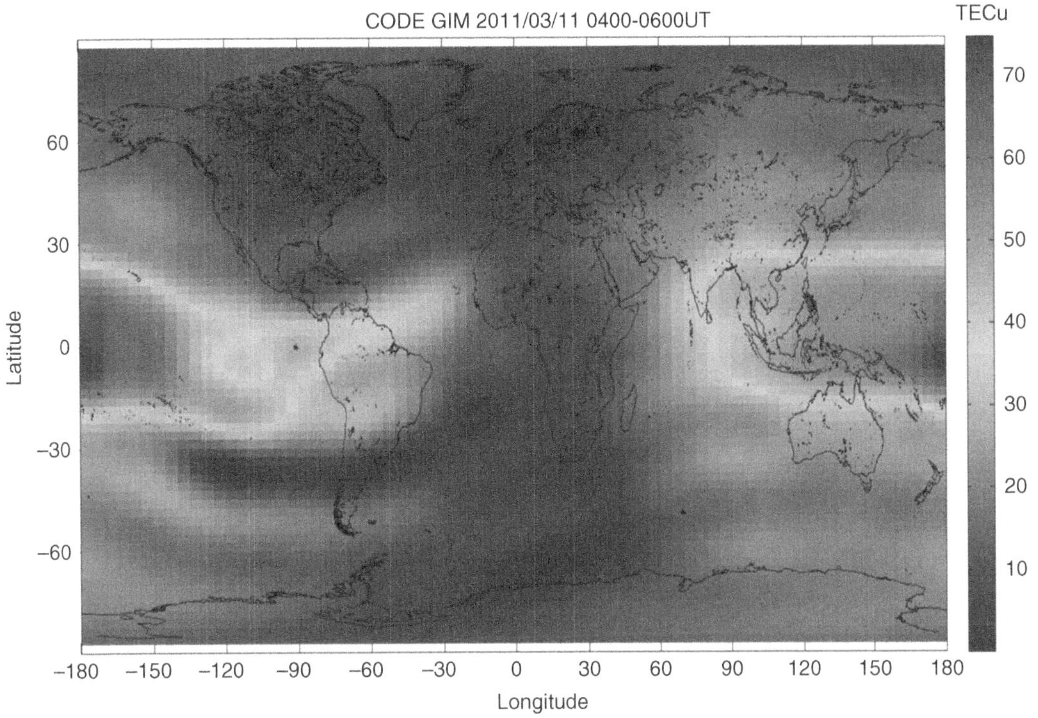

Figure 17.1 Snap shot of the total electron content in a global ionosphere map at 04:00–06:00 UT on 11 March 2011, the day of the Tohoku earthquake: the red star denotes the epicenter. (Source: Reproduced with permission of University of Berne.) (*See electronic version for color representation.*)

we declare an upper (increase) or lower (decrease) abnormal TEC signal. Since the GIM TEC time resolution is 2h, there are 12 data points per day. If more than one-third (4/12) of the upper or lower abnormal signals successively appear in one day, and the observed TEC is greater or smaller than the associated UB or LB, we then consider the upper or lower anomalous (positive or negative precursor) day as being detected. The probability of having a daily anomaly by observing four or more signals (negative or positive) is about 0.22, that of the successively appearing anomalies should be even less.

Figure 17.2 displays the GNSS TEC over the monitoring point at (38°N, 142°E) isolated from the GIM database 15 days before and after (from 25 February to 26 March 2011) the earthquake. Possibly due to the solar radiation enhancement (i.e. F10.7), positive anomalies in the GIM TEC appear during 1–13 March 2016. *Kon et al.* (2011) and *Liu et al.* [2013] conduct statistical analyses of TEC anomalies associated with $M \geq 6.0$ earthquakes during 1998–2011, and find significance for the positive TEC anomalies appearing 1–5 days before the earthquakes in Japan. Based on the these statistical results, the positive anomalies appearing 6–10 March, 1–5 days before, could be the temporal SIPs associated with the 2011 Tohoku Earthquake. However, the solar radio flux F10.7 index reaches a maximum of 160 sfu on 7 March, which could also result in the positive TEC anomalies appearing 1–5 days before the earthquakes. Figure 17.3 depicts the LTT plots within ±60°N along the monitoring longitude 140°E, again isolated from the GIM database 30 days before and 1 day after the earthquake (from 9 February to 13 March 2011). It can be seen in both Northern and Southern Hemispheres that the EIA crests reach their extreme maximum on 8 March 2011 the during the observation period.

To see if the positive anomalies of the GIM TEC in a 5-day period of 6–10 March mainly appear over the epicenter region or simply occur coincidently with the solar radiation enhancement on 7 March worldwide, a two-dimensional spatial analysis is conducted. Figure 17.4 displays that the distribution of the positive anomaly at each lattice repeatedly appears at various time points (1 point = 2 h) during the 5-day period. To quickly yield the distribution, a 1–30 days' moving reference for each lattice is constructed and utilized. The extreme maximum lattices with 4, 8, 12, and more repeating time points (times) occur worldwide, whereas those with 16, 20, 24, and more repeating times appear mainly in the Northern Hemisphere. It can be seen that the extreme maximum grids repeating up to 7–8 and even more times occur mainly around the epicenter and the conjugate point. The extreme maximum lattices with 28, 32, 36, and more repeats appear in relatively large latitudinal zones of the epicenter and its conjugate point. Finally, the extreme

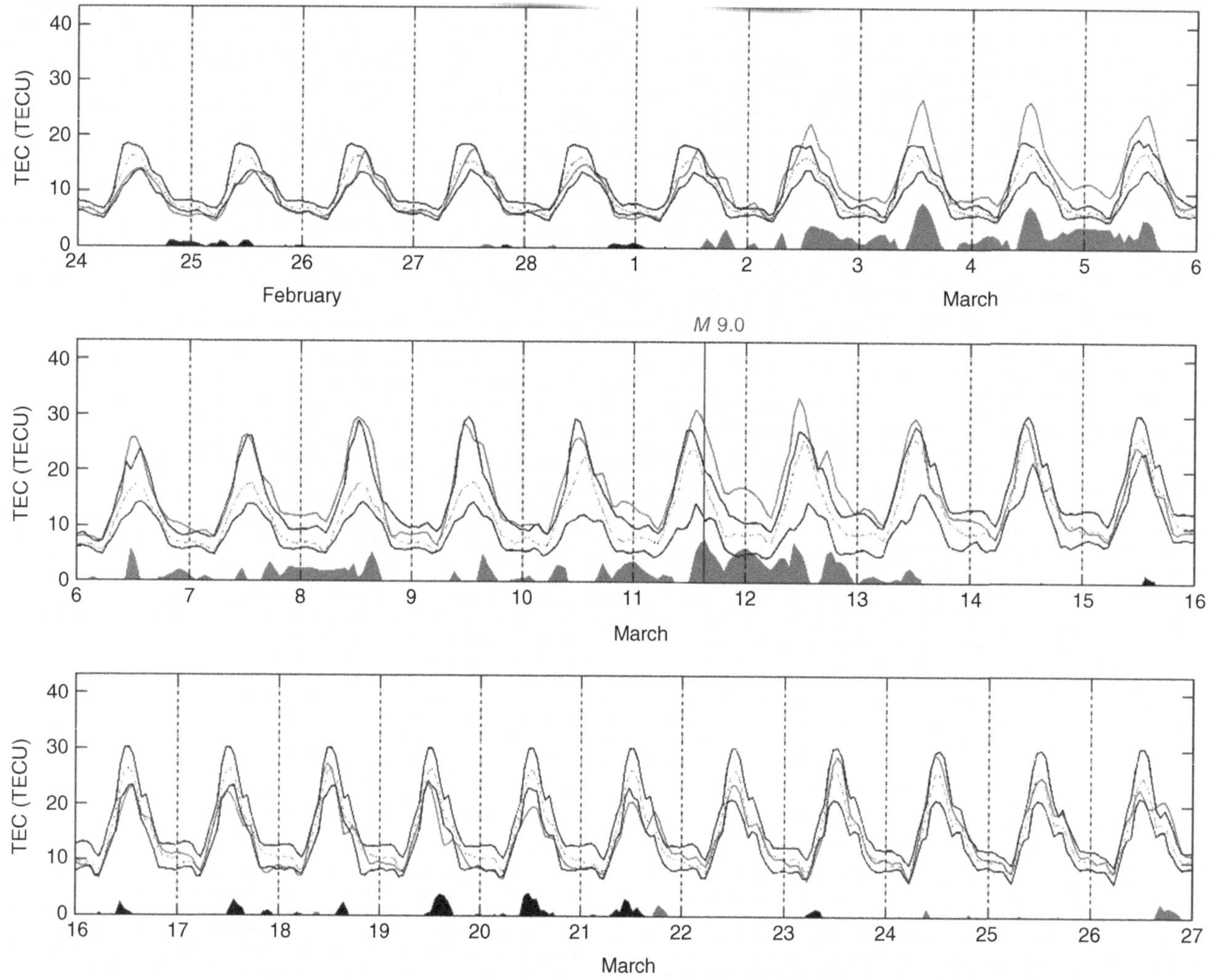

Figure 17.2 A time series of the total electron content (TEC) around the Tohoku epicenter extracted from the global ionosphere map (GIM) of 25 February to 26 March 2011. The red, gray-dashed, and two black curves denote the GIM TEC, the associated median value, and the upper and lower bounds (UB/LB), respectively. The LB and UB are constructed using the 1–15 previous days' moving median (M), lower quartile (LQ), and upper quartile (UQ). Here, $LB = \tilde{M} - 1.5(\tilde{M} - LQ)$ and $UB = \tilde{M} - LQ - 1.5(UQ - \tilde{M} - LQ)$. Red and black shaded areas denote differences of O–UB and LB–O, respectively, where O is the observed TEC. (*See electronic version for color representation.*)

maximums appearing specifically in a small area around the epicenter repeat up to 40 times and even in a sole lattice, which is about 2000 km northeast of the epicenter (i.e., 42 times; 70% = 42/60 of the 5-day period). Since the global 5183 lattices are simultaneously examined, the persistency (i.e. repeating times) can be used to discriminate anomalies induced by the local (earthquake) effect and those caused by the global (solar radiation, magnetic storm, etc.) effect. Results of this spatial analysis suggest that the positive anomalies are mostly likely the SIPs related to the 2011 Tohoku earthquake.

The probability of the anomaly successively (or continuously) appearing at a certain lattice should be even less than 3.3% (=1 day/30 day). Figure 17.4 reveals the global distribution of the 30-day extreme maximum, with various hours of continuous appearance (i.e. persistence) of some. It can be seen that three lattices do persist for 28 hours at about 800 km northeast from the epicenter during DOY 077-067 (7–8 March) of 2011. Figures 17.3 and 17.4 indicate that the positive anomalies of the GIM TEC specifically appearing around the epicenter are most likely associated with the 2011 Tohoku EQ.

17.3. THREE DIMENSIONAL TOMOGRAPHIC ANALYSIS

Understanding the behavior of ionospheric electron densities has recently become important for both academic purposes and practical applications. For investigation of the dynamics of ionospheric phenomena, three-dimensional ionospheric tomography is effective. The TEC observed by a ground-based receiver

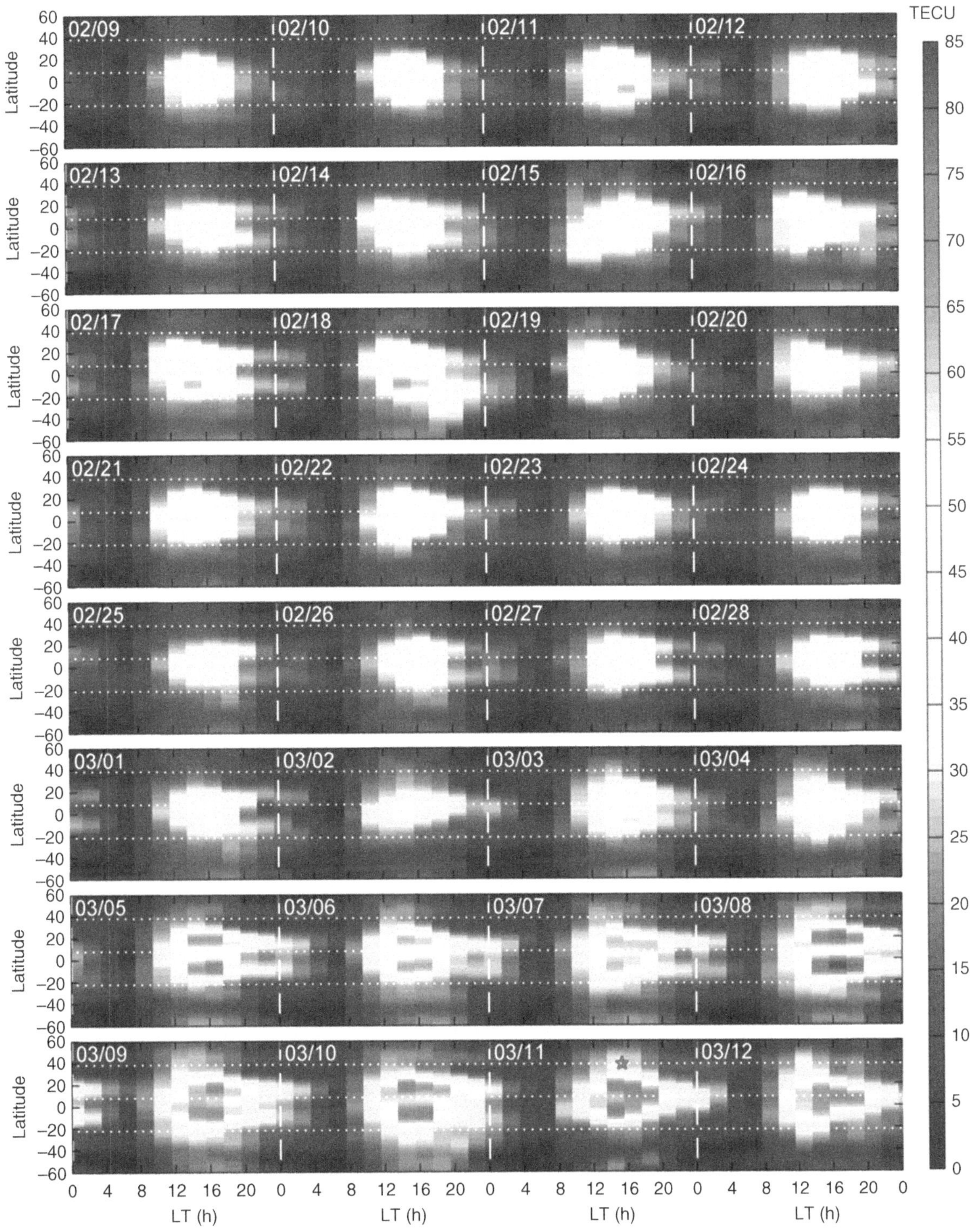

Figure 17.3 The latitude–time–TEC plots along 135°E extracted from the GIM during 9 February to 12 March 2011. The star symbol is the epicenter of the Tohoku earthquake. The three dashed horizontal white lines denote, from the top, magnetic latitudes of the epicenter, magnetic equator, and conjugate point, respectively.

is an integrated value of the ionospheric electron density along the raypath between the satellite and the receiver. The electron density distribution can be reconstructed based on a number of TEC data points. However, it is an ill-posed problem, and accurate reconstruction is difficult because of the small number of data points and the lack of horizontal raypaths. Especially, the lack of horizontal raypaths makes

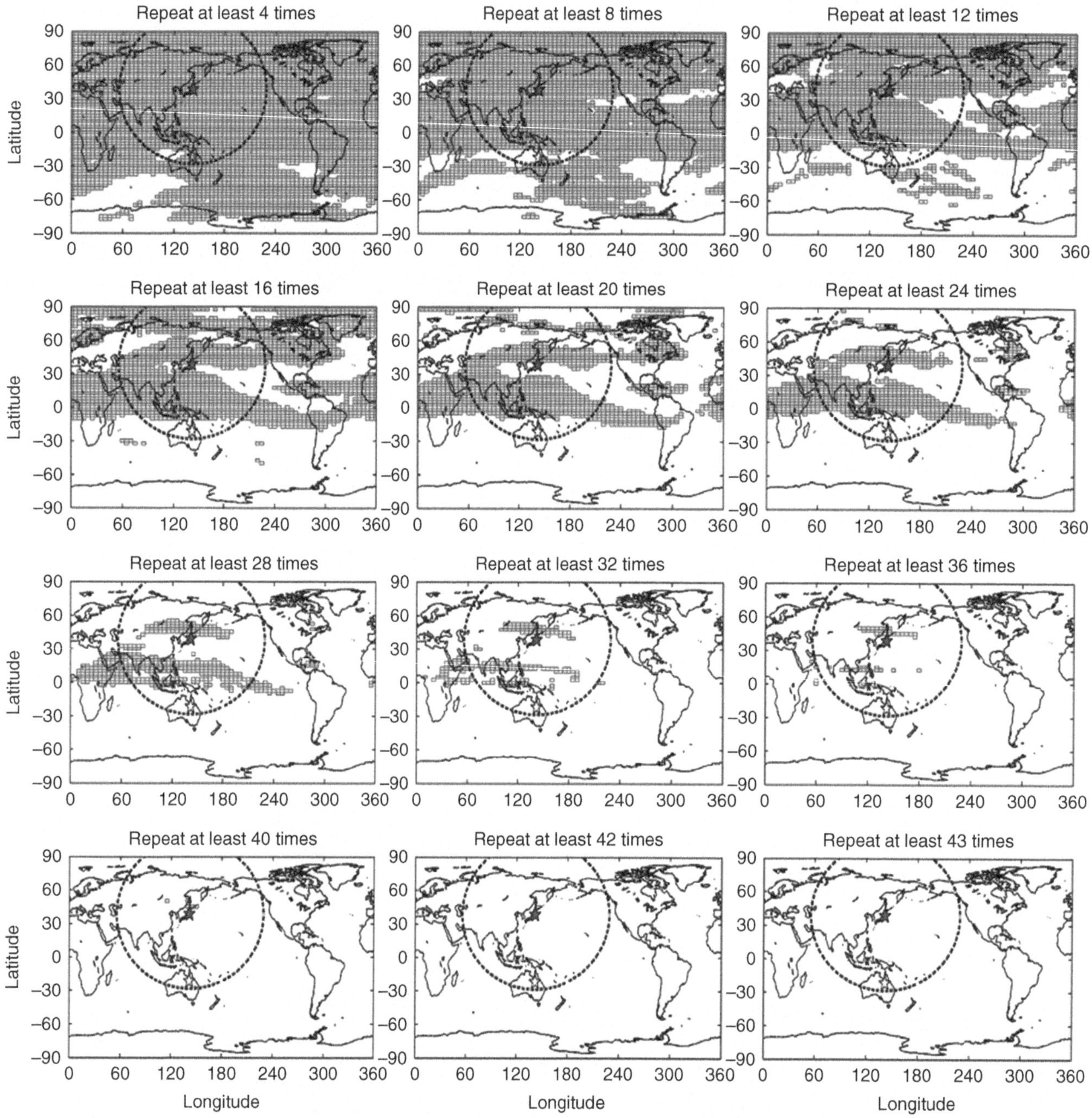

Figure 17.4 The distribution of the 30-day extreme maximum (the positive anomaly) at each lattice with various repeats (or occurrence frequency) during the 5-day period of 6–10 March 2011 (DOY 65–69; 5–1 days before the Tohoku earthquake, red star). The red star and the open star denote the epicenter and corresponding conjugate point of the Tohoku earthquake, respectively. The dashed circle with radius $R = 100.43M = 7413$ km indicates the earthquake preparation area of the lithosphere [*Dobrovolsky et al.*, 1979]. (Source: Reproduced with permission of Springer.) (*See electronic version for color representation.*)

reconstruction of vertical electron density distribution extremely difficult. Previous studies have proposed various algorithms for ionospheric tomography [*Austen et al.*, 1988; *Raymund et al.*, 1990; *Garcia-Fernandez et al.*, 2003; *Kunitsyn and Tereshchenko*, 2003; *Lee et al.*, 2008; *Yao et al.*, 2013].

In this section, the residual minimization training neural network (RMTNN) tomographic approach is explained [*Ma et al.*, 2005a; *Takeda and Ma*, 2007], using TEC data, including location and altitude, derived by ground-based GNSS receivers and ionosondes. The principle and application will be given in the following.

17.3.1. Principle

The slant TEC (STEC) [*Mannucci et al.*, 1998] along a raypath between a GNSS satellite and a ground-based receiver is defined as the integrated value of the ionospheric and plasmaspheric electron densities, including instrumental bias, as follows:

$$I_j^i = \int_{r_i}^{r^j} N(\vec{r})ds + B_i + B^j \tag{17.1}$$

where I_j^i is the STEC, $N(\vec{r})$ is the electron density, r_i and r^j are the position of the ith ground-based receiver and the jth satellite, and B_i and B^j are the instrumental bias of the ith ground-based receiver and the jth satellite. To determine $N(\vec{r})$, a neural network is constructed; r is the neural network's input parameter and $N(\vec{r})$ its output. In the neural network system, r is given by a geographical latitude, longitude, and altitude. For evaluation of the residuals, the equation is separated into the ionospheric and the plasmaspheric parts and discretized as

$$I_j^i \approx \sum_{q=i}^{Q} \alpha_q N(\vec{r}) + B_i + B^j + P_i^j \tag{17.2}$$

where q and α denote a sampling point and the corresponding weight for the numerical integration, respectively, Q is the total number of sampling points along a raypath in the ionosphere, and P is the contribution of the plasmaspheric electron density to the STEC I. We define altitudes from 100 to 700 km as the ionosphere and altitudes above 700 km as the plasmasphere, as modeled by the simple diffusive-equilibrium model [*Angerami and Thomas*, 1964]. Based on this model, the plasmaspheric contribution to the STEC value along the raypath is represented by P_i^j. In this model, we assume the altitude of the top of the ionosphere is 700 km and the scale height of density decay in the plasmasphere is 480 km. To estimate the electron density $N(\vec{r})$, we take the squares of the residuals of the integral equation (17.2) as the neural network's objective function. The objective function E1 is given as

$$E1 = \left(\sum_{q=1}^{Q} \alpha_q N(\vec{r}) + B_i + B^j + P_i^j - I_i^j \right)^2 \tag{17.3}$$

In order to estimate the $N(\vec{r})$, we minimize the objective function E1 in the learning process of the neural network. Although the instrumental biases B_i and B^j can be estimated during reconstruction by the RMTNN method using an additional single neuron [*Ma et al.*, 2005b], we assumed that these biases are removed before processing.

The disadvantage of using ionospheric tomography using ground-based GNSS receivers is the associated difficulty of obtaining sufficient vertical resolution because of the lack of horizontal raypaths. To solve this problem, we use information about the peak electron density (NmF2) and the corresponding height (hmF2) observed by an ionosonde station to constrain the problem. The neural network was trained using these data through the conventional supervised back-propagation algorithm [*Rumelhart et al.*, 1986]. The object function E2 for the ionosonde data is given by

$$E2 = \sum_{s=1}^{S} \left(N_S(r_S) - N_S^{\text{ion}} \right)^2, \tag{17.4}$$

where S is the number of ionosondes, N_S is the output of the neural network for the corresponding position that gives hmF2, and N_S^{ion} is the observed NmF2 value. The objective function E2 is evaluated from electron density at hmF2 height over the ionosonde locations using the same neural network. Therefore, the overall objective function E is

$$E = gE1 + E2, \tag{17.5}$$

where g is a balance parameter between the GNSS and ionosonde data. In this chapter, if the value of the objective function E is below 10^{-3}, we decided that the neural network was convergent; the criterion of E (10^{-3}) was empirically determined by numerical simulations. Consequently, $g = 1.0$ was adopted as the balance parameter in this study [*Hirooka et al.*, 2011].

Figure 17.5 shows a schematic diagram of the data flow for ionospheric tomography by RMTNN. The objective function E1 can be used only after all density values of the sampling points along a raypath have been evaluated. Therefore, conventional online updating (where the values of the weights are updated after every forward calculation) cannot be used. For the objective function E1, the weight updating process of the neural network is derived in clusters for the nth path. The weight update for E2 is executed after updating process E1 for all GNSS raypaths. In the following, we used data obtained during 15-min periods for the reconstruction. We assume that the electron density distribution is constant during these periods. Furthermore, the spatial resolution is 0.5° latitude × 0.5° longitude × 30 km in altitude [*Hirooka et al.*, 2011].

The map of the epicenter, the selected GNSS receivers, and the ionosonde is shown in Figure 17.6. Before applying the RMTNN method described above, we check the existence of possible TEC anomalies before the earthquake. Figure 17.7 indicates the two-dimensional maps of TEC* using local GNSS receivers (GEONET) during DOY66-67

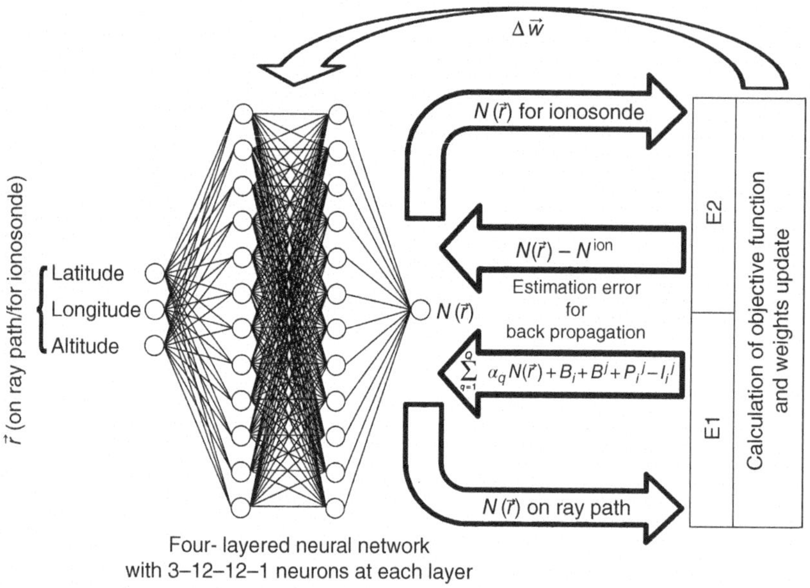

Figure 17.5 Schematic diagram of the data flow for ionospheric tomography using the residual minimization training neural network (RMTNN), where r is the three-dimensional coordinate, $N(\vec{r})$ is the output value of the neural network and Δw is the updated weight value.

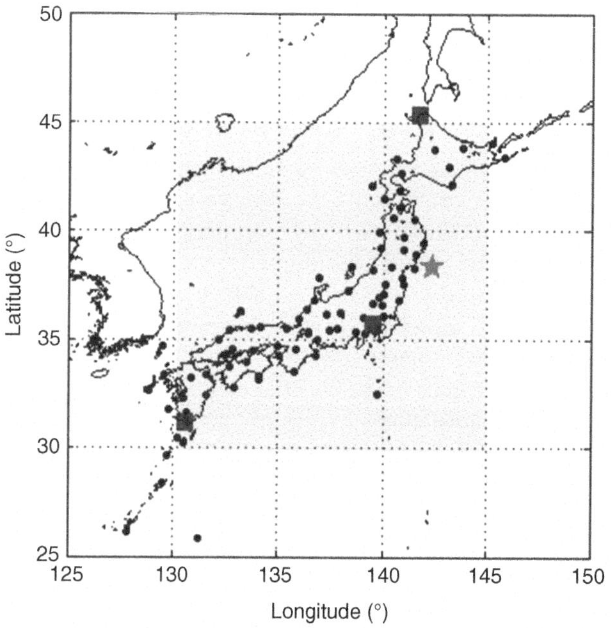

Figure 17.6 Map of the epicenter of the 2011 Tohoku earthquake, GNSS ground receivers, and Wakkanai, Kokubunji, and Yamakawa ionosonde stations. The red star represents the epicenter, black circles show GNSS stations, and blue squares indicate ionosonde stations. (*See electronic version for color representation.*)

(4–5 March), while the apparent GIM TEC enhancement in zero-dimensional, one-dimensional, and two-dimensional space are registered in Figures 17.2, 17.3, and 17.4, respectively. Here, TEC* indicates the normalized TEC using mean 15-day backward propagation and its standard deviation and has a unit of σ [*Nishihashi et al.*, 2009; *Ichikawa et al.*, 2011; *Kon et al.*, 2011; *Hirooka et al.*, 2013]. The GIM TEC has a wider view with the less resolution (Figure 17.4). The positive anomaly starts to appear at 01:00 UT on DOY66 around Hokkaido and by 07:00 UT appears over all Japan. The peak of the anomalous region is not located at the epicenter but it does stay in the area of Japan for more than one day. Note in Figure 17.4 the continuity of the positive anomaly area from DOY066 to DOY067 (3–4 days prior to the earthquake).

17.3.2. Electron Density Distribution Based on RMTNN Tomography

Eighty GNSS receivers were selected from the GEONET (GNSS Earth Observation Network System) to be used with three ionosonde stations (Wakkanai, Kokubunji, and Yamakawa) for the area shown in Figure 17.6: covering 26°N–42°N latitude, 128°E–142°E longitude and 100–700 km in altitude. To understand the anomalous changes in three-dimensional electron density distributions, the difference between the reconstructed data at 04:00 UT on DOY67 (8 March; 3 days before the Tohoku earthquake) and the model data constructed using the median value from the 15-day backward propagation computations [*Hirooka et al.*, 2013] is taken. The results are shown in Figures 17.7–17.9.

Figure 17.8 shows differential images of the tomographic results as meridional plane plots. Warm colors

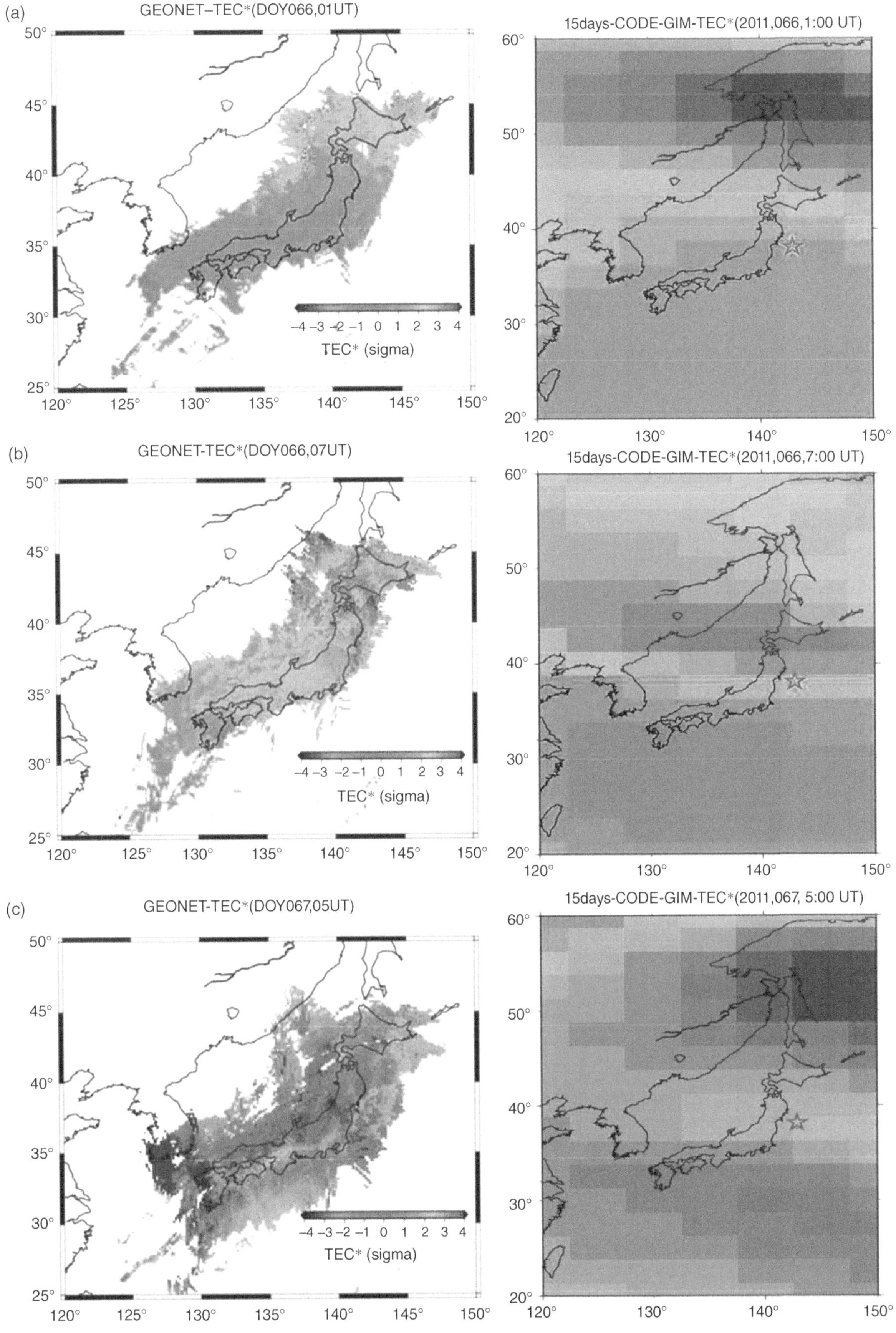

Figure 17.7 Spatial distributions of TEC* (left side, local map) and GIM-TEC* (right side, global map). (a) 01:00 UT on 7 March 2011, (b) 07:00 UT on 7 March 2011, (c) 05:00 UT on 8 March 2011, and (d) 23:00 UT on 8 March 2011.

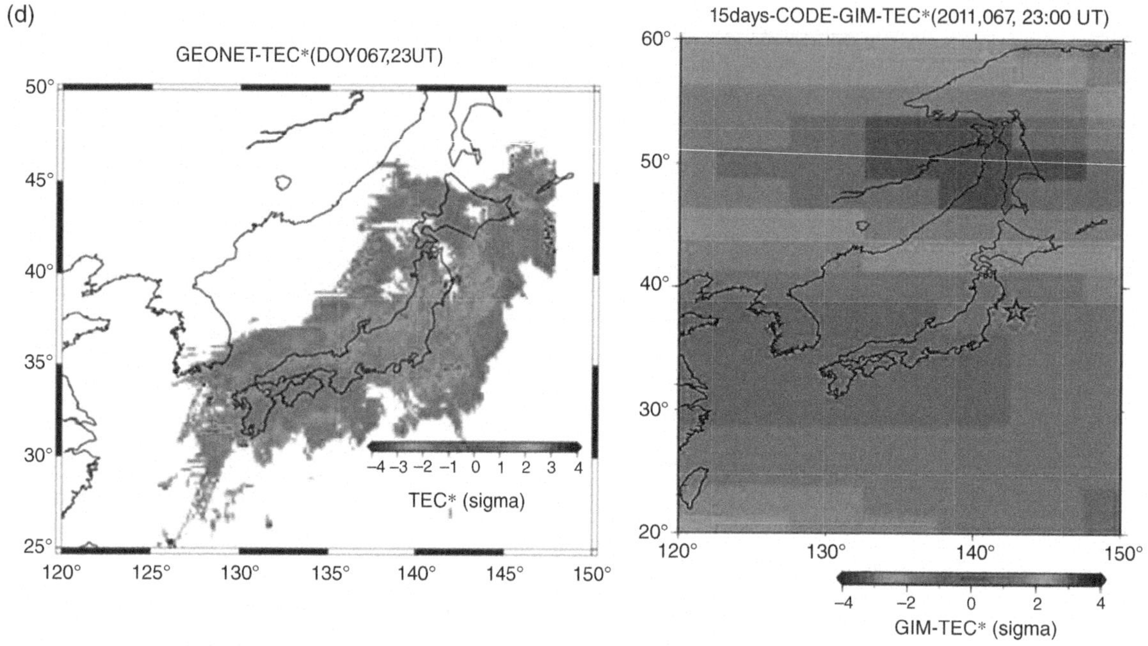

Figure 17.7 (Continued)

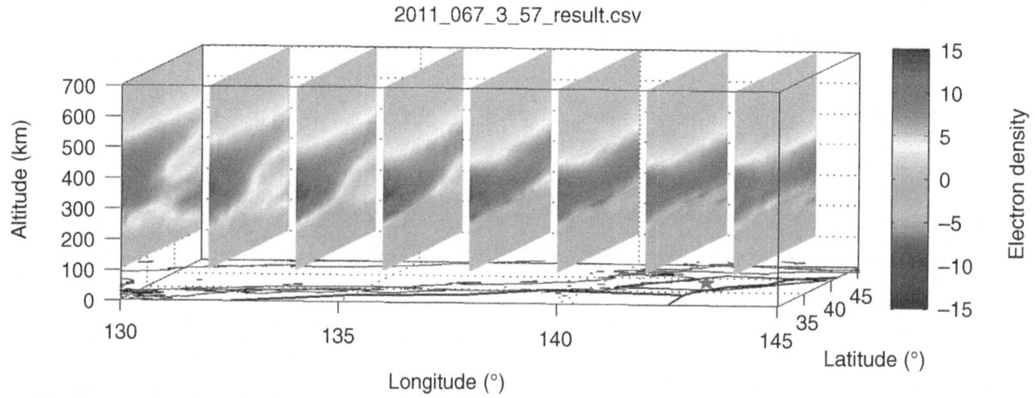

Figure 17.8 Differential images of the tomographic results for anomaly data on 8 March, 2011 (DOY67). Warm colors indicate increase and cold colours decrease of electron density against the 15-day backward propagation median value model. The red star in the map of Japan at the bottom shows the earthquake epicenter. (*See electronic version for color representation.*)

indicate increase and cold colours decrease of electron density against the 15-day backward propagation median value model. The positive enhancement near the F2 layer is dominant (see electronic version for color representation). It is generally consistent with the two-dimensional TEC anomaly shown in Figures 17.4 and 17.7. Over the epicenter, however, a negative anomaly region exists below the F2 layer (200-250 km height). Figure 17.9 presents an altitude map of electron density showing the above features more clearly. Figure 17.10 shows only cells with a difference quantity of <0 or $>5\times10^{11}$ (e m^{-3}). Decreases of local plasma density beyond -6×10^{11} (e m^{-3}) were around 250 km height over the epicenter and increases of plasma density of 10×10^{11} (e m^{-3}) around 400 km height are confirmed. The increase spreads over the whole area of Japan around this altitude. Fluctuation of the density reaches about −60% for the decrease and about 200% for the increase against the background density. Figure 17.11 shows an example of the differential images for the tomographic results of a quiet day. We can see there are no significant changes against the 15-day backward propagation median value. These results are highly suggestive of earthquake-related anomalies in electron densities in ionosphere, showing a positive enhanced region near the F2 layer and a negative anomaly region just below the F2 layer (200–250 km height) over the epicenter.

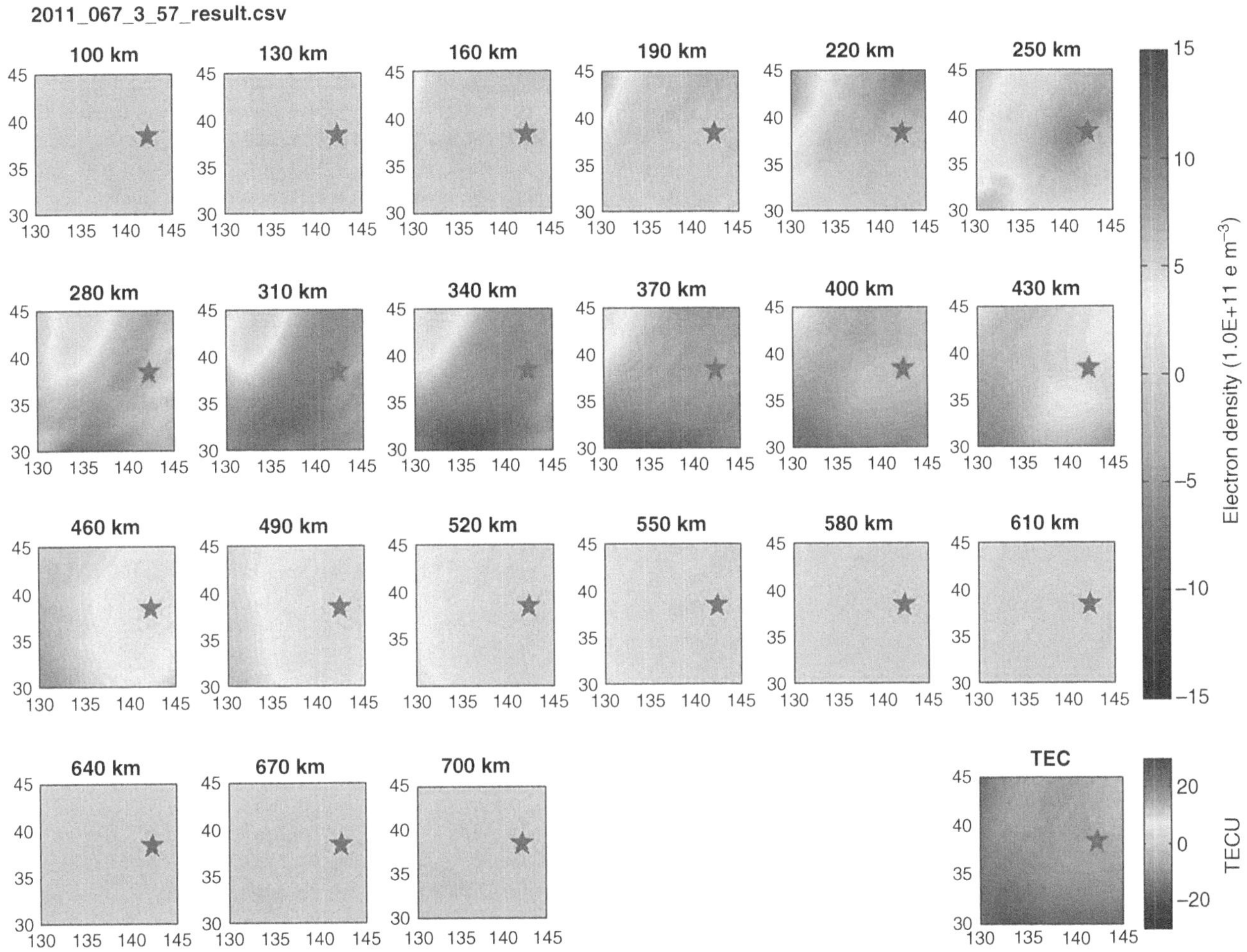

Figure 17.9 An altitude map of electron density. Horizontal and vertical axes are longitude and latitude, respectively. Red stars in each panel indicate the epicenter. At 200–250 km altitude over the epicenter region a 60% decrease of electron density is registered. (*See electronic version for color representation.*)

17.4. SUMMARY

Temporal anomalies at a specific location can be detected using the time series of GNSS TEC data, but statistical verification is required in order to detect/identify the anomalies. When a detected anomaly is statistically validated as characteristic of SIPs, taking into consideration the polarity (positive or negative), local time, duration period, lead day, etc., of previous earthquakes at the same region, it may be considered a reliable temporal SIP. When a temporal SIP is detected, a spatial analysis considering the global GNSS TEC is required in order to distinguish the frequency and persistency distributions of the anomalies, which will confirm whether or not the temporal SIPs related to an earthquake. Finally, in this study a dense GNSS receiver network was used to construct three-dimensional ionospheric tomography in order to understand the structure and dynamics of the SIPs.

ACKNOWLEDGMENTS

The authors would like to thank to the Geospatial Information Authority of Japan for GNSS data (GEONET), the Center for Orbit Determination in Europe (CODE) for GNSS data (GIM), and the Japanese National Institute of Information and Communications Technology for the ionosonde data. The D_{st} index and the K_p index were supplied by the World Data Center (WDC) for Geomagnetism, Kyoto. The F10.7 solar flux data were provided by the Solar Radio Monitoring Program operated jointly by the National Research Council and the Canadian Space Agency. We also thank the US Geological Survey for the earthquake data. This research is partly supported by the Ministry of Science and Technology MOST 104-2628-M-008-001 grant, Grant-in-Aid for Scientific Research of Japan Society for Promotion of Science (19403002, 26249060 and 26240004), National Institute of

Information and Communication Technology (R & D Promotion Funding International Joint Research), and 2015 International Team selected by ISSI-BJ of Validation of Lithosphere–Atmosphere–Ionosphere–Magnetosphere Coupling (LAIMC).

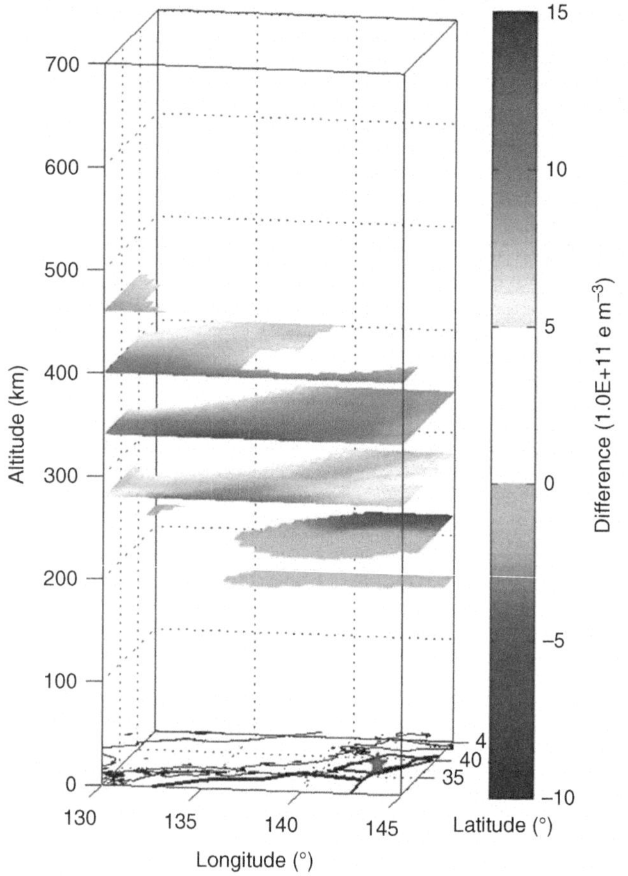

Figure 17.10 Three-dimensional difference image between observed data at 04:00 UT on 8 March 2011 and the 15-day backward propagation median value.

REFERENCES

Angerami, J. J. and Thomas, J. O. (1964), Studies of planetary atmospheres: The distribution of electrons and ions in the Earth's exosphere, *J. Geophys. Res.*, *69*, 4537–4560.

Austen, J. R., Franke, S. J., and Liu, C. H. (1988), Ionospheric image using computerized tomography, *Radio Sci.*, *23*, 299–307.

Chen, C. H., Liu, J. Y., Yen, H. Y., Zeng, X., and Yeh, Y. H. 2004. Changes of geomagnetic total field and occurrences of earthquakes in Taiwan, *Terr. Atmos. Oceanic Sci.*, *15*, 361–370.

Liu, J. Y., Chen, Y. I., Chuo, Y. J., and Chen, C. S. 2006. A statistical investigation of pre-earthquake ionospheric anomaly, *J. Geophys. Res.*, *111*, A05304, doi:10.1029/2005JA 011333

Chen, Y. I., Huang, C. S., and Liu, J. Y. (2015), Statistical evidences of seismo-ionospheric precursors applying receiver operating characteristic (ROC) curve on the GPS total electron content in China, *J. Asian Earth Sci.*, *114*, 393–402.

Dobrovolsky, I. P., Zubkov, S. I., and Miachkin, V. I. (1979), Estimation of the size of earthquake preparation zones, *Pure Appl. Geophys.*, *117*, 1025–1044.

Garcia-Fernandez, M., Hernandez-Pajares, M., Juan, M., Sanz, J., Orus, R., Coisson, P., Nava, B., and Radicella, S. M. (2003), Combining ionosonde with ground GNSS GPS data for electron density estimation, *J. Atmos. Solar Terr. Phys.*, *65*, 683–691.

Hayakawa, M. (1999), *Atmospheric and Ionospheric Electromagnetic Phenomena with Earthquakes*, Terra Scientific Publishing, Tokyo.

Hayakawa, M. (2000), *Seismo Electromagnetics, Monograph of International Workshop on Seismo Electromagnetics*, Terra Scientific Publishing, Tokyo.

Hayakawa, M. and Fujinawa, Y. (1994), *Electromagnetic Phenomena Relater to Earthquake Predication*, Terra Scientific Publishing, Tokyo.

Hayakawa, M. and Molchanov, O. A. (2002), *Seismo Electromagnetics (Lithospheric–Atmospheric–Ionospheric Coupling)*, Terra Scientific Publishing, Tokyo.

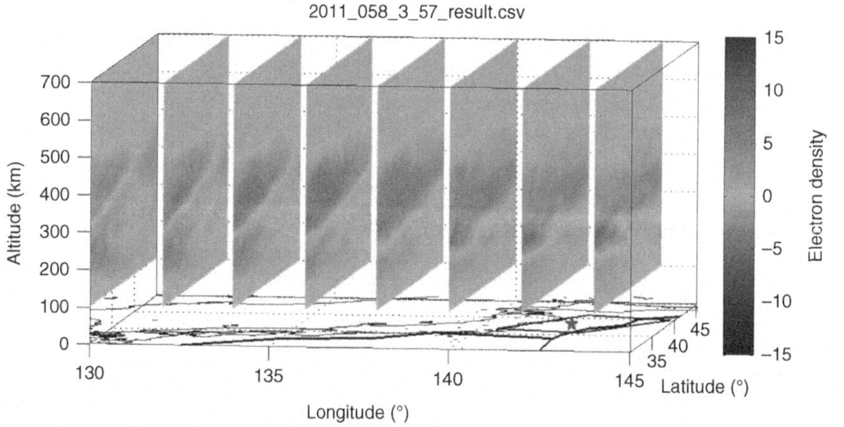

Figure 17.11 Differential images of the tomographic results for a quiet day. Warm colors indicate increase and cold colours decrease of electron density against the 15-day backward propagation median value model. (*See electronic version for color representation.*)

Hirooka, S., Hattori, K., and Takeda, T. (2011), Numerical validations of neural network based ionospheric tomography for disturbed ionospheric conditions and sparse data, *Radio Sci.*, *46*, RS0F05.

Hirooka, S., Hattori, K., Ichikawa, T., and Han, P. (2013), Ionospheric anomalies associated with large earthquakes: tomographic approach, *AGU Fall Meeting*, 318–1614.

Ichikawa, T., Hattori, K., Kon, S., Hirooka, S., and Liu, J. Y. (2011), Ionospheric anomalies possibly associated with M > 6.0 earthquakes in the Japan area during 1998–2010 and the 2011 off the Pacific Coast of Tohoku Earthquake (*Mw*9.0), *AGU Fall Meeting*, NH22A-08.

Klotz, S. and Johnson, N. L. (1983), *Encyclopedia of Statistical Sciences*, John Wiley & Sons, Inc., Hoboken, NJ.

Kon, S., Nishihashi, M., and Hattori, K. (2011), Ionospheric anomalies possibly associated with M ≥ 6 earthquakes in Japan during 1998–2010: Case studies and statistical study, *J. Asian Earth Sci.*, *41*(4): 410–420, doi: 10.1016/j.jseaes.2010.10.005.

Kunitsyn, V. E. and Tereshchenko, E. D. (2003), *Ionospheric Tomography*, Springer. Berlin, 260 pp.

Lee, J. K., Kamalabadi, F., and Makela, J. J. (2008), Three-dimensional tomography of ionospheric variability using a dense GNSS GPS receiver array, *Radio Sci.*, *43*, RS3001.

Lin, C. H., Yeh, Y. H., Chen, Y. I., Liu, J. Y., and Chen, K. J. (2003), Earthquake clustering relative to lunar phases in Taiwan, *Terr. Atmos. Oceanic Sci.*, *14*, 289–298.

Liu, J. Y., H. F. Tsai, and T. K. Jung (1996), Total electron content obtained by using the global positioning system, *Terr. Atmos. Oceanic Sci.*, *7*, 107–117.

Liu, J. Y., Tsai, H. F., Wu, C.-C., Tseng, C. L., Tsai, L.-C., Tasi, W. H., Liou, K., and Chao, J. K. (1999), The effect of geomagnetic storm on ionospheric total electron content at the equatorial anomaly region, *Adv. Space Res.*, *24*, 1491–1494.

Liu, J. Y., Chen, Y. I., Pulinets, S. A., Tsai, Y. B., and Chuo, Y. J. (2000), Seismo-ionospheric signatures prior to M > 6.0 Taiwan earthquakes, *Geophys. Res. Lett.*, *27*, 3113–3116.

Liu, J. Y., Chen, Y. I., Chuo, Y. J., and Tsai, H. F. (2001), Variations of ionospheric total electron content during the Chi-Chi earthquake, *Geophys. Res. Lett.*, *28*, 1383–1386.

Liu, J. Y., Chuo, Y. J., Shan, S. J., Tsai, Y. B., Chen, Y. I., Pulinets, S. A., and Yu, S. B. (2004a), Pre-earthquake ionospheric anomalies registered by continuous GNSS GPS TEC measurement, *Ann. Geophy.*, *22*, 1585–1593.

Liu, J. Y., Lin, C. H., Tsai, H. F., and Liou, Y. A. (2004b), Ionospheric solar flare effects monitored by the ground-based GNSS GPS receivers: Theory and observation, *J. Geophys. Res.*, *109*, doi: 10.1029/2003JA009931

Liu, J. Y., Chen, Y. I., Chen, C. H., Liu, C. Y., Chen, C. Y., Nishihashi, M., Li, J. Z., Xia, Y. Q., Oyama, K. I., Hattori, K., and Lin, C. H. (2009), Seismo-ionospheric GNSS GPS total electron content anomalies observed before the 12 May 2008 *Mw*7.9 Wenchuan earthquake, *J. Geophys. Res.*, *114*, A04320, doi: 10.1029/2008JA013698

Liu, J. Y., Chen, Y. I., Chen, C. H., and Hattori, K. (2010), Temporal and spatial precursors in the ionospheric global positioning system (GNSS GPS) total electron content observed before the 26 December 2004 *M*9.3 Sumatra-Andaman Earthquake, *J. Geophys. Res.*, *115*, A09312, doi: 10.1029/2010JA015313

Liu, J. Y., Le, H., Chen, Y. I., Chen, C. H., Liu, L., Wan, W., Su, Y. Z., Sun, Y. Y., Lin, C., and Chen, M. Q. (2011), Observations and simulations of seismo-ionospheric GNSS GPS total electron content anomalies before the 12 January 2010 *M*7 Haiti earthquake, *J. Geophys. Res.*, *116*, A04302, doi: 10.1029/2010JA015704.

Liu, J. Y., Yang, W. H., Lin, C. H., Chen, Y. I., and Lee, I. T. (2013), A statistical study on the characteristics of ionospheric storms in the equatorial ionization anomaly region: GNSS GPS-TEC observed over Taiwan, *J. Geophys. Res. Space Phys.*, *118*, 3856–3865, doi: 10.1002/jgra.50366.

Ma, X. F., Maruyama, T., Ma, G., and Takeda, T. (2005a), Three dimensional ionospheric tomography using observation data of GPS ground receivers and ionosonde by neural network, *J. Geophys. Res.*, *110*, A05308.

Ma, X. F., Maruyama, T., Ma, G., and Takeda, T. (2005b), Determination of GPS receiver differential biases by neural network parameter estimation method, *Radio Sci.*, *40*, RS1002.

Mannucci, A. J., Wilson, B. D., Yuan, D. N., Ho, C. H., Lindqwister, U. J., and Runge, T. F. (1998), A global mapping technique for GPS derived ionospheric electron content measurements, *Radio Sci.*, *33*, 565–582.

Nishihashi, M., Hattori, K., Jhuang, H. K., and Liu, J. Y. (2009), Spatial distribution of ionospheric GNSS GPS-TEC and NmF2 anomalies during the 1999 Chi-Chi and Chia-Yi Earthquakes in Taiwan, *Terr. Atmos. Oceanic Sci.*, *20*, 779–789.

Pulinets, S. and Boyarchuk, K. (2004), *Ionospheric Precursors of Earthquakes*, Springer, New York.

Raymund, T. D., Austen, J. R., Franke, S. J., Liu, C. H., Klobuchar, J. A., and Stalker, J. (1990), Application of computerized tomography to the investigation of ionospheric structures, *Radio Sci.*, *25*, 771–789.

Rumelhart, D. E., Hinton, G. E., and Williams, R. J. (1986), Learning internal representations by error propagation, in D. Rumelhart and J. Mclelland (eds), *Parallel Distributed Processing*, Vol. *1*, pp. 318–362, MIT Press, Cambridge, MA.

Takeda, T. and Ma, X. F. 2007. Ionospheric tomography by neural network collocation method, *Plasma Fusion Res.*, *2*, S1015, doi:10.1585/pfr.2. S1015

Yao, Y., Tang, J., Kong, J., Zhang, L., and Zhang, S. (2013), Application of hybrid regularization method for tomographic reconstruction of midlatitude ionospheric electron density, *Adv. Space Res.*, *52*(12), 2215–2225.

18

Statistical Analysis of the Ionospheric Density Recorded by the DEMETER Satellite During Seismic Activity

Michel Parrot[1] and Mei Li[2]

ABSTRACT

Over tens of years ionospheric phenomena linked to earthquakes have been analyzed from the ground with ionospheric sounders. These ionospheric perturbations can be considered as short-term precursors of earthquakes. The ionosphere is highly variable, however, and it is necessary to process many events in order to demonstrate the characteristics of a particular precursory phenomenon. The DEMETER (an acronym for detection of electro-magnetic emissions transmitted from earthquake regions) satellite was especially launched for the study of these events. This chapter presents a review of statistical analyses of unusual ionospheric perturbations. The analyses indicate that the ionosphere is perturbed above the epicenter a few days before the earthquake, mainly at nighttime. Only local measurements of the ionospheric density were performed at the altitude of the satellite (660 km), but study of the whistler-wave propagation revealed that there is also an excess of ionization at the bottom of the ionosphere (90 km) a few hours before an earthquake. The chapter goes on to present a statistical analysis of ionospheric data showing that it is possible to observe perturbations in the conjugated region of an earthquake. An attempt to identify precursors of the M 8.1 Samoa earthquake in 2009 indicates that an accurate prediction is not conceivable but that it is possible to give a warning.

18.1. INTRODUCTION

A substantial number of effects in the Earth's atmosphere and the ionosphere possibly associated with earthquakes has been revealed in the past 20 years. The ionosphere and the upper atmosphere phenomena correlated with seismic events have been examined in numerous experimental and theoretical studies (see the reviews by *Hayakawa et al.* [2018] and *Liu et al.* [2018]). The seismo-electromagnetic effects involve the electric and magnetic perturbations and include: electromagnetic emissions in a large frequency range, perturbations of ionospheric layers, anomalies in the records of very low frequency (VLF) transmitter signals, night airglow observations, etc. (see the review by *Pulinets et al.* [2018]). Such phenomena are of great interest, because they may start a few hours before the main shock and can be considered as short-term precursors.

This chapter will deal only with the study of the ionospheric density observed by the satellite DEMETER (an acronym for detection of electro-magnetic emissions transmitted from earthquake regions) before earthquakes. Wave emissions and electron density perturbations can be linked through various mechanisms in the ionosphere. For example, density fluctuations are often associated with electrostatic turbulence. Thus, the same hypotheses proposed for the generation mechanism of precursors are also valid for explaining changes in density fluctuations or in electrostatic turbulence. These hypotheses are mainly related to: wave production by compression of rocks, diffusion of water in the epicenter area, redistribution of

[1] LPC2E/CNRS, Orléans, France
[2] China Earthquake Networks Center, China Earthquake Administration, Beijing, China

Pre-Earthquake Processes: A Multidisciplinary Approach to Earthquake Prediction Studies, Geophysical Monograph 234, First Edition. Edited by Dimitar Ouzounov, Sergey Pulinets, Katsumi Hattori, and Patrick Taylor.

electric charges at the surface of Earth and then in Earth's atmospheric system, radon emission, and heating and gas release which can trigger acoustic gravity waves. These mechanisms are well described and discussed in relation to lithosphere–atmosphere–ionosphere coupling (LAIC) models by *Pulinets et al.* [2015, 2018], *De Santis et al.* [2015], and *Kuo et al.* [2018].

It has to be borne in mind, however, that all these ionospheric parameters also display variations in the absence of seismic activity because the mid-latitude and equatorial ionosphere is affected by a number of other sources of perturbations, but primarily by solar activity [*Liu et al.*, 2007]. Under these circumstances only a statistical study including many events will show the general behavior of such ionospheric perturbations and help to define characteristic ionospheric perturbations prior to earthquakes. In contrast to ground experiments, satellite experiments cover most seismic zones of the Earth, and statistical studies become meaningful because of the much larger number of recorded events.

The satellite DEMETER and its payload are briefly described in the following section.Section 18.3 presents results from observations at the time of large events, including the Wenchuan earthquake that occurred in 2008 in China. Statistical analyses including all events occurring during the DEMETER mission are presented in section 18.4, and discussion and conclusions follow in sections 18.5 and 18.6.

18.2. THE DEMETER SATELLITE

DEMETER is a microsatellite (130 kg) with a low-altitude (710 km) and nearly polar orbit. The satellite was launched by the CNES (French National Space Agency) in June 2004, and its scientific mission came to an end in December 2010. The orbit was nearly sun-synchronous (10.30–22.30 LT). The main scientific objectives of the DEMETER experiment were to study disturbances of the ionosphere due to seismo-electromagnetic effects and anthropogenic activities (power line harmonic radiation, VLF transmitters, and high-frequency broadcasting stations, etc.). In order to achieve these objectives, the payload of the microsatellite measured electromagnetic waves in different frequency ranges from ultralow to medium, and also some important plasma parameters (ion composition, electron density and temperature, and energetic particles). Description of the different experiments and of the raw data processing can be found in *Parrot* [2006].

18.3. THE OBSERVATIONS

18.3.1. Examples of Events

During the lifetime of DEMETER, many ionospheric perturbations have been observed in relation with earthquakes and more than 140 articles have been published to date (see the list in https://demeter.cnes.fr/en/DEMETER/A_publications.htm).

An example of recorded ionospheric perturbations is given in Figure 18.1. It is related to a *M* 6.8 earthquake that occurred on 5 January 2010 at 12:15:33 UT with an epicenter located at latitude 9.06°S and longitude 157.59°E, at a depth of 18.7 km. The electron and ion O^+ densities are plotted as a function of time along an orbit close to this epicenter, in the top and middle panel, respectively. A large increase in densities (more than 100% for the O^+ ion density) can be observed at the closest approach to the epicenter and at a time less than 40 min before the main shock. Owing to the difficulty of calibration between the two instruments, measuring on one hand the electron density and on the other the ion density (especially during night time), in Figure 18.1 only the relative variations of the two densities should be considered and the two absolute values should not be compared.

As the ionosphere also has variations not related to seismic activity it is important to show that the observations that we attribute to earthquakes are very uncommon. This has been done in published articles in which the ionospheric behavior has been studied at the same location a long time before the occurrence of powerful earthquakes. In the following we will summarize two statistics performed with the complete set of DEMETER data (see section 18.4).

18.3.2. The Wenchuan Earthquake

The *M* 8.0 Wenchuan earthquake occurred at 14:28:01 LT (06:28:01 UT) on 12 May 2008. The epicenter was located in Sichuan province at 31.021°N, 103.367°E with a depth of 19 km. This large earthquake is very well specified because there were many experiments recording various seismic parameters: reviews of many different precursors identified have been carried out by *Singh et al.* [2010], *Zhang and Shen* [2011], and *Ma and Wu* [2012]. Anomalous changes have been reported in ground, meteorological, and atmospheric parameters (air temperature and relative humidity) compared with other days. Studies have shown that electromagnetic anomalies started 2.5 years earlier and were recorded until 3 days before the event, and that the ionospheric anomalies are very short-term precursors with a peak appearance 5 days before the earthquake.

Concerning the DEMETER data, *Zhang et al.* [2009] found that the lowest value of the daily averaged ion densities measured in the epicenter area during the local nighttime occurred 3 days before the earthquake. With a short statistical analysis of electron density data before and after the quake, *He et al.* [2011a,b] found an anomalous increase centered close to the epicenter. *Yan et al.* [2013] confirmed these results by comparing data from

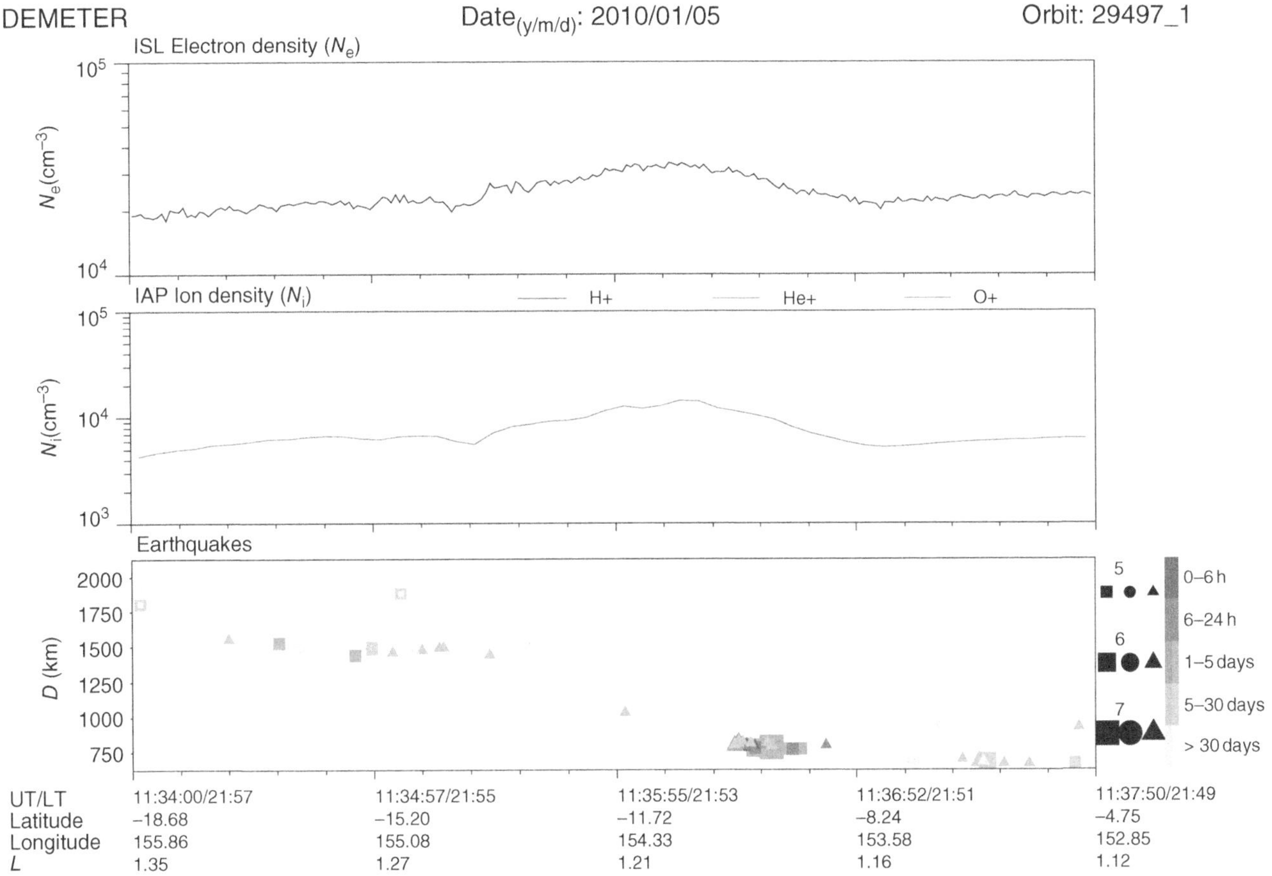

Figure 18.1 Ionospheric density variation recorded by the DEMETER satellite on 5 January 2010 between 11:34:00 and 11:37:50 UT as a function of the parameters displayed at the bottom of the figure (UT/LT, latitude, longitude, L (the McIlwain value)). The top panel is a record of the electron density, the middle panel a record of the ion O^+ density (the densities of the other ions are much lower), and the bottom panel displays the occurrence of the earthquakes along the corresponding orbit and the distance between the epicenter and this orbit. The bright red triangle at 11:36:28 UT indicates the closest approach to the epicenter of the earthquake being considered. The description of the other symbols can be found in *Parrot et al.* [2006]. (*See electronic version for color representation.*)

GPS receivers, the DEMETER satellite, and the Advanced Very High Resolution Radiometer (AVHRR) onboard NOAA satellite. *Ryu et al.* [2014] have shown that, during daytime, there is an enhancement of the equatorial ionospheric anomaly (EIA) near the epicenter longitude that began approximately 1 month before the earthquake and reached its maximum 8 days prior to the main shock. Normally DEMETER does not observe the EIA due to its high altitude in the ionosphere.

Walker et al. [2013] performed a statistical study using ULF data over several months around the time of the Wenchuan earthquake. All nighttime orbits close to the epicenter have been considered and individual spectra at 5.8 Hz were averaged using 30 s time bins and then compared against the average wave power for the whole orbit. They observed that the noise measured at this frequency exhibits large changes relative to the background levels when DEMETER flies over the region of the epicenter or over the region of its conjugated point. In particular, the greatest enhancements occurred around the end of March, mid- to late April, mid-May, and early and mid-June.

18.4. THE STATISTICAL ANALYSES

18.4.1. Classical Analysis

A statistical study on the variations of the electric field up to 10 kHz has been carried by *Němec et al.* [2008, 2009]. The intensity of the signal received when the satellite passed close to the epicenter of an earthquake some days before and after, was compared with its mean intensity at the same place and under the same conditions. More than 2.5 years of satellite data have been analyzed and about 2000 earthquakes with magnitudes larger than or equal to 4.8 have been taken into account. The data related to earthquakes (satellite orbits close to epicenters in space and in time) have been compared with usual data at the same place and at the same time when there was no earthquake. A superposed epoch method (all earthquake occurrences are adjusted to a zero time) is then used to display these ratios considering the time (before and after the earthquakes) and the distance between the projection of the satellite orbit and the epicenters. This study has shown that at night (22.30 LT), there is a statistically significant decrease by 4–6 dB of the measured wave intensity shortly (~4 h) before earthquakes with their epicenter at a depth less than or equal to 40 km. Aftershocks have been removed from the database in order not to mix pre- and post-seismic effects. The results are presented as a function of the frequency under the form of relative intensity normalized by the standard deviation. They show a decrease of the wave intensity measured by DEMETER (more intense when the magnitude is larger) in a frequency range between 1 and 2 kHz. It is important to note that the frequency band where the decrease is observed can be related to the cut-off frequency f_c of the first transverse magnetic (TM) mode in the Earth–ionosphere waveguide nighttime. This frequency is given by $f_c = nc/2h$, where n is the mode number, c is the light velocity, and h is the height of the Earth–ionosphere waveguide. For $n = 1$ and $h = 90$ km, one obtains $f_c = 1666$ Hz. If the intensity of the electric field decreases, it means that the cut-off frequency is increasing, which means that the height of the ionosphere is statistically lower above epicenters. Therefore this study statistically shows that there should be an excess of ionization at the bottom of the ionosphere above epicenters of future earthquakes. *Píša et al.* [2012, 2013] extended the work of Němec and colleagues by using the complete DEMETER data set at the end of the mission (9000 earthquakes). With more than 6 years of data, they confirm exactly the previous results (see Figure 18.2).

Hobara et al. [2013] presented another statistical study of the ionosphere by using electric field measurements, detailing results of ionospheric perturbations in relation with equatorial anomalies observed by DEMETER. In their study, weak correlation between nighttime ionospheric perturbations and land earthquakes was obtained.

Several statistical analyses have been performed with the complete sets of electron and ion density data. *He et al.* [2010] compared the density close to epicenters of earthquakes between and 1 and 30 days before the earthquake with the density recorded at the same location between 31 and 75 days before. They normalized this ratio by the variance and they averaged this quantity for all earthquakes as a function of the distance from the epicenters. They have statistically shown that there is an increase of the density close to the epicenters at nighttime. This increase does not occur if the locations of the

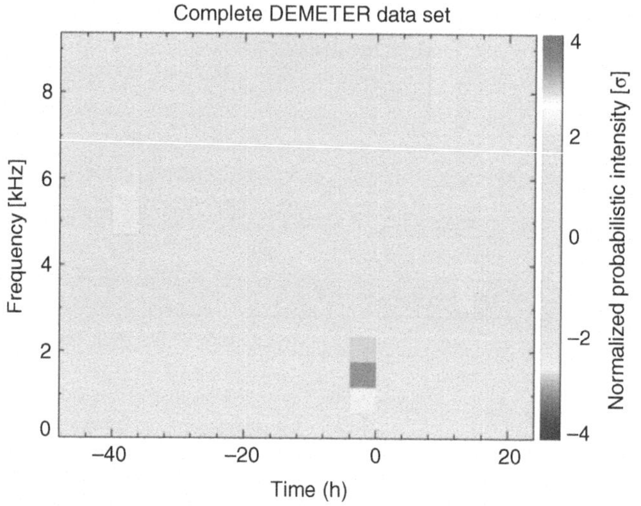

Figure 18.2 Time–frequency spectrogram of the normalized probability density obtained with night measurements of an electric antenna at a distance up to 330 km from earthquakes with a magnitude larger than 4.8. It shows a decrease of the wave intensity at ~1.7 kHz, 4 hours before the earthquakes. (Source: Reproduced with permission of Istituto Nazionale di Geofisica e Vulcanologia.)

earthquakes are chosen randomly. This increase also is more important with increasing magnitude of the earthquake, however, it is not present if the depth of the earthquake is > 60 km.

18.4.2. "Real Time" Analysis of Perturbations

Li and Parrot [2012, 2013] performed a two-step statistical analysis of the ion density. In the first step, they automatically searched the complete DEMETER data set (6.5 years) for ionospheric density peaks similar to that shown in Figure 18.1. They then eliminated perturbations during episodes of large geomagnetic activity ($K_p > 3$) and those not above seismic areas. Their outputs were the amplitude, time, and location of the remaining perturbations. In the second step they used a list of earthquakes (see Table 18.1) and they automatically checked if a given perturbation could be attributed to a given earthquake. The parameters in this step were: earthquake magnitude (*Mw*) and depth (*d*), maximum distance (*D*) between the earthquake epicenter and the perturbation location, maximum time (*T*) between the earthquake and the perturbation occurrence. Their outputs were the number of good detections (one perturbation corresponds to one earthquake), the number of false alarms (one perturbation but no earthquake), and the number of bad detections (no perturbation but one earthquake).

For the purpose of the present study the same data processing has been used but with a different earthquake database. There is a link between the two hemispheres of the Earth due to the magnetic field lines, and LF electromagnetic waves that are emitted in a hemisphere have a tendency to propagate into the opposite hemisphere along these magnetic field lines. In doing so they interact with particles present in the radiation belts (see, e.g., figure 2 of *Sauvaud et al.* [2008]). These particles precipitate in the ionosphere and trigger additional ionization [*Parrot et al.*, 2013]. In this way, a perturbation observed in one hemisphere may have its counterpart in the opposite hemisphere close to its magnetically conjugated point. Earthquakes with epicenters too close to the Equator ($|lat| < 10°$) must be removed (see Table 18.2), however, otherwise it would be difficult to define a clear conjugated region because the perturbations and the conjugated perturbations could be mixed. In section 18.4.2.1 we will verify whether earthquakes occur at the time of and close to ionospheric perturbations, as in *Li and Parrot* [2012, 2013], and in section 18.4.2.2 we will verify whether perturbations around the magnetically conjugated point of the earthquake epicenters can be observed.

18.4.2.1. "Real Time" Analysis of Perturbations Found Close to Epicenters

In comparison with *Li and Parrot* [2012, 2013] the software for anomaly detections has been slightly improved and the number of earthquakes considered is slightly changed (Table 18.2), but the latest results, given in Tables 18.3 and 18.4, are not significantly changed. Table 18.3 concerns the statistics of the perturbations occurring close to the epicenters and lists the data for three magnitude classes of earthquakes, the parameter r (the ratio between the number of earthquakes with good detections and the total number of earthquakes), and the parameter n (the average number of perturbations for each earthquake detected considering the values of D (distance in km) and T (time in days)). Table 18.4 presents some relevant statistical parameters. The number of right alarms N_p is less than the number of good detections N_g because a given earthquake can induce several perturbations at different times. The results for $D = 0$–1500 km and $T = 0$–15 days indicate that the number of false alarms and wrong detections are high, but the good points (the variations which are expected if there is a seismo-electromagnetic effect) of this statistical analysis are:

- the number of earthquakes detected increases with increasing earthquake magnitude;
- the number of perturbations increases with increasing earthquake magnitude;
- the number of perturbations starts increasing remarkably from 4–5 days before the earthquake and attains its maximum value the day before the earthquake (see Figure 18.3).
- the average amplitude of the perturbations increases with increasing earthquake magnitude (not shown here).

A similar type of study has also been carried out for shallow earthquakes ($d < 20$ km). The results are presented in Tables 18.5 and 18.6. As expected the parameters r and n increase relative to the results when all earthquakes are taken into account (cf. Table 18.3) because deep earthquakes have less interaction with Earth's surface.

These results must be considered with caution, however, because a good detection does not mean that we are able to predict the earthquake: there is great uncertainty about the position and even greater uncertainties about the time and the magnitude. Natural variations of electromagnetic signals in the ionosphere give rise to false alarms. Importantly, a number of wrong detections are made, which can be explained by the fact that the satellite is above a seismic area only a few minutes per day, and that we do not expect continuous perturbations before a given earthquake, therefore

Table 18.1 Number of earthquakes considered in past studies.

$4.8 \le Mw \le 5.0$	$5.1 \le Mw \le 6.0$	$Mw \ge 6.1$
12057	8953	853

Table 18.2 Number of earthquakes meeting the limitations in this chapter.

$4.8 \le Mw \le 5.0$	$5.1 \le Mw \le 6.0$	$Mw \ge 6.1$
8081	6110	573

Table 18.3 Statistical results concerning seismic ionospheric influences on the ion density close to the earthquake epicenters ($D = 1500$ km, $d = 0$–1000 km).

	Mw 4.8–5.0		*Mw* 5.1–6.0		*Mw* > 6.1	
T	*r* (%)	*n*	*r* (%)	*n*	*r* (%)	*n*
15–0	55.1	3.5	60.4	4.1	79.8	5.8

Table 18.4 Data list for Table 18.3.

	Mw 4.8–5.0			*Mw* 5.1–6.0			*Mw* > 6.1		
T	N_e	N_g	N_p	N_e	N_g	N_p	N_e	N_g	N_p
15–0	8081	4454	1555	6110	3690	14959	573	457	2649

N_e is the number of earthquakes; N_g is the number of good detections with perturbations close to the earthquake epicenters; N_p is the number of correct alarms close to the earthquake epicenters.

possible perturbations could be missed. It is feasible to reduce the number of wrong detections by simultaneously using several satellites to monitor perturbations above future epicenters for a much longer time period. At the moment the ESA (European Space Agency) SWARM mission has three identical satellites in the ionosphere.

18.4.2.2. "Real Time" Analysis of Perturbations Found Close to Conjugated Points of Epicenters

Using the International Geomagnetic Reference Field (IGRF) model, for each earthquake epicenter the position of its conjugated point at an altitude of 400 km has been considered because it is expected that, if there is a perturbation, it will be more important at the ionospheric height where the density is larger. The same anomaly file as before is considered as this file covers the complete Earth whatever the positions of the earthquakes are. We then search whether there are perturbations close to the conjugated points using the same criteria as before concerning the distance and the time ($D = 0$–1500 km and $T = 0$–15 days). An identical statistical study was performed concerning those perturbations that can be associated with conjugated points of earthquakes. The results are presented in Tables 18.7 and 18.8.

Table 18.7 (8) is similar to Table 18.3 (4) but concerns the statistics at the conjugated point of the epicenters. It can be seen that the ratio r between the number of earthquakes with good detections and the total number of earthquakes is almost similar to the ratio calculated in the epicenter hemisphere and this ratio also increases with increasing earthquake magnitude. Figure 18.4 indicates the variation of the number of conjugated

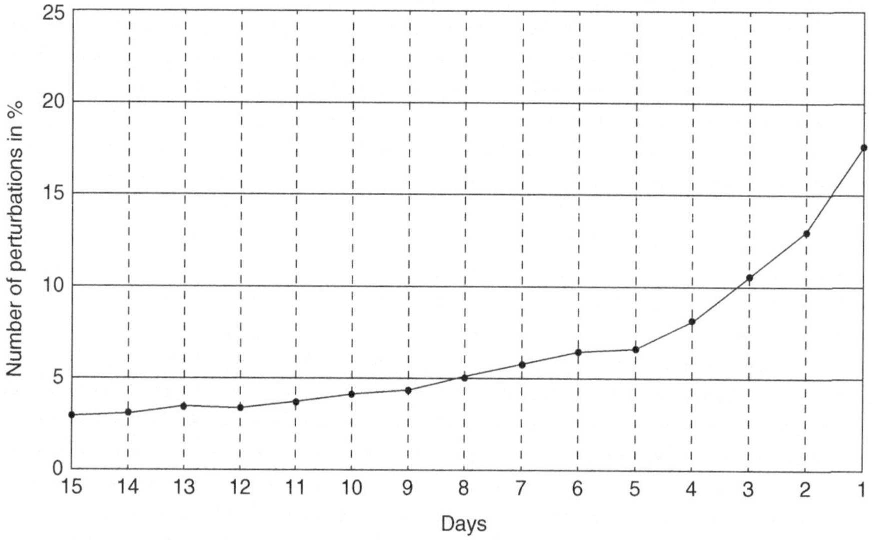

Figure 18.3 Variation of the number of perturbations as a function of the days before the earthquakes. The results are expressed as a percentage relative to the total number of perturbations.

Table 18.5 The statistical results concerning seismic ionospheric influences on the ion density (D = 1500 km, d = 0–20 km).

	Mw 4.8–5.0		Mw 5.1–6.0		Mw > 6.1	
T	*r* (%)	*n*	*r* (%)	*n*	*r* (%)	*n*
15–0	58.2	4.8	63.2	5.4	80.7	6.1

Table 18.7 Statistical results concerning seismic ionospheric influences on the ion density close to the conjugated points of the earthquake epicenters ($D = 1500$ km, $d = 0$–1000 km).

	Mw 4.8–5.0		Mw 5.1–6.0		Mw > 6.1	
T	*r* (%)	*n*	*r* (%)	*n*	*r* (%)	*n*
15–0	54.8	3.0	59.4	4.0	80.5	5.3

Table 18.6 Data list for Table 18.5.

	Mw 4.8–5.0			Mw 5.1–6.0			Mw > 6.1		
T	N_e	N_g	N_p	N_e	N_g	N_p	N_e	N_g	N_p
15–0	3511	2045	9739	2930	1851	9971	296	239	1462

N_e is the number of earthquakes; N_g is the number of good detections with perturbations close to the earthquake epicenters; N_p is the number of correct alarms close to the earthquake epicenters.

Table 18.8 Data list for Table 18.7.

	Mw 4.8–5.0			*Mw* 5.1–6.0			*Mw*>6.1		
T	N_e	N_g	N_p	N_e	N_g	N_p	N_e	N_g	N_p
15–0	8081	4431	16297	6110	3631	14493	573	461	2440

N_e is the number of earthquakes; N_g is the number of good detections with perturbations at the conjugated points; N_p is the number of perturbations at the conjugated points.

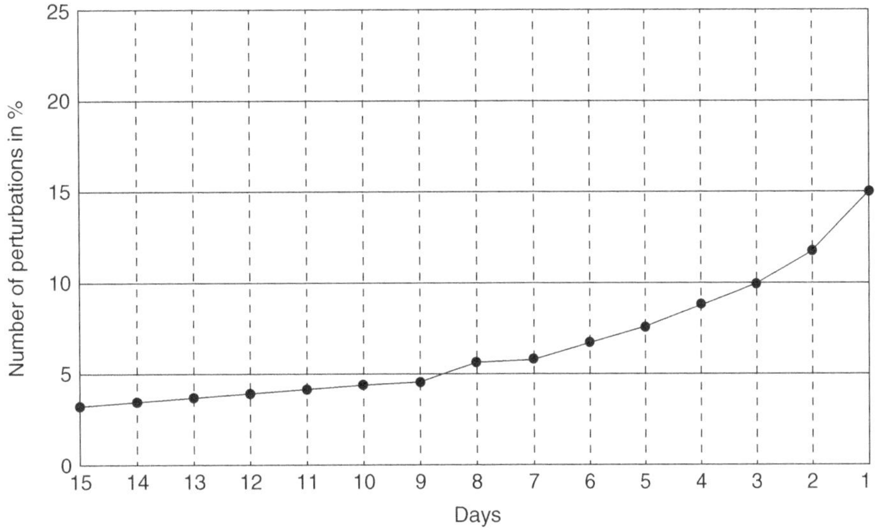

Figure 18.4 Variation of the number of conjugated perturbations as a function of the days before the earthquakes.

perturbations as a function of the days before the earthquakes. Compared to Figure 18.3, the number of perturbations for the day just before the earthquakes is a slightly lower but the overall variation is similar, i.e., the number of perturbations is maximum for days close to the day of the earthquake and smoothly decreases with increasing days from the earthquakes. Our statistical analysis means that it is also possible to check perturbations in the conjugated region of the seismic areas.

As previously undertaken for variations above epicenters, a similar type of study has been carried out in the conjugated areas considering only shallow earthquakes ($d<20$ km). The results are shown in Tables 18.9 and 18.10. The value of r is slightly larger than in Table 18.7, as expected.

Table 18.9 Statistical results concerning seismic ionospheric influences on the ion density close to the conjugated points of the earthquake epicenters ($D=1500$ km, $d=0-20$ km).

	Mw 4.8–5.0		*Mw* 5.1–6.0		*Mw*>6.1	
T	*r* (%)	*n*	*r* (%)	*n*	*r* (%)	*n*
15–0	58.6	4.6	62.3	4.6	81.8	5.2

Table 18.10 Data list for Table 18.9.

	Mw 4.8–5.0			*Mw* 5.1–6.0			*Mw*>6.1		
T	N_e	N_g	N_p	N_e	N_g	N_p	N_e	N_g	N_p
15–0	3511	2058	9370	2930	1826	8446	296	242	1261

N_e is the number of earthquakes; N_g is the number of good detections with perturbations at the conjugated points; N_p is the number of perturbations at the conjugated points.

18.4.3. Application to the Samoa Earthquake

The M8.1 Samoa earthquake occurred on 29 September 2009 with an epicenter located at 15.49°S, 172.10°W. The related ionospheric anomalies have been studied previously by *Akhoondzadeh et al.* [2010a,b] and *Akhoondzadeh* [2013]. Here we want to see whether it would have been possible to predict this earthquake using the database of ionospheric anomalies described in section 18.4.2 [*Li and Parrot*, 2012]. Again these anomalies were detected automatically. We check these anomalies day by day as if we were in real time, considering an area limited in latitude (between 30°S and 0°) and longitude (between 174°E and 157°W). We know nothing except the past. Using 2 months of data before the earthquake, density variations have been monitored in the vicinity of the epicenter by *Akhoondzadeh et al.* [2010a]. Their figures 1 and 2 indicate that increase of the plasma density is very uncommon at this location and at this time. In our database we select relative anomalies larger than 10%. The result of our process is shown in Figure 18.5. The days for each panel are running from the left to the right and from the top to the bottom. We start to plot an anomaly on the map on 16 September 2009, and then we automatically plotted other anomalies for the other days. On 20 September 2009 we have several anomalies in the area and we decided to average the positions of all anomalies in order to determine the position of a possible earthquake. This position is indicated by a blue triangle. The time proceeds we continue the automatic process. Every day we recalculate the position of a possible earthquake, but when we arrive at 29 September, 2009, the day of the earthquake, it is impossible to claim that it is the day of the earthquake and consequently impossible to predict the time of the earthquake. We can provide only a warning that anomalies are registered in the vicinity of a given location. Concerning the magnitude of this earthquake we can only say that it would be large, because anomalies started to appear 15 days previously (the delay between the occurrence of precursors and the occurrence of earthquakes increases with the magnitude of earthquakes).

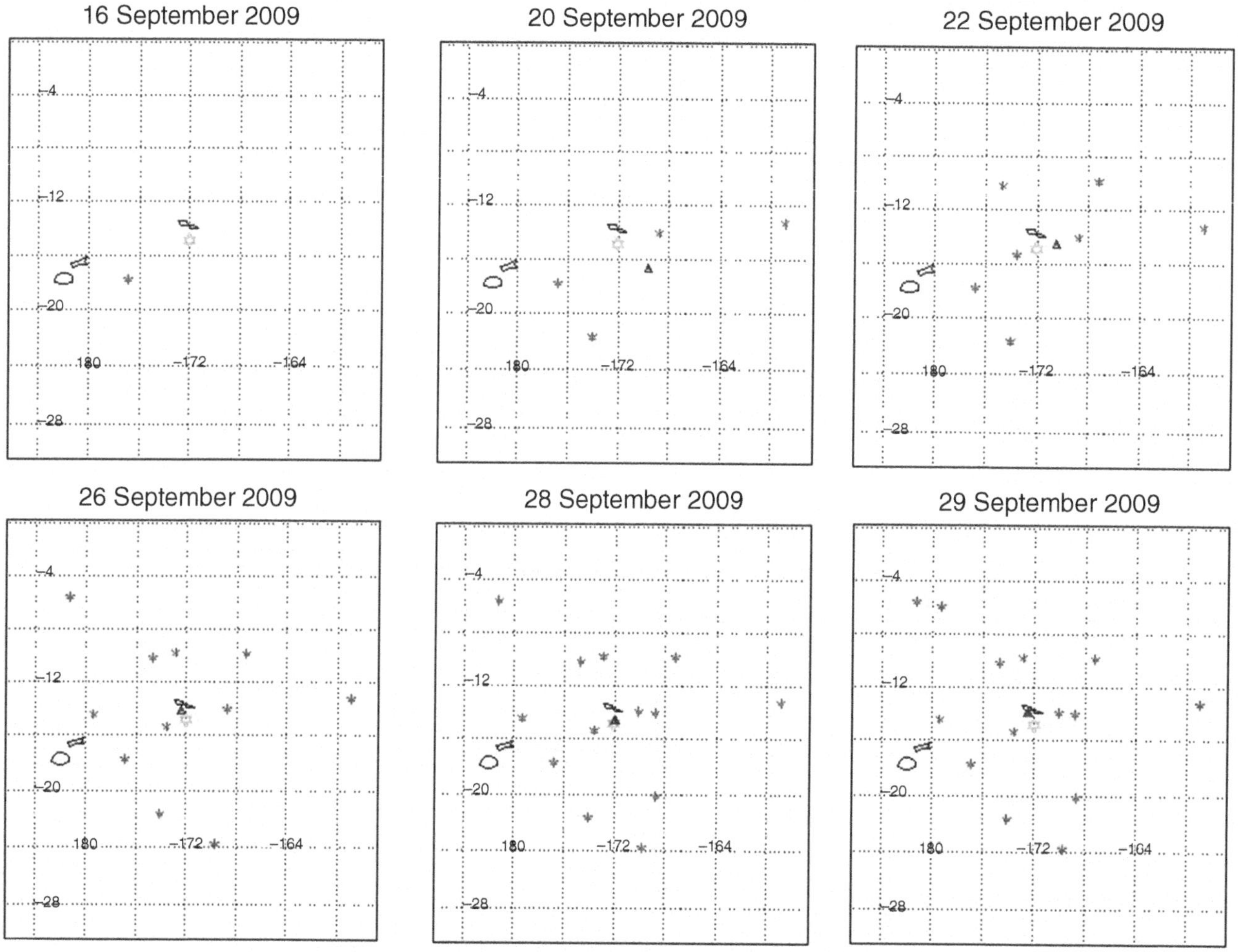

Figure 18.5 Maps of the Samoa region where latitude and longitude coordinates are indicated. The red crosses indicate locations of ionospheric perturbations recorded prior to the 2009 earthquake whose epicenter is indicated by the green star. The last panel corresponds to the day of the earthquake. The blue triangle corresponds to the automatic determination of the epicenter obtained by the average of the anomaly positions (see text for explanation). (*See electronic version for color representation.*)

18.5. DISCUSSION

Even though the mechanism to produce the ionospheric perturbations prior to earthquakes is not known at the moment, there are several hypotheses to explain these phenomena (see section 18.1). Depending on the crust configuration, it cannot be excluded that a mechanism could be efficient in a given seismic area and not in another one. It can, however, be said that there is always a fairweather electric field between the ground and the lower ionosphere [*Rycroft et al.*, 2008; *Pulinets*, 2009, and references therein]; this electric field is enhanced during thunderstorm activity, for example. If there is a change in the crust at the time of an earthquake, it is certain that it will induce changes at ionospheric levels.

18.6. CONCLUSIONS

Finally, this review underlines the important role of satellite observations in the study of lithosphere–atmosphere–ionosphere coupling. It has been shown statistically that the seismic activity induces perturbations in the ionosphere between a few hours and a few days before the earthquakes. It remains the case, however, that the information we can extract from the ionosphere with a single satellite measurement is far from accurate enough to be able to make earthquake prediction (i.e. to predict time, location, and magnitude). It also has been shown that it might be useful to combine observations close to the future epicenters with those at their magnetically conjugated locations. Measurements with several satellites with more frequent passes over a given seismic zone, in combination with ground-based experiments, will greatly improve surveys of the seismic areas. It is expected that new satellites in other countries will be launched soon to continue this survey of the ionospheric variations related to seismic activities. At the moment, another dedicated ionospheric satellite named CSES (China Seismo-Electromagnetic Satellite) has been developed in China and launched in February 2018 [*Shen*, 2014]. It must also be noted that large earthquakes occurring in areas well covered by experiments, such as the Wenchuan earthquake, provide useful data to validate LAIC models.

ACKNOWLEDGMENTS

The DEMETER satellite was supported by the Centre National d'Etudes Spatiales. ML acknowledges funding from NSFC (National Natural Science Foundation of China) under grant agreement n°41204057. The data presented in this paper are available at https://cdpp-archive.cnes.fr/.

REFERENCES

Akhoondzadeh, M. (2013), Novelty detection in time series of ULF magnetic and electric components obtained from the DEMETER satellite experiments above Samoa (29 September 2009) earthquake region, *Nat. Hazards Earth Syst. Sci.*, *13*, 15–25, doi: 10.5194/nhess-13-15-2013

Akhoondzadeh, M., Parrot, M., and Saradjian, M. R. (2010a), Electron and ion density variations before strong earthquakes ($M > 6.0$) using DEMETER and GPS data, *Nat. Hazards Earth Syst. Sci.*, *10*, 7–18.

Akhoondzadeh, M., Parrot, M., and Saradjian, M. R. (2010b), Investigation of VLF and HF waves showing seismo-ionospheric anomalies induced by the 29 September 2009 Samoa earthquake (Mw=8.1), *Nat. Hazards Earth Syst. Sci.*, *10*, 1061–1067, doi: 10.5194/nhess-10-1061-2010

De Santis, A., De Franceschi, G., Spogli, L., Perrone, L., Alfonsi, L., Qamili, E., Cianchini, G., Di Giovambattista, R., Salvi, S., Filippi, E., Pavon-Carrasco, F. J., Monna, S., Piscini, A., Battiston, R., Vitale, V., Picozza, P. G., Conti, L., Parrot, M., Pincon, J.-L., Balasis, G., Tavani, M., Argan, A., Piano, G., Rainone, M. L., Liu, W., and Tao, D. (2015), Geospace perturbations induced by the Earth: the state of the art and future trends. *Phys. Chem. Earth, Parts A/B/C*, *85–86*, 17–33.

Hayakawa, M., Asano, T., Rozhnoi, A., and Solovieva, M. (2018), Very-low- to low-frequency sounding of ionospheric perturbations and possible association with earthquakes, in D. Ouzounov, S. Pulinets, K. Hattori, and P. Taylor (eds), *Pre-Earthquake Processes: A Multi-disciplinary Approach to Earthquake Prediction Studies*, pp. 277–304, Geophysical Monograph 234, American Geophysical Union, Washington, DC, and John Wiley & Sons, Inc., Hoboken, NJ.

He, Y., Yang, D., Zhu, R., Qian, J., and Parrot, M. (2010), Variations of electron density and temperature in ionosphere based on the DEMETER ISL data, *Earthquake Sci.*, *23*(4), 349–355, doi: 10.1007/s11589-010-0732-8

He, Y., Yang, D., Qian, J., and Parrot, M. (2011a), Response of the ionospheric electron density to different types of seismic events, *Nat. Hazards Earth Syst. Sci.*, *11*, 2173–2180, doi: 10.5194/nhess-11-2173-2011

He, Y., Yang, D., Qian, J., and Parrot, M. (2011b), Anomaly of the ionospheric electron density close to earthquakes: Case studies of Pu'er and Wenchuan earthquakes, *Earthquake Sci.*, *24*, 549–555

Hobara, Y., Nakamura, R., Suzuki, M., Hayakawa, M., and Parrot, M. (2013), Ionospheric perturbations observed by the low altitude satellite DEMETER and possible relation with seismicity, *J. Atmos. Electr.*, *33*, 21–29, doi:/10.1541/jae.33.21.

Kuo, C. L., Ho, Y. Y., and Lee, L. C. (2018), Electrical Coupling Between the Ionosphere and Surface Charges in the Earthquake Fault Zone, in D. Ouzounov, S. Pulinets, K. Hattori, and P. Taylor (eds), *Pre-Earthquake Processes: A Multi-disciplinary Approach to Earthquake Prediction Studies*, pp. 99–124, Geophysical Monograph 234, American Geophysical Union, Washington, DC, and John Wiley and Sons, Inc., Hoboken, NJ.

Li, M. and Parrot, M. (2012), "Real Time Analysis"of the Ion Density measured by the satellite DEMETER in relation with the seismic activity, *Nat. Hazards Earth Syst. Sci.*, *12*, 2957–2963.

Li, M. and Parrot, M. (2013), Statistical analysis of an ionospheric parameter as a base for earthquake prediction, *J. Geophys. Res. Space Phys.*, *118*(6), 3731–3739.

Liu, L., Wan, W., Yue, X., Zhao, B., Ning, B., and Zhang, M. L. (2007), The dependence of plasma density in the topside ionosphere on the solar activity level, *Ann. Geophys.*, *25*(6), 1337–1343.

Liu, J.-Y, Hattori, K., and Chen, Y. (2018), Application of total electron content derived from the Global Navigation Satellite System for detecting earthquake precursors, in D. Ouzounov, S. Pulinets, K. Hattori, and P. Taylor (eds), *Pre-Earthquake Processes: A Multidisciplinary Approach to Earthquake Prediction Studies*, pp. 305–317, Geophysical Monograph 234, American Geophysical Union, Washington, DC, and John Wiley and Sons, Inc., Hoboken, NJ.

Ma, T. and Wu, Z. (2012), Precursor-like anomalies prior to the 2008 Wenchuan Earthquake: a critical-but-constructive review, *Int. J. Geophys.*, 2012, article ID 583097, doi: 10.1155/2012/583097

Němec, F., Santolík, O., Parrot, M., and Berthelier, J. J. (2008), Spacecraft Observations of Electromagnetic Perturbations connected with Seismic Activity, *Geophys. Res. Lett.*, *35*(5), L05109.

Němec, F., Santolík, O., and Parrot, M. (2009), Decrease of intensity of ELF/VLF Waves observed in the upper ionosphere close to earthquakes: a statistical study, *J. Geophys. Res. Space Phys.*, *114*(A4), A04303.

Parrot, M. (2006), First results of the DEMETER micro-satellite, *Planet. Space Sci.*, *54*(5), 411–558.

Parrot, M., Berthelier, J. J., Lebreton, J. P., Sauvaud, J. A., Santolik, O., and Blecki, J. (2006), Examples of unusual ionospheric observations made by the DEMETER satellite over seismic regions, *Phys. Chem. Earth, Parts A/B/C*, *31*(4), 486–495.

Parrot, M., Sauvaud, J. A., Soula, S., Pinçon, J. L., and van der Velde, O. (2013), Ionospheric density perturbations recorded by DEMETER above intense thunderstorms, *J. Geophys. Res. Space Phys.*, *118*, doi: 10.1002/jgra.50460

Píša, D., Němec, F., Parrot, M., and Santolík, O. (2012), Attenuation of electromagnetic waves at the frequency ~1.7 kHz in the vicinity of earthquakes observed in the upper ionosphere by the DEMETER satellite, *Ann. Geophys.*, *55*(1), 157–163.

Píša, D., Němec, F., Santolík, O., Parrot, M., and Rycroft, M. (2013), Additional attenuation of natural VLF electromagnetic waves observed by the DEMETER spacecraft resulting from pre-seismic activity, *J. Geophys. Res. Space Phys.*, *118*(8), 5286–5295.

Pulinets, S. A. (2009), Physical mechanism of the vertical electric field generation over active tectonic faults, *Adv. Space Res.*, *44*(6), 767–773, doi: 10.1016/j.asr.2009.04.038

Pulinets, S. A., Ouzounov, D. P., Karelin, A. V., and Davidenko, D. V. (2015), Physical bases of the generation of short-term earthquake precursors: A complex model of ionization-induced geophysical processes in the lithosphere-atmosphere-ionosphere-magnetosphere system. *Geomagn. Aeron.*, *55*(4), 521–538.

Pulinets, S., Ouzounov, D., Karelin, A., and Dmitry Davidenko, D. (2018), Lithosphere–atmosphere–ionosphere–magnetosphere coupling—a concept for pre-earthquake signals generation, in D. Ouzounov, S. Pulinets, K. Hattori, and P. Taylor (eds), *Pre-Earthquake Processes: A Multi-disciplinary Approach to Earthquake Prediction Studies*, pp. 79–98, Geophysical Monograph 234, American Geophysical Union, Washington, DC, and John Wiley and Sons, Inc., Hoboken, NJ.

Rycroft, M. J., Harrison R. G., Nicoll, K. A., and Mareev, E. A. (2008), An overview of Earth's global electric circuit and atmospheric conductivity, *Space Sci. Rev.*, *137*, 83–105.

Ryu, K., Parrot, M., Kim, S. G., Jeong, K. S., Chae, J. S., Pulinets, S., and Oyama, K.-I. (2014), Suspected seismo-ionospheric coupling observed by satellite measurements and GPS TEC related to the M7.9 Wenchuan earthquake of 12 May 2008, *J. Geophys. Res. Space Phys.*, *119*, doi: 10.1002/2014JA020613

Sauvaud, J.-A., Maggiolo, R., Jacquey, C., Parrot, M., Berthelier, J.-J., Gamble, R. J., and Rodger, C. J. (2008), Radiation belt electron precipitation due to VLF transmitters: Satellite observations, *Geophys. Res. Lett.*, *35*, L09101, doi: 10.1029/2008GL033194

Shen, X. (2014), The experimental satellite on electromagnetism monitoring, *Chin. J. Space Sci.*, *34*(5), 558–562.

Singh, R. P., Mehdi, W., Gautam, R., Kumar, J. S., Zlotnicki, J., and Kafatos, M. (2010), Precursory signals using satellite and ground data associated with the Wenchuan Earthquake of 12 May 2008, *Int. J. Remote Sens.*, *31*(13), 3341–3354.

Walker, S. N., Kadirkamanathan, V., and Pokhotelov, O. A. (2013), Changes in the ultra-low frequency wave field during the precursor phase to the Sichuan earthquake: DEMETER observations, *Ann. Geophys.*, *31*, 1597–1603, doi: 10.5194/angeo-31-1597–2013

Yan X., Shan, X., Zhang, X., Qu, C., Tang, J., Wang, F., and Wen, S. (2013), Multiparameter seismo-ionospheric anomaly observation before the 2008 Wenchuan, China, *Mw*7.9 earthquake, *J. Appl. Remote Sens.*, *7*(1), 073532, doi: 10.1117/1.JRS.7.0773532

Zhang, X. and Shen, X. (2011), Electromagnetic anomalies around the Wenchuan earthquake and their relationship with earthquake preparation, *Int. J. Geophys.*, article ID 904132, doi: 10.1155/2011/904132

Zhang, X., Shen, X., Liu, J., Ouyang, X., Qian, J., and Zhao, S. (2009), Analysis of ionospheric plasma perturbations before Wenchuan earthquake. *Nat. Hazards Earth Syst. Sci.*, *9*, 1259–1266.

Part VII: Interdisciplinary Approach to Earthquake Forecasts/Predictions

19

Significant Cases of Preseismic Thermal Infrared Anomalies

Valerio Tramutoli[1], Nicola Genzano[1], Mariano Lisi[1], and Nicola Pergola[2]

ABSTRACT

From appropriate identification and real-time integration of independent observations we expect to improve significantly our present capability to assess the seismic hazard in the short term (from weeks to days before an earthquake). One specific observation (e.g. anomaly in one parameter) sometimes can be used as a trigger or as a reference point (in the space and/or time domain) for activating/improving analysis on other independent parameters whose systematic computation otherwise could be computationally very expensive or impossible. In this chapter, we briefly exemplify these advantages, and the utility of the Robust Satellite Technique (RST) approach. The RST data analysis approach has been implemented on daily TIR satellite records collected over three different areas (Italy, Greece, and Turkey) by the geostationary satellite sensor Spinning Enhanced Visible and Infrared Imager (SEVIRI) on board Meteosat Second Generation (MSG) satellite, in order to evaluate its possible contribution to an improved multiparametric system for a time-dependent assessment of seismic hazard.

19.1. INTRODUCTION

Chapter 14 introduced and described the Robust Satellite Technique (RST) approach [*Tramutoli*, 1998, 2005, 2007], as well as demonstrating its application to the identification of preseismic near-surface thermal anomalies. In this chapter some illustrative examples are presented to better highlight the potential of integrating multiparametric (e.g., chemical, physical, biological, etc.) observations; Table 19.1 lists those earthquakes taken into account. Although individual techniques can identify anomalous variations that can be, to some extent, associated with the complex processes of earthquake preparation, a multidisciplinary approach can be expected to improve our present capability of assessing seismic hazard in the short-medium term.

[1]*School of Engineering, Basilicata University, Potenza, Italy*

[2]*Institute of Methodologies for Environmental Analysis of the National Research Council, Tito Scalo, Italy*

19.2. ILLUSTRATIVE CASES OF PRESEISMIC THERMAL INFRARED ANOMALIES

In the framework of the "S3—Short-Term Earthquake Prediction and Preparation" project, supported by the National Department of Civil Protection (DPC) and the National Institute for Geophysics and Volcanology (INGV), the RST approach was used to monitor the space–time fluctuations of Earth's thermal emission in possible relation to the 2012 Emilia (Northern Italy) seismic sequences. As reported by *Scognamiglio et al.* [2012], the Emilia region was hit by a seismic sequence at the end of May 2012. The main shock with a magnitude of *Mw* 5.8 occurred on 20 May 2012 at 6.3 km depth and epicenter at 44.89°N, 11.23°E. Another six earthquakes with *Mw* > 5.0 followed. The largest aftershock, considered as

Pre-Earthquake Processes: A Multidisciplinary Approach to Earthquake Prediction Studies, Geophysical Monograph 234, First Edition. Edited by Dimitar Ouzounov, Sergey Pulinets, Katsumi Hattori, and Patrick Taylor.

Table 19.1 List of analyzed seismic events.

Case studies	Date	Magnitude	Epicenter coordinates	Source
Emilia seismic sequence	20 May 2012	*Mw* 5.8	44.896°N, 11.264°E	Italian Seismic Instrumental and Parametric Database seismic catalog [*INGV*, 2016]
	29 May 2012	*Mw* 5.6	44.842°N, 11.066°E	
Turkish seismic events	22 July 2012	M_L 5.6	37.5740°N, 36.3707°E	DDA catalog [*AFAD*, 2016]
	16 October 2012	M_L 4.6	37.2710°N, 37.1372°E	
	16 October 2012	M_L 4.6	37.2773°N, 37.1484°E	
Aegean Sea seismic sequence	29August 2014	*M* 5.7	36.67°N, 23.67°E	National Observatory of Athens catalog [*NOA*, 2016]

the second main shock, was a *Mw* 5.6 earthquake that occurred on 29 May 2012 (07:00:03 UTC) [*Scognamiglio et al.*, 2012].

To study Earth's thermal emission during the different phases of the Emilia seismic sequence, all nighttime (i.e. 00.00–00.15 UTC time slot) Thermal InfraRed (TIR) records available over the Italian peninsula during the months of May and June measured by the MSG-SEVIRI (Meteosat Second Generation—Spinning Enhanced Visible and Infrared Imager) satellite sensor, from 2004 to 2012, were used. The TIR anomalies were identified using the RST methodology [*Tramutoli et al.*, 2005] in combination with the Robust Estimator of TIR Anomalies (RETIRA) index [*Filizzola et al.*, 2004].

Figure 19.1 shows a sequence of daily images with the distribution of the RETIRA values over northern Italy, from 10 to 20 May 2012 (day of main shock occurrence). Due to a cloud cover that was frequently present over the area in the same days, it was not possible to recognize the space–time persistence required to qualify the detected TIR anomalies as a preseismic Significant Sequence of Thermal Anomalies (SSTA). According to *Tramutoli et al.* [2005] and *Genzano et al.* [2015], a SSTA is defined as a sequence of TIR anomalies with a relative intensity of RETIRA index (in this case ≥ 3), not due to particular meteorological conditions (e.g. wide cloudy coverage and/or known spurious effects, such as cold spatial average effect; see *Aliano et al.* [2008], *Genzano et al.* [2009], and *Eleftheriou et al.* [2016]) and navigational errors, with a spatial extension at least 150 km^2 within an area of 1° × 1° and a temporal persistence within an area of 1° × 1° around *x*,*y* of at least one more time in the 7 days preceding/following their appearance at time *t* (see Chapter 14 and *Genzano et al.* [2015] and *Eleftheriou et al.* [2016]). Nevertheless, it is worth noting that the low-intensity TIR anomalies with 2 ≤ RETIRA < 3 that appear on 13 May 2012 in a quite evident spatial relationship with the main fault system of the Northern Apennine area that originated the Emilia earthquakes, occurred 7 days (*Mw* 5.8) and 16 days (*Mw* 5.6) after the first detection of the TIR anomalies.

Usually, low-intensity TIR anomalies (RETIRA < 3) are not considered significant, therefore for a meaningful evaluation we need to consider them in combination with other parameters. The same area affected by the presence of TIR anomalies was considered using the CN algorithm (named after its development application to California and Nevada) for Italy [*Peresan et al.*, 2005] as an area with an active state of Times of Increased Probability (TIP) for the occurrence of earthquakes with $M \geq 5.4$ during the period March–June 2012. Briefly, the CN algorithm [*Keilis-Borok and Rotwain*, 1990] is based on a pattern recognition scheme [*Keilis-Borok and Soloviev*, 2003, and references therein] devoted to determine the TIP for the occurrence of strong earthquakes. For specific regions (mainly identified on the basis of the spatial distribution of seismicity and the geometry of fault systems) and time windows, the CN algorithm indicates the probability of occurrence of earthquakes with magnitude greater than a fixed threshold M_0 (i.e. $M_0 = 5.4$ for northern Italy) on the basis of the quantification of seismicity patterns observed for the events that occurred in the analysed region, the level of seismic activity, the seismic quiescence, and space–time clustering of the events.

So it is possible that even low intensity (i.e. RETIRA < 3) TIR anomalies with nonappreciable space–time persistence (Figure 19.2) can assume significance when analysed in combination with other methods (like CN products). Moreover, the intersection of medium-term alert (until June 2012) areas identified using the CN algorithm with those identified by low-intensity TIR anomalies (RETIRA > 2), which usually occur from a few weeks to a few days before an earthquake (with $M > 4$, e.g. *Eleftheriou et al.* [2016]), allows the space-time window of an alert to be restricted (Figure 19.2). Attention can thereby be focused on this restricted area in terms of the precaution and prevention procedures [e.g. *Peresan et al.*, 2012] that are normally carried out during periods of

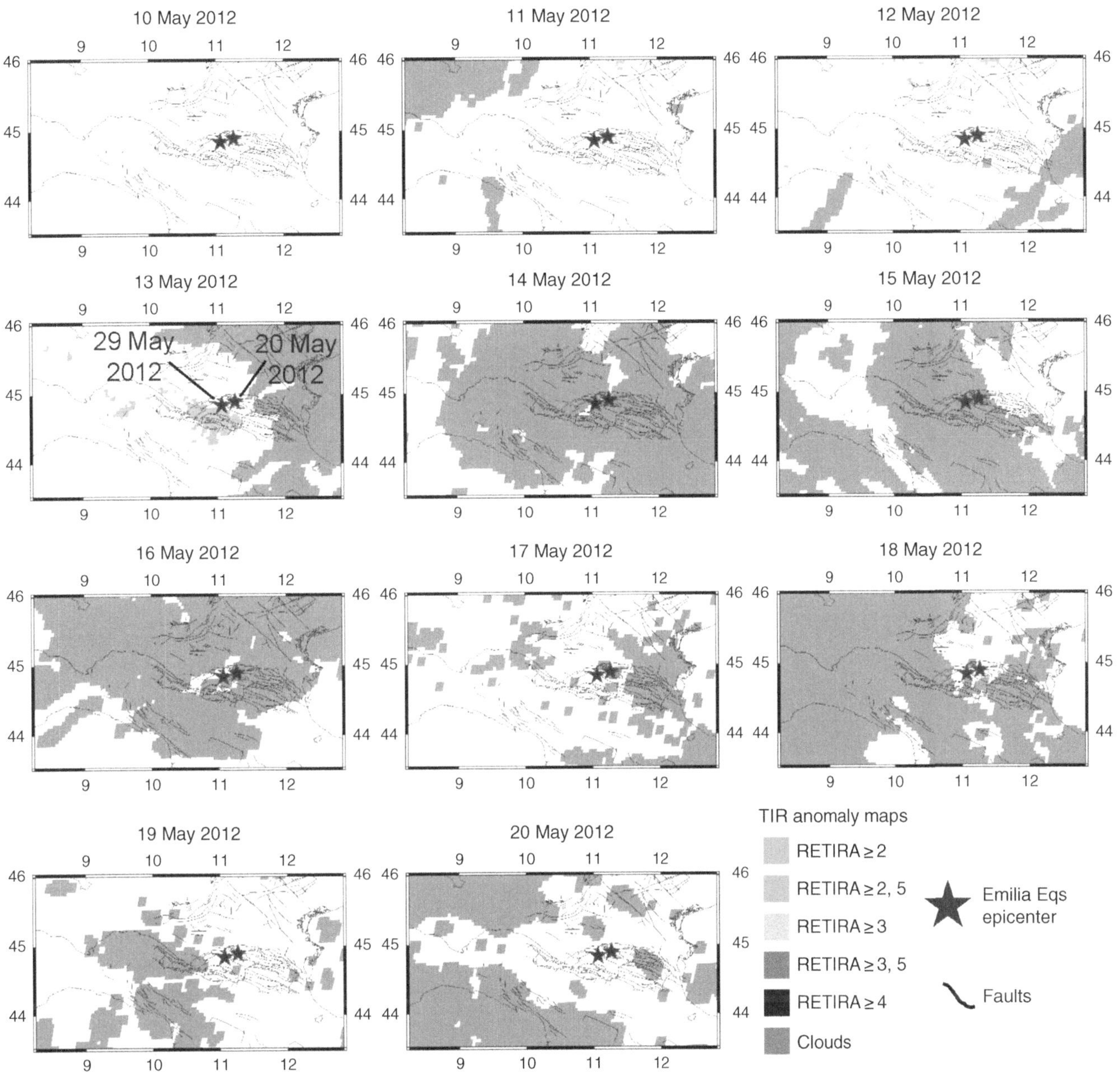

Figure 19.1 TIR anomaly maps in the days before the *Mw* 5.8 main shock (20 May 2012) of the Emilia seismic sequence. Low intensity (RETIRA≥2) TIR anomalies located near tectonic lineaments (black lines) of Northern Apennine appear in the epicentral area on May 13, 2012.

seismic crisis, and also on addressing additional analysis to reduce the uncertainty of the seismic hazard assessment provided.

The RST approach was also used during the Pre-earthquakes Real-time Integration and Monitoring Experiment (PRIME) performed in the framework of the Processing Russian and European EARTH observations for earthQUAKE precursor Studies (EC/FP7 PRE-EARTHQUAKES) project [http://www.pre-earthquakes.org/]. This approach was used,in order to evaluate the added value of the integration of independent candidate precursor signals in different seismogenic areas, including Turkey. From October 2011, based on TIR records collected by MSG/SEVIRI, a daily TIR anomaly map was generated systematically applying the RST methodology. Figures 19.3 and 19.4 show the areas of SSTAs located near the same tectonic lineaments present in the area of Maras triple junction [*Chorowicz et al.*, 1994], where African, Anatolian, and Arabian plates collide (central-southern part of Turkey).

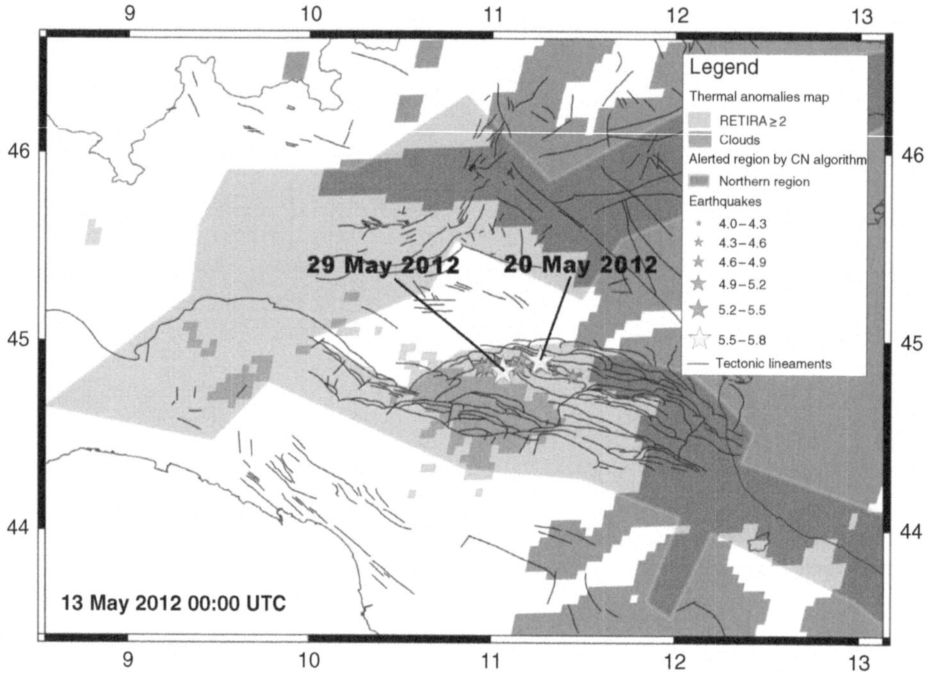

Figure 19.2 A view of the Po Plain region (northern Italy): the alerted area (purple) monitored by the CN algorithm for an earthquake with $M \geq 5.4$ during the period January–June 2012 is superimposed on TIR anomalies (green) highlighted by the RST approach on 13 May 2012. Stars indicate earthquakes with $M \geq 4$ that occurred in May 2012). (*See electronic version for color representation.*)

In particular:

• Figure 19.3 shows TIR anomalies with RETIRA ≥ 3 from 18 to 20 July 2012, located near the main faults present in the Maras area, that appeared 4 days before the occurrence of a $M > 5$ earthquake on 22 July 2012;

• the same region was affected by high-intensity TIR anomalies (i.e. RETIRA ≥ 3), having a shape similar to the above-mentioned TIR anomalies, from 28 to 30 September 2012 (Figure 19.4). The TIR anomalies appeared 18 days before two M 4.6 earthquakes on 16 October 2012.

In this particular case the evident similarity (in shape and space–time evolution) of the TIR anomalies observed in September 2012 with those observed in July, provides a further element that increases forecast strength. Moreover, this suggests that a typical shape/space-time evolution of some pre-seismic TIR anomalies could reflect a specific local tectonic setting and the main degassing regime associated with it. This also testifies to the importance, mainly in an Operational Earthquake Forecastcasting (OEF) context, of continuous observations, which allow accumulation of knowledge for improving and strengthening the quality of the seismic forecasts based on past activity and will better address the decision-making process of civil protection authorities. As was the case during the PRIME, the observation of a similar pre-seismic SSTA in July 2012 stimulated a program of further observations that confirmed the presence of anomalous variations detected using other parameters [see *PRE-EARTHQUAKES*, 2013].

Another interesting example that demonstrates how the integration of TIR anomalies identified by the RST method with other parameters could be used for improving seismic hazard assessment in the short–medium term, is that provided by a seismic sequence that occurred in the Aegean Sea (Greece) during August 2014. Based on TIR records collected by the MSG-SEVIRI satellite sensor, a SSTA was identified from 17 to 25August in that area a few days before the M 5.7 earthquake that occurred on 29August 2014 (Figure 19.5).

Like the nearby Turkey examples, the daily real-time analysis of TIR anomaly maps allowed recognition in real-time the particular space–time relation between the seismicity and TIR anomalies occurring in the area. In fact, as shown in Figure 19.5, from 22 to 24 August 2014 several seismic events with magnitude $M \geq 1.5$ surrounded the areas affected by the high-intensity TIR anomalies (RETIRA ≥ 3) where a few days later on 29August 2014 the M 5.7 earthquake occurred.

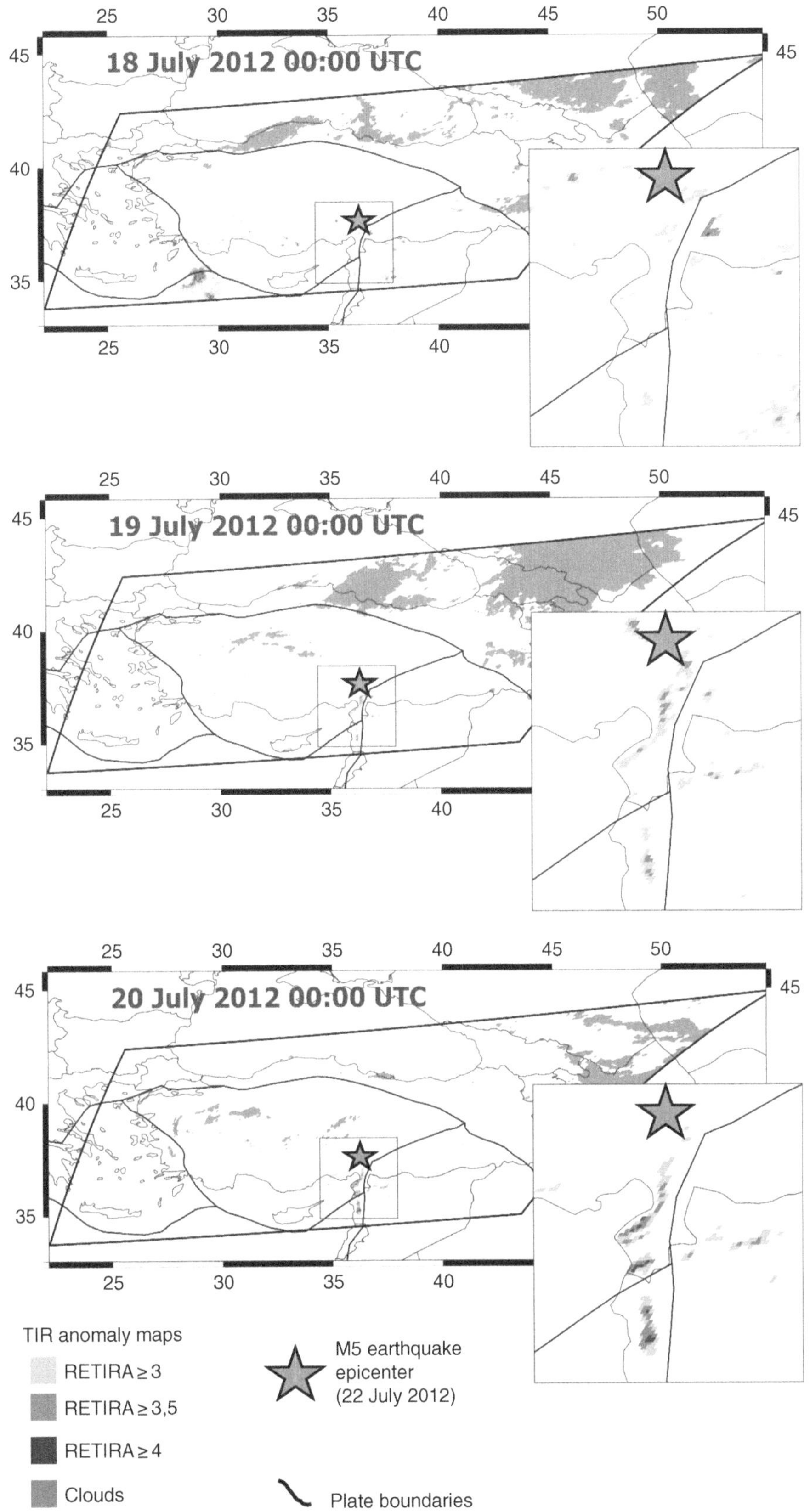

Figure 19.3 Significant Sequence of TIR Anomalies identified at Maras triple junction on 18–20 July 2012. Green star indicates the epicenter of the *M* 5 Kahramanmaras earthquake of 22 July 2012. Note that the linear shaped TIR anomalies close tectonic lineaments start to appear 4 days before the Kahramanmaras earthquake. (*See electronic version for color representation.*)

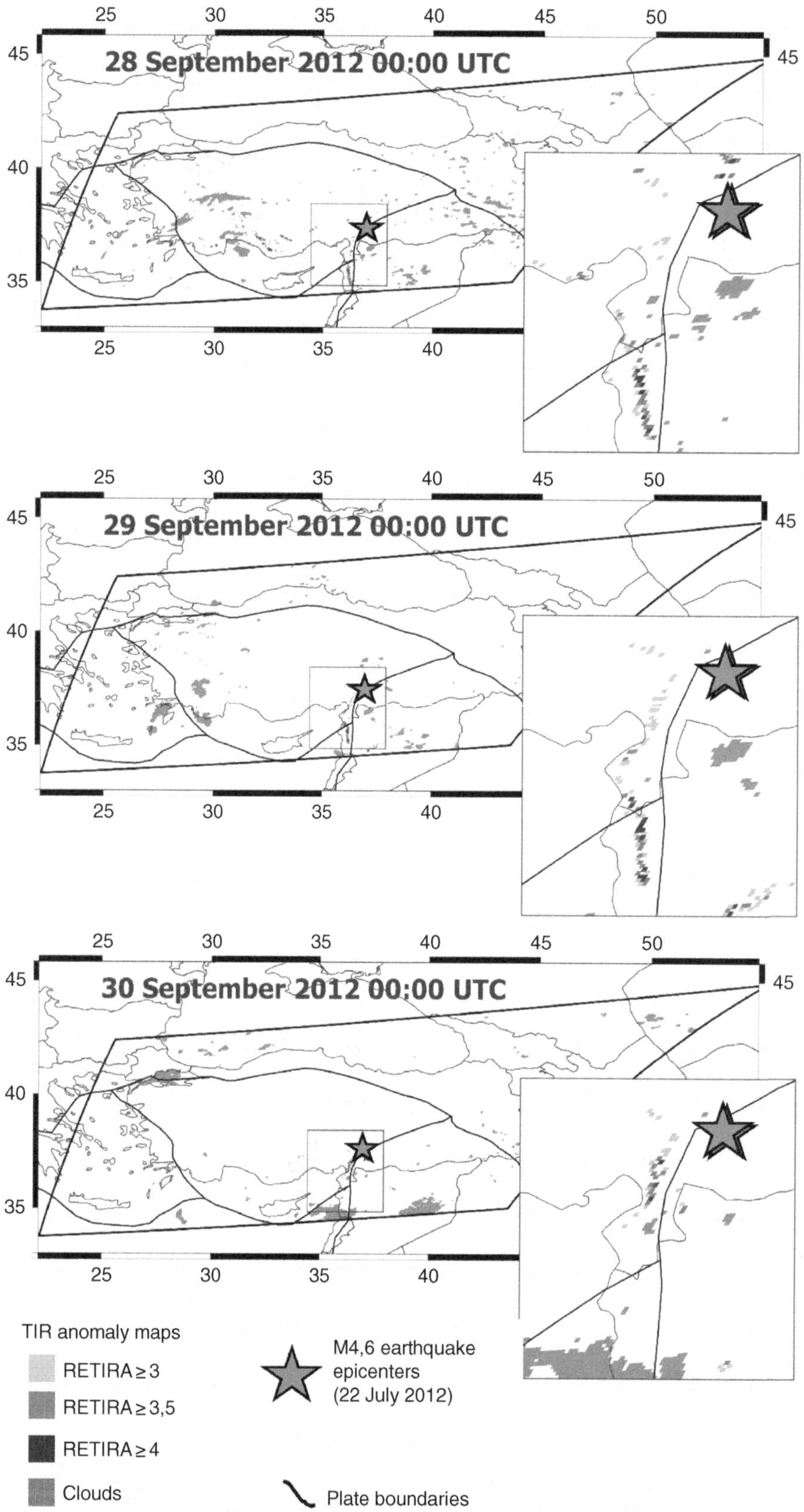

Figure 19.4 Significant Sequence of TIR Anomalies identified at Maras triple junction on 28–30 September 2012. Two *M* 4.6 earthquakes occurred on 16 October 2012 (double green stars). Note the similar shape of the TIR anomalies to those highlighted in July 2012 (Figure 19.3). The anomalies precede the two earthquakes that occurred in the same area. (*See electronic version for color representation.*)

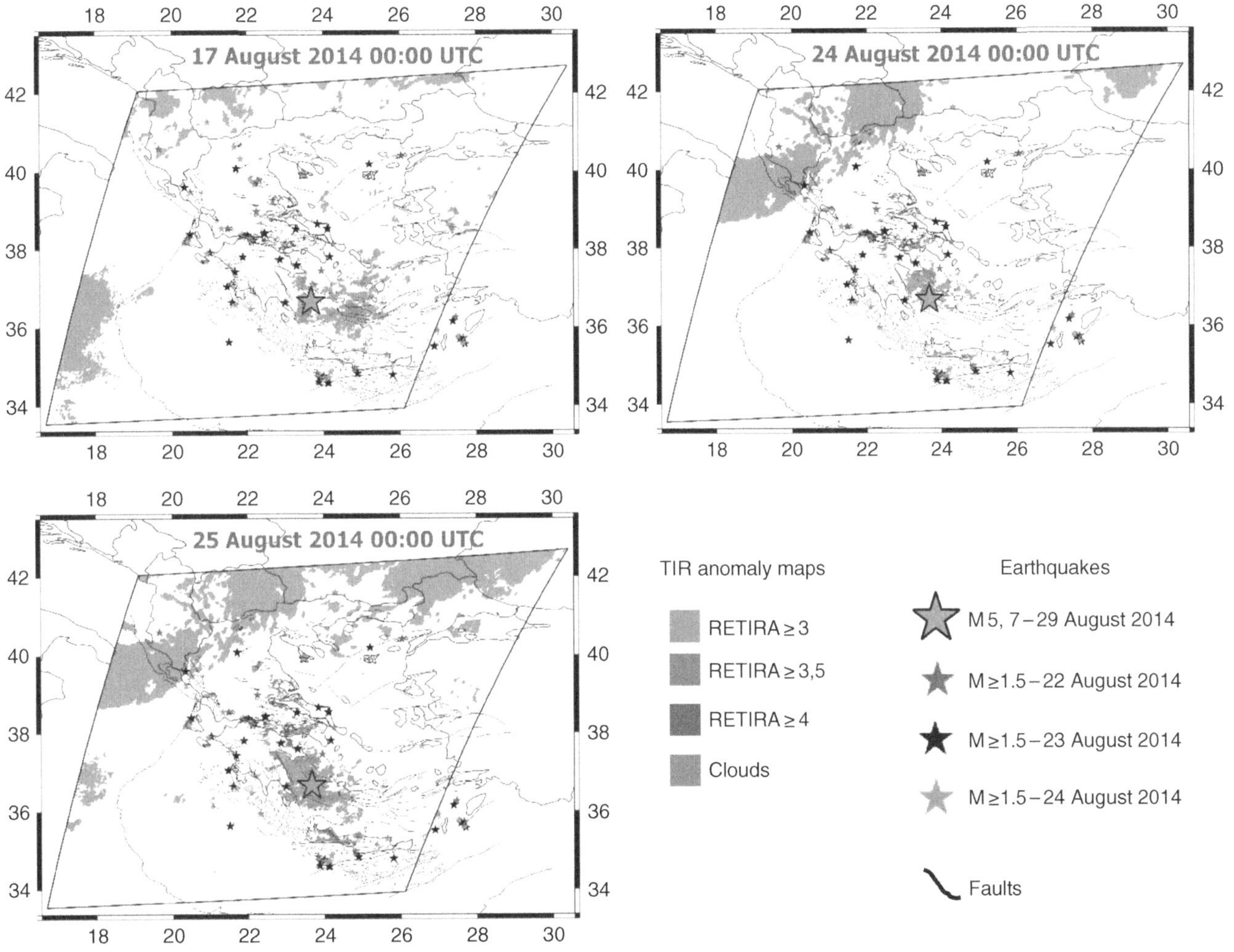

Figure 19.5 Significant Sequence of TIR Anomalies identified in the Aegean Sea on 17–25 August 2014. High-intensity TIR anomalies (RETIRA ≥ 3) appeared from 17 to 25 August 2014, few days before the *M* 5.7 earthquake of 29 August 2012 in the Aegean Sea (green star). Note that several earthquakes with $M \geq 1.5$ occurred in an area surrounding that affected by TIR anomalies on 22–24 August. (*See electronic version for color representation.*)

19.3. CONCLUSIONS

The three nontrivial examples presented in this chapter indicate the added value of integration using a multiparametric system for the time-Dependent Assessment of Seismic Hazard (t-DASH). These three examples demonstrate the potential of the parameter "satellite preseismic TIR anomalies highlighted by the RST approach in combination with the RETIRA index" and its possible use (also in a possible operational capacity) in a t-DASH system.

The first example concerned the seismic sequence that struck the Emilia region at the end of May 2012. For that case thermal anomalies were observed 1 week before the main shock (*Mw* 5.8) occurred on 20 May 2012, but it was impossible to confirm their temporal persistence due to the quite constant presence of clouds over the area as well as their low relative intensity (with RETIRA < 3). Following the prescriptions advocated in Chapter 14 (see *Tramutoli et al.* [2018] as well as *Genzano et al.* [2015] and *Eleftheriou et al.* [2016]), these circumstances would not allow seismic forecasting in any event. This weak, but not null, indication instead proved very useful in reducing the temporal (from 3 months to few weeks) and spatial (from tens to a few thousands of square kilometers) alert window established by application of the CN algorithm, and in turn strengthened the significance of the latter. Thus the combination of two independent observations provided information that could be utilized, but which independently would have been discarded.

The second example highlighted another not trivial advantage of the combination of multiple observations. In this case, thanks to the continuity of satellite TIR observations, the significance of a single parameter observation was strengthened not only by its combination with the observations of another independent parameter (as in the previous case) but also the "memory" of previous observations of the same parameter. In fact, the appearance on 28–30 September 2012 of SSTAs so similar to those observed on 18-20 July 2012 in the same area just before a *M* 5 earthquake, with the same linear shape and close to the same fault (Maras triple junction), strongly suggested that the later SSTA was announcing a similar seismic event, which turned out to be two as it happens.

The third example concerned combination of SSTA observations with the most commonly available data, i.e., seismicity. Not always does such a seismicity precede a strong earthquake. In the case of the Aegean Sea earthquake, however, the combination of seismic activity data with the SSTA appearing in the middle of the area affected by the foreshock, led to better identification not only of the area but also the timing (the temporal window of an SSTA alert ranges from few days up to few weeks) of the impending earthquake.

REFERENCES

AFAD (2016), *DDA Earthquake Catalogue*, Disaster and Emergency Management Authority, available at http://www.deprem.gov.tr/en/ddacatalogue

Aliano, C., Corrado, R., Filizzola, C., Genzano, N., Pergola, N., and Tramutoli, V. (2008), Robust TIR satellite techniques for monitoring earthquake active regions: limits, main achievements and perspectives, *Ann. Geophys.*, *51*, 303–317, doi 10.4401/ag-3050

Chorowicz, J., Luxey, P., Lyberis, N., Carvalhoa, J., and Parrot, J. F. (1994), The Maras Triple Junction (southern Turkey) based on digital elevation model and satellite imagery interpretation, *J. Geophys. Res.*, *99*(B10), 20.225-20-242, doi: 10.1029/94JB00321

Eleftheriou, A., Filizzola, C., Genzano, N., Lacava, T., Lisi, M., Paciello, R., Pergola, N., Vallianatos, F., and Tramutoli, V. (2016), Long term (2004–2013) RST analysis of anomalous TIR sequences in relation with earthquake occurrence in Greece, *Pure Appl. Geophys.*, *173*(1), 285–303, doi: 10.1007/s00024-015-1116-8

Filizzola, C., Pergola, N., Pietrapertosa, C., and Tramutoli, V. (2004), Robust satellite techniques for seismically active areas monitoring: a sensitivity analysis on September 7, 1999 Athens's earthquake, *Phys. Chem. Earth*, *29*, 517–527, doi: 10.1016/j.pce.2003.11.019

Genzano, N., Aliano, C., Corrado, R., Filizzola, C., Lisi, M., Mazzeo, G., Paciello, R., Pergola, N. and Tramutoli, V. (2009), RST analysis of MSG-SEVIRI TIR radiances at the time of the Abruzzo 6 April 2009 earthquake, *Nat. Hazards Earth Syst. Sci.*, *9*, 2073–2084.

Genzano, N., Filizzola, C., Paciello, R., Pergola, N., and Tramutoli, V. (2015), Robust satellite techniques (RST) for monitoring earthquake prone areas by satellite TIR observations: the case of 1999 Chi-Chi earthquake (Taiwan), *J. Asian Earth Sci.*, *114*, 289–298, doi: 10.1016/j.jseaes.2015.02.010

INGV (2016), *Italian Seismic Instrumental and Parametric Database*, Istituto Nazionale di Geofisica e Vulcanologia, available at http://iside.rm.ingv.it/iside/standard/index.jsp

Keilis-Borok, V. I. and Rotwain, I. M. (1990), Diagnosis of time of increased probability of strong earthquakes in different regions of the world: algorithm CN, *Phys. Earth Planet. Inter.*, *61*, 57–72.

Keilis-Borok, V. I. and Soloviev, A. (2003), *Nonlinear Dynamics of the Lithosphere and Earthquake Prediction*. Springer-Verlag, Berlin-Heidelberg.

NOA (2016), *Earthquake Catalogues*, National Observatory of Athens, available at http://www.gein.noa.gr/en/seismicity/earthquake-catalogs

Peresan, A., Kossobokov, V., Romashkova, L., and Panza, G. F. (2005), Intermediate-term middle-range earthquake predictions in Italy: a review, *Earth Sci. Rev.*, *69*(1–2), 97–132, doi: 10.1016/j.earscirev.2004.07.005

Peresan, A., Kossobokov, V., and Panza, G. F. (2012), Operational earthquake forecast/prediction, *Rend. Fis. Acc. Lincei 23*, 131–138, doi 10.1007/s12210-012-0171-7

PRE-EARTHQUAKES (2013), PRE-earthquakes Final Project Report, available online in the "Downloads" area of the Pre-earthquakes Project website (http://www.pre-earthquakes.org/).

Scognamiglio, L., Margheriti, L., Mele, F., Tinti, E., Bono, A., De Gori, P., Lauciani, V., Lucente, F., Mandiello, A., Marcocci, C., Mazza, S., Pintore, S., and Quintiliani, M. (2012), The 2012 Pianura Padana Emiliana seimic sequence: locations, moment tensors and magnitudes, *Ann. Geophys.*, *55*(4), 549–559, doi: 10.4401/ag-6159

Tramutoli, V. (1998), Robust AVHRR techniques (RAT) for environmental monitoring: theory and applications, in *Earth Surface Remote Sensing II*, pp. 101–113, SPIE Proceedings 3496.

Tramutoli, V. (2005), Robust satellite techniques (RST) for natural and environmental hazards monitoring and mitigation: ten years of successful applications, in *9th International Symposium on Physical Measurements and Signatures*, pp. 792–795, Remote Sensing XXXVI,7/W.

Tramutoli, V. (2007), Robust satellite techniques (RST) for natural and environmental hazards monitoring and mitigation: theory and applications, *International Workshop on the Analysis of Multi-Temporal Remote Sensing Images 1*, pp. 1–6, doi: 10.1109/Multitemp.2007.4293057

Tramutoli, V., Cuomo, V., Filizzola, C., Pergola, N., and Pietrapertosa, C. (2005), Assessing the potential of thermal infrared satellite surveys for monitoring seismically active areas. The case of Kocaeli (Izmit) earthquake, August 17th, 1999, *Remote Sens. Environ.*, *96* (3–4), 409–426, doi: 10.1016/j.rse.2005.04.006

Tramutoli, V., Filizzola, C., Genzano, N., and Lisi, M. (2018), Robust satellite techniques for detecting preseismic thermal anomalies, in D. Ouzounov, S. Pulinets, K. Hattori, and P. Taylor (eds), *Pre-Earthquake Processes: A Multidisciplinary Approach to Earthquake Prediction Studies*, pp. 243–258, Geophysical Monograph 234, American Geophysical Union, Washington, DC, and John Wiley and Sons, Inc., Hoboken, NJ.

20

Multiparameter Assessment of Pre-Earthquake Atmospheric Signals

Dimitar Ouzounov[1], Sergey Pulinets[2], Jann-Yenq (Tiger) Liu[3], Katsumi Hattori[4], and Peng Han[5]

ABSTRACT

We apply interdisciplinary observations to study earthquake processes, their physics, and the phenomena that precede their energy release. Our approach is based on multisensor observations of short-term pre-earthquake phenomena preceding large earthquakes ($M > 6$). The integrated satellite and terrestrial framework is our method for validation and is based on a sensor web of several physical and environmental parameters (satellite thermal infrared radiation (STIR), electron concentration in the ionosphere, air temperature, and relative humidity measurements) that were associated with earthquakes. The scientific rationale for multidisciplinary analysis is founded on the concept lithosphere–atmosphere–ionosphere coupling. To check the predictive potential of pre-earthquake signals we validate in retrospective and prospective modes. Our validation processes consist of two steps: (a) a retrospective analysis preformed over three different regions with high seismic activity (M 6.0 Napa of 2014, M 6.0 Taiwan of 2016, and M 7.0 Kumamoto, Japan of 2016); (b) testing of Molchan's error diagram (MED) for STIR and differential total electron content anomalous events over Japan and Taiwan. Our findings suggest that: (a) pre-earthquake signals (with 1–30 days time lag) follow a general temporal–spatial evolution pattern; (b) pre-earthquake atmospheric anomalies can provide short-term predictive information for the occurrence of major earthquakes in the tested regions.

"To come to the truth, you need one hundred years, and to begin to follow her, you need another hundred years."

Alexander von Humboldt

[1]*Center of Excellence in Earth Systems, Modeling and Observations (CEESMO), Chapman University, Orange, CA, USA*

[2]*Space Research Institute Russian Academy of Sciences Moscow, Russia*

[3]*Institute of Space Science, National Central University, Taoyuan, Taiwan*

[4]*Graduate School of Science, Chiba University, Chiba, Japan*

[5]*Southern University of Science and Technology, Shenzhen, China*

20.1. INTRODUCTION

The reporting of physical phenomena observations before large earthquakes covers a historical span of about twenty-five centuries [*Martinelli*, 2000]. Historically, Aristotle was the first scientist from ancient Greece to report comprehensively about earthquake-associated phenomena in his treatise "Meteorology" written in 340 BC. He presented the first description of water evaporation, earthquakes, and other weather phenomena, including his theory of "pneuma," a phenomenon emanating before earthquakes [*Aristotle*, 1951]. The basic meaning of "pneuma" is a wind, an air in motion [*Humboldt*, 1897]. An analogous idea known as "respiration of Earth" was developed at the beginning of the 20th century by Russian academician V. Vernadsky [*Vernadsky*,

Pre-Earthquake Processes: A Multidisciplinary Approach to Earthquake Prediction Studies, Geophysical Monograph 234, First Edition. Edited by Dimitar Ouzounov, Sergey Pulinets, Katsumi Hattori, and Patrick Taylor.

1912, 1945]. The Aristotle concept for pre-earthquake phenomena can be summarized in the following steps: "pneuma" arises from underground to the surface (gas and fluids), on reaching the surface (gas emanations of methane, radon, CO) an immediate calming of the wind is seen (because of temperature rise, as result of gas-release-triggered air ionization, a high pressure is created), producing fogs and making the atmosphere dusty (hydration leads to growth of aerosols). This is followed by large ion clusters rising to the upper layers of atmosphere over the active tectonic faults within convective flows that lead to the formation of linear cloud structures (see the review by *Pulinets et al.* [2018]). Shortly afterwards an underground hum can be heard, immediately prior to an earthquake [*Aristotle*, 1951].

For at least 12 centuries, the oracle at Delphi (near Athens, Greece) spoke on behalf of the gods, providing recommendations on all manner of things. The oracle was always a woman, her divine utterances were made in response to a petitioner's request. Science scholars have suggested that the oracle's inspiration most likely came from the vapors rising from the beneath the temple's floor. Recent studies [*Piccardi*, 2000 *De Boer et al.*, 2001] indicate the presence of two faults that cross exactly beneath the ruined temple, creating a path by which petrochemical fumes could rise to the surface to help induce visions. Ethylene, ethane, and methane were very likely to be present. Ethylene has been used for general anesthesia. It is hypothesized that the oracle's visions were stimulated by the gas fluxes emanating from the unique geodynamic setting. Proof of this hypothesis lies in the powerful 373 BC earthquake that destroyed Delphi. Although the epicenter was far to the south near the Gulf of Corinth, geochemical analysis has shown evidence that methane and ethylene were released in Delphi in advance of the devastating earthquake [*Piccardi*, 2000].

In recent years, many new studies have reported on the possible connection between atmospheric parameters and earthquake processes. Recently *Tronin* [2011] published a comprehensive catalog of associated earthquake phenomena based on historical data. This study includes 1500 earthquakes that occurred between 550 BC and AD 2000. Only factual historical documents were consulted from variety of cases with reports of thermal, atmospheric, geochemical hydrothermal, acoustic, animal behavior, and electromagnetic phenomenon being observed before and during the seismic events. Other observational facts that have been widely reported include: (a) deformation precursors [*Wyss et al.*, 1981; *Mogi*, 1985; *Neresov and Latynina*, 1992; *Kalenda et al.*, 2012]; (b) geochemical precursors [*Irwin and Barnes*, 1980; *Wakita*, 1982; *Zia*, 1984; *Gold and Soter*, 1985; *King*, 1986; *Thomas*, 1988; *Toutain and Baubron*, 1998; *Martinelli et al.*, 2000; *Cicerone et al.*, 2009; *Fu et al.*, 2015]; (c) thermal infrared [*Qiang et al.*, 1991; *Tronin*, 1996; *Tronin et al.*, 2002, 2004; *Tramutoli et al.*, 2001, 2005; *Ouzounov and Freund*, 2004; *Ouzounov et al.*, 2007; *Saraf and Choudhury*, 2005]; (d) latent heat [*Cervone et al.*, 2002; *Dey and Singh*, 2003]; (e) earthquake clouds [*Morozova*, 1996; *Guo and Xie*, 2007; *Doda et al.*, 2013]; (f) earthquake lights [*Stothers*, 2004; *St-Laurent et al.*, 2006]; (g) jet stream [*Wu*, 2004; *Wu et al.*, 2015]; (h) air temperature and humidity [*Mil'kis*, 1986; *Dunajecka and Pulinets*, 2005], and atmospheric pressure [*Bokov*, 2010]; (9) VHF signals [*Kushida and Kushida*, 2002; *Fujiwara et al.*, 2004]; (i) VLF signals [*Biagi et al.*, 2004; *Hayakawa*, 2004, 2011; *Rozhnoi et al.*, 2007]; (j) GPS associated total electron content (TEC) [*Liu et al.*, 2004; *Pulinets et al.*, 2006b; *Zakharenkova et al.*, 2006; *Kon et al.*, 2010]. Ari Ben-Menahem (1995) warned the seismological community that: "Seismology has reached a stage where its lofty goals will not be reached by seismologists alone and unless we launch an interdisciplinary research and observational effort, we shall always be surprised by the next major earthquake." Unfortunately he was right, now, 20 years later, about 21 destructive earthquakes have since claimed an aggregate count close to a million of lives lost and damage worth hundreds of billions US dollars: short–term prediction was not issued prior to any those earthquakes.

20.2. MULTIPARAMETER REGISTRATION OF PRE-EARTHQUAKE ANOMALIES

Taking into consideration Ben-Benahem's concerns, a framework for global studies of multiparameter satellite and in situ data for validation of physically based pre-earthquake signals has been proposed [*Ouzounov et al.*, 2006b; *Pulinets*, 2011; *Liu et al.*, 2016]. It was realized that the complex and dynamic nature of the pre-earthquake phenomena require spatial, spectral, and temporal coverage that is far beyond any existing single method and any satellite mission. The concept of a multiparameter approach to monitoring pre-earthquake signals has been explored lately using a variety of methods: atmospheric/ionospheric and lithospheric sensor web [*Ouzounov et al.*, 2006b]; multisatellite observations [*Singh et al.*, 2007]; multiparameter geochemical analysis [*Yuce et al.*, 2010]; data fusion [*Akhoondzadeh and Saradjian*, 2012]; atmospheric multiparameter analysis [*Jing et al.*, 2013]; electromagnetic (EM) multiprecursory phenomena [*Nagao et al.*, 2002, 2014; *Saraev et al.*, 2014]; multiparametric analysis [*Wu et al.*, 2015]. The result of these studies highlighted three common problems in the multiparameter analysis of pre-earthquake signals.

1. Lack of monitoring over an extended period and integration of different physical observations.
2. Lack of a physical mechanism explaining the possible link between the pre-earthquake signal and seismic cycle.
3. Lack of statistical validation of pre-earthquake signals.

With the aim of addressing these problems we present here some new information.

The authors are actively participating in several international projects focused on multiparameter pre-earthquake signals namely: iSTEP; PRE-EARTHQUAKE, and lithosphere–atmosphere–ionosphere–magnetosphere coupling (LAIMC).

1. iSTEP (integrated Search for Taiwan Earthquake Precursor) is a Taiwanese project concerned with earthquake precursors. Studies were undertaken in three different phases during 2001–2016. After the devastating *M* 7.6 earthquake of 21 September 1999, the iSTEP1 program was established, consisting of a main project and five subprojects. Research was conducted to search for credible precursors in seismological variations, geomagnetic and gravity fields, ground surface deformations, and ionospheric electron density anomalies, as well as evaluation of the statistical significance of observed precursors in Taiwan [*Tsai et al.*, 2006]. Results revealed that anomalies in P-wave velocity, ground surface deformation, geomagnetic field intensity, ionospheric electron density could appear a few years, months, and days before large earthquakes in Taiwan. An integrated ground-based seismo-electromagnetic observation system, including eight networks of magnetometers, electrode arrays, corona probes, FM tuners, Doppler sounding systems, ionosondes, GPS receivers, and all-sky cameras, was constructed and routinely monitors for earthquake precursors in the lithosphere, atmosphere, and ionosphere, as well as evidence for possible lithosphere–atmosphere–ionosphere coupling (LAIC) in the Taiwan area. Several statistical analyses were developed in order to validate that the observed anomalies are credible precursors (see *Liu et al.* [2016] and references listed).

During iSTEP2 (2006–2012), the project was extended so as to have a longer time period for data collection and analysis, as well as developing physical and statistical models. Many new observations possibly related to seismo-lithospheric precursors of the earth and space surface magnetic field were reported, including GPS detection of surface deformation and seismo-atmospheric precursors of the infrasound signal, as well as satellite observations of seismo-ionospheric precursors (SIPs) in the electron density profile, the electron temperature, ion density, and neutral temperature.

The iSTEP-3 (2012-2016) program was formally funded and focused on the study of SIPs, and comprised a main project and three subprojects. The main project continues to operate the integrated ground-based observation system, develop physical models, and compare model simulations with observed precursors, while the three subprojects aim to: develop a near real-time global ionosphere map (GIM) with a 4-h time delay for worldwide SIP monitoring; monitor lithospheric, atmospheric, and ionospheric precursors in order to find the precursor link; and conduct earthquake hazard assessment using observed precursors [*Liu et al.*, 2016].

2. The project PRE-EARTHQUAKE (Processing Russian and European Earth Observations for Earthquake Precursors Studies) was led by the University of Basilicata (Italy) and the Space Research Institute (Russia) and was developed under the European Union project EU-FP7cordis.europa.eu/result/rcn/57410_en.html. The intent of the project was to commit EU and Russian researchers to integrate different ground and satellite data in order to improve, by cross-validating, their methodologies, through: (a) substantially improving our knowledge of preparatory phases of earthquakes and their possible precursors; and (b) promoting a worldwide Earthquake Observation System as a dedicated component of the Global Earth Observation System of Systems (GEOSS). One of the main scientific objectives of the project was to investigate and demonstrate the extent to which a systematic integration of several independent observations (instead of a single parameter approach) can improve our present capabilities of short-term earthquake prediction. Three European (from Italy, Germany, and Turkey) and three Russian scientific institutions among the most quoted for these studies worldwide constituted the core partnership of the project. In this context the project investigated the extent to which combined use of different observations/parameters and the refinement of data analysis methods can reduce rates of false alarm and improve reliability and precision (in the space–time domain) of short-term prediction. Results achieved by applying a validation/confutation approach devoted to evaluating the presence/absence of anomalous space–time transients in the presence/absence of significant seismic events were discussed in comparison with results achieved independently by other studies [*Tramutoli et al.*, 2012].

3. Validation of LAIMC is the goal of international projects supported by the International Space Science Institute (ISSI) in Bern (2014–2016) and in Beijing (2015–2017) that involves an international team of scientists from the United States, Russia, France, Italy, Taiwan, Japan, and China (http://www.issibern.ch/teams/spaceborneobserve/?page_id=13). Projects were designed to study the complex chain of interactions between lithosphere–atmosphere and near Earth space plasma related to earthquake processes and other natural and anthropogenic processes. The LAIC concept initially was created to provide common understanding of the multiparameter pre-earthquake phenomena and to link different variations of geochemical, atmospheric, and ionospheric parameters in a logical chain of processes united by a common physical mechanism. The main goal of the projects was to validate LAIC with a wide range of space-based observations using experimental data from the recent major earthquakes worldwide [*Ouzounov et al.*, 2015, 2016]. From the experimental data and their detailed analysis some key atmosphere–ionosphere processes were studied, which modify Earth's plasma environment system

under various geophysical conditions. As input for analysis and modeling we used atmospheric, ionospheric, and magnetospheric data products from different satellites, including EOS (Terra, Aqua), NPOESS, ESA/EUM (METEOSAT, Swarm), DEMETER/CNES, and FORMOSAT-3/COSMIC, as well as ground observations of GPS/TEC, ground EM fields, meteorological monitoring data, geochemistry, etc. Initial results of LAIC validation demonstrated: (a) consistency and ability to reveal pre-earthquake patterns for all types of the earthquakes (coastal, intraplate, ocean), thereby confirming its universal character; and (b) a complex dissipative open system associated with earthquake processes and other natural and anthropogenic processes [*Pulinets et al.*, 2015; *Ouzounov et al.*, 2016].

The results of these projects show that no solitary existing method—the geochemical magnetic field, thermal infrared (TIR), ionospheric variation or GPS/TEC—can provide successful and consistent short-term forecasting on a global scale. This is most likely because of the local geology, tectonic terrane, and the possibility that strong earthquakes are a result of a long-lasting, ongoing, large-scale process with different stages involving different parts of a fault network [*Keilis-Borok et al.*, 2004]. However, the simultaneous use of selected standardized measurements within an integrated network of multiple sensing sources both on the ground and from space, and linked to a common physical mechanism, should provide the necessary information to determine pre-earthquake signals.

This was the principle goal behind proposing an integrated satellite and terrestrial framework (ISTF). The ISFT concept for Japan is shown in Figure 20.1 and the physical parameters used in ISTF analysis are listed in Table 20.1. The approach is a coordinated observation infrastructure using multiple sensors that are distributed on one or more platforms in order to cover large areas and provide temporal and spatial coherence in the monitoring. The advantage of ISTF approach is that it facilitates maximal use of existing multiple and previously validated physical measurements, e.g., see list in Table 20.1, to be

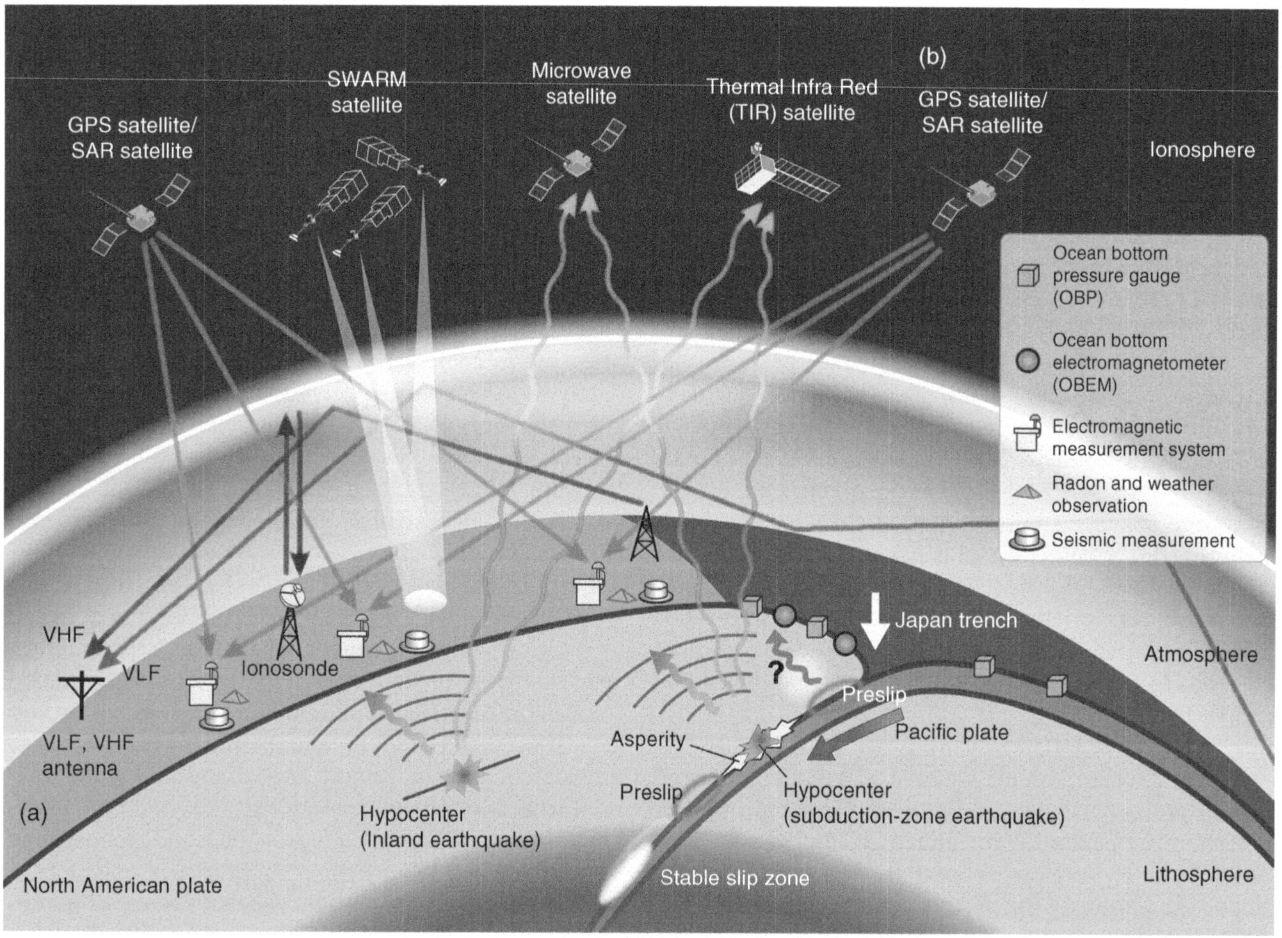

Figure 20.1 Conceptual diagram of the integrated satellite and terrestrial framework (ISTF) for multiparameter observations of pre-earthquake signals in Japan: (a) ground component includes seismic, EM observations, radon, weather, VLF–VHF, and ocean-bottom EM sensors; (b) satellite component includes GPS/TEC, SAR, Swarm, microwave, and thermal infrared (TIR) satellites.

Table 20.1 List of physical parameters used in ISTF analysis

Parameter	Lag time	Anomaly range	Sensors	Data source	Methodology	References
Ionospheric variations	5–1 days before the shock	Local irregularities 15–100%	GPS/TEC	IGS, NASA	Ionospheric perturbation detection	*Liu et al.*, 2000; *Pulinets and Boyarchuk*, 2004; *Zakharenkova et al.*, 2006; *Kon et al.*, 2010; *Hirooka et al.*, 2011
Outgoing longwave radiation	1–4 weeks	2–80 W m^{-2}	NPOESS/AVHRR EOS Aqua	AIRS spectrometer, AQUA, AVHRR spectrometer, NOAA	Spatial and temporal detection of hot spots at TOA	*Ouzounov et al.*, 2007, 2011b; *Xiong et al.*, 2010
Atmospheric chemical potential	One week	Sharp increase	EOS Assimilation models	NASA-GEOS5	Time series data analysis	*Pulinets et al.*, 2006b

integrated into one framework with the latest physical models of generation and propagation, and provides feedback on data gaps that may then be acquired from other sources [*Ouzounov et al.*, 2011b]. We are applying different observation techniques to study earthquake processes, their physics and the phenomena that precede their energy release. Our approach is based on multisensor observations of short-term pre-earthquake phenomena preceding large earthquakes ($M > 6$). Our method for validation is based on a web of several physical and environmental parameters—satellite thermal infrared radiation (STIR), electron concentration in the ionosphere (GPS/TEC), and air temperature measurements that were found to be associated with earthquakes. The scientific rationale for multidisciplinary analysis is based on the LAIC concept [*Pulinets and Ouzounov*, 2011], which explains the synergy of the different physical processes and anomalous variations usually associated with short-term pre-earthquake anomalies. To check the predictive potential of pre-earthquake signals we validate different anomalous signals in retrospective and prospective modes. Our validation processes comprise two steps: (a) case studies of retrospective analysis preformed in three different regions with high seismicity; and (b) statistical check of STIR and differential total electron content (dTEC) anomalies over Japan and Taiwan.

20.2.1. Temperature and Humidity Anomalies

Probably for the first time in the scientific literature John Milne reported atmospheric temperature changes ahead of earthquakes [*Milne*, 1913]. He observed that the sinusoidal curves of mean monthly temperature were generally a little in advances of the crest of the waves indicating the earthquake arrival for 387 earthquakes in northern Japan. Later *Mil'kis* [1986] demonstrated the presence of long-term thermal effects associated with seismic activity by studying thermal weather anomalies during the month (or season) of strong earthquakes in the former Soviet Union. He used the data from more than 120 meteorological stations in Turkmenistan, Uzbekistan, and other regions of Central Asia. His results show that within intervals of tens of years, the mean monthly temperature for the year of the earthquake in practically all cases is anomalously high, and is the local maximum for the multiyear interval.

Similar results were found in a study of short-term change in air temperate and relative humidity (RH) around the time of the Colima Earthquake in Mexico in 2003 and before the devastating M 7.6 Kashmir earthquake of 8 October 2005 in Pakistan [*Pulinets et al.*, 2006a]. To understand the day-by-day variability of the daytime and nighttime temperature during the time of earthquake events the hourly temperature and relative humidity near to the epicenter were analyzed.

The presence of ions in the atmosphere creates the possibility for molecules of water vapor to join them through the hydration process, which is different from condensation. In the process of condensation the chemical potential is equal to the latent heat, which is $Q = 40.683\,\text{kJ mol}^{-1}$ or $U_0 = 0.422\,\text{eV mol}^{-1}$. The process of evaporation/condensation, which is a first-order phase transition, always takes place during chemical potential equality, but newly formed ions have a different chemical potential that we should take into account. In this case the real chemical potential can be expressed as:

$$U(t) = U_0 + \Delta U \cos^2 t \tag{20.1}$$

where U_0 is the chemical potential for pure water, U(t) is the chemical potential where the ionization and hydration are taken into account, and $\cos^2 t$ takes into account the daily changes of solar radiation. The increase in chemical potential, ΔU, of the water molecules (which can be derived from equation (20.1) indicates the strength of nucleation process and can be used as an indicator of an impending earthquake, as shown by *Pulinets et al.* [2006a] and *Boyarchuk et al.* [2010].

20.2.2. Thermal Anomalies of Outgoing Radiation

Natural phenomena observations and availability of data stimulated the analysis of long time series of thermal images in relation to earthquake activity. Historically, the first application of thermal satellite observation of earthquake phenomena was performed in the 1980s and 1990s over Central Asia [*Gorny et al.*, 1988; *Qiang et al.*, 1991; *Salman et al.*, 1992; *Tronin*, 1996]. Similar studies were developed later in China, Japan, Italy, United States, and other countries. Thermal observations from satellites indicate significant changes in (a) Earth's surface temperature, (b) Earth radiation, (c) latent heat, and (d) outgoing radiation at the top of the atmosphere (TOA). Many thermal anomalies prior to earthquakes in areas of high seismic activity have been reported in Central Asia, Iran, China, Turkey, Japan, Kamchatka, India, Turkey, Italy, Greece, and Spain [*Qiang and Du*, 2001; *Saraf and Choudhury*, 2005; *Ouzounov et al.*, 2007, 2016; *Tronin*, 2011; *Tramutoli et al.*, 2015].

One of the main parameters used to characterize Earth's radiation environment is the outgoing longwave radiation (OLR, 8–12 μm). The OLR occurs at the TOA, and are integrated emissions from the ground, lower atmosphere, and clouds [*Ohring and Gruber*, 1982], and primarily OLR data have been used to study Earth's radiative budget and climate [*Gruber and Krueger*, 1984; *Mehta and Susskind*, 1999]. The National Oceanic and Atmospheric Administration (NOAA) Climate Prediction Center provides daily and monthly OLR databases on Polar Orbiting Environmental

Satellites (POES). The algorithm for analyzing the Advanced Very High Resolution Radiometer (AVHRR) OLR data is not a direct application, but is calculated from the raw data using a separate algorithm [*Gruber and Krueger*, 1984]. The total longwave radiation flux is obtained by applying a regression equation to the infrared(IR) window observations [*Mehta and Susskind*, 1999]:

$$T_f = T_w\left(a + bT_w\right) \qquad (20.2)$$

where T_w is the radiance equivalent of the brightest temperature (BT in K) in the IR window over nadir, T_f is the flux equivalent BT(K), and a and b are the regression coefficients.

The OLR is mainly sensitive to near surface and cloud temperatures. A daily mean data footprint covering a significant area (90°N–90°S, 0°E to 357.5°E) and with a spatial resolution of 2.5° by 2.5° was used to study the OLR variability in the zone of earthquake activity [*Ouzounov et al.*, 2007; *Xiong et al.*, 2010]. An increase in radiation and a transient change in OLR were recorded at the top of the atmosphere over seismically active regions and were proposed to be related to thermodynamic processes in Earth's surface. The OLR anomaly variations were defined [*Ouzounov et al.*, 2007] in similarly to the definition of an anomalous thermal field proposed by *Tramutoli et al.* [2004]. The anomaly was constructed as statically estimated variability in OLR values for specific locations and time periods:

$$Anomaly_{i,j}(t) = \left(S^*\left(x_{i,j}, y_{i,j}, t\right) - \bar{S}^*\left(x_{i,j}, y_{i,j}, t\right)\right)/\tau_{i,j} \qquad (20.3)$$

where $t = 1$, K days, $S^*(x_{i,j}, y_{i,j}, t)$ is the current OLR, $\bar{S}^*(x_{i,j}, y_{i,j}, t)$ is the computed mean of the OLR field, defined for multiple years of observations over the same location and same local time, and $\tau_{i,j}$ is the standard deviation.

We use the anomaly index, a modified version of equation (20.3), which represents the regional calibration of anomaly estimates as:

$$\text{Anomaly Index} = \left(A * \text{Anomaly}(t)\right)/B \qquad (20.4)$$

where A and B are regional calibration coefficients, mainly defined by the regional seismo-tectonic patterns, historical events in the area, and by the coverage and quality of the thermal satellite data for the region (for more details, see the review by *Ouzounov et al.* [2018]).

20.2.3. Seismo-Ionospheric Anomalies

Recently the presence of a special type of ionospheric variability caused by EM coupling between the seismically active area and the ionosphere was clearly demonstrated [*Pulinets et al.*, 1998; *Liu et al.*, 2000]. The possible physical mechanism of seismo-ionospheric anomalies (SEA) as part of LAIC [*Pulinets and Boyarchuk*, 2004; *Pulinets and Ouzounov*, 2011] is described here and more details are available in Chapter 6. Due to the wide availability of global GPS data we can derive the time delay and phase variations of the radio signal received from GPS satellites, and hence integratie TEC vertically. The ionosphere variability can be monitored by different techniques both from the ground and from satellites. The GIM of the TEC constructed using 200 GPS ground-based receivers worldwide, has been routinely published at a time interval of 2 h (now 1 h) with a time delay of 2–4 days.

To detect SEA of the GPS TEC variations we perform a quartile-based process. At each time point, we compute the median M of every successive 15 days of the GPS TEC, as well as find the deviation between the observed GPS TEC on the 16th day and the computed median M. To provide the information about the deviation, we also calculate the first (or lower) and the third (or upper) quartiles, denoted by LQ and UQ, respectively. Note that under the assumption of a normal distribution with mean m and standard deviation s for the GPS TEC, the expected value of M and LQ or UQ is m and $1.34s$, respectively [*Klotz and Johnson*, 1983; *Liu et al.*, 2018]. Since the GPS TEC time resolution is 2h, there are 12 data points per day. If more than one-third (=4/12) of the upper or lower abnormal signals appear in one day, and the observed GPS TEC is greater or smaller than the associated UB or LB, we then declare detection of a positive or negative SEA day [*Liu et al.*, 2009; *Kon et al.*, 2010).

20.2.4. Physical Connection with the Earthquake Processes

Natural variability of atmospheric chemical potential (ACP), OLR, and SEA phenomena present a fundamental problem for LAIC associated with earthquake processes (see more about LAIC in Chapter 6). The causes of pre-earthquake anomalies are related to the earthquake preparation process. The geological structures (faults, cracks, fractures, etc.) act as preferred conduits because of the convective flow of fluids and gas in the upper levels of the lithosphere. The ACP and OLR anomalies are typically observed above large faults and their intersections over the land and over the seas. Depending on the geological and tectonic setting, gases like Rn, H_2, He, CH_4, CO_2, and O_3 make their way to the atmosphere, but what is the possible cause of gas release before large earthquake events?

Thomas [1988] reviewed five possible mechanisms to account for geochemical pre-earthquake signals: ultrasonic seismic vibration, pressure-sensitive solubility, pore-volume collapse, fracture-induced increase in reactive surfaces, and aquifer breaching. All directly or indirectly require increasing crustal strain. However, all proposed mechanism face a common dilemma: crustal strain measurements often fail to detect any unusual changes prior to earthquakes, even when other kinds of pre-earthquake phenomena are very much in evidence [*Soter*, 1999]. Observations therefore are lacking in support of the interpretation of pre-earthquake signals being 'strain indicators'. *Gold* [1979] and *Gold and Soter* [1985] proposed a mechanism based on the episodic emission of radon. Accordingly, the lower crust and upper mantle generate gas at lithostatic pressure. At a particular concentration the gas could form large-scale domains in the rock and when reaching a certain vertical extent could become hydrostatically unstable, forcing its way upward through the lithosphere. On reaching the crust, the gas spreads out, inflating and depleting any accessible faults over a large volume. These processes can happen without affecting strain measurements. Mantle gases have been observed in the San Andreas Fault system [*Kennedy et al.*, 1997] and because the gas bears a large part of the overburden pressure, it evidently reduces the static friction across the walls of the faults, preparing the ground for an earthquake [*Gold and Soter*, 1985].

This hypothesis could be verified by the integration of several radon, GPS, and satellite-based observations. The rapid lithospheric degassing would trigger several atmospheric processes near Earth's surface, namely: air molecule ionization and plasma chemical reactions, processes that dominate formation of complex molecular ions and of large ion clusters up to aerosol size. Temperature differences between areas proximal to faults and those distal to faults leads to horizontal air movement, air mixing, and air temperature rise over the entire earthquake preparation area. A decrease in relative humidity often accompanies this process as water vapor attachment to the ions decreases the content of free water vapor in air. This primary chain leads to two major consequences: change of air conductivity and therefore change to the near-ground atmospheric electric field, and emanation of latent heat resulting from water molecule attachment (condensation) to ions. The ionosphere as a part of the global electric circuit immediately reacts to the changes of near-ground electric properties, and an electric field induced within the ionosphere causes ion drift and creation of irregularities in electron concentration. Anomalous electric field penetration into the ionosphere is mapped along the geomagnetic field lines in the magnetosphere up to the conjugated point of the opposite hemisphere. As the whole magnetosphere channel connected to the tectonically active area starts to be modified this leads to generation of different types of plasma instabilities, generation of extra low frequency (ELF) and very low frequency (VLF) emissions, and stimulated precipitation of energetic particles [*Pulinets et al.*, 2015].

Enomoto [2012] proposed a novel concept in support of lithosphere–ionosphere coupling, also involving deep-gas migration. This was suggested in order to provide an explanation of observed ionospheric electron enhancement prior to some recent offshore earthquakes in Japan. The hypothesis is based on interaction between earthquake nucleation and gases deep in the Earth. This interaction causes a negatively electrifiied gas as the gases pass through fractured asperities, and the mechanism could be summarized as a pressure-induced electric current generator. There is no assessment in the model, however, about the effect of the gas on the atmospheric physical properties, which makes it difficult to validate within the observations under the LAIC framework. *Hayakawa* [2011] recognized the important role of the atmosphere as part of LAIC processes in proposing a mechanism known as "atmospheric" channels, based on the key role of atmospheric oscillations (acoustic waves (AW) or atmospheric gravity waves (AGW)). Perturbation at Earth's surface, such as temperature and pressure changes in a seismo-active region, could produce atmospheric oscillations traveling up to the ionosphere and inducing ionospheric density changes. Despite the obvious link with thermal atmospheric anomalies, this mechanism still needs experimental proof of generation of AGW immediately before an earthquake.

20.3. STATISTICAL ASSESSMENT OF PRE-EARTHQUAKE ATMOSPHERIC ANOMALIES

20.3.1. Case Studies of Multiparameter Analysis

In our studies we integrate OLR, GPS/TEC, and ACP data for monitoring atmospheric pre-earthquake signals associated with three recent major earthquake events: (a) *M* 6.0 South Napa, California, earthquake of 24 August 2014; (b) *M* 6.4 southeast of Yujing, Taiwan, earthquake of 6 February 2016; and (c) *M* 7.0 Kumamoto, Japan, earthquake of 16 April 2016: earthquake catalog data are presented in Table 20.2.

We process in a controlled environment the different satellite and ground data.

1. Remote sensing data from the NPOESS and EOS AIRS sensors recording OLR in the infrared. We used daily OLR data calculated at the TOA to study Earth's radiation budget because it represents emissions from Earth's surface, the lower atmosphere, and clouds, and it is sensitive to near-surface and cloud temperatures. The OLR daily mean global data have a resolution of 2.5° by 2.5° and are available from NOAA's Advance Radiation Radiometer (AVHRR) in polar orbit.

2. GPS/TEC data from the Global Navigation Satellite System (GNSS). To the detriment of SEA a mask conception of ionospheric precursors is used [*Pulinets et al.*, 2003] which recognizes SEA based on their self-similarities. The definition of the mask for a given region that we

Table 20.2 List of earthquakes (US Geological Survey) studied

Location	Geographic coordinates	Date and time (UTC)	*M*	Depth (km)
Napa Valley, California	38.21°N/122.31°W	24 August 2014 10:24:44	6.0	11.11
Southeast of Yujing, Taiwan	22.938°N/120.601°E	6 February 2016 19:57:27	6.4	23
Kumamoto, Japan	32.791°N/130.754°E	16 April 2016 16:25:06	7.0	10

compute is usually presented in a two-dimensional distribution of ΔTEC based on single IGS (International GNSS Service) stations, where the days in relation to the main shock are on the horizontal axis, and local time on the vertical axis. The magnitude of ΔTEC is color coded.

3. ACP data from Earth Observing System assimilation models. A separate algorithm calculates ACP from the GOES-5 assimilation model. The 3 hourly data variations represent the correction of ACP, which indicates the deviation of the latent heat transformation in the boundary layer of the atmosphere in the presence of an ionization source. The anomalous value was defined based on equation (20.5). For revealing transitional ACP anomalous patterns in the ACP background, data for the period 2013–2016 have been used.

20.3.1.1. M 6.0 South Napa Valley California Earthquake of 24 August 2014

In 2014, the largest earthquake in over 25 years hit the San Francisco Bay Area, causing significant damage in California's famous Napa Valley. The *M* 6.0 earthquake occurred early in the morning of 24 August 2014, on the West Napa Fault (Figure 20.2b). A smaller magnitude earthquake on the same fault had damaged the city of Napa in 2000. We detected large anomalies in the OLR transient field at the TOA over northern California during two periods: 28 July to 2 August and 22–23 August. The location was about 150 km northeast of the 24 August earthquake epicenter. Compared to the reference field of August 2004 to August 2014 the hotspot anomalies record the largest values over the epicentral area for all of 2014 (Figure 20.2a). In comparison the same approach was computed for 2013, the previous year with no significant seismic activity, and no significant anomalies similar to the OLR signals of 2014 were observed (Figure 20.2a). Retrospective GPS/TEC data analysis using a dTEC mask show anomalious variations at nighttime 1–3 days before and 1 day after the earthquake, over the Station p222, south of the Napa earthquake epicenter (Figure 20.2c). We integrate the three parameters OLR, dTEC, and ACP for a 30 = day time period, spanning the day of the *M* 6 earthquake of 24 August 2014 (Figure 20.2d). Positive ACP anomalies and dTEC enhancement (03:00 LT) peaked during the major OLR anomaly (nighttime 23 August). Observations for GPS Station p222 indicate that the day before the earthquake (23August 2014) was the only day that witnessed coordinated enhancement between the calculated index of ACP, OLR at the TOA, and dTEC variations, when they reached their peak values, which probably indicates the presence of physical coupling between the atmosphere and ionosphere near the epicentral area (Table 20.3).

20.3.1.2. M 6.0 Southeast of Yujing, Taiwan, Earthquake of 6 February 2016

On 6 February 2016, a *Mw* 6.4 earthquake occurred northeast of Pingtung City in southern Taiwan (Figure 20.3b). The earthquake was shallow (depth 23 km) and intense reverberations on the surface caused widespread damage and 117 deaths. The earthquake was the deadliest in Taiwan since the Chi-Chi earthquake of 1999. We are conducting validation on the pre–earthquake temporal–spatial pattern of multiparameter precursor signals associated with the 6 February 2016 earthquake in Taiwan. The continuous analysis of OLR data obtained from the NPOESS satellite system shows rapid increase of OLR at the TOA (Figure 20.3a) on 29 January 2016 (2.5σ significance for 25 years of analysis). The significance of detected OLR anomalies in 2016 (Figure 20.3b) was estimated by comparison with the anomalies detected in the year of 2015, a year with no major seismic activity in the same region (Figure 20.3a). The full year comparison shows that the major OLR anomaly in 2016 was slightly ahead of the 6 February earthquake (Figure 20.3d) and that there is no similarity with the 2015 trend of OLR anomalies, which also probably means that the anomalies are not related to weather phenomena. The GPS/TEC analysis using a dTEC mask shows anomalous variations during local nighttime 1–3 days before and 1 day after the earthquake, over Station TWTF located to the north of the 6 February earthquake epicenter (Figure 20.3c). In Figure 20.3d we present integration of monthly data for the three parameters OLR, dTEC and ACP, spanning the day of the *M* 6 earthquake of 06 February 2016. The time series variations of ACP characterizing the ionization processes inside the atmospheric boundary layer, show rapid increases during 1–8 February (12:00 LT) synchronized with enhancements of dTEC (18:00 LT). In comparison, the major OLR anomaly appears ahead (29 January) of the ACP and dTEC anomalies, which is not common. According to *Pulinets and Ouzounov* [2011] and *Ouzounov et al.* [2011b] the

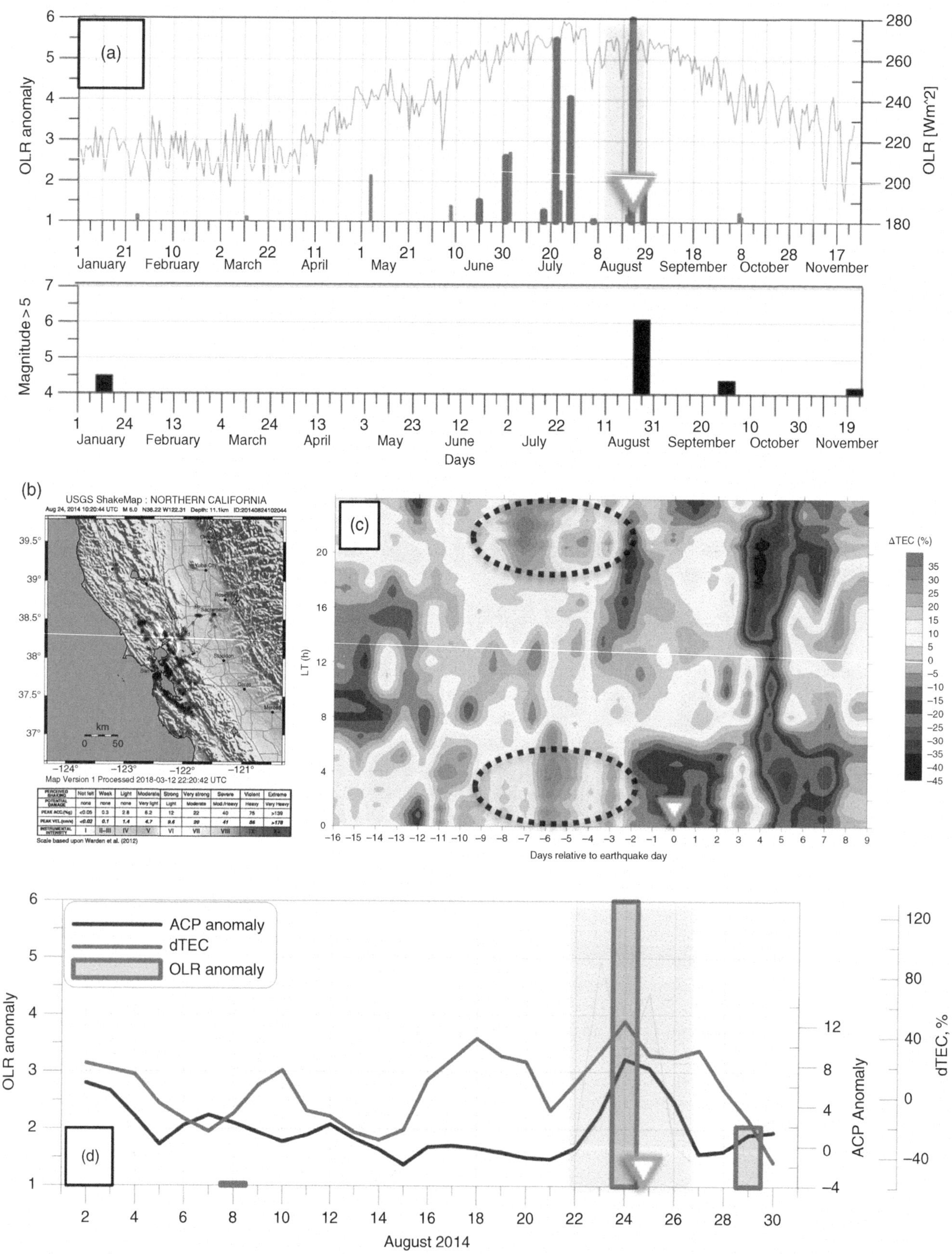

Figure 20.2 Different observations for the *M* 6 South Napa earthquake. (a) Time series of OLR nighttime data for 2014 (red columns) for location near to the earthquake epicenter. Includes OLR anomalies for 2013, a year with no major seismic activities (blue columns). The gray line denotes the 2014 OLR daily values. (Bottom panel) Seismic events with M≥4.5 (European–Mediterranean Seismological Centre) for 2014. (b) Shake map (US Geological Survey). (c) 30 days of dTEC mask data for receiver p222 spanning the day of the earthquake (24 August 2014). (d) 30-day integration graph for OLR, dTEC, and ACP: gray shaded area spans anomalous patterns; white triangle marks earthquake occurrence. (*See electronic version for color representation.*)

evolution pattern usually seen (as shown for the two other cases studies in Figures 20.2d and 20.4d) is an ACP anomaly followed by OLR and dTEC variations. The probable reason for the earlier appearance of the OLR anomaly in this case was the large areal integration (2.5 × 2.5°) used for the OLR data versus the single pixel analysis of ACP. As the OLR anomaly data encompass a large area it is probable that activity associated with additional faults close to the epicenter, and not represented in the single point ACP graph, contributed to the signal (Figure 20.3d). The outcome of our multi-instrument analyses of the *Mw* 6.4 earthquake of 6 February 2016 in Taiwan reveals a general temporal–spatial evolution pattern of earthquake precursor signals (Table 20.3).

20.3.1.3. M 7.0 Kumamoto, Japan, Earthquake of 16 April 2016

The 2016 Kumamoto series of earthquakes started with a *M* 6.2 foreshock (12:26 UTC) on 14 April 2016 and climaxed with a *M* 7.0 earthquake at 16:25 UTC on 15April, beneath Kumamoto City in the Kyushu Region of Japan (Figure 20.4b. The two earthquakes killed at least 50 people and injured about 3000 others in total (Wikipedia).We retrospectively analyzed the transient variation of three physical parameters characterizing the state of the atmosphere and ionosphere several days before the onset of this earthquake: OLR on the TOA; ACP, related to the thermodynamic proprieties in the lower atmosphere; and dTEC from GPS observations for the ionosphere. Satellite monitoring over Japan shows that beginning on 12April (3 days before the *M* 7 event) a rapid increase of OLR occurred near the Kumamoto epicentral area. Based on the OLR reference field (2004–2016) the hotspot anomalies were the largest values for all of 2016 (Figure 20.4a) over the Kumamoto epicentral area. To check for confutation (no earthquakes with no thermal anomalies) we utilized the same approach for the 2015 data, a year with no major seismic activities in the same area, and no significant anomalies were observed (Figure 20.4a). We integrated the three parameters OLR, dTEC, and ACP for a 30-day time period, spanning the day of the *M* 7 earthquake of 15 April 2016 (Figure 20.4d). Positive ACP anomalies (00:00 LT) were observed coinciding with the OLR anomaly (nighttime 15 April) and dTEC enhancement (21.00 LT). The ACP data demonstrate a weak type of anomaly in comparison with the two other cases (California and Taiwan), probably because of the volcanic setting of the Kyushu region. The GPS/TEC retrospective analysis using a dTEC mask for data from the GMSD (an IGS GPS station in Japan) receiver (Figure 20.4c) located south of the earthquake epicenter shows positive anomalies during nighttime 1–2 days before and negative anomalies 1 day after the quake. Our results show correlation between the appearances of pre-earthquake transient anomalies (OLR, ACP, dTEC) predominantly with a very short time-lag, from hours up to a few days, which appear to be characteristic earthquake precursors in the atmosphere and ionosphere associated with the Kumamoto earthquake series of 2016 (Table 20.3)

20.3.2. Statistical Evaluation

We present the latest developments in the validation of multisensor observations of short-term pre-earthquake phenomena preceding major earthquakes. To check the predictive potential of pre-earthquake signals we have started to validate different anomalous signals in retrospective and prospective modes. Our validation processes consist of: (a) a 9-month period of prospective testing for OLR in Japan 2014–2015; (b) continuous retrospective analysis of SEA for Japan for 1998–2010; and (c) continuous retrospective analysis of SEA for Taiwan for 1998–2010.

20.3.2.1. Outgoing Longwave Radiation Anomalies

For the period 1 June 2014 to 31 March 2015 we performed continuous prospective alerts for Japan for major events of ≥ *M* 5.5. We used the combination of the two parameters: OLR and ACP. Each alert contained four parameters: (a) location, comprising latitude and longitude coordinates and radius of confidence of 2°; (b) estimated magnitude ±0.5; (c) time interval, usually within 30 days; and (d) confidence level, between 0.5 and 1.0. Each alert was shared with three people in real time and posted online. We used the ISTF approach based on equations (20.1–20.3). For the testing period of 310 days, 22 alarms were issued for ≥ *M* 5.5 potential events. Table 20.4 lists contingencies for the OLR prospective tests.

In the area of analysis 22 earthquakes occurred and were forecast correctly, but one earthquake was missed, i.e., was not forecast (Figure 20.5a). The average lag time is 12 days (Figure 20.5b). Alerts were issued only for earthquakes ≥ *M* 5.5+, with radius of confidence of 250 km valid of period of 30 days. To assess the statistical significance of the alerts issued we used Molchan's error diagram (MED) [*Molchan*, 1997; *Zechar and Jordan*, 2008]. As our alarms include time and location, a two-dimensional time–space MED is used. The alarm unit in time domain is 1 day, and in space domain is 2° × 2°. As a result, there are 310 alarm units (301 days) in time and 61 alarm pixels in space for the Japan region. The total alarm units are 310 × 61 = 18910. In Table 20.4 we computed true positive (TP), false negative (FN), true negative (TN), Actual Positive (AP), Actual negative (AN), Predicted positive (PP) and predicted negative (PN). Because we have a fixed alarm threshold, we can produce only one point on the MED diagram. The alarm rate is

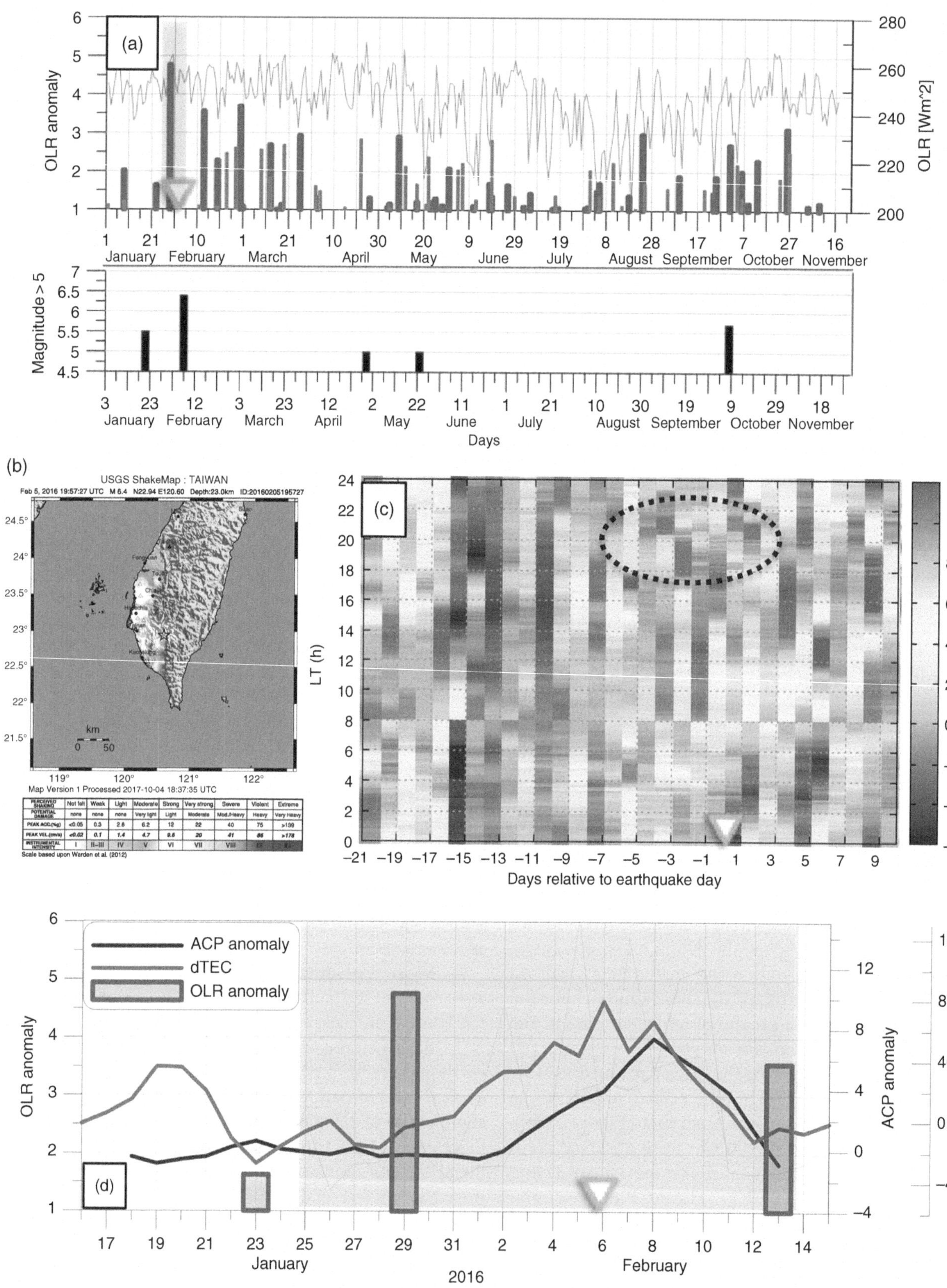

Figure 20.3 Different observations for the *M* 6 southeast of Yujing, Taiwan earthquake. (a) Time series of OLR nighttime data for 2016 (red columns) for location near the earthquake epicenter. Includes OLR anomalies for 2015, a year with no major seismic activities (blue columns). The gray line denotes the 2016 OLR daily values. (Bottom panel) Seismic events with $M \geq 4.5$ (European–Mediterranean Seismological Centre) for 2016. (b) Shake map (US Geological Survey). (c) 30 days of dTEC mask data for receiver TWTF spanning the day of the earthquake (6 February 2016). (d) 30-day integration graph for OLR, dTEC, and ACP: gray shaded area spans anomalous patterns; white triangle marks earthquake occurrence. (*See electronic version for color representation.*)

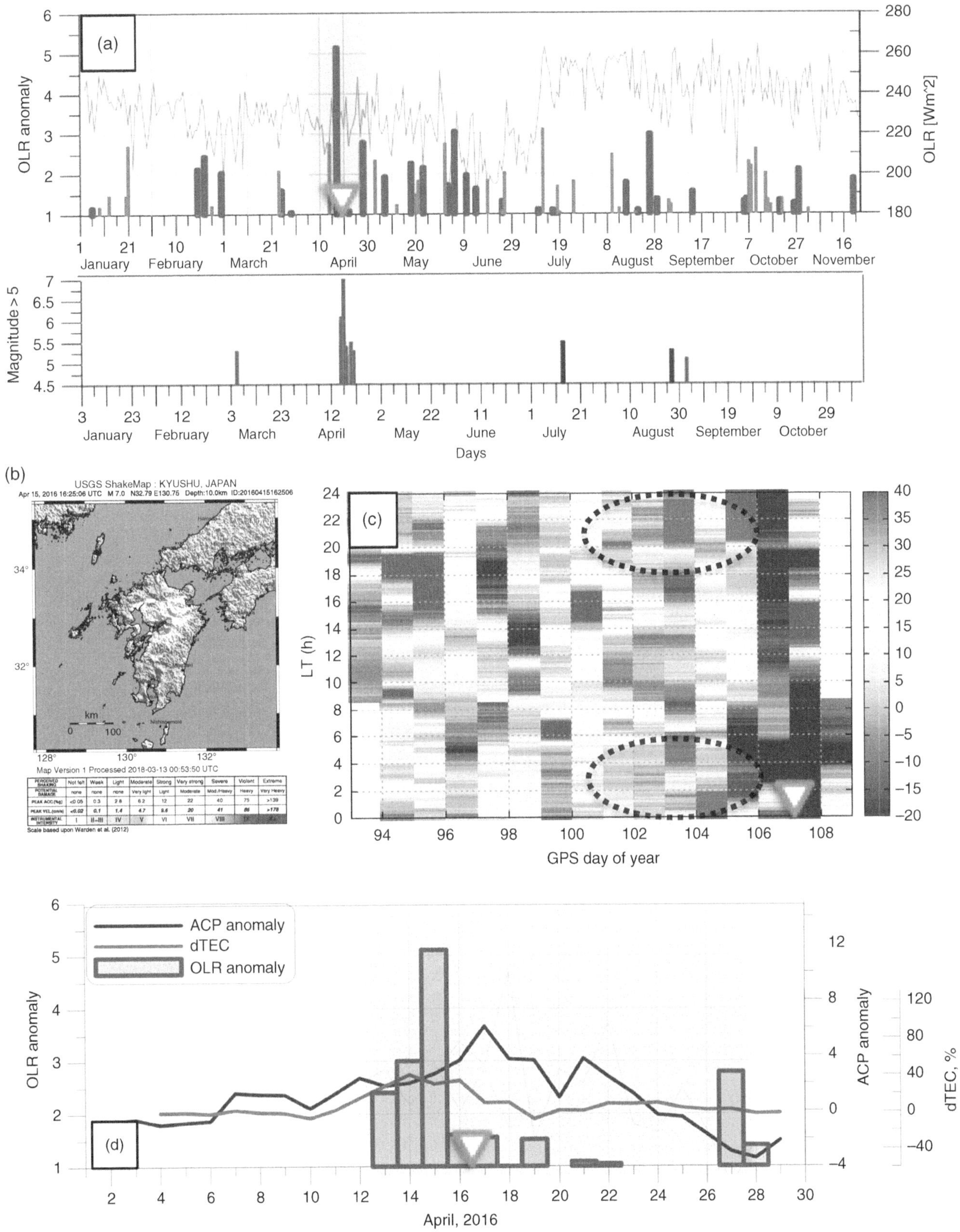

Figure 20.4 Different observations of the *M* 7 earthquake in Kumamoto, Japan. (a) Time series of OLR night ime data for 2016 (red columns) for location near to the earthquake epicenter. Includes OLR anomalies for 2015, a year with no major seismic activities (blue columns). The gray line denotes the 2016 OLR daily values. (Bottom panel) Seismic events with $M \geq 4.5$ (European–Mediterranean Seismological Centre) for 2014. (b) Shake map (US Geological Survey). (c) 30 days of dTEC mask data for receiver GMSD spanning the day of the earthquake (16 April 2016). (d) 30-day integration graph for OLR, dTEC, and ACP: gray shaded area spans anomalous patterns; white triangle marks earthquake occurrence. (*See electronic version for color representation.*)

Table 20.3 Summary of ACP, OLR, and dTEC anomalies observed in relation to the *M* 6 Napa Valley, California, *Mw* 6.4 Yujing, Taiwan, and *M* 7 Kumamoto, Japan, earthquakes

		ACP anomalies			OLR anomalies			dTEC anomalies		
Date and time	Earthquake	Date	Value	Lag (days)	Date	Value	Lag (days)	Date	Value (%)	Lag (days)
24 August 2014 10:24:44	*M* 6 Napa Valley, California	22–23 August	8	−2	22 August	6	−1	08.22-23	+40	−1–3
6 February 2016 19:57:27	*Mw* 6.4 SE of Yujing, Taiwan	1–8 February	8	−5	29 January	5	−7	02.04-06	+70	−1–3
16 April 2016 16:25:06	*M* 7.0 Kumamoto, Japan	12 April	8	−3	12 April	5	−3	04.14	+40	−1–2

Table 20.4 Contingency table for OLR analysis June 2014 to March 2015, Japan

	Alarm	Non-alarm	Total
Earthquake	True positive(TP) 21	False negative(FN) 1	Actual positive (TP+FN) 22
No earthquake	False positive(FP) 2535 − 21 = 2514	True negative (TN) 18910 − 21 − 1 − 2514 = 16374	Actual negative (FP+TN) 18888
Total	Predicted positive (TP+FP) 2535	Predicted negative (FN+TN) 16375	TP+FP+FN+TN 18910

about 0.134, and the detection rate is about 0.955 (Figure 20.5c). Then, the gain is about 7. Note that here we use a uniform assumption in which earthquakes occur completely at random in both time and space.

20.3.2.2. SEA Evaluation

On Figure 20.6 we present the concept of the MED analysis to evaluate retrospectively the effectiveness of the short-term earthquake forecast using the GPS/TEC approach over Japan. We use the data for 1998–2013. Figure 20.6a shows the distribution of earthquakes satisfy the following condition in the target area; $M \geq 6.0$ and depth less than 40 km in the target region. The SEA result is given in Figure 20.6b. Because of the 2011 Tohoku event, the SEA result is different from the result for 1998–2010 due to aftershocks. In this figure, the grey and black lines correspond to significance levels for 1 day and for 5-day accumulation of data, respectively. After adequately removing the aftershocks, the post-earthquake anomalies disappear. In order to examine the efficiency of the precursor signal, we use 5-day accumulation data for the MED analysis. The result is shown in Figure 20.6c and indicates that the strategy of using the GPS/TEC anomaly is statistically sound. In our case, the total number of earthquakes is 87 for 1998–2013 and the number of detected earthquakes is 24 for −1 to −5 days before the earthquake. Thus the detection rate is about 0.276, the alarm rate is about 0.18, and the gain is about 1.53.

To cross-compare with results of Japan shown in Figure 20.7 we also examine by applying the MED analysis to evaluate retrospectively the effectiveness of the short-term earthquake forecast using the GPS/TEC approach over Taiwan. We use data from 1998–2013. Figure 20.7a shows the distribution of earthquakes satisfy the following condition in the target area: 28 $M \geq 6.0$ and depth less than 40 km in target region. The TEC at Central Taiwan (24°N, 121°E) is retrieved from CODE GIM. We also calculate a 15-day running mean (m), and construct the associated upper bound $UB = 1.25(TEC/m)$ and lower bound $LB = 0.75(TEC/m)$. When the TEC on day 16 exceeds UB and LB it indicates an increase (positive) or decrease (negative) in SEA, respectively. Based on the previous results [*Liu et al.*, 2004], in Taiwan the negative SEA tends to appear in the afternoon period 1–5 days the earthquake. Thus, we focus on the negative anomaly. To remove the confounded effect of magnetic storms, we also eliminate the SEA appearing 1–2 days after the D_{st} magnetic index exceeds −100 nT.

The SEA result is given in Figure 20.7b. Because of the 1999 Chi-Chi event, the SEA result is different from the result of 1996–1999 due to aftershocks. In this figure, the grey and black lines correspond to the significance levels for 1-day and for 5-day accumulation data, respectively. After adequately removing the aftershocks, the post anomalies are found to disappear. In order to examine the efficiency of precursor, we use the 5-day accumulation data for MED analysis, shown in Figure 20.7c, where the lead-time is 2 days and the alarm window is 1 day. The result shows that the strategy of using the TEC anomaly is statistically sound. Note that the alarm rate cannot reach 1 as the anomalies on the storm days are excluded. In this case, the total number of earthquakes is 28 for 1998–2013 and the number of detected earthquakes is 21 for −1 to −5 days before the earthquake. Thus the detection rate is about 0.75, the alarm rate is about 0.6, and the gain is about 1.25.

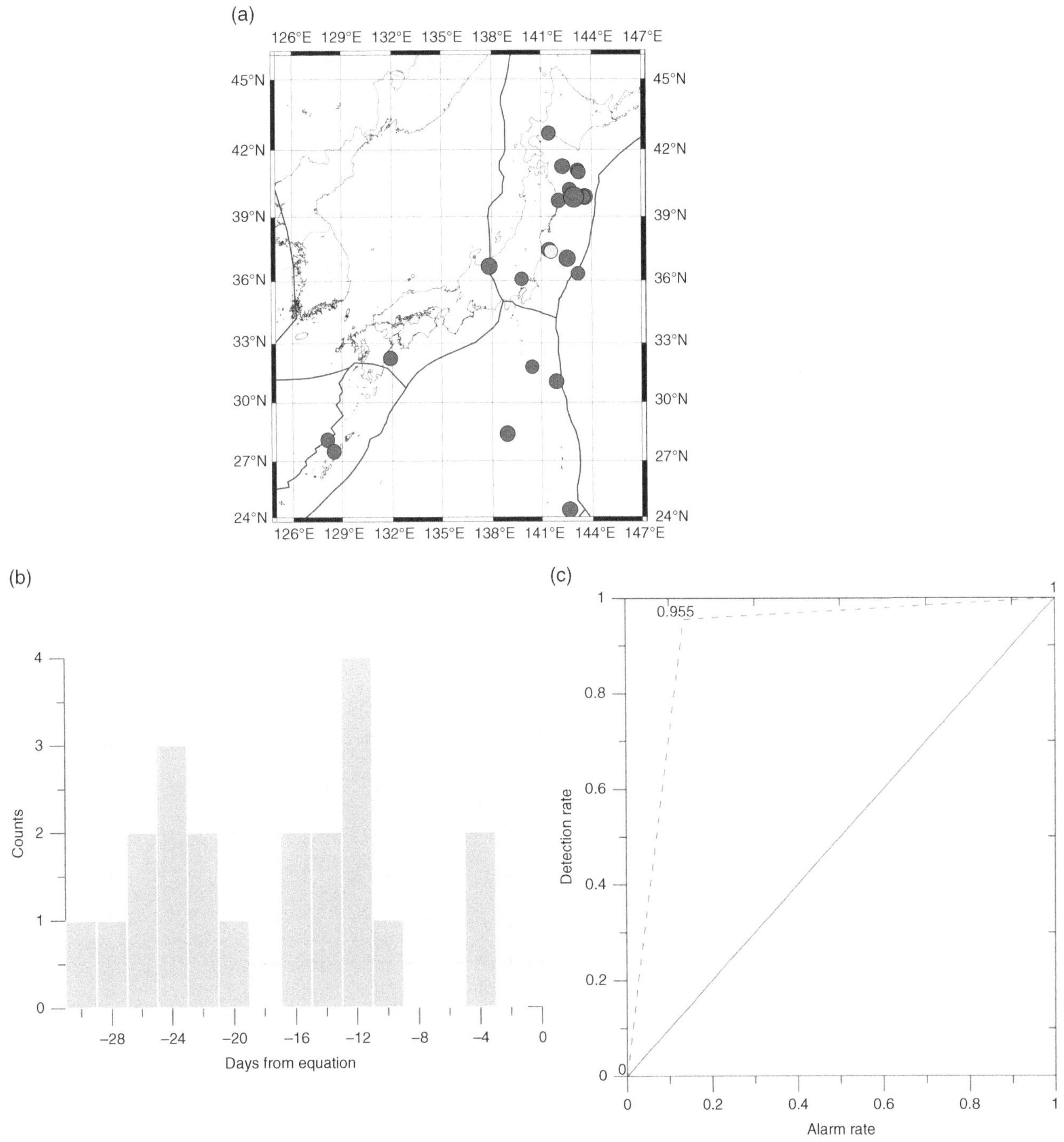

Figure 20.5 Evaluation of OLR earthquake anomalies using Molchan's error diagram (MED) for Japan, 2014–2015. (a) Distribution map for $M>5.5$ earthquakes, with dark color – successfully alerted and with light color-missed earthquakes. (b) Temporal distribution (time-lag) of OLR anomalies in relation to the timing earthquakes. (c) The MED diagram. (*See electronic version for color representation.*)

20.4. SUMMARY AND CONCLUSIONS

This chapter describes the latest progress in advancing knowledge concerning the connection between atmospheric–ionospheric variability and medium to strong earthquakes. Understanding this fundamental connection will contribute to earthquake hazard reduction. We apply interdisciplinary-based multisensor observations of short-term (days, weeks) pre-earthquake phenomena preceding large earthquakes ($M>6$). The ISTF method for validation is based on a web network of several physical and environmental parameters (satellite thermal infrared

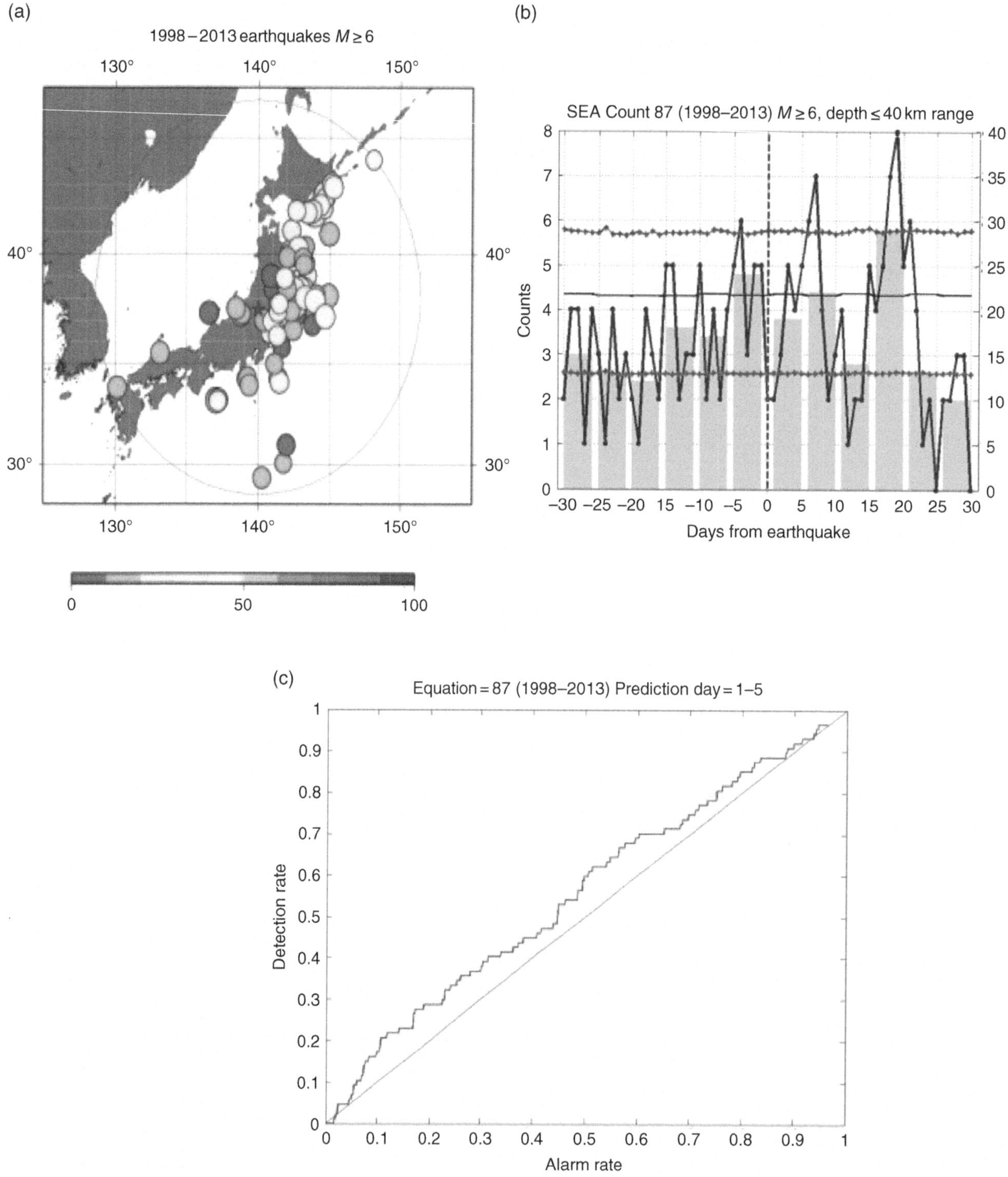

Figure 20.6 Evaluation of GPS/TEC seismic earthquake anomalies (SEA) with Molchan's error diagram (MED) for Japan, 1998–2013 (a) Distribution map for $M>6$ earthquakes, 1998–2013. (b) Distribution of anomalies, the green and black lines correspond to the significance levels for 1-day and for 5-day accumulation data, respectively. (c) The MED diagram for 5-day accumulation data. (*See electronic version for color representation.*)

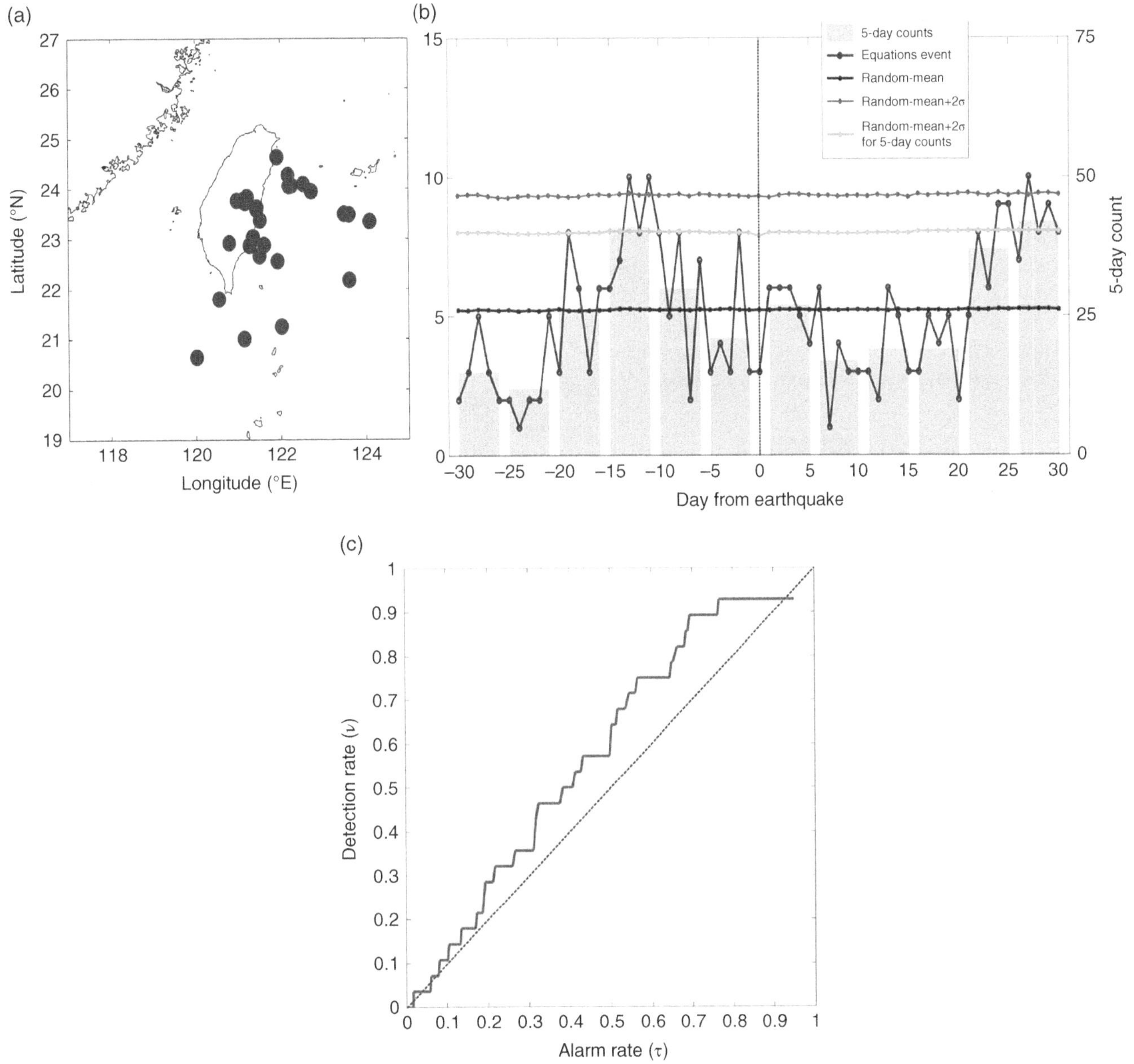

Figure 20.7 Evaluation of GIM/TEC Seismic Earthquake Anomalies (SEA) with Molchan's error diagram (MED) for Taiwan, 1998–2013. (a) Distribution map for $M>6$ earthquakes, 1998–2013. (b) Distribution of anomalies, the green and black lines correspond to the significance levels for 1-day and for 5-day accumulation data, respectively. (c) The MED diagram for 5-day accumulation data. *(See electronic version for color representation.)*

radiation (STIR), electron concentration in the ionosphere (GPS/TEC), air temperature/humidity measurements) that were found to be associated with earthquakes. The multi-sensors approach provides two observational advantages: (a) multiparameter observations reveal that the pre-earthquake phase follows a general temporal–spatial evolution pattern, and (b) multiparameter observations play a critical role in the understanding of LAIC associated with earthquake processes. These two features could be observed only by using multi-instrument observations from space and the ground. The scientific rationale for the proposed multidisciplinary analysis is based on the concept of LAIC [*Pulinets and Ouzounov*, 2011], which explains the synergy of different physical processes and anomalous variations, usually known as short-term pre-earthquake anomalies. The initial testing using MED for STIR and SEA anomalies over Japan and Taiwan shows results that are better than random guesses and suggests that physically based pre-earthquake atmospheric signals can provide short-term predictive information about the occurrence of major earthquakes in the tested regions. For the M 6 Napa earth quake of 2014, the M 6 Taiwan earthquake of 2016, and the M 7 Kumamoto earthquake of 2016 case studies, we combined independent analysis of OLR, dTEC, and ACP and found that pre-earthquake atmospheric signals follow a general temporal–spatial evolution pattern (within 1–30 days time-lag), which has been seen associated with other large earthquakes worldwide. The latest Earth observations satellites together with ground sensor networks provide a unique opportunity for comprehensive understanding of earthquake processes and this knowledge could eliminate the existing gap in our understanding of the interaction between solid Earth processes and the atmosphere in the global concept of Earth–Space system science.

ACKNOWLEDGMENTS

The authors thank to NOAA's Climate Prediction Center (CPR), NASA's Goddard Earth Sciences Data and Information Center (GES DISC), the International GNSS Service (IGS) and GEONET-GSI-Japan for providing access and services to the science data. Special thanks go to the US Geological Survey and European–Mediterranean Seismological Centre for the earthquake information services and data. The authors also thank Earth Scientific Inc. for their assistance and the International Space Science Institute (Bern, Beijing) for the international support of the team "Validation of Lithosphere-Atmosphere-Ionosphere-Magnetosphere Coupling (LAIMC) as a concept for geospheres interaction by utilizing space-borne multi-instrument observations".

REFERENCES

Akhoondzadeh, M. and Saradjian, M. R. (2012), Fusion of multi precursors earthquake parameters to estimate the date, magnitude and affected area of the forthcoming powerful earthquakes, *International Archives of the Photogrammetry, Remote Sensing and Spatial Information Sciences*, Vol. *XXXIX-B8*, 2012, XXII ISPRS Congress, 25 August – 01 September 2012, Melbourne, Australia.

Aristotle (1951), *Meteorolgica*, English Translation, H. F. P. Lee, Harvard.

Ben-Menahem, A. A. (1995), Concise history of mainstream seismology: Origina, legacy and perspectives. *Bull. Seismol. Soc. Am.*. *85*, 1202–1225.

Biagi, P. F., Piccolo, R., Castellana, L., Maggipinto, T., Ermi-ni, A., Martellucci, S., Bellecci, C., Perna, G., Capozzi, V., Molchanov, O. A., Hayakawa, M., and Ohta, K. (2004), VLF-LF radio signals collected at Bari (South Italy): a prelim-inary analysis on signal anomalies associated with earthquakes, *Nat. Hazards Earth Syst. Sci.*, *4*, 685–689.

Bokov, V. (2010), On the link between the atmospheric circulation and seismicity within the range of the seasonal variability, *Mem. Russian State Hydro-Meteorol. Univ.*, *14*, 89–100.

Boyarchuk, K. A., Karelin, A. V., and Nadolski, A. V. (2010), Statistical analysis of the chemical potential correction value of the water vapor in atmosphere on the distance from earthquake epicenter, *Issues Electromech.*, *116*, 39–46.

Cervone, G., Maekawa, S., Singh, R., Hayakawa, M., Kafatos, M., and Shavets, A. (2002), Surface latent heat flux and nighttime LF anomalies prior to the $Mw = 8.3$ Tokachi-Oki earthquake. *Nat. Hazards Earth Syst. Sci.*, *6*, 109–114.

Cicerone, R., Ebel, J., and Briton, J. (2009), A systematic compilation of earthquake precursors, *Tectonophysics*, *476*, 371–396.

De Boer, J. Z., Hale, J. R., and Chanton, J. (2001), New evidence of the geological origins of the ancient Delphic oracle (Greece), *Geology*, *29*, 707–710.

Dey, S. and Singh, R. P. (2003), Surface latent heat flux as an earthquake precursor, *Nat. Hazards Earth Syst.*, *3*, 749–755.

Doda, L., Natyaganov, V., and Stepanov, I. (2013), An empirical scheme of short-term earthquake prediction, *Dokl. Earth Sci.*, *453*(2), 1257–1263.

Dunajecka, M. A. and Pulinets, S. A. (2005), Atmospheric and thermal anomalies observed around the time of strong earthquakes in Mexico, *Atmosfera*, *18*, 235–247.

Enomoto, Y. (2012), Coupled interaction of earthquake nucleation with deep Earth gases: a possible mechanism for seismo-electromagnetic phenomena, *Geophys. J. Int.*, *191*, 1210–1214.

Fu, C. C., Wang, P. K., Lee, L. C., Lin, C. H., Chang, W. Y., Giuliani, G., and Ouzounov, D. (2015), Temporal variation of gamma rays as a possible precursor of earthquake in the Longitudinal Valley of eastern Taiwan, *J. Asian Earth Sci.*, *114*(2),362–372.

Fujiwara, H., Kamogawa, M., Ikeda, M., Liu, J. Y., Sakata, H., Chen, Y. I., Ofuruton, H., Muramatsu, S., Chuo, Y. J., and Ohtsuki, Y. H. (2004), Atmospheric anomalies observed during earthquake occurrences, *Geophys. Res. Lett.*, *31*, L17110.

Gold, T. (1979), Terrestrial sources of carbon and earthquake outgassing, *J. Petrol. Geol.*, *1*, 3–19.

Gold, T. and Soter, S. (1985), Fluid ascent through the solid lithosphere and its relation to earthquakes, *Pure Appl. Geophys.*, *122*, 492–530.

Gorny, V. I., Salman, A. G., Tronin, A. A., and Shilin, B. V. (1988), The earth's outgoing IR radiation as an indicator of seismic activity, *Proc. Acad. Sci. USSR.*, *301*, 67–69.

Gruber, A. and Krueger, A. (1984), The status of the NOAA outgoing longwave radiation dataset, *Bull. Am. Meteorol. Soc.*, *65*, 958–962.

Guo, G., and Xie, G. (2007), Earthquake cloud over Japan detected by satellite, *Int. J. Remote Sens.*, *28*, 5375–5376.

Hayakawa, M. (2004), Electromagnetic phenomena associated with earthquakes: A frontier in terrestrial electromagnetic noise environment, *Recent Res. Devel. Geophys.*, *6*, 81–112.

Hayakawa, M. (2011), Probing the lower ionospheric perturbations associated with earthquakes by means of subionospheric VLF/LF propagation, *Earthquake Sci.*, *24*, 609–637.

Hirooka, S., Hattori, K., and Takeda, T. (2011), Numerical validations of neural network based ionospheric tomography for disturbed ionospheric conditions and sparse data, *Radio Sci.*, *46*, RS0F05.

Humboldt, A. Von (1897), *COSMOS: A Sketch of the Physical Description of the Universe*, New York, 462 pp.

Irwin, W. P. and Barnes, I. (1980), Tectonic relations of carbon dioxide discharges and earthquakes. *J. Geophys. Res.*, *85*(B6), 3115–3121.

Jing, F., Shen, X. H., Kang, C. L., and Xiong, P. (2013), Variations of multi-parameter observations in atmosphere related to earthquake, *Nat. Hazards Earth Syst. Sci.*, *13*, 27–33, doi: 10.5194/nhess-13-27-2013

Kalenda, P., Neumann, L., Malek, J., and Wandrol, I. (2012), *Tilts, Global Tectonics And Earthquake Prediction*, Science Without Borders, London, 251 pp.

Keilis-Borok, V., Shebalin, P., Gabrielov, A., and Turcotte, D. (2004), Reverse tracing of short-term earthquake precursors. *Phys. Earth Planet. Inter.*, *145*(1–4), 75–85.

Kennedy, B. M., Kharaka, Y. K., Evans, W. C., Ellwood, A., Depaolo, D. J., Thordsen, J., Ambats, G., and Mariner, R. H. (1997), Mantle fluids in the San Andreas Fault System, *California, Science*, *278*, 1278–1281.

King, C.-Y. (1986), Gas geochemistry applied to earthquake prediction: An overview. *J. Geophys. Res.*, *91*(B12), 12,269–12,281.

Klotz, S. and Johnson, N. L. (1983), *Encyclopedia of Statistical Sciences*, John Wiley & Sons, Hoboken, NJ, 672 pp.

Kon, S., Nishihashi, M., and Hattori, K. (2010), Ionospheric anomalies possibly associated with $M \geq 6$ earthquakes in Japan during 1998–2010: Case studies and statistical study, *J. Asian Earth Sci.*, *41*(4), 410–420, doi: 10.1016/j.jseaes.2010.10.005

Kushida, Y. and Kushida, R. (2002), Possibility of earthquake forecast by radio observations in the VHF band, *J. Atmos. Electr.*, *22*, 239–255.

Liu, J. Y., Chen, Y. I., Pulinets, S. A., Tsai, Y. B., and Chuo, Y. J. (2000), Seismo-ionospheric signatures prior to M>6.0 Taiwan earthquakes, *Geophys. Res. Lett.*, *27*, 3113–3116.

Liu, J. Y., Chuo, Y. J., Shan, S. J., Tsai, Y. B., Chen, Y. I., Pulinets, S. A., and Yu, S. B. (2004), Pre-earthquake ionospheric anomalies registered by continuous GNSS GPS TEC measurement, *Ann. Geophys.*, *22*, 1585–1593.

Liu, J. Y., Chen, Y. I., Chen, C. H., Liu, C. Y., Chen, C. Y., Nishihashi, M., Li, J. Z., Xia, Y. Q., Oyama, K. I., Hattori, K., and Lin, C. H. (2009), Seismo-ionospheric GNSS GPS total electron content anomalies observed before the 12 May 2008 *Mw*7.9 Wenchuan earthquake, *J. Geophys. Res.*, *114*, A04320, doi: 10.1029/2008JA013698

Liu, J. Y., Tsai, Y., Chen, C., Chen, Y., and Yen, H. (2016), Integrated search for Taiwan earthquake precursors (ISTEP), *IEEJ Trans. Fund. Mater.*, *136*(5): 214–220.

Liu, J.-Y, Hattori, K., and Chen, Y. (2018), Application of total electron content derived from the Global Navigation Satellite System for detecting earthquake precursors, in D. Ouzounov, S. Pulinets, K. Hattori, and P. Taylor (eds), *Pre-Earthquake Processes: A Multidisciplinary Approach to Earthquake Prediction Studies*, pp., Geophysical Monograph 234, American Geophysical Union, Washington, DC, and John Wiley and Sons, Inc., Hoboken, NJ, 305–318.

Martinelli, G., (2000), Contributions to a history of earthquake prediction research, Seismological Research Letters *Seismol. Res. Lett.*, *vol. 71* no. (5), 583-5–588.

Mehta, A. and Susskind, J. (1999), Outgoing longwave radiation from the TOVS Pathfinder path A data set, *J. Geophys. Res.*, *104*(D10), 12193–12212.

Mil'kis, M. R. (1986). Meteorological precursors of earthquakes, *Izvest. Earth Phys.*, *22*, 195

Milne, J. (1913), *Earthquakes and Other Movements*, 2nd edn, London, 210 pp.

Mogi, K. (1985), Temporal variation of crustal deformation during the days preceding a thrust-type great earthquake—the 1944 Tonankai earthquake of magnitude 8.1. Japan, *Pure Appl. Geophys.*, *122*, 765–780.

Molchan, G. M. (1997), Earthquake prediction as a decision-making problem, *Pure Appl. Geophys*, *147*(1), 1–15.

Morozova, L. I. (1996), Features of atmo-lithmospheric relationships during periods of strong Asian earthquakes. Izvestiya, *Phys. Solid Earth*, *5*, 63–68.

Nagao, T., Enomoto, Y., Fujinawa, Y., Hata, M., Hayakawa, M., Huang, Q., Izutsu, J., Kushida, Y., Maeda, K., Oike, K., Uyeda, S., and Yoshino, T. (2002), Electromagnetic anomalies associated with 1995 KOBE earthquake, *J. Geodyn.*, *33*, 401–411.

Nagao, T., Orihara, Y., and Kamogawa, M. (2014), Precursory phenomena possibly related to the 2011 *M*9.0 off the Pacific coast of Tohoku earthquake, *J. Disaster Res.*, *9*(3), 303–310

Neresov, I. L. and Latynina, L. A. (1992), Strain processes before the Spitak earthquake, *Tectonophysics*, *202*, 221–225.

Ohring, G. and Gruber, A. (1982), Satellite radiation observations and climate theory. *Adv. Geophys.*, *25*, 237–304.

Ouzounov, D. and Freund, F. (2004), Mid-infrared emission prior to strong earthquakes analyzed by remote sensing data, *Adv. Space Res.*, *33*, 268–273.

Ouzounov, D., Pulinets, S., Cervone, G., Kafatos, M., and Taylor, P. (2006), New methodology for global earthquake monitoring using joint multi-parameter satellite and in-situ data, *36th COSPAR Scientific Assembly*, 16–23 July, Beijing, China

Ouzounov, D., Liu, D., Kang, C., Cervone, G., Kafatos, M., and Taylor, P. (2007), Outgoing long wave radiation variability from IR satellite data prior to major earthquakes, *Tectonophysics*, *431*, 211–220.

Ouzounov, D., Pulinets, S., Hattori, K., Parrot, M., Liu, J. Y., Yang, T. F., Arellano-Baeza, A. A., Kafatos, M., and Taylor, P. (2011a), Progress in multidisciplinary validation of earthquake atmospheric signals by joint satellite and ground based observations, *The XXV IUGG General Assembly*, Melbourne, 28 June to 7 July.

Ouzounov, D., Pulinets, S., Hattori, K., Kafatos, M., and Taylor, P. (2011b), Atmospheric signals associated with major earthquakes. a multi-sensor approach, in M. Hayakawa (ed.), *Frontier of Earthquake Short-Term Prediction Study*, pp. 510–531, Terra Scientific Publishing, Tokyo.

Ouzounov, D., Pulinets, S. A., Hernandez-Pajares, M., Alberto Garcia Rigo, A. G., Davidenko, D., Hatzopoulos, N., and Kafatos, M. (2015), Transient effects in atmosphere and ionosphere preceding the two 2015 *M*7.8 and *M*7.3 earthquakes in Nepal, *AGU Fall Meeting*, abstract #NH32B-05.

Ouzounov, D., Pulinets, S. A., Hernandez-Pajares, M., Alberto Garcia Rigo, A. G., Davidenko, D., Petrov, L., Hatzopoulos, N., and Kafatos, M. (2016), Pre-earthquake signatures in atmosphere/ionosphere and their potential for short-term earthquake forecasting, *Case Studies for 2015, EGU General Assembly*, 17–22 April, Vienna, p. 3496.

Ouzounov, D., Pulinets, S., Kafatos, M., and Taylor, P. (2018), Thermal radiation anomalies associated with major earthquakes, in D. Ouzounov, S. Pulinets, K. Hattori, and P. Taylor (eds), *Pre-Earthquake Processes: A Multidisciplinary Approach to Earthquake Prediction Studies*, pp., Geophysical Monograph 234, American Geophysical Union, Washington, DC, and John Wiley and Sons, Inc., Hoboken, NJ, 259–274.

Piccardi, L. (2000), Active faulting at Delphi, Greece: seismotectonic remarks and a hypothesis for the geologic environment of a myth,. *Geology*, *28*, 651 – 654.

Pulinets, S. (2011), The synergy of earthquake precursors, *Earthquake Sci.*, *24*, 535–548.

Pulinets, S. and Boyarchuk, K. (2004), *Ionospheric Precursors of Earthquakes*, Springer, New York, 289. pp.

Pulinets, S. and Ouzounov, D. (2011), Lithosphere–atmosphere–ionosphere coupling (LAIC) model–a unified concept for earthquake precursors validation, *J. Asian Earth Sci.*, *41*(4), 371–382.

Pulinets, S. A., Khegai, V. V., Boyarchuk, K. A., and Lomonosov, A. M. (1998), Atmospheric electric field as a source of ionospheric variability. *Phys. Uspekhi*, *41*(5), 515–522.

Pulinets, S., Legen'Ka, A. D., Gaivoronskaya, T. V., and Depuev, V. K. (2003), Main phenomenological features of ionospheric precursors of strong earthquakes, *J. Atmos. Solar Terr. Phys.*, *65*(16), 1337–1347.

Pulinets, S., Ouzounov, D., Ciraolo, L., Singh, R., Cervone, G., Leyva, A., Dunajecka, M., Karelin, A. V., Boyarchuk, K. A., and Kotsarenko, A. (2006a), Thermal, atmospheric and ionospheric anomalies around the time of the Colima M7.8 earthquake of 21 January 2003, *Ann. Geophys.*, *24*, 835–849.

Pulinets, S., Ouzounov, D., Karelin, A. V., Boyarchuk, K. A., and Pokhmelnykh, L. A. (2006b), The physical nature of the thermal anomalies observed before strong earthquakes, *Phys. Chem. Earth*, *31*, 143–153.

Pulinets, S., Ouzounov, D., Karelin, A., and Davidenko, D. (2015), Physical bases of the generation of short-term earthquake precursors: A complex model of ionization-induced geophysical processes in the lithosphere–atmosphere–ionosphere–magnetosphere system, *Geomagn. Aeron.*, *55*(4), 540–558.

Pulinets, S., Ouzounov, D., Karelin, A., and Dmitry Davidenko, D. (2018), Lithosphere–atmosphere–ionosphere–magnetosphere coupling—a concept for pre-earthquake signals generation, in D. Ouzounov, S. Pulinets, K. Hattori, and P. Taylor (eds), *Pre-Earthquake Processes: A Multi-disciplinary Approach to Earthquake Prediction Studies*, pp. Geophysical Monograph 234, American Geophysical Union, Washington, DC, and John Wiley and Sons, Inc., Hoboken, NJ, 79–98.

Qiang, Z.; Xu, X.; Dian, C.(1991a) Thermal infrared anomaly precursors of impending earthquakes. *Chin. Sci. Bull. 1991*, 36, 319–323.

Qiang, Z.; and Du, L.-T. (2001), Earth degassing, forest fire and seismic activities,. *Earth Science Sci. FrontiersFront.*, *2001, 8*, 235-2–245.

Qiang, Z., Xu, X., and Dian, C. (1991b), Thermal infrared anomaly precursors of impending earthquakes, *Chin. Sci. Bull.*, *36*, 319–323.

Rozhnoi A, Molchanov O, Solovieva M, Gladyshev V, Akentieva O, Berthelier J J, Parrot M, Lefeuvre F, Hayakawa M, Castellana L and Biagi P F (2007a). Possible seismo-ionosphere perturbations revealed by VLF signals col-lected on ground and on a satellite. *Nat Hazards Earth Syst Sci* *7*: 617–624.

Rozhnoi, A., Molchanov, O., Solovieva, M., Gladyshev, V., Aken-tieva, O., Berthelier, J. J., Parrot, M., Lefeuvre, F., Hayakawa, M., Castellana, L., and Biagi, P. F. (2007b), Possible seismo-ionosphere perturbations revealed by VLF signals col-lected on ground and on a satellite, *Nat. Hazards Earth Syst. Sci.*, *7*, 617–624.

Salman, A., Egan, W. G., and Tronin, A. A. (1992), Infrared remote sensing of seismic disturbances, in *Polarization and Remote Sensing, SPIE*, 1747, San Diego, CA.

Saraev, A. K., Antaschuk, K. M., Simakov, A. E., et al. (2014), Multiparameter monitoring of electromagnetic earthquake precursors in the frequency range 0.1 Hz–1 MHz, *Seismol. Instr.*, *50*, 52, doi: 10.3103/S0747923914010071

Saraf, A. K., Rawat, V., Banerjee, P., Choudhury, S., Panda, S. K., Dasgupta, S., and Das, J. D. (2008,) Satellite detection of earthquake thermal precursors in Iran, *Nat. Hazard*, *47*, 119–135.

Singh, R. P., Cervone, G., Kafatos, M., Prasad, A. K., Sahoo, A. K., Sun, D., Tang, D. L., and Yang, R. (2007), Multi-sensor studies of the Sumatra earthquake and tsunami of 26 December 2004, *Int. J. Remote Sens.*, *28*, 2885–2896.

Soter, S. (1999), Macroscopic seismic anomalies and submarine pockmarks in the Corinth–Patras rift, *Greece, Tectonophysics*, *308*, 275–290.

St-Laurent, F., Derr, J., and Freund, F. (2006), Earthquake lights and the stress-activation of positive hole charge carriers in rocks, *Phys. Chem. Earth*, *31*,305–312.

Stothers, R. (2004), Ancient and modem earthquake lights in Northwestern Turkey, *Seismol. Res. Lett.*, *75*(2), 199–204.

Thomas, D. (1988), Geochemical Precursors to Seismic Activity, *Pure Appl. Geophys.*, *125*(2–4), 241–266.

Toutain, J.-P. and Baubron, J.-C. (1998), Gas geochemistry and seismotectonics: a review, *Tectonophysics*, *304*, 1–27.

Tramutoli, V., Bello, G. D., Pergola, N., and Piscitelli, S. (2001), Robust satellite techniques for remote sensing of seismically active areas, *Ann. Geofis.*, *44*, 295–312.

Tramutoli, V., Cuomo, V., Filizzola, C., Pergola, N., and Pietrapertosa, C. (2005), Assessing the potential of thermal infrared satellite surveys for monitoring seismically active areas. The case of Kocaeli (İzmit) earthquake, August 17, 1999. *Remote Sens. Environ.*, *96*, 409–426.

Tramutoli, V., Inan, S., Jakowski, N., Pulinets, S. A., Romanov, A., Filizzola, C., Shagimuratov, I., Pergola, N., Ouzounov, D. P., Papadopoulos, G. A., Parrot, M., Genzano, N., Lisi, M., Alparlsan, E., Wilken, V., Tsybukia, K., Romanov, A., Paciello, R., Zakharenkova, I., and Romano, G. (2012), Dynamic Assessment of Seismic Risk (DASR) by Multi-parametric Observations: Preliminary Results of PRIME experiment within the PRE-EARTHQUAKES EU-FP7 Project, *AGU Fall Meeting Abstracts*, #NH44A-02.

Tramutoli, V., Corrado, R., Filizzola, C., Genzano, N., Lisi, M., and Pergola, N. (2015), From visual comparison to robust satellite techniques: 30 years of thermal infrared satellite data analyses for the study of earthquake preparation phase, *Boll. Geofis. Teor. Appl.*, *56*(2), 167–202.

Tronin, A. A. (1996), Satellite thermal survey—a new tool for the studies of seismoactive regions, *Int. J. Remote Sens.*, *17*, 1439–1455.

Tronin, A. A. (2011), *Catalog of Thermal and Atmospheric Phenomena Associated with Earthquakes*, Sankt Petersburg, Russia, 260 pp. (In Russian.)

Tronin, A. A., Hayakawa, M., and Molchanov, O. A. (2002), Thermal IR satellite data application for earthquake research in Japan and China, *J. Geodyn.*, *33*, 519–534.

Tronin, A. A., Biagi, P. F., Molchanov, O. A., Khatkevich, Y. M., and Gordeev, E. I. (2004), Temperature variations related to earthquakes from simultaneous observation at the ground stations and by satellites in Kamchatka area, *Phys. Chem. Earth*, *29*, 501–506.

Tsai, Y. B., Liu, J. Y., Ma, K. F., Yen, H. Y., Chen, K. S., Chen, Y. I., and Lee, C. P. (2006), Precursory phenomena associated with the 1999 Chi-Chi earthquake in Taiwan as identified under iSTEP Program, *Phys. Chem. Earth*, *31*, 365–377.

Vernadsky, V. (1912), About the gas exchange of Earth crust, *Russ. Acad. Sci. St Petersburg*, *6*(2), 141–162. (In Russian.)

Vernadsky, V. (1945), The biosphere and the noösphere, *Am. Sci.*, *33*(1),1–12.

Wakita, H. (1982), Changes in groundwater level and chemical composition, in T. Asada (ed.), *Earthquake Prediction Techniques*. pp. 175–216, University of Tokyo Press.

Wu, H. C. (2004), Preliminary findings on perturbation of jet stream prior to earthquakes, *Eos (Trans. Am. Geophys. Union)*, *85*(47), T51B-0455.

Wu, H. C., Tikhonov, I., and C′esped, A. R. (2015), Multi-parametric analysis of earthquake precursors, *Russ. J. Earth. Sci.*, *15*, ES3002.

Wyss, M., Klein, F. W., and Johnston, A. C. (1981), Precursors to the Kalapana *M*7.2 earthquake, *J. Geophys. Res.*, *86*, 3881–3900.

Xiong, P., Shen, X. H., Bi, Y. X., Kang, C. L., Chen, L. Z., Jing, F., and Chen, Y. (2010), Study of outgoing longwave radiation anomalies associated with Haiti earthquake, *Nat. Hazards Earth Syst. Sci.*, *10*, 2169–2178.

Yuce, G., Ugurluoglu, D. Y., Adar, N., Yalcin, T., Yaltirak, C., Streil, T., and Oeser, V. (2010), Monitoring of earthquake precursors by multi-parameter stations in Eskisehir region (Turkey), *Appl. Geochem.*, *25*(4), 572–579.

Zakharenkova, I. E., Krankowski, A., and Shagimuratov, I. I. (2006), Modification of the low-latitude ionosphere before December 26, 2004 Indonesian earthquake, *Nat. Hazards Earth Syst. Sci.*, *6*, 817–823.

Zechar, J. and Jordan, T. (2008), Testing alarm-based earthquake predictions, *Geophys. J. Int.*, *172*, 715–724.

Zia, R. (1984), Forewarning phenomena for the earthquake that took place in the Basilicata and Campania regions of South Italy on November 23rd 1980, *International Symposium on Continental Seismicity and Earthquake Prediction*, pp. 534–540, Seismological Press.

INDEX

Pre-Earthquake Processes: A Multidisciplinary Approach to Earthquake Prediction Studies, Geophysical Monograph 234, First Edition. Edited by Dimitar Ouzounov, Sergey Pulinets, Katsumi Hattori, and Patrick Taylor.

Wiley The manufacturer's authorized representative according to the EU General Product Safety Regulation is Wiley-VCH GmbH, Boschstr. 12, 69469 Weinheim, Germany, e-mail: Product_Safety@wiley.com.

Printed and bound by CPI Group (UK) Ltd, Croydon, CR0 4YY

01/10/2026

02225815-0001